Chapter 4

Chapter 5

Chapter 6

Chapter 7

Chapter 8

Chapter 9

ELEMENTARY ALGEBRA
with Applications

ELEMENTARY ALGEBRA
with Applications

FOURTH EDITION

TERRY H. WESNER
Henry Ford Community College

HARRY L. NUSTAD

WCB **Wm. C. Brown Publishers**

Dubuque, IA Bogota Boston Buenos Aires Caracas Chicago
Guilford, CT London Madrid Mexico City Sydney Toronto

Book Team

Developmental Editor *Theresa Grutz*
Production Editor *Eugenia M. Collins*
Designer *K. Wayne Harms*

 Wm. C. Brown Publishers

President and Chief Executive Officer *Beverly Kolz*
Vice President, Publisher *Earl McPeek*
Vice President, Director of Sales and Marketing *Virginia S. Moffat*
Vice President, Director of Production *Colleen A. Yonda*
National Sales Manager *Douglas J. DiNardo*
Marketing Manager *Keri L. Witman*
Advertising Manager *Janelle Keeffer*
Production Editorial Manager *Renée Menne*
Publishing Services Manager *Karen J. Slaght*
Royalty/Permissions Manager *Connie Allendorf*

A Times Mirror Company

Copyedited by *Patricia Steele*

Production, illustrations, photo research, and composition by
Publication Services, Inc.

Cover credit © Kerrick James Photography

The credits section for this book begins on page 663 and is considered an extension of the
copyright page

To my parents
Harold and Nina,
for everything

Terry

To my parents
Harry and Cordelia,
for their guidance throughout my life

Harry

Chapter 1

Operations with Real Numbers

Chapter 2

Solving Equations and Inequalities

Chapter 3

Polynomials and Exponents

Chapter 4

Factoring and Solution of Quadratic Equations by Factoring

Chapter 5

Rational Expressions and Equations

Chapter 6

Linear Equations in Two Variables

Chapter 7

Systems of Linear Equations

Chapter 8

Roots and Radicals

Chapter 9

Quadratic Equations

Applications

Business and Economics

Electrical

General

Geometry

Chemistry and Physics

Nature is in a constant state of change, with each new development building on what already existed. *Elementary Algebra with Applications,* Fourth Edition, follows the same pattern— it develops concepts and ideas as natural extensions of previously learned material. This problem-solving approach leads today's students beyond memorization to a true conceptual understanding.

Embark on an algebraic exploration through Wesner and Nustad's innovative learning environment. Ideal for lecture-discussion, or self-paced classes, *Elementary Algebra with Applications,* Fourth Edition, successfully guides students without previous algebra experience through the unknowns of beginning algebra.

Problem Solving

To navigate safely through beginning algebra, a firm grasp of problem solving is essential. In chapter 1, students are asked to translate English phrases into simple algebra. They are then asked to form and solve equations from word problems that have simple arithmetic solutions. Continuing throughout the rest of the text, real-world applications are featured to further develop problem solving.

Readability

Students won't need a map and compass to follow along in *Elementary Algebra with Applications,* Fourth Edition. Wesner and Nustad's friendly and accessible writing style is modeled after the way you would interact with your own students in the classroom.

Functional Use of Color

Nature uses color to warn or attract attention, and so do Wesner and Nustad. Color is used to guide students throughout the text:

Green identifies headings, notes, and mastery points within the text, and core problems in the exercise sets that can form the basis of an assignment or student review session.

Blue is used for example headings, side notes, and step-by-step explanations within examples, and to identify procedures.

Red highlights the Quick Checks that provide student interaction within each group of examples. The step-by-step solutions of Quick Checks are shown in the exercise sets to provide convenient reinforcement of the mathematics necessary to complete the exercises.

Explore to Learn

"Explore to Learn" exercises fine-tune students' creative problem-solving skills and can be used as cooperative learning activities for classroom discussion, or as extension problems.

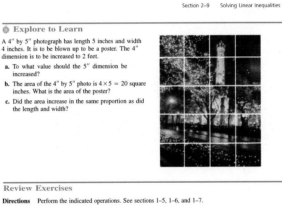

⊛ Explore to Learn

A 4″ by 5″ photograph has length 5 inches and width 4 inches. It is to be blown up to be a poster. The 4″ dimension is to be increased to 2 feet.

a. To what value should the 5″ dimension be increased?

b. The area of the 4″ by 5″ photo is $4 \times 5 = 20$ square inches. What is the area of the poster?

c. Did the area increase in the same proportion as did the length and width?

Review Exercises

Directions Perform the indicated operations. See sections 1–5, 1–6, and 1–7.

1. $-14 + 8$ **4.** $7 - (-7)$ **7.** $(-8)(-4)$ **9.** $(-4)^2$

2. $5 + (-11)$ **5.** $4(-3)$ **8.** -4^2 **10.** $(-4)^3$

3. $-4 - (-4)$ **6.** $(-2)(5)$

2–9   Solving Linear Inequalities

Many of the everyday problems that need to be solved are **inequalities**. For example, an elevator has a sign stating that the maximum safe load is 4,000 pounds, or a company says it must sell 8,500 units before making a profit, or a computer requires an operating environment of from 17 to 27 degrees Celsius; these are common situations that are expressed using inequalities.

Inequality Symbols

In chapter 1, we studied the meaning of the inequality symbols[3]

< "is less than"
≤ "is less than or equal to"
> "is greater than"
≥ "is greater than or equal to."

These symbols define the *sense* or *order* of an inequality. For example, if we wish to denote that the variable x represents 3 or any number greater than 3, we write $x \geq 3$.

[3]The symbols > and < were first used in 1631 in the *Artis analyticae praxis.* The symbols ≥ and ≤ were first used by the Frenchman P. Bouguer in 1734.

Chapter Lead-in Problems

Students will use important skills introduced in each chapter to solve real-world problems. Photographs illustrate real-world applications of the mathematics developed in the chapter. Complete solutions to the lead-in applications are shown at the end of each chapter.

Applications

Real-world applications, word problems, and exercises have been updated and expanded to cover a broad range of relevant and interesting real-life applications. Numerous applications emphasize the important role algebra plays in today's computerized world.

Footnotes

Footnotes offer points of interest, while labeled tables, figures, centered equations, and bold-faced terms and phrases emphasize ideas of importance.

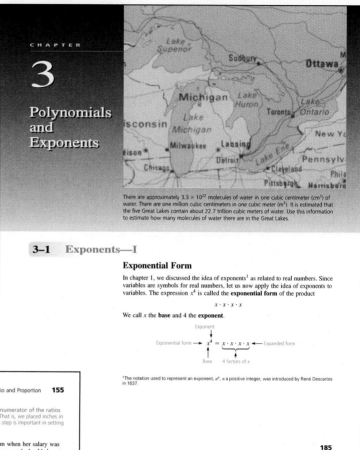

CHAPTER

3

Polynomials and Exponents

There are approximately 3.3×10^{22} molecules of water in one cubic centimeter (cm^3) of water. There are one million cubic centimeters in one cubic meter (m^3). It is estimated that the five Great Lakes contain about 22.7 trillion cubic meters of water. Use this information to estimate how many molecules of water there are in the Great Lakes.

3–1   Exponents—I

Exponential Form

In chapter 1, we discussed the idea of exponents[1] as related to real numbers. Since variables are symbols for real numbers, let us now apply the idea of exponents to variables. The expression x^4 is called the **exponential form** of the product

$$x \cdot x \cdot x \cdot x$$

We call x the **base** and 4 the **exponent**.

Exponent

Exponential form → $x^4 = \underbrace{x \cdot x \cdot x \cdot x}_{\text{4 factors of } x}$ ← Expanded form

Base

[1]The notation used to represent an exponent, a^n, n a positive integer, was introduced by René Descartes in 1637.

185

Section 2–8   Ratio and Proportion   **155**

Note In example 1, the same units of measure are in the numerator of the ratios and the same units of measure are in the denominators. That is, we placed inches in the numerator and miles in the denominator of each ratio. This step is important in setting up the proportion you will use to solve for the unknown.

2. Cheryl set aside $20 per week for her savings program when her salary was $200 per week. If her salary is now $250 per week, how much should she set aside for her weekly savings to be proportional to what she saved before?

   Let x represent the amount to be set aside when Cheryl earns $250 per week. Then, $20 is to $200 as x is to $250.

$$\frac{20}{200} = \frac{x}{250} \qquad \text{Set up a proportion using } \frac{\text{savings}}{\text{salary}}$$
$$200 \cdot x = 20 \cdot 250 \qquad \text{Property of proportions}$$
$$200x = 5{,}000 \qquad \text{Multiply as indicated}$$
$$x = \frac{5{,}000}{200} \qquad \text{Divide each side by 200}$$
$$x = 25$$

Cheryl should set aside $25 when making $250 per week.

   Proportions can be used to convert one unit of measure to another unit of measure. To set up the proportion we will need to know the relationship between the two units of measure.

3. Convert 6 inches to centimeters.

   Let x represent the number of centimeters that is equal to 6 inches. The relationship is 1 inch = 2.54 centimeters. We use this relationship to write the ratio on the left side of the equation.

$$\frac{1 \text{ inch}}{2.54 \text{ centimeters}} = \frac{6 \text{ inches}}{x \text{ centimeters}} \qquad \text{Set up a proportion using } \frac{\text{inches}}{\text{centimeters}}$$
$$\frac{1}{2.54} = \frac{6}{x} \qquad \text{Drop the units}$$
$$1 \cdot x = 6 \cdot 2.54 \qquad \text{Property of proportions}$$
$$x = 15.24 \qquad \text{Multiply}$$

   Therefore, 6 inches = 15.24 centimeters.

4. Convert 5.3 quarts to liters.

   Let x represent the number of liters that is equal to 5.3 quarts. The relationship is 1 liter = 1.06 quarts.

$$\frac{1 \text{ liter}}{1.06 \text{ quarts}} = \frac{x \text{ liters}}{5.3 \text{ quarts}} \qquad \text{Set up a proportion using } \frac{\text{liters}}{\text{quarts}}$$
$$\frac{1}{1.06} = \frac{x}{5.3} \qquad \text{Drop the units}$$
$$5.3 = 1.06x \qquad \text{Property of proportions}$$
$$5 = x \qquad \text{Divide both sides by 1.06}$$

   Therefore, 5.3 quarts = 5 liters.

   ☞ **Quick check** On a map, 1 inch represents 9 miles. How many inches are needed to represent 42 miles.

Examples

Wesner and Nustad are there every step of the way with a proven step-by-step approach to all examples. This approach makes it easier for students to follow the problem-solving process and encourages students to develop a clear understanding of how the problem is worked.

Notes to the Student

Notes to the student highlight important ideas and point out common errors. These ideas are revisited in the end-of-chapter error analysis.

Properties and Definition Boxes

Properties and definition boxes include "concepts"—restatements of the property or definition in everyday language—to facilitate the learning process and to help students understand and remember the main idea.

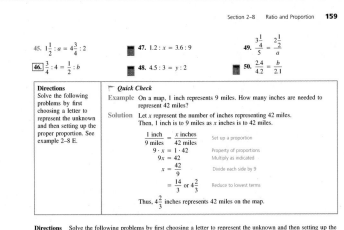

Section 2–8   Ratio and Proportion   **159**

45. $1\frac{1}{2} : a = 4\frac{3}{4} : 2$ ■ 47. $1.2 : x = 3.6 : 9$ 49. $\dfrac{3\frac{1}{4}}{5} = \dfrac{2\frac{1}{2}}{a}$

46. $\dfrac{3}{4} : 4 = \dfrac{1}{2} : b$ ■ 48. $4.5 : 3 = y : 2$ ■ 50. $\dfrac{2.4}{4.2} = \dfrac{b}{2.1}$

Directions
Solve the following problems by first choosing a letter to represent the unknown and then setting up the proper proportion. See example 2–8 E.

⌐ *Quick Check*

Example On a map, 1 inch represents 9 miles. How many inches are needed to represent 42 miles?

Solution Let x represent the number of inches representing 42 miles. Then, 1 inch is to 9 miles as x inches is to 42 miles.

$$\frac{1 \text{ inch}}{9 \text{ miles}} = \frac{x \text{ inches}}{42 \text{ miles}}$$ Set up a proportion

$$9 \cdot x = 1 \cdot 42$$ Property of proportions

$$9x = 42$$ Multiply as indicated

$$x = \frac{42}{9}$$ Divide each side by 9

$$= \frac{14}{3} \text{ or } 4\frac{2}{3}$$ Reduce to lowest terms

Thus, $4\frac{2}{3}$ inches represents 42 miles on the map.

Directions Solve the following problems by first choosing a letter to represent the unknown and then setting up the proper proportion. See example 2–8 E.

51. A man earns $180 per week. How many weeks must he work to earn $1,260?

52. An automobile uses 8 liters of gasoline to travel 84 kilometers. How many liters are needed to travel 1,428 kilometers?

53. If 24 grams of water will yield 4 grams of hydrogen, how many grams of hydrogen will there be in 216 grams of water?

54. The operating instructions for a gasoline chain saw call for a 16 gallons : 1 pint fuel-to-oil mixture. How many *pints* of oil are needed to mix with 88 gallons of fuel?

55. The power-to-weight ratio of a given engine is 5 : 3. What is the weight in pounds of the engine if it produces 650 horsepower?

■ 56. If a 20-pound casting costs $1.50, at this same rate, how much would a 42-pound casting cost?

57. American television signals are broadcast in a 4 : 3 aspect ratio (the ratio of width to height of the picture). A television screen is to be adjusted so the width is 14 inches; to what value should the height be adjusted?

■ 58. In a hydraulic press, the force on the output piston is to the force on the input piston as the area of the output piston is to the area of the input piston. That is, $\dfrac{F_o}{F_i} = \dfrac{A_o}{A_i}$ or $F_o : F_i = A_o : A_i$. Find the area of the input piston if F_o is 15.2 pounds, F_i is 6.5 pounds, and A_o is 10.4 inches[2].

59. If the ratio of the wins to the losses of the Chicago Cubs in a given season is 6 : 5, how many games did they lose if they won 90 games?

60. A rectangular picture that is 10 inches long and 8 inches wide is to be enlarged so that the enlargement will be 36 inches wide. What should be the length of the enlargement?

Quick Checks

Quick checks provide convenient reinforcement of the mathematics necessary for the related exercises and support each group of similar problems in the exercise sets. These exercises directly parallel the development and examples in the text. Immediately after each group of examples is a Quick Check problem for students to work thereby getting immediate feedback that they understood that group of examples. Complete solutions are incorporated in the Section Exercises, allowing students to see a line-by-line check of their Quick Check solution. These solutions also provide a convenient quick reference while working the section exercises.

Section Exercises

Section exercises provide abundant opportunities for students to check their understanding of the concepts being presented. The problems in the exercise sets are carefully paired and graded by level of difficulty to guide students easily from straightforward computations to more challenging, multistep problems.

Core Problems

Core problems identify the variety of problem types on which to base an assignment. These problems reflect the range of concepts and skills learned in each section and are easily identified by green type.

Trial Exercise Problems

Identified by a box around the problem number, these helpful problems appear throughout the exercise set and represent the various problem types in each exercise set. Completely worked-out solutions in the answer appendix give students the opportunity to check their work, step-by-step, as they do their homework.

Review Exercises

Review exercises help students refresh their understanding of concepts and skills needed for success in the next section and continually reinforce previously developed material.

Graphing Calculators

Modern-day algebra exploration benefits from the use of graphing calculators. Wesner and Nustad recognize the usefulness of both scientific and graphing calculator technology. Throughout the text, examples and exercises for the TEXAS INSTRUMENTS and Casio families of graphing calculators are integrated.

Procedure Boxes

Procedure boxes provide a student-friendly, step-by-step summary of each of the problem-solving processes.

Mastery Points

Mastery points alert students to particular skills they will need to successfully complete their journey through each section. The tutorial software and the test/quiz items in the WCB Test Writer program are directly tied to the Mastery Points, as are the core of the course materials in the *Instructor's Resource Manual*. The Mastery Points are also listed at the beginning of each section in the Annotated Instructor's Edition to set forth the objectives of the section.

Error Analysis Problems

With "Error Analysis Problems" Wesner and Nustad skillfully help students develop the ability to find errors when checking solutions and prevent future errors. Error analysis problems contain common mathematical errors that students must identify. The student must then state in their own words what error has been made.

Critical Thinking Exercises

These exercises promote a greater depth of understanding and a higher level of analysis skills. The *Instructor's Resource Manual* outlines methods and techniques for incorporating critical thinking into the course using these exercises.

---

5. Combining polynomials
   Example: $(3x^2 - 2x + 1) - (x^2 - x + 2)$
   $$= 3x^2 - 2x + 1 - x^2 - x + 2$$
   $$= 2x^2 - 3x + 3$$
   Correct answer: $(3x^2 - 2x + 1) - (x^2 - x + 2)$
   $$= 2x^2 - x - 1$$
   What error was made? (*see page 106*)

6. Reciprocal of a number
   Example: The reciprocal of 0 is $\frac{1}{0}$.
   Correct answer: 0 has no reciprocal.
   What error was made? (*see page 123*)

7. Graphing linear inequalities
   Example: The graph of $x \le 3$ is

   0   3   5

   Correct answer: The graph of $x \le 3$ is

   0   3   5

   What error was made? (*see page 162*)

8. Multiplying sides of an inequality
   Example: If $3 < 4$, then $3 \cdot -2 < 4 \cdot -2$.
   Correct answer: If $3 < 4$, then $3 \cdot -2 > 4 \cdot -2$.
   What error was made? (*see page 164*)

9. Multiplication of negative numbers
   Example: $(-5)(7) = 35$
   Correct answer: $(-5)(7) = -35$
   What error was made? (*see page 62*)

10. Division using zero
    Example: $\frac{7}{0} = 0$

    Correct answer: $\frac{7}{0}$ is undefined.
    What error was made? (*see page 65*)

11. Solving proportion problems
    Example: On a map, 1 inch represents 10 miles. How many inches represent 24 miles? Let $x =$ inches represented by 24 miles.
    $$\frac{1}{10} = \frac{24}{x}$$
    $x = 240$ inches
    Correct answer: 2.4 inches
    What error was made? (*see page 154*)

12. Solving inequalities
    Example: $\frac{x}{-3} \le 6$, so $x \le -18$
    Correct answer: $x \ge -18$.
    What error was made? (*see page 165*)

13. Solving problems with percent
    Example: 5% of a number is 20; what is the number?
    $0.5x = 20$, so $x = 40$
    Correct answer: $x = 400$
    What error was made? (*see page 28*)

Chapter 2 Critical Thinking

1. If you add any three consecutive odd integers, the sum will be a multiple of 3. Why is this true?

2. If you add any three consecutive even integers, is the sum a multiple of 3? Justify your answer.

3. The distributive property: $a(b + c) = ab + ac$ is assumed to be true for any real numbers a, b, c. It cannot be proven true. Recall than the area of a rectangle is the product of its length and width.

Thus the area of figure a is $3 \cdot 7 = 21$. Show how figure b could be used to justify the distributive property.

---

4. A parent left $17,000 to her three children. The will stipulated that half the money be left to the first child, a third to the second child, and a ninth to the third child. Because the arithmetic wasn't working, the administrator of the estate consulted a wise accountant. The accountant loaned the administrator of the estate $1,000. This solved the problem. Explain.

5. The Gunning Fog index for prose is 0.4 times the sum of "average sentence length" and "percentage of words with three or more syllables." It approximates the grade level of education needed to understand written material. Suppose this index is to be computed for a newspaper article. Suppose the following variables are defined as indicated: L_s = average length of sentences (number of words in the sentence) and W_3 = percentage of words with three or more syllables. Create a formula for the Gunning Fog index. Apply this formula to this problem.

6. It takes 365 days, 5 hours, 48 minutes, and 46 seconds for the earth to go around the sun once. If there was no adjustment to the calendar by leap years the seasons would drift, what is considered

summer would eventually become the vice versa. The adjustment made in by adding a day every 4 years, does problem exactly. (a) What is the err after such an adjustment? The years a century are not leap years unless the divisible by 400. For example, 1900 year, but 2000 will be. (b) How much occurred since the year 1600 at the en 2000?

7. Count the number of males and fema three of the classes in your school. F many total students there are in the s your data to predict how many males are attending the school. Ask the registrar if your estimates are correct. What factors should govern the classes you choose for the counting to obtain the most accurate results? Do you think two or three classes is a large enough sample to obtain accurate results?

8. How many people are awake while you are eating your lunch?

Chapter 2 Review

[2–1]

Directions   Specify the number of terms in each expression.

1. $4x^2 + 3x + 2$ 2. $5a^2b$ 3. $7xy + 5$ 4. $ab + cd + xy$

Directions   Determine which of the following algebraic expressions are polynomials. If it is a polynomial, state the degree. If it is not a polynomial, state why not.

5. $2x^2 - 3x + 4$ 8. $\frac{a+b}{c}$

6. $x^2y - xy$

7. $4a^2$

Directions   Write an algebraic expression for each of the following.

9. 5 times x 11. 4 more than z

10. 7 less than y 12. 2 times a number, plus 6

Chapter Summaries, Reviews, Tests, and Cumulative Tests

These end-of-chapter materials thoroughly review and test student knowledge—preparing students for the next chapter and upcoming exams. These features help students determine if they need further work on a particular section. The answers in the answer appendix are keyed to refer students back to the section from which they are drawn.

Discover These Features New to the Fourth Edition

- A new section (1–1) on **study skills for success in algebra** offers specific strategies for improving study skills and maximizing students' efforts. These skills are reinforced by the authors in the Student Study Skills audio tape, supplied free with the textbook.
- The two chapters on **rational expressions** have been combined.
- The section on the **quotient of two polynomials** has been moved to chapter 3.
- The two sections on **multiplication and division of real numbers** have been combined.
- The section on **evaluating algebraic expressions** has been moved to chapter 1 to provide immediate use of order of operations. Variables and subscripts give students the feeling of the transition from arithmetic to algebra.
- The **operations of addition and subtraction of algebraic expressions** is covered at the beginning of chapter 2, and **multiplication and division of algebraic expressions** is covered in chapter 3. This separation provides time to solidify the procedures for each process without the possibility of confusing the two.
- A new section on representation of data shows students not only how to read and interpret different ways that data can be presented, but also how to construct circle graphs, bar graphs, and broken-line graphs.
- **New geometry coverage** is integrated throughout the text.
- The **concept of function**, and in particular linear functions, is introduced at the end of the chapter on linear equations in two variables.
- **Scientific calculator problems** have been increased.
- **Applications and word problems** have been expanded, improved, and updated.
- The **list of applications** can be found on page xiii. To illustrate the range of applied problems, the applications are listed by category. Page references are given to make locating particular applications easy.
- End-of-section **"Explore to Learn" problems** can be used as cooperative learning problems, for classroom discussion, or as extension problems. "Explore to Learn" problems have been carefully developed to support and extend the ideas presented in each section and enhance student understanding. Detailed suggestions for cooperative learning groups are available in *Cooperative Learning for the College Mathematics Classroom.*
- **Error analysis** has been expanded.
- **Separate chapter and cumulative tests** are at the end of each chapter.
- **Graphing calculator examples and exercises** have been included. A brief introduction to scientific and graphing calculators is found at the end of section 1–4. More comprehensive graphing calculator coverage is developed in the appendix.
- **Critical thinking has been expanded.** These problems can be used for cooperative learning, classroom discussion, or as extension problems. The critical thinking problems in each chapter are complimented by a series of hints in the *Instructor's Resource Manual.*
- The **Chapter Summaries** give students a comprehensive overview of each chapter. Each summary provides a convenient review of the chapter content through the glossary of terms, list of new notation, properties and definitions, and procedures.
- The **endsheet procedure lists** give students a quick guide to the location of all the procedures developed in the textbook.

Ancillaries to Help Guide You Through Elementary Algebra

The Wesner and Nustad ancillary package is part of a strongly integrated instructional and support package. New additions include the **Annotated Instructor's Edition**, text-specific tutorial software, and the **Discovering Mathematics** video series developed specifically for the Wesner and Nustad textbooks. Specifically created for *Elementary Algebra with Applications*, Fourth Edition, these supplements are carefully crafted to ensure the highest level of quality and integration.

The **Annotated Instructor's Edition** contains all of the content found in the student text with answers and teaching tips inserted for all exercise sets.

The *Instructor's Resource Manual* features reproducible chapter tests and section quizzes with answers keys. Suggested uses of the support material are also included, as well as a complete listing of mastery points (objectives) outlining the chapter, an organizational chart and time line, detailed critical thinking lessons, and guides with solutions for the critical thinking and Explore to Learn exercises.

WCB Testwriter software is text-specific, **algorithm-based** testing software for DOS, Windows, and Macintosh. Test items are coded to the Mastery Points in the text.

The *Cooperative Learning for the College Mathematics Classroom* guide provides detailed guidance for incorporating learning groups in your course. Reproducibles, forms, and activities related to *Elementary Algebra with Applications*, Fourth Edition, are included.

The **Discovering Mathematics** video series provides text-specific reinforcement of all major topics in each chapter of *Elementary Algebra with Applications*, Fourth Edition. Each concept is introduced with a real-world problem, followed by careful explanation and solved examples. These tapes are ideal for use in the math lab, remediation, or lecture enhancement.

Tutorial Software for IBM and Macintosh computers has been specifically developed for *Elementary Algebra with Applications*, Fourth Edition. This interactive software follows each section of the book, providing numerous examples, practice exercises, on-line testing, and diagnostic feedback. A content-sensitive help system provides students with suggestions based on the specific error. This software can also be used to monitor student progress and test scores.

An **audio study tape** by Terry Wesner provides practical ideas for improving study skills and performance in mathematics. It also introduces students to all of the features of the textbook, which will allow them to get the full benefit of the learning system and not view the textbook as just a set of exercises. It is provided *free* with the textbook.

Plotter Software is a graphing calculator emulator available in IBM and Macintosh formats. This menu-driven software graphs and analyzes functions and includes a manual describing operations and recommended student exercises.

The *Student's Solutions Manual* contains chapter introductions, solutions to every other odd-numbered problem in the section exercises with step-by-step explanations, hints and warnings, solutions to all odd-numbered chapter review problems, and self-tests at the end of each chapter. This manual is available for students to purchase.

Study Cards reinforce important concepts from the text—all in a flash card format. Students receive a "starter set" that gives them approximately fifty review cards for the first few chapters and teaches them how to create their own cards for the rest of the course. These are provided *free* with the textbook. The process of creating their own study cards also helps students understand and remember the concepts.

Acknowledgments

We wish to express our heartfelt thanks and grateful appreciation for the many comments and suggestions given to us during the preparation of the first edition. In particular, we wish to thank George Gullen III, Lynne Hensel, Lisa Miyazaki, Terry Baker, Harry Datsun, and Robert Olsen for their excellent effort in reviewing each stage of the book and supplying us with the numerous valuable comments, suggestions, and constructive criticisms.

Throughout the development, writing, and production of this text, two people have been of such great value that we are truly indebted to them for their excellent work on our behalf. We wish to express our utmost thanks to Theresa Grutz and Eugenia M. Collins.

We are most grateful for the hard work and dedication to this project from Philip Mahler and Suresh Ailawadi. They truly helped make this book possible.

We are grateful to our "book team," for without them there would be no book. In particular, we would like to express our sincere thanks to Earl McPeek, Gene Collins, Theresa Grutz, K. Wayne Harms, Keri Witman, and Pat Steele.

The following people deserve special thanks for their excellent work in the following areas: Signe Kastberg, Abraham Baldwin College, for *Cooperative Learning for the College Mathematics Classroom*; Suresh Ailawadi, Henry Ford Community College, *Student's Solutions Manual*; Jeanne M. Draper, Solano Community College, for the *Instructor's Resource Manual*; Vince McGarry, Austin Community

College, and Irene Doo, Austin Community College, for the *Instructor's Solutions Manual*; Pat Foard, South Plains College, for the Algebra Study Cards and Elementary Explore to Learn Tutorial Software; Cynthia Fleck, Wright State University, for contributions to the Annotated Instructor's Edition; Kathy Struve, Columbus State Community College, for contributions to Explore to Learn; Brenda Lackey, University of Tennessee–Martin, for graphing calculator accuracy check; William Radulovich, Florida Community College–Jacksonville, for the videos; Mary Wilber, Pasco-Hernando Community College, Dade City, Florida, for contributions to the exercises; and Pat Martin for supplying so many wonderful photographs.

A very special thanks goes out to the accuracy checkers. In particular, we wish to thank: Suresh Ailawadi, Henry Ford Community College; Oiyin Pauline Chow, Harrisburg Area Community College; Pat Foard, South Plains College; Shaun Kleitsch; Rita B. Sowell, Volunteer State Community College; Faye W. White, Canton College; Ruth Ann Henke, Southern Illinois University at Edwardsville; and Carroll Wells, Western Kentucky University.

We would like to express our appreciation for the valuable contributions to the fourth edition that were made by the following people: Bobby Avila, Mount San Jacinto College; Ondis Bible, Volunteer State Community College; Philip E. Buechner, Cowley County Community College; Susan J. Decker, Lackawanna Junior College; Diane K. Downie, Pierce College; Edward A. Gallo, Indiana University East; Kay Haralson, Austin Peay State University; Joe Howe, St. Charles County Community College; Joseph G. Karnowski, St. Louis Community College–Florrisant Valley; Patrick Keeler, Corning Community College; Rebecca Kitto, Antelope Valley College; Harvey W. Lambert, University of Nevada; Giles Wilson Maloof, Boise State University; Laurie K. McManus, St. Louis Community College; Beverly K. Michael, University of Pittsburgh; Katherine Muller, Cisco Junior College; Shai Neumann, Brevard Community College–Melbourne; A. J. Perrotto, El Paso Community College; Mary V. Portman, State University of New York–Cobleskill; Dr. Leon F. Sagan, Anne Arundel Community College; Patricia Stone, Tomball College; Diane Tischer, Metropolitan Community College; Wesley W. Tom, Chaffey College; Kevin Trutna, Arizona Western College; and George Witt, Glendale Community College.

We would like to thank the following reviewers of the Fourth Edition of *Elementary Algebra with Applications*. Their classroom experience and concern for their students positively influenced the development of this textbook.

Barbara Allen
Southeastern Louisiana University

James Blackburn
Tulsa Junior College

Dave Busekist
Southeastern Louisiana University

John Coburn
St. Louis Community College

Natalie Creed
Belmont Abbey College

John Garlow
Tarrant County Junior College

Peg Hovde
Grossmont College

Joyce Huntington
Walla Walla Community College

Judy Kasabian
El Camino College

Sharon Killian
Asheville Buncombe Technical Community College

Gail Kingrey
Pueblo Community College

Jill McKenney
Lane Community College

Ralph Merah
University of Louisville

Vicki Minor
Columbus State Community College

Michael Montemuro
West Chester University

Gerry Moultine
Northwood University

Joanne Peeples
El Paso Community College

Sarah Percy Jones
San Jacinto College North

Angela Peterson
Portland Community College

Richard Rockwell
Pacific Union College

Joel Spring
Broward Community College

Judith Stahl
Lake Sumter Community College

Ron Staszkow
Ohlone College

Allyson Stewart
Napa Valley College

Mary Jane Still
Palm Beach Community College

Katharine R. Struve
Columbus State Community College

Carole Sutphen
Muskegon Community College

Froy Tiscareno
Mt. San Antonio College

Rebecca Wong
West Valley College

Neil Aiken
Milwaukee Area Technical College

Joe Albree
Austin University at Montgomery

Ann Anderson
Broward Community College

Robert Baer
Miami University–Hamilton Branch

Pat Barbalich
Jefferson Community College

Charles Beals
Hartnell College

Don Bellairs
Grossmont College

Marybeth Beno
South Suburban College

John P. Bibbo
Southwestern College

Nancy Bray
San Diego Mesa College

Daniel Burns
Sierra College

Sharlene Cadwallader
Mount San Antonio College

P. M. Commons
Florida Junior College–South Campus

Ben Cornelius
Oregon Institute of Technology

Lena Dexter
Faulkner State Junior College

Louis Dyson
Clark College

Gail Earles
St. Cloud University

Pat Foard
South Plains College

Alice Grandgeorge
Manchester Community College

Michele Greenfield
Middlesex County College

George Gullen III
Henry Ford Community College

Ray Haertel
Central Oregon Community College

Pam Hager
College of the Sequoias

Harry Hayward
Westmoreland County Community College

Lynne Hensel
Henry Ford Community College

Angela Hernandez
University of Montevallo

Marty Hodges
Colorado Technical College

Tom Householder
Muskingham Area Technical College

Roe Hurst
Central Virginia Community College

Elizabeth Huttenlock
Pennsylvania State University

T. Henry Jablonski, Jr.
East Tennessee Sate University

Martha Jordan
Okaloosa-Walton Junior College

Glen Just
Mount St. Clare

Judy Kasabian
El Camino College

Margaret A. Kimbell
Texas State Technical Institute Waco Campus

Joanne F. Korsmo
New Mexico State University

Henry Kubo
West Los Angeles College

Theodore Lai
Hudson County Community College

Special thanks go to the reviewers of the study tips audiotape for their helpful comments. In particular, we wish to thank Roy D. Frisinger, *Harrisburg Area Community College;* John McIntosh, *St. Louis Community College at Meramac;* Richard D. Edie, *Southeastern Louisiana University;* David Pitrie, *Cypress College;* Mitzi Chaffer, *Central Michigan University;* James E. Coleman, *Baltimore City Community College;* Jennie Preston-Sabin, *Austin Peay State University;* Mary Jean Brod, *University of Montana;* Joe Karnowski, *St. Louis Community College–Florissant Valley;* Kathryn Pletsch, *Antelope Valley College;* and David Harvey, *University of Texas at El Paso.*

In addition, we would like to thank the reviewers of *Principles of Elementary Algebra with Applications,* First and Second Editions and *Elementary Algebra with Applications,* First, Second, and Third Editions, whose comments have positively influenced this edition.

Howard B. Lambert
East Texas State University

Calvin Latham
Monroe Community College

Wanda J. Long
*St. Charles County
Community College*

Jeri Vorwerk Love
Florida Junior College

Thomas McGannon
Chicago City College

Vincent McGarry
Austin Community College

Phil Mahler
*Middlesex Community
College*

J. Robert Malena
*Community College of
Allegheny County*

Gerald Marlette
*Cuyahoga Community
College*

Hank Martel
*Broward Community
College Samuels Campus*

Jerry J. Maxwell
Olney Central College

Donald Mazukelli
Los Angeles Valley College

Michael Montemuro
Westchester University

Robert Olsen
Dearborn Public Schools

Rita B. Sowell
*Volunteer State Community
College*

Gerry C. Vidrine
Louisiana State University

Keith L. Wilson
*Oklahoma City Community
College*

Kelly Wyatt
*Umpaqua Community
College*

Elementary Algebra
with Applications

CHAPTER

1

Operations with Real Numbers

While on a trip to Canada, Tonya heard on the radio that the temperature today will be 20° Celsius. Will she need her winter coat? What will the temperature be in degrees Fahrenheit? We can determine what 20° Celsius is in degrees Fahrenheit by the following formula.

$$F = \frac{9}{5} \cdot C + 32$$

1–1   Study Skills for Success in Algebra

Studying Mathematics

There are certain study skills that you as an algebra student need to have, or develop, to ensure your success in this course. Being successful in mathematics is different from being successful in other classes. In mathematics you not only need to learn and understand the material, but you must also be able to apply what you have learned.

In some other classes, not understanding the previous chapter's material may not affect your ability to understand the material in the present chapter, but in mathematics, not understanding the previous chapter will probably prevent you from learning new material. Mathematics constantly builds on previous material. Practice is the key. You must review and practice regularly.

You should not expect to understand every new topic the first time you see it. Understanding algebra takes time and effort. It requires that you attend all of your classes. You should prepare before going to class, take notes, and listen carefully in class. Read the textbook, do your homework, and ask questions about what you do not understand.

Put forth the necessary effort. You must be willing to give this course and yourself a fair chance. Always maintain a positive attitude.

The Classroom—Attendance, Preparation, and Note Taking

Attending every class is extremely important to your success in the course. The classroom is an essential ingredient in understanding the material and working the problems. If you have missed a class, make every effort to immediately get the assignment and notes from a classmate.

Before going to class, read the material to be covered. By reading the material before class, you will have seen the procedures that will be explained in class. You will have read definitions and properties that will be covered in the class. You will have introduced yourself to any new vocabulary. All of these things will make the classroom lecture more meaningful and easier to understand.

Where you sit in the classroom should be chosen with at least as much thought as choosing your seat at a movie. Get to the class early and choose a seat that allows you to clearly see the board and/or overhead projector and be seen by the instructor. The following diagram represents where you would want to sit in a classroom of four rows with ten seats per row.

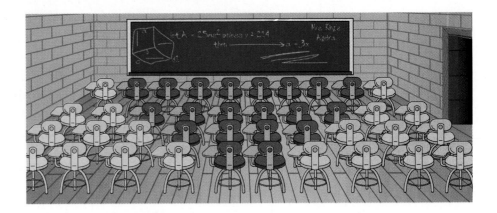

When you are taking notes in class, remember that notes are just notes. They do not have to be grammatically correct or in complete sentences. Get the ideas down and revise your notes as soon after class as possible. If you have read the material before class, you will be able to take more meaningful notes. You will have already read the definitions and properties and be familiar with any new vocabulary. You can watch and understand more of the presentation without feeling that you have to copy down every word.

A tape recorder can be a useful tool. (Always ask your instructor if you are permitted to use one.) Start with a tape fully rewound and the tape counter set to zero. If you do not understand something or if you missed something, write down the number on the tape counter and you will be able to quickly reference that part of the lecture. If you feel that you were able to take complete notes, you would not need to listen to the tape and it could be erased.

Listen for clues from the instructor about possible test questions. Statements such as "This is important" or "This is not in the textbook" indicate probable test questions. Keep a list of the probable test questions and make sure you can answer them on the next test.

Time Management

A general guideline in college is that students study and do homework for at least two hours for each hour of class time. The amount of time necessary to master the material is different for each person. If a course seems difficult (time consuming), you need to adjust the rest of your schedule to allow adequate time to study.

You should study as soon as possible after class, and you should study your most difficult course first. Make sure that you take a 10-minute break each hour, because learning decreases if you do not take any breaks.

You should make a weekly schedule that shows everything you normally do. Shown here is a possible chart. The schedule should show class time, travel time, sleep time, free time, etc. Start by filling in those things that you must do such as class time and work time. Study time should be next. Continue filling out the chart according to how important something is to you. Free time is what is left. A template of this chart is in the appendix for you to photocopy and use.

Make sure that your family and friends know your schedule and allow you to stick to it.

Time	SUN	MON	TUES	WED	THURS	FRI	SAT
6 am – 7 am							
7 am – 8 am							
8 am – 9 am							
9 am – 10 am							
10 am – 11 am							
11 am – 12 pm							
12 pm – 1 pm							
1 pm – 2 pm							
2 pm – 3 pm							
3 pm – 4 pm							
4 pm – 5 pm							
5 pm – 6 pm							
6 pm – 7 pm							
7 pm – 8 pm							
8 pm – 9 pm							
9 pm – 10 pm							
10 pm – 11 pm							
11 pm – 12 am							
12 am – 6 am							
Special Notes							

WEEKLY SCHEDULE

Doing Homework

You should have a quiet place to study. It should have good lighting, a good work area, and everything you need should be within reach.

Before you start the homework, you should review your notes, read the textbook, and work through the examples from class and the textbook. You will not save any time and will probably get off to a bad start if you jump into exercises before reviewing your notes and reading the textbook.

When you are reading the book, your lecture notes should supplement the textbook's explanations. Write notes while you are reading the material. Write all the important facts, definitions, properties, and some key examples on note cards. Anytime you have a few extra minutes, you can take these out and review. Highlight only the most important material for quick review later. Always look up unfamiliar words.

Read the book slowly and carefully. Pay attention to detail. Mathematics textbooks try to present the information with as few words as possible, therefore, careful reading is important. If you do not understand a paragraph or sentence, read it again.

Examples must be worked through step-by-step. You do not read an example, you work through it. Working the example on a separate piece of paper is helpful in that you can compare your solution step-by-step to the one in the book.

Always read the preface of a textbook. The different features of the textbook and how to best use the book are presented there.

When you get stuck on a problem, check to see if it is copied correctly. Look at your notes and the textbook for an example that is similar to the problem. If you still cannot work it, come back to it later. Sometimes working other problems will show you how to work a problem with which you were previously unsuccessful.

A study partner or a study group can make learning mathematics easier. You can discuss things you know and work together on things that you do not understand. In a study group, you can ask questions that you might feel uncomfortable asking in class.

This textbook comes with a large variety of supplements that provide extra help or additional exercises. Your should ask your instructor what extra materials are available and which ones might benefit you.

Taking an Exam

Start early, if you have an exam next week, start studying this week. Spread out the study time. Do not stay up late cramming the night before a test. You should get a good night's sleep. Do not try to learn any new material the day of the test. Avoid classmates who are cramming at the last minute because discussing the test with them may cause you to become uncertain about what you know. Just before the test, review your note cards of important facts and examples. Get to class early, sit in your usual seat, and relax for a few minutes until the test is handed out.

After you get your exam:

1. Write down useful formulas and facts on the test paper. By doing this you will not worry that you might not be able to remember them when you needed them. If you remember other facts as you are doing the test, write them down also.

2. Look the exam over to get a sense of the time you will need, and pace yourself to make sure you complete the test.

3. Read the directions carefully and understand how your answers are to be given.

4. Answer the questions that are easiest for you first. Then come back to the more difficult ones. Do not spend too much time on one problem.

5. Attempt every problem. Never leave a question blank. Try to identify the type of problem and what information is given. This may jog your memory enough to get you going on the problem. Make sure that you write down every step. You might get partial credit even though you did not get the correct answer.

6. Use the entire time given for the exam. After you have worked all the problems, check the answers.

7. If the time is running out, try not to get anxious. Take a few deep breaths and continue working at your usual pace. Finish what you can.

When you get the exam back, you should correct any errors. Because mathematics is cumulative, you must take time when you get the exam back to learn what you have missed before moving on to new material. It is a mistake to think that since the exam is over, you can forget about this material.

You must learn the material, know the vocabulary, and practice. This may seem like a lot of work for one course, and it is. You should give the course your best effort from the beginning. Anything less may result in a low grade or your repeating the course.

MASTERY POINTS

Can you
- Prepare for class?
- Take notes?
- Make a weekly schedule?
- Read the text book and understand all of its special features?
- Prepare for an exam?

Exercise 1–1

Directions   Answer the following questions. See section 1–1 and the preface.

1. If a course seems difficult, what should you do?

2. How many hours should you spend studying and doing homework for each hour of class time?

3. Fill out a copy of the weekly schedule template from the appendix.

4. What three things should you do before you start the homework?

5. How can using note cards be useful in learning mathematics?

6. When you are reading the textbook and come to an example, what should you do?

7. What area of the textbook describes the different features of the book and how to best use them?

8. What are some of the key things you should do when taking an exam?

9. What should you do when you get an exam back?

10. The following are some key features of this book. You should be able to describe what each of these are. If you are unclear about any of these features, you should read the preface again.

a. Chapter Lead-in Problem and Solution
b. Explanations
c. Examples
d. Quick Checks
e. Procedure Boxes
f. Definitions
g. Concepts
h. Notes
i. Problem Solving
j. Mastery Points
k. Section Exercises
l. Quick-Check Examples
m. Trial Problems
n. Core Exercise Problems
o. Section Review Exercises
p. Chapter Summary
q. Error Analysis
r. Critical Thinking
s. Chapter Review Exercises
t. Cumulative Test
u. Explore to Learn—cooperative learning problems
v. Functional use of color
w. Historical notes

1–2   Operations with Fractions

Fractions

In day-to-day living, the numbers we use most often are the whole numbers,

$$0, 1, 2, 3, 4, \text{ and so on}$$

and the fractions[1], such as

$$\frac{1}{2}, \frac{3}{4}, \frac{9}{10}, \text{ and so on}$$

In a fraction, the top number is called the **numerator** and the bottom number is called the **denominator**.

$$\frac{9}{10} \quad \begin{array}{l} \longleftarrow \text{ Numerator} \\ \longleftarrow \text{ Denominator} \end{array}$$

There are two types of fractions:

1. **Proper fractions** where the numerator is less than the denominator; for example, $\frac{9}{10}$.

2. **Improper fractions** where the numerator is greater than or equal to the denominator; for example, $\frac{10}{9}$ or $\frac{9}{9}$.

Prime Numbers and Factorization

Any whole number can be stated as a **product**[2] of two or more whole numbers, called **factors** of the number. For example,

---

[1] The notation $\frac{a}{b}$ to denote division was introduced about 1202 by the Arab author al-Hassar.

[2] The dot symbol · to denote multiplication was introduced by Thomas Harriot in his book *Artis Analyticae Praxis* (*Practice of the Analytic Art*) in 1631.

Product Factors	
$12 = 2 \cdot 6$	2 and 6 are factors
$12 = 1 \cdot 12$	1 and 12 are factors
$12 = 4 \cdot 3$	4 and 3 are factors
$12 = 2 \cdot 2 \cdot 3$	2, 2, and 3 are factors

Note The raised dot means multiply. Some other ways to indicate multiplication are the multiplication cross 2×6, or with no symbol between parentheses (2)(6), or with no symbol between a factor and a set of parentheses 2(6). Thus, $2 \cdot 6 = 2 \times 6 = (2)(6) = 2(6)$ all mean 2 times 6.

To factor a whole number is to write the number as a product of factors. In future work, it will be necessary to factor whole numbers such that the factors are **prime numbers**.

Learning System
"Notes" are used to highlight important ideas and facts, and to point out potential errors.

Prime Numbers

A prime number is any whole number greater than 1 whose only factors are the number itself and 1.

The first ten prime numbers are

2, 3, 5, 7, 11, 13, 17, 19, 23, and 29

Thus, when we factored 12 as $12 = 2 \cdot 2 \cdot 3$, the number was stated as a product of prime factors. 12 is called a **composite number**. Whole numbers greater than 1 that are not prime are called composite numbers.

● **Example 1–2 A**
Factoring numbers

Write each number as a product of prime factors.

1. 36

$2 \cdot 18$ Divide $36 \div 2 = 18$

$2 \cdot 2 \cdot 9$ Divide $18 \div 2 = 9$

$2 \cdot 2 \cdot 3 \cdot 3$ Divide $9 \div 3 = 3$

Thus $36 = 2 \cdot 2 \cdot 3 \cdot 3$

Note This could have been done in the following way.

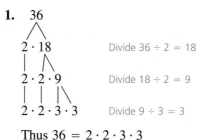

② | 36   Divide $36 \div 2 = 18$
② | 18   Divide $18 \div 2 = 9$
③ | 9   Divide $9 \div 3 = 3$
③

We successively divide by prime numbers, starting with 2 if possible, until the quotient is a prime number.

2.

$$
\begin{array}{c|l}
& 90 \\
② & 90 \\
③ & 45 \\
③ & 15 \\
⑤ &
\end{array}
\qquad
\begin{array}{l}
\text{Divide } 90 \div 2 = 45 \\
\text{Divide } 45 \div 3 = 15 \\
\text{Divide } 15 \div 3 = 5
\end{array}
$$

Thus $90 = 2 \cdot 3 \cdot 3 \cdot 5$

Reducing Fractions to Lowest Terms

A fraction is **reduced to lowest terms** when the only factor common to the numerator and the denominator is 1.

To Reduce a Fraction to Lowest Terms

1. Write the numerator and the denominator as a product of prime factors.
2. Divide the numerator and the denominator by all common prime factors. The product of all common prime factors is called the **greatest common factor, GCF.**

● **Example 1–2 B**

Reducing a fraction to lowest terms

Perform the indicated operations.

1. Reduce the following fraction to lowest terms.

$$
\begin{aligned}
\frac{14}{21} &= \frac{2 \cdot 7}{3 \cdot 7} && \text{Write as a product of prime factors} \\
&= \frac{2}{3} && \text{Divide numerator and denominator by common factor 7}
\end{aligned}
$$

$\dfrac{2}{3}$ is the answer since 2 and 3 have only 1 as a common factor.

2. In a class of 60 students, 15 students are absent. Describe the part of the class present as a reduced fraction.

$60 - 15 = 45$ students are present, so 45 out of 60, or $\dfrac{45}{60}$ of the class is present.

$$
\begin{aligned}
\frac{45}{60} &= \frac{3 \cdot 3 \cdot 5}{2 \cdot 2 \cdot 3 \cdot 5} && \text{Write as a product of prime factors} \\
&= \frac{3}{2 \cdot 2} && \text{Divide numerator and denominator by } 3 \cdot 5 \\
&= \frac{3}{4} && \text{Multiply in denominator}
\end{aligned}
$$

Thus we could say that three fourths of the class is present, or three out of every four students are present.

Note If you are able to determine the greatest common factor (GCF) by inspection, you can reduce the fraction to lowest terms by dividing the numerator and the denominator by the GCF. In example 1–2 B number 2 we could divide by 15, the greatest common factor, instead of $3 \cdot 5$.

⚐ *Quick check*   Reduce $\frac{25}{45}$ to lowest terms.

Products and Quotients of Fractions

So that the discussion can be more general, we will now introduce the concept of a **variable. A variable is a symbol (generally a lowercase letter) that represents an unspecified number.** A variable holds a position for a number.

Multiplication of Fractions

If $\frac{a}{b}$ and $\frac{c}{d}$ are fractions, then

$$\frac{a}{b} \cdot \frac{c}{d} = \frac{a \cdot c}{b \cdot d}$$

To multiply two or more fractions, we use the following procedure.

To Multiply Fractions

1. Write the numerator and the denominator as an indicated product (do not multiply).
2. Reduce the resulting fraction to lowest terms.
3. Multiply the remaining factors.

● **Example 1–2 C**
Multiplying fractions

Perform the indicated operations.

1. Multiply the following fractions and reduce to lowest terms.

$$\frac{2}{3} \cdot \frac{5}{7} = \frac{2 \cdot 5}{3 \cdot 7} \qquad \text{Multiply numerators}$$
$$\text{Multiply denominators}$$
$$= \frac{10}{21} \qquad \text{Perform multiplications}$$

2. A forester estimates that five out of every six trees in a wooded area are hardwood trees, and that three fourths of these hardwood trees are oak trees. Describe the number of oak trees in this wooded area as a fraction of the total number of trees.

   Five sixths, $\left(\frac{5}{6}\right)$ of the trees are hardwood. We want three fourths $\left(\frac{3}{4}\right)$ of this number. When discussing fractions the word "of" almost always means multiplication. We compute $\frac{5}{6} \cdot \frac{3}{4}$.

$$\frac{5}{6} \cdot \frac{3}{4} = \frac{5 \cdot 3}{6 \cdot 4} \qquad \text{Multiply numerators}$$
$$\text{Multiply denominators}$$
$$= \frac{5 \cdot 3}{2 \cdot 3 \cdot 2 \cdot 2} \qquad \text{Factor } 6 = 2 \cdot 3,\ 4 = 2 \cdot 2$$
$$= \frac{5}{2 \cdot 2 \cdot 2} \qquad \text{Divide numerator and denominator by 3}$$
$$= \frac{5}{8} \qquad \text{Multiply in the denominator}$$

Learning System
"Quick Check" problems are worked step-by-step within the exercise set. You should check your solution of this problem line-by-line and use this solution as a quick reference while doing the problems within the exercise set.

Thus, five eighths of all the trees, or five out of every eight trees, are oak trees. ⬤

Suppose we multiply the fractions

$$\frac{5}{6} \cdot \frac{6}{5} = \frac{5 \cdot 6}{6 \cdot 5}$$
$$= \frac{30}{30}$$
$$= 1$$

When the product of two numbers is 1, we call each number the **reciprocal** of the other number. Thus

$$\frac{5}{6} \text{ and } \frac{6}{5} \text{ are reciprocals,}$$

$$\frac{2}{7} \text{ and } \frac{7}{2} \text{ are reciprocals,}$$

$$\frac{14}{13} \text{ and } \frac{13}{14} \text{ are reciprocals,}$$

$$\frac{1}{2} \text{ and } 2 \text{ are reciprocals.}$$

We can see that the reciprocal of any fraction is obtained by interchanging the numerator and the denominator. The reciprocal of a fraction is used to divide fractions.

Division of Fractions

If $\frac{a}{b}$ and $\frac{c}{d}$ are fractions, then

$$\frac{a}{b} \div \frac{c}{d} = \frac{a}{b} \cdot \frac{d}{c} = \frac{a \cdot d}{b \cdot c}$$

To Divide Two Fractions

1. Multiply the first fraction by the **reciprocal** of the second fraction.
2. Reduce the resulting product to lowest terms.

⬤ **Example 1–2 D**

Dividing fractions and mixed numbers

Divide[3] the following fractions and reduce to lowest terms.

1. $\dfrac{7}{8} \div \dfrac{6}{7} = \dfrac{7}{8} \cdot \dfrac{7}{6}$ Multiply by the reciprocal of $\frac{6}{7}$

$= \dfrac{7 \cdot 7}{8 \cdot 6}$ Multiply numerators
   Multiply denominators

$= \dfrac{49}{48}$

---

[3]The ÷ symbol was introduced in 1659 by the Swiss mathematician Johann Heinrich Rahn in his book *Teutsche Algebra*.

<u>Note</u> The improper fraction $\frac{49}{48}$ can be written as the **mixed number** $1\frac{1}{48}$, which is the sum of a whole number and a proper fraction. This is obtained by dividing the numerator by the denominator.

$$
\begin{array}{r}
1 \\
48\overline{)49} \\
\underline{48} \\
1
\end{array}
\quad = 1\frac{1}{48}
$$

Quotient ←
Remainder ←
Original denominator ←
Remainder

2. 12 ounces of uranium is to be divided into equal parts, each part weighing $\frac{1}{20}$ pound. How many of these parts can be made?

There are 16 ounces in a pound, so there are $\frac{12}{16}$ pound of uranium to be divided into parts that are $\frac{1}{20}$ pound each. We want to divide the total amount into equal parts, so use division.

$$\frac{\frac{12}{16}}{\frac{1}{20}} = \frac{12}{16} \div \frac{1}{20} = \frac{12}{16} \cdot \frac{20}{1}$$ Multiply by the reciprocal of $\frac{1}{20}$

$$\frac{2 \cdot 2 \cdot 3 \cdot 2 \cdot 5}{2 \cdot 2 \cdot 2 \cdot 1}$$ Factor the numerators
Factor the denominators

$$\frac{3 \cdot 5}{1} = \frac{15}{1}$$ Reduce the resulting fraction and multiply the remaining factors

$$= 15$$ Simplify

Thus there will be 15 parts, each weighing $\frac{1}{20}$ pound.

The improper fraction answer is usually the one preferred in algebra. The mixed number form is usually preferred in an application problem.

3. $3\frac{1}{4} \div 5\frac{2}{3}$

We change the mixed numbers to improper fractions.

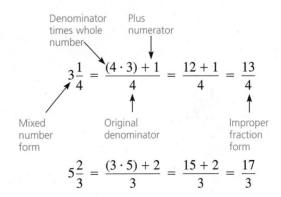

$$3\frac{1}{4} = \frac{(4 \cdot 3) + 1}{4} = \frac{12 + 1}{4} = \frac{13}{4}$$

Denominator times whole number
Plus numerator
Mixed number form
Original denominator
Improper fraction form

$$5\frac{2}{3} = \frac{(3 \cdot 5) + 2}{3} = \frac{15 + 2}{3} = \frac{17}{3}$$

Learning System
Examples have short phrase statements next to most steps stating exactly what has been done. These will enable you to develop a clear understanding of how a problem is worked without having to guess what went on in a particular step.

We now divide as indicated.

$$3\frac{1}{4} \div 5\frac{2}{3} = \frac{13}{4} \div \frac{17}{3}$$

$$= \frac{13}{4} \cdot \frac{3}{17} \qquad \text{Multiply by the reciprocal of } \frac{17}{3}$$

$$= \frac{13 \cdot 3}{4 \cdot 17} \qquad \text{Multiply numerators}$$
$$\qquad\qquad\quad \text{Multiply denominators}$$

$$= \frac{39}{68} \qquad \text{Perform indicated operations}$$

Quick check Divide $\dfrac{\frac{5}{6}}{\frac{5}{8}}$ and reduce to lowest terms.

4. The area of a rectangle is found by multiplying the length of the rectangle by the width of the rectangle. Find the area of a rectangle that is $2\frac{1}{2}$ feet long and $1\frac{5}{6}$ feet wide.

$$\text{Area} = 2\frac{1}{2} \cdot 1\frac{5}{6} \qquad \text{Multiply the given dimensions}$$

$$= \frac{5}{2} \cdot \frac{11}{6} \qquad \text{Change mixed numbers to improper fractions}$$

$$= \frac{5 \cdot 11}{2 \cdot 6} \qquad \text{Multiply numerators}$$
$$\qquad\qquad\quad \text{Multiply denominators}$$

$$= \frac{55}{12} \text{ or } 4\frac{7}{12} \qquad \text{Perform indicated operations}$$

The area of the rectangle is $4\frac{7}{12}$ square feet.

Sums and Differences of Fractions

To add or subtract fractions, the fractions must have a *common* (same) *denominator.*

To Add or Subtract Fractions with Common Denominators

1. Add or subtract the numerators.
2. Place the sum or difference over the common denominator.
3. Reduce the resulting fraction to lowest terms.

● **Example 1–2 E**

Addition and subtraction of fractions with common denominators

Add or subtract[4] the following fractions as indicated. Reduce to lowest terms.

1. $\dfrac{3}{8} + \dfrac{1}{8} = \dfrac{3+1}{8}$ Add numerators

$\phantom{\dfrac{3}{8} + \dfrac{1}{8}} = \dfrac{4}{8}$ Combine in numerator

$\phantom{\dfrac{3}{8} + \dfrac{1}{8}} = \dfrac{1}{2}$ Reduce to lowest terms

2. $\dfrac{7}{16} - \dfrac{5}{16} = \dfrac{7-5}{16}$ Subtract numerators

$\phantom{\dfrac{7}{16} - \dfrac{5}{16}} = \dfrac{2}{16}$ Combine in numerator

$\phantom{\dfrac{7}{16} - \dfrac{5}{16}} = \dfrac{1}{8}$ Reduce to lowest terms

When the fractions have different denominators, we must rewrite all of the fractions with a new common denominator. Many numbers can satisfy the condition for any set of denominators, but we want the *least* of these numbers, called the **least common denominator** (denoted by LCD). For example, 24 is the least common denominator of the fractions

$$\frac{7}{8} \text{ and } \frac{5}{6}$$

since 24 is the least (smallest) number that can be divided by 6 and 8 exactly. The procedure for finding the LCD is outlined next.

To Find the Least Common Denominator (LCD)

1. Express each denominator as a product of prime factors.
2. List all the *different* prime factors.
3. Write each prime factor the *greatest* number of times it appears in any of the prime factorizations in step 1.
4. The least common denominator is the product of all factors from step 3.

● **Example 1–2 F**

Finding the LCD

Find the least common denominator (LCD) of the fractions with the following denominators.

1. 24 and 18

   a. Write 24 and 18 as products of prime factors.

$$24 = 2 \cdot 2 \cdot 2 \cdot 3$$
$$18 = 2 \cdot 3 \cdot 3$$

   b. The different prime factors are 2 and 3.

[4]The symbols + and − to denote addition and subtraction were first found in manuscripts in Germany during the late fifteenth century.

c. 2 is a factor three times in 24 and 3 is a factor two times in 18 (the greatest number of times).
d. The LCD is $2 \cdot 2 \cdot 2 \cdot 3 \cdot 3 = 72$.

2. 6, 8, and 14

a. $6 = 2 \cdot 3$
   $8 = 2 \cdot 2 \cdot 2$
   $14 = 2 \cdot 7$
b. The different prime factors are 2, 3, and 7.
c. 2 is a factor three times in 8, 3 is a factor once in 6, and 7 is a factor once in 14.
d. The LCD is $2 \cdot 2 \cdot 2 \cdot 3 \cdot 7 = 168$.

☛ *Quick check*   Find the LCD for fractions with denominators of 6, 9, and 12.

Building Fractions

To write the fraction $\dfrac{5}{6}$ as an equivalent fraction with new denominator 24, we find the number that is multiplied by 6 to get 24. Since

$$6 \cdot 4 = 24$$

we use the factor 4. Now multiply the given fraction $\dfrac{5}{6}$ by the fraction $\dfrac{4}{4}$. The fraction $\dfrac{4}{4}$ is equal to 1 and is called a **unit fraction**. Multiplying by the unit fraction $\dfrac{4}{4}$ will not change the value of $\dfrac{5}{6}$, only its form. Thus

$$\frac{5}{6} = \frac{5}{6} \cdot \frac{4}{4} = \frac{5 \cdot 4}{6 \cdot 4} = \frac{20}{24}$$

————— Multiplication by 1

We use the following procedure to write equivalent fractions.

To Find Equivalent Fractions

1. Divide the original denominator into the new denominator.
2. Multiply the numerator and the denominator of the given fraction by the number obtained in step 1.

● **Example 1–2 G**
Changing the denominator

Write equivalent fractions having the new denominator.

1. $\dfrac{3}{5} = \dfrac{?}{30}$

Since $30 \div 5 = 6$, multiply $\dfrac{3}{5}$ by $\dfrac{6}{6}$.

$$\dfrac{3}{5} = \dfrac{3}{5} \cdot \dfrac{6}{6} \qquad \text{Multiply by } \dfrac{6}{6}$$

$$= \dfrac{3 \cdot 6}{5 \cdot 6} \qquad \begin{array}{l}\text{Multiply numerators}\\ \text{Multiply denominators}\end{array}$$

$$= \dfrac{18}{30}$$

2. $\dfrac{7}{9} = \dfrac{?}{72}$

Since $72 \div 9 = 8$, multiply $\dfrac{7}{9}$ by $\dfrac{8}{8}$.

$$\dfrac{7}{9} = \dfrac{7}{9} \cdot \dfrac{8}{8} \qquad \text{Multiply by } \dfrac{8}{8}$$

$$= \dfrac{7 \cdot 8}{9 \cdot 8} \qquad \begin{array}{l}\text{Multiply numerators}\\ \text{Multiply denominators}\end{array}$$

$$= \dfrac{56}{72}$$

To add or subtract fractions having different denominators, we use the following procedure.

> ### To Add or Subtract Fractions Having Different Denominators
>
> 1. Find the LCD of the fractions.
> 2. Write each fraction as an equivalent fraction with the LCD as the new denominator.
> 3. Perform the addition or subtraction.
> 4. Reduce the resulting fraction to lowest terms.

● **Example 1–2 H**

Addition and subtraction of fractions and mixed numbers with different denominators

Add or subtract the following fractions as indicated. Reduce the resulting fraction to lowest terms.

1. $\dfrac{7}{8} + \dfrac{5}{6}$

　　a. The LCD of the fractions is 24.
　　b. Since $24 \div 8 = 3$, then

$$\dfrac{7}{8} = \dfrac{7}{8} \cdot \dfrac{3}{3} = \dfrac{7 \cdot 3}{8 \cdot 3} = \dfrac{21}{24} \qquad \text{Multiply by } \dfrac{3}{3}$$

and since $24 \div 6 = 4$, then

$$\dfrac{5}{6} = \dfrac{5}{6} \cdot \dfrac{4}{4} = \dfrac{5 \cdot 4}{6 \cdot 4} = \dfrac{20}{24} \qquad \text{Multiply by } \dfrac{4}{4}$$

c. $\dfrac{7}{8} + \dfrac{5}{6} = \dfrac{21}{24} + \dfrac{20}{24}$ Add fractions with LCD

$= \dfrac{21 + 20}{24}$ Add numerators

$= \dfrac{41}{24} \text{ or } 1\dfrac{17}{24}$

2. $\dfrac{7}{8} - \dfrac{1}{3}$

a. The LCD of the fractions is 24.

b. Since $24 \div 8 = 3$, then

$\dfrac{7}{8} = \dfrac{7}{8} \cdot \dfrac{3}{3} = \dfrac{21}{24}$ Multiply by $\dfrac{3}{3}$

Since $24 \div 3 = 8$, then

$\dfrac{1}{3} = \dfrac{1}{3} \cdot \dfrac{8}{8} = \dfrac{8}{24}$ Multiply by $\dfrac{8}{8}$

c. $\dfrac{7}{8} - \dfrac{1}{3} = \dfrac{21}{24} - \dfrac{8}{24}$ Subtract fractions with LCD

$= \dfrac{21 - 8}{24} = \dfrac{13}{24}$ Subtract numerators

3. A metal bar was $3\dfrac{7}{8}$ feet long before $2\dfrac{3}{4}$ feet was cut off of it. Ignoring the thickness of the cut, how long is the remaining part of the bar?

$$3\dfrac{7}{8} - 2\dfrac{3}{4}$$

Change each of the mixed numbers to an improper fraction.

$$3\dfrac{7}{8} = \dfrac{(8 \cdot 3) + 7}{8} = \dfrac{31}{8}; \qquad 2\dfrac{3}{4} = \dfrac{(4 \cdot 2) + 3}{4} = \dfrac{11}{4}$$

a. The LCD of the fractions is 8.

b. $\dfrac{31}{8}$ already has the LCD in its denominator.

Since $8 \div 4 = 2$, then

$\dfrac{11}{4} = \dfrac{11 \cdot 2}{4 \cdot 2} = \dfrac{22}{8}$ Multiply by $\dfrac{2}{2}$

c. $3\dfrac{7}{8} - 2\dfrac{3}{4} = \dfrac{31}{8} - \dfrac{11}{4}$ Subtract improper fractions

$= \dfrac{31}{8} - \dfrac{22}{8}$ Subtract fractions with LCD

$= \dfrac{31 - 22}{8}$ Subtract numerators

$= \dfrac{9}{8} \text{ or } 1\dfrac{1}{8}$

Thus the remaining part is $1\dfrac{1}{8}$ feet long.

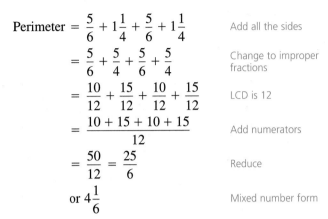

4. The perimeter (distance around) of a rectangle is found by *adding* the four sides of the rectangle. Find the perimeter of a rectangle that is $1\frac{1}{4}$ yards long and $\frac{5}{6}$ yard wide.

$$
\begin{aligned}
\text{Perimeter} &= \frac{5}{6} + 1\frac{1}{4} + \frac{5}{6} + 1\frac{1}{4} && \text{Add all the sides} \\
&= \frac{5}{6} + \frac{5}{4} + \frac{5}{6} + \frac{5}{4} && \text{Change to improper fractions} \\
&= \frac{10}{12} + \frac{15}{12} + \frac{10}{12} + \frac{15}{12} && \text{LCD is 12} \\
&= \frac{10 + 15 + 10 + 15}{12} && \text{Add numerators} \\
&= \frac{50}{12} = \frac{25}{6} && \text{Reduce} \\
&\text{or } 4\frac{1}{6} && \text{Mixed number form}
\end{aligned}
$$

The perimeter of the rectangle is $4\frac{1}{6}$ yards.

⚑ *Quick check*   Subtract and reduce to lowest terms $4\frac{1}{2} - 2\frac{3}{4}$.

Learning System
"The mastery points" form a checklist of the skills you must master from this section.

MASTERY POINTS

Can you
- Reduce a fraction to lowest terms?
- Multiply and divide fractions and mixed numbers?
- Find the least common denominator (LCD) of two or more fractions?
- Add and subtract fractions and mixed numbers?

Exercise 1–2

Directions
Reduce the following fractions to lowest terms. See example 1–2 B.

⚑ *Quick Check*

Example $\dfrac{25}{45}$

Solution
$$
\begin{aligned}
\frac{25}{45} &= \frac{5 \cdot 5}{3 \cdot 3 \cdot 5} && \text{Factor numerator} \\
&&& \text{Factor denominator} \\
&= \frac{5}{3 \cdot 3} && \text{Divide numerator and denominator by 5} \\
&= \frac{5}{9} && \text{Multiply in denominator}
\end{aligned}
$$

Learning System
"Quick Check examples" are a step-by-step solution of the problems that follow each set of examples within the text.

Directions   Reduce the following fractions to lowest terms. See example 1–2 B.

1. $\dfrac{4}{8}$ 　　　　　**2.** $\dfrac{3}{9}$ 　　　　　**3.** $\dfrac{10}{12}$ 　　　　　**4.** $\dfrac{8}{14}$

5. $\dfrac{16}{18}$

6. $\dfrac{14}{21}$

7. $\dfrac{28}{36}$

8. $\dfrac{50}{75}$

9. $\dfrac{64}{32}$

10. $\dfrac{96}{48}$

11. $\dfrac{100}{85}$

12. $\dfrac{120}{84}$

Directions
Multiply or divide as indicated. Reduce to lowest terms.
See examples 1–2 C and D.

🚩 *Quick Check*

Example $\dfrac{\frac{5}{6}}{\frac{5}{8}}$

Solution $\dfrac{\frac{5}{6}}{\frac{5}{8}} = \dfrac{5}{6} \div \dfrac{5}{8} = \dfrac{5}{6} \cdot \dfrac{8}{5}$ Multiply by the reciprocal of $\frac{5}{8}$

$= \dfrac{5 \cdot 8}{6 \cdot 5}$ Multiply numerators
Multiply denominators

$= \dfrac{8}{6}$ Reduce by common factor 5

$= \dfrac{2 \cdot 2 \cdot 2}{2 \cdot 3}$ Write numerator and denominator as the product of prime factors

$= \dfrac{2 \cdot 2}{3}$ Reduce by common factor 2

$= \dfrac{4}{3}$ or $1\dfrac{1}{3}$ Multiply in numerator

Directions Multiply or divide as indicated. Reduce to lowest terms. See examples 1–2 C and D.

13. $\dfrac{5}{6} \cdot \dfrac{3}{5}$

14. $\dfrac{2}{3} \cdot \dfrac{5}{6}$

15. $\dfrac{7}{8} \cdot \dfrac{7}{12}$

16. $\dfrac{7}{5} \cdot \dfrac{3}{2}$

17. $\dfrac{7}{9} \cdot \dfrac{3}{4}$

18. $\dfrac{3}{4} \cdot 6$

19. $\dfrac{3}{7} \div \dfrac{4}{5}$

20. $\dfrac{12}{25} \div \dfrac{8}{15}$

21. $\dfrac{6}{7} \div 3$

22. $4 \div \dfrac{3}{8}$

23. $\dfrac{15}{17} \div \dfrac{3}{5}$

24. $4 \div \dfrac{7}{2}$

25. $17 \div 2\dfrac{1}{3}$

26. $12 \cdot 1\dfrac{5}{6}$

27. $7\frac{1}{3} \cdot 2\frac{4}{7}$

28. $1\frac{1}{5} \cdot 2\frac{1}{2}$

29. $4\frac{4}{5} \cdot 2\frac{1}{2}$

30. $7\frac{1}{2} \div 5\frac{1}{4}$

31. $\dfrac{\frac{8}{2}}{3}$

32. $\dfrac{17}{\frac{3}{4}}$

33. $\dfrac{\frac{7}{8}}{\frac{4}{3}}$

34. $\dfrac{15}{64} \div \dfrac{45}{8}$

35. $\dfrac{4}{5} \cdot \dfrac{2}{3} \cdot \dfrac{3}{8}$

36. $\dfrac{9}{8} \cdot \dfrac{2}{3} \cdot \dfrac{3}{8}$

37. $\dfrac{8}{3} \cdot \dfrac{4}{7}$

38. $\dfrac{8}{3} \div \dfrac{15}{14}$

39. $\dfrac{8}{3} \cdot \dfrac{15}{14}$

Directions   Work the following problems. See example 1–2 D numbers 2 and 4.

40. What is the total length of 25 pieces of steel, each $5\frac{1}{2}$ inches long?

41. The volume of a rectangular block is found by multiplying the length times the width times the height.

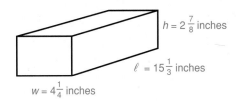

$h = 2\frac{7}{8}$ inches

$\ell = 15\frac{1}{3}$ inches

$w = 4\frac{1}{4}$ inches

a. What is the volume in cubic inches of a rectangular block of wood $15\frac{1}{3}$ inches long, $4\frac{1}{4}$ inches wide, and $2\frac{7}{8}$ inches high?

b. What is the volume in cubic inches of a block of steel $8\frac{1}{2}$ inches long, $2\frac{1}{8}$ inches wide, and $1\frac{3}{4}$ inches high?

42. A wire $61\frac{1}{2}$ inches long is divided into 14 equal parts. What is the length of each part?

Directions Find the LCD of the fractions with the following groups of denominators. See example 1–2 F.	☛ *Quick Check*
	Example 6, 9, and 12
	Solution **1.** Write 6, 9, and 12 as a product of prime factors.
	$$6 = 2 \cdot 3$$ $$9 = 3 \cdot 3$$ $$12 = 2 \cdot 2 \cdot 3$$
	2. The different prime factors are 2 and 3.
	3. 2 is a factor twice in 12 and 3 is a factor twice in 9.
	4. The LCD is $2 \cdot 2 \cdot 3 \cdot 3 = 36$.

Directions   Find the LCD of the fractions with the following groups of denominators. See example 1–2 F.

43. 3, 8, 10

44. 9, 15, 21

45. 6, 14, 18

46. 5, 10, 12

47. 16, 24, 36

48. 12, 16, 24

49. 5, 7, 11

50. 10, 20, 30

51. 9, 12, 54

52. 10, 14, 18

53. 10, 15, 20

54. 10, 15, 24

Directions	☛ *Quick Check*
Add or subtract as indicated. Reduce to lowest terms. See examples 1–2 E, G, and H.	**Example** $4\frac{1}{2} - 2\frac{3}{4}$

Solution We first change the mixed numbers to improper fractions.

$$4\frac{1}{2} = \frac{(4 \cdot 2) + 1}{2} = \frac{9}{2}; \quad 2\frac{3}{4} = \frac{(4 \cdot 2) + 3}{4} = \frac{11}{4}$$

$$4\frac{1}{2} - 2\frac{3}{4} = \frac{9}{2} - \frac{11}{4} \qquad \text{Replace mixed numbers with improper fractions}$$

The LCD is 4.

$$= \frac{18}{4} - \frac{11}{4} \qquad \text{Write } \frac{9}{2} \text{ as } \frac{18}{4}$$

$$= \frac{18 - 11}{4} \qquad \text{Subtract numerators}$$

$$= \frac{7}{4} \text{ or } 1\frac{3}{4}$$

Directions   Add or subtract as indicated. Reduce to lowest terms. See examples 1–2 E, G, and H.

55. $\dfrac{1}{3} + \dfrac{1}{3}$

56. $\dfrac{2}{5} + \dfrac{3}{10}$

57. $\dfrac{1}{3} + \dfrac{1}{4}$

58. $\dfrac{5}{6} - \dfrac{1}{6}$

59. $\dfrac{4}{5} - \dfrac{2}{10}$

60. $\dfrac{5}{6} - \dfrac{3}{8}$

61. $1 + \dfrac{5}{8}$

62. $3 + \dfrac{5}{6}$

63. $4 - \dfrac{3}{5}$

64. $\dfrac{2}{3} + \dfrac{3}{4}$

65. $\dfrac{3}{5} + \dfrac{7}{15}$

66. $\dfrac{5}{6} - \dfrac{1}{3}$

67. $\dfrac{3}{8} - \dfrac{1}{12}$

68. $\dfrac{7}{24} - \dfrac{3}{16}$

69. $\dfrac{7}{54} + \dfrac{19}{45}$

70. $\dfrac{1}{2} + \dfrac{1}{5} + \dfrac{1}{10}$

71. $\dfrac{7}{15} + \dfrac{5}{6} - \dfrac{3}{4}$

72. $\dfrac{9}{16} + \dfrac{5}{18} - \dfrac{2}{15}$

73. $8\dfrac{3}{16} - 4\dfrac{5}{8}$

74. $7\dfrac{1}{2} + 2\dfrac{3}{4}$

75. $\dfrac{2}{7} + \dfrac{2}{3} + \dfrac{5}{7}$

Directions   Work the following problems. See example 1–2 H numbers 3 and 4.

76. Althea owed Jasmine some money. If she paid Jasmine $\dfrac{1}{4}$ of the original debt on June 15, $\dfrac{1}{3}$ of the original debt on July 1, and $\dfrac{3}{8}$ of the original debt on August 10, how much of her original debt had Althea paid by August 10?

77. A flower garden in the form of a rectangle has two sides that are $24\dfrac{1}{2}$ feet long and two sides that are $18\dfrac{3}{4}$ feet long. Find the perimeter (total distance around) of the rectangle.

78. On a given day, Alicia purchased $\frac{5}{6}$ yard of one material, $\frac{3}{4}$ yard of another material, and $\frac{2}{3}$ yard of a third material. How many yards of material did she purchase altogether?

79. Butcher Angelo has $32\frac{1}{4}$ pounds of pork chops. If he sells $21\frac{1}{3}$ pounds of the pork chops on a given day, how many pounds of pork chops does he have left?

80. A machinist has a piece of steel stock that weighs $12\frac{7}{8}$ ounces. If he cuts off $5\frac{1}{5}$ ounces, how many ounces does he have left?

81. Five eighths of the houses in a certain town are painted white. Of these, three fourths are more than one story high. Describe the part of the houses in the town that are white and more than one story high as a fraction.

82. In a certain city 16 out of every 22 streets are named after a famous person. Of these people it is estimated that three fourths lived before the year 1940. Describe the portion of the streets in this town that are named after famous people who lived *after* the year 1940.

83. A garden is $42\frac{1}{2}$ feet long. Plants are to be planted in a row spaced $1\frac{1}{4}$ feet apart, beginning at one end. How many plants will be needed?

84. A garden is 41 feet long. Plants are to be planted in a row equally spaced $1\frac{1}{4}$ feet apart from each other, with an equal distance from the plants to each end. As many plants as possible are to be planted. How many plants can be planted, and how far in from each end will they be placed?

85. The instructions on a do-it-yourself table kit direct that a hole be drilled $2\frac{1}{2}'' \pm \frac{1}{8}''$ from a certain edge. (The symbol $\pm$ means plus or minus.) Compute the largest and smallest distance from the edge that is described by the measurements.

86. In the stock market, prices are quoted as multiples of $\frac{1}{8}$ of a dollar. Suppose a stock is $\$37\frac{1}{4}$ on Monday. On Tuesday the price goes down $1\frac{1}{8}$. On Wednesday it goes up $\frac{5}{8}$. On Thursday it goes up $1\frac{1}{4}$, and on Friday it goes down $\frac{1}{2}$. What is the price of the stock when the market closes on Friday?

⚜ Explore to Learn

1. You are a parent driving your child to school, when your child announces there is a test today on reducing fractions, and that your child doesn't know how to do this. The child says that the fractions will be like "ten over twenty five," "eighteen over fifty," and "three over nine." Use complete sentences to describe a way to do these kinds of problems. Remember, you're driving to school, so you can't make any drawings or actually write any fractions. Only write what you could say while (safely) driving an automobile.

2. If the test were going to cover finding the least common denominator, use complete sentences to describe how to add one fourth and one sixth.

1–3 Operations with Decimals and Percents

In section 1–2, we studied fractions. A **decimal number** is a special fraction with a denominator that is 10, 100, 1,000, and so on.

In a number such as 23, the digits 2 and 3 have place value as follows:

$$23 = (2 \cdot 10) + (3 \cdot 1)$$

Now consider the same two digits with a dot, called the **decimal point**, in front of them, .23. We call this number a **decimal fraction**, or just plain **decimal**. It is standard procedure to place a zero to the left of the decimal point if the decimal number is less than 1. The zero helps the reader see the decimal point and emphasizes the fact that we are dealing with a decimal fraction. The decimal point is placed so that the number of digits to the right of it indicates the number of zeros in the denominator of the fraction. If there is one digit to the right of the decimal point, the denominator is 10, read "tenths." If there are two digits to the right of the decimal point, the denominator is 100, read "hundredths." If three digits are to the right of the decimal point, the denominator is 1,000, read "thousandths." And so on.

To read a decimal fraction: Read the whole number (if any); next read "and" for the decimal point. Then read the portion after the decimal point as a whole number. Finally, read the name of the decimal place of the last digit on the right.

For example, 0.27, which is read "twenty-seven *hundredths*," is written

$$0.27 = \frac{27}{100} \leftarrow \text{Hundred}$$

while 0.149, which is read "one hundred forty-nine *thousandths*," is written

$$0.149 = \frac{149}{1,000} \longleftarrow \text{Thousand}$$

The last digit in the number is the key to the denominator of the fraction.

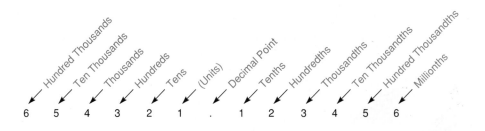

● Example 1–3 A

Changing a decimal to a fraction

Write the following decimal numbers as fractions reduced to lowest terms.

1. 0.57 (read "fifty-seven *hundredths*")

$$0.57 = \frac{57}{100} \longleftarrow \text{Hundred}$$

2. 0.1234 (read "one thousand two hundred thirty-four *ten thousandths*")

$$0.1234 = \frac{1,234}{10,000} \longleftarrow \text{Ten thousand}$$

$$= \frac{617}{5,000} \qquad \text{Reduce to lowest terms}$$

<u>Note</u> A decimal number that is written as a fraction will reduce *only if* the numerator is divisible by 2 or 5. This was the case in example 2.

☞ *Quick check*   Write 0.42 as a fraction reduced to lowest terms. ●

Addition and Subtraction of Decimal Numbers

To add or subtract decimal numbers, we place the numbers under one another so that the decimal points line up vertically and then proceed as in adding or subtracting whole numbers. The decimal point will appear in the answer directly below where it is lined up in the problem.

● **Example 1–3 B**

Addition and subtraction of decimal numbers

Add or subtract the following numbers as indicated.

1. $5.67 + 32.046 + 251.7367 + 0.92$

Decimal points lined up vertically

$$
\begin{array}{r}
5.67 \\
32.046 \\
251.7367 \\
+\ \ 0.92 \\
\hline
290.3727
\end{array}
$$

Arrange numbers in columns

2. Subtract (a) 18.7 from 39.62, (b) 4.38 from 19.2

a.
$$
\begin{array}{r}
\overset{8}{3\cancel{9}}.62 \\
-18.7 \\
\hline
20.92
\end{array}
$$

b.
$$
\begin{array}{r}
\overset{8\ 1}{1\cancel{9}.\cancel{2}} \\
-4.38 \\
\hline
14.82
\end{array}
$$

3. What is the perimeter of the figure in the diagram? (Recall that the perimeter is the total distance around the figure.)

$$
\begin{array}{r}
3.97 \text{ m} \\
7.39 \text{ m} \\
3.18 \text{ m} \\
+7.83 \text{ m} \\
\hline
22.37 \text{ m}
\end{array}
$$

7.39 m

3.97 m

3.18 m

7.83 m

☞ *Quick check*   Subtract 14.9 from 83.42 ●

> ### To Multiply Decimal Numbers
>
> 1. Multiply the numbers as if they are whole numbers (ignore the decimal points).
> 2. Count the number of decimal places in both factors. That is, count the number of digits to the right of the decimal point in each factor. This total is the number of decimal places the product must have.
> 3. Beginning at the right in the product, count off to the left the number of decimal places from step 2. Insert the decimal point. If necessary, zeros are inserted so there are enough decimal places.

● **Example 1–3 C**

Multiplication of decimal numbers

Multiply the following.

1. 2.36×0.413

$$
\begin{array}{r}
2.36 \quad \longleftarrow \text{2 decimal places} \\
0.413 \quad \longleftarrow \text{3 decimal places} \\
\hline
708 \\
236 \\
944 \\
\hline
0.97468 \quad \longleftarrow \text{5 decimal places}
\end{array}
$$

$(2 + 3 = 5)$

2. $(18.14)(116.4)$

$$
\begin{array}{r}
18.14 \quad \longleftarrow \text{2 decimal places} \\
116.4 \quad \longleftarrow \text{1 decimal place} \\
\hline
7256 \\
10884 \\
1814 \\
1814 \\
\hline
2{,}111.496 \quad \longleftarrow \text{3 decimal places}
\end{array}
$$

$(2 + 1 = 3)$

☛ *Quick check* $(206.1)(9.36)$ ●

Learning System
"Procedure Boxes" clearly state the step-by-step process for working problems.

> ### To Divide Decimal Numbers
>
> 1. Change the *divisor* to a whole number by moving the decimal point to the *right* as many places as is necessary.
> 2. Move the decimal point in the *dividend* to the right this same number of places. If necessary, zeros are inserted so there are enough decimal places.
> 3. Insert the decimal point in the *quotient* directly above the new position of the decimal point in the dividend.
> 4. Divide as with whole numbers.

● **Example 1–3 D**
Division of decimal numbers

Divide the following.

1. $360.5 \div 1.03$

   a. Write the problem $1.03\overline{)360.5}$

   b. Move the decimal point *two* places to the right in 1.03 and 360.5

$$103\overline{)36{,}050.}\quad \longleftarrow \text{ Zero inserted as placeholder}$$

   c. Now divide as with whole numbers.

$$
\begin{array}{r}
350. \quad\longleftarrow \text{ Quotient}\\
\text{Divisor} \longrightarrow 103\overline{)36{,}050.} \quad\longleftarrow \text{ Dividend}\\
\underline{30\ 9}\\
5\ 15\\
\underline{5\ 15}\\
0
\end{array}
$$

The quotient is 350.

2. If an automobile travels 429.76 miles and uses 15.8 gallons of gas, how many miles per gallon did the automobile achieve?

To determine the miles per gallon, we divide the total number of miles traveled by the amount of gasoline used.

$429.76 \div 15.8$

$$
\begin{array}{r}
27.2\\
15.8\overline{)4{,}297.6}\\
\underline{316}\\
1137\\
\underline{1106}\\
316\\
\underline{316}\\
0
\end{array}
$$

The automobile achieved 27.2 miles per gallon.

➥ *Quick check* $4{,}950.3 \div 5.69$

To Change a Fraction to a Decimal Number

Divide the denominator into the numerator.

● **Example 1–3 E**

Changing a fraction to a decimal number

Convert each fraction to a decimal number.

1. $\frac{3}{4}$

We divide 3 by 4. To do this, we must add zero placeholders.

$$
\begin{array}{r}
0.75 \\
4\overline{)3.00} \quad \longleftarrow \text{ Add zeros} \\
\underline{2\ 8} \\
20 \\
\underline{20} \\
0
\end{array}
$$

Thus $\frac{3}{4} = 0.75$

2. $\frac{1}{3}$

We divide $1 \div 3$.

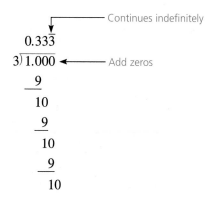

$$
\begin{array}{r}
\text{Continues indefinitely} \\
0.33\overline{3} \\
3\overline{)1.000} \quad \longleftarrow \text{ Add zeros} \\
\underline{9} \\
10 \\
\underline{9} \\
10 \\
\underline{9} \\
10
\end{array}
$$

We can see that no matter how many zero placeholders we add, the quotient will continue to add digits of 3. This is called a **repeating decimal** (denoted by the bar placed over the last digit, or digits, that are repeating). Therefore, $\frac{1}{3} = 0.\overline{3}$. We can round a repeating decimal to as many places as are needed. We can say that $\frac{1}{3}$ is *approximately equal* to 0.333, denoted by

$$\frac{1}{3} \approx 0.333$$

☞ *Quick check*   Convert $\frac{3}{8}$ to a decimal number.

Percent

We use decimal numbers extensively in our work with **percent**. The word percent means "per one hundred."

Definition of Percent

Percent is defined to be parts per one hundred.

We use the symbol "%" to represent percent. Thus

$$3\% \text{ means "three parts per one hundred"}$$

or

$$3\% \text{ means "three one hundredths."}$$

From the above discussion,

$$3\% = \frac{3}{100} = 0.03$$

To Write a Percent as a Decimal Number

Move the decimal point two places to the *left* and drop the % symbol.

To Write a Percent as a Fraction

Drop the % symbol and write the number over a denominator of 100.

To write a fraction or decimal number as a percent, we reverse the procedure.

To Write a Decimal Number as a Percent

Move the decimal point two places to the *right* and affix the % symbol.

To Write a Fraction as a Percent

Find the decimal number equivalent of the fraction and change this decimal number to a percent.

● **Example 1–3 F**
Changing between fractions, decimals, and percents

Write the following as decimal numbers, fractions, and percents.

1. 0.9

$$0.9 = 90\%$$
Move the decimal point two places to the right and affix % symbol (Add a zero placeholder)

$$\text{Since } 0.9 = \frac{9}{10}$$
Write as a fraction

$$\text{then } 0.9 = \frac{9}{10} = 90\%$$

2. 1.25

1.25 = 125%   Move the decimal point two places to the right and affix % symbol

Since $1.25 = \dfrac{125}{100}$ Write as a fraction

$= \dfrac{5}{4}$ Reduce to lowest terms

then $1.25 = \dfrac{5}{4} = 125\%$

3. $\dfrac{7}{8}$

Divide $7 \div 8$ to obtain the decimal equivalent. Doing this we find that

$\dfrac{7}{8} = 0.875$

Then $\dfrac{7}{8} = 0.875 = 87.5\%$ Move the decimal point two places to the right and affix % symbol

4. 3.9%

3.9% = 0.039   Move the decimal point two places to the left and drop the % symbol

Since $0.039 = \dfrac{39}{1,000}$ Write as a fraction

then $3.9\% = 0.039 = \dfrac{39}{1,000}$

☞ **Quick check** Write 241% as a fraction and as a decimal. Write 1.75 as a fraction and a percent. ●

Percentage

When we find 60% of 500, we find the **percentage**. In the language of mathematics, "of" usually means the operation multiplication. Thus

60% of 500 means 60% · 500

However, we cannot multiply 60% times 500. We must first change 60% to a decimal number (or a fraction) before we can perform the multiplication.

60% of 500 = 0.60 · 500   Change 60% to 0.60
= 300   Percentage

Therefore   60%   of 500 is   300

(percent) · (base) = (percentage)

● **Example 1–3 G**
Finding percentages

Find the following percentages.

1. 8% of 35

8% = 0.08   Change percent to a decimal number
8% of 35 = (0.08)(35)   Multiply
= 2.8

Thus 8% of 35 = 2.8

2. 224% of 50

224% = 2.24   Change percent to a decimal number

224% of 50 = (2.24)(50)

= 112   Multiply

Thus 224% of 50 = 112

3. $3\frac{1}{2}\%$ of 270

$3\frac{1}{2}\% = 3.5\%$ $\frac{1}{2}$ = 0.5 as a decimal number

$= 0.035$ Change percent to a decimal number

$3\frac{1}{2}\%$ of 270 = (0.035)(268)

$= 9.45$ Multiply

Thus $3\frac{1}{2}\%$ of 270 = 9.45

📭 *Quick check*   236% of 20

MASTERY POINTS

Can you
- Write decimal numbers as fractions?
- Add and subtract decimal numbers?
- Multiply and divide decimal numbers?
- Write fractions as decimal numbers?
- Change a percent to a decimal number?
- Change a decimal number to a percent?
- Change a fraction to a percent?
- Change a percent to a fraction?
- Find a percentage?

Exercise 1–3	📭 *Quick Check*
Directions	**Example** 0.42
Write each decimal number as a fraction reduced to lowest terms. See example 1–3 A.	**Solution** 0.42 is read "forty-two *hundredths*."

Example 0.42

Solution   0.42 is read "forty-two *hundredths*."

$0.42 = \dfrac{42}{100}$ ◀— Hundred

$= \dfrac{21}{50}$ Reduce to lowest terms

Directions Write each decimal number as a fraction reduced to lowest terms. See example 1–3 A.

1. 0.4   **3.** 0.15   **5.** 0.125   **7.** 0.875

2. 0.8   **4.** 0.36   **6.** 0.248   **8.** 0.625

Directions Add or subtract the following as indicated. See example 1–3 B.	☛ *Quick Check* **Example** Subtract 14.9 from 83.42 **Solution** We want $83.42 - 14.9$ $$\begin{array}{r} 7\,2 \\ 8\!\!\!/\,3\!\!\!/.42 \\ -14.9 \\ \hline 68.52 \end{array}$$

Directions Add or subtract the following as indicated. See example 1–3 B.

9. $6.8 + 0.354 + 2.78 + 7.083 + 2.002$ **14.** $27.376 - 14.007$

10. $4.76 + 0.573 + 3.57 + 40.09 + 13$ **15.** $367.0076 - 210.02$

11. $8.0007 + 360.01 + 25.72 + 6.362 + 140.2$ **16.** $836 - 0.367$

12. $7.0001 + 8 + 7.067 + 803.1 + 5.25$ **17.** $1.07 - 0.00036$

13. $10.03 + 3.113 + 0.3342 + 0.0763 + 0.005$ **18.** $4,563.2 - 274.063$

Directions Multiply the following. See example 1–3 C.	☛ *Quick Check* **Example** $(206.1)(9.36)$ **Solution** $$\begin{array}{r} 206.1 \quad \longleftarrow \text{1 decimal place} \\ 9.36 \quad \longleftarrow \text{2 decimal places} \\ \hline 12366 \\ 6183 \\ 18549 \\ \hline 1,929.096 \quad \longleftarrow \text{3 decimal places} \end{array}$$ $(1 + 2 = 3)$

Directions Multiply the following. See example 1–3 C.

19. $(7.006)(1.36)$ **22.** 703.6×1.7

20. $(42.6)(73)$ **23.** 30.0303×0.030303

21. $(56.37)(0.0076)$ **24.** 2.456×0.00012

Directions	
Divide the following. See example 1–3 D.	**⚑ Quick Check**
	Example 4,950.3 ÷ 5.69
	Solution **1.** Write the problem $5.69\overline{)4,950.3}$
	2. Move the decimal point *two* places to the right in 5.69 and 4,950.3

$$569\overline{)495,030.} \longleftarrow \text{Add zero placeholder}$$

3. Divide as whole numbers.

$$
\begin{array}{r}
870. \longleftarrow \text{Add zero placeholder} \\
569\overline{)495,030.} \\
\underline{455\ 2} \\
39\ 83 \\
\underline{39\ 83} \\
0
\end{array}
$$

Thus 4,950.3 ÷ 5.69 = 870

Directions   Divide the following. See example 1–3 D.

25. 0.84 ÷ 0.7

26. 0.525 ÷ 0.5

27. 10.4 ÷ 0.26

28. 21.681 ÷ 8.03

29. 6,125.1 ÷ 60.05

30. 166.279 ÷ 64.7

31. 31.50 ÷ 0.0126

32. 2.9868 ÷ 0.057

Directions	
Convert each fraction to a decimal number. See example 1–3 E.	**⚑ Quick Check**
	Example $\dfrac{3}{8}$
	Solution We divide 3 ÷ 8, adding zero placeholders where necessary.

$$
\begin{array}{r}
0.375 \\
8\overline{)3.000} \longleftarrow \text{Zero placeholders} \\
\underline{2\ 4} \\
60 \\
\underline{56} \\
40 \\
\underline{40} \\
0
\end{array}
$$

The decimal equivalent of $\dfrac{3}{8}$ is 0.375

Directions Convert each fraction to a decimal number. See example 1–3 E.

33. $\dfrac{3}{20}$ **35.** $\dfrac{13}{20}$ **37.** $\dfrac{2}{9}$

34. $\dfrac{5}{8}$ **36.** $\dfrac{17}{50}$ **38.** $\dfrac{5}{9}$

Directions Work the following problems. See examples 1–3 B and D.

39. Heating oil costs 89.9 cents per gallon. What is the total cost of 14.36 gallons, correct to the nearest cent?

40. A carpenter has three pieces of wood that are 24.5 inches, 35.25 inches, and 62.375 inches long, respectively. How many inches of wood does she have all together?

41. A wood craftsman has 74.75 inches of a particular stock. He needs 5.75 inches of the stock to carve out a cardinal bird. How many cardinals can he make?

42. A rectangular field is 21.3 yards long and 15.75 yards wide. Find the area (length × width) of the field.

43. The record for the mile run in 1895 was 257 seconds. The record in 1993 was 224.39 seconds. How much faster was the 1993 time?

44. A student bought a book for $41.68. If she gave the cashier $50, how much change did she receive?

45. On a 4-day trip, the Adams family used 32.5 gallons, 28.36 gallons, 41.87 gallons, and 19.55 gallons of gasoline. How many gallons of gasoline did they use all together?

46. An airline pilot flew distances of 210.6 kilometers, 504.3 kilometers, 319.6 kilometers, 780.32 kilometers, and 421.75 kilometers on five flights. How many kilometers did she fly all together?

47. A rectangular field is 43.3 yards long and 25.34 yards wide. Find the area (length × width) of the field.

48. If a cubic foot of water weighs 62.4 pounds, how many pounds of water are there in a tank containing 10.4 cubic feet?

Directions
Write each percent as a decimal number and as a fraction. See example 1–3 F.

☞ *Quick Check*

Example 241%

Solution 241% means "two hundred forty-one one hundredths."

$$241\% = \frac{241}{100} = 2.41$$

Directions Write each percent as a decimal number and as a fraction. See example 1–3 F.

49. 5%   **51.** 12%   **53.** 135%   **55.** 325%

50. 1%   **52.** 64%   **54.** 150%   **56.** 570%

Directions	☛ *Quick Check*
Write exercises 57–62 as a fraction and a percent and exercises 63–66 as a decimal and a percent. See example 1–3 F.	**Example** Write 1.75 as a fraction and a percent.

Solution $1.75 = 175\%$ Move decimal point two places to the right and affix % symbol

$\qquad\qquad 1.75 = \dfrac{175}{100}$ Write as a fraction

$\qquad\qquad\quad = \dfrac{7}{4}$ Reduce fraction to lowest terms

Thus $1.75 = \dfrac{7}{4} = 175\%$

Directions   Write exercises 57–62 as a fraction and a percent and exercises 63–66 as a decimal and a percent. See example 1–3 F.

57. 0.8

58. 0.9

59. 0.54

60. 0.08

61. 1.15

62. 2.40

63. $\dfrac{3}{4}$

64. $\dfrac{5}{2}$

65. $\dfrac{3}{8}$

66. $\dfrac{5}{8}$

Directions	☛ *Quick Check*
Find the following percentages. See example 1–3 G.	**Example** 236% of 20

Solution $236\% = 2.36$ Change percent to a decimal number

$\qquad\qquad 236\% \text{ of } 20 = 2.36 \cdot 20$

$\qquad\qquad\qquad\qquad\quad = 47.2$ Multiply

Thus 236% of $20 = 47.2$

Directions   Find the following percentages. See example 1–3 G.

67. 5% of 40

68. 8% of 45

69. 26% of 130

70. 78% of 900

71. 110% of 500

72. 240% of 60

Directions   Work the following problems. See example 1–3 G.

73. City Bank pays 3.7% interest per year on its savings accounts. What is the annual interest on a savings account that has $4,500? (*Hint:* Annual Interest = Percent · Amount in savings.)

74. The sales tax on retail sales is 4%. How much sales tax does John pay on a purchase of $250?

75. If Micah pays 5% of her weekly salary in state income tax, how much state tax does she pay if her weekly salary is $460?

76. A local retailer predicts his profit in a given year will be 116% of the previous year. What is his predicted profit for this year if last year's profit was $42,500?

77. The local shoe store is giving a 25% discount on clearance items. How much discount is there on a pair of shoes costing $34? What is the price of the shoes *after* the discount excluding sales tax?

78. A company charges $4\frac{1}{2}\%$ shipping and handling charges on all items shipped. What are the shipping and handling charges on goods that cost $70? What is the total cost for the goods?

79. A bottle of solution is 4% salt. How much salt is there in a 24-fluid ounce bottle of solution?

80. Self-employed persons must pay a Social Security tax of 12%. What is the Social Security tax on earnings of $25,000? If the person is in the 28% federal income tax bracket, how much federal income tax does the person pay?

⊕ Explore to Learn

Teacher A, bragging: "12 of my students got A's last semester."
Teacher B: "Well 16 of mine got A's."
Teacher C (the math teacher): "A, you had 40 students, and B, you had 50 students, so it's hard to compare your statements. Now, using the concept of percent I can see that . . ."
Finish C's statement in a complete sentence or two.

1–4   The Set of Real Numbers and the Real Number Line

Set Symbolism

Algebra is often referred to as "a generalized arithmetic." The operations of arithmetic and algebra differ only in respect to the symbols we use in working with each of them. Therefore, we will begin our study of algebra by dealing first with numbers and their properties. From this work, we will see algebra develop naturally as a generalized arithmetic.

To begin our study, we will start with a very simple, but important, mathematical concept—the idea of the **set**.[5] **A set is any collection of things.** In mathematics, we use the idea of a set primarily to denote a group of numbers. Any one of the things that belong to the set is called a **member** or an **element** of the set. One way we write a set is by listing the elements, separating them by commas, and including this listing within a pair of braces, { }.

---

[5]Georg Cantor (1845–1918) is credited with the development of the ideas of set theory. He described a set as a grouping together of single objects into a whole.

● **Example 1–4 A**
Using set notation

1. Using set notation, write the set of months that have exactly 30 days.

{April,June,September,November}

2. Using set notation, write the set of seasons of the year.

{spring,summer,fall,winter}

<u>Note</u> When we form a set, the elements within the set are never repeated and they can appear in any order.

☞ *Quick check*   Write the set of letters in the word "mathematics."
Write the set of odd numbers between 5 and 10.   ●

We use capital letters A, B, C, D, and so on, to represent a set. The symbol used to show that an element belongs to a set is the symbol $\in$, which we read "is an element of" or "is a member of." Consider the set $A = \{1, 2, 3, 4\}$, which is read "the set A whose elements are 1, 2, 3, and 4." If we want to say that 2 is an element of the set A, this can be written symbolically as $2 \in A$.

A slash mark is often used in mathematics to negate a given symbol. Therefore if $\in$ means "is an element of the set," then $\notin$ would mean "is *not* an element of the set." To express the fact that 7 is not an element of set A, we could write $7 \notin A$.

Subset

Suppose that P is the set of people in a class and M is the set of men in the same class. It is obvious that the members of M are also members of P, so we say that M is a **subset** of P.

Definition
The set A is a subset of the set B if every element in A is also an element of B.

The symbol for subset is $\subseteq$, which we read "is a subset of." Therefore $A \subseteq B$ is read "the set A is a subset of the set B." Consider the following sets: $A = \{1, 2, 4\}$, $B = \{1, 2, 3, 4, 5\}$, and $C = \{1, 3, 5, 7\}$. We observe that $A \subseteq B$ since every element in A is also an element of B. $C \nsubseteq B$ is read "the set C is *not* a subset of the set B," because not every element of C is an element of B. The set C contains the element 7, which is not an element of the set B.

Natural Numbers and Whole Numbers

The most basic use of our number system is that of counting. We use 1, 2, 3, 4, 5, and so on, as symbols to represent the **natural** or **counting numbers**. The set of natural numbers will be denoted by N, as follows:

$$N = \{1, 2, 3, 4, 5, \cdots\}$$

The three dots, called ellipsis, tell us to continue this counting pattern indefinitely.

If we include 0 with the set of natural numbers, we have the set of **whole numbers**, W.

$$W = \{0, 1, 2, 3, 4, 5, \cdots\}$$

Learning System
Color is used to guide you throughout the text. When you are reviewing for an exam, the functional use of color will allow you to quickly locate key information.

We can use the set of whole numbers to represent physical quantities such as profit (100 dollars), room temperature (72 degrees), and distance (1,250 feet above sea level). However, with the set of whole numbers, we are not able to represent such things as losses of money, temperatures below zero, and distances below sea level. Therefore to represent such situations, we define a new set of numbers called the set of **integers**. We will denote this set by J.

Integers

We shall start by giving the natural numbers another name, the **positive integers**. We then form the *opposites*, or *negatives*,[6] of the positive integers as follows: $-1, -2, -3, \cdots$. Combining the positive integers, the negative integers, and 0, we have the set of integers, J.

$$J = \{\cdots, -3, -2, -1, 0, 1, 2, 3, \cdots\}$$

● **Example 1–4 B**
Using integers

Use integers to represent each of the following.

1. Bromine melts at seven degrees below zero Celsius.
Answer: -7 degrees Celsius   Less than zero is a negative value

2. Bromine boils at fifty-nine degrees Celsius.
Answer: 59 degrees Celsius   Greater than zero is a positive value

⚑ *Quick check*   Use integers to represent a debt of nine dollars.

Rational Numbers

The set of integers is sufficient to represent many physical situations, but it is unable to provide an answer for the following problem. If we want to determine the miles per gallon (mpg) that our car is getting, and we find that 8 gallons of gas enable us to travel 325 miles, then our miles per gallon can be computed by dividing the number of miles by the number of gallons used.

$$\frac{325 \text{ miles}}{8 \text{ gallons}} = \frac{325}{8} \text{ mpg or } 40\frac{5}{8} \text{ mpg}$$

This value is not in the set of integers. Therefore to represent such situations, we define a new set of numbers called the set of **rational numbers**, which is denoted by Q. Recall that a quotient is an answer to a division problem. Hence $\frac{325}{8}$ is called a quotient of two integers since 325 and 8 are both integers.

> **Definition**
>
> A rational number is any number that can be expressed as a quotient of two integers in which the divisor is not zero.[7]

[6]The Babylonians of over 3,500 years ago knew about negative values, although they were considered as not applying to real-world situations, and therefore usually ignored. This attitude did not change until the Middle Ages.
[7]Division involving zero will be discussed in section 1–7.

Other examples of rational numbers would be

$$\frac{2}{3}, -\frac{1}{2}, \frac{6}{1}, \frac{19}{5}, -\frac{23}{7}, \frac{15}{3}, \frac{0}{8}, \frac{-5}{1}$$

The decimal representation of a rational number is either a terminating or a repeating decimal. Some examples of terminating or repeating decimals are

$$\frac{1}{2} = 0.5, \quad \frac{1}{3} = 0.\overline{3}, \quad -\frac{1}{6} = -0.1\overline{6}, \quad -\frac{5}{4} = -1.25, \quad \frac{4}{33} = 0.\overline{12}$$

where a bar placed over a number or group of numbers indicates that the number(s) repeat indefinitely.

Irrational Numbers

At this point, we might feel that we now have numbers that will answer all possible physical situations. However, that is not the case. Consider the following question.

What is the exact length of the side of a square whose area is 10 square units (figure 1–1)? To be able to answer this question, we need to find that number such that when it is multiplied with itself, the product is 10. If we use 3.16, the result would be $(3.16) \times (3.16) = 9.9856$. This is close to 10 but is not equal to 10.

It can be shown that there is no rational value that when multiplied with itself has a product of 10. The answer to this question and many others cannot be found in the set of rational numbers. In chapter 8, we will see that the answer to this question is $\sqrt{10}$ (read "the square root of 10"). Such numbers that cannot be expressed as the quotient of two integers belong to the set of **irrational numbers**, which is denoted by H.

Real Numbers

Since a rational number can be expressed as the quotient of two integers and an irrational number cannot, we should realize that a number can be rational or irrational, but it cannot be both. The set that contains all of the rational numbers and all of the irrational numbers is called the set of **real numbers**, which is denoted by R. Whenever we encounter a problem and a specific set of numbers is not indicated, it will be understood that we are dealing with real numbers.

All of the sets that we have examined thus far are subsets of the set of real numbers. Figure 1–2 shows the relationship.

The Real Number Line

To picture the set of real numbers, we shall use a **real number line**. We begin by drawing a line where the arrowhead at each end of the line indicates that the line continues on indefinitely in both directions. Next we choose any point on the line to represent 0. This point is called the **origin** of the number line. Numbers to the right of zero are positive and to the left of zero are negative (figure 1–3).

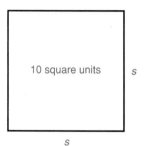

10 square units   s

s

Figure 1–1

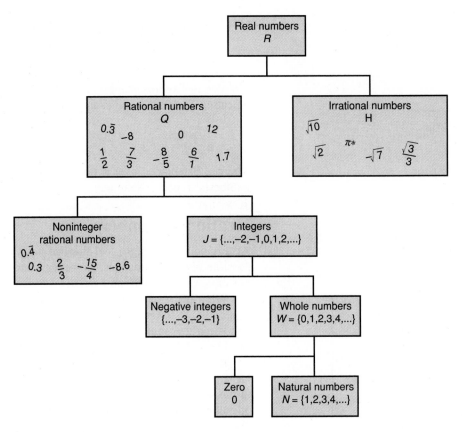

*A well-known irrational number is π (pi), which represents the distance around a circle (circumference) divided by the distance across the circle through its center (diameter). Common rational number approximations for π are 3.14 and $\frac{22}{7}$.

Figure 1–2

Figure 1–3

Any real number can now be located on the number line. Consider the number line in figure 1–4.

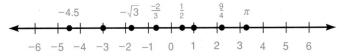

Figure 1–4

The number that is associated with each point on the line is called the **coordinate** of the point. The solid circle that is associated with each number is called the **graph** of that number. In figure 1–4, the numbers -4.5, -3, $-\sqrt{3}$,

$\dfrac{-2}{3}, \dfrac{1}{2}, 1, \dfrac{9}{4}$, and π are the *coordinates* of the points indicated on the line by solid circles. The solid circles are the *graphs* of these numbers.

Note The coordinates $-\sqrt{3}$ and π represent irrational numbers. To graph these points, we would use a calculator to find a rational approximation for the coordinate of the point. $-\sqrt{3} \approx -1.732$, $\pi \approx 3.142$ ($\approx$ is read "is approximately equal to").

Order on the Number Line

The direction in which we move on the number line is also important. If we move to the right, we are moving in a positive direction and the numbers are *increasing*. If we move to the left, we are moving in a negative direction and the numbers are *decreasing* (figure 1–5).

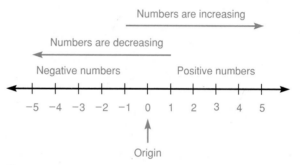

Figure 1–5

If we choose any two points on the number line and represent them by a and b, where a and b represent some *unspecified* numbers, we observe that there is an *order* relationship between a and b (figure 1–6). Since the point associated with a is to the *left* of the point associated with b, we say that a is **less than** b, which in symbols is $a < b$. We might also say that b is **greater than** a, which in symbols is $b > a$. The symbols $<$ (is less than) and $>$ (is greater than) are inequality symbols called **strict inequalities** and they denote an **order relationship** between numbers.

Figure 1–6

● **Example 1–4 C**
Using inequality symbols

Replace the ? with the proper inequality symbol ($<$ or $>$).

1. $-5 ? -3$ <—●——————●—→ Answer: $-5 < -3$ Because -5 is to
   $\quad\quad\quad\quad -6\;\; -5\;\; -4\;\; -3\;\; -2$ the left of -3

Note If we have difficulty deciding which of two numbers is greater, think of the numbers as representing temperature readings. The -5 would be thought of as 5 degrees below zero and the -3 would be 3 degrees below zero. It is easy to realize that -3 is the greater (warmer) temperature and the inequality would be $-5 < -3$.

2. $0 ? -4$ <—————●————————●——→ Answer: $0 > -4$ Because 0 is to
   $\quad\quad\quad\quad -5\; -4\; -3\; -2\; -1\;\; 0\;\; 1$ the right of 4

3. $0 ? 3$ <——●————————●——→ Answer: $0 < 3$ Because 0 is to the left
   $\quad\quad\quad\quad -1\;\; 0\;\; 1\;\; 2\;\; 3\;\; 4$ of 3

Note No matter which inequality symbol we use, the arrow _always_ points at the lesser number.

☛ _Quick check_   Replace ? with < or >.

2 ? 4

-3 ? -6

There are two other inequality symbols, called **weak inequalities**. They are **is less than or equal to**, ≤, and **is greater than or equal to**, ≥. The weak inequality symbol ≥, is greater than or equal to, denotes that one number could either be greater than a second number or equal to that second number. The inequality $x \geq 3$ means that x is _at least_ 3. That is, x represents all numbers that are 3 or more.

The other weak inequality symbol, ≤, is less than or equal to, denotes that one number could either be less than a second number or equal to that second number. The inequality $x \leq 5$ means that x is _at most_ 5. That is, x represents all numbers that are 5 or less.

Absolute Value

As we study the number line, we observe a very useful property called **symmetry**. The numbers are symmetrical with respect to the origin. That is, if we go four units to the right of 0, we come to the number 4. If we go four units to the left of 0, we come to the _opposite_ of 4, which is −4 (figure 1–7).

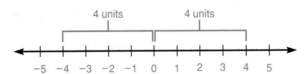

Figure 1–7

Each of these numbers is four units away from the origin. How far a given number is from the origin is called the **absolute value** of the number. **The absolute value of a number is the undirected distance that the number is from the origin.** The symbol for absolute value is | |.

● **Example 1–4 D**

Finding absolute value

Evaluate the following expressions.

1. $|-7| = 7$ **3.** $|0| = 0$ **5.** $|-3.7| = 3.7$

2. $|3| = 3$ **4.** $\left|\dfrac{2}{3}\right| = \dfrac{2}{3}$ **6.** $-|-6| = -(6) = -6$

Note The absolute value of a number is _never_ negative; that is, $|x| \geq 0$, for every $x \in R$, however the absolute value bars are only applied to the symbol contained within them. The − sign in front of the absolute value bars is not affected by the absolute value bars. Example 6 would be read "the opposite of the absolute value of −6," and the answer is −6.

☞ *Quick check*   Evaluate the following expressions.

$$|-3|; \qquad |12|; \qquad -|-4|$$

Visualizing our number system on the number line demonstrates the fact that each number possesses two important properties.

1. The sign of the number denotes a direction from zero. (The absence of a sign indicates a positive number.)
2. The absolute value represents a distance from zero.

MASTERY POINTS

Can you
- Write sets?
- Draw a number line?
- Graph a number on the number line?
- Tell which of two real numbers is greater?
- Find absolute values?
- Approximate the value of the coordinate of a graph on the number line?

Exercise 1–4

Directions
Write each set by listing the elements. See example 1–4 A.

☞ *Quick Check*

Examples   The letters in the word "mathematics"   The odd numbers between 5 and 10

Solutions   {m,a,t,h,e,i,c,s}   The letters *m, a,* and *t* are not repeated within the set   {7,9}   The word between means that we do not include the 5 or 10

Directions   Write each set by listing the elements. See example 1–4 A.

1. The days of the week

2. The days of the week that begin with "T"

3. The first 3 months of the year

4. The months of the year that begin with "J"

5. The months with 31 days

6. The letters in the word "repeat"

7. The letters in the word "algebra"

8. The letters in the word "elementary"

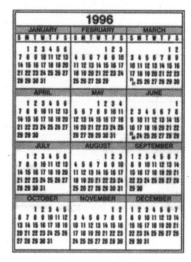

9. The letters in the word "intermediate"

10. The even numbers between 5 and 11

11. The odd numbers between 2 and 10

12. The months of the year that begin with "A"

13. The days of the week that begin with "S"

14. The months of the year that begin with "M"

Directions Use integers to represent each of the following. See example 1–4 B.	☛ ***Quick Check***
	Example A debt of nine dollars
	Solution −9 dollars A debt is a negative value

Directions   Use integers to represent each of the following. See example 1–4 B.

15. Ten dollars overdrawn in a checking account; 150 dollars in a savings account

16. Mercury's melting point is 39 degrees below zero Celsius. Its boiling point is 357 degrees Celsius.

17. A loss of 10 yards on a football play; a gain of 16 yards

18. Mt. Everest rises 29,028 feet above sea level; the Dead Sea has a depth of 1,290 feet below sea level.

19. The Dow Jones Industrial Stock Average fell 14 points; it rose 8 points.

20. Hydrogen's melting point is 259 degrees below zero Celsius. Water boils at 100 degrees Celsius. Water freezes at zero degrees Celsius.

Directions Plot the graph of the following numbers, using a different number line for each exercise. See figure 1–4.	**Example** $-3, -2, 1, 1\frac{1}{2}, 3$
	Solution 

Directions   Plot the graph of the following numbers, using a different number line for each exercise. See figure 1–4.

21. $-3, -1, \frac{1}{2}, 3, 4.5$

22. $-2, -1, \frac{3}{4}, 2, 4$

23. $-5, -2.5, 0, 1, \pi, 3\frac{1}{2}$

24. $-1.5, 0, \frac{2}{3}, 1, 3$

25. $-4, -2, 1, 2\frac{1}{2}, 4$

27. $-5, -1.5, 0, \frac{1}{2}, \sqrt{2}, 6$

26. $-3, -1, \frac{1}{2}, 1, 2$

28. $-4, -\sqrt{4}, \frac{1}{2}, \sqrt{3}, \pi$

Directions Approximate the values of the set of coordinates of the graphs on the following number lines to the nearest $\frac{1}{4}$ of a unit. See figure 1–4.	**Example** **Solution** $-4, -2, 1, 2, 4$

Directions   Approximate the values of the set of coordinates of the graphs on the following number lines to the nearest $\frac{1}{4}$ of a unit. See figure 1–4.

29.

30.

31.

32.

33.

34.

35.

36.

Directions Replace the ? with the proper inequality symbol, $<$ or $>$. See example 1–4 C.	🏳 *Quick Check* **Examples** 2 ? 4 **Solutions** $2 < 4$ Because 2 is to the left of 4 on the number line	-3 ? -6 $-3 > -6$ Because -3 is to the right of -6 on the number line

Directions Replace the ? with the proper inequality symbol, $<$ or $>$. See example 1–4 C.

37. 4 ? 8   **40.** -2 ? -4 **43.** -10 ? -5 **46.** -3 ? 0

38. 6 ? 3   **41.** -3 ? -8 **44.** 0 ? 2   **47.** 0 ? -6

39. 9 ? 2   **42.** -9 ? -6 **45.** 0 ? 4

Directions Evaluate the following expressions. See example 1–4 D.	☞ *Quick Check*
	Examples $\|-3\|$ $\|12\|$ $-\|-4\|$
	Solutions 3 −3 is 3 units from the origin 12 12 is 12 units from the origin −4 −4 is 4 units from the origin. The problem is to find the opposite of the absolute value

---

Directions Evaluate the following expressions. See example 1–4 D.

48. $\|0\|$ **53.** $\|4\|$ **56.** $\left\|-\dfrac{3}{4}\right\|$ **59.** $-\left\|\dfrac{5}{8}\right\|$

49. $\|2\|$ **54.** $\left\|\dfrac{2}{3}\right\|$ **60.** $-\|-2\|$

50. $\|8\|$ **57.** $\left\|1\dfrac{1}{2}\right\|$ **61.** $-\|6\|$

51. $\|-5\|$ **55.** $\left\|-\dfrac{1}{2}\right\|$

52. $\|-7\|$ **58.** $-\left\|-2\dfrac{3}{4}\right\|$

Directions Replace the ? with the proper inequality symbol, $<$ or $>$. See examples 1–4 C and D.	**Examples** $\|-6\|$? $\|-3\|$ $\|4\|$? $\|-7\|$
	Solutions 6 ? 3 $\|-6\|$ is 6 and $\|-3\|$ is 3 4 ? 7 $\|4\|$ is 4 and $\|-7\|$ is 7
	6 > 3 6 is to the right of 3 on the number line 4 < 7 4 is to the left of 7 on the number line
	then $\|-6\| > \|-3\|$ then $\|4\| < \|-7\|$

Directions Replace the ? with the proper inequality symbol, $<$ or $>$. See examples 1–4 C and D.

62. $\|-2\|$? $\|-4\|$ **65.** $\|0\|$? $\|-2\|$ **68.** $\|-9\|$? $\|7\|$ **71.** 7 ? $\|-2\|$

63. $\|5\|$? $\|-7\|$ **66.** $\|-3\|$? $\|4\|$ **69.** $\|-6\|$? $\|-2\|$ **72.** 4 ? $\|-8\|$

64. $\|-3\|$? $\|-4\|$ **67.** $\|-8\|$? $\|-5\|$ **70.** $\|-5\|$? 3   **73.** $\|-4\|$? 6

Directions Use absolute value to write each of the following. See figure 1–7.	**Example** The distance between -8 and 0
	Solution $\|-8\|$ The absolute value of a number is the distance from the number to the origin.

Directions Use absolute value to write each of the following. See figure 1–7.

74. The distance between 14 and 0   **77.** The distance between -9 and 0

75. The distance between -27 and 0   **78.** The distance between -19 and 0

76. The distance between 18 and 0

✦ Explore to Learn

When mathematics is applied to the rest of the world it is sometimes called modeling. In this modeling, mathematical terms represent real-world terms, and what is true in mathematics is supposed to be true in the real-world situation, and vice versa. Show how the idea of debt and wealth, and phrases like "richer than" and "poorer than," can be modeled by inequalities and signed numbers. Describe a related real-world situation that might be modeled by each of the following three mathematical statements: $5 < 10, 10 > 5, -5 < 2, -10 < -5$

Calculators

Modern calculators can do all of the arithmetic required in this chapter. As indicated in the error analysis section, this can lead to error if not used properly. The following table and the notes on p. 46 shows how to perform some typical calculations on several representative types of calculator. Work through each problem using the steps for your type of calculator.

Problem	Scientific calculator	Graphing calculators			RPN	Answer
		TI-81/82	TI-85	Casio 7700/9700	Hewlett-Packard	
$\dfrac{6+9}{3}$	(6 + 9) ÷ 3 =	(6 + 9) ÷ 3 ENTER	(6 + 9) ÷ 3 ENTER	(6 + 9) ÷ 3 EXE	6 ENTER 9 + 3 ÷	5
$6+\dfrac{9}{3}$	6 + 9 ÷ 3 =	6 + 9 ÷ 3 ENTER	6 + 9 ÷ 3 ENTER	6 + 9 ÷ 3 EXE	6 ENTER 9 ENTER 3 ÷ +	9
$(5+2)(8-3)$	(5 + 2) × (8 - 3) =	(5 + 2) (8 - 3) ENTER	(5 + 2) (8 - 3) ENTER	(5 + 2) (8 - 3) EXE	5 ENTER 2 + 8 ENTER 3 - ×	35
$\sqrt{10}$	10 √	2nd √ 10 ENTER	2nd √ 10 ENTER	√ 10 EXE or SHIFT √ 10 EXE	10 √	3.162*
$8+(3)(2)$	8 + 3 × 2 =	8 + 3 × 2 ENTER	8 + 3 × 2 ENTER	8 + 3 × 2 EXE	8 ENTER 3 ENTER 2 × +	14
$\dfrac{1}{2}+\dfrac{1}{3}$	1 ÷ 2 + 1 ÷ 3 =	1 ÷ 2 + 1 ÷ 3 ENTER	1 ÷ 2 + 1 ÷ 3 ENTER	1 ÷ 2 + 1 ÷ 3 EXE	1 ENTER 2 ÷ 1 ENTER 3 ÷ +	$0.8\overline{3}$
$\dfrac{1}{2}+\dfrac{1}{3}$ on calculators with a fraction key (see note 1) or feature (see note 2)	1 $a^b/_c$ 2 + 1 $a^b/_c$ 3 =	TI-82 only 1 ÷ 2 + 1 ÷ 3 MATH 1:▶Frac ENTER	1 ÷ 2 + 1 ÷ 3 ENTER 2nd MATH F5 MORE F1 ENTER	1 $a^b/_c$ 2 + 1 $a^b/_c$ 3 EXE		$\dfrac{5}{6}$

* Rounded to three decimal places.

1. Fraction key: Many calculators have a key that computes numeric fractions exactly, up to certain limits. To enter $5\frac{1}{3} \div 2\frac{3}{8}$, enter 5 $\boxed{a^b/_c}$ 1 $\boxed{a^b/_c}$ 3 $\boxed{\div}$ 2 $\boxed{a^b/_c}$ 3 $\boxed{a^b/_c}$ 8 $\boxed{=}$ to see the result $2\frac{14}{57}$. To see the answer as the improper fraction $\frac{128}{57}$, select $\boxed{\text{SHIFT}}$ $\boxed{\text{d/c}}$ (or $\boxed{\text{2nd}}$ $\boxed{\text{d/c}}$).

2. Fraction feature: The TI-85 will convert a result into fractional form when possible, up to certain limits. To select the $\boxed{\blacktriangleright\text{Frac}}$ feature, use $\boxed{\text{2nd}}$ $\boxed{\text{MATH}}$ $\boxed{\text{F5}}$ $\boxed{\text{MORE}}$ $\boxed{\text{F1}}$ as shown in the example in the table. If you want to use this feature often, add it to the custom menu. Do this as follows.

$\boxed{\text{2nd}}$ $\boxed{\text{CATALOG}}$ $\boxed{.}$ This gets one near the bottom of the catalog menu.
Select $\boxed{\text{F1}}$ seven times.   Use $\boxed{\text{PAGE}\downarrow}$ to find ▶Frac in the menu.
Select ▶Frac by using the down arrow key $\boxed{\blacktriangledown}$.
$\boxed{\text{F3}}$ Select the $\boxed{\text{CUSTOM}}$ menu item.
Select whichever of $\boxed{\text{F1}}$ to $\boxed{\text{F5}}$ corresponds to an empty slot in the custom menu. Use $\boxed{\text{MORE}}$ if all are already filled. $\boxed{\text{EXIT}}$ $\boxed{\text{EXIT}}$

Now, to convert, say, 0.2 to fraction form, select $\boxed{\text{CUSTOM}}$ and whichever of F1 to F5 contains the ▶Frac menu item. Assuming this is $\boxed{\text{F1}}$, use 0.2 $\boxed{\text{CUSTOM}}$ $\boxed{\text{F1}}$ $\boxed{\text{ENTER}}$. The result is $\frac{1}{5}$.

1–5   Addition of Real Numbers

Addition of Two Positive Numbers

When we perform the operations of addition and subtraction with integers, we will refer to this as operations with **signed numbers**. We use the minus sign (−) to indicate a negative number and the plus sign (+) to indicate a positive number. We should realize that the minus sign is identical to the symbol used for subtraction, and the plus sign is identical to the symbol used for addition. The meanings of these symbols will depend on their use in the context of the problem. In the case of a positive number, the plus sign need only be used if we wish to emphasize the fact that the number is positive. *When there is no sign, the number is understood to be positive.*

To visualize the idea of addition of signed numbers, we will use the number line to represent a checking account in which the origin represents a zero balance. We will let moves in the positive direction represent deposits and moves in the negative direction represent checks that we write, withdrawals. If we have a zero balance and deposit 5 dollars and 4 dollars, represented by (+5) and (+4), the balance in the account would be as shown in figure 1–8.

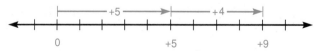

Figure 1–8

Performing the addition, we have $(+5) + (+4) = (+9)$. Notice that the sum, $(+9)$, has the same sign as the 5 and the 4.

So the discussion can be more general, we will use variables to state the process of addition on the number line. Recall that a variable represents an unspecified number. It is a placeholder for a number.

> ## Addition on the Number Line
>
> To add *a* and *b*, that is $a + b$, we locate *a* on the number line and move from there according to the value of *b*.
> 1. If *b* is positive, we move to the right *b* units.
> 2. If *b* is negative, we move to the left the absolute value of *b* units.
> 3. If *b* is 0, we stay at *a*.

● **Example 1–5 A**

Addition of positive numbers

Add the following numbers.

1. $(+4) + (+5) = +9$ 4 and 5 are called addends, 9 is the sum

2. $(+3) + (+8) = +11$ The sum of 3 and 8 is 11

3. $(+6) + (0) = +6$ 6 plus 0 equals 6

We have used the plus (+) sign in front of a number to emphasize the fact that the number was positive. In future examples, we will omit the plus sign from a positive number and it will be understood that 3 means +3.

From example 3, we see that when zero and a given number are added, the sum is the given number. For this reason, zero is called the **identity element of addition**. We now state this property.

> ## Identity Property of Addition
>
> For every real number *a*,
> $$a + 0 = 0 + a = a$$
>
> ### Concept
> Adding zero to a number leaves the number unchanged.

Observe that our example in figure 1–8 and number 1 in example 1–5 A are the same addition problem with the order of the numbers reversed. That is, $5 + 4 = 9$ and $4 + 5 = 9$. This observation illustrates an important mathematical principle called the **commutative property of addition**.

Learning System
"Concept" is a translation of the mathematical property into everyday language.

> ## Commutative Property of Addition
>
> For every real number *a* and *b*,
> $$a + b = b + a$$
>
> ### Concept
> This property says that when we are *adding* numbers, changing the order in which the numbers are added will not change the answer (sum).

Addition of Two Negative Numbers

We now examine addition of two negative numbers. If we have a zero balance and write a check for 6 dollars, expressed as (−6), and another for 5 dollars, expressed as (−5), the loss to our checking account would be as shown in figure 1–9.

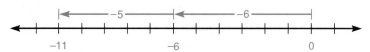

Figure 1–9

Our total withdrawal would be (−6) dollars + (−5) dollars = −11 dollars.

We can summarize what we have done by saying: *When we add two negative numbers, we add their absolute values and prefix the sum with their common sign, −.*

● **Example 1–5 B**

Addition of negative numbers

Add the following numbers.

1. $(-2) + (-7) = -9$

2. $(-9) + (-4) = -13$

3. $(-20) + (-30) = -50$

4. $(-6) + (-11) = -17$

$(-7) + (-8)$

Scientific 7 $\boxed{+/-}$ $\boxed{+}$ 8 $\boxed{+/-}$ $\boxed{=}$

Graphing $\boxed{(-)}$ 7 $\boxed{+}$ $\boxed{(-)}$ 8 $\boxed{\text{ENTER}}$
Result −15

☞ *Quick check* $(-5) + (-7)$

We see from our examples that when we add two signed numbers and their signs are the same, we add their absolute values and prefix the sum with their common sign.

Addition of Two Numbers with Different Signs

To consider the addition of two numbers with different signs, we again refer to our checking account. Suppose we make a deposit to and a withdrawal from our checking account. If we deposit more money than we withdraw, we will have a positive balance in our account. For example, if we have a zero balance and deposit 15 dollars, represented by (+15) dollars, and withdraw 10 dollars, represented by (−10) dollars, the result to our checking account will be as shown in figure 1–10.

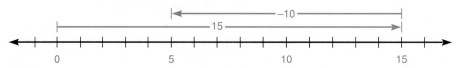

Figure 1–10

The balance in the checking account would be $(+15)$ dollars $+ (-10)$ dollars $= (+5)$ dollars.

We see that our answer is the absolute value of the difference of 15 and 10, prefixed by the sign of the number with the greater absolute value, $(+15)$.

Consider a second example in which we have a zero balance and deposit 10 dollars, $(+10)$ dollars, and write a check for 15 dollars, (-15) dollars. The result to our checking account this time would be as shown in figure 1–11.

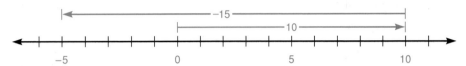

Figure 1–11

The balance would be $(+10)$ dollars $+ (-15)$ dollars $= (-5)$ dollars.

We again see that our answer is the absolute value of the difference of the two numbers prefixed by the sign of the number with the greater absolute value, (-15).

We now summarize the procedure for adding two numbers of different signs. *The sum of a positive number and a negative number is found by subtracting the lesser absolute value from the greater absolute value. The answer has the sign of the number with the greater absolute value.*

● **Example 1–5 C**

Addition of real numbers

Add the following numbers.

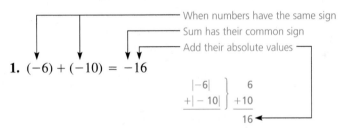

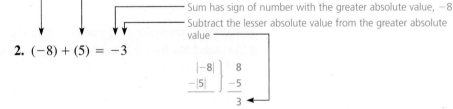

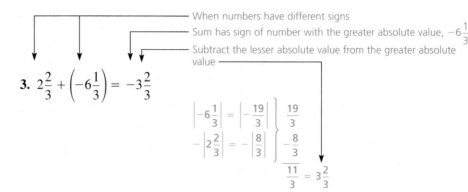

When numbers have different signs

Sum has sign of number with the greater absolute value, $-6\frac{1}{3}$

Subtract the lesser absolute value from the greater absolute value

3. $2\frac{2}{3} + \left(-6\frac{1}{3}\right) = -3\frac{2}{3}$

$$\left|-6\frac{1}{3}\right| = \left|-\frac{19}{3}\right| \left.\begin{array}{c} \frac{19}{3} \\ -\frac{8}{3} \end{array}\right\}$$

$$-\left|2\frac{2}{3}\right| = -\left|\frac{8}{3}\right|$$

$$\frac{11}{3} = 3\frac{2}{3}$$

  $2\frac{2}{3} + \left(-6\frac{1}{3}\right)$ Exact answer only possible on some calculators.

Scientific with $\boxed{a^b/_c}$ key $2 \boxed{a^b/_c} 2 \boxed{a^b/_c} 3 \boxed{+} 6 \boxed{a^b/_c} 1 \boxed{a^b/_c}$
$3 \boxed{+/-} \boxed{=}$

TI-82 $2 \boxed{+} 2 \boxed{\div} 3 \boxed{+} \boxed{(-)} \boxed{(} 6 \boxed{+} 1 \boxed{\div} 3 \boxed{)}$
$\boxed{\text{MATH}} \boxed{1:\blacktriangleright\text{Frac}} \boxed{\text{ENTER}}$

Casio 7700GE/9700GE $2 \boxed{a^b/_c} 2 \boxed{a^b/_c} 3 \boxed{+} \boxed{(-)} 6 \boxed{a^b/_c} 1 \boxed{a^b/_c} 3 \boxed{\text{EXE}}$
Result $-3\frac{2}{3}$

4. $(-4) + (6) = 2$ Difference of the absolute values. The sign comes from the +6

5. $(10) + (-10) = 0$ The sum of a number and its opposite is zero

Note $(-8) + (5)$ will often be written $-8 + 5$, since the leading $-$ must mean the opposite of 8, as opposed to subtraction, and the $+$ must mean addition, just as it would in $8 + 5$.

➤ *Quick check* $(3) + (-8)$ ●

We observe from example 5 that $(10) + (-10) = 0$. That is, a number added to its opposite gives a sum equal to zero. The opposite of a number is also called the **additive inverse** of that number. We now state this property.

Additive Inverse Property

For every real number *a*,

$$a + (-a) = 0$$

and

$$(-a) + a = 0$$

Concept

The sum of a number and its opposite is zero. The opposite of a number is also called its additive inverse.

We can summarize our procedure for addition of real numbers as follows:

Addition of Two Real Numbers

1. If the signs are the same, add their absolute values and prefix the sum by their common sign.
2. If the signs are different, subtract the lesser absolute value from the greater absolute value. The answer has the sign of the number with the greater absolute value.
3. The sum of a number and its opposite (additive inverse) is zero.

In many problems, there will be more than two numbers being added together. In those situations, as long as the operation involved is *strictly addition*, we can add the numbers in any order we wish.

● **Example 1–5 D**

Addition of more than two real numbers

Find the sum.

1. $-2 + (-7) + 12 = -9 + 12$ First $(-2) + (-7)$ is -9
$= 3$ Then $(-9) + (12)$ is 3

Note For convenience, we simply add the numbers as they appear, reading them from left to right.

2. $14 + (-4) + 4 = 10 + 4$ First $(14) + (-4)$ is 10
$= 14$ Then $(10) + (4)$ is 14

Since the numbers in the preceding examples can be added in any order, we might feel in example 2 that it would be easier to add the -4 and 4 first, as follows:

$$14 + (-4) + 4 = 14 + 0 = 14$$

Therefore we observe the following:

$$[14 + (-4)] + 4 = 14 + [-4 + 4] = 14$$

This illustrates a mathematical principle called the **associative property of addition**.

Associative Property of Addition

For every real number a, b, and c,

$$(a + b) + c = a + (b + c)$$

Concept
Changing the grouping of the numbers will not change the sum.

Problem Solving

To solve the following word problems, we must find the sum of the given quantities. Represent gains by positive integers and losses by negative integers.

● **Example 1–5 E**
Problem solving

Choose a variable to represent the unknown quantity and find its value.

1. A pipe that is 4 feet long is joined with a pipe that is 6 feet long. What is the total length of the pipe?

Let ℓ represent the total length of the pipe. To find the total length, we must *add* the individual lengths.

total length	is	4-foot pipe	joined with	6-foot pipe
ℓ	$=$	4	$+$	6

$$\ell = 4 + 6$$
$$\ell = 10$$

The total length of the pipe is 10 feet.

2. The quarterback of the Detroit Lions attempted four passes with the following results: a 12-yard gain, an incomplete pass, a 5-yard loss (tackled behind the line), and a 15-yard gain. What was his total gain (or loss)?

Let t represent the total gain (or loss). To find the total gain (or loss), we must *add* the results of the four plays. Then

total gain (or loss)	is	12-yard gain		incomplete pass		5-yard loss		15-yard gain
t	$=$	(12)	$+$	(0)	$+$	(-5)	$+$	(15)

$$t = (12) + (0) + (-5) + (15)$$
$$t = (12) + (-5) + (15) = (7) + (15) = 22$$

The total *gain* was 22 yards after the 4 plays. ●

MASTERY POINTS

Can you
- Add real numbers on the number line?
- Add real numbers mentally?
- Use the commutative and associative properties of addition and the identity property of addition?

Exercise 1–5	☛ *Quick Check*
Directions Find each sum. See examples 1–5 A, B, C, and D.	**Examples** $(-5) + (-7)$ $\qquad\qquad$ $(3) + (-8)$ **Solutions** -12 Signs are the same, add their absolute values and prefix the sum by their common sign $\qquad$ -5 Signs are different, subtract lesser absolute value from greater, sign comes from number with the greater absolute value

Directions Find each sum. See examples 1–5 A, B, C, and D.

1. $(-9) + (-4)$

2. $(+3) + (-5)$

3. $(+7) + (-2)$

4. $(-14) + (-10)$

5. $-8 + 3$

6. $12 + (-8)$

7. $4 + (-9)$

8. $-11 + 7$

9. $4 + (-4)$

10. $(-3) + 3$

11. $-8.7 + (-4.9)$

12. $(-12.1) + 8.6$

13. $(-3.7) + (-7.4)$

14. $-8.3 + 15.8$

15. $-\dfrac{1}{6} + \left(-\dfrac{1}{3}\right)$

16. $\dfrac{3}{4} + \left(-\dfrac{7}{8}\right)$

17. $\dfrac{1}{5} + \left(-\dfrac{1}{10}\right)$

18. $-1\dfrac{1}{2} + \left(-1\dfrac{1}{4}\right)$

19. $2\dfrac{1}{2} + \left(-3\dfrac{1}{4}\right)$

20. $5\dfrac{3}{8} + \left(-2\dfrac{1}{4}\right)$

21. $10 + (-5) + (-2)$

22. $3 + (-4) + 1$

23. $-12 + (-10) + 8 + 24$

24. $-24 + 12 + 12$

25. $-30 + 14 + (-8) + (-20)$

26. $-2 + 3 + (-4) + (-5)$

27. $(-25) + 4 + (-32) + 28 + 3$

28. $11 + (-12) + (-14) + (-9)$

Directions	Example	The sum of -6, 2, and -9
Find the sum.	Solution	$(-6) + 2 + (-9) = -4 + (-9) = -13$ First add -6 and 2 / Then add -4 and -9

Directions   Find the sum.

29. The sum of 7, -11, and -6

30. The sum of -16, -6, and -5

31. 18 plus -14 plus -4

32. -9 plus 15 plus -17

33. The sum of -5 and 12 increased by 4

34. The sum of 15 and -18 increased by 10

35. 15 added to the sum of -9 and 9

36. 9 added to the sum of -6 and -11

Directions	Example	If a man borrows \$1,800 and has \$700 in savings, what is his net worth?
Work the following problems. See example 1–5 E.	Solution	We represent the borrowed amount as $(-1,800)$ dollars and his savings as $(+700)$ dollars. His net worth is then represented by $(-1,800)$ dollars + $(+700)$ dollars = $(-1,100)$ dollars.

Directions   Work the following problems. See example 1–5 E.

37. A temperature of $(-18)°$ C is increased by $25°$ C. What is the resulting temperature?

38. In accounting negative values are sometimes indicated by parentheses. In an informal accounting ledger the following data are entered. Compute the balance.

Initial balance	\$568.50
Deposit	61.72
Monday sales	211.43
Wages paid	(312.50)
Postage	(13.78)
Tuesday sales	189.13
Wednesday sales	225.58
Sales tax paid	(23.54)
Balance	

39. (See exercise 38.) In an informal accounting ledger the following data are entered. Determine the amount that was paid for materials.

Initial balance	\$568.50
Deposit	61.72
Monday sales	211.43
Wages paid	(312.50)
Postage	(13.78)
Tuesday sales	189.13
Wednesday sales	225.58
Sales tax paid	(23.54)
Paid for materials	_____
Balance	(125.44)

40. The stock market rose by 23 points on Monday, fell by 10 points on Tuesday, rose by 8 points on Wednesday, rose by 31 points on Thursday, and fell by 19 points on Friday. What was the total gain (or loss) during the 5 days?

41. The barometric pressure rose 6 mb (millibars), then dropped 9 mb. Later that day the pressure dropped another 3 mb and then rose 8 mb. What was the total gain (or loss) in barometric pressure that day?

42. A small business showed profits of $157.25, $250.12, $109.18, $47.96, and $312.20 on five consecutive days. What was the total profit?

43. Armand's body temperature was 103.2 degrees. It changed by −2.9 degrees. Find his present body temperature.

44. The temperature in Sault Ste. Marie was −13° F at 8 A.M. By 1 P.M. that day, it had risen 39° F. What was the temperature at 1 P.M.?

45. The temperature on a given day in Anchorage, Alaska, was −19° F. The temperature then went down 22° F. What was the final temperature that day?

46. An American Airlines plane is flying at an altitude of 33,000 feet. It suddenly hits an air pocket and drops 4,200 feet. What is its new altitude?

47. Jamal has $35 in his checking account. He deposits $52, $25, and $32; he then writes checks for $18 and $62. What is the final balance in his checking account?

48. A football team has the ball on its 25-yard line. On three successive plays, the team gains 6 yards, loses 3 yards, and then gains 4 yards. Where does the ball rest for the fourth play?

⚜ Explore to Learn

1. Suppose numbers represent temperature in degrees Fahrenheit (as indeed they do on thermometers). What would 70 mean? What would −5 mean? What sentence might $25 + (−6) = 19$ mean? What might $−12 + 45 = 33$ mean?

2. Make up a problem modeling the equations in number 1 that does not involve temperature.

1–6   Subtraction of Real Numbers

Subtraction of Two Real Numbers

We are already familiar with the operation of subtraction in problems such as $15 − 10 = 5$. From the definition of subtraction, we know that $15 − 10 = 5$ since $5 + 10 = 15$. In figure 1–10 of section 1–5, we see that $15 + (−10)$ is also equal to 5. That is,

$$\underset{\text{Subtraction}}{15 − 10 = 5} \quad \text{and} \quad \underset{\text{Addition}}{15 + (−10) = 5}$$

From this example, we see that we obtain the same results if we change the operation from subtraction to addition and change the sign of the number that we are subtracting. We would have an addition problem and could proceed as we did in section 1–5.

To summarize the procedure for subtracting real numbers, we can state algebraically:

Definition of Subtraction

For any two real numbers, *a* and *b*,

$$a - b = a + (-b)$$

Concept

"*a* minus *b*" means the same as "*a* plus the opposite of *b*."

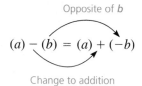

Opposite of *b*

$$(a) - (b) = (a) + (-b)$$

Change to addition

The steps to carry out the subtraction would be as follows:

Subtraction of Two Real Numbers

1. Change the operation from subtraction to addition.
2. Change the sign of the number that follows the subtraction symbol.
3. Perform the addition, using the rules for adding real numbers.

● **Example 1–6 A**

Subtraction of real numbers

Subtract the following numbers.

			Step 1 Subtraction to addition		Step 2 Change sign of number being subtracted		Step 3 Add
1.	$9 - 5$	=	9	+	(-5)	=	4
2.	$4 - 11$	=	4	+	(-11)	=	-7
3.	$-9 - 5$	=	-9	+	(-5)	=	-14
4.	$6 - (-8)$	=	6	+	8	=	14
5.	$-12 - (-8)$	=	-12	+	8	=	-4
6.	$-5 - (-14)$	=	-5	+	14	=	9
7.	$8 - 5$	=	8	+	(-5)	=	3
8.	$5 - 8$	=	5	+	(-8)	=	-3

Note From examples 7 and 8, we see that the operation of subtraction is *not* commutative. That is,

$$8 - 5 \neq 5 - 8$$

$-2 - (-6)$

Scientific $2 \boxed{+/-} \boxed{-} 6 \boxed{+/-} \boxed{=}$

Graphing $\boxed{(-)} 2 \boxed{-} \boxed{(-)} 6 \boxed{\text{ENTER}}$
Result 4

☛ *Quick check* $(4) - (-6);$ $(-2) - (-8)$

Addition and Subtraction of More Than Two Real Numbers

When several numbers are being added and subtracted in a horizontal line, do the problem in order from left to right. For example, in

$$9 - 3 + 4 + 3 - 6 - 1 + 4$$

	Operation being performed
$= \underline{9 - 3} + 4 + 3 - 6 - 1 + 4$	$9 - 3 = 6$
$= \underline{6 + 4} + 3 - 6 - 1 + 4$	$6 + 4 = 10$
$= \underline{10 + 3} - 6 - 1 + 4$	$10 + 3 = 13$
$= \underline{13 - 6} - 1 + 4$	$13 - 6 = 7$
$= \underline{7 - 1} + 4$	$7 - 1 = 6$
$= \underline{6 + 4}$	$6 + 4 = 10$
$= 10$	

Note If we had changed each of the indicated subtractions to addition, $9 + (-3) + 4 + 3 + (-6) + (-1) + 4$, then the order in which the problem was carried out would not change the answer. For example, $9 - 3 \neq 3 - 9$, but $9 + (-3) = (-3) + 9$.

Grouping Symbols

Many times, part of the problem will have a group of numbers enclosed with grouping symbols, such as parentheses ( ), brackets [ ], or braces { }. *If any quantity is enclosed with grouping symbols, we treat the quantity within as a single number.* Thus in

$$9 - (3 + 2) + (6 - 2) - (5 - 4)$$

we perform operations within parentheses first to get

$$9 - 5 + 4 - 1$$

which gives

$$4 + 4 - 1 = 8 - 1 = 7$$

● **Example 1–6 B**

Addition and subtraction of real numbers

Perform the indicated operations.

1. $8 - 3 + 2 - 5 - 1 = 5 + 2 - 5 - 1$
$= 7 - 5 - 1$
$= 2 - 1$
$= 1$

2. $(6 + 1) - (2 - 5) + 7 + (9 - 6) = 7 - (-3) + 7 + 3$
$= 10 + 7 + 3$
$= 17 + 3$
$= 20$

3. $(14 - 7) - 2 = 7 - 2 = 5$

4. $14 - (7 - 2) = 14 - 5 = 9$

Note We observe from examples 3 and 4 that the operation of subtraction is _not_ associative. That is, order of grouping does make a difference in subtraction.

$$(14 - 7) - 2 \neq 14 - (7 - 2)$$

☞ _Quick check_ $14 - 11 + 18 - (7 - 12) + 2$

Problem Solving

To solve the following word problems, we must find the difference between two quantities. To find the difference we must _subtract_.

● **Example 1–6 C**

Problem solving

Choose a letter for the unknown and find its value by subtracting.

1. On a given winter's day in Detroit, Michigan, the temperature was $31°$ Fahrenheit (F) in the afternoon. By 9 P.M. the temperature was $-12°$ Fahrenheit. How many degrees did the temperature drop from afternoon to 9 P.M.?

Let t represent the number of degrees fall in temperature. We must find the difference between $31°$ and $-12°$. Thus

degrees fall is difference between
$31°$ and $-12°$

$$t \quad = \quad 31 - (-12)$$

$$t = 31 - (-12) = 31 + 12 = 43$$

There was a $43°$ F drop in temperature.

2. From a board that is 16 feet long, Kesia must cut a board that is 7 feet long. How much is left of the original board?

Let f represent the number of feet of board left. We must find the difference between 16 and 7. Thus

feet left is difference between
16 and 7

$$f \quad = \quad 16 - 7$$

$$f = 16 - 7 = 9$$

Kesia has 9 feet of the original board left.

☞ _Quick check_   A temperature of $14°$ Celsius is decreased by $18°$ Celsius. What is the resulting temperature?

MASTERY POINTS

Can you
• Subtract real numbers?
• Add and subtract in order from left to right?
• Remember that subtraction is _not_ commutative or associative?
• Remember that quantities within grouping symbols represent a single number?

Exercise 1–6	☛ *Quick Check*
Directions Find each sum or difference. See examples 1–6 A and B.	**Examples** **Solutions**

$$\begin{array}{cccccccc} & & & \text{Step 1} & & \text{Step 2} & & \text{Step 3} \\ 4 & - & (-6) & = & 4 & + & 6 & = & 10 \\ -2 & - & (-8) & = & -2 & + & 8 & = & 6 \end{array}$$

Directions   Find each sum or difference. See examples 1–6 A and B.

1. $4 - 5$

2. $-6 - 2$

3. $4 - (-2)$

4. $-3 - (-7)$

5. $-8 - 4$

6. $4 - (-8)$

7. $4 - 9$

8. $-8 - (-5)$

9. $-8 - (-4)$

10. $-12 - (-16)$

11. $8 - (-6)$

12. $14 - (4)$

13. $-6 + 0$

14. $9 - (11)$

15. $7 - 0$

16. $6 + (-10)$

17. $-\dfrac{1}{2} - \left(-\dfrac{1}{4}\right)$

18. $-\dfrac{2}{3} - \left(-\dfrac{1}{4}\right)$

19. $1\dfrac{3}{8} - \left(-1\dfrac{1}{4}\right)$

20. $5\dfrac{5}{6} - \left(-2\dfrac{1}{3}\right)$

21. $-18.7 - (-9.3)$

22. $107.4 - (-12.6)$

23. $-215.8 - 96.2$

24. $-119.1 - 218.8$

25. $-512.7 - (-814.5)$

26. $-12 - (-10) - 8$

27. $-30 + 14 - 8$

28. $-25 + 4 - 32 + 28$

29. $24 - (-12) - 12 + (-13)$

30. $-2 - 3 + (-4) - (-5) + (-6)$

31. $-15 - 13 - (-7) - 32$

32. $-17 - 11 - (-12) - (-5)$

Directions Find each sum or difference. See example 1–6 B.	☛ *Quick Check*

Example $14 - 11 + 18 - (7 - 12) + 2$

Solution
$$\begin{aligned} &= 14 - 11 + 18 - (-5) + 2 \\ &= 3 + 18 - (-5) + 2 \\ &= 21 - (-5) + 2 \\ &= 26 + 2 \\ &= 28 \end{aligned}$$

Perform operations within parentheses first, then add and subtract from left to right

Directions   Find each sum or difference. See example 1–6 B.

33. $17 + 4 - (7 - 2)$

34. $(25 - 2) - (12 - 3)$

35. $-6 - 4 + 8 - (8 - 7)$

36. $32 - 5 + 7 - 4 - (11 - 8)$

37. $10 - 10 + (10 + 10) - 10$

38. $12 + 3 - 16 - 10 - (12 + 5)$

39. $10 + (2 - 21) - (7 - 8)$

40. $(12 + 3) - 16 - 10 + (5 - 12)$

41. $(14 - 18) - (17 - 12) - 16$

42. $8 - 4 + 7 - (2 - 5) - 3$

Directions Work the following problems. See example 1–6 C.	☛ *Quick Check*

Example A temperature of 14° C (Celsius) is decreased by 18° C. What is the resulting temperature?

Solution $14 - 18 = 14 + (-18)$
$$ = -4$$

The resulting temperature is −4° C.

Directions   Work the following problems. See example 1–6 C.

43. A temperature of $(-6)°$ C is decreased by $32°$ C. What is the resulting temperature?

44. An electronics supply house has 432 resistors of a certain type. If 36 are sold during the first week, 72 during the second week, 29 during the third week, and 58 during the fourth week, how many are left at the end of the month?

45. Tim owes Tom and Rob $343.79 and $205.43, respectively, and Terry owes Tim $176.62. In terms of positive and negative symbols, how does Tim stand monetarily?

46. If a person has $78.81 after paying off a debt of $23.12, how much money did he have before paying off the debt? Write a statement involving the operation of subtraction of integers to show your answer.

47. A piece of wood 12 feet long is cut into three pieces so that two of the pieces measure $3\frac{1}{2}$ feet and $4\frac{1}{4}$ feet. What is the length of the third piece?

Directions Find the difference.	**Example** 12 diminished by 20
	Solution $12 - 20$
	$\quad = 12 + (-20)$ Diminished by 20 means subtract 20
	$\quad = -8$ Change to addition

Directions   Find the difference.

48. -8 diminished by 11

49. -15 diminished by 7

50. -6 diminished by -21

51. -18 diminished by -9

52. Subtract -26.3 from -18.1.

53. Subtract -19.6 from 41.7.

54. Subtract $-17\frac{1}{2}$ from $28\frac{1}{4}$.

55. From $-26\frac{1}{3}$ subtract $-45\frac{5}{6}$.

56. From -43 subtract -16.

57. 5 less than -8

58. 4 less than -12

59. 8 less than 3

60. 10 less than 5

Directions   Choose a letter for the unknown quantity and find the indicated difference. See example 1–6 C.

61. Chan has $23.12 in his savings account. If he spends $15.97 to buy a CD, how much is left in his savings account?

62. Ali has $44.84 in his savings account. He wishes to buy a baseball glove for $36.14. How much is left in his savings account?

63. The temperature was $-15°$ at 6 A.M. but by noon the temperature has risen to $23°$. How many degrees did it rise from 6 A.M. to noon?

64. The temperature dropped $22°$ from $-7°$ at midnight to just before daybreak at 7 A.M. What was the temperature at 7 A.M.?

65. Erin Nustad was born in 1986. How old will she be in the year 2000?

66. A chemist has 100 ml of acid and she needs 368 ml of the acid. How much more is needed?

67. Amy has $451.63 in assets and she wants to borrow enough money to buy a stereo system for $695. How much will she owe?

68. The top of Mt. Everest in Asia is 29,028 feet above sea level and the top of Mt. McKinley in Alaska is 20,320 feet above sea level. How much higher is the top of Mt. Everest than the top of Mt. McKinley?

69. Death Valley in California is 282 feet below sea level (-282). What is the difference in the altitude between Mt. McKinley and Death Valley? See exercise 68.

70. Mt. Whitney in California is 14,494 feet above sea level and the Salton Sea in California is 235 feet below sea level. What is the difference between the altitude of Mt. Whitney and Salton Sea?

71. Chad owes John $25.25. He paid back $13.50 and then had to borrow another $6.75. How much does Chad owe John now?

72. On 4 successive hands in a poker game, Sheila won $10, then lost $9, then lost $14, and finally lost $6. What is her financial position after the 4 hands?

❀ Explore to Learn

1. Suppose that signed numbers represent money in a checking account. It is allowed to have a negative balance. Then $200 - 10$ might mean that there is $200 in the account and a check for $10 is cashed against the account. Write a sentence that $-200 - 10$ might mean. Write a sentence that $-200 - (-500)$ could mean. For a logical interpretation you should think about the fact that money has to get into an account too!

2. Let a be a positive number and b be a negative number.

$$\text{Is } a - b > 0? \quad \text{Is } b - a > 0?$$
$$\text{Is } a + b > 0? \quad \text{Is } b + a > 0?$$

1–7   Multiplication and Division of Real Numbers

Multiplication of Two Positive Numbers

We are already familiar with the fact that the product of two positive numbers is positive. We can see this fact by considering multiplication as repeated addition. For example, if we wish to add four 3s, then $3 + 3 + 3 + 3 = 12$. Another way of expressing this repeated addition is $4 \cdot 3 = 12$, in which case the raised dot, $\cdot$, means multiply or times. We could also have added three 4s: $4 + 4 + 4 = 12$, which could be written as $3 \cdot 4 = 12$. This observation illustrates an important mathematics principle called the **commutative property of multiplication**.

Commutative Property of Multiplication

For every real number a and b,

$$a \cdot b = b \cdot a$$

Concept

This property tells us that changing the order of the numbers when we multiply will not change the answer (product).

In the previous paragraph, the number 12 is called the **product** of 4 and 3, and 4 and 3 are called **factors** of 12. *The numbers or variables in an indicated multiplication are referred to as the factors of the product.*

In our example, we used a raised dot to indicate the operation of multiplication. The cross, $\times$, is used in arithmetic to indicate multiplication.

We avoid using it in algebra because it may become confused with the variable x. Another way to indicate multiplication is the absence of any operation symbol between factors.

Multiplication of Two Numbers with Different Signs

As an illustration of multiplying a positive number times a negative number, consider the following pattern:

$$3 \cdot 3 = 9$$
$$2 \cdot 3 = 6$$
$$1 \cdot 3 = 3$$
$$0 \cdot 3 = 0$$
$$(-1) \cdot 3 = -3$$
$$(-2) \cdot 3 = -6$$
$$(-3) \cdot 3 = -9$$

The product decreases by 3

The product of a negative and a positive is a negative

We observe from this pattern that our product decreases by 3 each time. It logically follows that the product of a negative number and a positive number is a negative number.

A second observation from this pattern is that zero times a number has a product of zero. This is called the **zero factor property**.

Zero Factor Property

For every real number a,

$$a \cdot 0 = 0 \cdot a = 0$$

Concept

Multiplying any number by zero always gives zero as the answer. That is, whenever we are multiplying and zero is one of the factors, the product will be zero.

A third observation from this pattern is that 1 times a number is equal to the number. For this reason, 1 is called the **identity element of multiplication**.

Identity Property of Multiplication

For every real number a,

$$a \cdot 1 = 1 \cdot a = a$$

Concept

Multiplying a number by 1 leaves the number unchanged.

Multiplication of Two Negative Numbers

We will observe another pattern when we consider the following.

$$3(-3) = -9$$
$$2(-3) = -6$$
$$1(-3) = -3$$
$$0(-3) = 0$$
$$(-1)(-3) = +3$$
$$(-2)(-3) = +6$$
$$(-3)(-3) = +9$$

The product increases by 3

The product of two negatives is a positive

From this pattern, we can see that our product increases by 3 each time. It logically follows that *the product of two negative numbers is a positive number.*

We can summarize our procedures for multiplication of real numbers as follows:

Multiplication of Two Real Numbers

To multiply two real numbers, multiply their absolute values and
1. the product will be positive if the numbers have the same sign;
2. the product will be negative if the numbers have different signs.

● **Example 1–7 A**

Multiplication of real numbers

Multiply the following numbers.

1. $(-2) \cdot 3 = -6$

Product of their absolute values: $2 \cdot 3 = 6$
Product is negative because the numbers have different signs

2. $(-2)(-4) = 8$

Product of their absolute values: $2 \cdot 4 = 8$
Product is positive because the numbers have the same sign

3. $4(-4) = -16$ Negative because signs are different

4. $(-5)(-5) = 25$ Positive because signs are the same

☞ *Quick check* $(-4)(-3)$

Multiplication of More than Two Real Numbers

1. If in the numbers being multiplied there is an **odd** number of negative factors, the answer will be negative.
2. If in the numbers being multiplied there is an **even** number of negative factors, the answer will be positive.

● **Example 1–7 B**

Multiplication of more than two real numbers

Multiply the following numbers.

1. $(-7)(-2)(5) = (14)(5) = 70$ Even number of negative factors

2. $(-6)(2)(-4) = (-12)(-4) = 48$ Even number of negative factors

3. $[(-3)(5)](4) = (-15)(4) = -60$ Odd number of negative factors

4. $(-3)[(5)(4)] = (-3)(20) = -60$ Odd number of negative factors

$(-9)(-2)(5)$

Scientific   9 [+/−] [×] 2 [+/−] [×] 5 [=]

Graphing   [(−)] 9 [×] [(−)] 2 [×] 5 [ENTER]
Result 90

☛ *Quick check* $(-3)(-2) \cdot 4$

Examples 3 and 4 illustrate an important mathematical principle called the **associative property of multiplication**.

Associative Property of Multiplication

For every real number *a*, *b*, and *c*,

$$(a \cdot b)c = a(b \cdot c)$$

Concept
Changing the grouping of the numbers will not change the product.

Problem Solving

To solve the following problems, we must multiply the quantities.

● **Example 1–7 C**

Problem solving

Choose a letter for the unknown and find the indicated product.

1. What is the cost of 7 VHS tapes if each tape costs $5.95?

Let *c* represent the cost of the 7 tapes. We must multiply to find the total cost of all 7 tapes. Thus

total cost	is	7 tapes	at	$5.95 each
c	=	7	·	(5.95)

$$c = 7 \cdot (5.95) = 41.65$$

The 7 tapes cost $41.65.

2. On 4 successive days, the stock market dropped 9 points (represented by a negative number) each day. How many points did the market change in the 4 days?

Let *d* represent the total drop. Represent the 9-point drop by -9. We multiply to obtain

total drop	is	4 days	at	9-point drop each day
d	=	4	·	(−9)

$$d = 4 \cdot (-9) = -36$$

The stock market dropped 36 points (-36) in the 4 days.

Division of Two Real Numbers

Recall that when we divide a number (called the *dividend*) by another number (called the *divisor*), we compute an answer (called the *quotient*). We define the operation of division as follows:

Definition of Division

If $b \neq 0$,* $\dfrac{a}{b} = q$ provided that $b \cdot q = a$, where a is the dividend, b is the divisor, and q is the quotient.

The second part of this definition of division shows how to check the answer to the problem. We multiply the divisor by the quotient to get the dividend ($b \cdot q = a$).

By the above definition, the quotient of the two negative numbers $(-20) \div (-5)$ or $\dfrac{-20}{-5}$ must be that number which multiplied by -5 gives -20. That number is 4, since $(-5)(4) = -20$. Therefore $\dfrac{-20}{-5} = 4$. We observe that *the quotient of two negative numbers is a positive number.*

To divide a positive number by a negative number, or a negative number by a positive number, consider the following divisions:

$$(-14) \div (2) = \frac{-14}{2} = -7$$

since

$$(2)(-7) = -14$$

and

$$(24) \div (-6) = \frac{24}{-6} = -4$$

because

$$(-6)(-4) = 24$$

We find that *the quotient of a positive number and a negative number is always a negative number.*

Division of Two Real Numbers

To divide two real numbers, perform the division using the absolute values of the numbers and
1. the quotient will be positive if the numbers have the same sign.
2. the quotient will be negative if the numbers have different signs.

*The reason for this restriction will be explained on page 65.

● **Example 1–7 D**

Division of real numbers

Divide the following numbers.

1. $\dfrac{-14}{-7} = 2$, since $(-7)(2) = -14$

2. $\dfrac{-36}{-6} = 6$, since $(-6)(6) = -36$

3. $\dfrac{-24}{3} = -8$, since $(3)(-8) = -24$

4. $\dfrac{15}{-5} = -3$, since $(-5)(-3) = 15$

☞ *Quick check* $\dfrac{-18}{-9}$; $\dfrac{-15}{3}$

Division Involving Zero

In section 1–4, we defined a rational number to be any number that can be expressed as a quotient of two integers in which the divisor is not zero. The number zero, 0, is the only number that we cannot use as a divisor. To see why we exclude zero as a divisor, recall that we check a division problem by multiplying the divisor times the quotient to get the dividend. If we apply this idea in connection with zero as a divisor, we observe the following situations. Suppose there was a number q such that $3 \div 0 = q$. Then $q \cdot 0$ would have to be equal to 3 for our answer to check, but this product is zero regardless of the value of q. Therefore we cannot find an answer for this problem. We say that division by zero is *undefined*. If we try to divide zero by zero and again call our answer q, we have $0 \div 0 = q$. When we check our work, $0 \cdot q = 0$, we see that any value for q will work. Since any value for q will work, there is no single quotient. We therefore decide that **division by zero is not allowed and whenever a division problem involves division by zero, we will write undefined**.

It is important to note that although division by zero is not allowed, this does not extend to the division of zero by some other number. We can see that $\dfrac{0}{-4} = 0$ since $(-4) \cdot 0 = 0$. Thus, **the quotient of zero divided by any number other than zero is always zero**.

● **Example 1–7 E**

Division involving zero

Perform the division, if possible.

1. $\dfrac{0}{5} = 0$ **3.** $\dfrac{-7}{0}$ is undefined

2. $\dfrac{2}{0}$ is undefined   **4.** $\dfrac{0}{-7} = 0$

☞ *Quick check*   Divide $\dfrac{11}{0}$, if possible.

Problem Solving

To solve the following problems, we will have to divide the given quantities.

● **Example 1–7 F**
Problem solving

Choose a letter for the unknown quantity and find the indicated quotient.

1. If $7.68 is spent on 6 three-way light bulbs, how much did each light bulb cost?

Let c represent the cost of each light bulb. We must divide $7.68 by 6. Thus

cost per bulb	is equal to	total cost	divided	by 6 identical items
c	=	(7.68)	÷	6

$$c = (7.68) \div 6 = 1.28.$$

Each light bulb cost $1.28.

2. There are 400 people seated in a full auditorium. If there are 25 identical rows of seats, how many people are there in each row?

Let x represent the number of people in each row. We must divide 400 by 25. Thus

people per row	is equal to	total number of people	divided	by 25 identical rows
x	=	400	÷	25

$$x = 400 \div 25 = 16$$

There are 16 people seated in each row.

MASTERY POINTS

Can you
- Use the commutative and associative properties of multiplication, the zero factor property, and the identity property of multiplication?
- Multiply real numbers?
- Perform division with real numbers?
- Remember the results of division involving zero?

Exercise 1–7

Directions
Perform the indicated operations. See examples 1–7 A and B.

⚑ *Quick Check*

Examples $(-4)(-3)$

Solutions $= 12$ Product is positive because the numbers have the same sign

$(-3)(-2) \cdot 4$

$= 6 \cdot 4$ $(-3)(-2) = 6$ because the numbers have the same sign

$= 24$ 6 times 4 is 24

Directions   Perform the indicated operations. See examples 1–7 A and B.

1. $(-3)(-5)$
2. $0 \cdot (-6)$
3. $4 \cdot (-7)$
4. $(-8) \cdot 3$
5. $4 \cdot (-3) \cdot 5$
6. $(-2)(2)(-2)$
7. $4 \cdot (-9)$
8. $(-3)(-2)(-8)$
9. $(-1)(-4)(5)$

10. $(-5)(2)(4)(3)$

11. $7 \cdot (-1)(-3)(-5)$

12. $2 \cdot (-3)(-1)(2)(-2)(3)$

13. $(-1.8)(2.4)$

14. $(-5.7)(-6.12)$

15. $(0.49)(-28.1)$

16. $(-8.9)(-8.9)$

17. $(-27)(0.08)$

18. $\left(-\dfrac{1}{3}\right)\left(\dfrac{3}{5}\right)$

19. $\left(-\dfrac{3}{4}\right)\left(-\dfrac{3}{4}\right)$

20. $\left(-\dfrac{3}{4}\right)\left(\dfrac{8}{9}\right)$

21. $\left(-\dfrac{5}{8}\right)\left(-\dfrac{2}{5}\right)$

22. $\left(\dfrac{5}{12}\right)\left(-\dfrac{9}{10}\right)$

23. $(-5)(-4)(-3)(2)$

24. $(-2)(-7)(7)(4)$

25. $(-3)(3)(-4)(4)$

26. $(-1)(-1)(-1)(-1)$

27. $(-2)(0)(3)(-4)$

28. $(-3)(-2)(4)(0)$

29. $(-5)(0)(-4)$

Directions   Choose a letter for the unknown and multiply to find the value. See example 1–7 C.

30. A man acquires a debt of $6 each day for 5 days. If we represent a $6 debt by (-6), write a statement of the change in his assets after 5 days. What is the change?

31. An auditorium contains 42 rows of seats. If each row contains 25 seats, how many people can be seated in the auditorium?

32. There are seven rows of desks in a classroom. If each row contains eight desks, how many students will the classroom hold?

33. A clothier ordered 15 suits, each costing him $93.28. What was the total cost of the 15 suits?

34. Richelle purchased two dozen (24) cans of frozen orange juice concentrate that was on sale for 57¢ per can. How much did the orange juice cost her?

35. Nassar lost an average of $23 in four successive poker games. What were his total losses? (Represent this by a negative answer.)

36. If a bank advertises that a person can double her investment in a savings account in 9 years, how much will $3,125.63 grow to in 9 years?

37. Tara sold 35 tee shirts at the concert. If she charges $15.50 each, what was the total sales?

38. Gerardo deposits $27.50 per week in a savings account. How much has he deposited in 3 years? (Use 52 weeks = 1 year.)

39. A grocer averages selling 25 gallons of milk each day. How many gallons of milk does he sell in 4 weeks? (Assume the grocery is open 7 days per week.)

40. A department store averages a loss of $75, due to thefts, each day. How much is lost in a month of 31 days?

Directions Perform the indicated operations, if possible. See examples 1–7 D and E.	☛ **Quick Check**		
	Examples $\dfrac{-18}{-9}$	$\dfrac{-15}{3}$	$\dfrac{11}{0}$
	Solutions $= 2$ Quotient of two negatives is a positive	$= -5$ Quotient of a positive and a negative is a negative	is undefined

Directions   Perform the indicated operations, if possible. See examples 1–7 D and E.

41. $\dfrac{-14}{-7}$

42. $\dfrac{-15}{5}$

43. $\dfrac{32}{-4}$

44. $\dfrac{18}{3}$

45. $\dfrac{-22}{-11}$

46. $\dfrac{18}{-3}$

47. $\dfrac{-16}{2}$

48. $\dfrac{-25}{-5}$

49. $\dfrac{7}{0}$

50. $\dfrac{-4}{0}$

51. $\dfrac{0}{-9}$

52. $\dfrac{0}{5}$

53. $\dfrac{10}{0}$

54. $\dfrac{-24}{-6}$

55. $\dfrac{49}{-7}$

56. $\dfrac{36}{-6}$

57. $\dfrac{-25}{5}$

58. $\dfrac{-64}{8}$

59. $\dfrac{(-4)(-3)}{-6}$

60. $\dfrac{(-18)(2)}{-4}$

61. $\dfrac{(16)(2)}{-8}$

62. $\dfrac{(-4)(0)}{-8}$

63. $\dfrac{(-16)(0)}{-8}$

64. $\dfrac{(-5)(-2)}{(-1)(-10)}$

65. $\dfrac{(-18)(3)}{(-2)(-9)}$

66. $\dfrac{(-2)(-4)}{(0)(4)}$

67. $\dfrac{(-3)(6)}{(0)(-2)}$

68. $\dfrac{8-8}{3+4}$

69. $\dfrac{(-6)}{(-3)(0)}$

70. $\dfrac{6+6}{6-6}$

Directions Choose a letter for the unknown quantity and find the indicated quotient. See example 1–7 F.	**Example** A football player carried the ball eight times, making the following yardages: gain of 6 yards (yd), loss of 3 yd, loss of 4 yd, gain of 4 yd, gain of 3 yd, loss of 1 yd, loss of 2 yd, gain of 5 yd. Show his gains and losses by positive and negative integers. What was his average gain or loss per carry?
	Solution 6 yd, −3 yd, −4 yd, 4 yd, 3 yd, −1 yd, −2 yd, 5 yd. To find an average, we add together all of the values and divide by the total number of values. $$\dfrac{6 \text{ yd} + (-3) \text{ yd} + (-4) \text{ yd} + 4 \text{ yd} + 3 \text{ yd} + (-1) \text{ yd} + (-2) \text{ yd} + 5 \text{ yd}}{8}$$ $$= \dfrac{8 \text{ yd}}{8} = 1 \text{ yd}$$

Directions   Choose a letter for the unknown quantity and find the indicated quotient. See example 1–7 F.

71. The temperature at 1 P.M. for seven consecutive days in January was 5° C, −8° C, −7° C, −1° C, 10° C, −6° C, and 0° C. What was the average temperature for the seven days?

72. If the stock market showed the following gains and losses during 6 consecutive hours of trading on a given day, determine the average gain or loss during that 6-hour period. Gain 36 points, loss 23 points, loss 72 points, gain 25 points, loss 31 points, loss 21 points.

73. Between Chicago and Detroit, a distance of 282 miles, a driver averages 47 miles per hour. How long will it take her to make the trip?

74. A trip of 369 miles takes 9 hours to complete. What was the average rate of speed?

75. Light travels at a rate of 186,000 miles per second. How long will it take to travel 1,674,000 miles?

76. How long does it take light from the sun to reach earth if the sun is approximately 93,000,000 miles away? (Refer to exercise 75.)

77. Benita paid $36 for nine crates of peaches for her fruit market. How much did each crate cost her?

78. A carpenter wishes to cut a 10-foot board into four pieces that are all the same length. Find the length of each piece.

79. A man drove 384 miles and used 15 gallons of gasoline. How many miles per gallon did he get?

80. Irene drove 420 miles in 8 hours. How many miles did she travel each hour (in miles per hour) if she drove at a constant speed?

81. During a recent cold wave, the temperature fell 28° over a 7-day period. What was the average change per day?

82. Terrell typed 1,350 words in 30 minutes. How many words did he type per minute?

83. The college bookstore purchased 480 math textbooks. If the books came in 15 boxes of the same size, how many books were in each box?

84. Four girls worked together painting a barn. If they received $159 to split evenly among them, how much did each girl receive?

85. Hiroko took part in a 26-mile marathon run. If she ran the marathon in 5 hours and 12 minutes, how long did it take her to run 1 mile (in minutes) if she ran at a constant speed?

86. A farmer got 2,592 bushels of wheat from a 30-acre field. How many bushels did he get per acre?

✵ Explore to Learn

1. We seem to use a lot of parentheses in algebra. Consider $8 - (-9)$, or $(-8)(-9)(3)$. This is mainly because we use the same symbol, $-$, for the subtraction operation and to indicate negative values. Experiment with a different symbol to represent the opposite (negative) of a value and, assuming we keep the $\cdot$ symbol for multiplication, do we need all those parentheses? Some books use something like $^-3$ to mean (-3). Describe in writing some advantages and some disadvantages of a new symbol.

2. Revisit why we do not define division by zero by looking at one example. Suppose we consider $10 \div 0$, or $\dfrac{10}{0}$. Assume this gives some other number.

If so it must be something logical like 0, 1, or 10. Remembering that $\dfrac{10}{2} = 5$ because $10 = 2 \cdot 5$, explore each of $\dfrac{10}{0} = 0$, $\dfrac{10}{0} = 1$, and $\dfrac{10}{0} = 10$. Could any of these be correct? Why or why not?

3. Use $a = 15$ and $b = 5$ to show that $\dfrac{-a}{b} = \dfrac{a}{-b} = -\dfrac{a}{b}$. Use $a = 15$ and $b = 5$ to show that $\dfrac{a}{b} = a \cdot \dfrac{1}{b}$.

1–8   Properties of Real Numbers and Order of Operations

In the previous four sections, we introduced some of the properties of real numbers. We also saw how these properties are used when performing fundamental operations with numbers. Since variables represent numbers, we will be using these and other properties throughout our study of algebra. The properties that we have covered so far are listed and the page number where the property was first introduced is given for reference.

Properties of Real Numbers

If a, b, and c are any real numbers, then

$a + b = b + a$, commutative property of addition (page 47)
$a \cdot b = b \cdot a$, commutative property of multiplication (page 60)
$(a + b) + c = a + (b + c)$, associative property of addition (page 51)
$(a \cdot b)c = a(b \cdot c)$, associative property of multiplication (page 63)
$a + 0 = 0 + a = a$, identity property of addition (page 47)
$a \cdot 1 = 1 \cdot a = a$, identity property of multiplication (page 61)
$a + (-a) = 0$, additive inverse property (page 50)
$a \cdot 0 = 0 \cdot a = 0$, zero factor property (page 61)

Exponents

Consider the indicated products

$$4 \cdot 4 \cdot 4 = 64$$

and

$$3 \cdot 3 \cdot 3 \cdot 3 = 81$$

A more convenient way of writing $4 \cdot 4 \cdot 4$ is 4^3, which is read "4 to the third power" or "4 cubed." We call the number 4 the **base** of the expression and the number 3, to the upper right of 4, the **exponent**.

Thus

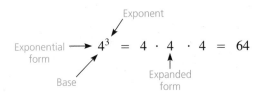

Note The exponent tells us how many times the base is used as a factor in an indicated product. Also, the exponent is understood to be 1 when a number or a variable has no exponent. That is, $5 = 5^1$.

Remember that when we have a negative number, we place it inside parentheses. With this fact in mind, we can see that there is a definite difference between $(-2)^4$ and -2^4. In the first case, the parentheses denote that this is a negative number to a power: $(-2)^4 = (-2)(-2)(-2)(-2) = +16$. In the second case, since there are no parentheses around the number, we understand that this is *not* (-2) to a power. It is, rather, the opposite of the answer when we raise 2^4: $-2^4 = -(2)^4 = -(2 \cdot 2 \cdot 2 \cdot 2) = -(16) = -16$.

● **Example 1–8 A**

Powers involving negative numbers

Perform the indicated multiplication.

1. $(-3)^3 = (-3)(-3)(-3) = -27$ Odd number of negative factors

2. $-3^3 = -(3 \cdot 3 \cdot 3) = -27$ Opposite of the value of 3^3

3. $(-3)^4 = (-3)(-3)(-3)(-3) = 81$ Even number of negative factors

4. $-3^4 = -(3 \cdot 3 \cdot 3 \cdot 3) = -81$ Opposite of the value of 3^4

☛ *Quick check* -3^2

When we are performing several different types of arithmetic operations within an expression, we need to agree on an order in which the operations will be performed. To show that this is necessary, consider the following numerical expression.

$$3 + 4 \cdot 5 - 3$$

More than one answer is possible, depending on the order in which we perform the operations. To illustrate,

$$3 + 4 \cdot 5 - 3 = 7 \cdot 2 = 14$$

if we add and subtract as indicated before we multiply. However,

$$3 + 4 \cdot 5 - 3 = 3 + 20 - 3 = 20*$$

if we multiply before we add or subtract. A third possibility would be

$$3 + 4 \cdot 5 - 3 = 3 + 4 \cdot 2 = 3 + 8 = 11$$

if we subtract, then multiply, and finally add. To standardize the answer, we agree to the following order of operations, or priorities.

Order of Operations, or Priorities

1. **Groups**   Perform any operations within a grouping symbol such as ( ) parentheses, [ ] brackets, { } braces, | | absolute value, and above or below the fraction bar.
2. **Exponents**   Perform operations indicated by exponents.
3. **Multiply and divide**   Perform multiplication and division in order from left to right.
4. **Add and subtract**   Perform addition and subtraction in order from left to right.

Note

   a. Within a grouping symbol, the order of operations will still apply.
   b. If there are several grouping symbols intermixed, remove them by starting with the innermost one and working outward.

● **Example 1–8 B**

Order of operations

Perform the indicated operations in the proper order and simplify.

1. $7 + 8 \cdot 3 \div 2 = 7 + 24 \div 2$ Multiply
$$= 7 + 12 \qquad\qquad \text{Divide}$$
$$= 19 \qquad\qquad\quad \text{Add}$$

2. $(7 - 1) \div 2 + 3 \cdot 4 = 6 \div 2 + 3 \cdot 4$ Parentheses
$$= 3 + 12 \qquad\qquad \text{Divide and multiply}$$
$$= 15 \qquad\qquad\quad \text{Add}$$

*This is the correct answer

3. $\dfrac{1}{2} + \dfrac{3}{4} \div \dfrac{5}{8} = \dfrac{1}{2} + \dfrac{3}{2 \cdot 2} \cdot \dfrac{2 \cdot 2 \cdot 2}{5}$ Invert, factor, divide out common factors

$= \dfrac{1}{2} + \dfrac{6}{5}$ Multiply remaining factors

$= \dfrac{5}{10} + \dfrac{12}{10}$ Least common denominator

$= \dfrac{5 + 12}{10}$ Add

$= \dfrac{17}{10}$ or $1\dfrac{7}{10}$ Add

4. $2^2 \cdot 3 - 3 \cdot 4 = 4 \cdot 3 - 3 \cdot 4$ Exponent

$= 12 - 12$ Multiply

$= 0$ Subtract

  $2^2 \cdot 3 - 3 \cdot 4$

Scientific or Graphing $2 \boxed{x^2} \boxed{\times} 3 \boxed{-} 3 \boxed{\times} 4 \boxed{=}$ (or $\boxed{\text{ENTER}}$)
Result 0

5. $\dfrac{3}{4} - \dfrac{1}{2} \cdot \dfrac{2}{3} = \dfrac{3}{4} - \dfrac{1}{3}$ Divide out common factors and multiply

$= \dfrac{9}{12} - \dfrac{4}{12}$ Least common denominator

$= \dfrac{9 - 4}{12}$ Subtract

$= \dfrac{5}{12}$ Subtract

6. $(7.28 + 1.6) \div 2.4 - (6.1)(3.8)$

$= (8.88) \div 2.4 - (6.1)(3.8)$ Parentheses

$= 3.7 - 23.18$ Division and multiplication

$= -19.48$ Subtract

7. $\left(\dfrac{2}{3} + \dfrac{7}{8}\right) \div \dfrac{5}{6} = \left(\dfrac{16}{24} + \dfrac{21}{24}\right) \div \dfrac{5}{6}$ Parentheses

$= \left(\dfrac{16 + 21}{24}\right) \div \dfrac{5}{6}$ Parentheses

$= \dfrac{37}{24} \div \dfrac{5}{6}$ Parentheses

$= \dfrac{37}{2 \cdot 2 \cdot 2 \cdot 3} \cdot \dfrac{2 \cdot 3}{5}$ Invert, factor, divide out common factors

$= \dfrac{37}{20}$ or $1\dfrac{17}{20}$ Multiply remaining factors

8. $(5.4)^2 - 4(3.1)(2.8)$

$= 29.16 - 4(3.1)(2.8)$ Exponent

$= 29.16 - 34.72$ Multiply

$= -5.56$ Subtract

9. $\dfrac{3(2+4)}{4-2} - \dfrac{4+6}{5} = \dfrac{3(6)}{4-2} - \dfrac{4+6}{5}$ Groups: numerator and denominator

$$= \dfrac{18}{2} - \dfrac{10}{5}$$ Numerator and denominator

$$= 9 - 2$$ Divide

$$= 7$$ Subtract

    $\dfrac{3(2+4)}{4-2} - \dfrac{4+6}{5}$

Scientific or Graphing $3\;\boxed{\times}\;\boxed{(}\;2\;\boxed{+}\;4\;\boxed{)}\;\boxed{\div}\;\boxed{(}\;4\;\boxed{-}\;2\;\boxed{)}\;\boxed{-}$
$\boxed{(}\;4\;\boxed{+}\;6\;\boxed{)}\;\boxed{\div}\;5\;\boxed{=}\;(\text{or }\boxed{\text{ENTER}}\;)$
Result 7

10. $5[7 + 3(10 - 4)]$

We first evaluate within the grouping symbol, applying the order of operations.

$$5[7 + 3(6)] = 5[7 + 18]$$ Groups

$$= 5[25]$$ Groups

$$= 125$$ Multiply

☛ Quick check $18 \div 6 \cdot 3 + 10 - (4 + 5)$

Problem Solving

Solve the following word problems using the order of operations.

● **Example 1–8 C**
Problem solving

Choose a variable to represent the unknown quantity and find its value by performing the indicated operations.

1. Asha purchased 6 boxes of cereal at $3.25 per box and 7 cans of tuna at $1.49 per can. What was her total bill?

Let t represent Asha's total bill. 6 boxes at $3.25 per box cost $6 \cdot \$3.25$, 7 cans at $1.49 per can cost $7 \cdot \$1.49$. The total bill is given by

total bill	is equal to	6 boxes of cereal	at	$3.25 per box	and	7 cans of tuna	at	$1.49 per can
t	$=$	6	$\cdot$	(3.25)	$+$	7	$\cdot$	(1.49)

$$t = 6 \cdot (3.25) + 7 \cdot (1.49)$$

$$= 19.50 + 10.43$$ Multiply

$$= 29.93$$ Add

Asha's total bill was $29.93.

2. A man works a 40-hour week at $12 per hour. If he works 11 hours of overtime at time and a half, how much will he receive for the 51 hours of work?

Let w represent the man's total wages for the week. 40 hours at $12 per hour is $40 \cdot \$12$. Hourly rate at time and a half is $\left(1\frac{1}{2} \cdot 12 = 18\right)$ and 11 hours at time and a half is $11 \cdot 18$. Thus

| total | is equal | 40 | at $12 per | and | 11 | at $18 per |
| wages | to | hours | hour | | hours | hour |

$$w = 40 \cdot 12 + 11 \cdot 18$$

$$w = 40 \cdot 12 + 11 \cdot 18$$
$$w = 480 + 198 \quad \text{Multiply}$$
$$w = 678 \quad \text{Add}$$

The man will receive $678 for 51 hours of work.

MASTERY POINTS

Can you
- Use the order of operations?
- Use exponents?

Exercise 1–8

Directions
Perform the indicated operations. See example 1–8 A.

⚑ Quick Check

Example -3^2

Solution $= -(3^2)$
$= -9$

-3^2 is not the same as $(-3)^2$, it is the opposite of the value of 3^2

Directions   Perform the indicated operations. See example 1–8 A.

1. $(-4)^2$
2. $(-5)^4$
3. $(-3)^3$
4. -4^2
5. -6^2
6. -2^4
7. -1^2
8. -2^2
9. $(-1)^2$
10. $(-2)^2$

Directions
Perform the indicated operations and simplify. See example 1–8 B.

⚑ Quick Check

Example $18 \div 6 \cdot 3 + 10 - (4 + 5)$

Solution
$= 18 \div 6 \cdot 3 + 10 - 9 \quad \text{Parentheses}$
$= 3 \cdot 3 + 10 - 9 \quad \text{Division}$
$= 9 + 10 - 9 \quad \text{Multiplication}$
$= 19 - 9 \quad \text{Addition}$
$= 10 \quad \text{Subtraction}$

Directions   Perform the indicated operations and simplify. See example 1–8 B.

11. $\dfrac{4+2}{3} + 2$
12. $-6 \cdot 7 + 8$
13. $6 + 5 \cdot 4$
14. $\dfrac{1}{5} \cdot 5 + 6$
15. $-2 + 10 \cdot \dfrac{1}{5}$
16. $4(3 - 2)(2 + 1)$
17. $0(5 + 2) + 3$
18. $\dfrac{24 \cdot 3}{9} - 6$
19. $7 - \left(\dfrac{7-2}{2-7}\right)$
20. $\left(\dfrac{4-8}{8-4}\right) - (-3)4$
21. $\left(\dfrac{10-5}{5-10}\right)\left(\dfrac{2-6}{6-2}\right)$

22. $\left(\dfrac{3-9}{9-3}\right)\left(\dfrac{8-2}{2-8}\right)$

23. $(24-6)\div 3$

24. $(37-4)\div 11$

25. $\dfrac{2}{3}\div\left(\dfrac{5}{6}-\dfrac{4}{9}\right)$

26. $12\cdot 4+2$

27. $2+3(8-5)$

28. $5+2(11-6)$

29. $6+4(8+2)$

30. $7+3^2(9-4)$

31. $8-3^2(6-4)$

32. $10-2(7-11)$

33. $15\cdot 3^2-14$

34. $(8-3)(5+3)^2$

35. $\dfrac{7}{8}-\dfrac{1}{2}\div\dfrac{3}{4}$

36. $\dfrac{3}{8}+\dfrac{7}{12}\cdot\dfrac{3}{14}$

37. $3(6-2)(7+1)$

38. $12+3\cdot 16\div 4^2-2$

39. $9-3(12+3)-4\cdot 3$

40. $15-2(8+1)-6\cdot 4$

41. $50-4(6-8)+5\cdot 4$

42. $18-5(7+3)-6^2$

43. $10-3\cdot 4\div 6-5^2$

44. $8-(12+3)-4\cdot 3$

45. $4(2-5)^2-2(3-4)$

46. $6(-8+10)^2-5(4-7)$

47. $\dfrac{5(3-5)}{2}-\dfrac{27}{-3}$

48. $\dfrac{3(8-6)}{2}-\dfrac{8}{-2}$

49. $\dfrac{5(6-3)}{3}-\dfrac{(-14)}{2}$

50. $(14.13+11.4)\div 3.7-(2.4)(7.8)$

51. $(5.1+2.2)(4.8)-(6.3)(8.1)$

52. $(5.1)^2\cdot 3-(14.64)\div(6.1)$

53. $(1.9)^2+4(3.3)^2-8.7$

54. $5[10-2(4-3)+1]$

55. $18+[14-5(6-4)+7]$

56. $(8-2)[16+4(5-7)]$

57. $(9-6)[21+5(4-6)]$

58. $\left(\dfrac{6-3}{7-4}\right)\left(\dfrac{14+2\cdot 3}{5}\right)$

59. $\left(\dfrac{3}{12}-\dfrac{1}{6}\right)\left(\dfrac{2}{3}+\dfrac{1}{8}\right)$

60. $\left(\dfrac{1}{4}-\dfrac{1}{6}\right)\div\left(\dfrac{2}{3}-\dfrac{1}{8}\right)$

Directions   Perform the indicated operations and simplify. See example 1–8 B.

61. To convert 74° Fahrenheit (F) to Celsius (C), we use the expression

$$C=\dfrac{5}{9}(F-32);$$ thus, in this case,

$$C=\dfrac{5}{9}(74-32).$$ Find C.

62. A Murray Loop is used to determine the point at which a telephone line is grounded. The unknown distance to the point of the ground, x, for a length of the loop of 32 miles and resistances of 222 and 384 ohms is given by

$$x=\dfrac{384}{222+384}\cdot 32$$

Find x.

63. The surface area in square inches of a flat ring whose inside radius is 2 inches and whose outside radius is 3 inches is approximately

$$\dfrac{22}{7}\cdot 3^2-\dfrac{22}{7}\cdot 2^2$$

Find the area of this surface in square inches.

64. The surface area in square inches of a ring section whose inside diameter is 18 inches and whose outside diameter is 26 inches is approximately

$$\dfrac{22}{7}\cdot\dfrac{26+18}{2}\cdot\dfrac{26-18}{2}$$

Find the area of this surface in square inches.

65. To find the pitch diameter, D, of a gear with 36 teeth and an outside diameter of 8 inches, we use

$$D=\dfrac{36(8)}{38+3}$$

Find D in inches.

Directions   Choose a letter for the unknown quantity and use the order of operations to find its value. See example 1–8 C.

66. A woman purchased a case of soda (24 bottles) at 35¢ per bottle, 5 pounds of candy at 89¢ per pound, and 20 jars of baby food at 75¢ per jar. What was her total bill (a) in cents and (b) in dollars and cents?

67. Colleen Meadow is a typist in a law firm. Her base pay is $7 per hour for a 40-hour week and she receives time and a half for every hour she works over 40 hours in a week. How much will she earn if she works 49 hours in one week?

68. In a series of poker games, Ace McGee won $4,000 in each of three games, lost $1,500 in each of four games, and won $2,000 in each of two games. How much did Ace win (or lose) in the nine games?

69. A carpenter must cut a 16-foot long board into 4-foot lengths and a 12-foot long board into 3-foot lengths. How many pieces of lumber will he have?

70. Jane, David, and Mary are typists in an office. David can type 75 words per minute, Jane can type 80 words per minute, and Mary can type 95 words per minute. How many words can they type together in 15 minutes?

71. The stock market opened at 3,325 points on a given day. If it lost nine points per hour during the first 3 hours after opening and then gained six points per hour during the next 5 hours, what did the stock market close at?

⟡ Explore to Learn

1. Why do we need the order of operations? Obviously we need to all agree on the same order, or a problem like $8 + 2 \cdot 5$ could give two answers. But, do we need the order as currently accepted? Consider just always doing operations from left to right, except where parentheses are used, of course. Wouldn't this work? Consider $20 + \dfrac{10}{2}$. How would you do this from left to right? Does this make any sense? Isn't this the same as $20 + 10 \div 2$, which clearly could be done from left to right? Does this cause a problem?

2. Use your calculator to do the following problems. Determine the necessary key strokes to answer the problem without writing down any intermediate results.

   a. $18 + 3 \cdot 24 \div 2^3 - 5 \cdot 2^2 = 7$

   b. $\dfrac{6^2 - 10}{5^2 - 13} = 2.1\overline{6}$

   c. $\dfrac{48 - 2 \cdot 3^2}{40 + 2^2} = 0.68\overline{1}$

1–9   Evaluating Expressions

Substitution Property

An extremely important process in algebra is that of calculating the numerical value of an expression when we are given specific replacement values for the variables. This process is called **evaluation**. To perform evaluation, we need the following **property of substitution**.

Property of Substitution

If $a = b$, then a may be replaced by b or b may be replaced by a in any expression without altering the value of the expression.

Concept

When two expressions are equal, they can replace each other anywhere.

To find the distance (d) traveled when the rate (r) and time (t) are known, we use

$$d = rt$$

If the rate is 45 miles per hour and the time is 3 hours, we can substitute these values into the expression as follows:

$$d = (45)(3) = 135$$

The distance traveled is 135 miles. We replaced the respective variables representing rate and time with their values. We then carried out the indicated arithmetic by using the order of operations.

Note When replacing variables with the numbers they represent, it is a good procedure to put each of the numbers inside parentheses.

● **Example 1–9 A**

Evaluating expressions

Evaluate the following expressions for the given real number replacement for the variable or variables.

1. $x^2 + 2x - 7$, when $x = 4$

The expression would be $(\ )^2 + 2(\ ) - 7$ without the x. Substituting 4 for each x, we have $(4)^2 + 2(4) - 7$. Using the order of operations, we have

$= 16 + 2(4) - 7$	Exponents
$= 16 + 8 - 7$	Multiply
$= 24 - 7$	Add
$= 17$	Subtract

Therefore the expression $x^2 + 2x - 7$ evaluated for $x = 4$ is 17.

2. $5a - b + 2(c + d)$, when $a = 2$, $b = 3$, $c = -2$, and $d = -3$.

$5a - b + 2(c + d)$	Original expression
$= 5(\ ) - (\ ) + 2[(\ ) + (\ )]$	Expression ready for substitution
$= 5(2) - (3) + 2[(-2) + (-3)]$	Substitute
$= 5(2) - (3) + 2[-5]$	Order of operations, groups
$= 10 - (3) + (-10)$	Multiply
$= 7 + (-10)$	Subtract
$= -3$	Add

3. $4ab - c^2 + 3d$, when $a = 2$, $b = 3$, $c = -2$, and $d = -3$.

$4ab - c^2 + 3d$ Original expression
$= 4(\ )(\ ) - (\ )^2 + 3(\ )$ Expression ready for substitution
$= 4(2)(3) - (-2)^2 + 3(-3)$ Substitute
$= 4(2)(3) - (4) + 3(-3)$ Order of operations, exponents
$= 8(3) - 4 + 3(-3)$ Multiply
$= 24 - 4 + (-9)$ Multiply
$= 20 + (-9)$ Subtract
$= 11$ Add

 We can use a calculator to check these calculations. To check number 3 of example 1–9 A.

Scientific calculator   4 $\times$ 2 $\times$ 3 $-$ $($ 2 $+/-$ $)$ x^2 $+$ 3 $\times$ 3 $+/-$ $=$

Graphing   4 $\times$ 2 $\times$ 3 $-$ $($ $(-)$ 2 $)$ x^2 $+$ 3 $\times$ $(-)$ 3 **ENTER**
Result 11

☞ *Quick check*   Evaluate $3a - 2(c - d) + b$ when $a = 2$, $b = 3$, $c = -2$, and $d = -3$.   ●

To Evaluate an Expression

1. Write parentheses in place of each variable.
2. Place the value that the variable is representing inside the parentheses.
3. Perform the indicated operations according to the order of operations.

Formulas

A **formula** expresses a relationship between quantities in the physical world, for example, $d = rt$ (distance equals rate times time).

● **Example 1–9 B**
Evaluating formulas

Evaluate the following formulas for the real number replacements of the variables.

1. A formula in electricity is $I = \dfrac{E}{R}$, where I represents the current measured in amperes in a certain part of a circuit, E represents the potential difference in volts across that part of the circuit, and R represents the resistance in ohms in that part of the circuit. Find I in amperes if $E = 110$ volts and $R = 44$ ohms.

$I = \dfrac{E}{R}$ Original formula

$I = \dfrac{(\ )}{(\ )}$ Formula ready for substitution

$I = \dfrac{(110)}{(44)}$ Substitute

$I = \dfrac{5}{2} = 2\dfrac{1}{2}$ Reduce and change to a mixed number

The current is $2\dfrac{1}{2}$ amperes.

2. If we know the temperature in degrees Fahrenheit (F), the temperature in degrees Celsius (C) can be found by the formula $C = \dfrac{5}{9}(F - 32)$. Find the temperature in degrees Celsius if the temperature is 86 degrees Fahrenheit.

$$C = \frac{5}{9}(F - 32) \qquad \text{Original formula}$$

$$C = \frac{5}{9}[(\ ) - 32] \qquad \text{Formula ready for substitution}$$

$$C = \frac{5}{9}[(86) - 32] \qquad \text{Substitute}$$

$$C = \frac{5}{9}(54) \qquad \text{Order of operations, groups first}$$

$$C = 30 \qquad \text{Multiply}$$

The temperature is 30 degrees Celsius.

☛ *Quick check*   Evaluate $I = \dfrac{E}{R}$ when $E = 220$ volts and $R = 11$ ohms.   ●

Subscripts

In some formulas, two or more measurements of the same unit may be given. It is customary to label these by using **subscripts**. To illustrate, given two different measurements of pressure in a science experiment, we might label them

$$P_1 \text{ and } P_2$$

The 1 and 2 are the subscripts. Subscripts are always written to the lower right of the letter. The symbols above are read "P sub-one" and "P sub-two."

Note Do not confuse a subscript, such as P_2, that helps distinguish between different measurements, and an exponent, such as P^2, that indicates the number of times a given base is used as a factor in an indicated product. Subscripts are written to the lower right of the symbol. Exponents are written to the upper right of the symbol.

● **Example 1–9 C**

Evaluating formulas with subscripts

Evaluate the following formulas for the real number replacements of the variables.

1. In a science problem $V_1 = \dfrac{V_2 T_1}{T_2}$, where V_1 and V_2 represent different measurements of volume and T_1 and T_2 represent different measurements of temperature.

Find V_1 in liters if $V_2 = 8.7$ liters, $T_1 = 18$ degrees Celsius, and $T_2 = 45$ degrees Celsius.

$$V_1 = \frac{V_2 T_1}{T_2} \qquad \text{Original formula}$$

$$V_1 = \frac{(\ )(\ )}{(\ )} \qquad \text{Formula ready for substitution}$$

$$V_1 = \frac{(8.7)(18)}{(45)} \qquad \text{Substitute}$$

$$V_1 = \frac{156.6}{45} \qquad \text{Multiply}$$

$$V_1 = 3.48 \qquad \text{Divide}$$

V_1 is 3.48 liters.

2. In $R_t = \dfrac{R_1 \cdot R_2}{R_1 + R_2}$, R_1 and R_2 represent two different measurements of resistance and R_t represents the total resistance in the electrical circuit. Find R_t in ohms if $R_1 = 3.1$ ohms and $R_2 = 6.9$ ohms.

$$R_t = \frac{R_1 \cdot R_2}{R_1 + R_2} \qquad \text{Original formula}$$

$$R_t = \frac{(\ ) \cdot (\ )}{(\ ) + (\ )} \qquad \text{Formula ready for substitution}$$

$$R_t = \frac{(3.1) \cdot (6.9)}{(3.1) + (6.9)} \qquad \text{Substitute}$$

$$R_t = \frac{21.39}{10} \qquad \begin{array}{l}\text{Perform the operations in the numerator}\\ \text{and the denominator}\end{array}$$

$$R_t = 2.139$$

The total resistance, R_t, is 2.139 ohms.

☞ **Quick check**   Evaluate $C_t = \dfrac{C_1 \cdot C_2}{C_1 + C_2}$ when $C_1 = 4$ and $C_2 = 6$.   ●

MASTERY POINTS

Can you
• Evaluate an expression?
• Evaluate a formula?
• Use subscripts?

Exercise 1–9	☞ *Quick Check*	
Directions	**Example** $3a - 2(c - d) + b$	
Evaluate the following expressions if $a = 2$, $b = 3$, $c = -2$, $d = -3$. See example 1–9 A.	**Solution** $= 3(\ ) - 2[(\ ) - (\ )] + (\ )$	Expression ready for substitution
	$= 3(2) - 2[(-2) - (-3)] + (3)$	Substitute
	$= 3(2) - 2(1) + 3$	Order of operations, groups
	$= 6 - 2 + 3$	Multiply
	$= 7$	Subtract and add

Directions   Evaluate the following expressions if $a = 2$, $b = 3$, $c = -2$, $d = -3$. See example 1–9 A.

1. $2a + b - c$
2. $a + b$
3. $2a - 2d$
4. $3d - 5a$
5. $ad - bc$
6. $ac - bd$
7. $a^2 + b^2$
8. $b^2 - d^2$
9. $a^2 - c^2$
10. $b^2 + 2d^2$
11. $3a - 2b - (c + d)$
12. $a - 3(c + d)$

13. $5a + 7b - 3c(a - d)$
14. $3c - 2(3a + b)$
15. $2ab(c + d)$
16. $3ab - 4c^2 + d$
17. $(3a + 2b)(a - c)$
18. $(3a - 5c)(2b - 4d)$
19. $(5c - 3a)(4d - 2b)$
20. $(c - d)^2$
21. $(c + d)^2$
22. $(3d - 5c)^2$
23. $(4a - 2b)^2$
24. $(4a + b) - (3a - b)(c + 2d)$

25. $(2a - b) - (a + 2d)(b - c)$
26. $(b - d) - (b - 2c)(a + d)$
27. $a^2 - 2a + 3$
28. $3a^2 + 2a + 1$
29. $2c^2 - 3c + 6$
30. $3c^2 + 2c - 5$
31. $3ac - 2a^2c^2$
32. $a^2b^2 + c^2d^2$
33. $(ab)^2 - (cd)^2$
34. $(bd)^2 - (cd)^2$
35. $(c - d)^2(a + b)$
36. $(a - b)^2(c + d)$

Directions
Evaluate the following formulas. See examples 1–9 B and C.

🏳 *Quick Check*

Examples $I = \dfrac{E}{R}$,

$E = 220$ and $R = 11$

$C_t = \dfrac{C_1 \cdot C_2}{C_1 + C_2}$,

$C_1 = 4$ and $C_2 = 6$

Solutions $I = \dfrac{(\ )}{(\ )}$

$= \dfrac{(220)}{(11)}$

$= 20$

$C_t = \dfrac{(\ )(\ )}{(\ ) + (\ )}$ Formulas ready for substitution

$= \dfrac{(4)(6)}{(4) + (6)}$ Substitute

$= \dfrac{24}{10}$ Order of operations

$= \dfrac{12}{5}$ Reduce

Directions   Evaluate the following formulas. See examples 1–9 B and C.

37. $I = \dfrac{E}{R}$, $E = 220$ and $R = 33$
    (Electronics)

38. $I = \dfrac{E}{R}$, $E = 180$ and $R = 21$ (Electronics)

39. $I = prt$, $p = 1{,}000$; $r = 0.08$; and $t = 2$
    (Business)

40. $I = prt$, $p = 2{,}000$; $r = 0.04$; and $t = 3$
    (Business)

41. $W = I^2R$, $I = 12$ and $R = 2$ (Electronics)

42. $W = I^2R$, $I = 4$ and $R = 3$ (Electronics)

43. $V = \ell wh$, $\ell = 11.2$, $w = 7.1$, and $h = 6.8$
    (Geometry)

44. $V = \ell wh$, $\ell = 9.3$, $w = 4.5$, and $h = 5.2$
    (Geometry)

45. $F = ma$, $m = 18$ and $a = 6$ (Physics)

46. $F = ma$, $m = 12$ and $a = 6$ (Physics)

47. $V = k + gt$, $k = 24$, $g = 9$, and $t = 4$
    (Physics)

48. $V = k + gt$, $k = 20$, $g = 8$, and $t = 4$
    (Physics)

49. $\ell = a + (n - 1)d$, $a = 2$, $n = 14$, and $d = 3$
    (Mathematics)

50. $\ell = a + (n - 1)d$, $a = 3$, $n = 10$, and $d = 2$
    (Mathematics)

51. $H = \dfrac{D^2N}{2}$, $D = 4$ and $N = 6$ (Automotive)

52. $H = \dfrac{D^2N}{2}$, $D = 3$ and $N = 4$ (Automotive)

53. $A = \dfrac{1}{2}h(b_1 + b_2)$, $h = 6$, $b_1 = 8$, and $b_2 = 10$
(Geometry)

54. $A = \dfrac{1}{2}h(b_1 + b_2)$, $h = 5$, $b_1 = 6$, and $b_2 = 8$
(Geometry)

55. $A = p + pr$, $p = 2{,}000$ and $r = 0.07$
(Business)

56. $A = p + pr$, $p = 500$ and $r = 0.05$
(Business)

57. $p = 2\ell + 2w$, $\ell = 8.2$ and $w = 6.1$
(Geometry)

58. $p = 2\ell + 2w$, $\ell = 12.3$ and $w = 8.9$
(Geometry)

59. $A = \dfrac{I^2R - 120E^2}{R}$, $E = 5$, $I = 12$, and $R = 100$
(Electronics)

60. $A = \dfrac{I^2R - 120E^2}{R}$, $E = 3$, $I = 6$, and $R = 50$
(Electronics)

61. $S = \dfrac{1}{2}gt^2$, $g = 32$ and $t = 4$ (Physics)

62. $S = \dfrac{1}{2}gt^2$, $g = 32$ and $t = 6$ (Physics)

63. $V_1 = \dfrac{V_2P_2}{P_1}$, $V_2 = 18$, $P_2 = 12$, and $P_1 = 36$
(Chemistry)

64. $V_1 = \dfrac{V_2P_2}{P_1}$, $V_2 = 12.6$, $P_2 = 26$, and $P_1 = 30$
(Chemistry)

65. $C_t = \dfrac{C_1 \cdot C_2}{C_1 + C_2}$, $C_1 = 2.8$ and $C_2 = 7.2$
(Electronics)

66. $C_t = \dfrac{C_1 \cdot C_2}{C_1 + C_2}$, $C_1 = 6$ and $C_2 = 12$ (Electronics)

67. $R_t = \dfrac{R_1 \cdot R_2}{R_1 + R_2}$, $R_1 = 5.1$ and $R_2 = 4.9$
(Electronics)

68. $R_t = \dfrac{R_1 \cdot R_2}{R_1 + R_2}$, $R_1 = 8$ and $R_2 = 12$
(Electronics)

69. $V_2 = \dfrac{V_1T_2}{T_1}$, $V_1 = 6.3$, $T_1 = 16$, and $T_2 = 24$
(Chemistry)

70. $V_2 = \dfrac{V_1T_2}{T_1}$, $V_1 = 9.7$, $T_1 = 15$, and $T_2 = 45$
(Chemistry)

Directions   Evaluate the following formulas. See examples 1–9 B and C.

71. Find the horsepower (h) required by a hydraulic pump when it needs to pump 10 gallons per minute (g) and the pounds per square inch (p) equal 3,000. Use $h = \dfrac{g \cdot p}{1{,}714}$.

72. The required ratio of gearing (R) of a milling machine is given by $R = (A - N) \cdot \dfrac{40}{A}$, where $N =$ required number of divisions and $A =$ approximate number of divisions. Find R when $A = 280$ and $N = 271$.

73. In a gear system, the velocity (V) of the driving gear is defined by $V = \dfrac{vn}{N}$, where $v =$ velocity of follower gear, $n =$ number of teeth of follower gear, and $N =$ number of teeth of driving gear. Find V when $v = 90$ revolutions per minute, $n = 30$ teeth, $N = 65$ teeth.

74. The tap drill size (T) of a drill needed to drill threads in a nut is given by $T = D - \dfrac{1}{N}$, where $D =$ diameter of the tap and $N =$ number of threads per inch. Find T with a $\dfrac{1}{2}$-inch diameter tap and 13 threads per inch.

75. It is necessary to drag a box 600 feet across a level lot in 3 minutes. The force required to pull the box is 2,000 pounds. What is the horsepower (h) needed to do this if horsepower is defined by $h = \dfrac{\ell \cdot w}{33,000 \cdot t}$, where $\ell = $ length to be moved, $w = $ force exerted through distance ℓ, and $t = $ time in minutes required to move the box through ℓ?

76. A pulley 12 inches in diameter that is running at 320 revolutions per minute is connected by a belt to a pulley 9 inches in diameter. How many revolutions per minute will the smaller pulley make if $s = \dfrac{SD}{d}$, where $s = $ speed of smaller pulley and $d = $ diameter of smaller pulley, $S = $ speed of larger pulley and $D = $ diameter of larger pulley?

✸ Explore to Learn

1. Consider the following expressions:

a. $x^2 - y^2$ **b.** $(x - y)^2$ **c.** $x^2 y^2$ **d.** $(xy)^2$ **e.** $\dfrac{1}{6} x^2 y$

f. $\dfrac{x^2 y}{6}$ **g.** $\dfrac{x^2}{6} \cdot \dfrac{y}{6}$ **h.** $(y - x)^2$ **i.** $y^2 - x^2$

Now suppose that x represents 5 and y represents 2. If these expressions were evaluated, the same answer would sometimes appear for different expressions. (a) Make a prediction as to which expressions will give the same answer (for example, e, f, g perhaps?) and then (b) compute each expression and see how well you did.

2. In the formula $A = \ell \cdot w$, the area of a rectangle, suppose that the value of ℓ was doubled. What would happen to the value of A? Suppose that the values of *both* ℓ and w were doubled. What is the effect on A? Use specific values for ℓ and w to find the answer.

Chapter 1 Lead-in Problem

While on a trip to Canada, Tonya heard on the radio that the temperature today will be 20° Celsius. Will she need her winter coat? What will the temperature be in degrees Fahrenheit? We can determine what 20° Celsius is in degrees Fahrenheit by the following formula.

$$F = \frac{9}{5} \cdot C + 32$$

Solution

$F = \dfrac{9}{5} \cdot C + 32$	Original formula
$F = \dfrac{9}{5} \cdot (\ \ ) + 32$	Formula ready for substitution
$F = \dfrac{9}{5}(20) + 32$	Substitute
$F = 36 + 32$	Order of operations: multiply
$F = 68$	Add

The temperature is 68 degrees Fahrenheit. She will not need her winter coat.

Chapter 1 Summary

• Glossary

absolute value (page 40) the undirected distance that a number is from the origin.

composite number (page 7) a whole number greater than one that is not prime.

denominator (page 6) the bottom number of a fraction.

evaluation (page 76) calculating the numerical value of an expression when given specific replacement values for the variables.

exponent (page 70) indicates how many times a base is used as a factor in an indicated product.

factor (page 60) the numbers or variables in an indicated multiplication.

integers (page 36) $\{\cdots, -3, -2, -1, 0, 1, 2, 3, \cdots\}$

irrational numbers (page 37) real numbers that cannot be expressed as the quotient of two integers.

least common denominator (page 13) the smallest number that can be divided by two or more denominators.

natural numbers (page 35) $\{1, 2, 3, 4, 5, \cdots\}$

numerator (page 6) the top number of a fraction.

percent (page 27) parts per one hundred.

prime number (page 7) any whole number greater than 1 whose only factors are the number itself and 1.

rational numbers (page 36) any number that can be expressed as the quotient of two integers in which the divisor is not zero.

real numbers (page 37) the set of numbers composed of all the rational numbers and all of the irrational numbers.

real number line (page 37) a line on which we visually represent the set of real numbers.

reduced to lowest terms (page 8) when the only factor common to the numerator and the denominator of a fraction is 1.

set (page 34) any collection of objects or things.

undefined (page 65) division by zero is undefined.

variable (page 9) a symbol that represents an unspecified number.

whole numbers (page 35) $\{0, 1, 2, 3, 4, 5, \cdots\}$

• New Symbols and Notation

$=$ is equal to

$\neq$ is not equal to

$\approx$ is approximately equal to

$>$ is greater than

$<$ is less than

$\geq$ is greater than or equal to

$\leq$ is less than or equal to

$|\ |$ absolute value

$2 \times 6,\ (2)(6),\ 2(6),\ 2 \cdot 6,\ 2$ times 6

$\%$ percent

$(\ )$ parentheses (grouping symbol)

$[\ ]$ brackets (grouping symbol)

$\{\ \}$ braces (encloses a set or grouping symbol)

$\{a, b, c\}$ the set whose elements are a, b, and c

N the set of natural numbers

W the set of whole numbers

J the set of integers

Q the set of rational numbers

H the set of irrational numbers

R the set of real numbers

$\cdots$ three dots, called ellipsis, indicating that an established pattern continues

$\in$ is an element of or is a member of

$\subseteq$ is a subset of

$^{-}$ a bar placed over a number or group of numbers indicates that the number(s) repeat indefinitely

π pi

$4^3 = 4 \cdot 4 \cdot 4$, exponent (3 factors of 4)

P_1 subscript (P-sub-one)

$\dfrac{a}{b}$, a/b, a divided by b

• Properties and Definitions

Prime number (page 7)
A prime number is any whole number greater than 1 whose only factors are the number itself and 1.

Multiplication of fractions (page 9)
If $\dfrac{a}{b}$ and $\dfrac{c}{d}$ are fractions, then

$$\frac{a}{b} \cdot \frac{c}{d} = \frac{a \cdot c}{b \cdot d}$$

Division of fractions (page 10)
If $\dfrac{a}{b}$ and $\dfrac{c}{d}$ are fractions, then

$$\frac{a}{b} \div \frac{c}{d} = \frac{a}{b} \cdot \frac{d}{c} = \frac{a \cdot d}{b \cdot c}$$

Definition of percent (page 27)
Percent is defined to be parts per one hundred.

Definition (page 35)

The set A is a subset of the set B if every element in A is also an element of B.

Identity property of addition (page 47)

For every real number a,

$$a + 0 = 0 + a = a$$

Commutative property of addition (page 47)

For every real number a and b,

$$a + b = b + a$$

Additive inverse property (page 50)

For every real number a,

$$a + (-a) = 0$$

and

$$(-a) + a = 0$$

Associative property of addition (page 51)

For every real number a, b, and c,

$$(a + b) + c = a + (b + c)$$

Definition of subtraction (page 55)

For any two real numbers, a and b,

$$a - b = a + (-b)$$

Commutative property of multiplication (page 60)

For every real number a and b,

$$a \cdot b = b \cdot a$$

Zero factor property (page 61)

For every real number a,

$$a \cdot 0 = 0 \cdot a = 0$$

Identity property of multiplication (page 61)

For every real number a,

$$a \cdot 1 = 1 \cdot a = a$$

Associative property of multiplication (page 63)

For every real number a, b, and c,

$$(a \cdot b)c = a(b \cdot c)$$

Definition of division (page 64)

If $b \neq 0$, $\dfrac{a}{b} = q$ provided that $b \cdot q = a$, where a is the dividend, b is the divisor, and q is the quotient.

Division involving zero (page 65)

1. Division by zero is not allowed, and whenever a division problem involves division by zero, we will write undefined.

2. The quotient of zero divided by any number other than zero is always zero.

Property of substitution (page 77)

If $a = b$, then a may be replaced by b or b may be replaced by a in any expression without altering the value of the expression.

• Procedures

[1–2]

To reduce a fraction to lowest terms (page 8)

1. Write the numerator and the denominator as a product of prime factors.

2. Divide the numerator and the denominator by all common prime factors. The product of all common prime factors is called the greatest common factor, GCF.

To multiply fractions (page 9)

1. Write the numerator and the denominator as an indicated product (do not multiply).

2. Reduce the resulting fraction to lowest terms.

3. Multiply the remaining factors.

To divide two fractions (page 10)

1. Multiply the first fraction by the reciprocal of the second fraction.

2. Reduce the resulting product to lowest terms.

To add or subtract fractions with common denominators (page 12)

1. Add or subtract the numerators.

2. Place the sum or difference over the common denominator.

3. Reduce the resulting fraction to lowest terms.

To find the least common denominator (LCD) (page 13)

1. Express each denominator as a product of prime factors.

2. List all the different prime factors.

3. Write each prime factor the greatest number of times it appears in any of the prime factorizations in step 1.

4. The least common denominator is the product of all factors from step 3.

To find equivalent fractions (page 14)

1. Divide the original denominator into the new denominator.

2. Multiply the numerator and the denominator of the given fraction by the number obtained in step 1.

To add or subtract fractions having different denominators (page 15)

1. Find the LCD of the fractions.

2. Write each fraction as an equivalent fraction with the LCD as the new denominator.

3. Perform the addition or subtraction.

4. Reduce the resulting fraction to lowest terms.

[1–3]

Addition and subtraction of decimal numbers (page 23)
To add or subtract decimal numbers, we place the numbers under one another so that the decimal points line up vertically and then proceed as in adding or subtracting whole numbers. The decimal point will appear in the answer directly below where it is lined up in the problem.

To multiply decimal numbers (page 24)

1. Multiply the numbers as if they are whole numbers (ignore the decimal points).

2. Count the number of decimal places in both factors. That is, count the number of digits to the right of the decimal point in each factor. This total is the number of decimal places the product must have.

3. Beginning at the right in the product, count off to the left the number of decimal places from step 2. Insert the decimal point. If necessary, zeros are inserted so there are enough decimal places.

To divide decimal numbers (page 24)

1. Change the divisor to a whole number by moving the decimal point to the right as many places as is necessary.

2. Move the decimal point in the dividend to the right this same number of places. If necessary, zeros are inserted so there are enough decimal places.

3. Insert the decimal point in the quotient directly above the new position of the decimal point in the dividend.

4. Divide as with whole numbers.

To change a fraction to a decimal number (page 25)
Divide the denominator into the numerator.

To write a percent as a decimal number (page 27)
Move the decimal point two places to the left and drop the % symbol.

To write a percent as a fraction (page 27)
Drop the % symbol and write the number over a denominator of 127.

To write a decimal number as a percent (page 27)
Move the decimal point two places to the right and affix the % symbol.

To write a fraction as a percent (page 27)
Find the decimal number equivalent of the fraction and change this decimal number to a percent.

[1–5]

Addition on the number line (page 47)
To add a and b, that is $a + b$, we locate a on the number line and move from there according to the value of b.

1. If b is positive, we move to the right b units.

2. If b is negative, we move to the left the absolute value of b units.

3. If b is 0, we stay at a.

Addition of two real numbers (page 51)

1. If the signs are the same, add their absolute values and prefix the sum by their common sign.

2. If the signs are different, subtract the lesser absolute value from the greater absolute value. The answer has the sign of the number with the greater absolute value.

3. The sum of a number and its opposite (additive inverse) is zero.

[1–6]

Subtraction of two real numbers (page 55)

1. Change the operation from subtraction to addition.

2. Change the sign of the number that follows the subtraction symbol.

3. Perform the addition, using the rules for adding real numbers.

[1–7]

Multiplication of two real numbers (page 62)

To multiply two real numbers, multiply their absolute values and

1. the product will be positive if the numbers have the same sign;

2. the product will be negative if the numbers have different signs.

Multiplication of more than two real numbers (page 62)

1. If in the numbers being multiplied there is an odd number of negative factors, the answer will be negative.

2. If in the numbers being multiplied there is an even number of negative factors, the answer will be positive.

Division of two real numbers (page 64)

To divide two real numbers, perform the division using the absolute values of the numbers and

1. the quotient will be positive if the numbers have the same sign;

2. the quotient will be negative if the numbers have different signs.

[1–8]

Order of operations, or priorities (page 71)

1. Groups: Perform any operations within a grouping symbol such as ( ) parentheses, [ ] brackets, { } braces, | | absolute value, and above or below the fraction bar.

2. Exponents: Perform operations indicated by exponents.

3. Multiply and divide: Perform multiplication and division in order from left to right.

4. Add and subtract: Perform addition and subtraction in order from left to right.

Note

a. Within a grouping symbol, the order of operations will still apply.

b. If there are several grouping symbols intermixed, remove them by starting with the innermost one and working outward.

[1–9]

To evaluate an expression (page 78)

1. Write parentheses in place of each variable.

2. Place the value that the variable is representing inside the parentheses.

3. Perform the indicated operations according to the order of operations.

☼ Chapter 1 Error Analysis

1. Determining order between numbers
   Example: $-3 < -5$
   Correct answer: $-3 > -5$
   What error was made? (*see page 39*)

2. Evaluate absolute value
   Example: $-|-3| = 3$
   Correct answer: $-|-3| = -3$
   What error was made? (*see page 40*)

3. Adding real numbers
   Example: $(-3) + 4 = 7$
   Correct answer: $(-3) + 4 = 1$
   What error was made? (*see page 49*)

Learning System
"Error Analysis" provides a group of problems where a common error has been made. A page reference is provided so that you can refer to relevant examples and notes for assistance in finding and correcting the error.

4. Subtracting real numbers
Example: $(-9) - (-4) = -13$
Correct answer: $(-9) - (-4) = -5$
What error was made? (*see page 55*)

5. Combining using grouping symbols
Example: $4 - (5 - 2) = 4 - 5 - 2 = -3$
Correct answer: $4 - (5 - 2) = 1$
What error was made? (*see page 56*)

6. Exponents
Example: $-3^2 = 9$
Correct answer: $-3^2 = -9$
What error was made? (*see page 70*)

7. Multiplication of negative numbers
Example: $(-2)(-6) = -12$
Correct answer: $(-2)(-6) = 12$
What error was made? (*see page 62*)

8. Division of real numbers
Example: $\dfrac{-15}{3} = 5$

Correct answer: $\dfrac{-15}{3} = -5$

What error was made? (*see page 64*)

9. Division by zero
Example: $\dfrac{-5}{0} = 0$

Correct answer: $\dfrac{-5}{0}$ is undefined.

What error was made? (*see page 65*)

10. Exponents
Example: $3^3 = 9$
Correct answer: $3^3 = 27$
What error was made? (*see page 70*)

11. Calculators
The following keystrokes are entered on a calculator
to compute $\dfrac{6 + 8}{2}$:

6 $\boxed{+}$ 8 $\boxed{\div}$ 2 $\boxed{=}$ ($\boxed{=}$ may be $\boxed{\text{ENTER}}$ on some
calculators). The result is 10, but the correct answer
is 7.
What error was made? (*see page 45*)

12. Calculators
To compute $(3 + 5)(2 + 3)$ the following keystrokes
are entered on a calculator:
3 $\boxed{+}$ 5 $\boxed{\times}$ 2 $\boxed{+}$ 3 $\boxed{=}$ ($\boxed{=}$ may be $\boxed{\text{ENTER}}$ on
some calculators). Although the result should be 40,
the calculator displays 16.
What error was made? (*see page 45*)

Chapter 1 Critical Thinking

1. A watch is started at 12 noon. Each time the watch reaches the next hour, it is stopped for 10 minutes. How long will it take the watch to go from 12 noon to 12 midnight?

2. A 6-oz size of laundry soap costs $1.23; the 10-oz size costs $1.98, the 1-pound size costs $3.45, and the 1-pound 8-ounce size costs $4.95. Which size is the most economical?

3. The order of operations says that multiplications and divisions are to be done from left to right. Is it really necessary that these operations be done in this order?

4. A student was given a very inexpensive credit card calculator for free. The student took it to algebra class and calculated the expression $2 + 3 \cdot 5$ by entering $\boxed{2}$ $\boxed{+}$ $\boxed{3}$ $\boxed{\times}$ $\boxed{5}$ $\boxed{=}$. The answer 25 appeared. Is this correct? Explain.

5. The following is an expense report form with some entries filled in. Explain what this system of bookkeeping implies about the operation of addition. What properties of the addition of real numbers are involved? The report has an error. Describe how the layout of the report helps to quickly find mathematical errors.

| | | | Mileage at 0.28/mile | | | |
Date	Reason	Hotel	Meals	Miles	Amount	Phone	Total
June 4	Visit vendor	85.96	41.54	78	21.84	4.51	153.85
June 23	Site visit	56.97	38.12	49	13.72		108.81
June 24	Site visit	71.40	39.88	56	16.24	3.45	130.97
Total		214.33	119.54	183	51.80	7.96	393.63

6. Describe how figures (a), (b), and (c) all describe the fraction $\frac{1}{2}$. Some would say that (b) is better than (c) for this purpose. Why?

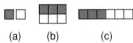

   (a)   (b)   (c)

7. Describe how figures (a), (b), (c), and (d) all describe the fraction $\frac{1}{3}$. Some would say that (d) is better than (b) or (c) for this purpose. Why?

  (a)   (b)   (c)   (d)

8. Using the previous two statements as a guide, what statement about fractions could the figure represent?

9. How can you arrange the numbers from 1 to 9 to form a 3 by 3 square array, such that each row, each column, and the two diagonals will all total exactly 15?

10. If three days ago was the day before Friday, what will the day after tomorrow be?

Chapter 1 Review

[1–1]

1. What are your instructor's office number and office hours?

2. What are your instructor's policies for this course? For example: grading scale, attendance, homework, etc.

[1–2]

Directions   Reduce each fraction to lowest terms.

3. $\frac{10}{14}$

4. $\frac{36}{48}$

5. $\frac{120}{180}$

Directions   Multiply or divide the following as indicated. Reduce to lowest terms.

6. $\frac{6}{7} \cdot \frac{5}{3}$

7. $\frac{2}{3} \cdot \frac{9}{10}$

8. $\frac{7}{8} \div \frac{5}{6}$

9. $\frac{5}{12} \div \frac{10}{21}$

10. $3\frac{3}{4} \div 1\frac{1}{5}$

11. $2\frac{1}{2} \cdot 3\frac{1}{3}$

12. Hannah rents $\frac{3}{4}$ of a plot of land. If the plot is $\frac{5}{6}$ of an acre in size, how many acres does Hannah rent?

13. A recipe calls for $\frac{4}{5}$ of a cup of sugar. If George wishes to make $\frac{1}{2}$ of the recipe, how many cups of sugar should he use?

Directions   Find the LCD of the fractions with the following groups of denominators.

14. 4, 8, 14

15. 8, 10, 12

Directions   Add or subtract the following fractions as indicated. Reduce to lowest terms.

16. $\frac{3}{7} + \frac{5}{7}$

18. $\frac{11}{12} - \frac{1}{12}$

20. $4\frac{1}{4} + 2\frac{3}{5}$

17. $\frac{5}{8} + \frac{1}{6}$

19. $\frac{8}{9} - \frac{2}{3}$

21. $\frac{1}{5} - \frac{2}{3}$

22. Marta paid $\frac{1}{3}$ of her debt one week and $\frac{1}{4}$ of her debt the second week. At the end of the second week, how much of her debt had she paid off?

23. Darrell owns $3\frac{1}{8}$ acres of land. If he sells $2\frac{1}{4}$ acres to his friend, how many acres does he have left?

[1–3]

Directions   Perform the indicated operations on decimal numbers.

24. 20.6 + 1.373 + 210.42 + 0.027 + 31.09

25. 42.5 − 10.705

26. 213.4 × 6.35

27. 316.03 ÷ 22.1

28. It is estimated that 24% of a certain population of deer is less than 2 years old. If there are about 4,800 deer in this population, how many are less than 2 years old?

29. A grocery bill contains two parts. The first is $89.36 for food, which is not taxed. The second is $8.79 for nonfood items, which is taxed at 5%. What is the total grocery bill?

30. An automobile uses 15.2 gallons of gasoline to travel 188.8 miles. How many miles per gallon did the automobile average? (Round the answer to the tenths.)

Directions   Fill in the missing parts.

	Fraction	Decimal	Percent		Fraction	Decimal	Percent
31.	$\dfrac{3}{5}$			**34.**			187%
32.		0.2		**35.**		0.22	
33.			10%	**36.**	$\dfrac{7}{9}$		

Directions   Find the following percentages.

37. 4% of 250

38. 57% of 120

39. 62.5% of 40

40. 131.2% of 60

[1–4]

Directions   List the elements of the following sets.

41. Integers between 49 and 56

42. Natural numbers less than 5

43. Whole numbers that are not natural numbers

44. Integers between -4 and 4

Directions   Plot the graphs of the following numbers, using a different number line for each problem.

45. $-2, -\dfrac{1}{2}, 0, 3$

46. $-\dfrac{3}{4}, 1, \dfrac{3}{2}, \dfrac{5}{2}$

47. $-4, -1, \sqrt{2}, 4$

48. $-3, -2, \dfrac{1}{2}, \pi$

Directions   Replace the ? with the proper inequality symbol ($<$ or $>$) to get a true statement.

49. 4 ? 8

50. -5 ? 0

51. -10 ? -20

52. $|-10|$? $|-20|$

53. $|-5|$? $|0|$

54. $|-8|$? $|4|$

[1–5, 1–6]

Directions   Find the sum or difference.

55. $-1 + (-3)$

56. $6 - 3$

57. $7 - 13$

58. $-4 + 5$

59. $-8 + 2$

60. $7 - (-8)$

61. $-8 - (-4)$

62. $-3 - 6$

63. $0 + (-3) - (-7) + 3 - 4$

64. $4 - 3 + 7 - 8 + 12 - (-3)$

65. $1\dfrac{1}{2} + \left(-2\dfrac{1}{4}\right)$

66. $\left(-\dfrac{3}{8}\right) + 4\dfrac{1}{2}$

67. $-10.9 + 14.3$

68. $5.6 + (-8.1)$

69. $(12 - 16) - (20 - 15)$

70. $9 - 5 + 6 - (3 - 8)$

71. The temperature on a given day in Montreal was $-2°$ C. The temperature went down $8°$ C. What was the final temperature that day?

72. The sales on five consecutive days at a card shop were $119.08, $156.41, $201.31, $518.62, and $86.43. What was the total sales for those days?

73. A man suffers successive financial losses of $3,000, $2,560, and $3,300 on three business transactions. A loss is denoted by a negative number, and the man originally had $52,000.

    a. Write a statement using negative numbers representing his assets after the losses.

    b. Find the total assets after the losses.

74. At 7 A.M. the temperature was −17°. At noon that same day the reading was 23°. How much of a rise in temperature was there from 7 A.M. to noon?

75. The temperature readings during a 5-hour period were 63°, 72°, 80°, 75°, and 69°.

    a. Represent by positive and negative integers how much rise (+) and fall (−) there was from hour to hour.

    b. Was the numerical value of the total rise greater than, equal to, or less than that of the total fall? How much? (Represent by a positive or negative integer.)

    c. If the sixth hour showed a drop of 11°, what was the temperature during the sixth hour? Write a statement involving a negative integer representing this answer.

[1–7]

Directions Find the product or quotient. If a quotient does not exist, so state.

76. $3 \cdot (-7)$

77. $3\frac{1}{5} \cdot \left(-2\frac{1}{3}\right)$

78. $(-8) \cdot 3 \cdot (-1)$

79. $8 \cdot (-9) \cdot (-1) \cdot (-2)$

80. $(-4) \cdot 3 \cdot (-5) \cdot 0$

81. $\dfrac{-14}{2}$

82. $\dfrac{-8}{-4}$

83. $24 \div (-4)$

84. $\dfrac{7}{0}$

85. $\dfrac{0}{-8}$

86. $(0.3)(-4.1)$

87. $\dfrac{(-2)(-3)}{-6}$

88. An auditorium contains 52 rows of seats. If each row contains 80 seats, how many people can be seated in the auditorium?

89. In a classroom, there are 5 rows of desks. If each row contains 7 desks, how many students will the classroom hold?

90. Mariko paid $20.40 for six watermelons for her fruit market. How much did each watermelon cost her?

91. A man drove 398.2 miles and used 11 gallons of gasoline. How many miles did he drive on each gallon of gasoline?

[1–8]

Directions Perform the indicated operations and simplify.

92. $(-5)^2$

93. -4^3

94. -4^2

95. -3^3

96. $100 - 4 \cdot 5 + 18$

97. $-7 + 14 \div 7 + 2$

98. $18 + 3 \cdot 12 \div 2^2 - 7$

99. $19 - (14 - 6) + 7^2 - 11$

100. $\dfrac{8(2 - 4)}{4} - \dfrac{35}{7}$

101. $4[8 - 2(5 - 3) + 1]$

102. $\left[\dfrac{8 + (-2)}{-3}\right]\left[\dfrac{14 \div (-2)}{-1}\right]$

103. $\left[\dfrac{(-12) + (-6)}{4}\right]\left[\dfrac{(-18)(-3)}{-9}\right]$

[1–9]

Directions   Evaluate the following expressions if $a = 3$, $b = 4$, $c = -4$, and $d = -3$.

104. $3a - b + c$

105. $d - 2(a + c)$

106. $a^2b - a^2c$

107. $(2a + c)(b + 2d)$

108. $(c - 2d)^2$

109. $c^2 - d^2$

110. Evaluate R when $R = \dfrac{P \cdot L}{D^2}$ given (a) $P = 6$, $L = 8$, $D = 4$; (b) $P = 7$, $L = 3$, $D = \dfrac{2}{3}$.

111. The volume of a gas V_2 is given by $V_2 = \dfrac{P_1 V_1}{P_2}$. Find V_2 when $P_1 = 780$, $V_1 = 80$, and $P_2 = 60$.

Chapter 1 Test

Directions   Perform the indicated operations, if possible.

1. $\dfrac{1}{4} + \dfrac{1}{6}$

2. $(2)(0)(-3)$

3. $5\dfrac{3}{8} - 2\dfrac{1}{4}$

4. $(1.8)(-3.2)$

5. $-\dfrac{28}{4}$

6. $\left(2\dfrac{2}{5}\right)\left(3\dfrac{3}{4}\right)$

7. $\dfrac{4 - 4}{11}$

8. $20 - 2 \cdot 3$

9. -5^2

10. $\dfrac{-7}{0}$

11. $(9 - 12) - (16 - 14)$

12. $15 + 8 \cdot 12 \div 2^2 - 4 \cdot 3$

Directions   Fill in the missing parts.

Fraction	Decimal	Percent
13.	0.85	
14. $\dfrac{9}{20}$		

Directions   Find the following percentages.

15. 23% of 200

16. 6% of 86

17. List the elements of the set of integers between -2 and 3.

18. Replace the ? with the proper inequality symbol ($<$ or $>$) to get a true statement.
$|-7| \,?\, |4|$

19. Evaluate $(2a - b)(c + 2d)$ if $a = 3$, $b = -2$, $c = 2$, and $d = -4$.

20. Evaluate the following formula.
$P = 2\ell + 2w$, $\ell = 6.9$ and $w = 3.4$
(Geometry).

21. Rico can type 2,146 words in 29 minutes. How many words can he type per minute?

22. A chemist now has 125 ml (milliliters) of acid and she needs 350 ml of the acid. How much more is needed?

23. Kelsey has $175 and wants to buy a television for $349. How much more does she need?

24. If a woman has $27.43 just after paying off a debt of $16.12, how much money did she have before paying off the debt?

25. Mary Ann has $175.46 in her checking account. She deposits $19.52, $14, and $23.12. She then writes checks for $52 and $78.48. What is her final balance in the checking account?

2

Solving Equations and Inequalities

On a road map, 3 inches represents a distance of 45 miles. If the distance between Detroit and Sault Ste. Marie is 23 inches on the map, how far is it from Detroit to Sault Ste. Marie?

2–1 Algebraic Notation and Terminology

Algebraic Terminology

Mathematics is a language and as with any language has a vocabulary that you must learn so that you can understand a lecture, read the textbook, and follow directions. In this section we will introduce some of the vocabulary that will be used in this and other math classes that you take.

In section 1–2, we defined a **variable** to be a symbol that represents an unspecified number. A variable is able to take on any one of the different values that it represents. In the relationship

$$y = x + 2$$

y and x are variables since they both can assume various numerical values.

A constant is a symbol that does not change its value, such as 5, -3, $\dfrac{29}{7}$, or π. In the relationship

$$y = x + 2$$

2 is a constant. A number is a constant. If a symbol represents only one value, that symbol is a constant.

In algebra the symbols $+$, $-$, $\cdot$, $\div$, ^{2}, etc. signify the same operations as in arithmetic. One innovation, due to René Descartes in 1637, is that we do not always write the $\cdot$ for multiplication. For example, xy means $x \cdot y$. This is because variables are understood to be one letter only, not two or more. If we write $5xy$ we mean $5 \cdot x \cdot y$. As in arithmetic, $(5)(x)(y)$ would mean the same thing.

The only other place we do not have to write the operation is in mixed numbers, such as $2\frac{1}{3}$. This means $2 + \frac{1}{3}$, so omitting the operation here does not mean multiplication. In fact the product of 2 and $\frac{1}{3}$ must show or indicate the operation, as in $2 \cdot \frac{1}{3}$ or $2\left(\frac{1}{3}\right)$.

Any meaningful collection of variables, constants, grouping symbols, and signs of operations is called an algebraic expression. Examples of algebraic expressions would be

$$prt, \quad \ell w, \quad 5xy, \quad \frac{xy}{z}, \quad 2\ell + 2w, \quad \frac{x^2 - 1}{x^2 + 1}, \quad 3x^2 + 2x - 1, \quad 5(a + 2b), \quad \pi r^2$$

In an algebraic expression, terms are any constants, variables, or products or quotients of these. *Terms are separated by plus or minus signs.* The $+$ and $-$ signs that separate the algebraic expression into terms are part of the term.

● **Example 2–1 A**

Determining the number of terms

Determine the number of terms in the algebraic expression.

The plus and minus signs separate the algebraic expression into three terms

1. $5x^2 + 2x - 1$ — There are three terms $5x^2$, $2x$, and -1
  1st   2nd   3rd

2. $x^2 + y^2$ — There are two terms
  1st   2nd

3. $4x^5 y^2 z^4$ — There is one term
  1st

4. $a^2 + \dfrac{b + c^2}{d}$ — There are two terms since the fraction bar forms a grouping. Observe that the second term has two terms in the numerator
  1st   2nd

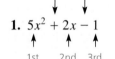

▶ *Quick check* Determine the number of terms in $5 + x^2 y - z$ and $4x^2 - 2x$. ●

Learning System
"Quick Check" problems are worked step-by-step within the exercise set. You should check your solution of this problem line-by-line and use this solution as a quick reference while doing the problems within the exercise set.

In the expression $5xy$, *each factor or grouping of factors is called the* **coefficient** *of the remaining factors.* That is, 5 is the coefficient of xy; x is the coefficient of $5y$; $5x$ is the coefficient of y; and so on. The 5 is called the **numerical coefficient**, and it tells us how many xy's we have in the expression.

Since we often talk about the numerical coefficients of a term, we will eliminate the word "numerical" and just say "coefficient." It will be understood that we are referring to the numerical coefficient. If no numerical coefficient appears in a term, the coefficient is *understood* to be 1, because x means $1 \cdot x$.

● **Example 2–1 B**

Determining the numerical coefficient

The algebraic expression $6x - 3y + z$ is thought of as the sum of terms $6x + (-3y) + z$, therefore 6 is the coefficient of x, -3 is the coefficient of y, and 1 is understood to be the coefficient of z.

☞ *Quick check*   What are the numerical coefficients in the algebraic expression $a^2 - 2a + 4b$?   ●

A special kind of algebraic expression is a **polynomial**. The following are characteristics of a polynomial.

1. It has real number coefficients.
2. All variables in a polynomial are raised to only natural number powers.
3. The operations performed by the variables are limited to addition, subtraction, and multiplication.

A polynomial that contains just one term is called a **monomial**; a polynomial that contains two terms is called a **binomial**; and a polynomial that contains three terms is called a **trinomial**. No special names are given to polynomials that contain more than three terms.

● **Example 2–1 C**

Identifying polynomials

Determine if each of the following algebraic expressions is a polynomial. If it is a polynomial, what name best describes it? If it is not a polynomial, state why it is not.

1. x, $4x$, 3, and $5x^2y$ are monomials.

2. $3x + 1$, $x + y$, and $81W^2 - 9T^2$ are binomials.

3. $5x^3 + 2y - 1$ and $z^2 + 9z - 10$ are trinomials.

4. $6x^3 - 2x^2 + 4x + 1$ is a polynomial of 4 terms.

5. $\dfrac{4}{x + 2}$ is not a polynomial since it contains a variable in the denominator.

☞ *Quick check*   Determine if each is a polynomial. If it is, what name best describes it? If it is not, state why it is not.

$$5x^2 + 2x; \qquad 5x^2 + \frac{2}{x}$$ ●

Another way that we identify different types of polynomials is by the **degree** of the polynomial. **The degree of a term is the sum of the exponents of the variables. The degree of a polynomial is the highest degree of any nonzero term of the polynomial.**

● **Example 2–1 D**

Determining the degree

Determine the degree of the polynomial.

1. $5x^3$ Degree is 3 because the exponent of *x* is 3

2. $x^4 - 2x^3 + 3x - 5$ Degree is 4 because the greatest exponent of *x* in any one term is 4

Note   In example 2, the polynomial has been arranged in **descending powers** of the variable. This is the form that we will use when we write polynomials in one variable.

3. $6a^2b^3$ Degree is 5 because the sum of the exponents of the variables (2 + 3) is 5

4. $4x^5 - 2x^3y^4 + y^3$ Degree is 7 because the highest degree of any term is 7

5. 3   Degree is 0. The reason for this will be explained in section 3–3.

<u>Note</u> Any nonzero constant *a* has degree 0. The number 0 has no degree, since 0 times a variable to any power is 0.

Algebraic Notation

Many problems that we encounter will be stated verbally. These will need to be translated into algebraic expressions. While there is no standard procedure for changing a verbal phrase into an algebraic expression, the following guidelines should be of use.

Changing a Verbal Phrase into an Algebraic Expression

1. Read the phrase carefully, determining useful prior knowledge. Prior knowledge is knowing that increased means add, diminished means subtract, product means multiply, and so on.
2. If there is no variable used in the phrase, choose one to represent the unknown value.
3. Write an algebraic expression using the variable to express the conditions stated in the phrase.

When translating verbal phrases into algebraic expressions, we should be looking for phrases that involve the basic operations of addition, subtraction, multiplication, and division. Table 2–1 shows some examples of phrases that are commonly encountered. We will let *x* represent the unknown number.

Table 2–1

Phrase	Algebraic expression	Phrase	Algebraic expression
Addition		*Multiplication*	
6 more than a number		a number multiplied by 6	
the sum of a number and 6		6 times a number	
6 plus a number	$x + 6$	the product of a number and 6	$6x$
a number increased by 6			
6 added to a number			
Subtraction		*Division*	
6 less than a number		a number divided by 6	
a number diminished by 6		the quotient of a number and 6	$\dfrac{x}{6}$
the difference of a number and 6		$\dfrac{1}{6}$ of a number	
a number minus 6	$x - 6$		
a number less 6			
a number decreased by 6			
6 subtracted from a number			
a number reduced by 6			

● **Example 2–1 E**

Problem solving

Write an algebraic expression for each.

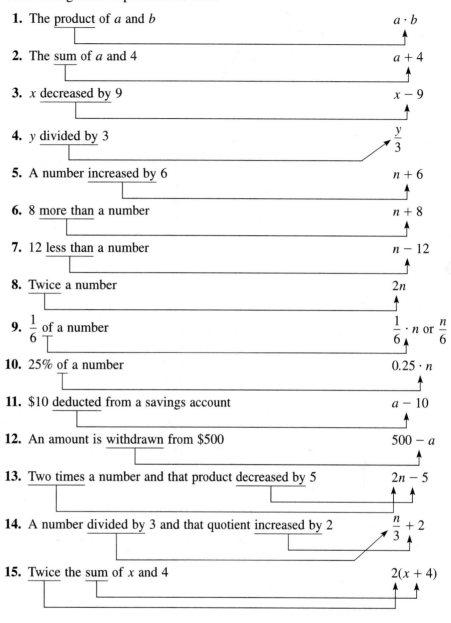

1. The product of a and b $a \cdot b$

2. The sum of a and 4 $a + 4$

3. x decreased by 9 $x - 9$

4. y divided by 3 $\dfrac{y}{3}$

5. A number increased by 6 $n + 6$

6. 8 more than a number $n + 8$

7. 12 less than a number $n - 12$

8. Twice a number $2n$

9. $\dfrac{1}{6}$ of a number $\dfrac{1}{6} \cdot n$ or $\dfrac{n}{6}$

10. 25% of a number $0.25 \cdot n$

11. \$10 deducted from a savings account $a - 10$

12. An amount is withdrawn from \$500 $500 - a$

13. Two times a number and that product decreased by 5 $2n - 5$

14. A number divided by 3 and that quotient increased by 2 $\dfrac{n}{3} + 2$

15. Twice the sum of x and 4 $2(x + 4)$

Learning System
"The mastery points" form a checklist of the skills you must master from this section.

📐 ***Quick check***   Write an algebraic expression for the product of x and y. Write an algebraic expression for a number increased by 6.   ⬤

MASTERY POINTS

Can you
- Identify terms in an algebraic expression?
- Identify a polynomial?
- Write an algebraic expression?

Learning System
"Quick Check examples" are a step-by-step solution of the problems that follow each set of examples within the text.

Exercise 2–1

Directions
Specify the number of terms in each algebraic expression and list the terms in each expression. See example 2–1 A.

📐 ***Quick Check***

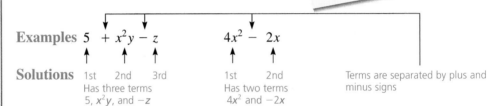

Examples   5 $+$ x^2y $-$ z $4x^2$ $-$ $2x$

Solutions   1st   2nd   3rd   1st   2nd   Terms are separated by plus and
            Has three terms   Has two terms   minus signs
            5, x^2y, and $-z$ $4x^2$ and $-2x$

Directions   Specify the number of terms in each algebraic expression and list the terms in each expression. See example 2–1 A.

1. $3x + 4y$

2. $5xyz$

3. $4x^2 + 3x - 1$

4. $x^3 - 4x + 7$

5. $\dfrac{6x}{5}$

6. $\dfrac{x}{3}$

7. $8xy + \dfrac{5y}{2} - 6x$

Learning System
"Trial exercise problems" are denoted by a ☐ around the problem number indicating that the solution is shown step-by-step in the answer appendix. You can use the problem as an example or as a line-by-line check of your solution.

8. $15x^2 + y$

9. $5x^3 + 3x^2 - 4$

10. $x^2 + a^2y^2 - z$

11. $x + y + z$

12. 7

13. $a^2b + c - x^2y + z$

14. $x^2 + y - z + c$

Directions Determine the numerical coefficients of the following algebraic expressions. See example 2–1 B.	☞ *Quick Check* Example $a^2 - 2a + 4b$ Solution 1 is understood to be the coefficient of a^2, -2 is the coefficient of a, 4 is the coefficient of b.

Directions   Determine the numerical coefficients of the following algebraic expressions. See example 2–1 B.

15. $5x^2 + x - 4z$

16. $a^2b + 4ab^2 - ab$

17. $x - y - 3z$

18. $3x^4 - x^3 + x^2$

19. $-2a - b + c$

Directions Determine if each of the following algebraic expressions is a polynomial. If it is a polynomial, what name best describes it and what is its degree? If it is not a polynomial, state why it is not. See examples 2–1 C and D.	☞ *Quick Check* Examples $5x^2 + 2x$ $\qquad\qquad\qquad\qquad 5x^2 + \dfrac{2}{x}$ Solutions It is a polynomial. Since there are two terms, it is a binomial. The greatest exponent is 2, so its degree is 2. $\qquad$ Not a polynomial because a variable is used as a divisor (appears in the denominator).

Directions   Determine if each of the following algebraic expressions is a polynomial. If it is a polynomial, what name best describes it and what is its degree? If it is not a polynomial, state why it is not. See examples 2–1 C and D.

20. $3x^2 + 2x + 5$

21. $7x + 4$

22. $5x^2 + 2x$

23. $x + \dfrac{1}{x}$

24. $x^5 + x^3 - x$

25. $3x^4y^2$

26. $4x^5y^3 - 7x^3y^2 + 3xy - 2$

27. $9x^6y + 2x^2y^2 + 4y^4$

Directions Write an algebraic expression for each of the following. See example 2–1 E.	☞ *Quick Check* Examples The product of x and y $\qquad\qquad$ A number increased by 6 Solutions $x \cdot y$ $\qquad\qquad\qquad\qquad\qquad$ Let x represent the number; hence $x + 6$

Directions   Write an algebraic expression for each of the following. See example 2–1 E.

28. The sum of a and b

29. 3 times a, subtracted from b

30. 7 less than x

31. 5 more than y

32. The sum of x and y, divided by z

33. x times the sum of y and z

34. a decreased by 5

35. a decreased by b

36. 6 more than a number

37. 9 less than a number

38. 15% of a number

39. 34% of a number

40. $\dfrac{3}{4}$ of a number

41. $\dfrac{1}{3}$ of a number

42. $40 withdrawn from a savings account

43. $50 deposited into a savings account

44. A savings account increased by $80

45. $100 deducted from a saving account

46. $\dfrac{1}{2}$ of x, decreased by 2 times x

47. A number decreased by 12

48. A number added to 4

49. 3 times a number and that product increased by 1

50. A number divided by 5

51. 2 times the sum of a number and 4

52. A number decreased by 6 and that difference divided by 11

53. 6 less than twice a number

54. **a.** Twice a number, less 6
   b. Twice the quantity represented by a number less 6

55. **a.** One third of the sum of a number and 5

   b. The sum of one half of a number and 5

56. The sum of one half a number and one third of the number

57. The product of a number and that same number less 3

58. 30% of a number

59. 5% of the difference between a number and 10,000

60. 25% of what remains when an amount is deducted from $12,000

61. The sum of 6% of an amount and what remains when that amount is deducted from $30,000

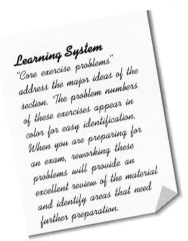

Learning System
"Core exercise problems" address the major ideas of the section. The problem numbers of these exercises appear in color for easy identification. When you are preparing for an exam, reworking these problems will provide an excellent review of the material and identify areas that need further preparation.

Directions   Multiplication and division are related. For example, $\dfrac{x}{6}$ means the same as $\dfrac{1}{6}x$. Rewrite the following as multiplication of the variable(s) by a fraction.

62. $\dfrac{x}{2}$

63. $\dfrac{y}{7}$

64. $\dfrac{3y}{5}$

65. $\dfrac{2x}{9}$

66. $\dfrac{3xy}{4}$

67. $\dfrac{2x^2 y}{5}$

✦ Explore to Learn

1. The symbol 2 is a constant because it is used to represent exactly one value. Think of a combination of symbols that represents this same value.

2. Suppose it is understood that x represents the age of any one person in your classroom. Is x a variable or constant?

3. Suppose x represents your age today. Is x a variable or constant?

4. The symbol π is a letter (from the Greek alphabet). Is it a variable or a constant?

Review Exercises

Directions   Perform the indicated operations. See sections 1–5 and 1–6.

1. $(-12) + 6$

2. $4 + (-8)$

3. $10 - 18$

4. $9 - (-9)$

5. $-6 - (-6)$

6. $(-14) + (-7)$

Directions   Perform the indicated operations. See section 1–8.

7. -5^2

8. $(-8)^2$

9. $10 - 6 \cdot 2$

10. $25 - 5 \cdot 2$

11. $100 \div 10 \cdot 2 + 2$

12. $28 - (8 - 12) - 3^2$

2–2   Algebraic Addition and Subtraction

Since algebraic expressions (including polynomials) represent real numbers when the variables are replaced by real numbers, the ideas and properties that apply to operations with real numbers also apply to algebraic expressions.

The Distributive Property

The distributive property is the only property that establishes a relationship between addition (or subtraction) and multiplication in the same expression. The distributive property allows us to change certain multiplication problems into sums or differences.

Distributive Property

For every real number a, b, and c,

$$a(b + c) = ab + ac \text{ and}$$
$$a(b - c) = ab - ac$$

Concept

If a number is being used to multiply the sum or difference of two others, it is "distributed" to them both. That is, it multiplies them both.

● **Example 2–2 A**
Distributive property

The following are applications of the distributive property.

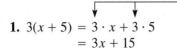
Each term inside the parentheses is multiplied by 3

1. $3(x + 5) = 3 \cdot x + 3 \cdot 5$
$\quad\quad\quad\quad = 3x + 15$

2. $2(3 - a) = 2 \cdot 3 - 2 \cdot a$
$\quad\quad\quad\quad = 6 - 2a$

We are now going to use the distributive property to carry out addition and subtraction of algebraic expressions and to remove grouping symbols.

Like Terms

We first need to define the types of quantities that can be added or subtracted. *We can add or subtract only like, or similar, quantities.* **Like terms, or similar terms, are terms whose variable factors are the same.**

Learning System
"Notes" are used to highlight important ideas and facts and to point out potential errors.

Note For two or more terms to be called **like terms**, the variable factors of the terms, along with their respective exponents, must be identical. However, the numerical coefficients of these identical variable factors may be different.

● **Example 2–2 B**
Determining like terms

1. $3a^2b^3$ and $-2a^2b^3$ are like terms because the variables are the same (a and b) and the respective exponents are the same (a is to the second power in each term and b is to the third power).

2. $2x^2y$ and $2xy^2$ both contain the same variables but are *not* like terms because the exponents of the respective variables are not the same.

☛ *Quick check*   Are $4a^2$ and $4a^3$ like terms?

Addition and Subtraction

Using the definition of like terms and the distributive property, we are ready to carry out addition and subtraction of algebraic expressions. Consider the following example:

$$3a + 4a$$

Using the distributive property, the expression can be written

$$3a + 4a = (3 + 4)a = 7a$$

Note The process of addition or subtraction is performed only with the numerical coefficients. *The variable factor and its exponent remain unchanged.*

Learning System
"Procedure Boxes" clearly state the step-by-step process for working problems.

Combining Like Terms

1. Identify the like terms.
2. If necessary, use the commutative and associative properties to group together the like terms.
3. Combine the numerical coefficients of the like terms and multiply that by the variable factor.
4. Remember that y is the same as $1 \cdot y$ and $-y$ is the same as $-1 \cdot y$.

● **Example 2–2 C**

Addition and subtraction of like terms

Perform the indicated addition and subtraction.

1. $5x + 7x$ Identify like terms
$$= (5 + 7)x$$ Distributive property
$$= 12x$$ Add numerical coefficients

2. $y + 3y - 2y$ Identify like terms
$$= (1 + 3 - 2)y$$ Distributive property
$$= 2y$$ Combine numerical coefficients

3. $6x^2 - 4x + 3x - 2x^2$
$$= (6x^2 - 2x^2) + (-4x + 3x)$$ Use the commutative and associative properties to group together like terms

$$= (6 - 2)x^2 + (-4 + 3)x$$ Distributive property
$$= 4x^2 - 1x = 4x^2 - x$$ Combine numerical coefficients

Note   After sufficient practice, we should be able to carry out the addition and subtraction by grouping mentally.

4. $a^2b + 5ab^2 - 4a^2b + 3ab^2 = (1 - 4)a^2b + (5 + 3)ab^2$
$$= -3a^2b + 8ab^2$$

Like terms

⚑ *Quick check*   Perform the indicated addition and subtraction:
$$3a - 2b + a + 5b; \qquad 3a^2 + 5a - 2a^2 - a$$

Learning System
Examples have short phrase statements next to most steps indicating exactly what has been done. These will enable you to develop a clear understanding of how a problem is worked without having to guess what went on in a particular step.

Grouping Symbols

If any quantity enclosed with grouping symbols, we treat the quantity within as a single number. We are now going to use the distributive property to remove grouping symbols such as ( ), [ ], and { }. Consider the following examples:

1. The quantity $(2a + 3b)$ can be written as $1 \cdot (2a + 3b)$. Applying the distributive property, we have

$$1(2a + 3b) = 1 \cdot 2a + 1 \cdot 3b = 2a + 3b$$

2. The quantity $+(2a + 3b)$ can be written as $(+1) \cdot (2a + 3b)$ giving

$$(+1)(2a + 3b) = (+1) \cdot 2a + (+1) \cdot 3b = 2a + 3b$$

3. The quantity $-(2a + 3b)$ can be written as $(-1) \cdot (2a + 3b)$ giving

$$(-1)(2a + 3b) = (-1) \cdot 2a + (-1) \cdot 3b = -2a - 3b$$

Removing Grouping Symbols

1. If a grouping symbol is preceded by no symbol or by a "+" sign, the grouping symbol can be dropped and the enclosed terms remain unchanged.
2. If a grouping symbol is preceded by a "−" sign, when the grouping symbol is dropped, we change the sign of **each** enclosed term.

● **Example 2–2 D**
Removing grouping symbols

Remove all grouping symbols and perform the indicated addition or subtraction.

1. $(3x^2 + 2x + 5) + (4x^2 + 3x + 6)$ Remove grouping symbols
$= 3x^2 + 2x + 5 + 4x^2 + 3x + 6$ Enclosed terms remain unchanged
$= (3x^2 + 4x^2) + (2x + 3x) + (5 + 6)$ Associative and commutative properties
$= (3 + 4)x^2 + (2 + 3)x + 11$ Distributive property
$= 7x^2 + 5x + 11$ Combine numerical coefficients

2. $(3x^2 - x + 4) - (2x^2 - 5x - 7)$ Remove grouping symbols
$= 3x^2 - x + 4 - 2x^2 + 5x + 7$ Change the sign of each term contained in the second set of parentheses
$= (3x^2 - 2x^2) + (-x + 5x) + (4 + 7)$ Associative and commutative properties
$= (3 - 2)x^2 + (-1 + 5)x + 11$ Distributive property
$= 1x^2 + 4x + 11$ Combine numerical coefficients
$= x^2 + 4x + 11$ x^2 is the same as $1x^2$

3. $3(2x^2 + x + 3) + 2(x - 4)$
$= 6x^2 + 3x + 9 + 2x - 8$ Distributive property
$= 6x^2 + (3x + 2x) + (9 - 8)$ Associative and commutative properties
$= 6x^2 + (3 + 2)x + 1$ Distributive property
$= 6x^2 + 5x + 1$ Combine like terms

<u>Note</u> In the following examples, we will mentally add or subtract the like terms.

4. $(8R^2 - 2R + 3) - (6R^2 + 6R - 1)$ Remove grouping symbols

$= 8R^2 - 2R + 3 - 6R^2 - 6R + 1$ Change the sign of each term in the second parentheses

$= 2R^2 - 8R + 4$ Combine like terms

5. $-2(2x - 7) + 3(4x + 1) - (x - 8)$

$= -4x + 14 + 12x + 3 - x + 8$ Distributive property

$= 7x + 25$ Combine like terms

▶ *Quick check* Remove all grouping symbols and perform the indicated addition or subtraction: $(5x^2 + 2x - 1) - (3x^2 - 4x + 3)$; $-2(5x - 4) + 3(6x - 2)$

Problem Solving

The following sets of word problems are designed to help us interpret word phrases and write expressions for them in algebraic symbols.

● **Example 2–2 E**

Problem solving

Write an algebraic expression for each of the following word phrases.

1. Nancy can type 90 words per minute. How many words can she type in n minutes?

If Nancy can type 90 words in one minute, then we multiply

$$90 \cdot n \text{ or } 90n$$

to obtain the number of words she can type in n minutes.

2. If John has n dollars in his savings account and on successive days he deposits $15 and then withdraws $34 to make a purchase, write an expression for the balance in his savings account.

We *add* the deposits and *subtract* the withdrawals. Thus

$$n + 15 - 34 = n - 19$$

represents the balance in John's savings account after the two transactions.

3. A woman paid d dollars for a 30-pound bag of dog food. How much did the dog food cost her per pound?

The price per pound is found by dividing the total cost by the number of pounds. Thus the price per pound of the dog food is represented by $\dfrac{d}{30}$ dollars.

☛ *Quick check*   Express the cost in terms of dollars and in terms of cents for x cassette tapes if each tape costs $2.95.   ●

MASTERY POINTS

Can you
- Identify like terms?
- Add and subtract algebraic expressions?
- Remove grouping symbols?
- Write an algebraic expression?

Exercise 2–2

Directions
For the groups of terms, determine if they are like or unlike terms. See example 2–2 B.

📌 **Quick Check**

Example $4a^2$ and $4a^3$

Solution Both contain the same variable but are **unlike** because the exponent of the variable is not the same.

Directions For the groups of terms, determine if they are like or unlike terms. See example 2–2 B.

1. $3a, -2a$

2. $5x, 7x$

3. $4a^2, a^2$

4. $b^3, -2b^3$

5. $2a^2, 2a^3$

6. $4x, 4x^2$

Directions
Perform the indicated addition and subtraction. See examples 2–2 A, B, and C.

📌 **Quick Check**

Examples $3a - 2b + a + 5b$

Solutions $= (3a + a) + (-2b + 5b)$

$= (3 + 1)a + (-2 + 5)b$
$= 4a + 3b$

$3a^2 + 5a - 2a^2 - a$

$= (3a^2 - 2a^2) + (5a - a)$ Commutative and associative properties

$= (3 - 2)a^2 + (5 - 1)a$ Distributive property
$= 1a^2 + 4a$ Combine numerical coefficients

$= a^2 + 4a$

Directions Perform the indicated addition and subtraction. See examples 2–2 A, B, and C.

7. $2x + x + 6x$

8. $8y - y + 2y$

9. $4a - 2b + 9a + 4b$

10. $a + 4b + 6a - 8b$

11. $3x + 4x + 7x$

12. $2a^2b - 4a^2b + 6a^2b$

13. $4ab + 11ab - 10ab - 8ab$

14. $d^2 + d - 3d^2 + d^4 + 4d^2$

15. $5x + x^2 - x + 6x^2$

16. $x + 2x^2 - 5 + x^3 - 2x - 2x^2$

17. $3a + b + 2a - 5c - b - 2x^2 + 8a$

18. $3a + 8a - 6a + 9a$

19. $x^2 + 5x - 8x + 2x^2$

20. $5a + 4a^2 - 2a - a^2$

21. $x^2 + 5x - 6 + 7x^2 - 3x + 7$

22. $6a^2 - 5a + 3 - 2a^2 - 4a + 8$

Directions
Remove all grouping symbols and combine like terms. See example 2–2 D.

📌 **Quick Check**

Example $(5x^2 + 2x - 1) - (3x^2 - 4x + 3)$

Solution $= 5x^2 + 2x - 1 - 3x^2 + 4x - 3$ Change the sign of each term in the second parentheses

$= (5x^2 - 3x^2) + (2x + 4x) + (-1 - 3)$ Commutative and associative properties
$= 2x^2 + 6x - 4$ Combine like terms

Example $-2(5x - 4) + 3(6x - 2)$

Solution

Like terms

$= -10x + 8 + 18x - 6$ Distributive property

Like terms

$= 8x + 2$ Combine like terms

Directions   Remove all grouping symbols and combine like terms. See example 2–2 D.

23. $(2x + 3y) + (x + 5y)$

24. $(4a - b) + (3a + 2b)$

25. $(5x + y) - (3x - 2y)$

26. $(7x - 3y) - (5x - 6y)$

27. $(3a - b + 4c) - (a - 2b - c)$

28. $(4x - 3y - 2z) - (3x - 4y - z)$

29. $(8x + 3y - 4z) - (6x - y - 4z)$

30. $(7a - b - 3c) - (5a - 4b + 3c)$

31. $(2x^2y - xy^2 + 7xy) + (xy^2 - 5x^2y + 8xy)$

32. $(5x^2 - y^2) - (6x^2 - 3y^2) - (8x^2 + 2y^2)$

33. $(48a + 3b) - (-22a - 6b)$

34. $(8xy + 9y^2z) - (13xy - 14yz)$

35. $(a - 3b + 2) - (a + 5b - 8)$

36. $(2x + 6z - 10y) - (8y + 3z - 6x + 4)$

37. $(3a - 2b) - (a + 4b) - (-a + 3b)$

38. $(5xy - y) - (3yz + 2xy) + (3y - 4xy)$

39. $(7x^2 - 2y) + (3z - 4y) - (4x^2 - 6y)$

40. $3(2x - 4) + 2(3x - 1)$

41. $5(x + 3) + 4(x - 3)$

42. $4(2a - 1) - 2(a - 3)$

43. $3(3a + 2) - 4(a - 1)$

44. $-3(2a - 4) + 5(a - 3) - (a - 7)$

45. $2(3x - 2) - 3(2x + 3) + (3x - 1)$

46. $4(x + 3) - (x - 6) - 2(x - 4)$

47. $\dfrac{1}{2}x + \dfrac{1}{3}x$

48. $\dfrac{5}{8}x - \dfrac{1}{8}x$

49. $\left(\dfrac{3}{5}x - \dfrac{2}{3}y\right) - \left(\dfrac{1}{6}x + \dfrac{1}{6}y\right)$

50. $2y - \left(3x - \dfrac{1}{2}y\right) + \dfrac{2}{3}x$

51. $(400 - 0.3x) + 0.2x$

52. $(0.25x + 0.3y) - (0.15x - 0.7y)$

53. $t_1 - t_2 - (3t_1 + 2t_2)$

54. $5a_1 - (a_1 + 2b_1) + (b_1 - a_2 - b_2)$

55. $(b_2 - b_1) - (a_2 - a_1) + 5a_1 - 2a_2 + (b_1 + b_2)$

56. $(x_1 + 3x_2) + (5x_1 - 2x_2) + x_1$

Directions   Write and simplify an expression for the perimeter (distance around) each of the figures.

57.

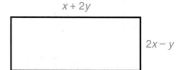

58.

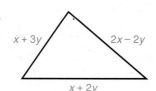

59. In the series electronic circuit shown, the total resistance r is the sum of the resistances. Compute the resistance.

Directions   For problems 60–63, an industrial robot can only move right or left along a straight line. Its motion is accomplished by two types of stepper motors. One moves the robot a distance d_1, and the second moves it a distance d_2. Moves to the right are considered positive, and moves to the left are considered negative. An expression like $2d_1 - 3d_2$ would mean the robot moved to the right in two steps of distance d_1, then moved left three steps of distance d_2.

61. Suppose d_1 is set to 4 centimeters (cm), and d_2 is set to 7 cm. (a) Find a combination of moves that leaves the robot 5 cm to the right of its home position. (b) Find a combination of moves that leaves the robot 6 cm to the left of its home position.

62. Suppose d_1 is set to 4 cm and d_2 is set to 6 cm. Discuss why this might not be a good combination of settings.

60. Suppose the robot makes the following sequence of moves starting at its "home" position: $3d_1 - d_2$, $d_1 + 5d_2$, $2d_1 - 2d_2$, $4d_1 + d_2$, $6d_1$, $d_1 - 4d_2$. (a) Write an expression that describes this sequence of moves and simplify the expression. (b) Interpret the result in terms of the robot's final position.

63. Referring to the robot's settings from problem 61, suppose the robot is 18 cm to the right of its home position. Give a series of moves that will bring the robot to a position 9 cm to the right of its home position.

Directions	⚐ *Quick Check*
Write an algebraic expression for the following word statements. See example 2–2 E.	**Example** Express the cost in terms of dollars and in terms of cents for x cassette tapes if each tape costs $2.95.
	Solution If we are buying x tapes at $2.95 each, then we must multiply x by $2.95. The algebraic expression in terms of dollars would be $2.95 \cdot x$ and in terms of cents, it would be $295 \cdot x$.

Directions   Write an algebraic expression for the following word statements. See example 2–2 E.

64. Lin enters 85 keystrokes per minute on the computer. How many keystrokes can he enter in m minutes?

65. A 10-pound box of candy costs y dollars. How much does the candy cost per pound?

66. Carlos paid $25 for a ticket to a play. If the play lasted h hours, what did it cost him per hour to see the play?

67. Arlene has n nickels and d dimes in her purse. Express in cents the amount of money she has in her purse. (*Hint:* n nickels is represented by $5n$.)

68. Jack has q quarters, d dimes, and n nickels. Express in cents the amount of money Jack has.

69. Maria is p years old now. Express her age (a) 12 years from now, (b) 5 years ago.

70. Ann is 3 years old. If Jan is n times as old as Ann, express Jan's age. Express Jan's age 8 years ago.

71. Bill's savings account has a current balance of $258. He makes a withdrawal of n dollars and then makes a deposit of m dollars. Express his new balance in terms of n and m.

72. Carmen has a balance of n dollars in her checking account. She makes a deposit of $36 and then writes 3 checks for m dollars each. Express her new balance in terms of n and m.

73. Pete has c cents, all in half-dollars. Write an expression for the number of half-dollars Pete has.

74. If x represents a whole number, write an expression for the next greater whole number.

75. If y represents an even integer, write an expression for the next greater even integer.

76. If z represents an odd integer, write an expression for the next greater odd integer.

77. If Larry is f feet and t inches tall, how tall is Larry in inches?

78. John earns $1,000 more than twice what Terry earns in a year. If Terry earns d dollars, write an expression for John's annual salary.

79. Jean's annual salary is $2,000 less than n times Lisa's salary. If Lisa earns $25,000 per year, express Jean's annual salary.

80. Express the total cost of purchasing x cans of tuna at 69¢ per can on Friday and y cans of the same tuna at 57¢ per can on Saturday.

81. A gallon of primer paint costs $9.95 and a gallon of latex-base paint costs $12.99. Express the cost in dollars of p gallons of primer and q gallons of latex-base paint.

82. Paula enters x calculations per minute on the calculator and Leigh enters 7 calculations per minute less than Paula. Write an expression for the number of calculations Leigh enters in 35 minutes.

✵ Explore to Learn

1. We said that in an expression like $-(?)$ we would change the sign of every term inside the parentheses if we remove the parentheses. If this is true, then the following should give the same results. Verify that they do by performing the operations according to the order of operations.

   a. $(6 - 4) - (8 - 13 + 6 + 2 - 10)$
   b. $(6 - 4) - 8 + 13 - 6 - 2 + 10$

2. A landscape architect designs a brick patio to be 3 feet longer than twice the width. The width is represented by x. Label the rectangle to show the length. How would the expression for the length be different if the patio were 3 feet less than twice the width? Is there ever an instance when you would label the width $2x$?

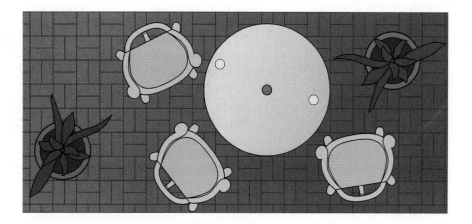

Review Exercises

Directions   Write an algebraic expression for each of the following. See section 2–1.

1. The product of x and 3

2. 6 times the sum of a and 7

3. y decreased by 2 and that difference divided by 4

4. A number multiplied by 5

5. A number diminished by 12

6. A number divided by 8 and that quotient decreased by 9

2–3   The Addition and Subtraction Property of Equality

Equations

An **equation** is a statement of equality. If two expressions represent the same number, then placing an equality sign, $=$,[1] between them forms an equation. The following diagram is used to show the parts of an equation.

$$\underbrace{3x - 7}_{} \quad = \quad \underbrace{2x + 5}_{}$$

Left side (also called left member of the equation)

Equality sign

Right side (also called right member of the equation)

A **mathematical statement** is a sentence that can be labeled true or false. $2 + 3 = 5$ is a true statement, and $3 + 4 = 8$ is a false statement. In this chapter we will learn procedures used to solve linear equations. An equation is said to be a **linear equation**[2] if the variable is raised only to the first power.

Linear Equation

An equation is linear if it can be written in the form

$$ax + b = 0 \qquad a \neq 0$$

where a and b are real numbers.

Solution Set

A replacement value for the variable that forms a true statement is called a **root**, or a **solution**, of the equation. The **solution set** is the set of all replacement values for the variable that cause the equation to be a true statement.

To Check an Answer of an Equation

1. **Substitute:** Replace the variable in the original equation with the answer.
2. **Order of operations:** Perform the indicated operations.
3. **True statement:** If step 2 produces a true statement, the answer is correct.

---

[1]The "$=$" symbol was first used by Robert Recorde in *The Whetstone of Witte* published in 1557. "I will sette as I doe often in woorke use, a paire of paralleles, or Gemowe [twin] lines of one lengthe, thus: ====, bicause noe 2 thynges, can be moare equalle."

[2]The algebra of ancient Egypt (some 4,000 years ago) was much concerned with linear equations.

● **Example 2–3 A**

Determining if a value is a solution

Determine if the given value is a solution of the equation.

1. $4 - x = 7$ when $x = -3$
If we replace x by -3 in the equation and simplify,

$$4 - x = 7$$
$$4 - (-3) = 7$$
$$7 = 7$$

the equation is true, and -3 is a solution of the equation.

2. $2 + x = 8$ when $x = 5$
If we replace x by 5 in the equation and simplify,

$$2 + x = 8$$
$$2 + (5) = 8$$
$$7 = 8$$

The statement is false, and 5 is not a solution of the equation.

☞ *Quick check*   Determine if the equation $3x + 3 = 6$ is true when $x = 1$.
Determine if the equation $2x - 1 = 3$ is true when $x = 4$. ●

Equivalent Equations

To solve an equation, we go through a series of steps whereby we form equations that are equivalent to the original equation until we have the equation in the form $x = n$, n being some real number. These equations that we form are called **equivalent equations**. *Equivalent equations are equations whose solution set is the same.*

● **Example 2–3 B**

Equivalent equations

The following are equivalent equations whose solution set is {6}.

1. $4x - 2 = 3x + 4$

2. $x - 2 = 4$

3. $x = 6$ ●

Addition and Subtraction Property of Equality

Since an equation is a statement of equality between two expressions, identical quantities added to or subtracted from each expression will produce an equivalent equation. We can state this property as follows:

Learning System
Color is used to guide you throughout the text. When you are reviewing for an exam, the functional use of color will allow you to quickly locate key information.

Addition and Subtraction Property of Equality

For any algebraic expressions a, b, and c,

$$\text{if } a = b, \text{ then } a + c = b + c \text{ and}$$
$$a - c = b - c$$

Concept

We can add or subtract the same quantity on each side of an equation and the result will be an equivalent equation.

● **Example 2–3 C**

Using the addition and subtraction property of equality

Find the solution set and check the answer.

1. $x - 5 = 7$

$x - 5 + 5 = 7 + 5$ Add 5 to both sides

$x + 0 = 12$ Additive inverse

$x = 12$ Solution

Check

$(12) - 5 = 7$ Substitute

$7 = 7$ True

The solution set is {12}.

2. $x + 4 = 12$

$x + 4 - 4 = 12 - 4$ Subtract 4 from both sides

$x = 8$ Solution

Check

$(8) + 4 = 12$ Substitute

$12 = 12$ True

The solution set is {8}.

Note Subtracting 4 is the same as adding −4. Either method may be used, but we must remember to perform the operation to *both* sides of the equation.

3. $2 = x + 7$

$2 - 7 = x + 7 - 7$ Subtract 7 from both sides

$-5 = x$ Solution

Check

$2 = (-5) + 7$ Substitute

$2 = 2$ True

The solution set is {−5}.

☛ *Quick check* Find the solution set for $x - 7 = 12$ and check the answer. ●

Symmetric Property of Equality

The **symmetric property of equality** is also useful in finding the solution set of equations.

Learning System
"Concept" is a translation of the mathematical property into everyday language.

Symmetric Property of Equality

If $a = b$, then $b = a$

Concept

This property allows us to interchange the right and left sides of the equation.

In example 2–3 C–3, instead of leaving the equation as $-5 = x$, we should use the symmetric property and write the equation as $x = -5$.

We can see that our goal in solving an equation is to **isolate** the unknown on one side of the equation and to place everything else on the other side. This forms an equation of the type $x = n$. When the unknown appears on both sides of the equation, we use the addition and subtraction property of equality to form an equivalent equation where the unknown appears only on one side of the equation.

● Example 2–3 D

Solving equations when the variable appears on both sides

Find the solution set and check the answer.

1.

$$6x - 4 = 7x + 2$$
$$6x - 6x - 4 = 7x - 6x + 2 \qquad \text{Subtract } 6x \text{ from both sides}$$
$$-4 = x + 2$$
$$-4 - 2 = x + 2 - 2 \qquad \text{Subtract 2 from both sides}$$
$$-6 = x$$
$$x = -6 \qquad \text{Symmetric property}$$

$$\text{Check:} \quad 6(-6) - 4 = 7(-6) + 2 \qquad \text{Substitute}$$
$$-36 - 4 = -42 + 2 \qquad \text{Order of operations}$$
$$-40 = -40 \quad \text{(True)} \qquad \text{Checks}$$

The solution set is $\{-6\}$.

2.

$$-2x - 5 = -3x + 4$$
$$-2x + 3x - 5 = -3x + 3x + 4 \qquad \text{Add } 3x \text{ to both sides}$$
$$x - 5 = 4$$
$$x - 5 + 5 = 4 + 5 \qquad \text{Add 5 to both sides}$$
$$x = 9$$

<u>Note</u> A good habit for us to develop is to form equivalent equations in which the unknown appears only on the side of the equation that has the greater coefficient of the unknown. This will ensure a positive coefficient for the unknown.

$$\text{Check:} \quad -2(9) - 5 = -3(9) + 4 \qquad \text{Substitute}$$
$$-18 - 5 = -27 + 4 \qquad \text{Order of operations}$$
$$-23 = -23 \quad \text{(True)} \qquad \text{Answer checks}$$

The solution set is $\{9\}$.

☞ ***Quick check***   Find the solution set for $4x - 2 = 3x + 5$ and check the answer. ●

Sometimes it is necessary to use the associative, commutative, and distributive properties to perform indicated operations on one or both sides of an equation. This will *simplify* the equation before the addition and subtraction property of equality is used. Consider the examples that follow.

● Example 2–3 E

Simplifying and solving equations

Find the solution set.

1.
$$5x - 4 + 2x = 6x - 6 + 11$$
$$7x - 4 = 6x + 5 \qquad \text{Simplify}$$
$$7x - 6x - 4 = 6x - 6x + 5 \qquad \text{Subtract } 6x \text{ from both sides}$$
$$x - 4 = 5$$
$$x - 4 + 4 = 5 + 4 \qquad \text{Add 4 to both sides}$$
$$x = 9$$

The solution set is $\{9\}$.

2. $$3x = 2(2x - 4)$$
$$3x = 4x - 8 \qquad \text{Simplify}$$
$$3x - 3x = 4x - 3x - 8 \qquad \text{Subtract } 3x \text{ from both sides}$$
$$0 = x - 8$$
$$0 + 8 = x - 8 + 8 \qquad \text{Add 8 to both sides}$$
$$8 = x$$
$$x = 8 \qquad \text{Symmetric property}$$

The solution set is {8}.

☛ *Quick check* Find the solution set for $5x + 2x - 4 = 6x + 7$ and for $3(2x + 1) = x + 4x - 2$.

A calculator can be used to check a numeric answer to an equation. There are several ways to do this. We illustrate several of these in example 2–3 F, where we check the answer to number 1 of example 2–3 E.

● **Example 2–3 F**

Checking a solution with a calculator

Verify that 9 is a solution to the equation $5x - 4 + 2x = 6x - 6 + 11$. One method is to evaluate the left side and evaluate the right side using the calculator; wherever x appears we use 9.

Scientific/graphing
Left side:
$$5 \; \boxed{\times} \; 9 \; \boxed{-} \; 4 \; \boxed{+} \; 2 \; \boxed{\times} \; 9 \; \boxed{=} \; (\text{or } \boxed{\text{ENTER}}) \qquad \text{Result: 59}$$
Right side:
$$6 \; \boxed{\times} \; 9 \; \boxed{-} \; 6 \; \boxed{+} \; 11 \; \boxed{=} \; (\text{or } \boxed{\text{ENTER}}) \qquad \text{Result: 59}$$

The same value, 59, is obtained for both sides. Thus 9 is a solution. Graphing calculators permit storage of a value in a variable. We show this method for a typical graphing calculator.

$$9 \; \boxed{\text{STO}\triangleright}^* \; \boxed{\text{X|T}} \; \boxed{\text{ENTER}} \qquad \text{Store 9 in the variable } x$$
Left side:
$$5 \; \boxed{\text{X|T}}^{**} \; \boxed{-} \; 4 \; \boxed{+} \; 2 \; \boxed{\text{X|T}} \; \boxed{\text{ENTER}} \qquad \text{Result: 59}$$
Right side:
$$6 \; \boxed{\text{X|T}} \; \boxed{-} \; 6 \; \boxed{+} \; 11 \; \boxed{\text{ENTER}} \qquad \text{Result: 59}$$

Problem Solving

We are now ready to combine our ability to write an expression and our ability to solve an equation and apply them to solve a word problem. While there is no standard procedure for solving a word problem, the following guidelines should be useful.

---

*$\boxed{\rightarrow}$ on the Casio 7700GE/9700GE

**$\boxed{\text{X,T,}\theta}$ on the TI-82, and Casio 7700GE/9700GE, and $\boxed{\text{x-VAR}}$ on the TI-85.

Solving Word Problems

Read **Analyze-Visualize**	1. Read the problem carefully, usually several times. Determine what information is given, and what information you are asked to find. If possible, draw a picture, make a diagram, or construct a table to help visualize the information.
Choose a variable	2. Choose a variable to represent one of the unknowns and write a sentence stating exactly what that variable represents. Use this variable to write algebraic expressions that represent any other unknowns in the problem. Label any picture, diagram, or chart using the chosen variable.
Write an equation	3. Use the unknowns from step 2 to translate the word problem into an equation. There may be some underlying relationship or formula that you need to know to write the equation. If it is a geometry problem, the formulas inside the front cover may be useful. Other commonly used formulas are: interest equals principal times rate times time ($I = prt$) or distance equals rate times time ($d = rt$). If there is no applicable formula, then the words in the problem give the information necessary to write the equation.
Solve the equation	4. Solve the equation.
Answer the **question(s)**	5. Answer the question or questions in the problem. If there is more than one unknown in the problem, substitute the solution of the equation into the algebraic expressions that represent the other unknowns to determine the other unknown values in the problem.
Check your results	6. Check your results in the original statement of the problem. Make sure your answer makes sense. Remember that in a geometry problem the dimensions of the figure cannot be negative. If you were determining how fast a car was traveling, 1,000 miles per hour would not make sense.

● **Example 2–3 G**

Problem solving

Solve the following word problems by setting up an equation and solving it.

1. A number increased by 16 gives 24. Find the number.

Let n represent the number we are looking for. The key words to use are "increased by," which means *add*, and "gives," which means *equals*. The equation is then

a number increased by 16 gives 24
$$n \quad + \quad 16 \quad = \quad 24$$

$$n + 16 = 24$$
$$n = 8 \qquad \text{Subtract 16 from each side}$$

The number is 8.

2. Jared skipped a day of work and was penalized $52. If his pay for that week was $208, what is his weekly salary?

Let s represent his weekly salary. The key words are "penalized," which means subtract, and "was," which means equals. The equation is

weekly   penalized $52   was $208
salary
s $-$ 52 $=$ 208

$$s - 52 = 208$$
$$s = 260 \qquad \text{Add 52 to both sides}$$

Jared's weekly salary is $260.

MASTERY POINTS

Can you
- Determine if a given number is a solution of an equation?
- Use the addition and subtraction property of equality?
- Simplify equations?
- Solve for an unknown?
- Check your answer?
- Solve word problems?

Exercise 2–3	⚑ *Quick Check*
Directions Determine if the given value is a solution of the equation. See example 2–3 A.	**Examples** $3x + 3 = 6$, when $x = 1$ $\qquad$ $2x - 1 = 3$, when $x = 4$ **Solutions** $3(1) + 3 = 6$ Substitute $\qquad$ $2(4) - 1 = 3$ Substitute $\qquad 3 + 3 = 6$ Order of operations $\qquad 8 - 1 = 3$ Order of operations $\qquad 6 = 6$ Checks $\qquad 7 = 3$ Does not check $\qquad$ 1 is a solution $\qquad\qquad$ 4 is not a solution

Directions   Determine if the given value is a solution of the equation. See example 2–3 A.

1. $4 + x = 8$; $x = 4$

2. $3 - x = 4$; $x = -1$

3. $x + 7 = 10$; $x = 3$

4. $3x - 2 = 4$; $x = 2$

5. $8x + 6 = 2x - 6$; $x = -2$

6. $\dfrac{4}{5}x + 2 = 10$; $x = 10$

7. $7x - 3 = 2x + 2$; $x = -2$

8. $3x + 2 = 5x - 1$; $x = \dfrac{3}{2}$

9. $2(x - 1) = 4x + 5$; $x = -\dfrac{7}{2}$

10. $5x - 1 = 11x - 1$; $x = 0$

11. $\dfrac{x}{5} - 2 = 3x + 1$; $x = 1$

12. $\dfrac{2x}{3} - 1 = \dfrac{x}{4} + 3$; $x = 2$

Directions Find the solution set by using the addition and subtraction property of equality. Check each answer. See examples 2–3 C and D.	☞ *Quick Check*

Examples $x - 7 = 12$ $\qquad\qquad$ $4x - 2 = 3x + 5$

Solutions $x - 7 + 7 = 12 + 7$ Add 7 $\qquad$ $4x - 3x - 2 = 3x - 3x + 5$ Subtract 3x

$\qquad\qquad\qquad x = 19$ $\qquad\qquad\qquad\qquad x - 2 = 5$

$\qquad\qquad\qquad\qquad\qquad\qquad\qquad x - 2 + 2 = 5 + 2$ Add 2

$\qquad\qquad\qquad\qquad\qquad\qquad\qquad\qquad x = 7$

Check: $\qquad\qquad\qquad\qquad\qquad$ *Check:*

$\qquad (19) - 7 = 12$ Substitute $\qquad$ $4(7) - 2 = 3(7) + 5$ Substitute

$\qquad\qquad 12 = 12$ (True) Checks $\qquad$ $28 - 2 = 21 + 5$ Order of operations

$\qquad\qquad\qquad\qquad\qquad\qquad\qquad\qquad 26 = 26$ (True)   Checks

The solution set is {19}. $\qquad\qquad$ The solution set is {7}.

Directions   Find the solution set by using the addition and subtraction property of equality. Check each answer. See examples 2–3 C and D.

13. $x - 4 = 12$ $\qquad$ **20.** $a - 5 = -2$ $\qquad$ **27.** $b + 4 = 2b + 5$

14. $y - 7 = 11$ $\qquad$ **21.** $-10 = x - 4$ $\qquad$ **28.** $-y - 6 = -2y + 1$

15. $a + 5 = 2$ $\qquad$ **22.** $a - 18 = -14$ $\qquad$ **29.** $-z - 8 = -2z - 4$

16. $b + 5 = 7$ $\qquad$ **23.** $b + 7 = 0$ $\qquad$ **30.** $5 - 3x = 7 - 4x$

17. $y - 6 = -8$ $\qquad$ **24.** $y - 14 = 0$ $\qquad$ **31.** $9 - 7a = 14 - 6a$

18. $5 = x + 7$ $\qquad$ **25.** $3x - 4 = 2x + 10$ $\qquad$ **32.** $3a - 5 = 2a - 2$

19. $9 = x + 14$ $\qquad$ **26.** $6x - 5 = 5x + 11$

Directions Find the solution set. See example 2–3 E.	☞ *Quick Check*

Examples $5x + 2x - 4 = 6x + 7$ $\qquad\qquad$ $3(2x + 1) = x + 4x - 2$

Solutions $7x - 4 = 6x + 7$ Combine like terms $\qquad$ $6x + 3 = 5x - 2$ Simplify

$\qquad\qquad\qquad\qquad\qquad\qquad\qquad\qquad\qquad$ $6x - 5x + 3 = 5x - 5x - 2$ Subtract 5x

$\qquad 7x - 6x - 4 = 6x - 6x + 7$ Subtract 6x $\qquad\qquad\qquad x + 3 = -2$

$\qquad\qquad\qquad x - 4 = 7$ $\qquad\qquad\qquad\qquad\qquad$ $x + 3 - 3 = -2 - 3$ Subtract 3

$\qquad x - 4 + 4 = 7 + 4$ Add 4 $\qquad\qquad\qquad\qquad\qquad x = -5$

$\qquad\qquad\qquad\qquad x = 11$

The solution set is {11}. $\qquad\qquad$ The solution set is {−5}.

Directions   Find the solution set. See example 2–3 E.

33. $6a - 3a + 7 = 9a - 5a + 2$ $\qquad\qquad$ **37.** $-4 - x = 4x + 2 - 6x$

34. $-4x - 2x + 1 = -5x + 7$ $\qquad\qquad$ **38.** $5(x + 2) = 4(x - 1)$

35. $7b - 2b + 5 - 4b = 11$ $\qquad\qquad$ **39.** $2(2y - 1) = 3(y + 2)$

36. $12 = 6x + 3 - 4x - x$ $\qquad\qquad$ **40.** $5(3x + 2) = 7(2x + 3)$

41. $5x - 4 + x = 5(x - 2)$

42. $3(2x + 1) - 7 = 5x - 4$

43. $(4a + 5) - (2 + 3a) = 8$

44. $(9b + 7) - (8b + 2) = -4$

45. $3(z + 7) - (8 + 2z) = 6$

46. $4(x - 5) - (3x + 4) = -2$

47. $2(a - 3) - (a - 2) = 8$

48. $5(a + 1) - (4a + 3) = 14$

49. $2(3x - 1) + 3(x + 2) = 4(2x + 5)$

50. $3(4x - 5) + 2(x - 4) = 3(5x + 2)$

51. $-2(b + 1) + 3(b - 4) = -5$

52. $-3(x - 2) + 4(x - 5) = -7$

Directions Solve the following word problems by setting up an equation and solving for the unknown. See example 2–3 G.

53. A number increased by 11 yields 37. Find the number.

54. If a number is decreased by 16, the result is 52. Find the number.

55. If Gary's age is increased by 4 years, he is 37 years old. How old is Gary now?

56. Harry is 6 years older than Dene. If Dene is 54 years old, how old is Harry?

57. If Jake withdraws $340 from his savings account, his balance will be $395. How much does Jake have in his savings account now?

58. Pam deposits $42.50 in her checking account. If her new balance is $125.30, how much did she have in her account originally?

59. Mr. Johnson took in $560 on a given day in his grocery store. If he paid out $195 to his employees in wages, how much profit did he realize?

60. Marsha can groom 11 more dogs per day than Margaret can. If Marsha can groom 24 dogs per day, how many dogs per day can Margaret groom?

Business application The basic principle of double-entry accounting relies on the addition and subtraction property of equality as well as the additive inverse property. The basic accounting model is expressed by the equation

$$A = L + E$$

where A means assets (things owned), L means liabilities (things owed), and E means owner's equity (net worth of the enterprise). This equation must always be true—if not there is a mistake in accounting and the books are said to be unbalanced.

As an example of the basic model, if the assets of a company are $50,000, the liability is $40,000, and the equity is $10,000, then for this business,

$$50,000 = 40,000 + 10,000$$

If the company buys a desk by promising to pay $450 then the value of the desk is added to the assets and the value of the debt is added to the liability:

$$
\begin{array}{rcl}
50,000 & = & 40,000 + 10,000 \\
+\ 450 & & +\ 450 \\
\hline
50,450 & = & 40,450 + 10,000
\end{array}
$$

This is an example of the addition and subtraction property of equality.

If the company buys the desk by paying cash then the assets are increased by the value of the desk, $450, but the cash part of the assets are decreased by the same amount:

$$
\begin{array}{rcl}
50{,}000 &=& 40{,}000 + 10{,}000 \\
+\ 450 & & \\
-\ 450 & & \\
\hline
50{,}000 &=& 40{,}000 + 10{,}000
\end{array}
$$

This is an example of the additive inverse property.

Directions   In each of the following problems assume the initial values $A = 50{,}000$, $L = 40{,}000$, $E = 10{,}000$. For each problem (a) Show the effect on the basic accounting model equation and (b) state whether the principle involved is the addition and subtraction property of equality or the additive inverse property.

61. The company pays $2,000 cash for a computer.

62. The company buys a $1,500 computer by agreeing to pay the amount next month.

63. The company is bought by another company, which gives it $5,000 cash.

64. The company sells a building for $12,000 cash.

65. The company takes $4,000 from its cash and pays $4,000 towards a debt it owes.

⚛ Explore to Learn

1. "If I subtract $-6\frac{2}{3}$ from a certain number the result is $-\frac{1}{2}$. What is the number?" Try to solve this problem without using an equation, and then using an equation. Any comments?

2. How many solutions does the following equation have? Explain.

$$2(x + 5) = 4x + 6 - 2x + 4$$

3. How many solutions does the following equation have? Explain.

$$2(x + 1) = 4x + 6 - 2x + 4$$

Review Exercises

Directions   Perform the indicated operations. See section 1–7.

1. $(-2)(-8)$

2. $(-4)(3)$

3. $\dfrac{-8}{-8}$

4. $\dfrac{6}{6}$

5. $\left(\dfrac{1}{3}\right)(3)$

6. $\left(-\dfrac{1}{4}\right)(-4)$

2–4   The Multiplication and Division Property of Equality

Multiplication and Division Property of Equality

In section 2–3, we used the associative, commutative, and distributive properties to simplify equations. We then used the addition and subtraction property of equality to solve for the unknown. These properties are sufficient to solve many of the equations that we encounter. However we cannot use them to solve such equations as

$$3x = 21 \qquad \text{or} \qquad \frac{2}{3}x = 12$$

Recall that we want our equation to be of the form $x = n$. This means that the coefficient of x must be 1. To achieve this, we make use of the multiplication and division property of equality.

The Multiplication and Division Property of Equality

For any algebraic expressions $a, b,$ and c ($c \neq 0$)

$$\text{if } a = b, \text{ then } a \cdot c = b \cdot c \text{ and}$$
$$a \div c = b \div c$$

Concept

An equivalent equation is obtained when we multiply or divide both sides of an equation by the same nonzero quantity.

The multiplication and division property of equality enables us to multiply or divide both sides of an equation by the same nonzero quantity. In the equation $3x = 21$, we use the multiplication and division property of equality to divide both sides of the equation by 3. This forms an equivalent equation where x has a coefficient of 1, that is, $x = n$.

$$3x = 21$$
$$\frac{3x}{3} = \frac{21}{3} \qquad \text{Divide both sides by 3}$$
$$x = 7$$

The solution set is $\{7\}$.

Multiplicative Inverse

For the equation $\frac{2}{3}x = 12$, recall that when we divide by a fraction, we invert and multiply. Therefore, if the coefficient is a fraction, we will multiply both sides of the equation by the multiplicative inverse (reciprocal) of the coefficient. The **multiplicative inverse** of a number, also called the **reciprocal** of the number, is such that when we multiply a number times its reciprocal, the answer will be 1.

Multiplicative Inverse Property

For every real number a, $a \neq 0$,

$$a \cdot \frac{1}{a} = 1$$

Concept

Every real number except zero has a multiplicative inverse, and the product of a number and its multiplicative inverse is always 1.

Note Zero is the only number that does not have a reciprocal. From the zero factor property, we know that zero times any number is zero. Therefore there can be no number such that zero times that number gives 1 as an answer.

$$a \cdot 0 = 0$$

● **Example 2–4 A**
Reciprocals

The following examples are illustrations of the multiplicative inverse property, where the second number can be considered the reciprocal of the first, and the first can be considered the reciprocal of the second.

1. $5 \cdot \frac{1}{5} = 1$

2. $\frac{1}{2} \cdot 2 = 1$

3. $b \cdot \frac{1}{b} = 1, b \neq 0$

4. $\frac{3}{4} \cdot \frac{4}{3} = 1$

5. $\left(-\frac{5}{7}\right)\left(-\frac{7}{5}\right) = 1$

We will now use the multiplicative inverse property to solve the equation $\frac{2}{3}x = 12$.

$$\frac{2}{3}x = 12$$

$$\frac{3}{2} \cdot \frac{2}{3}x = \frac{3}{2} \cdot 12 \qquad \text{Multiply both sides by the reciprocal of the coefficient}$$

$$x = 18$$

The solution set is {18}.

<u>Note</u> In the earlier example $3x = 21$, we could have multiplied by the reciprocal of 3 to solve the equation. That is,

$$3x = 21$$

$$\frac{1}{3} \cdot 3x = \frac{1}{3} \cdot 21 \qquad \text{Multiply both sides by the reciprocal } \frac{1}{3}$$

$$x = 7$$

Remember that to divide by a number is the same operation as to multiply by the reciprocal of that number.

● **Example 2–4 B**

Using the multiplication and division property of equality

Find the solution set.

1. $$\frac{3}{4}x = 9$$

$$\frac{4}{3} \cdot \frac{3}{4}x = \frac{4}{3} \cdot 9 \qquad \text{Multiply both sides by the reciprocal } \frac{4}{3}$$

$$x = 12$$

The solution set is {12}.

2. $$-x = -10$$
$$-1 \cdot x = -10 \qquad \text{-1 is the coefficient}$$
$$\frac{-1 \cdot x}{-1} = \frac{-10}{-1} \qquad \text{Divide both sides by the coefficient -1}$$
$$x = 10$$

The solution set is {10}.

3. $6x = 10$
$$\frac{6x}{6} = \frac{10}{6} \qquad \text{Divide both sides by the coefficient 6}$$
$$x = \frac{5}{3} \qquad \text{Reduce the fraction}$$

The solution set is $\left\{\dfrac{5}{3}\right\}$.

4. $$\frac{x}{4} = 6 \qquad \text{We can rewrite the left side to show that the coefficient is } \frac{1}{4}$$

$$\frac{1}{4}x = 6$$

$$4 \cdot \frac{1}{4}x = 4 \cdot 6 \qquad \text{Multiply both sides by the reciprocal 4}$$

$$x = 24$$

The solution set is {24}.

☞ *Quick check*   Find the solution set of the equations $5a = 35$, $\frac{2}{3}y = 8$, and $1.7x = 10.2$ ●

Problem Solving

Now we will translate some word statements into equations and solve the resulting equations.

● **Example 2–4 C**
Problem solving

Write an equation for each problem and then solve the equation.

1. When a number is multiplied by -6, the result is 48. Find the number.

Let n represent the number for which we are looking. The formation of the equation would be as follows:

$$\underset{n}{\underset{\text{a number}}{}} \quad \underset{\cdot \,(-6)}{\underset{\text{multiplied by } (-6)}{}} \quad \underset{=}{\underset{\text{result is}}{}} \quad \underset{48}{\underset{48}{}}$$

$$n \cdot (-6) = 48$$
$$\frac{n(-6)}{-6} = \frac{48}{-6} \qquad \text{Divide both sides by } -6$$
$$n = -8$$

The number is -8.

2. Alice makes \$4.50 per hour. If her pay was \$108, how many hours did she work?

Let n represent the number of hours that she worked. The formation of the equation would be as follows:

$$\underset{(4.50)}{\underset{\text{hourly rate}}{}} \quad \underset{\cdot}{\underset{\text{times}}{}} \quad \underset{n}{\underset{\substack{\text{number of} \\ \text{hours worked}}}{}} \quad \underset{=}{\underset{\text{gives}}{}} \quad \underset{108}{\underset{\text{total pay}}{}}$$

$$(4.50) \cdot n = 108$$
$$\frac{(4.50)n}{4.50} = \frac{108}{4.50} \qquad \text{Divide both sides by 4.50}$$
$$n = 24$$

Alice worked 24 hours.   ●

MASTERY POINTS

Can you
- Use the multiplication and division property of equality to form equivalent equations where the coefficient of the unknown is 1?
- Check your answer?
- Solve word problems?

Exercise 2–4

Directions

Find the solution set by using the multiplication and division property of equality. Check each answer. See example 2–4 B.

⚑ *Quick Check*

Examples

$$5a = 35 \qquad\qquad \frac{2}{3}y = 8 \qquad\qquad 1.7x = 10.2$$

Solutions

$$\frac{5a}{5} = \frac{35}{5} \;\text{Divide by 5} \qquad \frac{3}{2} \cdot \frac{2}{3}y = \frac{3}{2} \cdot 8 \;\text{Multiply by } \frac{3}{2} \qquad \frac{1.7x}{1.7} = \frac{10.2}{1.7} \;\text{Divide by 1.7}$$
$$a = 7 \qquad\qquad y = 12 \qquad\qquad x = 6$$

The solution set is {7}.   The solution set is {12}.   The solution set is {6}.

Check: *Check:*   *Check:*

$$5(7) = 35 \;\text{Substitute} \qquad \frac{2}{3}(12) = 8 \;\text{Substitute} \qquad 1.7(6) = 10.2 \;\text{Substitute}$$
$$35 = 35 \;\text{(True) Checks} \qquad\qquad\qquad 10.2 = 10.2 \;\text{(True) Checks}$$
$$8 = 8 \;\text{(True) Checks}$$

Directions   Find the solution set by using the multiplication and division property of equality. Check each answer. See example 2–4 B.

1. $2x = 8$

2. $3a = 18$

3. $6y = 36$

4. $9x = 45$

5. $\frac{3}{4}x = 12$

6. $\frac{2}{5}x = 10$

7. $\frac{1}{7}x = 5$

8. $\frac{1}{5}a = 9$

9. $\frac{3}{2}x = 18$

10. $14 = \frac{7}{3}n$

11. $5y = -15$

12. $-8 = 2n$

13. $-24 = 6a$

14. $-5x = 30$

15. $-4x = -28$

16. $-30 = -6x$

17. $-x = 4$

18. $-y = -11$

19. $6x = 14$

20. $5x = 9$

21. $4x = 6$

22. $3a = -8$

23. $5x = 0$

24. $0 = 7x$

25. $-3x = 0$

26. $-2n = 0$

27. $\frac{x}{3} = 5$

28. $\frac{x}{4} = 8$

29. $\frac{n}{-2} = 7$

30. $-2 = \frac{a}{-3}$

31. $2.6x = 10.4$

32. $3.1x = 21.7$

33. $-4.8x = 33.6$

34. $-7.1y = 35.5$

35. $-42.9 = -3.9x$

36. $(0.4)x = 7.2$

37. $(0.3)y = -7.8$

38. $\frac{5}{7}x = 8$

39. $\frac{3}{8}x = 14$

40. $\frac{2}{9}a = 11$

Directions   Write an equation for each exercise and then solve the equation. See example 2–4 C.

41. When a number is multiplied by 6, the result is 54. Find the number.

42. When a number is multiplied by -4, the result is 36. Find the number.

43. When a number is divided by 9, the result is -7. Find the number.

44. When a number is divided by -8, the result is -8. Find the number.

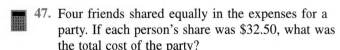 **45.** Nancy worked for 30 hours and received $135. Find her hourly wage.

46. Adam worked for 14 hours and received $52.50. Find his hourly wage.

47. Four friends shared equally in the expenses for a party. If each person's share was $32.50, what was the total cost of the party?

48. Six friends shared equally in the cost of dinner. If the cost of the dinner was $51, what was each person's share?

49. If $\frac{3}{4}$ of a number is 48, find the number.

50. If $\frac{2}{3}$ of a number is 26, find the number.

✦ Explore to Learn

Consider solving $\frac{2x}{3} = -\frac{5}{9}$. This could be solved by multiplying each side by $\frac{3}{2}$. Another way would be to first multiply each side by the least common denominator, 9, and then divide each side by 2. Try this and see if you prefer one way to the other.

Review Exercises

Directions   Perform all indicated operations. See section 2–2.

1. $3x + 2x + 1 - 3$

2. $7x - 5x - 3 + 4$

3. $8x - 5 + 4x + 7$

4. $6x + 3 - 3x - 8$

5. $2(3x + 1) + 4x - 3$

6. $3(x - 1) + 2(x + 2)$

2–5   Solving Linear Equations

Review of Properties

We now are ready to combine the properties from the previous sections to help us solve more involved equations. The process consists of forming equivalent equations until we have our equation in the form of $x = n$. The properties that we will use are the following:

    1. We can add or subtract the same number on both sides of the equation.

    2. We can multiply or divide both sides of the equation by the same nonzero number.

If we use these two properties, making sure that both sides of the equation are treated in exactly the same manner as we apply each of the properties, we will be forming equivalent equations.

Procedure for Solving a Linear Equation

Using these properties, there are the following basic steps to solve a linear equation. We shall now apply the properties to the equation $6(x + 1) = 4x + 10$.

Solving a Linear Equation

$$6(x + 1) = 4x + 10$$

1. *Simplify each side of the equation.* Perform all indicated addition, subtraction, multiplication, and division. Remove all grouping symbols. In our example, step 1 would be to carry out the indicated multiplication on the left side as follows:

$$6(x + 1) = 4x + 10$$
$$6x + 6 = 4x + 10$$

2. *Use the addition and subtraction property of equality to form an equivalent equation where all the terms involving the unknown are on one side of the equation.* By subtracting 4x from *both* sides of the equation, we have

$$6x + 6 = 4x + 10$$
$$6x - 4x + 6 = 4x - 4x + 10$$
$$2x + 6 = 10$$

3. *Use the addition and subtraction property of equality to form an equivalent equation where all the terms not involving the unknown are on the other side of the equation.* Subtracting 6 from *both* sides of the equation, we have

$$2x + 6 = 10$$
$$2x + 6 - 6 = 10 - 6$$
$$2x = 4$$

4. *Use the multiplication and division property of equality to form an equivalent equation where the coefficient of the unknown is 1.* That is, x = n. By dividing *both* sides of the equation by 2, we have

$$2x = 4$$
$$\frac{2x}{2} = \frac{4}{2}$$
$$x = 2$$

The solution set is denoted by {2}.

5. To check the answer, we substitute the answer in place of the unknown in the original equation. If we get a true statement, we say that the answer "satisfies" the equation.

In the equation $6(x + 1) = 4x + 10$, we found that $x = 2$. We can check the answer by substituting 2 in place of x in the original equation.

$$6[(2) + 1] = 4(2) + 10 \qquad \text{Substitute}$$
$$6[3] = 8 + 10 \qquad \text{Order of operations}$$
$$18 = 18 \quad \text{(True)} \qquad \text{Answer checks}$$

We see that $x = 2$ satisfies the equation.

● **Example 2–5 A**

Solving linear equations

Find the solution set.

1. $6y + 5 - 7y = 10 - 2y + 3$

$\qquad\qquad 5 - y = 13 - 2y$ $\qquad$ Simplify each side by combining like terms

$\qquad 5 - y + 2y = 13 - 2y + 2y$ $\qquad$ Add $2y$ to both sides

$\qquad\qquad 5 + y = 13$

$\qquad 5 + y - 5 = 13 - 5$ $\qquad$ Subtract 5 from both sides

$\qquad\qquad\qquad y = 8$

Check: $6(8) + 5 - 7(8) = 10 - 2(8) + 3$ $\qquad$ Substitute 8 for y

$\qquad\qquad 48 + 5 - 56 = 10 - 16 + 3$ $\qquad$ Order of operations

$\qquad\qquad\qquad 53 - 56 = -6 + 3$

$\qquad\qquad\qquad\qquad -3 = -3$ $\qquad$ (True) $\qquad$ Answer checks

The solution set is {8}.

2. $8y + 5 - 7y = 10 - 2y + 3$

$\qquad\qquad 5 + y = 13 - 2y$ $\qquad$ Combine like terms

$\qquad 5 + y + 2y = 13 - 2y + 2y$ $\qquad$ Add $2y$

$\qquad\qquad 5 + 3y = 13$ $\qquad$ Combine like terms

$\qquad 5 + 3y - 5 = 13 - 5$ $\qquad$ Subtract 5

$\qquad\qquad\qquad 3y = 8$ $\qquad$ Combine like terms

$\qquad\qquad\qquad \dfrac{3y}{3} = \dfrac{8}{3}$ $\qquad$ Divide by 3

$\qquad\qquad\qquad y = \dfrac{8}{3}$

Check: $8\left(\dfrac{8}{3}\right) + 5 - 7\left(\dfrac{8}{3}\right) = 10 - 2\left(\dfrac{8}{3}\right) + 3$ $\qquad$ Substitute $\dfrac{8}{3}$ for y

$\qquad \dfrac{64}{3} + \dfrac{15}{3} - \dfrac{56}{3} = \dfrac{30}{3} - \dfrac{16}{3} + \dfrac{9}{3}$ $\qquad$ Multiply, change to common denominator

$\qquad \dfrac{64 + 15 - 56}{3} = \dfrac{30 - 16 + 9}{3}$ $\qquad$ Add and subtract in numerators

$\qquad\qquad\qquad \dfrac{23}{3} = \dfrac{23}{3}$ $\qquad$ (True) $\qquad$ Answer checks

The solution set is $\left\{\dfrac{8}{3}\right\}$.

At this point, we will no longer show the check of our answer, but you should realize that a check of your answer is an important final step.

3. $4(5x - 2) + 7 = 5(3x + 1)$

$\qquad 20x - 8 + 7 = 15x + 5$ $\qquad$ Distributive property

$\qquad\qquad 20x - 1 = 15x + 5$ $\qquad$ Combine like terms

$\qquad 20x - 15x - 1 = 15x - 15x + 5$ $\qquad$ Subtract $15x$

$\qquad\qquad 5x - 1 = 5$

$\qquad 5x - 1 + 1 = 5 + 1$ $\qquad$ Add 1

$\qquad\qquad\qquad 5x = 6$

$\qquad\qquad\qquad \dfrac{5x}{5} = \dfrac{6}{5}$ $\qquad$ Divide by 5

$\qquad\qquad\qquad x = \dfrac{6}{5}$

The solution set is $\left\{\dfrac{6}{5}\right\}$.

Conditional Equations, Identities, and Contradictions

An equation that is true for some values of the variable and false for other values of the variable is called a **conditional equation**. All of the equations solved thus far have been first degree conditional equations whose solution set contains exactly one solution.

If an equation is true for every permissible replacement value of the variable, it is called an **identity**. The following example illustrates this type of equation.

$$2(3x + 4) = 6x + 8$$
$$6x + 8 = 6x + 8 \qquad \text{Distributive property}$$
$$6x - 6x + 8 = 6x - 6x + 8 \qquad \text{Subtract } 6x$$
$$8 = 8 \qquad \text{True statement}$$

When the variable has been eliminated from the equation and the resulting statement is true, the equation is an identity. The example above is an identity and is true for all real numbers. We represent the solution set with R, which denotes the set of real numbers.

An equation that has no solution is called a **contradiction** and its solution set is the empty set. **A set that contains no elements is called the empty set or the null set and is denoted by the symbol $\varnothing$.** The following example illustrates this type of equation.

$$3x + 2x + 7 = 5x + 5$$
$$5x + 7 = 5x + 5 \qquad \text{Combine like terms}$$
$$5x - 5x + 7 = 5x - 5x + 5 \qquad \text{Subtract } 5x$$
$$7 = 5 \qquad \text{False statement}$$

When the variable has been eliminated from the equation and the resulting statement is false, the equation is a contradiction and has no solution. The solution set is the empty set and we represent that with $\varnothing$. We could also write the empty set as a pair of braces with nothing inside, { }.

From the preceding examples we can see that a first degree equation's solution set will have exactly one solution, no solution, or infinitely many solutions. We will be primarily concerned with those equations that have a single value in the solution set.

Equations Containing Fractions

The following equations contain several fractions. When this occurs, it is usually easier to *clear the equation of all fractions*. We do this by multiplying each term on both sides of the equation by the least common denominator of all the fractions. Clearing all fractions is considered a means of simplifying the equation and will be done as a first step when necessary. Equations containing fractions will be studied more completely in chapter 5.

Example 2–5 B

Solving linear equations involving fractions

Find the solution set.

1.
$$\frac{1}{4}x + 2 = \frac{1}{2}$$

$$4\left(\frac{1}{4}x + 2\right) = 4\left(\frac{1}{2}\right)$$ The least common denominator of the fractions is 4, multiply both sides by 4

$$4\left(\frac{1}{4}x\right) + 4(2) = 4\left(\frac{1}{2}\right)$$ Simplify (distributive property)

$$x + 8 = 2$$ All fractions have been cleared

$$x + 8 - 8 = 2 - 8$$ Subtract 8

$$x = -6$$

The solution set is $\{-6\}$.

2.
$$\frac{5}{6}x - \frac{2}{3} = \frac{3}{4}x + 2$$

$$12\left(\frac{5}{6}x - \frac{2}{3}\right) = 12\left(\frac{3}{4}x + 2\right)$$ The least common denominator of the fractions is 12; multiply by 12

$$12\left(\frac{5}{6}x\right) - 12\left(\frac{2}{3}\right) = 12\left(\frac{3}{4}x\right) + 12(2)$$ Simplify (distributive property)

$$10x - 8 = 9x + 24$$

$$10x - 9x - 8 = 9x - 9x + 24$$ Subtract 9x

$$x - 8 = 24$$

$$x - 8 + 8 = 24 + 8$$ Add 8

$$x = 32$$

The solution set is $\{32\}$.

👉 **Quick check**   Find the solution set for $5x + 2(x - 1) = 4 - 3x$ and check. ●

Approximate solutions to problems that involve decimal values can be most easily found with a calculator. The arithmetic in the next example is best done by calculator.

Example 2–5 C

Finding approximate solutions

Find the solution set for the following equation; round the result to two decimal places.

$$3.51(2.67x - 3) = 1.05x$$

$$9.3717x - 10.53 = 1.05x$$ Distributive property

$$9.3717x - 1.05x = 10.53$$ Subtract 1.05x, add 10.53

$$8.3217x = 10.53$$ $9.3717 - 1.05 = 8.3217$ combine like terms

$$x = \frac{10.53}{8.3217} \approx 1.27$$ Perform the division on a calculator and round to two decimal places

The solution set is $\{1.27\}$. ●

Problem Solving

Example 2–5 D shows that it is often best to translate from English to algebra starting at the word "is" and working backward and forward from it.

● **Example 2–5 D**

Problem solving

Solve the following problems by setting up an equation and solving it.

1. Six less than three fourths of a number is 36. What is the number?

Let x represent the number. Translate the sentence as follows.

Six less than three fourths of a number $= 36$ Is means equals

Six less than three fourths of $x = 36$ x represents the number

Six less than $\dfrac{3}{4}x = 36$

$$\frac{3}{4}x - 6 = 36 \qquad \text{Observe how "six less than" is translated}$$

$$\frac{3}{4}x = 42 \qquad \text{Add 6}$$

$$4 \cdot \frac{3}{4}x = 4 \cdot 42 \qquad \text{Clear denominators}$$

$$3x = 168 \qquad \text{Multiply}$$

$$x = \frac{168}{3} = 56 \qquad \text{Divide by 3}$$

Thus the number is 56.

Check: Three fourths of 56 is 42, and six less than 42 is 36.

2. The sum of 12% of a number and 6% of the amount represented by 20,000 less that number is 1,620. Find the number.

Let x represent the number. The first translation is as follows:

The sum of 12% of x and 6% of the amount represented by 20,000 less $x = 1,620$

Working backward from the "$=$":

The sum of 12% of x and 6% of $(20,000 - x) = 1,620$

$$0.12x + 0.06(20,000 - x) = 1,620 \qquad \text{Change \% to its decimal form}$$

$$0.12x + 1,200 - 0.06x = 1,620 \qquad \text{Distributive property}$$

$$0.12x - 0.06x = 1,620 - 1,200 \qquad \text{Subtract 1,200}$$

$$0.06x = 420 \qquad \text{Combine like terms}$$

$$x = \frac{420}{0.06} = 7,000 \qquad \text{Divide by 0.06}$$

Thus the number is 7,000.

Check: 12% of 7,000 is 840
20,000 less 7,000 is 13,000
6% of 13,000 is 780
$840 + 780 = 1,620$

MASTERY POINTS

Can you
- Solve linear equations?
- Check your answers?
- Identify an equation as an identity?
- Identify an equation as a contradiction?
- Find an approximate solution to a linear equation?
- Write an equation for a word problem and then solve the problem?

Exercise 2–5

Directions
Find the solution set of the following equations, and check the answer. See examples 2–5 A and B.

☛ *Quick Check*

Example $5x + 2(x - 1) = 4 - 3x$

Solution

$5x + 2x - 2 = 4 - 3x$	Simplify (distributive property)
$7x - 2 = 4 - 3x$	Combine like terms
$7x + 3x - 2 = 4 - 3x + 3x$	Add $3x$
$10x - 2 = 4$	Combine like terms
$10x - 2 + 2 = 4 + 2$	Add 2
$10x = 6$	Combine like terms
$\dfrac{10x}{10} = \dfrac{6}{10}$	Divide by 10 and reduce
$x = \dfrac{3}{5}$	

Check:

$$5\left(\frac{3}{5}\right) + 2\left[\left(\frac{3}{5}\right) - 1\right] = 4 - 3\left(\frac{3}{5}\right) \quad \text{Substitute } \tfrac{3}{5} \text{ for } x$$

$$5\left(\frac{3}{5}\right) + 2\left[\frac{3}{5} - \frac{5}{5}\right] = 4 - \frac{9}{5} \quad \text{Order of operations}$$

$$5\left(\frac{3}{5}\right) + 2\left[\frac{-2}{5}\right] = \frac{20}{5} - \frac{9}{5}$$

$$\frac{15}{5} + \frac{-4}{5} = \frac{11}{5}$$

$$\frac{11}{5} = \frac{11}{5} \quad \text{(True)   Answer checks}$$

The solution set is $\left\{\dfrac{3}{5}\right\}$.

Directions   Find the solution set of the following equations, and check the answer. See examples 2–5 A and B.

1. $2x = 4$

2. $3x = 11$

3. $5x = -10$

4. $-2x = 8$

5. $\dfrac{x}{2} = 18$

6. $\dfrac{x}{4} = 24$

7. $\dfrac{3x}{2} = 8$

8. $\dfrac{5x}{3} = 18$

9. $x + 7 = 11$

10. $x - 4 = 9$

11. $x + 5 = 5$

12. $x - 4 = -4$

13. $3x + 1 = 10$

14. $5x - 2 = 13$

15. $4x + 7 = 7$

16. $6x + 2 = 2$

17. $5x + 2x = x + 6$

18. $2x + (3x - 1) = 4 - x$

19. $2x + 3x - 6x = 4x - 8$

20. $3x + 4x + 5 = 7x + 5$

21. $6x - 4 = 2(3x - 2)$

22. $3(2x + 3) = 6x + 9$

23. $x + 5x + 6 = 4x + 6 + 2x$

24. $\dfrac{x}{2} + 7 = 14$

25. $5 - \dfrac{3x}{5} = 11$

26. $\dfrac{5x + 2x}{6} = 10$

27. $\dfrac{1}{2}x + 3 = \dfrac{3}{4}$

28. $\dfrac{1}{5}x - 1 = \dfrac{7}{10}$

29. $\dfrac{1}{3}x + 2 = \dfrac{1}{2}x - 1$

30. $\dfrac{1}{4}x - 3 = \dfrac{1}{8}x + 1$

31. $\dfrac{2}{3}x + 5 = \dfrac{3}{4}$

32. $\dfrac{3}{5}x - 3 = \dfrac{3}{10}$

33. $\dfrac{3}{8}x + \dfrac{1}{2} = \dfrac{1}{4}x + 2$

34. $\dfrac{7}{12}x + 1 = \dfrac{2}{3}x - 1$

35. $3(3x - 2) = 2x + 7x + 5$

36. $10x - 3x + 4 = 5x + 2x + 6$

37. $x + 3x - 5 = 8x - 4x - 2$

38. $5(2x + 3) = 7x - 3 + 3x$

39. $3(2x - 1) = 4x + 3$

40. $5(7x - 3) = 30x + 11$

41. $12x - 8 = 5x + 2$

42. $3(2x + 5) = 4(x - 3)$

43. $8 - 2(3x + 4) = 5x - 16$

44. $(3x + 2) - (2x - 5) = 7$

45. $(7x - 6) - (4 - 3x) = 27$

46. $2(x - 4) - 3(5 - 2x) = 16$

47. $3(2x + 3) = 5 - 4(x - 2)$

48. $6(3x - 2) = 7(x - 3) - 2$

49. $2(x + 5) = 16$

50. $6 = 2(2x - 1)$

51. $2x - (3 - x) = 0$

52. $3(7 - 2x) = 30 - 7(x + 1)$

Directions Solve the following equations, and check the answer. See examples 2–5 A and B.

53. To convert Celsius temperature to Fahrenheit, we use $F = \dfrac{9}{5}C + 32$. Find C when (a) $F = 18$, (b) $F = -27$, (c) $F = 2$.

54. The Stefan-Boltzmann law in metallurgy, which is the temperature scale of radiation pyrometers, is given by $W = KT^4$. Find K when (a) $W = 36$, $T = 2$; (b) $W = 243$, $T = -3$.

55. The total creep of a metal (E_P) at time t is given by $E_P = E_0 + V_0 t$, where $E_0 =$ original creep, $t =$ time, and $V_0 =$ the original volume. Find V_0 when $E_P = 16$, $E_0 = 9$, and $t = 3$.

56. In a gear system, the speed, in number of revolutions, of two gears and the number of teeth in the gears are related by $S_D \cdot T_D = S_d \cdot T_d$, where S_D is the speed of the driver, T_D is the number of teeth in the driver, S_d is the speed of the driven gear, and T_d is the number of teeth in the driven gear. If (a) $S_D = 240$, $T_D = 40$, and $S_d = 360$, find T_d; (b) $S_D = 120$, $S_d = 90$, and $T_d = 18$, find T_D.

Directions Find the solution set for the following equations. Round answers to two decimal places if necessary. See example 2–5 C.

57. $2.65x = 9.31$

58. $-3.11x + 5.33 = 9.08$

59. $16.5 - 2.77x = 5.11x$

60. $17.5x + 12.65 = 3.84x$

61. $3.75x + 2.25(6.80x - 3) = 51.82$

62. $2.55(3.21x - 10.5) = 17 - 34.6x$

63. $6.28 + \dfrac{3.60x - 15}{12.6} = 8.5$

64. $\dfrac{3x}{5} - 12.6x = -10$

Directions Solve the following problems by setting up an equation and solving it. See example 2–5 D.

65. Three fourths of a number is 600. What is the number?

66. One half of six less than a number is 43. Find the number.

67. Two thirds of 10 more than a number is 312. What is the number?

68. One third of a number, less 7, is $19\frac{1}{2}$. What is the number?

69. Five more than two thirds of a number is 59. Find the number.

70. Five eighths of a number, decreased by 11, is 49. Find the number.

71. The sum of 8% of a number and 50 is 99.6. Find the number.

72. Thirty percent of a number is 156. Find the number.

73. The sum of 10% of a number and 8% of the amount represented by 10,000 less that number is 870. Find the number.

74. The sum of 3% of a number and 6% of the amount represented by 500 less that number is 26.4. Find the number.

75. 23% of a number less 6% of the amount represented by 2,500 less that number is 140. Find the number.

76. The sum of 18% of 100 more than a number and 6% of the number is 238.8. Find the number.

✺ Explore to Learn

The Alpha Auto Rental Agency charges $75 per week and $0.15 per mile. The Beta Car Rental Agency charges $50 per week and $0.25 per mile. For a certain number of miles, the charge for a rental car from Alpha Auto Rental will be the same as the charge from Beta Car Rental. Find the number of miles that must be driven for the charges to be equal.

Review Exercises

Directions Evaluate the following formulas. See section 1–8.

1. $W = I^2R$, $I = 6$ and $R = 3$

2. $S = \dfrac{1}{2}gt^2$, $g = 32$ and $t = 3$

3. $A = \dfrac{1}{2}h(b_1 + b_2)$, $h = 8$, $b_1 = 10$, and $b_2 = 12$

4. $I = prt$; $p = 2{,}000$; $r = 0.06$; and $t = 3$

5. $V = \ell wh$, $\ell = 10$, $w = 4$, and $h = 7$

6. $V = k + gt$, $k = 12$, $g = 16$, and $t = 5$

2–6   Applications of Geometry and Solving Literal Equations and Formulas

Applications of Geometric Formulas

Many word problems that you will encounter can be solved by using an existing formula. In sections 1–8, 1–9, and 2–5 there were problems in the exercise sets involving formulas for finding horsepower requirements, revolutions per minute of a pulley, converting Celsius temperature to Fahrenheit, etc. In this section we are going to apply some of the many formulas dealing with geometry. Inside the back cover are some of the more common geometric formulas.

When the values of all but one of the variables in a formula are given, the known values can be substituted into the formula and we can solve for the remaining variable.

● **Example 2–6 A**

Solving for the remaining value

Rectangle
Area $A = \ell w$

Solve for the value of the remaining variable.

1. $A = \ell w$, $A = 16$, $\ell = 5$

This formula gives the area of a rectangle having length ℓ and width w. We will substitute 16 for A and 5 for ℓ and solve for w.

$$\begin{aligned}
A &= \ell w &\quad& \text{Original formula} \\
(16) &= (5)w &\quad& \text{Substitute 16 for } A \text{ and 5 for } \ell \\
\frac{16}{5} &= w &\quad& \text{Divide by 5} \\
w &= 3.2 &\quad& \text{Symmetric property}
\end{aligned}$$

Therefore the width is 3.2. Usually in a geometry problem there will be a specific unit of measure such as feet or meters. We will not concern ourselves with specific units of measure here but rather our focus will be on solving formulas for the remaining variable.

2. $P = 2a + 2b$, $P = 56$, $a = 12$

This formula gives the perimeter P of a parallelogram (a four-sided figure with opposite sides equal and parallel) with sides a and b.

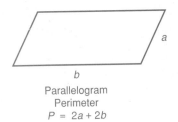

Parallelogram
Perimeter
$P = 2a + 2b$

$$\begin{aligned}
P &= 2a + 2b &\quad& \text{Original formula} \\
(56) &= 2(12) + 2b &\quad& \text{Substitute 56 for } P \text{ and 12 for } a \\
56 &= 24 + 2b &\quad& \text{Multiply} \\
32 &= 2b &\quad& \text{Subtract 24} \\
16 &= b &\quad& \text{Divide by 2} \\
b &= 16 &\quad& \text{Symmetric property}
\end{aligned}$$

The value of side b is 16.

☞ *Quick check*   Solve for the value of the remaining variable. $P = 4s$, $P = 36$

Literal Equations and Formulas

Equations that contain two or more variables are called **literal equations**. In a literal equation, we generally solve the equation for one variable in terms of the remaining variables and constants. *The procedure for solving a literal equation is the same as the procedure for solving linear equations.*

A formula is a literal equation that states the relationship between two or more physical conditions. Consider the formula $d = rt$, which expresses the fact that distance (d) is equal to rate (r) multiplied by the time (t). If we knew how far it was between two cities (d) and we wanted to travel this distance in a certain amount of time (t), then the equation could be solved for the necessary rate (r) to achieve this.

$$d = rt \qquad \text{Divide each side by } t$$
$$\frac{d}{t} = r$$

The equation is now solved for r in terms of d and t. If the distance and rate were known, the equation could be solved for time as follows:

$$d = rt \qquad \text{Divide each side by } r$$
$$\frac{d}{r} = t$$

The equation is now solved for t in terms of d and r.

● **Example 2–6 B**

Solving for a specified variable

Solve for the specified variable.

Geometry

1. The volume of a rectangular solid is found by multiplying length (ℓ) times width (w) times height (h), $V = \ell wh$. Solve the equation for h.

$$
\begin{aligned}
V &= \ell wh & &\text{Original equation} \\
V &= (\ell w)h & &\text{Coefficient of } h \text{ is } \ell w. \\
\frac{V}{\ell w} &= \frac{\ell wh}{\ell w} & &\text{Divide by } \ell w \\
\frac{V}{\ell w} &= h & &\text{Equation is solved for } h \text{ in terms} \\
& & &\text{of } V, \ell, \text{ and } w \\
h &= \frac{V}{\ell w} & &\text{Symmetric property}
\end{aligned}
$$

Business

2. The simple interest (I) earned on the principal (P) over a time period (t) at an interest rate (r) is given by $I = Prt$. Solve for r.

$$
\begin{aligned}
I &= Prt & &\text{Original equation} \\
I &= (Pt)r & &Pt \text{ is the coefficient of } r \\
\frac{I}{Pt} &= \frac{Ptr}{Pt} & &\text{Divide by } Pt \\
\frac{I}{Pt} &= r & &\text{Equation is solved for } r \text{ in terms} \\
& & &\text{of } I, P, \text{ and } t \\
r &= \frac{I}{Pt} & &\text{Symmetric property}
\end{aligned}
$$

Literal 3.

$$x = 2y + z \qquad \text{Solve for } y$$

$$x - z = 2y \qquad \text{Subtract } z$$

$$\frac{x - z}{2} = y \qquad \text{Divide by 2, the coefficient of } y$$

$$y = \frac{x - z}{2} \qquad \text{Symmetric property}$$

Literal 4.

$$\frac{3}{4}y = \frac{1}{3}x - 2 \qquad \text{Solve for } x$$

$$12 \cdot \frac{3}{4}y = 12\left(\frac{1}{3}x - 2\right) \qquad \text{The LCD of the fractions is 12, multiply by 12}$$

$$9y = 4x - 24 \qquad \text{Distributive property}$$

$$9y + 24 = 4x \qquad \text{Add 24}$$

$$\frac{9y + 24}{4} = x \qquad \text{Divide by 4, the coefficient of } x$$

$$x = \frac{9y + 24}{4} \qquad \text{Symmetric property}$$

☞ *Quick check* Solve $P = 2\ell + 2w$ for ℓ.

Whether we are solving for x in a linear equation or a literal equation, the procedure is the same.

Linear equation	**Literal equation**	
$5(x + 1) = 2x + 7$	$5(x + y) = 2x + 7y$	Original equation
$5x + 5 = 2x + 7$	$5x + 5y = 2x + 7y$	Simplify (distributive property)
$3x + 5 = 7$	$3x + 5y = 7y$	All x's on one side
$3x = 2$	$3x = 2y$	Terms not containing x on the other side
$x = \dfrac{2}{3}$	$x = \dfrac{2y}{3}$	Divide by the coefficient

In the linear equation, we have a solution for x, and in the literal equation, we have solved for x in terms of y.

MASTERY POINTS

Can you
- Solve for the value of a remaining variable?
- Solve literal equations and formulas for a specified variable?

Exercise 2–6	☞ *Quick Check*
Directions Solve for the value of the remaining variable. See example 2–6 A.	**Example** $P = 4s$, $P = 36$
	Solution

$$P = 4s \qquad \text{Original formula}$$
$$(36) = 4s \qquad \text{Substitute 36 for } P$$
$$9 = s \qquad \text{Divide by 4}$$
$$s = 9 \qquad \text{Symmetric property}$$

Directions   Solve for the value of the remaining variable. See example 2–6 A.

1. $A = bh$, $A = 112$, $b = 14$ (Area of a parallelogram)

2. $A = \ell w$, $A = 76$, $w = 8$ (Area of a rectangle)

3. $P = 2\ell + 2w$, $P = 34$, $w = 6.5$ (Perimeter of a rectangle)

4. $P = 2a + 2b$, $P = 29$, $b = 6$ (Perimeter of a parallelogram)

5. $A = \frac{1}{2}bh$, $A = 42$, $b = 14$ (Area of a triangle)

6. $A = \frac{1}{2}bh$, $A = 55$, $h = 10$ (Area of a triangle)

7. $P = a + b + c$, $P = 35$, $a = 5$, $c = 18$ (Perimeter of a triangle)

8. $P = 4s$, $P = 30$ (Perimeter of a square)

9. $A = \frac{1}{2}h(b_1 + b_2)$, $A = 48$, $b_1 = 7.5$, $h = 6$ (Area of a trapezoid)

10. $A = \frac{1}{2}h(b_1 + b_2)$, $A = 55$, $b_2 = 4.8$, $h = 5$ (Area of a trapezoid)

11. $C = 2\pi r$, $C = 31.4$, $\pi^* = 3.14$ (Circumference of a circle)

12. $C = \pi D$, $C = 47.1$, $\pi = 3.14$ (Circumference of a circle)

13. $V = \ell w h$, $V = 192$, $\ell = 8$, $w = 4$ (Volume of a rectangular solid)

14. $V = \pi r^2 h$, $V = 628$, $\pi = 3.14$, $r = 5$ (Volume of a right circular cylinder)

15. $V = \frac{1}{3}\pi r^2 h$, $V = 376.8$, $\pi = 3.14$, $r = 6$ (Volume of a right circular cone)

16. $V = \pi r^2 h$, $V = 628$, $\pi = 3.14$, $r = 10$ (Volume of a right circular cylinder)

Directions
Solve for the specified variable. See example 2–6 B.

☞ Quick Check

Example $P = 2\ell + 2w$, for ℓ (Geometry)

Solution $P - 2w = 2\ell$ Subtract $2w$

$\dfrac{P - 2w}{2} = \ell$ Divide by 2

$\ell = \dfrac{P - 2w}{2}$ Symmetric property

Directions   Solve for the specified variable. See example 2–6 B.

17. $V = \ell w h$, for w (Geometry)

18. $V = \ell w h$, for ℓ (Geometry)

19. $I = Prt$, for P (Business)

20. $I = Prt$, for t (Business)

21. $F = ma$, for m (Physics)

22. $E = IR$, for R (Electronics)

23. $K = PV$, for V (Chemistry)

24. $E = mc^2$, for m (Physics)

25. $W = I^2 R$, for R (Electronics)

26. $A = \ell w$, for w (Geometry)

27. $P = 2\ell + 2w$, for w (Geometry)

28. $C = \pi D$, for π (Geometry)

29. $P = a + b + c$, for a (Geometry)

30. $A = \frac{1}{2}bh$, for b (Geometry)

31. $x = y + 4$, for y (Literal)

*3.14 is a common approximation for π.

32. $x = y - 7$, for y (Literal)

33. $y = 3x - 2$, for x (Literal)

34. $y = 2x + 5$, for x (Literal)

35. $3x - y = 5$, for x (Literal)

36. $2x - y = 6$, for x (Literal)

37. $4x - 3y = 2$, for x (Literal)

38. $2x - 5y = 3$, for x (Literal)

39. $ay - 3 = by + c$, for a (Literal)

40. $ay - 3 = by + c$, for b (Literal)

41. $V = k + gt$, for k (Physics)

42. $V = k + gt$, for t (Physics)

43. $A = \dfrac{1}{2}h(b_1 + b_2)$, for b_1 (Geometry)

44. $A = \dfrac{1}{2}h(b_1 + b_2)$, for h (Geometry)

45. $\ell = a + (n - 1)d$, for a (Discrete mathematics)

46. $\ell = a + (n - 1)d$, for d (Discrete mathematics)

47. $A = P(1 + r)$, for P (Business)

48. $\ell = a + (n - 1)d$, for n (Discrete mathematics)

49. $T = 2f + g$, for f (Optics)

50. $i = \dfrac{prm}{12}$, for r (Business)

51. $D = dq + R$, for q (Business)

52. $y = 2(3x - 1)$, for x (Literal)

53. $y = 3(2x - 3)$, for x (Literal)

54. $\dfrac{1}{2}y = \dfrac{1}{4}x + 2$, for x (Literal)

55. $\dfrac{1}{3}y = \dfrac{1}{2}x - 1$, for x (Literal)

56. $\dfrac{3}{4}y - \dfrac{2}{3}x = 2$, for y (Literal)

57. $\dfrac{5}{6}x - \dfrac{1}{2}y = 3$, for y (Literal)

58. $\dfrac{1}{4}x - \dfrac{1}{3}y = \dfrac{1}{2}$, for y (Literal)

59. $\dfrac{2}{3}y - \dfrac{3}{4}x = \dfrac{5}{12}$, for x (Literal)

60. $M = -P(\ell - x)$, for x (Optics)

61. $R = W - b(2c + b)$, for c (Physics)

62. $F = k\dfrac{m_1 m_2}{d^2}$, for k (Physics)

63. $A = P(1 + rt)$, for r (Business)

64. $A = P(1 + rt)$, for P (Business)

65. $V = r^2(a - b)$, for a (Geometry)

66. $P = n(P_2 - P_1) - c$, for P_2 (Business)

67. $3x - y = 4x + 5y$, for x (Literal)

68. $3x - y = 4x + 5y$, for y (Literal)

69. $2S = 2vt - gt^2$, for g (Physics)

70. $ax + by = c$, for y (Literal)

71. The distance s that a body projected downward with an initial velocity of v will fall in t seconds because of the force of gravity is given by $s = \dfrac{1}{2}gt^2 + vt$. Solve for g.

72. Solve the formula in exercise 71 for v.

73. The net profit P on sales of n identical tape decks is given by $P = n(S - C) - e$, where S is the selling price, C is the cost to the dealer, and e is the operating expense. Solve for S.

74. Solve the formula in exercise 73 for C.

76. Solve the formula in exercise 73 for e.

75. If we know the temperature in degrees Fahrenheit (F), the temperature in degrees Celsius (C) can be found by the equation $C = \dfrac{5}{9}(F - 32)$. Solve the formula for F.

Directions   A device to help solve equations of the form $a = bc$ for any letter is a triangle like that shown. To see what the solution is for a given symbol, cover it up. For example, covering a shows $b \mid c$, which implies $a = bc$. Covering b shows a up top, c below, indicating that $b = \dfrac{a}{c}$. Use this device to help solve the following equations for the symbol indicated.

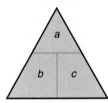

79. $d = rt$ for r (distance = rate · time, from physics)

80. $I = Pr$ for r (interest = principal · rate, from finance)

81. Try to develop a geometric aid like that above for solving problems like $V = \ell wh$ (exercises 17 and 18) and $I = Prt$ (exercises 19 and 20) for certain letters. Apply the aid to exercises 17 through 20.

77. $E = IR$ for I (voltage = current · resistance, Ohm's law from electronics)

78. $F = ma$ for a (force = mass · acceleration, from physics)

Directions   Match the correct sentence with the given equation.

82. $x + 7 = 3x$

   a. A number added to 7 is three times the number.
   b. Three times a number is 7 less than the number.
   c. A number diminished by 7 equals three times the number.

83. $2y + 1 = y$

   a. A number increased by the number is twice the number.
   b. The sum of a number and 1 is equal to twice the number.
   c. A number is equal to the sum of one and twice the number.

84. $4x - 3 = 3x$

   a. Three diminished by four times a number is equal to triple the number.
   b. Four times a number diminished by three is three times the number.
   c. Four less than three times a number is equal to the number.

85. $8y = 7 - y$

   a. A number diminished by seven is equal to eight.
   b. Eight times a number is seven less than the number.
   c. Seven diminished by a number is equal to eight times the number.

86. $3y - 2 = y + 7$

   a. Two less than triple a number is the sum of the number and seven.
   b. Three times a number added to seven is the number less two.
   c. The difference between two and three times a number is the number.

87. $4x + 3 = x - 9$

   a. The sum of a number and three is the number less nine.
   b. Four times a number diminished by nine is the sum of the number and three.
   c. The sum of four times a number and three is the difference between the number and nine.

88. $5 - 2y = y + 9$

   a. Twice a number less five is the sum of the number and nine.

   b. A number added to nine is five more than twice the number.

   c. Five diminished by twice a number is nine more than the number.

89. $2y + 7 = 9 - y$

   a. Nine less than a number is the sum of twice the number and seven.

   b. Twice a number increased by seven is nine less than the number.

   c. The difference between nine and a number is the sum of seven and twice the number.

90. $\dfrac{x}{2} = 3x - 1$

   a. The quotient of a number and two is triple the number diminished by 1.

   b. The quotient of two and a number equals the difference between triple the number and 1.

   c. Three times a number increased by 1 is equal to the quotient of the number and two.

91. $\dfrac{2y - 3}{4} = y$

   a. A number is equal to the quotient of three diminished by twice the number and four.

   b. Twice a number less three divided by the number is four.

   c. A number is twice the number diminished by three and that divided by four.

92. $\dfrac{x + 7}{2} = 3x - 4$

   a. Triple a number less four equals the sum of the number and seven divided by two.

   b. Triple a number less four is the quotient of seven more than the number and two.

   c. The sum of seven and a number when divided by two gives triple the number less four.

⚜ Explore to Learn

Sometimes a formula can be solved for a certain symbol by undoing the steps that the formula implies. For example, consider solving $C = \dfrac{5}{9}(F - 32)$ for F (exercise 75). This formula calculates C starting with F. It says to do the following steps: (1) Take F, (2) subtract 32, (3) multiply this by 5, (4) divide this by 9, (5) the result is C. If we "undo" these steps, we get: (5) Take C, (4) multiply this by 9, (3) divide this by 5, (2) add 32, (1) the result is F. As a formula this is $F = \dfrac{9C}{5} + 32$, which is another form for that given in problem 75.

As another example consider $I = Prt$. If we wish to solve this for t, we should start with it to calculate I. (1) Take t. (2) Multiply by r. (3) Multiply this by P. (4) The result is I. Backward this is (4) take I, (3) divide this by P, to get $\dfrac{I}{P}$, (2) divide this by

$$r \left(\dfrac{I}{P} \div r = \dfrac{I}{P} \cdot \dfrac{1}{r} = \dfrac{I}{Pr} \right).$$ (1) The result is $t \left(t = \dfrac{I}{Pr} \right).$

Try this idea on some of the exercises, such as 17, 24, 27, 42, 43, 45, and 48.

Review Exercises

Directions Perform the indicated operations. See section 1–8.

  1. -5^2 **2.** $(-5)^2$ **3.** -3^4 **4.** $(-3)^3$

Directions Write an algebraic expression for each of the following. See section 2–1.

  5. x raised to the fourth power

  6. A number squared

  7. The product of a and b

  8. x multiplied by y

2–7   Word Problems

Many problems that you will encounter will be written or verbally stated. These will need to be translated into algebraic equations and solved for the unknown values. In chapter 1, we solved arithmetic word problems. Throughout chapter 2 we have seen how to take a word phrase and write an algebraic expression for it and how to transform a word problem into an equation. On page 117 some useful guidelines for solving word problems were given. You should review them at this time.

● **Example 2–7 A**

Problem solving

Write an equation for the problem and solve for the unknown quantities.

Number Problem   1. One number is 4 more than a second number. If their sum is 38, find the two numbers.

Note In problems where we are finding more than one value, it is usually easiest to let the unknown represent the smallest unknown value.

Let x represent the smaller number (the second number). Then $x + 4$, which is 4 more than the smaller number, represents the other number. The parts that make up the equation are

smaller number	their sum	larger number	is	38
x	$+$	$(x + 4)$	$=$	38

$$
\begin{aligned}
x + (x + 4) &= 38 && \text{Original equation} \\
x + x + 4 &= 38 && \text{Remove grouping symbol} \\
2x + 4 &= 38 && \text{Combine like terms} \\
2x &= 34 && \text{Subtract 4} \\
x &= 17 && \text{Divide by 2}
\end{aligned}
$$

Therefore the smaller number is 17 and the larger number is 4 more than the smaller number: $x + 4$ and $(17) + 4 = 21$.

Number Problem   2. One number is 6 times a second number and their sum is 21. Find the numbers.

Let x represent the second number. Then six times the second number or $6x$ represents the other number. The parts that make up the equation are

second number	their sum	other number	is	21
x	$+$	$6x$	$=$	21

$$
\begin{aligned}
x + 6x &= 21 && \text{Original equation} \\
7x &= 21 && \text{Combine like terms} \\
x &= 3 && \text{Divide by 7}
\end{aligned}
$$

Hence the second number is 3 and the other number is 6 times the second number and is $6 \cdot (3) = 18$.

Consecutive Integer Problem   3. If the first of two consecutive integers is multiplied by 3, this product is 4 more than the sum of the two integers. Find the integers.

Note Prior knowledge that is needed for this problem is that consecutive integers differ by 1. Therefore we add 1 to the first to get the second, we would add 2 to the first to get the third, and so on.

first	second	third	fourth	fifth
x	$x + 1$	$x + 2$	$x + 3$	$x + 4$

first integer	second integer
x	$x + 1$

The parts that make up the equation are

three times the first integer	this product is	4	more than	the sum
$3x$	$=$	4	$+$	$[x + (x + 1)]$

$3x = 4 + [x + (x + 1)]$	Original equation
$3x = 4 + [x + x + 1]$	Remove innermost grouping symbols
$3x = 4 + [2x + 1]$	Combine like terms
$3x = 4 + 2x + 1$	Remove brackets
$3x = 2x + 5$	Combine like terms
$x = 5$	Subtract $2x$

Therefore the first consecutive integer is 5 and the second integer is one more than the first $x + 1$ and is $(5) + 1 = 6$.

Geometry Problem   **4.** The length of a rectangle is 3 times its width and its perimeter is 40 feet. Find the dimensions.

Let x represent the width, then 3 times the width or $3x$ represents the length.

$P = 2w + 2\ell$	Formula for perimeter
$40 = 2(x) + 2(3x)$	Substitute
$40 = 2x + 6x$	Multiply
$40 = 8x$	Combine like terms
$5 = x$	Divide by 8

Therefore the width is $x = 5$ feet and the length is $3x = 3(5) = 15$ feet.

Interest Problem   **5.** A man has $10,000, part of which he invests at 11% and the rest at 8%. If his total interest from the two investments for one year is $980, how much does he invest at each rate?

All interest problems in this textbook will be simple interest. The prior knowledge that is needed for this problem is that Interest = Principal · Rate · Time. Time will be equal to 1 year for all problems in this textbook.

If we have a total amount of $10,000 to invest and we invest x dollars at 11%, then the amount left to invest at 8% would be the total amount minus what we have already invested, $10,000 - x$. We can use a table to summarize the information.

	Investment earning 11%	Investment earning 8%	Total
Amount invested	x	$10,000 - x$	10,000
Interest received	$(0.11)x$	$(0.08)(10,000 - x)$	980

We get the equation for the problem from the bottom row of the table.

amount of interest at 11%   total   amount of interest at 8%   is   total interest

$$(0.11)x \quad + \quad (0.08)(10,000 - x) \quad = \quad 980$$

Solving for x,

$$x(0.11) + (0.08)(10,000 - x) = 980 \qquad \text{Original equation}$$
$$0.11x + 800 - 0.08x = 980 \qquad \text{Distributive property}$$
$$0.03x + 800 = 980 \qquad \text{Combine like terms}$$
$$0.03x = 180 \qquad \text{Subtract 800}$$
$$x = 6,000 \qquad \text{Divide by 0.03}$$

Hence he invests $x = 6,000$ dollars at 11% and $10,000 - x = 10,000 - (6,000) = 4,000$ dollars at 8%.

Note When we know the sum of two numbers, we let x equal one number, then the sum minus x(sum − x) will be the other number.

Mixture Problem

6. A trucking firm keeps antifreeze (alcohol/water) on hand in two concentrations, 30% and 70%. How many liters of each should be mixed to obtain 150 liters of 45% alcohol, to the nearest liter?

A process similar to that used for interest problems is used to solve problems about mixtures. It is important to understand the physical setting for such problems. For example, what does it mean to say that a 20-liter container is full of 30% alcohol/water solution? It means that 30% of the 20 liters is alcohol and the rest, 70%, is water. Thus, $(0.30)(20) = 6$ liters are alcohol, and $(0.70)(20) = 14$ liters are water. If we could separate the alcohol and water we could represent the situation as shown in the figure.

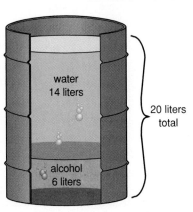

Let x represent the amount of 30% alcohol. Then, since a total of 150 liters is wanted, we need $150 - x$ liters of the 70% alcohol. To write an equation, focus on the actual amount of alcohol; this is 30% of x and 70% of $(150 - x)$. It is 45% of 150 liters.

As in the interest problem, we can use a table to summarize this information.

	30% alcohol mixture	70% alcohol mixture	45% alcohol mixture
Number of liters	x	$150 - x$	150
Amount of alcohol	$(0.3)x$	$(0.7)(150 - x)$	$(0.45)150$

30% of x and   70% of 150 − x equals   45% of 150

$$0.3x \quad + \quad 0.7(150 - x) \quad = \quad 0.45(150)$$
$$0.3x \quad + \quad 105 - 0.7x \quad = \quad 67.5$$
$$-0.4x \quad = \quad -37.5$$

$$x = \frac{-37.5}{-0.4} = 93.75 \approx 94 \text{ liters of 30\% alcohol and } 150 - (94) = 56 \text{ liters of}$$
70% alcohol.

☛ *Quick check*   One natural number is 5 times another natural number and their sum is 36. Find the numbers.

The sum of three consecutive integers is 36. Find the integers.

MASTERY POINTS

Can you
• Write an equation for a word problem?
• Solve for the unknown quantities?

Exercise 2–7

Directions
Write an equation for the problem and solve for the unknown quantities. See example 2–7 A.

☛ *Quick Check*

Example   One natural number is 5 times another natural number and their sum is 36. Find the numbers.

Solution   Let x represent the smaller natural number. Then five times the smaller natural number or $5x$ represents the larger natural number. The parts that make up the equation are

smaller natural number	sum	larger natural number	is	36
x	$+$	$5x$	$=$	36

$$x + 5x = 36 \qquad \text{Original equation}$$
$$6x = 36 \qquad \text{Combine like terms}$$
$$x = 6 \qquad \text{Divide by 6}$$

Therefore the smaller natural number is 6 and the larger natural number is 5 times the smaller natural number $5x$ and is $5 \cdot (6) = 30$.

Number problems
Directions   See example 2–7 A numbers 1, 2, and 3.

1. What number added to its double gives 63?

2. Six times a number, increased by 10, gives 94. Find the number.

3. If a number is divided by 4 and that result is then increased by 6, the answer is 13. Find the number.

4. If a number is decreased by 14 and that result is then divided by 5, the answer is 15. Find the number.

5. Nine times a number is decreased by 4, leaving 59. What is the number?

6. One third of a number is 8 less than one-half of the number. Find the number.

7. Find a number such that twice the sum of that number and 7 is 44.

8. If three times a number is increased by 11 and the result is 47, what is the number?

9. The difference between one half of a number and one third of the number is 9. Find the number.

10. One half of a number minus one third of the number is 8. Find the number.

11. A number plus one half of the number plus one third of the number equal 44. Find the number.

12. One number is 18 more than a second number. If their sum is 62, find the two numbers.

13. One number is 9 less than another number. If their sum is 47, find the two numbers.

14. The difference of two numbers is 17. Find the numbers if their sum is 87.

15. One number is 7 times a second number and their sum is 96. Find the numbers.

16. One number is 4 times a second number and their sum is 65. Find the numbers.

17. Find two numbers whose sum is 63 and whose difference is 5.

18. One number is 11 more than twice a second number. If their sum is 35, what are the numbers?

19. One number is 9 times a second number and their sum is 120. Find the numbers.

20. One number is 7 more than another number. Find the two numbers if three times the larger number exceeds four times the smaller number by 5.

21. One number is 4 more than another number. Find the two numbers if two times the larger number is 7 less than five times the smaller number.

22. One number is seven times another. If their difference is 18, what are the numbers?

23. The sum of three consecutive integers is 93. Find the integers.

24. The sum of three consecutive even integers is 72. Find the integers.

Geometry problems
Directions   See example 2–7 A number 4.

34. An equilateral triangle is a triangle in which all three sides have the same length. If the perimeter of an equilateral triangle is 52.8 centimeters, what is the length of a side?

35. If the perimeter of an equilateral triangle is 74.1 inches, what is the length of a side? (See exercise 34.)

36. A city ordinance requires that signs advertising garage sales must be 120 square inches. Cassandra has a poster-board that is 16 inches long, how wide must the poster-board be to comply with the ordinance?

25. The sum of three consecutive odd integers is 51. Find the integers.

26. The sum of three consecutive integers is 69. Find the integers.

27. The sum of three consecutive even integers is 66. Find the integers.

28. The sum of three consecutive odd integers is 75. Find the integers.

29. The sum of three numbers is 63. The first number is twice the second number and the third number is three times the first number. Find the three numbers.

30. The sum of three numbers is 44. The second number is three times the first number and the third number is 6 less than the first number. Find the three numbers.

31. Four times the first of three consecutive integers is 27 less than three times the sum of the second and third. Find the three integers.

32. Five times the first of three consecutive even integers is 2 less than twice the sum of the second and third. Find the three integers.

33. One fourth of the middle integer of three consecutive even integers is 27 less than one half of the sum of the other two integers. Find the three integers.

37. If you have 270 meters of fence and want to make a horse corral in the shape of a square, what would be the length of the side of the square?

38. An isosceles triangle is a triangle having two sides with the same length. If the perimeter of an isosceles triangle is 44 inches and the third side is 5 inches more than either of the equal sides, find the length of the three sides.

39. If the perimeter of an isosceles triangle is 21 centimeters and the third side is 4.5 centimeters shorter than either of the equal sides, find the length of the three sides. (See exercise 38.)

40. The length of a rectangle is 9 feet more than its width. The perimeter of the rectangle is 58 feet. Find the dimensions.

41. The width of a rectangle is 3 feet less than its length. The perimeter of the rectangle is 70 feet. Find the dimensions.

42. The width of a rectangle is $\frac{1}{3}$ of its length. If the perimeter is 96 feet, find the dimensions.

43. The length of a rectangle is 1 inch less than three times the width. Find the dimensions if the perimeter is 70 inches.

44. The width of a rectangle is 3 meters less than the length. If the perimeter of the rectangle is 142 meters, find the dimensions of the rectangle.

45. The length of a rectangle is 5 feet more than its width. If the perimeter is 82 feet, find the length and width.

46. The length of a rectangle is $2\frac{1}{2}$ times the width. If the perimeter is 84 centimeters, find the length and width.

Interest problems
Directions See example 2–7 A number 5.

47. Juan has $20,000, part of which he invests at 8% interest and the rest at 6%. If his total income for one year was $1,460 from the two investments, how much did he invest at each rate?

48. Rita has $18,000. She invests part of her money at $7\frac{1}{2}$% interest and the rest at 9%. If her income for one year from the two investments was $1,560, how much did she invest at each rate?

49. Charmain has $15,000. She invests part of this money at 8% and the rest at 6%. Her income for one year from these investments totals $1,120. How much is invested at each rate?

50. Alanzo invested $26,000, part at 10% and the rest at 12%. If his income for one year from these investments is $2,720, how much was invested at each rate?

51. Tyrone has $18,000, part of which he invests at 10% interest and the rest at 8%. If his income from each investment was the same, how much did he invest at each rate?

52. Amy invests a total of $12,000, part at 10% and part at 12%. Her total income for one year from the investments is $1,340. How much is invested at each rate?

53. Bina has $30,000, part of which she invests at 9% interest and the rest at 7%. If her income from the 7% investment was $820 more than that from the 9% investment, how much did she invest at each rate?

54. Paul invested a total of $18,000, part at 5% and part at 9%. If his income for one year from the 9% investment was $200 less than his income from the 5% investment, how much was invested at each rate?

55. Lynne made two investments totaling $25,000. On one investment she made an 18% profit, but on the other investment she took an 11% loss. If her net profit was $2,180, how much was each investment?

56. Grace made two investments totaling $18,000. She made a 14% profit on one investment, but she took a 9% loss on the other investment. If her net profit was $220, how much was each investment?

57. Kalil made two investments totaling $21,000. One investment made him a 13% profit, but on the other investment, he took a 9% loss. If his net loss was $196, how much was each investment?

58. Jeff made two investments totaling $34,000. One investment made him a 12% profit, but on the other investment he took a 21% loss. If his net loss was $2,940, how much was each investment?

Mixture problems
Directions   See example 2–7 A number 6.

62. A trucking firm has two mixtures of antifreeze; one is 35% alcohol, and the other is 65% alcohol. How much of each must be mixed to obtain 80 gallons of 50% alcohol?

63. A perfume manufacturer has two mixtures of toilet water; one is 15% alcohol, and the other is 35% alcohol. How much of each must be mixed to obtain 40 liters of 25% alcohol?

64. A company has 2.5 tons of material that is 30% copper. How much material that is 75% copper must be mixed with this to obtain a material that is 50% copper?

65. A plastic recycling firm has 8,000 lb of material that is 28% recycled plastic. How much material that is 70% recycled plastic must be mixed with this to obtain a material that is 50% recycled?

59. Dale has invested $5,000 at an 8% rate. How much more must he invest at 10% to make the total income for one year from both sources a 9% rate?

60. Fouad has $9,000 invested at 6%; how much more must he invest at 10% to realize a net return of 9%?

61. Chandra has $14,000 invested at 7% and is going to invest an additional amount at 11% so that her total investment will make 9%. How much does she need to invest at 11% to achieve this?

66. A company has 3000 gallons of a 10% pesticide solution. It also can obtain as much of a 4% pesticide solution as it needs. How much of this 4% solution should it mix with the 3000 gallons of 10% solution so that the result is an 8% solution?

67. A drug firm has on hand 400 lb of a mixture that is 3% sodium. The firm can buy as much of a 0.8% sodium mixture as it needs. It wishes to sell a mixture that is 2% sodium. How much of the 0.8% sodium mixture must it mix with the 400 lb of old material to obtain a 2% mixture?

68. A drug firm has an order for 300 liters of 40% hydrogen peroxide. It only stocks 20% and 55% solutions. How much of each should be mixed to fill the order?

69. A paint recycler has 450 gallons of recycled white paint on hand. How much new paint must be mixed with this to obtain a product that is 25% recycled material?

✿ Explore to Learn

The basic formula for interest for one year is $I = Pr$, where r is the rate. Suppose \$10,000 is invested, part in an account paying a 4% interest rate and the remaining amount in an account paying an 8% interest rate. Note that 4% of 10,000 is \$400, and 8% is \$10,000 is \$800.

a. Even not knowing how much of the \$10,000 is in each account, what can you say about the total interest earned from both accounts?

b. If the amount of interest earned was \$600, then the effective rate of the \$10,000 total is $\dfrac{600}{10,000} = 0.06$ or 6%. Even not knowing how much of the \$10,000 is in each account, what can you say about the effective rate of interest on the \$10,000 total?

Review Exercises

Directions   Evaluate the following formulas. See section 2–6.

1. $I = prt$, $p = 2,000$; $r = 0.05$; $t = 1$

2. $V = \ell wh$, $\ell = 7$, $w = 4$, $h = 3$

3. $F = ma$, $m = 34$, $a = 6$

4. $V = k + gt$, $k = 12$, $g = 32$, $t = 3$

5. $A = p + pr$, $p = 3,000$; $r = 0.06$

6. $A = \dfrac{1}{2}(b_1 + b_2)$, $b_1 = 20$, $b_2 = 12$

7. $S = \dfrac{1}{2}gt^2$, $g = 32$, $t = 4$

8. $\ell = a + (n - 1)d$, $a = 4$, $n = 10$, $d = 4$

2–8   Ratio and Proportion

A Ratio

We learned the fraction $\dfrac{a}{b}$ represents the indicated quotient of a divided by b. A **ratio** compares two numbers, or quantities, in the same way.

> ### Ratio
>
> A ratio is the comparison of two numbers (or quantities) by division.

The ratio of the number a to the number b is written

$$a \text{ to } b, \quad \frac{a}{b}, \quad \text{or} \quad a : b$$

We read $a : b$ as "the ratio of a to b," where a and b are called the *terms* of the ratio. The first number given is always the numerator and the second number is the denominator of the fraction representing the ratio.

● **Example 2–8 A**
Writing a ratio

Write each ratio statement in the forms $a : b$ and $\dfrac{a}{b}$ reduced to lowest terms.

1. If the quantities have the same unit of measure, the ratio will be expressed by a fraction without any unit designation required.

   45 minutes to 60 minutes

   $45 \text{ min} : 60 \text{ min} = \dfrac{45 \text{ min}}{60 \text{ min}}$ Write as fraction $\dfrac{a}{b}$

   $= \dfrac{3}{4}$ or $3 : 4$ Reduce by dividing each term by 15

2. When the compared quantities are not of the same unit of measure but *can be stated* in the same unit, it may be desirable to do so. The ratio again becomes only a fraction, as in example 1.

   3 feet to 4 inches
   Since 1 ft = 12 in., then 3 ft = 36 in., we have

   $36 \text{ in.} : 4 \text{ in.} = \dfrac{36 \text{ in.}}{4 \text{ in.}}$ Write as fraction $\dfrac{a}{b}$

   $= \dfrac{9}{1}$ or $9 : 1$ Reduce by dividing each term by 4

   We reduced to $\dfrac{9}{1}$ to demonstrate the comparison that is present.

   Note When we change to a common unit of measure, it is easiest to change to the *smaller* unit of measure, as we did in the previous example. Changing to the larger unit of measure usually involves fractions that are more difficult to reduce.

3. When you compare two measurable quantities by a ratio, it is not necessary for them to have the same unit of measure. If the units are not the same, you *must include the units* when you are expressing the ratio. These ratios represent rates of change.

   350 miles to 7 hours

   $350 \text{ mi} : 7 \text{ hr} = \dfrac{350 \text{ mi}}{7 \text{ hr}}$ Write as a fraction $\dfrac{a}{b}$

   $= \dfrac{50 \text{ mi}}{1 \text{ hr}}$ (stated 50 miles per hour)   Reduce by dividing each term by 7

🏴 *Quick check* Write the ratio of 16 minutes to 2 hours in the forms $a : b$ and $\dfrac{a}{b}$ reduced to lowest terms.

Express 64 pounds to 8 square inches as a ratio in lowest terms.

Unit Pricing

A **unit price** is the ratio of the price to the number of units. An important use of unit pricing is being able to determine which item is a better value. It allows us to compare things that are not the same size.

● **Example 2–8 B**
Finding unit prices

Find the unit price rounded to three decimal places and determine which has the lower unit price.

1. Brand A napkins at 120 napkins for 89¢ or Brand B napkins at 150 napkins for $1.09.

Brand A	Brand B	
$\dfrac{89¢}{120 \text{ napkins}}$	$\dfrac{\$1.09}{150 \text{ napkins}}$	Ratio of price to units
	$= \dfrac{109¢}{150 \text{ napkins}}$	Change to cents
$= 0.742$	$= 0.727$	Rounded to three decimal places

Brand B has the lower unit price at 0.727 cents per napkin.

2. Johnson's Corn Flakes at 18 ounces for $2.39 or Washington's Corn Flakes at 24 ounces for $3.29.

Johnson's	Washington's	
$\dfrac{\$2.39}{18 \text{ ounces}}$	$\dfrac{\$3.29}{24 \text{ ounces}}$	Ratio of price to units
$= 0.133$	$= 0.137$	Rounded to three decimal places

Johnson has the lower unit price at 0.133 dollars per ounce.

☞ *Quick check*   Find the unit price rounded to three decimal places and determine which one has the lower unit price. Brand A catsup at 9 ounces for 89¢ or Brand B catsup at 12 ounces for $1.19.   ●

A Proportion

A **proportion** establishes a relationship between two ratios.

Definition of a Proportion

A **proportion** is a statement of equality of two ratios.

Given the ratios a to b and c to d,

$$\frac{a}{b} = \frac{c}{d} \quad \text{or} \quad a : b = c : d$$

is a proportion. We read the statement $a : b = c : d$ "a is to b as c is to d." The numbers a, b, c, and d are called the *terms* of the proportion.

Given the proportion $\dfrac{a}{b} = \dfrac{c}{d}$,

$$bd \cdot \frac{a}{b} = bd \cdot \frac{c}{d} \qquad \text{Multiply each side by } bd$$
$$ad = bc \qquad \text{Reduce by } b \text{ on the left and by } d \text{ on the right}$$

Property of Proportions

If $\dfrac{a}{b} = \dfrac{c}{d}$, then $ad = bc$ $(b, d \neq 0)$

Also if $ad = bc$, then $\dfrac{a}{b} = \dfrac{c}{d}$

<u>Note</u> The products *ad* and *bc* are found by multiplying diagonally.

This process is frequently called *cross-multiplying,* especially by persons in applied fields, and *ad* and *bc* are called the *cross products*.

● **Example 2–8 C**

Determining when a proportion is true

Determine if the following proportions are true.

1. $\dfrac{3}{5} = \dfrac{12}{20}$

Using the property of proportions, we obtain

$5 \cdot 12 = 60$ and $3 \cdot 20 = 60$

The cross products are both 60, so the proportion is true.

2. $\dfrac{5}{6} = \dfrac{16}{18}$

Using the property of proportions, we obtain

$6 \cdot 16 = 96$ and $5 \cdot 18 = 90$

The cross products are not the same so the proportion is false.

We use the property of proportions to find the unknown term of a proportion if three of the four terms are known.

To Solve a Proportion

1. Use the property of proportions to form the cross products.
2. Solve for the unknown.

● **Example 2–8 D**

Solving a proportion

Find the unknown term of the given proportion. Check your solution.

1. $\dfrac{x}{8} = \dfrac{16}{64}$

$64 \cdot x = 8 \cdot 16$ Property of proportions

$64x = 128$ Multiply as indicated

$x = 2$ Divide each side by 64

Check: $\dfrac{2}{8} = \dfrac{16}{64}$ Then

$2 \cdot 64 = 8 \cdot 16$

$128 = 128$

2. $\dfrac{49}{y} = \dfrac{35}{5}$

$49 \cdot 5 = 35 \cdot y$ Property of proportions

$35y = 245$ Multiply as indicated

$y = 7$ Divide each side by 35

Check: $\dfrac{49}{7} = \dfrac{35}{5}$ Then

$5 \cdot 49 = 7 \cdot 35$

$245 = 245$

☛ *Quick check*   Find the value of z in the proportion $\dfrac{72}{z} = \dfrac{30}{6}$.

Problem Solving

Proportions are used in solving many applied problems.

To Solve a Problem Using a Proportion

1. Choose a variable to represent the unknown.
2. Make a proportion using the given information to set up a ratio on the left side of the equation. Set up the ratio on the right side of the equation using the unknown and the other given quantity. The ratio on the right side of the equation must be set up with the quantities occupying the same respective positions as in the ratio on the left side of the equation.

   For example: $\dfrac{\text{miles}}{\text{gallon}} = \dfrac{\text{miles}}{\text{gallon}}$ or $\dfrac{\text{gallon}}{\text{miles}} = \dfrac{\text{gallon}}{\text{miles}}$

   would be the correct way to set up a proportion involving miles and gallons.

   The units used in the problem are useful in correctly setting up the proportion, but are dropped once the proportion is set up.
3. Solve the proportion.
4. Answer the question(s) asked.

● **Example 2–8 E**

Using a proportion to solve a word problem

Set up a proportion for each problem and solve.

1. On a map, 1 inch represents 6 miles. How many inches are needed to represent 28 miles?

Let x represent the number of inches representing 28 miles. Now, 1 inch is to 6 miles as x inches is to 28 miles.

$\dfrac{1 \text{ in.}}{6 \text{ mi}} = \dfrac{x \text{ in.}}{28 \text{ mi}}$ Set up a proportion using $\dfrac{\text{inches on the map}}{\text{actual miles}}$

$\dfrac{1}{6} = \dfrac{x}{28}$ Drop the units

$6 \cdot x = 1 \cdot 28$ Property of proportions

$x = \dfrac{1 \cdot 28}{6} = \dfrac{28}{6}$ Divide each side by 6

$= \dfrac{14}{3}$ or $4\dfrac{2}{3}$ Reduce to lowest terms

Therefore, 28 miles are represented by $4\dfrac{2}{3}$ inches on the map.

Note In example 1, the same units of measure are in the numerator of the ratios and the same units of measure are in the denominators. That is, we placed inches in the numerator and miles in the denominator of each ratio. This step is important in setting up the proportion you will use to solve for the unknown.

2. Cheryl set aside $20 per week for her savings program when her salary was $200 per week. If her salary is now $250 per week, how much should she set aside for her weekly savings to be proportional to what she saved before?

   Let x represent the amount to be set aside when Cheryl earns $250 per week. Then, $20 is to $200 as x is to $250.

$$\frac{20}{200} = \frac{x}{250} \qquad \text{Set up a proportion using } \frac{\text{savings}}{\text{salary}}$$

$$200 \cdot x = 20 \cdot 250 \qquad \text{Property of proportions}$$

$$200x = 5{,}000 \qquad \text{Multiply as indicated}$$

$$x = \frac{5{,}000}{200} \qquad \text{Divide each side by 200}$$

$$x = 25$$

   Cheryl should set aside $25 when making $250 per week.

   Proportions can be used to convert one unit of measure to another unit of measure. To set up the proportion we will need to know the relationship between the two units of measure.

3. Convert 6 inches to centimeters.

   Let x represent the number of centimeters that is equal to 6 inches. The relationship is 1 inch = 2.54 centimeters. We use this relationship to write the ratio on the left side of the equation.

$$\frac{1 \text{ inch}}{2.54 \text{ centimeters}} = \frac{6 \text{ inches}}{x \text{ centimeters}} \qquad \text{Set up a proportion using } \frac{\text{inches}}{\text{centimeters}}$$

$$\frac{1}{2.54} = \frac{6}{x} \qquad \text{Drop the units}$$

$$1 \cdot x = 6 \cdot 2.54 \qquad \text{Property of proportions}$$

$$x = 15.24 \qquad \text{Multiply}$$

   Therefore, 6 inches = 15.24 centimeters.

4. Convert 5.3 quarts to liters.

   Let x represent the number of liters that is equal to 5.3 quarts. The relationship is 1 liter = 1.06 quarts.

$$\frac{1 \text{ liter}}{1.06 \text{ quarts}} = \frac{x \text{ liters}}{5.3 \text{ quarts}} \qquad \text{Set up a proportion using } \frac{\text{liters}}{\text{quarts}}$$

$$\frac{1}{1.06} = \frac{x}{5.3} \qquad \text{Drop the units}$$

$$5.3 = 1.06x \qquad \text{Property of proportions}$$

$$5 = x \qquad \text{Divide both sides by 1.06}$$

   Therefore, 5.3 quarts = 5 liters.

☞ _**Quick check**_   On a map, 1 inch represents 9 miles. How many inches are needed to represent 42 miles?

MASTERY POINTS

Can you
- Write ratios?
- Reduce ratios?
- Solve proportions for the unknown?
- Set up proportions to solve problems?

Exercise 2–8

Directions
Find the indicated ratios reduced to lowest terms expressed in two forms. See example 2–8 A.

☞ *Quick Check*

Example 16 minutes to 2 hours

Solution Since 2 hours = 120 minutes (1 hour = 60 minutes),

$$16 \text{ min} : 2 \text{ hr} = 16 \text{ min} : 120 \text{ min} \qquad \text{Replace 2 hr with 120 min}$$
$$= 16 : 120 \qquad \text{Eliminate unit of measure}$$
$$= 2 : 15 \text{ or } \frac{2}{15} \qquad \text{Reduce to lowest terms (Divide by 8)}$$

The ratio of 16 min to 2 hr is $2 : 15$ or $\frac{2}{15}$.

Example 64 pounds to 8 square inches

Solution $64 \text{ lb to 8 sq in.} = \dfrac{64 \text{ lb}}{8 \text{ sq in.}}$ Write ratio as a fraction

$$= \frac{8 \text{ lb}}{1 \text{ sq in.}} \qquad \text{Divide by 8 to reduce to lowest terms}$$
$$= 8 \text{ lb/sq in.} \qquad \text{Write as a rate}$$

64 lb to 8 sq in. is 8 lb/sq in. or 8 lb : 1 sq in.

Directions Find the indicated ratios reduced to lowest terms expressed in two forms. See example 2–8 A.

1. 6 in. to 14 in.
2. 4 ft to 18 ft
3. 25 cm to 10 cm
4. 35 lb to 5 lb
5. 36 km to 24 km
6. 48 lb to 16 lb
7. 15 in. to 3 ft
8. 10 ft to 4 yd
9. $3 to 35¢

10. 5 days to 15 weeks
11. 30 min to 13 hr
12. 16 lb to 8 oz
13. 48 lb to 24 ft^3
14. 50 oz to 5 in.^{3}
15. 16 grams to 2 cm^3
16. 300 mi to 10 gal
17. 1,020 mi to 17 hr
18. 105 kg to 35 m^3

Directions
Find the unit price rounded to three decimal places and determine which one has the lower unit price. See example 2–8 B.

☞ *Quick Check*

Example   Brand A catsup: 9 ounces for 89¢
          Brand B catsup: 12 ounces for $1.19

Solution   Brand A   Brand B
             89¢ $1.19
           ─────   ──────   Ratio of price to units
           9 ounces   12 ounces
                            119¢
                        = ──────   Change to cents
                          12 ounces
           = 9.889   = 9.917   Rounded to three decimal places

Brand A has the lower unit price at 9.889 cents per ounce.

Directions   Find the unit price rounded to three decimal places and determine which one has the lower unit price. See example 2–8 B.

19. Raisin bran cereal

   A: 16 ounces for $3.19
   B: 24 ounces for $4.39

20. Potato chips

   A: 14 ounces for $2.79
   B: 9 ounces for $1.89

21. Pretzels

   A: 10 ounces for $1.39
   B: 12 ounces for $1.79

22. Tea bags

   A: 50 bags for $2.73
   B: 80 bags for $3.99

23. Small garbage bags

   A: 30 bags for $1.65
   B: 40 bags for $2.29

24. Canola oil

   A: 32 ounces for $2.99
   B: 28 ounces for $2.49

25. Lasagna

   A: 16 ounces for $1.29
   B: 12 ounces for 99¢

26. Sugar

   A: 5 pounds for $2.15
   B: 4 pounds for $1.89

27. Flour

   A: 5 pounds for $1.45
   B: 8 pounds for $2.19

28. Peanut butter

   A: 1 pound 4 ounces for $2.39
   B: 15 ounces for $1.89

29. Cracker snacks

   A: 7 ounces for $2.39
   B: 1 pound 2 ounces for $5.19

Directions   Solve the following applied problems.

30. The *output* in horsepower is the useful energy delivered *by an engine* and the *input* in horsepower is the amount of energy delivered *to an engine.* The *mechanical efficiency* of the engine is given by the ratio

$$\text{mechanical efficiency} = \frac{\text{output}}{\text{input}}$$

Find the mechanical efficiency of an engine rated to deliver 425 horsepower (input) when it delivers only 375 horsepower.

31. An automobile engine is rated at 350 horsepower. When the engine is tested, it produces only 325 horsepower. What is the mechanical efficiency of the engine? (Refer to exercise 30.)

32. An electric motor uses 10 watts of electricity to produce an equivalent output of 8 watts. What is the mechanical efficiency of the motor? (Refer to exercise 30.)

33. A particular stock costing $63 paid an earnings of $6. What is the cost : earnings ratio?

34. A room is 24 feet long and 18 feet wide. What is the ratio of its length to its width?

35. The magnification (M) of an object by a lens is given by the ratio

$$M = \frac{q}{p}$$

where $q = $ the image distance and $p = $ the object distance from the lens. Find the magnification of an object whose image distance is 27 feet and object distance is 12 feet.

36. A mathematics class contains 32 male students and 10 female students. What is the ratio of the male students to the female students?

37. The *pitch* of a roof is the ratio of the *rise* of a rafter to the *span* of the roof.

$$\text{pitch} = \frac{\text{rise of rafter}}{\text{span of roof}}$$

Find the pitch if the roof rises 7 feet in a span of 21 feet.

38. A cement block weighs 120 pounds and a steel block weighs 1,860 pounds. What is the ratio of the weight of the steel block to the weight of the cement block?

Directions
Find the value of the unknown that makes the proportion true and check the answer. See example 2–8 D for how to solve the proportion and example 2–8 C for how to check the answer.

☞ *Quick Check*

Example   Find the value of z in the proportion $\dfrac{72}{z} = \dfrac{30}{6}$

Solution
$$\frac{72}{z} = \frac{30}{6}$$
$$30 \cdot z = 72 \cdot 6 \qquad \text{Property of proportions}$$
$$30z = 432 \qquad \text{Perform indicated multiplication}$$
$$z = \frac{432}{30} \qquad \text{Divide by 30}$$
$$= \frac{72}{5} \text{ or } 14\frac{2}{5} \qquad \text{Reduce to lowest terms}$$

Check:
$$\frac{72}{\frac{72}{5}} = \frac{30}{6}$$
$$72 \cdot 6 = \frac{72}{5} \cdot 30$$
$$432 = 432$$

Thus, $z = 14\frac{2}{5}$ makes the proportion true.

Directions   Find the value of the unknown that makes the proportion true and check the answer. See example 2–8 D for how to solve the proportion and example 2–8 C for how to check the answer.

39. $\dfrac{9}{x} = \dfrac{36}{5}$

40. $\dfrac{y}{7} = \dfrac{30}{42}$

41. $\dfrac{5}{9} = \dfrac{p}{20}$

42. $\dfrac{14}{10} = \dfrac{21}{z}$

43. $6 : 15 = x : 8$

44. $R : 12 = 15 : 100$

45. $1\frac{1}{2} : a = 4\frac{3}{4} : 2$

47. $1.2 : x = 3.6 : 9$

49. $\dfrac{3\frac{1}{4}}{5} = \dfrac{2\frac{1}{2}}{a}$

46. $\dfrac{3}{4} : 4 = \dfrac{1}{2} : b$

48. $4.5 : 3 = y : 2$

50. $\dfrac{2.4}{4.2} = \dfrac{b}{2.1}$

Directions

Solve the following problems by first choosing a letter to represent the unknown and then setting up the proper proportion. See example 2–8 E.

☞ *Quick Check*

Example On a map, 1 inch represents 9 miles. How many inches are needed to represent 42 miles?

Solution Let x represent the number of inches representing 42 miles. Then, 1 inch is to 9 miles as x inches is to 42 miles.

$$\frac{1 \text{ inch}}{9 \text{ miles}} = \frac{x \text{ inches}}{42 \text{ miles}}$$ Set up a proportion

$$9 \cdot x = 1 \cdot 42$$ Property of proportions

$$9x = 42$$ Multiply as indicated

$$x = \frac{42}{9}$$ Divide each side by 9

$$= \frac{14}{3} \text{ or } 4\frac{2}{3}$$ Reduce to lowest terms

Thus, $4\frac{2}{3}$ inches represents 42 miles on the map.

Directions   Solve the following problems by first choosing a letter to represent the unknown and then setting up the proper proportion. See example 2–8 E.

51. A man earns $180 per week. How many weeks must he work to earn $1,260?

52. An automobile uses 8 liters of gasoline to travel 84 kilometers. How many liters are needed to travel 1,428 kilometers?

53. If 24 grams of water will yield 4 grams of hydrogen, how many grams of hydrogen will there be in 216 grams of water?

54. The operating instructions for a gasoline chain saw call for a 16 gallons : 1 pint fuel-to-oil mixture. How many *pints* of oil are needed to mix with 88 gallons of fuel?

55. The power-to-weight ratio of a given engine is 5 : 3. What is the weight in pounds of the engine if it produces 650 horsepower?

56. If a 20-pound casting costs $1.50, at this same rate, how much would a 42-pound casting cost?

57. American television signals are broadcast in a 4 : 3 aspect ratio (the ratio of width to height of the picture). A television screen is to be adjusted so the width is 14 inches; to what value should the height be adjusted?

58. In a hydraulic press, the force on the output piston is to the force on the input piston as the area of the output piston is to the area of the input piston. That is, $\dfrac{F_o}{F_i} = \dfrac{A_o}{A_i}$ or $F_o : F_i = A_o : A_i$. Find the area of the input piston if F_o is 15.2 pounds, F_i is 6.5 pounds, and A_o is 10.4 inches^2.

59. If the ratio of the wins to the losses of the Chicago Cubs in a given season is 6 : 5, how many games did they lose if they won 90 games?

60. A rectangular picture that is 10 inches long and 8 inches wide is to be enlarged so that the enlargement will be 36 inches wide. What should be the length of the enlargement?

61. The corresponding sides of the triangles in the diagram are in proportion. Find the dimensions of the missing sides, x and y. (*Hint:* The corresponding sides are $6''$ and x, $7''$ and $5''$, $8''$ and y.)

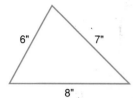 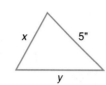

62. Ann is operating a machine that can produce 14 parts in 20 minutes. How long will it take for her to produce 224 parts?

63. A punch machine can make 72 holes in 4 minutes. How many holes can the machine make in 3 hours?

64. A roof rises $6\frac{1}{2}$ feet in a rafter span of 9 feet. At this rate, what would be the rise in a 15-foot span?

65. A man can type 3 pages of an English paper in 15 minutes. How long would it take him to type 54 pages? (State the answer in hours and minutes.)

66. On a draftsman scale, $\frac{1}{8}$ inch represents 1 foot. What length will a measurement of $2\frac{5}{8}$ inches on the scale represent?

67. An automobile engine uses $\frac{3}{4}$ quart of oil in 900 miles. How much oil will it take in 3,000 miles?

68. A copper wire 300 feet long has a resistance of 1,024 ohms. What is the resistance of 2,000 feet of copper wire?

69. Statistical sampling assumes the concept of proportion. The following is an example. A company makes a product called Woopi Cola. It normally sells about 5 million bottles per month nationwide. The company wants to introduce diet Woopi Cola, but has no idea how much to make nationwide. Before investing in more bottling equipment it uses its current equipment on weekends and places the new product in three cities it feels are typical of the company's national market. It sells about 250,000 bottles of Woopi per month in these cities. After a few months of test marketing in these cities, about 85,000 bottles of diet Woopi are being sold per month, with no change in sales of regular Woopi. Based on this data how many bottles of diet Woopi can the company expect to sell per month nationwide?

70. Refer to exercise 69. Suppose that the sales of regular Woopi Cola went down in these three cities to 230,000 per month after the introduction of diet Woopi, while national sales remained steady. What conclusion could the company make?

Directions   For problems 71–80 use the following relationships: 1 inch = 2.54 centimeters, 1 liter = 1.06 quarts, 1 kilometer = 0.621 mile, 1 pound = 454 grams, 1 kilogram = 2.2 pounds. Use a proportion to make the following conversions. See example 2–8 E.

71. 38.1 centimeters to inches

72. 8 inches to centimeters

73. 2.5 liters to quarts

74. 15.9 quarts to liters

75. 5 kilometers to miles

76. 31.05 miles to kilometers

77. 6.6 pounds to grams

78. 363.2 grams to pounds

79. 26.4 pounds to kilograms

80. 33 kilograms to pounds

✤ Explore to Learn

A 4″ by 5″ photograph has length 5 inches and width 4 inches. It is to be blown up to be a poster. The 4″ dimension is to be increased to 2 feet.

a. To what value should the 5″ dimension be increased?

b. The area of the 4″ by 5″ photo is $4 \times 5 = 20$ square inches. What is the area of the poster?

c. Did the area increase in the same proportion as did the length and width?

Review Exercises

Directions   Perform the indicated operations. See sections 1–5, 1–6, and 1–7.

1. $-14 + 8$

2. $5 + (-11)$

3. $-4 - (-4)$

4. $7 - (-7)$

5. $4(-3)$

6. $(-2)(5)$

7. $(-8)(-4)$

8. -4^2

9. $(-4)^2$

10. $(-4)^3$

2–9   Solving Linear Inequalities

Many of the everyday problems that need to be solved are **inequalities**. For example, an elevator has a sign stating that the maximum safe load is 4,000 pounds, or a company says it must sell 8,500 units before making a profit, or a computer requires an operating environment of from 17 to 27 degrees Celsius; these are common situations that are expressed using inequalities.

Inequality Symbols

In chapter 1, we studied the meaning of the inequality symbols[3]

$<$ "is less than"
$\leq$ "is less than or equal to"
$>$ "is greater than"
$\geq$ "is greater than or equal to."

These symbols define the *sense* or *order* of an inequality. For example, if we wish to denote that the variable x represents 3 or any number greater than 3, we write $x \geq 3$.

---

[3]The symbols $>$ and $<$ were first used in 1631 in the *Artis analyticae praxis*. The symbols $\geq$ and $\leq$ were first used by the Frenchman P. Bouguer in 1734.

<u>Note</u> In $x \geq 3$, x represents *any* real number that is greater than or equal to 3, and not just any integer greater than or equal to 3. Remember that 3.1, 3.004, and so on, are all greater than 3.

Linear Inequalities

When we replace the equal sign in a conditional linear equation with one of these inequality symbols, we form a *conditional linear inequality*.

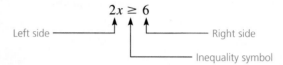

A major difference between the linear equation and the linear inequality is the solution. Consider the inequality $2x \geq 6$. We can, by inspection, see that if we substitute 3, $3\frac{1}{2}$, 4, or 5 for x, the inequality will be true. In fact, we see that if we were to substitute any number greater than or equal to 3, the inequality would be true. The inequality has an unlimited number of solutions. The values for x that would satisfy the inequality would be $x \geq 3$.

Another way to indicate the solution of an inequality is by graphing. To graph the solution, we simply draw a number line (as we did in chapter 1), place a solid circle at 3 on the number line to signify that 3 is in the solution, and draw an arrow extending from the solid circle to the right (figure 2–1). The solid line indicates that *all* numbers greater than or equal to 3 are part of the graph.

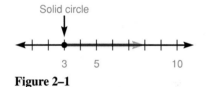

Figure 2–1

● **Example 2–9 A**

Graphing an inequality

Graph the following linear inequalities.

1. $x < 4$ Here x represents all real numbers less than 4, but not 4 itself. To denote the fact that x cannot equal 4, we put a **hollow circle** at 4.

2. $x \geq -3$ The is greater than or equal to symbol, $\geq$, indicates that the graph will contain the point -3, and we place a **solid circle** at -3.

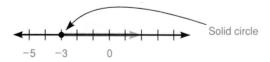

3. $-3 \le x < 4$ This statement is called a *compound inequality.* It is read "-3 is less than or equal to x and x is less than 4." We place a solid circle at -3 to denote that -3 is included and place a hollow circle at 4 to show that 4 is not included. We then draw a line segment between the two circles.

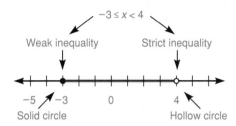

Note   When we graph inequalities, a strict inequality ($<$ or $>$) is represented by a hollow circle at the number. A weak inequality ($\le$ or $\ge$) is represented by a solid circle at the number.

☞ *Quick check*   Graph $-3 < x \le 2$ ●

Solving Linear Inequalities

The properties that we will be using to solve linear inequalities are similar to those that we used to solve linear equations.

Addition and Subtraction Property of Inequalities

For all real numbers *a*, *b*, and *c*, if $a < b$, then

$$a + c < b + c \text{ and}$$
$$a - c < b - c$$

Concept

The same number can be added to or subtracted from both sides of an inequality without changing the direction of the inequality symbol.

Multiplication and Division Property of Inequalities

For all real numbers a, b, and c, if $a < b$, and

1. If $c > 0$ (c represents a positive number), then

$$a \cdot c < b \cdot c \text{ and}$$

$$\frac{a}{c} < \frac{b}{c}$$

2. If $c < 0$ (c represents a negative number), then

$$a \cdot c > b \cdot c \text{ and}$$

$$\frac{a}{c} > \frac{b}{c}$$

Concept

We can multiply or divide *both* sides of the inequality by the same positive number without changing the direction of the inequality symbol.

We can multiply or divide *both* sides of an inequality by the same *negative* number, provided that we **reverse** the direction of the inequality symbol.

<u>Note</u> The two properties were stated in terms of the "is less than" ($<$) symbol. The properties apply for any of the other inequality symbols ($>$, $\leq$, or $\geq$).

To demonstrate these properties, consider the inequality $8 < 12$.

1. If we add or subtract 4 on each side, we still have a true statement.

$8 < 12$	or	$8 < 12$	Original true statement
$8 + 4 < 12 + 4$		$8 - 4 < 12 - 4$	Add or subtract 4
$12 < 16$		$4 < 8$	New true statement

2. If we multiply or divide by 4 on each side, we still have a true statement.

$8 < 12$	or	$8 < 12$	Original true statement
$8 \cdot 4 < 12 \cdot 4$		$\dfrac{8}{4} < \dfrac{12}{4}$	Multiply or divide by 4
$32 < 48$			
		$2 < 3$	New true statement

3. But if we multiply or divide by -4 on each side, we have to reverse the direction of the inequality to get a true statement.

$8 < 12$	or	$8 < 12$	Original true statement
$8(-4) > 12(-4)$		$\dfrac{8}{-4} > \dfrac{12}{-4} \longleftarrow$	Multiply or divide by -4 and reverse direction of the inequality symbol
$-32 > -48$		$-2 > -3$	New true statement

<u>Note</u> When we reverse the direction of the inequality symbol, we say that we **reversed the sense or order** of the inequality.

To summarize our properties, we see that they are the same as the properties for linear equations, with one exception. **Whenever we multiply or divide both**

sides of an inequality by a negative number, we must reverse the direction of the inequality symbol.

We shall now solve some linear inequalities. The procedure for solving a linear inequality uses the same four steps that we used to solve a linear equation.

Solving a Linear Inequality

1. Simplify on each side, where necessary, by performing the indicated operations.
2. Add, or subtract, to get all terms containing the unknown on one side of the inequality.
3. Add, or subtract, to get all terms *not* containing the unknown on the other side of the inequality.
4. Multiply, or divide, to obtain a coefficient of 1 for the unknown. *Remember, when multiplying or dividing by a negative number, always change the direction (order) of the inequality symbol.*

● **Example 2–9 B**

Solving a linear inequality

Find the solution and graph the solution.

1. $2x + 5x - 1 < 4x + 2$

   Step 1 *We simplify* the inequality by carrying out the indicated addition on the left side.

   $$2x + 5x - 1 < 4x + 2$$
   $$7x - 1 < 4x + 2$$

   Step 2 *We want all the terms containing the unknown,* x, *on one side of the inequality.* Therefore we subtract $4x$ from both sides of the inequality.

   $$7x - 1 < 4x + 2$$
   $$7x - 4x - 1 < 4x - 4x + 2$$
   $$3x - 1 < 2$$

   Note A negative coefficient of the unknown can be avoided if we form equivalent inequalities where the unknown appears only on the side of the inequality that has the greater coefficient of the unknown.

   Step 3 *We want all the terms not involving the unknown on the other side of the inequality.* Therefore we add 1 to both sides of the inequality.

   $$3x - 1 < 2$$
   $$3x - 1 + 1 < 2 + 1$$
   $$3x < 3$$

   Step 4 *We form an equivalent inequality where the coefficient of the unknown is 1.* Hence we divide both sides of the inequality by 3.

   $$3x < 3$$
   $$\frac{3x}{3} < \frac{3}{3}$$
   $$x < 1$$

   We can also graph the solution.

   $-5 \qquad 0 \ \ 1$

Note We should be careful to observe in step 4 whether we are multiplying or dividing by a positive or negative number so that we will form the correct inequality.

2. $-2x \leq 4$

The only operation we need to perform to solve the inequality is to divide by -2. Since we are dividing by a negative number, we must remember to **reverse** the direction of the inequality symbol.

$$-2x \leq 4$$
$$\frac{-2x}{-2} \geq \frac{4}{-2} \qquad \text{Reverse the direction of the inequality symbol}$$
$$x \geq -2$$

Graph

3. $\quad 5(2x + 1) \leq 7x - 4x + 3$
$$10x + 5 \leq 3x + 3 \qquad \text{Simplify by multiplying on left side and combining like terms on right side}$$
$$10x - 3x + 5 \leq 3x - 3x + 3 \qquad \text{Subtract } 3x \text{ from both sides}$$
$$7x + 5 \leq 3$$
$$7x + 5 - 5 \leq 3 - 5 \qquad \text{Subtract 5 from both sides}$$
$$7x \leq -2$$
$$\frac{7x}{7} \leq \frac{-2}{7} \qquad \text{Divide both sides by 7}$$
$$x \leq \frac{-2}{7}$$

Graph

$\frac{-2}{7}$

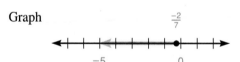

4. $-3 \leq 2x + 1 < 5$

When solving a compound inequality of this type, the solution must be such that the unknown appears only in the middle part of the inequality. We can still use all of our properties, if we apply them to all three parts, and we must reverse the direction of _all_ inequality symbols when multiplying or dividing by a negative number.

$$-3 \leq 2x + 1 < 5$$
$$-3 - 1 \leq 2x + 1 - 1 < 5 - 1 \qquad \text{Subtract 1 from all three parts}$$
$$-4 \leq 2x < 4$$
$$\frac{-4}{2} \leq \frac{2x}{2} < \frac{4}{2} \qquad \text{Divide all three parts by 2}$$
$$-2 \leq x < 2$$

Graph

5.
$$-6 < -3x - 3 \le 9$$
$$-6 + 3 < -3x - 3 + 3 \le 9 + 3 \qquad \text{Add 3 to all three parts}$$
$$-3 < -3x \le 12$$
$$\frac{-3}{-3} > \frac{-3x}{-3} \ge \frac{12}{-3} \qquad \text{Divide all three parts by } -3, \text{ reversing}$$
$$\qquad\qquad\qquad\qquad\qquad \text{the direction of } \textbf{\textit{both}} \text{ inequality symbols}$$
$$1 > x \ge -4$$

Graph
$$\qquad\qquad -5\ -4 \qquad\quad 0\ \ 1$$

<u>Note</u> From our discussions in chapter 1, we could have written the solution in the previous example as $-4 \le x < 1$. This is usually the preferred form.

☞ ***Quick check***   Find the solution for $4x + 5x - 4 < 6x - 1$ ●

Problem Solving

We are now ready to combine our abilities to write an expression and to solve an inequality and apply them to solve word problems. The guidelines for solving a linear inequality are the same as those for solving a linear equation in section 2–5. The following table shows a number of different ways that an inequality symbol could be written with words.

Symbol	<	≤	>	≥
In words	is less than is fewer than is almost	is at most is no more than is no greater than is less than or equal to	is greater than is more than exceeds	is at least is no less than is no fewer than is greater than or equal to

● **Example 2–9 C**
Problem solving

1. Write an inequality for the following statement: A student's test grade, G, must be at least 75 to have a passing grade.

If the student's test grade must be *at least* 75, the grade must be 75 or greater. Thus,

$$G \ge 75$$

2. Four times a number less 5 is to be no more than three times the number increased by 2. Find all numbers that satisfy these conditions.

Let x represent the number.

4 times a number	less	5	is no more than	3 times the number	increased by	2
$4x$	$-$	5	$\le$	$3x$	$+$	2

The inequality is $4x - 5 \leq 3x + 2$

$4x - 5 \leq 3x + 2$	Subtract $3x$ from each side
$4x - 5 - 3x \leq 3x + 2 - 3x$	Combine on each side
$x - 5 \leq 2$	Combine on each side
$x - 5 + 5 \leq 2 + 5$	Add 5 to each side
$x \leq 7$	Combine on each side

The number is any real number x such that $x \leq 7$.

Quick check   To complete an order for cement, a company will need at least 3 trucks. Write an inequality for the number of trucks needed.

If 4 is subtracted from three times a number, the result is greater than 2 more than twice the number. Find all numbers that satisfy this condition.

MASTERY POINTS

Can you
- Graph linear inequalities and compound linear inequalities?
- Solve linear inequalities and compound linear inequalities?
- Solve word problems involving linear inequalities?

Exercise 2–9

Directions
Graph the following.
See example 2–9 A.

Quick Check

Example $-3 < x \leq 2$

Solution

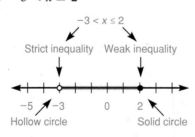

Directions   Graph the following. See example 2–9 A.

1. $x > 2$

2. $x > -2$

3. $x \geq 1$

4. $x \geq 4$

5. $x < 0$

6. $x < -2$

7. $x > 0$

8. $x \leq 3$

9. $x \leq -4$

10. $-2 < x < 0$

11. $-1 < x < 2$

12. $-3 \leq x \leq 4$

13. $0 \leq x \leq 5$

14. $1 \leq x < 4$

15. $-1 < x \leq 3$

Directions	Quick Check
Find the solution and graph the solution. See example 2–9 B.	**Example** $4x + 5x - 4 < 6x - 1$

Solution

$$9x - 4 < 6x - 1 \quad \text{Combine like terms on left side}$$
$$9x - 4 - 6x < 6x - 1 - 6x \quad \text{Subtract } 6x \text{ from each side}$$
$$3x - 4 < -1 \quad \text{Combine like terms}$$
$$3x - 4 + 4 < -1 + 4 \quad \text{Add 4 to each side}$$
$$3x < 3 \quad \text{Combine like terms}$$
$$x < 1 \quad \text{Divide each side by 3}$$

Thus $x < 1$ and the solutions are all real numbers less than 1.

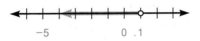

Directions   Find the solution and graph the solution. See example 2–9 B.

16. $4x > 10$

17. $2x \leq 5$

18. $3x \geq 15$

19. $5x < 30$

20. $\dfrac{3}{4}x < 9$

21. $\dfrac{2}{3}x \geq 12$

22. $-4x < 12$

23. $-3x \leq 27$

24. $-6x > 18$

25. $-2x < 10$

26. $4x + 3x \geq 2x + 7$

27. $8x - 2x > 4x - 5$

28. $3x + 2x < x + 6$

29. $\dfrac{3x}{2} > 8$

30. $\dfrac{4x}{3} > 12$

31. $x + 7 \leq 9$

32. $x + 4 \geq -12$

33. $x - 5 < 6$

34. $x - 12 < -9$

35. $2x + (3x - 1) > 5 - x$

36. $2(3x + 1) < 7$

37. $3(2x - 1) \geq 4x + 3$

38. $4(5x - 3) \leq 25x + 11$

39. $12x - 8 > 5x + 2$

40. $9x + 4 > x - 11$

41. $2 + 5x - 16 < 6x - 4$

42. $3(2x + 5) > 4(x - 3)$

43. $8 - 2(3x + 4) > 5x - 16$

44. $(3x + 2) - (2x - 5) > 7$

45. $(7x - 6) - (4 - 3x) \le 27$

46. $2(x - 4) - 16 \le 3(5 - 2x)$

47. $3(1 - 2x) \ge 2(4 - 4x)$

48. $4(5 - x) > 7(2 - x)$

49. $3(2x + 3) \ge 5 - 4(x - 2)$

50. $6(3x - 2) \le 7(x - 3) - 2$

51. $-1 < 2x + 3 < 4$

52. $-3 < 3x - 4 < 6$

53. $-2 \le 5x + 2 \le 3$

54. $0 \le 7x - 1 \le 7$

55. $-5 < 4x + 3 \le 8$

56. $-2 < -x \le 3$

57. $-1 \le -x < 4$

58. $-4 \le 2 - x < 3$

59. $-3 < 4 - x \le 5$

60. $1 < 3 - 4x < 6$

61. $0 \le 1 - 3x < 7$

62. $-4 \le 3 - 2x \le 0$

Directions Write an inequality to represent the following statements. See example 2–9 C.	☞ **Quick Check**
	Example To complete an order for cement, a company will need at least 3 trucks.
	Solution The words *at least* 3 trucks means that the company will need *3 or more* trucks. If *x* is the number of trucks needed, then
	$$x \ge 3$$

Directions   Write an inequality to represent the following statements. See example 2–9 C.

63. Mark's score must be at least 72 on the final exam to pass the course.

64. The temperature today will be less than 38.

65. An automobile parts company needs to order at least 8 new lift trucks.

66. An accounting company will hire at least 2 new employees, but not more than 7.

67. The selling price (P) must be at least twice the cost (C).

Directions	🏴 *Quick Check*
Write an inequality using the given information and solve. See example 2–9 C.	**Example** If 4 is subtracted from three times a number, the result is greater than 2 more than twice the number. Find all the numbers that satisfy this condition.

Solution Let x represent the number. Then

$3x - 4$ is "4 subtracted from three times the number,"
$2x + 2$ is "2 more than twice the number."

Since the two expressions are related by "is greater than," the inequality is $3x - 4 > 2x + 2$

$$3x - 4 > 2x + 2$$
$$3x - 4 - 2x > 2x + 2 - 2x \qquad \text{Subtract } 2x \text{ from each side}$$
$$x - 4 > 2 \qquad \text{Combine on each side}$$
$$x - 4 + 4 > 2 + 4 \qquad \text{Add 4 to each side}$$
$$x > 6 \qquad \text{Combine on each side}$$

The number is any real number that is greater than 6.

Directions   Write an inequality using the given information and solve. See example 2–9 C.

68. When 7 is subtracted from two times a number, the result is greater than or equal to 9. Find all numbers that satisfy this condition.

69. Five times a number minus 11 is less than 19. Find all numbers that satisfy this condition.

70. The product of 6 times a number added to 2 is greater than or equal to 1 subtracted from five times the number. What are the numbers that satisfy this condition?

71. Twice a number increased by 7 is no more than three times the number decreased by 5. Find the numbers that satisfy this condition.

72. The zoning bylaws of Carlisle, Massachusetts, require, among other things, that building lots have a ratio of area, A, to perimeter, P, conforming to the relation $\dfrac{16A}{P^2} \geq 0.4$. If the perimeter of a building lot is 20,000 feet, what are the values for the area for the lot that will satisfy the code?

73. If one third of a number is added to 23, the result is greater than 30. Find all numbers that satisfy this condition.

74. Two times a number plus 4 is greater than 6 but less than 14. Find all numbers that satisfy these conditions.

75. Eugenia has scores of 7, 6, and 8 on three quizzes. What must she score on the fourth quiz to have an average of 7 or higher?

76. Sam has scores of 72, 67, and 81 on three tests. If an average of 70 is required to pass the course, what is the minimum score he must have on the fourth test to pass?

77. To make the honor society at a college, a student must have at least a 3.2 grade point average. After five semesters, Angelina has grade point averages of 3.1, 3.2, 2.9, 2.7, and 4.0. What must she at least attain in the sixth semester to be accepted into the honor society? (*Note:* The grade point average cannot exceed 4.0.)

78. Crystal has grades of 76, 82, 95, and 92 on four computer science tests. What grade must she attain on the fifth test to have an average test score of at least 88?

79. Luis wanted to earn an average of at least $450 per month during his first semester at school. If he earned $500 in September, $350 in October, and $390 in November, what must he earn in December to achieve his goal?

80. Two sides of a triangle are 10 ft and 12 ft long, respectively. If the perimeter must be at least 31 ft, what are the possible values for the length of the third side?

81. The perimeter (the sum of the sides) of a triangle is more than 52 cm. If two sides of the triangle are 18 cm and 16 cm, respectively, what are the possible values for the length of the third side?

82. The perimeter of a rectangle must be less than 100 feet. If the length is known to be 30 feet, find all numbers that the width could be. (*Note:* The width of a rectangle must be a positive number.)

83. The perimeter of a square must be greater than 20 inches but less than 108 inches. Find all values of a side that satisfy these conditions.

84. A rectangular garden plot must have a perimeter no greater than 240 meters. If the width must be 30 meters, what are the possible lengths of the garden?

85. The area of a triangle is given by $A = \frac{1}{2}bh$, where b is the length of the base and h is the altitude. If a triangle has a base of 8 inches, what altitudes can the triangle have if the area is to be between 32 and 48 square inches?

86. The area of a parallelogram is given by $A = bh$, where b is the length of a base and h is the altitude to that base. If you wish to construct a parallelogram having an altitude of 8 inches, what are the possible values of the base if the area is to be at most 70 square inches and at least 48 square inches?

87. In a motor control circuit the voltage across a motor has to be between 12 and 21 volts. If current, I, is 1.5 amperes and voltage, E, is given by $E = IR$, where R is shunt resistance, what should the value of the shunt resistance be? (*Note:* the answer will be in ohms.)

88. In exercise 87 what would the shunt resistance be if the voltage has to be between 8 and 18 volts?

Explore to Learn

1. As stated in this section, a linear inequality is solved the same way a linear equality (equation) is solved with one major exception. What is this exception? Give a numeric example of why this exception is true.

2. Determine whether each number is a solution of the inequality $|x| < 4$. Then graph the solution to $|x| < 4$ on a number line.
   a. 4
   b. 3.25
   c. 0
   d. −3
   e. 1
   f. −2.8
   g. −5.2
   h. −1.6

Review Exercises

Directions   Perform the indicated operations. See section 1–8.

1. -4^2
2. $(-4)^2$
3. -2^4
4. $(-2)^4$

Directions   Write an algebraic expression for each of the following. See section 2–1.

5. x raised to the fifth power

6. A number cubed

7. A number squared

8. The product of x and y

Chapter 2 Lead-in Problem

On a road map, 3 inches represent a distance of 45 miles. If the distance between Detroit and Sault Ste. Marie is 23 inches on the map, how far is it from Detroit to Sault Ste. Marie?

Solution   Let x represent the distance from Detroit to Sault Ste. Marie. Then, since 3 inches represents 45 miles on the map, we use the relationship

3 in. is to 45 mi as 23 in. is to x mi

which we write as the proportion

$$\frac{3}{45} = \frac{23}{x}$$

$3 \cdot x = 45 \cdot 23$ Property of proportions

$x = \dfrac{45 \cdot 23}{3}$ Divide each side by 3

$x = 15 \cdot 23$ Reduce by 3

$x = 345$ Multiply on the right side

The distance from Detroit to Sault Ste. Marie is 345 miles.

Chapter 2 Summary

• Glossary

algebraic expression (page 96)   any meaningful collection of variables, constants, grouping symbols, and signs of operations.

binomial (page 97)   a polynomial of two terms.

conditional equation (page 130)   an equation that is true for some values of the variable and false for other values of the variable.

constant (page 95)   a symbol that does not change its value.

contradiction (page 130)   an equation that has no solution.

degree of a polynomial (page 97)   the highest degree of any of the nonzero terms of the polynomial.

degree of a term (page 97)   the sum of the exponents on the variables of the term.

empty set or null set (page 130)   the set that contains no elements.

equation (page 112)   a statement of the equality of two expressions.

equivalent equations (page 113)   equations whose solution set is the same.

identity (page 130)   an equation that is true for every permissible replacement value of the variable.

inequality (page 161)   a mathematical statement containing one or more inequality symbols.

like terms or similar terms (page 104)   terms whose variable factors are the same.

literal equation (page 137)   an equation that contains two or more variables.

mathematical statement (page 112)   a sentence that can be labeled true or false.

monomial (page 97)   a polynomial of one term.

numerical coefficient (page 96)   the numerical factor of a term.

polynomial (page 97)   an algebraic expression with real number coefficients, variables raised to only natural number powers, and operations performed by the variables limited to addition, subtraction, and multiplication.

solution or root (page 112)   a replacement value for the variable that forms a true statement.

solution set (page 112)   the set of all replacement values for the variable that cause the equation to be a true statement.

trinomial (page 97)   a polynomial of three terms.

unit price (page 151)   the ratio of the price to the number of units.

• New Symbols and Notation

a to b, $\dfrac{a}{b}$, or $a : b$ the ratio of a to b

$\varnothing$ or $\{\ \}$ the empty set or the null set

hollow circle denotes that the end point is not part of the graph

solid circle denotes that the end point is part of the graph

• Properties and Definitions

Distributive property (page 103)
For every real number a, b, and c,

$$a(b + c) = ab + ac \text{ and}$$
$$a(b - c) = ab - ac$$

Linear Equation (page 112)
An equation is linear if it can be written in the form

$$ax + b = 0 \qquad a \neq 0$$

where a and b are real numbers.

Addition and subtraction property of equality
(page 113)
For any algebraic expressions a, b, and c,

if $a = b$, then $a + c = b + c$ and
$$a - c = b - c$$

Symmetric property of equality (page 114)

If $a = b$, then $b = a$

The multiplication and division property of equality
(page 122)
For any algebraic expressions a, b, and c ($c \neq 0$)

if $a = b$, then $a \cdot c = b \cdot c$ and
$$a \div c = b \div c$$

Multiplicative inverse property (page 123)
For every real number a, $a \neq 0$,

$$a \cdot \frac{1}{a} = 1$$

Ratio (page 150)
A ratio is the comparison of two numbers (or quantities) by division.

Definition of a proportion (page 152)
A proportion is a statement of equality of two ratios.

Property of proportions (page 153)

$$\text{If } \frac{a}{b} = \frac{c}{d}, \text{ then } ad = bc \ (b, d \neq 0)$$

Also if $ad = bc$, then $\dfrac{a}{b} = \dfrac{c}{d}$

Addition and subtraction property of inequalities
(page 163)
For all real numbers a, b, and c, if $a < b$, then

$$a + c < b + c \text{ and}$$
$$a - c < b - c$$

Multiplication and division property of inequalities
(page 164)
For all real numbers a, b, and c, if $a < b$, and

1. If $c > 0$ (c represents a positive number), then

$$a \cdot c < b \cdot c \text{ and}$$
$$\frac{a}{c} < \frac{b}{c}$$

2. If $c < 0$ (c represents a negative number), then

$$a \cdot c > b \cdot c \text{ and}$$
$$\frac{a}{c} > \frac{b}{c}$$

• Procedures
[2–1]

Changing a verbal phrase into an algebraic expression (page 98)

1. Read the phrase carefully, determining useful prior knowledge. Prior knowledge is knowing that increased means add, diminished means subtract, product means multiply, and so on.

2. If there is no variable used in the phrase, choose one to represent the unknown value.

3. Write an algebraic expression using the variable to express the conditions stated in the phrase.

[2–2]

Combining like terms (page 104)

1. Identify the like terms.

2. If necessary, use the commutative and associative properties to group together the like terms.

3. Combine the numerical coefficients of the like terms and multiply that by the variable factor.

4. Remember that y is the same as $1 \cdot y$ and $-y$ is the same as $-1 \cdot y$.

Removing grouping symbols (page 105)

1. If a grouping symbol is preceded by no symbol or by a "+" sign, the grouping symbol can be dropped and the enclosed terms remain unchanged.

2. If a grouping symbol is preceded by a "−" sign, when the grouping symbol is dropped, we change the sign of *each* enclosed term.

[2–3]

To check an answer of an equation (page 112)

1. Substitute: Replace the variable in the original equation with the answer.

2. Order of operations: Perform the indicated operations.

3. True statement: If step 2 produces a true statement, the answer is correct.

Solving Word Problems (page 117)

Read Analyze-Visualize	1. Read the problem carefully, usually several times. Determine what information is given, and what information you are asked to find. If possible, draw a picture, make a diagram, or construct a table to help visualize the information.
Choose a variable	2. Choose a variable to represent one of the unknowns and write a sentence stating exactly what that variable represents. Use this variable to write algebraic expressions that represent any other unknowns in the problem. Label any picture, diagram, or chart using the chosen variable.
Write an equation	3. Use the unknowns from step 2 to translate the word problem into an equation. There may be some underlying relationship or formula that you need to know to write the equation. If it is a geometry problem, the formulas inside the front cover may be useful. Other commonly used formulas are: interest equals principal times rate times time ($I = Prt$) or distance equals rate times time ($d = rt$). If there is no applicable formula, then the words in the problem give the information necessary to write the equation.
Solve the equation	4. Solve the equation.
Answer the question(s)	5. Answer the question or questions in the problem. If there is more than one unknown in the problem, substitute the solution of the equation into the algebraic expressions that represent the other unknowns to determine the other unknown values in the problem.
Check your results	6. Check your results in the original statement of the problem. Make sure your answer makes sense. Remember that in a geometry problem the dimensions of the figure cannot be negative. If you were determining how fast a car was traveling, 1,000 miles per hour would not make sense.

[2–5]

Solving a linear equation (page 128)

1. Simplify each side of the equation. Perform all indicated addition, subtraction, multiplication, and division. Remove all grouping symbols.

2. Use the addition and subtraction property of equality to form an equivalent equation where all the terms involving the unknown are on one side of the equation.

3. Use the addition and subtraction property of equality to form an equivalent equation where all the terms not involving the unknown are on the other side of the equation.

4. Use the multiplication and division property of equality to form an equivalent equation where the coefficient of the unknown is 1. That is, $x = n$.

5. To check the answer, we substitute the answer in place of the unknown in the original equation. If we get a true statement, we say that the answer "satisfies" the equation.

[2–8]

To solve a proportion (page 153)

1. Use the property of proportions to form the cross products.

2. Solve for the unknown.

To solve a problem using a proportion (page 154)

1. Choose a variable to represent the unknown.

2. Make a proportion using the given information to set up a ratio on the left side of the equation. Set up the ratio on the right side of the equation using the unknown and the other given quantity. The ratio on the right side of the equation must be set up with the quantities occupying the same

respective positions as in the ratio on the left side of the equation.

For example: $\dfrac{\text{miles}}{\text{gallon}} = \dfrac{\text{miles}}{\text{gallon}}$ or $\dfrac{\text{gallon}}{\text{miles}} = \dfrac{\text{gallon}}{\text{miles}}$

would be the correct ways to set up a proportion involving miles and gallons.

The units used in the problem are useful in correctly setting up the proportion, but are dropped once the proportion is set up.

3. Solve the proportion.

4. Answer the question(s) asked.

[2–9]

Solving a linear inequality (page 165)

1. Simplify on each side, where necessary, by performing the indicated operations.

2. Add, or subtract, to get all terms containing the unknown on one side of the inequality.

3. Add, or subtract, to get all terms not containing the unknown on the other side of the inequality.

3. Add, or subtract, to get all terms not containing the unknown on the other side of the inequality.

4. Multiply, or divide, to obtain a coefficient of 1 for the unknown. Remember, when multiplying or dividing by a negative number, always change the direction (order) of the inequality symbol.

☀ Chapter 2 Error Analysis

1. Degree of a polynomial
   Example: $2x - 3x^2 + x^3 - 1$ has degree 6.
   Correct answer: $2x - 3x^2 + x^3 - 1$ has degree 3.
   What error was made? (*see page 97*)

2. Terms in an algebraic expression
   Example: $x^2 + \dfrac{2x - 3}{5}$ has 3 terms.

   Correct answer: $x^2 + \dfrac{2x - 3}{5}$ has 2 terms.

   What error was made? (*see page 96*)

3. Applying the distributive property
   Example: $5(4 + b) = 20b$
   Correct answer: $5(4 + b) = 20 + 5b$
   What error was made? (*see page 104*)

4. Combining like terms
   Example: $3a^2 + 4a = 7a^3$
   Correct answer: $3a^2 + 4a = 3a^2 + 4a$
   What error was made? (*see page 105*)

Learning System
"Error Analysis" provides a group of problems where a common error has been made. A page reference is provided so that you can refer to relevant examples and notes for assistance in finding and correcting the error.

5. Combining polynomials
Example: $(3x^2 - 2x + 1) - (x^2 - x + 2)$
$$= 3x^2 - 2x + 1 - x^2 - x + 2$$
$$= 2x^2 - 3x + 3$$
Correct answer: $(3x^2 - 2x + 1) - (x^2 - x + 2)$
$$= 2x^2 - x - 1$$
What error was made? (*see page 106*)

6. Reciprocal of a number
Example: The reciprocal of 0 is $\dfrac{1}{0}$.
Correct answer: 0 has no reciprocal.
What error was made? (*see page 123*)

7. Graphing linear inequalities
Example: The graph of $x \le 3$ is

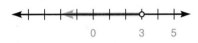

Correct answer: The graph of $x \le 3$ is

What error was made? (*see page 162*)

8. Multiplying sides of an inequality
Example: If $3 < 4$, then $3 \cdot -2 < 4 \cdot -2$.
Correct answer: If $3 < 4$, then $3 \cdot -2 > 4 \cdot -2$.
What error was made? (*see page 164*)

9. Multiplication of negative numbers
Example: $(-5)(7) = 35$
Correct answer: $(-5)(7) = -35$
What error was made? (*see page 62*)

10. Division using zero
Example: $\dfrac{7}{0} = 0$
Correct answer: $\dfrac{7}{0}$ is undefined.
What error was made? (*see page 65*)

11. Solving proportion problems
Example: On a map, 1 inch represents 10 miles. How many inches represent 24 miles? Let $x =$ inches represented by 24 miles.
$$\frac{1}{10} = \frac{24}{x}$$
$$x = 240 \text{ inches}$$
Correct answer: 2.4 inches
What error was made? (*see page 154*)

12. Solving inequalities
Example: $\dfrac{x}{-3} \le 6$, so $x \le -18$
Correct answer: $x \ge -18$.
What error was made? (*see page 165*)

13. Solving problems with percent
Example: 5% of a number is 20; what is the number?
$0.5x = 20$, so $x = 40$
Correct answer: $x = 400$
What error was made? (*see page 28*)

Chapter 2 Critical Thinking

1. If you add any three consecutive odd integers, the sum will be a multiple of 3. Why is this true?

2. If you add any three consecutive even integers, is the sum a multiple of 3? Justify your answer.

3. The distributive property: $a(b + c) = ab + ac$ is assumed to be true for any real numbers a, b, c. It cannot be proven true. Recall than the area of a rectangle is the product of its length and width.

Thus the area of figure a is $3 \cdot 7 = 21$. Show how figure b could be used to justify the distributive property.

4. A parent left $17,000 to her three children. The will stipulated that half the money be left to the first child, a third to the second child, and a ninth to the third child. Because the arithmetic wasn't working, the administrator of the estate consulted a wise accountant. The accountant loaned the administrator of the estate $1,000. This solved the problem. Explain.

5. The Gunning Fog index for prose is 0.4 times the sum of "average sentence length" and "percentage of words with three or more syllables." It approximates the grade level of education needed to understand written material. Suppose this index is to be computed for a newspaper article. Suppose the following variables are defined as indicated: L_s = average length of sentences (number of words in the sentence) and W_3 = percentage of words with three or more syllables. Create a formula for the Gunning Fog index. Apply this formula to this problem.

6. It takes 365 days, 5 hours, 48 minutes, and 46 seconds for the earth to go around the sun once. If there was no adjustment to the calendar by leap years the seasons would drift, what is considered

summer would eventually become the winter, and vice versa. The adjustment made in a leap year, by adding a day every 4 years, does not fix this problem exactly. (a) What is the error remaining after such an adjustment? The years at the end of a century are not leap years unless the number is divisible by 400. For example, 1900 was not a leap year, but 2000 will be. (b) How much drift will have occurred since the year 1600 at the end of the year 2000?

7. Count the number of males and females in two or three of the classes in your school. Find out how many total students there are in the school, and use your data to predict how many males and females are attending the school. Ask the registrar if your estimates are correct. What factors should govern the classes you choose for the counting to obtain the most accurate results? Do you think two or three classes is a large enough sample to obtain accurate results?

8. How many people are awake while you are eating your lunch?

Chapter 2 Review

[2–1]

Directions   Specify the number of terms in each expression.

1. $4x^2 + 3x + 2$
2. $5a^2b$
3. $7xy + 5$
4. $ab + cd + xy$

Directions   Determine which of the following algebraic expressions are polynomials. If it is a polynomial, state the degree. If it is not a polynomial, state why not.

5. $2x^2 - 3x + 4$
6. $x^2y - xy$
7. $4a^2$

8. $\dfrac{a + b}{c}$

Directions   Write an algebraic expression for each of the following.

9. 5 times x

10. 7 less than y

11. 4 more than z

12. 2 times a number, plus 6

13. 7 less than twice y

14. The sum of one third a number and 10

15. 25% of the sum of a number and 6,000

[2–2]

Directions Write an algebraic expression for the following word statements.

16. If 20 pounds of coffee cost c dollars, how much does the coffee cost per pound?

17. If the height of a doorway is f feet and t inches, how high is the doorway in inches?

Directions Remove all grouping symbols and combine like terms.

18. $(4a^2 - b^2) - (3a^2 + 2b^2) - (7a^2 - 3b^2)$

19. $(a^2 - 3a + 4) - (2a^2 - 4a - 7)$

20. $(4ab + 7b^2c) - (15ab - 11bc)$

21. $(x - 2y + 7) - (x + 4y + 6)$

22. $(4ab - 2ac) - (6bc - 5ac) + (ab + 2bc)$

23. $3(2a - 1) + 4(a + 2)$

24. $5(x + 1) + 2(2x - 3)$

25. $4(2x + 3) - 3(3x - 1)$

26. $4(a - 3) - 2(a - 3)$

27. $-3(x - 1) + 5(x + 1)$

[2–3]

Directions Determine if the given value is a solution of the equation.

28. $x + 7 = 11; x = 4$

29. $2x + 1 = 9; x = 2$

[2–5]

Directions Solve the following problems by setting up an equation and solving it.

30. Four more than $\frac{1}{3}$ of a number is 20. What is the number?

31. Seven percent of a number is 15.4. Find the number.

[2–3, 2–4, 2–5]

Directions Find the solution set.

32. $x + 5 = 12$

33. $x - 4 = 17$

34. $a + 7 = -4$

35. $b - 3 = -9$

36. $5z + 3z - 7z + 3 = 7$

37. $2(3x - 4) - 5x = 11$

38. $3(2y + 3) = 7 + 5y$

39. $3(x - 1) - 2(x + 1) = 4$

40. $3x = 9$

41. $4x = 12$

42. $-2x = 14$

43. $-3x = 21$

44. $\frac{x}{3} = 4$

45. $\frac{x}{2} = 7$

46. $\frac{3x}{5} = 9$

47. $\frac{2x}{7} = 6$

48. $\frac{1}{3}x - 1 = \frac{3}{4}$

49. $\frac{1}{3}x + 1 = \frac{1}{6}x - 2$

50. $\frac{3}{4}x + 4 = \frac{5}{8}$

51. $\frac{3}{5}x + \frac{1}{2} = \frac{7}{10}x - 3$

52. $3x = 0$

53. $-x = -4$

54. $3.7a = 22.2$

55. $32.8 = -4.1x$

56. $2x + 5 = 11 - 6 + 2x$

57. $3b - 8 = 6$

58. $y + (2y - 1) = 6$

59. $x + 3x = 5 + 7$

60. $3(2a - 1) = 4a - 2$

61. $5(x + 3) = 5x - 7$

62. $(3x - 2) - (4x - 1) = 3x$

63. $2a + 5a - 4 = 3(1 - 2a)$

64. $8 - 3x + 7 = 5(x + 7)$

65. $2y - 3(y + 1) = 11$

66. $7x - 4(2x + 3) = 12$

67. $8x - 14 = 14 - 8x$

68. $5b + 4 = 4 - 2b$

69. $-3(2x + 1) = 4x - 5$

70. $3(c + 2) - 2(c + 1) = 5c + 11$

71. $4x - 2(1 - 3x) = 8x + 2$

[2–6]

Directions   Solve for the specified variable.

72. $F = ma$, for a (Physics)

73. $E = IR$, for I (Electronics)

74. $k = PV$, for P (Chemistry)

75. $V = k + g + t$, for g (Physics)

76. $A = \frac{1}{2}h(b_1 + b_2)$, for b_2 (Geometry)

77. $5x - y = 2x + 3y$, for x (Literal)

[2–7]

79. If a number is divided by 9 and that result is then increased by 7, the answer is 11. Find the number.

80. The difference between one third of a number and one fifth of a number is 6. Find the number.

81. The sum of three consecutive odd integers is 297. Find the integers.

82. Part of $12,000 was invested at 4% and the rest at 5%. If the total return on the investments was $545, how much was invested at each rate?

83. John invested part of $20,000 at 8% and the rest at 7%. If his income from the 8% investment was $250 more than that from the 7% investment, how much was invested at each rate?

84. Anne made two investments totaling $25,000. On one investment she made a 12% profit but on the other she took a 19% loss. If her net loss was $1,030, how much was in each investment?

78. Match the correct sentence with the equation $2x - 5 = 4x$.

    a. Five diminished by two times a number is four times the number.

    b. Twice a number diminished by five is four times the number.

    c. Twice the difference of a number and five is four times the number.

85. A certain fertilizer contains 12% phosphorous. A second fertilizer contains 40% phosphorous. To the nearest pound, how many pounds of each must be mixed to obtain 250 pounds of a 30% phosphorous mixture?

[2–8]

Directions   Find the indicated ratios in two different forms reduced to lowest terms.

86. 15 meters to 35 meters

87. 36 pounds to 16 pounds

88. 12 inches to $2\frac{1}{2}$ feet

89. 450 miles to 15 gallons

90. In business, the current ratio compares current assets to current liabilities and represents the measure of the firm's ability to pay off the liabilities over a time period. What is the current ratio, reduced to lowest terms, if the firm's total current assets are $4,386 and total current liabilities are $1,762?

91. The May company has 42 sales representatives who are meeting their sales quota. If another 18 sales representatives have fallen short of their quota, what is the ratio of success to failure?

Directions   Find the unit price rounded to three decimal places and determine which one has the lower unit price.

92. Cheese puffs

    A: 12 ounces for $1.89
    B: 10 ounces for $1.59

93. Grape jelly

    A: 1 pound 4 ounces for $2.29
    B: 12 ounces for $1.29

Directions   Find the value of the unknown that makes the proportion true.

94. $\dfrac{8}{x} = \dfrac{9}{36}$

95. $\dfrac{5.4}{3.6} = \dfrac{a}{2.4}$

96. $\dfrac{y}{18} = \dfrac{15}{25}$

97. $\dfrac{\frac{5}{6}}{\frac{1}{2}} = \dfrac{\frac{2}{3}}{p}$

98. If a blueprint is drawn to the scale $\frac{1}{8}$ inch = 1 foot, what is the size of the corresponding part of a final product if the blueprint measurement is $4\frac{3}{8}$ inches?

99. An automobile has a 16-quart cooling system. If the ratio of antifreeze to water is 3 to 1, how much of each does the system have? (*Hint:* Let x be the amount of antifreeze. Then $16 - x$ is the amount of water.)

[2–9]

Directions   For problems 100–112, find the solution and graph the solution.

100. $3x > 12$

101. $5x \le 15$

102. $-2x < 14$

103. $-4x > 16$

104. $2x + 1 < 5$

105. $7x - 4 > 11$

106. $3x + 7 < 5x - 2$

107. $9x + 13 \ge 4x + 7$

108. $6(2x - 1) \le 3x - 4$

109. $-4 < 5x + 7 < 10$

110. $0 \le 1 - 5x < 6$

111. $5 < 4x + 3 < 12$

112. $-8 \le 3x + 5 \le 4$

113. Pam received scores of 52, 70, 80, 65, and 52 on five of six chemistry lab tests. To have a passing grade she needs an average of at least 65. What must she score on the remaining test to have a passing grade?

114. A company's profits are given by the following equation: $21x - 420 = P$, where P represents profit in dollars, x represents the number of items

sold, and 420 is a fixed overhead cost. How many items must be sold if their profits are at least $378?

115. Four times a number minus 7 is greater than 17 but less than 25. Find all numbers that satisfy these conditions.

Chapter 2 Test

Directions Determine if each of the following algebraic expressions is a polynomial. If it is a polynomial, what name best describes it and what is its degree? If it is not a polynomial, state why not.

1. $2x^2 + 3x - 4$

2. $a^3b^2 - a^2b^2 + 2a^4b^3$

3. $x^2 - \dfrac{2}{x}$

4. Write an algebraic expression for: two more than twice a number.

Directions Remove all grouping symbols and combine like terms.

5. $(2x^2 - x) + (x^2 - 3x) - (x^2 - 4x)$

6. $2(3a - 2) - 3(a + 1)$

Directions Find the solution set.

7. $5x - 3 = 12$

8. $2(3x - 1) = 2x + 4x - 2$

9. $\dfrac{1}{2}x - 2 = \dfrac{1}{3}x + 2$

10. $\dfrac{z}{2} = \dfrac{14}{5}$

11. $7.6x + 18.4 = 3.2x + 66.8$

12. $9y - 7 = -7$

13. $6(y + 3) - 2y = 2(2y + 1)$

14. $7x + 1 - 2x = 4(3 - 2x) + 3$

Directions Solve for the specified variable.

15. $A = P + Pr$ for r business

16. $3x - 2 = 2y - 7$ for y literal

Directions Find the solution for the following inequalities.

17. $(2x + 5) - (x - 4) \geq 8$

18. $-4x < 24$

19. $-7 \leq 3x + 2 \leq 20$

20. Find the unit price rounded to three decimal places and determine which brand of potato chips has the lower unit price.

    A: 14 ounces for $2.99
    B: 18 ounces for $3.89

Directions Solve the following problems.

21. If 6 pounds of nails cost $4.92, at this same rate, how much would 25 pounds of nails cost?

22. A chemist wishes to make 100 liters of a 4% acid solution by mixing a 10% acid solution with a 2% acid solution. How many liters of each solution is necessary?

23. If a number is increased by 9 and that result is divided by 3, the answer is 7. Find the number.

24. Karleen has $4,520 to invest. Stock A is selling for $57 per share and stock B is selling for $28 per share. If she wishes to purchase twice as many shares of stock B as stock A, how many shares of each will she purchase?

25. Zachary scored 76, 83, 72, and 86 on four tests in algebra. What must he get on the fifth test to get a B or better if a B or better requires an average of at least 80?

26. Dwala made two investments totaling $17,000. On one investment she made a 12% profit but on the other she took a 19% loss. If her net loss was $1,215, how much was in each investment?

Chapter 2 Cumulative Test

Directions   Perform the indicated operations, if possible, and simplify.

1. $(-10) - (-14)$

2. $\dfrac{8}{0}$

3. -6^2

4. $6 + 4(10 - 2)$

5. $14 + 2 \cdot 15 \div 6 - 3 + 4$

6. $(3x^2y - 2xy^2) - (5xy^2 - x^2y)$

7. $6(2a - 3) + 5(a - 4)$

Directions   Evaluate the following if $a = -2$, $b = -3$, $c = 4$, and $d = 5$.

8. $(a + 2d)^2$

9. $(a - 4c)(b - 2d)$

10. There are 6 rows of desks in a classroom. If each row contains 7 desks, how many desks are in the classroom?

11. Write an algebraic expression for x decreased by y

Directions   Find the solution set for 12–16 and the solution for 17–21.

12. $10x - 7 = 4x + 3$

13. $5x + 6 = 6$

14. $\dfrac{1}{3}x + 4 = \dfrac{5}{6}$

15. $3(2x - 1) + 2(5x - 3) = 8$

16. $16 - 2(4x - 1) = 3x - 12$

17. $-2x \geq 12$

18. $5x + 3x < 6x - 14$

19. $3x + (x - 1) > 7 - x$

20. $-1 < 2x + 3 < 11$

21. $-16 \leq 8 - 4x \leq 12$

Directions   Solve for the specified variable.

22. $P = a + b + c$, for b geometry

23. $x = a(y + z)$, for y literal

Directions   Find the value of the unknown that makes the proportion true.

24. $\dfrac{4.5}{2.7} = \dfrac{a}{2.4}$

Directions   Solve the following word problems.

25. Phil has $10,000, part of which he invests at 6% and the rest at 5%. If his total income from the two investments was $560, how much did he invest at each rate?

26. The sum of three consecutive even integers is 48. Find the three integers.

27. Twice a number decreased by 2 is at most 10. Find all numbers that satisfy this condition.

28. How many liters of a 5% salt solution must be mixed with 20 liters of a 10% salt solution to obtain a 7% salt solution?

29. An automobile uses 8.6 gallons of gasoline to travel 262.3 miles. How many gallons are needed to travel 1830 miles?

30. Find the unit price rounded to three decimal places and determine which brand of coffee has the lower unit price.

A: 13 ounces for $3.89
B: 16 ounces for $4.59

3

Polynomials and Exponents

There are approximately 3.3×10^{22} molecules of water in one cubic centimeter (cm^3) of water. There are one million cubic centimeters in one cubic meter (m^3). It is estimated that the five Great Lakes contain about 22.7 trillion cubic meters of water. Use this information to estimate how many molecules of water there are in the Great Lakes.

3–1   Exponents—I

Exponential Form

In chapter 1, we discussed the idea of exponents[1] as related to real numbers. Since variables are symbols for real numbers, let us now apply the idea of exponents to variables. The expression x^4 is called the **exponential form** of the product

$$x \cdot x \cdot x \cdot x$$

We call x the **base** and 4 the **exponent**.

Exponent
↓
Exponential form ⟶ $x^4 = \underbrace{x \cdot x \cdot x \cdot x}_{\text{4 factors of } x}$ ⟵ Expanded form
↑
Base

[1] The notation used to represent an exponent, a^n, n a positive integer, was introduced by René Descartes in 1637.

Definition of Exponents

$$a^n = \underbrace{a \cdot a \cdot a \cdots a}_{n \text{ factors of } a}, \text{ where } n \text{ is a positive integer.}$$

Concept

The exponent tells us how many times the base is used as a factor in an indicated product.

Note An exponent acts only on the symbol immediately to its left. That is, in ab^4 the exponent 4 applies only to b and expands to $a \cdot b \cdot b \cdot b \cdot b$, whereas $(ab)^4$ means the exponent applies to both a and b and expands to $ab \cdot ab \cdot ab \cdot ab$.

● **Example 3–1 A**

Determining form, base, and exponent

Change the following from exponential form to expanded form and determine the base and the exponent.

Variables

	Exponential form	Expanded form	Base	Exponent
1.	a^4	$= a \cdot a \cdot a \cdot a$	a	4
2.	$(x - y)^3$	$= (x - y)(x - y)(x - y)$	$(x - y)$	3

Real Numbers

			Product	Base	Exponent
3.	5^3	$= 5 \cdot 5 \cdot 5$	$= 125$	5	3
4.	$(-3)^4$	$= (-3)(-3)(-3)(-3)$	$= 81$	-3	4
5.	-3^4	$= -(3 \cdot 3 \cdot 3 \cdot 3)$	$= -81$	3	4

Note Examples 4 and 5 review the ideas from section 1–8 on exponents related to real numbers. Recall that $(-3)^4 = 81$, whereas $-3^4 = -81$.

☞ *Quick check* Write $y \cdot y \cdot y \cdot y$ in exponential form. Write c^5 in expanded form. ●

Multiplication of Like Bases

Consider the indicated product of $x^2 \cdot x^3$. If we rewrite x^2 and x^3 by using the definition of exponents, we have

$$x^2 \cdot x^3 = \overbrace{x \cdot x}^{x^2} \cdot \overbrace{x \cdot x \cdot x}^{x^3}$$

and again using the definition of exponents, this becomes

$$x^2 \cdot x^3 = \overbrace{x \cdot x \cdot x \cdot x \cdot x}^{5 \text{ factors}} = x^5$$

This leads us to the observation that

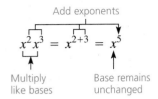

Add exponents

$$x^2 x^3 = x^{2+3} = x^5$$

Multiply
like bases

Base remains
unchanged

Thus we have the following **product property of exponents**.

Product Property of Exponents

$$a^m \cdot a^n = a^{m+n}$$

Concept
When multiplying expressions with like **bases**, add the exponents to get the exponent of the common base.

Note The base stays the same throughout the process. It is by adding the exponents that the multiplication is carried out.

● **Example 3–1 B**

Using the product property

Find the product.

1. $x^3 \cdot x^5 = x^{3+5} = x^8$

2. $3^2 \cdot 3^4 = 3^{2+4} = 3^6 = 729$

Note A common error in multiplying $3^2 \cdot 3^4$ is to multiply the bases $3 \cdot 3 = 9$ and add the exponents, getting the incorrect answer of 9^6. The correct way is to say $3^2 \cdot 3^4 = 3^6$, not 9^6.

3. $y^2 \cdot y^3 \cdot y^4 = y^{2+3+4} = y^9$

4. $a^2 \cdot a \cdot a^3 = a^{2+1+3} = a^6$

Note The variable a means the same as a^1. Likewise, 3 means the same as 3^1. If there is no exponent written with a numeral or a variable, the exponent is understood to be 1.

5. $(a + b)^3 (a + b)^4 = (a + b)^{3+4} = (a + b)^7$

6. $(-2)^3 (-2)^2 = (-2)^{3+2} = (-2)^5 = -32$

☛ *Quick check* Find the product. $x^4 \cdot x^5$

Group of Factors to a Power Property of Exponents

Several additional properties of exponents can be derived using the definition of exponents and the commutative and associative properties of multiplication. Observe the following:

$$3 \text{ factors of } xy$$
$$(xy)^3 = \overbrace{xy \cdot xy \cdot xy}$$

$$\begin{array}{cc} 3 \text{ factors of} & 3 \text{ factors of} \\ x & y \end{array}$$
$$= \overbrace{x \cdot x \cdot x} \cdot \overbrace{y \cdot y \cdot y}$$
$$= x^3 y^3$$

This leads us to the following property of exponents.

Group of Factors to a Power Property of Exponents

$$(ab)^n = a^n b^n$$

Concept

When a group of factors is raised to a power, raise each of the factors in the group to this power.

● **Example 3–1 C**

Using the group of factors to a power property

Simplify.

1. $(ab)^4 = a^4 b^4$ Both a and b are raised to the 4th power

Groups of factors to a power   Raise each factor to the power

2. $(2ab)^3$ $=$ $2^3 a^3 b^3$ $=$ $8a^3 b^3$

Note In example 2, the number 2 is a factor in the group. Therefore it is also raised to the indicated power. Also, when asked to simplify an expression involving exponents, we will write an equivalent expression with bases and exponents occurring as few times as possible.

3. $(3 \cdot 4)^3 = 3^3 \cdot 4^3 = 27 \cdot 64 = 1{,}728$ $(3 \cdot 4)^3$ also is $(12)^3 = 1{,}728$ ●

Note The expression $(a + b)^3 \neq a^3 + b^3$ because a and b are *terms,* not factors as the property specified. If we consider $(a + b)$ to be a single factor, then by the definition of exponents we have

$$(a + b)^3 = (a + b)(a + b)(a + b)$$

We will see the method of multiplying this later in this chapter.

Power to a Power

Consider the expression $(x^4)^3$. Applying the definition of exponents and the product property of exponents, we have

$$3 \text{ factors of } x^4 \qquad \text{Add the exponents}$$
$$(x^4)^3 = \overbrace{x^4 \cdot x^4 \cdot x^4} = \overbrace{x^{4+4+4}} = x^{12}$$

In chapter 1, we reviewed the idea that multiplication is repeated addition of the same number. Therefore adding the exponent 4 three times is the same as $4 \cdot 3$. Thus

Power to a power   Multiply exponents

$$(x^4)^3 = x^{4 \cdot 3} = x^{12}$$

Therefore we have the following property of exponents.

Power to a Power Property of Exponents

$$(a^m)^n = a^{m \cdot n}$$

Concept

A power to a power is found by multiplying the exponents.

● **Example 3–1 D**

Using the power of a power property

Simplify.

1. $(y^3)^2 = y^{3 \cdot 2} = y^6$

2. $(4^2)^5 = 4^{2 \cdot 5} = 4^{10} = 1{,}048{,}576$

3. $(x^5)^4 = x^{5 \cdot 4} = x^{20}$

☞ *Quick check*   Simplify. $(a^4)^3$

Products of Monomials

To multiply the monomials

$$3x^2 \cdot 5x$$

we apply the commutative and associative properties of multiplication along with the properties of exponents. We then write this expression as a product of the numerical coefficients times the product of the variables. That is,

$$3x^2 \cdot 5x = (3 \cdot 5)(x^2 \cdot x) = 15x^3$$

To find the product of

$$5a \cdot 4b$$

we apply the same properties to get

$$5a \cdot 4b = (5 \cdot 4)(a \cdot b) = 20ab$$

Note It is a good procedure to write the variable factors of any term in alphabetical order. This makes identifying like terms much simpler. For example, $3a^2c^3b$ and $4bc^3a^2$ are like terms, but recognizing that fact would have been easier if they had been written as $3a^2bc^3$ and $4a^2bc^3$.

● **Example 3–1 E**
Multiplying monomials

Perform the indicated multiplication.

1. $4x \cdot 3xy = (4 \cdot 3) \cdot (x \cdot x) \cdot y = 12x^2y$

2. $8a^3 \cdot 4a^3 \cdot 3a = (8 \cdot 4 \cdot 3) \cdot (a^3 \cdot a^3 \cdot a) = 96a^7$

3. $(-2a^2) \cdot (3ab) = (-2 \cdot 3) \cdot (a^2 \cdot a) \cdot b = -6a^3b$

Note The product of a^3 and b can *only* be written as a^3b since a and b are not like bases.

4. $(5x^2y^3z)(4x^3yz^4) = (5 \cdot 4)(x^2x^3)(y^3y)(zz^4) = 20x^5y^4z^5$ ●

Problem Solving

The following problems require us to write algebraic expressions involving the use of exponents.

● **Example 3–1 F**
Problem solving

Write an algebraic expression for each of the following verbal statements.

1. The volume of a cube is found by using the length of the edge, e, as a factor 3 times. Write an expression for the volume of a cube.

   We write e as a factor 3 times as $e \cdot e \cdot e = e^3$. Then the volume, V, of a cube is given by

   $$V = e^3.$$

2. Write an expression for 5 less than the square of a number.

   Let n represent the number, then the square of the number is given as n^2, and since "less than" means to subtract, the expression is given by

   $$n^2 - 5.$$ ●

MASTERY POINTS

Can you
- Change between exponential form and expanded form?
- Use the product property of exponents?
- Raise a group of factors to a power?
- Raise a power to a power?
- Multiply monomials?

Exercise 3–1	☞ *Quick Check*
Directions Write the following expressions in exponential form. See example 3–1 A.	**Example** $y \cdot y \cdot y \cdot y$ **Solution** $= y^4$ *y to the fourth power*

Directions   Write the following expressions in exponential form. See example 3–1 A.

1. *aaaaa*

2. *bbbb*

3. $(-2)(-2)(-2)(-2)$

4. $-(2 \cdot 2 \cdot 2 \cdot 2)$

5. *xxxxxx*

6. $(2a)(2a)(2a)$

7. $(xy)(xy)(xy)(xy)$

8. $(a + b)(a + b)$

9. $(x - y)(x - y)(x - y)$

10. $(2a - b)(2a - b)(2a - b)$

Directions Write in expanded form and perform the multiplication where possible. See example 3–1 A.	☞ **Quick Check** **Example** c^5 **Solution** $= c \cdot c \cdot c \cdot c \cdot c$ *c* written as a factor 5 times

Directions   Write in expanded form and perform the multiplication where possible. See example 3–1 A.

11. x^4

12. y^5

13. $(-2)^3$

14. -2^4

15. 5^3

16. $(5x)^3$

17. $(4y)^4$

18. $(a + b)^3$

19. $(x - y)^2$

20. $(2x + y)^3$

Directions Simplify by using the properties of exponents. See examples 3–1 B, C, D, and E.	☞ **Quick Check** **Examples** $x^4 \cdot x^5$ **Solutions** $= x^{4+5}$ Like bases $= x^9$ Add exponents	$(a^4)^3$ $= a^{4 \cdot 3}$ Power to a power $= a^{12}$ Multiply exponents

Directions   Simplify by using the properties of exponents. See examples 3–1 B, C, D, and E.

21. $x^4 \cdot x^7$

22. $a^5 \cdot a^5$

23. $R^2 \cdot R$

24. $a \cdot a^4$

25. $a^2 \cdot a^3 \cdot a^4$

26. $x^5 \cdot x \cdot x^3$

27. $5^2 \cdot 5^3$

28. $6 \cdot 6^3$

29. $4 \cdot 4^2 \cdot 4^4$

30. $(a + b)^2(a + b)^5$

31. $(x - 2y)^4(x - 2y)^6$

32. $(3a + b)^2(3a + b)^3$

33. $(a - b)^4(a - b)^7$

34. $(ab)^5$

35. $(xy)^4$

36. $(2abc)^3$

37. $(4xyz)^3$

38. $(a^2)^4$

39. $(x^5)^3$

40. $(y^2)^2$

41. $(b^5)^5$

42. $(c^9)^3$

43. $(2xy^2)(3x^3y)$

44. $(4x^2y^3)(5xy^4)$

45. $(a^2b^3)(a^5b^2)$

46. $(x^2y^3)(x^4y^3)$

47. $(6x^3)(5x^2)$

48. $(4a)(3a^4)$

49. $(2a^3b^4c)(6a^4b^3)$

50. $(5xy)(xy)$

51. $(3a^2b)(4a^3b^2)$

52. $(a^3b^4)(5a^2b^5)$

53. $(-2a^2b)(3ab^4)$

54. $(-5x^2y^5)(-2x^2y)$

55. The formula for finding the volume of a cube is $V = e^3$, where V represents volume in some cubic unit of measure and e represents the length of the edge of the cube. Determine the number of cubic units in the figure for each of the following values of e: (a) $e = 5$, (b) $e = 4$, (c) $e = 6$.

Directions Using exponents, write an expression for each of the following verbal statements. See example 3–1 F.

56. The area, *A,* of a square is found by using the length of the side, *s,* as a factor twice. Write an expression for the area of a square.

57. The distance, *s,* a falling object will fall in time, *t,* seconds is found by multiplying $\frac{1}{2}$ times the gravity, *g,* times the square of *t.* Write an expression for the distance the object will fall.

58. The area of a circle is found by multiplying the constant π times the length of the radius, *r,* used as a factor 2 times. Write an expression for the area of a circle.

59. The volume, *V,* of a sphere is found by multiplying $\frac{4}{3}\pi$ times the radius, *r,* used as a factor three times. Write an expression for the volume of a sphere.

60. Theo is *n* years old. His mother says that she is 6 years more than the cube of Theo's age. Write an expression for his mother's age.

61. Christina is *m* years old. Her father is 8 years less than Christina's age used as a factor four times. Write an expression for her father's age.

62. Write an expression for two times the square of *t.*

63. Write an expression for twice the square of *x* less the cube of *y.*

64. A number can be written in the form *a* times 10 used as a factor 8 times, where *a* is a number between 1 and 10. Write an expression for the number in terms of *a.*

65. Write an expression for the quotient of the cube of *p* divided by the square of *q.*

66. Write algebraically "the quantity *a* plus *b* squared" and "*a* plus *b* squared."

67. Write algebraically "the quantity *x* plus *y* cubed" and "*x* plus *y* cubed."

68. Explain why $-5^2 = -25$ and $(-5)^2 = 25$.

⚜ Explore to Learn

One of the properties in this section states that $(a \cdot b)^n = a^n \cdot b^n$. We are stressing that the operation in the parentheses is multiplication.

a. Does a similar property apply for division? That is, is it true that $(a \div b)^n = a^n \div b^n$? Try an example, such as letting $a = 18$, $b = 6$, and $n = 3$.

b. Does a similar property apply for addition? That is, is it true that $(a + b)^n = a^n + b^n$? Try an example, such as letting $a = 2$, $b = 3$, and $n = 2$.

Review Exercises

Directions Perform the indicated addition and subtraction. See section 2–2.

1. $2a + 3a + 4a$

2. $5x + x + 2x$

3. $3ab - 2ab + 5ab$

4. $9xy + 4xy - 6xy$

5. $4a^2 + 3a^2 - 2a + 7a$

6. $6x^2 + 3x - x^2 + 2x$

7. $2x^2y - x^2y + 3xy^2 + 4xy^2$

8. $5ab^2 + 3a^2b - 2ab^2 - a^2b$

3–2   Multiplication of Polynomials

Product of a Monomial and a Polynomial

To multiply a monomial and a polynomial of more than one term, we use the distributive property. For example, to multiply

$$3x^2y(x^2 + 2xy - y^2)$$

we multiply each term in the trinomial by the monomial $3x^2y$. We use arrows to indicate the process.

$$3x^2y(x^2 + 2xy - y^2)$$
$$3x^2y \cdot x^2 + 3x^2y \cdot 2xy - 3x^2y \cdot y^2$$
$$3x^4y + 6x^3y^2 - 3x^2y^3$$

In each indicated product, note that we multiplied like bases by using the properties of exponents. For example, in the first term,

$$3x^2y \cdot x^2 = 3 \cdot (x^2 \cdot x^2) \cdot y = 3 \cdot x^{2+2} \cdot y = 3x^4y$$

● **Example 3–2 A**

Multiplying a monomial and a polynomial

Perform the indicated multiplication.

1. $5y(2y + 3) = 5y \cdot 2y + 5y \cdot 3$ Distribute $5y$ times each term in the parentheses

   $= 10y^2 + 15y$ Multiply

2. $x^3(x^2 + xy - y^2) = x^3 \cdot x^2 + x^3 \cdot xy - x^3 \cdot y^2$

   $= x^5 + x^4y - x^3y^2$

Note In example 2, when we multiplied x^3 times the third term of the trinomial, y^2, the subtraction sign remained, giving $-x^3y^2$.

3. $4x^2y(2x^3 - 3x^2y^2 + y^4) = 4x^2y \cdot 2x^3 - 4x^2y \cdot 3x^2y^2 + 4x^2y \cdot y^4$

   $= 8x^5y - 12x^4y^3 + 4x^2y^5$

☞ *Quick check*   Perform the indicated multiplication. $3ab^2(2a - 3b)$ ●

Product of Two Polynomials

The product of two polynomials requires the use of the distributive property several times. That is, in the product

$$(x + 2y)(x + y)$$

we consider $(x + 2y)$ a single number and apply the distributive property.

$$(x + 2y)(x + y) = (x + 2y) \cdot x + (x + 2y) \cdot y$$

We now apply the distributive property again.

$$(x + 2y) \cdot x + (x + 2y) \cdot y = x \cdot x + 2y \cdot x + x \cdot y + 2y \cdot y$$
$$= x^2 + 2xy + xy + 2y^2$$

The last step in the problem is to combine like terms, if there are any.

$$x^2 + (2xy + xy) + 2y^2 = x^2 + 3xy + 2y^2$$

Notice that in this product, each term of the first factor is multiplied by each term of the second factor. We can generalize our procedure as follows:

Multiplying Two Polynomials

When we are multiplying two polynomials, we multiply each term in the first polynomial by each term in the second polynomial. We then combine like terms.

● **Example 3–2 B**
FOIL

Perform the indicated multiplication and simplify.

1. $(a + 3)(a - 4) = a \cdot a - a \cdot 4 + 3 \cdot a - 3 \cdot 4$ Distribute multiplication
$= a^2 - 4a + 3a - 12$ Multiply
$= a^2 - a - 12$ Combine like terms

Note We have drawn arrows to indicate the multiplication that is being carried out. This should be a convenient way for us to indicate the multiplication to be performed.

2. $(2x + 3)(5x - 2) = 10x^2 - 4x + 15x - 6$ Distribute and multiply
$= 10x^2 + 11x - 6$ Combine like terms

Note A word that is useful for remembering the multiplication to be performed when multiplying two binomials is **FOIL**. Foil is an abbreviation signifying **F**irst times first, **O**uter times outer, **I**nner times inner, and **L**ast times last.

3. $(3a - 2b)(2a - 5b) = 6a^2 - 15ab - 4ab + 10b^2$ Distribute and multiply
$= 6a^2 - 19ab + 10b^2$ Combine like terms

☞ **Quick check**   Perform the indicated multiplication and simplify.
$(2x + y)(x - 3y)$

In arithmetic, we multiply numbers stated vertically. We can use this same procedure to multiply two polynomials. To perform the example 3 multiplication vertically, we would proceed as follows:

$$2a - 5b$$
$$\underline{3a - 2b}$$

Multiply $-2b$ and $2a - 5b$.

$$2a - 5b$$
$$3a - 2b$$
$$-4ab + 10b^2$$

Multiply $3a$ and $2a - 5b$. Place any like terms in the same columns.

$$2a - 5b$$
$$3a - 2b$$
$$-4ab + 10b^2$$
$$6a^2 - 15ab$$

Add like terms.

$$2a - 5b$$
$$3a - 2b$$
$$-4ab + 10b^2$$
$$\underline{6a^2 - 15ab}$$
$$6a^2 - 19ab + 10b^2$$

Special Products

Three special products appear so often that the form of the answers can be computed by a special formula. Consider the product

$$(x + 6)^2 = (x + 6)(x + 6)$$

which becomes

$$x^2 + 6x + 6x + 36$$

When we combine the second and third terms, we get

$$x^2 + 12x + 36$$

This is called the **square of a binomial** or a **perfect square trinomial** and has certain characteristics. Inspection shows us that in

$$(x + 6)^2 = x^2 + 12x + 36$$

the three terms of the product can be obtained in the following manner:

The Square of a Binomial

1. The first term of the product is the *square of the first term* of the binomial $[(x)^2 = x^2]$.
2. The second term of the product is *two times the product of the two terms of the binomial* $[2(x \cdot 6) = 12x]$.
3. The third term of the product is the *square of the second term* of the binomial $[(6)^2 = 36]$.

If we apply this to

$$(x - 7)^2 = [x + (-7)]^2$$

we get

$$= x^2 + [2 \cdot x \cdot (-7)] + (-7)^2$$

and so

$$(x - 7)^2 = x^2 + (-14x) + 49$$
$$= x^2 - 14x + 49$$

The Square of a Binomial

In general, for real numbers a and b,

$$(a + b)^2 = a^2 + 2ab + b^2$$

and

$$(a - b)^2 = a^2 - 2ab + b^2$$

Note $(a + b)^2 = a^2 + 2ab + b^2$, not $a^2 + b^2$. This is a common error. *The square of a binomial is always a trinomial.*

● **Example 3–2 C**

Square of a binomial

Perform the indicated multiplication and simplify.

1. $(2x + 3)^2 = (2x)^2 + (2 \cdot 2x \cdot 3) + (3)^2$ Apply special products property
$$= 4x^2 + 12x + 9$$ Multiply

2. $(5a - 4b)^2 = (5a)^2 - [2 \cdot 5a \cdot (4b)] + (4b)^2$ Special products property
$$= 25a^2 - [40ab] + 16b^2$$ Multiply
$$= 25a^2 - 40ab + 16b^2$$ Remove the grouping symbol   ●

The third special product is obtained by multiplying the sum and the difference of the same two terms. Consider the following:

$$(x + 3)(x - 3) = x^2 - 3x + 3x - 9$$
$$= x^2 - 9$$

Special characteristics are evident in this product also.

The Difference of Two Squares

For real numbers a and b,

$$(a + b)(a - b) = a^2 - b^2$$

Concept

1. The product is obtained by first squaring the first term of the factors, and then
2. subtracting the square of the second term of the factors.

● **Example 3–2 D**

Difference of two squares

Perform the indicated multiplication and simplify.

1. $(x + 7)(x - 7) = (x)^2 - (7)^2 = x^2 - 49$

2. $(a + 2b)(a - 2b) = (a)^2 - (2b)^2 = a^2 - 4b^2$

3. $(3x - 2y)(3x + 2y) = (3x)^2 - (2y)^2 = 9x^2 - 4y^2$

    In all the examples that we have looked at, whether they were special products or not, a single procedure is sufficient. When multiplying two polynomials, we multiply each of the terms in the first polynomial times each of the terms in the second polynomial and then combine like terms.

● **Example 3–2 E**

Multiplying polynomials

Perform the indicated multiplication and simplify.

1. $(3x - y)(2x + 3y)$ $= 6x^2 + 9xy - 2xy - 3y^2$ Distribute multiplication

                                     $= 6x^2 + 7xy - 3y^2$ Combine like terms

2. $(a - 2)(2a^2 + 3a + 2) = 2a^3 + 3a^2 + 2a - 4a^2 - 6a - 4$ Distribute multiplication

                                       $= 2a^3 - a^2 - 4a - 4$ Combine like terms

Note   Although there are three terms in the second parentheses, we still follow the procedure of every term of the first polynomial times every term of the second polynomial.

3. $(a + 6)(a - 2)(a - 1)$. When there are three polynomials to be multiplied, we apply the associative property to multiply two of them together first and take that product times the third.

$$[(a + 6)(a - 2)](a - 1) = [a^2 - 2a + 6a - 12](a - 1)$$
$$= [a^2 + 4a - 12](a - 1)$$
$$= a^3 - a^2 + 4a^2 - 4a - 12a + 12$$
$$= a^3 + 3a^2 - 16a + 12$$

MASTERY POINTS

Can you
- Multiply a monomial and a polynomial?
- Multiply polynomials?
- Use the special products for the square of a binomial and the difference of two squares?

Exercise 3–2

Directions
Perform the indicated multiplication and simplify. See examples 3–2 A, B, C, D, and E.

☞ *Quick Check*

Examples $3ab^2(2a - 3b)$ $\qquad$ $(2x + y)(x - 3y)$

Solutions $= 3ab^2 \cdot 2a - 3ab^2 \cdot 3b$ $\;$ Distributive property $\qquad$ $= 2x \cdot x - 2x \cdot 3y + y \cdot x$ $\;$ Distributive property
$\qquad\qquad\qquad\qquad\qquad\qquad\qquad\qquad\quad -y \cdot 3y$
$\qquad\qquad = 6a^2b^2 - 9ab^3$ $\qquad$ Multiply $\qquad\qquad = 2x^2 - 6xy + xy - 3y^2$ $\;$ Multiply
$\qquad\qquad\qquad\qquad\qquad\qquad\qquad\qquad\qquad\quad = 2x^2 - 5xy - 3y^2$ $\;$ Combine like terms

Directions Perform the indicated multiplication and simplify. See examples 3–2 A, B, C, D, and E.

1. $2ab(a^2 - bc + c^2)$
2. $6x(4y + 7z)$
3. $3a(5b^2 - 7c^2)$
4. $-ab(a^4 - a^2b^2 - b^4)$
5. $-5ab^2(3a^2 - ab + 4b^2)$
6. $6x^2(4x^2 - 2x + 3)$
7. $3ab(a^2 - 2ab - b^2)$
8. $(2x)(x - y + 5)(5y)$
9. $(3a)(2a - b)(2b^2)$
10. $(x^2y)(x^2 + y^2)(xy^2)$
11. $(x + 3)(x + 4)$
12. $(a + 5)(a - 3)$
13. $(y - 9)(y - 4)$
14. $(z + 7)(z - 11)$
15. $(a + 1)(a + 1)$
16. $(b - 1)(b - 1)$
17. $(R - 3)^2$
18. $(R + 2)(R - 2)$
19. $(a + 3)(a - 3)$
20. $(3x + 2)(x - 4)$
21. $(c - 8)(c + 8)$
22. $(z + 5)(z - 5)$
23. $(2a - 1)(2a + 1)$
24. $(3x + 2)(3x - 2)$
25. $(3a - 5)(2a - 7)$
26. $(3 - 2y)(2 - y)$
27. $(7 + 2x)(2x - 7)$
28. $(4r + 3)(r - 12)$
29. $(3k + w)(k - 6w)$
30. $(a - 6bc)(5a + 4bc)$

31. $(2x + y)(2x - y)$
32. $(3a - 2b)(3a + 2b)$
33. $(x - 0.4)(x + 0.4)$
34. $(a + 0.5)(a - 0.5)$
35. $(0.3b + 0.4)(0.3b - 0.4)$
36. $(0.2y + 0.7)(0.2y - 0.7)$
37. $(x + 5)^2$
38. $(x + 6)^2$
39. $(2a + b)^2$
40. $(a + 3b)^2$
41. $(a + 6b)^2$
42. $(2a + 3b)^2$
43. $(x - 4)^2$
44. $(y - 3)^2$
45. $(x - 3y)^2$
46. $(3x - 3y)^2$
47. $(2a + 3b)(2a - 3b)$
48. $(4x - y)(4x + y)$
49. $(x + 0.4)^2$
50. $(a + 0.3)^2$
51. $(a + 0.5)^2$
52. $(x + 0.7)^2$
53. $(0.2y - z)^2$
54. $(0.3x - y)^2$
55. $(0.5a - b)^2$
56. $(0.7a - b)^2$
57. $(a + 4b)(a^2 - 2ab + b^2)$
58. $(x - 2y)(2x^2 - 3xy + y^2)$

59. $(x + 4)(6x^2 - 3x + 7)$

60. $(x - y)(x^2 - 2xy + y^2)$

61. $(x^2 - 2x - 3)(x^2 + x + 4)$

62. $(a^2 - 3a + 6)(a^2 + 2a - 5)$

63. $(a - 6)(a - 2)(a + 1)$

64. $(2b - 1)(b + 2)(2b + 1)$

65. $(a - b)(a + b)(2a - 3b)$

66. $(a + b)^3$

67. $(a - b)^3$

68. $(2a + b)^3$

69. $(a - 2b)^3$

70. The area of the shaded region between the two circles is $\pi(R + r)(R - r)$. Perform the indicated multiplication.

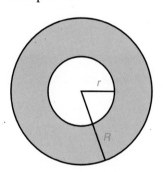

71. When squares of c units on a side are cut from the corners of a square sheet of metal x units on a side, and the metal sheet is then folded up into a tray, the volume is $c(x - 2c)(x - 2c)$. Perform the indicated multiplication.

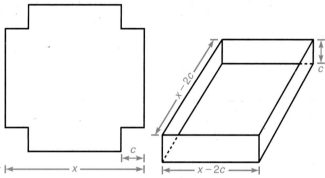

72. Explain why $(a + b)^2 \neq a^2 + b^2$

Explore to Learn

1. Consider the operations defined in mathematics. Most are called binary, such as addition, because they affect two expressions. Some are called unary, such as squaring, because they affect one value. Sometimes students, like mathematicians, look for short cuts for operating on three or more values at once. Consider the two examples: (a) $(xy)(xy)(xy)$ and (b) $(a - b)(a - b)(a - b)$. Example (a) can be done by multiplying all the x's then all the y's. Can you find a short cut to do (b) (b is exercise 67)?

2. How many terms are there in the product $(a - b)(a - b)$, in the product $(a - b)(a - b)(a - b)$? Guess how many terms there will be in the product $(a - b)(a - b)(a - b)(a - b)$. Confirm your guess by carrying out the multiplication.

3. Use the difference of two squares to find the product $18 \cdot 22$ by finding the product $(20 + 2)(20 - 2)$. Then use another special product to **mentally** find 21^2 and 59^2.

Review Exercises

Directions   Perform the indicated addition or subtraction. See sections 1–5 and 1–6.

1. $(-3) + (-2)$

2. $8 - (-4)$

3. $(-7) - (-10)$

4. $2 + (-5) + (-6)$

Directions   Simplify by using the properties of exponents. See section 3–1.

5. $x^4 \cdot x^8$

6. $(a^3)^5$

7. $(3ab)^3$

8. $(2x)^3$

3–3   Exponents—II

Fraction to a Power Property of Exponents

In section 3–1, we learned several useful properties of exponents. Now we shall learn several more.

Our next property of exponents can be derived from the definition of exponents. Consider the expression $\left(\dfrac{a}{b}\right)^3$.

$$\left(\frac{a}{b}\right)^3 = \underbrace{\frac{a}{b} \cdot \frac{a}{b} \cdot \frac{a}{b}}_{\text{3 factors of } \frac{a}{b}} = \frac{\overbrace{a \cdot a \cdot a}^{\text{3 factors of } a}}{\underbrace{b \cdot b \cdot b}_{\text{3 factors of } b}} = \frac{a^3}{b^3}$$

Thus

$$\text{Fraction raised to a power} \longrightarrow \left(\frac{a}{b}\right)^3 = \frac{a^3}{b^3} \longleftarrow \begin{array}{l}\text{Numerator raised to the power} \\ \text{Denominator raised to the power}\end{array}$$

Fraction to a Power Property of Exponents

$$\left(\frac{a}{b}\right)^n = \frac{a^n}{b^n}, b \neq 0$$

Concept

Whenever a fraction is raised to a power, the numerator and the denominator are *both* raised to that power.

● **Example 3–3 A**

Fraction to a power

Perform the indicated operations and simplify.

1. $\left(\dfrac{3}{4}\right)^3 = \dfrac{3^3}{4^3} = \dfrac{27}{64}$

2. $\left(\dfrac{a}{b}\right)^5 = \dfrac{a^5}{b^5}$

3. $\left(\dfrac{2a}{b}\right)^3 = \dfrac{(2a)^3}{b^3} = \dfrac{2^3 a^3}{b^3} = \dfrac{8a^3}{b^3}$

Division of Expressions with Like Bases

Consider the expression

$$\frac{x^6}{x^2}$$

We can use the definition of exponents to write the fraction as

$$\frac{x^6}{x^2} = \frac{x \cdot x \cdot x \cdot x \cdot x \cdot x}{x \cdot x}$$

We reduce the fraction as follows:

$$\frac{x \cdot x \cdot x \cdot x \cdot x \cdot x}{x \cdot x} = \frac{x \cdot x \cdot x \cdot x}{1} = \frac{x^4}{1} = x^4$$

In our example, we reduced by two factors of x, leaving $6 - 2 = 4$ factors of x in the numerator. Therefore

$$\frac{x^6}{x^2} = x^{6-2} = x^4$$

Thus we have the following property of exponents.

Quotient Property of Exponents

$$a^m \div a^n = \frac{a^m}{a^n} = a^{m-n}, a \neq 0$$

Concept

To divide expressions having *like* bases, subtract the exponent of the denominator from the exponent of the numerator to get the exponent of the given base in the quotient.

Note If the base a is zero, $a = 0$, we have an expression that has no meaning. Therefore $a \neq 0$ indicates that we want our variables to assume no values that would cause the denominator to be zero.

● **Example 3–3 B**

Using the quotient property of exponents

Simplify. Assume that no variable is equal to zero.

1. $\dfrac{x^7}{x^5} = x^{7-5} = x^2$

2. $a^{11} \div a^4 = a^{11-4} = a^7$

3. $\dfrac{5^4}{5} = 5^{4-1} = 5^3 = 125$

Note Remember that when we are dividing like bases, their exponents are subtracted, but the *base is not changed*.

☞ *Quick check*   Simplify $a^{11} \div a^7$. Assume that a is not equal to zero.   ●

Negative Exponents

To this point, we have considered only those problems where the exponent of the numerator is greater than the exponent of the denominator. Consider the example

$$\frac{x^2}{x^6}$$

By the definition of exponents, this becomes

$$\frac{x^2}{x^6} = \frac{x \cdot x}{x \cdot x \cdot x \cdot x \cdot x \cdot x}$$

and reducing the fraction,

$$\frac{x \cdot x}{x \cdot x \cdot x \cdot x \cdot x \cdot x} = \frac{1}{x \cdot x \cdot x \cdot x} = \frac{1}{x^4}$$

Again, we reduced by two factors of x, leaving $6 - 2 = 4$ factors of x in the denominator. Hence

$$\frac{x^2}{x^6} = \frac{1}{x^4}$$

However using the quotient property of exponents to carry out the division, we would have

$$\frac{x^2}{x^6} = x^{2-6} = x^{-4}$$

Since we should arrive at the same answer regardless of which procedure we use, then x^{-4} must be $\frac{1}{x^4}$, thus $x^{-4} = \frac{1}{x^4}$. This leads us to the definition of negative exponents.[2]

Definition of Negative Exponents

$$a^{-n} = \frac{1}{a^n}, a \neq 0$$

Concept

A negative exponent on any base (except zero) can be written as 1 over that base with a positive exponent.

● **Example 3–3 C**

Using negative exponents

Simplify. Leave the answer with only positive exponents. Assume that no variable is equal to zero.

1. $a^{-9} = \frac{1}{a^9}$ Rewritten as 1 over a to the positive 9th

Note From the definition of negative exponents, if a **factor** is moved from either the numerator to the denominator or from the denominator to the numerator, the sign of its exponent will change. The sign of the base will not be affected by this change.

2. *Alternative Procedure*

$$a^{-9} = \frac{a^{-9}}{1}$$ Rewrite as a fraction

$$= \frac{1}{a^9}$$ Sign of the exponent is changed as the **factor** is moved from the numerator to the denominator

[2]It was understood by 1553, by M. Stifel, that a quantity with an exponent zero had the value one. Newton introduced our modern notation for negative (and fractional) exponents in a letter in 1676. He wrote "Since algebraists write a^2, a^3, a^4, etc. for *aa*, *aaa*, *aaaa*, etc., so I write ... a^{-1}, a^{-2}, a^{-3}, etc. for $\frac{1}{a}$, $\frac{1}{aa}$, $\frac{1}{aaa}$, etc."

3. $\dfrac{1}{b^{-4}} = \dfrac{b^4}{1}$ Sign of the exponent is changed as the **factor** is moved from the denominator to the numerator

   $= b^4$ Standard form is to leave only positive exponents

4. $(-3)^{-3} = \dfrac{1}{(-3)^3}$ Sign of the exponent is changed as the **factor** is moved from the denominator to the numerator

   $= \dfrac{1}{-27} \text{ or } -\dfrac{1}{27}$ $(-3)^3 = -27$

☞ *Quick check*   Write b^{-2} with positive exponents.   ●

Zero as an Exponent

Now consider the situation involving the division of like bases that are raised to the same power.

$$\frac{x^3}{x^3}, x \neq 0$$

By the definition of exponents, we have

$$\frac{x^3}{x^3} = \frac{x \cdot x \cdot x}{x \cdot x \cdot x} = \frac{1}{1} = 1$$

By the quotient property of exponents,

$$\frac{x^3}{x^3} = x^{3-3} = x^0$$

Since $\dfrac{x^3}{x^3} = 1$ and $\dfrac{x^3}{x^3} = x^0$, then x^0 must be equal to 1. This leads us to the definition of zero as an exponent.

Definition of Zero as an Exponent

$$a^0 = 1, a \neq 0$$

Concept

Any number other than zero raised to the zero power is equal to 1.

Note In section 2–1, it was stated that the degree of any nonzero constant is zero. For example, the constant 3 has degree 0, because 3 can be written as $3x^0$.

● **Example 3–3 D**

Using zero as an exponent

Simplify. Assume that no variable is equal to zero.

1. $b^0 = 1$ **5.** $(a + b)^0 = 1$

2. $r^0 = 1$ **6.** $3x^0 = 3 \cdot 1 = 3$

3. $5^0 = 1$ **7.** $(3x)^0 = 1$

4. $(-2)^0 = 1$

<u>Note</u> The exponent acts only on the symbol immediately to its left. In example 6, only the *x* is raised to the zero power. The exponent of 3 is understood to be 1. In example 7, the parentheses indicate that both the 3 and the *x* are raised to the zero power.

☛ **Quick check** Simplify C^0. Assume that C is not equal to zero. ●

● **Example 3–3 E**

Using a combination of properties

Simplify. Leave the answer with only positive exponents. Assume that no variable is equal to zero.

1. $\dfrac{x^5}{x^{11}} = x^{5-11}$ Division of like bases

        $= x^{-6}$ Subtract exponents

        $= \dfrac{1}{x^6}$ Rewrite with positive exponents

2. $a^{-7} \cdot a^5 = a^{-7+5}$ Multiplication of like bases

        $= a^{-2}$ Add exponents

        $= \dfrac{1}{a^2}$ Rewrite with positive exponents

3. $(b^{-2})^{-4} = b^{(-2)\cdot(-4)}$ Power of a power

        $= b^8$ Multiply exponents

4. $\dfrac{a^3 b^5}{a^7 b^2} = a^{3-7} b^{5-2}$ Division of like bases

        $= a^{-4} b^3$ Subtract exponents

        $= \dfrac{b^3}{a^4}$ Rewrite with positive exponents

5. $\dfrac{a^3 b^2 c^4}{a b^5 c^4} = a^{3-1} b^{2-5} c^{4-4}$ Division of like bases

        $= a^2 b^{-3} c^0$ Subtract the exponents

        $= \dfrac{a^2 \cdot 1}{b^3}$ The *a*'s remain in the numerator, the *b*'s drop to the denominator, and c^0 is 1

        $= \dfrac{a^2}{b^3}$ Rewrite with positive exponents

6. $\dfrac{a^{-2} b^4}{a^{-5} b^6} = a^{-2-(-5)} b^{4-6}$ Division of like bases

        $= a^3 b^{-2}$ Subtract exponents

        $= \dfrac{a^3}{b^2}$ Rewrite with positive exponents

☛ **Quick check** Simplify $\dfrac{b^4}{b^{10}}$. Leave the answer with only positive exponents.

Assume that b is not equal to zero. ●

MASTERY POINTS

Can you
- Raise a fraction to a power?
- Perform division on expressions having like bases?
- Perform operations involving negative exponents?
- Perform operations involving zero as an exponent?

Exercise 3–3

Directions

Write each expression with only positive exponents. Assume that no variable is equal to zero. See examples 3–3 C and D.

📌 *Quick Check*

Examples C^0 $\qquad\qquad\qquad\qquad b^{-2}$

Solutions $= 1$ By definition is equal to 1 $\qquad = \dfrac{1}{b^2}$ Rewritten as 1 over b to the positive 2nd

---

Directions Write each expression with only positive exponents. Assume that no variable is equal to zero. See examples 3–3 C and D.

1. x^0

2. $(2y)^0$

3. $5a^0$

4. $7x^0$

5. $(3B)^0$

6. S^{-2}

7. R^{-5}

8. $(2x)^{-3}$

9. $(3P)^{-2}$

10. $4z^{-2}$

11. $9C^{-4}$

12. $\dfrac{5}{x^{-4}}$

13. $\dfrac{1}{2y^{-3}}$

14. $\dfrac{1}{3x^{-2}}$

15. $2x^{-4}y^2$

16. $x^{-2}y^4$

17. $p^0r^{-2}t^5$

18. $x^{-3}y^2z^{-4}$

Directions

Perform all indicated operations and leave your answer with only positive exponents. Assume that no variable is equal to zero. See examples 3–3 A, B, and E.

📌 *Quick Check*

Examples $a^{11} \div a^7$ $\qquad\qquad\qquad\qquad \dfrac{b^4}{b^{10}}$

Solutions $= a^{11-7}$ Division of like bases $\qquad = b^{4-10}$ Division of like bases

$\qquad\qquad = a^4$ Subtract exponents $\qquad\qquad = b^{-6}$ Subtract exponents

$\qquad\qquad\qquad\qquad\qquad\qquad\qquad\qquad = \dfrac{1}{b^6}$ Rewrite with positive exponent

---

Directions Perform all indicated operations and leave your answer with only positive exponents. Assume that no variable is equal to zero. See examples 3–3 A, B, and E.

19. $\left(\dfrac{a}{b}\right)^6$

20. $\left(\dfrac{x}{y}\right)^4$

21. $\left(\dfrac{2}{3}\right)^3$

22. $\left(\dfrac{1}{2}\right)^4$

23. $\left(\dfrac{2x}{y}\right)^4$

24. $\left(\dfrac{2ab}{c}\right)^3$

25. $\left(\dfrac{3a}{b}\right)^3$

26. $x^{12} \div x^6$

27. $y^4 \div y^2$

28. $\dfrac{a^5}{a^3}$

29. $\dfrac{b^9}{b^7}$

30. $\dfrac{c^6}{c^9}$

31. $\dfrac{R^4}{R^8}$

32. $\dfrac{3^4}{3^2}$

33. $\dfrac{2^5}{2^3}$

34. $\dfrac{4^2}{4^5}$

35. $\dfrac{6}{6^3}$

36. $\dfrac{x^4x^3}{x^2}$

37. $\dfrac{y^5y}{y^2}$

38. $\dfrac{a^4a^2}{a^5}$

39. $\dfrac{a^4}{a^2a}$

40. $\dfrac{x^7}{x^2x^3}$

41. $\dfrac{y^3}{y^4y^5}$

42. $\dfrac{b^2}{bb^4}$

43. $\dfrac{a^7b^5}{a^4b^2}$

44. $\dfrac{x^9y^7}{x^4y}$

45. $\dfrac{2^3x^3y^7}{2xy^5}$

46. $\dfrac{3^3 a^4 b^5}{3^2 a^2 b^3}$

47. $\dfrac{3a^2 b^5}{3^4 a^5 b^5}$

48. $\dfrac{5^2 a^3 b}{5^3 a^7 b^3}$

49. $x^{-4} x^7$

50. $y^{-2} y^{10}$

51. $a^5 a^{-11}$

52. $R^{-2} R^{-5}$

53. $x^{-2} x^4 x^0$

54. $x^5 x^0 x^{-2}$

55. $a^0 a^{-5} a^3$

56. $a^{-7} a^4 a^0$

57. $(-5)^{-3}$

58. $(-2)^{-4}$

59. $\dfrac{3^{-2}}{3^{-5}}$

60. $\dfrac{2^{-6}}{2^{-3}}$

61. $(a^{-2})^3$

62. $(b^4)^{-4}$

63. $(x^5)^{-2}$

64. $(y^{-3})^4$

65. $(a^{-2})^{-3}$

66. $(z^{-4})^{-4}$

67. $(x^0)^{-2}$

68. $(a^{-3})^0$

69. $(x^{-2})^0$

70. $(b^0)^{-4}$

71. $\dfrac{R^2 S^{-4}}{R^{-3} S^5}$

72. $\dfrac{4 y^{-3}}{4^{-1} y^2}$

73. Explain why $2x^{-2} \neq \dfrac{1}{2x^2}$.

⚙ **Explore to Learn**

Suppose $a = 2, b = 3$. Are any of the following statements true?

a. $(ab)^{-1} = \dfrac{1}{ab}$

b. $(a^2 b)^{-2} = \dfrac{1}{a^4 b^2}$

c. $(a + b)^{-1} = \dfrac{1}{a + b}$

d. $(a + b)^{-1} = \dfrac{1}{a} + \dfrac{1}{b}$

Review Exercises

Directions   Perform the indicated operations. See sections 1–5 to 1–8.

1. $(-4) + (-6)$ **2.** $(-3)(-7)$ **3.** $(-4) - (-8)$ **4.** -4^2

Directions   Simplify by using the properties of exponents. See section 3–1.

5. $a^3 a^5$ **6.** $(x^3)^4$ **7.** $x x^2 x^3$ **8.** $(2ab)^2$

3–4   Exponents—III

Properties and Definitions of Exponents

The following is a summary of the properties and definitions of exponents that we have studied so far.

> **Definitions**
>
> n factors
> $$a^n = a \cdot a \cdot a \cdots a, \text{ where } n \text{ is a positive integer}$$
> $$a^{-n} = \frac{1}{a^n}, a \neq 0$$
> $$a^0 = 1, a \neq 0$$

Properties

$a^m \cdot a^n = a^{m+n}$	Product
$(ab)^n = a^n b^n$	Group of factors to a power
$(a^m)^n = a^{m \cdot n}$	Power to a power
$\left(\dfrac{a}{b}\right)^n = \dfrac{a^n}{b^n}, b \neq 0$	Fraction to a power
$a^m \div a^n = \dfrac{a^m}{a^n} = a^{m-n}, a \neq 0$	Quotient

The following examples illustrate some more problems in which more than one property of exponents is applied within the same problem.

● **Example 3–4 A**

Using a combination of properties

Simplify. Leave the answer with only positive exponents. Assume that no variable is equal to zero.

1. $(2a^2 b^3)^3 = 2^3 (a^2)^3 (b^3)^3$ — Each factor in the group is raised to the third power

$= 2^3 a^6 b^9$ — Power to a power, multiply exponents

$= 8a^6 b^9$ — 2^3 is 8

2. $(5a^4 b^2)^4 = 5^4 (a^4)^4 (b^2)^4$ — Each factor is raised to the power four

$= 5^4 a^{16} b^8$ — Power to a power

$= 625 a^{16} b^8$ — 5^4 is 625

3. $(3a^{-2} b^3)^{-3} = 3^{-3} (a^{-2})^{-3} (b^3)^{-3}$ — Each factor is raised to the power

$= 3^{-3} a^6 b^{-9}$ — Power to a power

$= \dfrac{a^6}{3^3 b^9}$ — The 3's and the b's move to the denominator

$= \dfrac{a^6}{27 b^9}$ — 3^3 is 27

4. $(-3a^2)(2ab^3)(-4a^3 b^5)$ — Multiply like bases using the commutative and associative properties

$= [(-3)(2)(-4)](a^2 a a^3)(b^3 b^5)$

$= 24 a^6 b^8$ — Signed numbers and multiplication of like bases, add exponents

5. $(3a^4 b)^2 (3^2 ab^5)^2 = 3^2 (a^4)^2 b^2 \cdot (3^2)^2 a^2 (b^5)^2$ — Group of factors to a power

$= 3^2 a^8 b^2 \cdot 3^4 a^2 b^{10}$ — Power to a power

$= (3^2 \cdot 3^4)(a^8 a^2)(b^2 b^{10})$ — Multiply like bases

$= 3^6 a^{10} b^{12}$ — Add exponents

$= 729 a^{10} b^{12}$ — 3^6 is 729

6. $\left(\dfrac{2a^2 b^3}{c^5}\right)^3 = \dfrac{(2a^2 b^3)^3}{(c^5)^3}$ — Both the numerator and the denominator are raised to the third power

$= \dfrac{2^3 (a^2)^3 (b^3)^3}{(c^5)^3}$ — Each factor in the numerator is raised to the third power

$= \dfrac{8a^6 b^9}{c^{15}}$ — Power to a power, multiply exponents

7. $\dfrac{a^{-2}b^3}{a^{-4}b^6} = a^{(-2)-(-4)}b^{3-6}$ Division of like bases

$\qquad = a^2 b^{-3}$ Subtract exponents

$\qquad = \dfrac{a^2}{b^3}$ Rewrite with positive exponents

8. $\left(\dfrac{a^{-2}b}{c^3}\right)^{-2} = \dfrac{(a^{-2}b)^{-2}}{(c^3)^{-2}}$ Numerator and denominator are raised to the power

$\qquad = \dfrac{(a^{-2})^{-2}b^{-2}}{(c^3)^{-2}}$ Numerator has a group of factors to a power

$\qquad = \dfrac{a^{(-2)(-2)}b^{-2}}{c^{(3)(-2)}}$ Power to a power

$\qquad = \dfrac{a^4 b^{-2}}{c^{-6}}$ Multiply exponents

$\qquad = \dfrac{a^4 c^6}{b^2}$ Factors raised to a negative power are moved to the other side of the fraction bar

☛ *Quick check*   Simplify. Leave the answer with only positive exponents. Assume that no variable is equal to zero. $(2a^{-2}b^3)^3$

MASTERY POINTS

Can you
• Apply the definitions and properties of exponents?

Exercise 3–4

Directions

Simplify by using the properties and definitions of exponents. Leave the answer with only positive exponents. Assume that no variable is equal to zero. See example 3–4 A.

☛ *Quick Check*

Example $(2a^{-2}b^3)^3$

Solution $= 2^3(a^{-2})^3(b^3)^3$ Groups of factors to a power
$= 2^3 a^{-6} b^9$ Power to a power
$= \dfrac{8b^9}{a^6}$ Rewrite with positive exponents

Directions   Simplify by using the properties and definitions of exponents. Leave the answer with only positive exponents. Assume that no variable is equal to zero. See example 3–4 A.

1. $(2a^2)^3$

2. $(3x^4)^2$

3. $(2x^2y)^3$

4. $(4ab^3)^2$

5. $(3x^2y)^3$

6. $(2^2 a^3 b)^2$

7. $(3^2 xy^4)^2$

8. $(4x^5 y^6)^2$

9. $(x^4 y^3 z)^4$

10. $(2a^5 b^2 c)^3$

11. $(5a^5 b^2 c^4)^2$

12. $(a^3 b^2)^3$

13. $(2a^2)^{-2}$

14. $(5x^{-3})^{-2}$

15. $(4^{-1}x^2)^{-2}$

16. $(2a^2 b^{-3})^{-2}$

17. $(2^{-1}x^{-1}y)^{-2}$

18. $(3^{-1}a^{-2}b^3)^{-3}$

19. $(a^2 b^{-3} c)^{-3}$

20. $(a^4 b^{-2} c)^{-2}$

21. $(3xy^{-4})^{-3}$

22. $(3x^{-2} y^{-3})^2$

23. $(x^2 y^{-5} z^3)^{-2}$

24. $(a^{-3} b^2 c^{-4})^{-3}$

25. $(3x^2)(2x^0 y^2)(x^5 y)$

26. $(a^2 b)(-3a^0 b^2)(a^3 b)$

27. $(-2x^2 y)(3x^3 y^2)(x^5 y)$

28. $(a^2 bc)(-2a^2 b^2 c^2)(3abc)$

29. $(x^3 yz^4)(-3xyz^2)(-2x^2 yz)$

30. $(a^2 b)(-3b^2 c^2)(2a^2 c^2)$

31. $\left(\dfrac{2x}{y^2}\right)^3$

32. $\left(\dfrac{x^2 y}{z^2}\right)^3$

33. $\left(\dfrac{3a^2 c^0}{b^3}\right)^2$

34. $\left(\dfrac{x^3}{y^0 z^4}\right)^3$

35. $\left(\dfrac{2x^2}{y^3}\right)^3$

36. $\left(\dfrac{ab^2}{c^4}\right)^4$

37. $\left(\dfrac{2x^3 y^3}{z^5}\right)^2$

38. $\left(\dfrac{a^5 bc^4}{d^2 e}\right)^5$

39. $\dfrac{a^{-2} b^3}{a^3 b^{-5}}$

40. $\dfrac{x^{-5} y^2}{x^3 y^{-4}}$

41. $\dfrac{3R^{-1} S^{-2}}{9R^{-3} S^2}$

42. $\dfrac{R^2 S^{-4}}{R^{-3} S^5}$

43. $\dfrac{2^{-1} x^0}{2^{-2} x^{-2}}$

44. $\dfrac{3^{-2} a^{-2} b^3}{3a^0 b^{-2}}$

45. $\dfrac{6^{-1} a^3 b^{-2}}{3^{-2} a^{-2} b^0}$

46. $\dfrac{4^{-1} x^{-2} y^3}{2^{-3} x^{-2} y^0}$

47. $\dfrac{8a^{-2} b^{-5}}{2a^{-1} b^4}$

48. $\dfrac{2x^{-1} y^{-2}}{3x^{-2} y^2}$

49. $\dfrac{6R^{-2} S^0}{2R^2 S^{-3}}$

50. $\dfrac{2a^{-1} b^0 c^2}{5a^3 b^{-1} c^{-3}}$

51. $(a^2 b^3)^3 (ab^2)^4$

52. $(xy^2)^3 (x^2 y^2)^2$

53. $(2a^3)^2 (2a^2)^3$

54. $(3x^5)^3 (3x^3)^2$

55. $(2x^2 y)^3 (2x^4 y^5)^2$

56. $(3r^2 s^4)^3 (r^5 s^6)^2$

57. $\left(\dfrac{xy^{-2}}{z^{-4}}\right)^{-1}$

58. $\left(\dfrac{x^{-3} y}{z^5}\right)^{-2}$

59. $\left(\dfrac{2a^{-3}}{b^5}\right)^{-2}$

60. $\left(\dfrac{4^{-1} a^{-2}}{b^{-5}}\right)^{-2}$

61. $\left(\dfrac{ab^{-2}}{c^{-1}}\right)^{-3}$

62. $\left(\dfrac{2^{-2} x^3}{y^{-2}}\right)^{-3}$

63. Explain why $2^3 \cdot 2^4 \neq 4^7$.

⚜ Explore to Learn

1. Write any multiplication problem whose answer is $\dfrac{3x}{y}$. You must use at least one negative exponent in your problem.

2. Write any division problem whose answer is $\dfrac{24x^2 y}{5z}$. You must use at least one negative exponent in your problem.

Review Exercises

Directions   Perform the indicated multiplication. See section 1–2.

1. $(6.2) \cdot (5.7)$

2. $(2.8) \cdot (3.7)$

3. $(1.9) \cdot (8.8)$

4. $(4.2) \cdot (6.9)$

5. $(9.9) \cdot (1.9)$

6. $(7.5) \cdot (6.6)$

Directions   Perform the indicated multiplication. Leave the answer in exponential form. See section 3–3.

7. $10^3 \cdot 10^5$

8. $10^6 \cdot 10^6$

9. $10^{-2} \cdot 10^{-4}$

10. $10^5 \cdot 10^{-8}$

3–5   Scientific Notation

Scientific Notation

An important use of integer exponents is in scientific, engineering, and technical fields where we deal with very large or very small numbers. For example, the mass of a hydrogen atom is 0.000 000 000 000 000 000 000 001 67 gram; the mass of an electron is 0.000 000 000 000 000 000 000 000 000 91 gram; the half-life of lead-204 is 14,000,000,000,000,000,000 years. To work with such numbers on the calculator, they must often be entered in **scientific notation.** We define the scientific notation of a positive number X to be the product

$$X = a \times 10^n$$

where $1 \le a < 10$ and n is an integer. To achieve this form of the decimal number X, use the following steps.

To Write a Number in Scientific Notation

1. Move the decimal point in the original number to the right of the first nonzero digit.
2. Count the number of places, n, the decimal point has been moved. This number is the exponent of 10.
3. If
   a. the decimal point is moved to the *left, n* is *positive.*
   b. the decimal point is moved to the *right, n* is *negative.*
   c. the decimal point already follows the first nonzero digit, n is *zero.*

● **Example 3–5 A**

Using scientific notation

Express the following numbers in scientific notation.

1. 250

$$250 = 2.50. \times 10^2 = 2.5 \times 10^2$$

2. 45,000,000

$$45,000,000 = 4.5000000. \times 10^7 = 4.5 \times 10^7$$

3. 0.000152

$$0.000152 = 0.0001.52 \times 10^{-4} = 1.52 \times 10^{-4}$$

☛ *Quick check*   Express the following numbers in scientific notation.
4,380   0.00592

Standard Form

Sometimes it is necessary to convert a number in scientific notation to its standard form. To do this, we apply the procedure in reverse.

> ### To Convert a Number in Scientific Notation to its Standard Form
>
> When the power of 10, n, is
> 1. *positive,* the decimal point is moved to the *right n* places.
> 2. *negative,* the decimal point is moved to the *left* $|n|$ places.
> 3. *zero,* the decimal point is not moved.

● **Example 3–5 B**

Writing numbers in standard form

Express the following numbers in standard form.

1. 1.45×10^4

Since the exponent of 10 is positive 4, we move the decimal point 4 places to the *right* to get

$$1.45 \times 10^4 = 1.4500. = 14,500$$

2. 5.23×10^{-3}

The *negative* exponent, -3, tells us to move the decimal point 3 places to the *left* to get

$$5.23 \times 10^{-3} = 0.005.23 = 0.00523$$

Note In each example, it was necessary to insert zeros to properly locate the decimal point.

☞ **Quick check**   Express the following numbers in standard form.
9.98×10^{-4} 5.63×10^4 ●

Computation Using Scientific Notation

Scientific notation can be used to simplify numerical calculations when the numbers are very large or very small. We first change the numbers to scientific notation and use the properties of exponents to help perform the indicated operations.

● **Example 3–5 C**

Computation using scientific notation

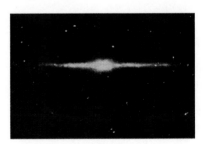

Perform the indicated operations using scientific notation. Leave the answer in standard form.

1. $(349,000,000)(0.0816)$

$$\begin{aligned}
&= (3.49 \times 10^8)(8.16 \times 10^{-2}) &&\text{Scientific notation} \\
&= (3.49 \cdot 8.16) \times (10^8 \cdot 10^{-2}) &&\text{Commutative and associative properties} \\
&= 28.4784 \times 10^6 &&\text{Multiply} \\
&= 28,478,400 &&\text{Standard form}
\end{aligned}$$

2. Our sun, with its family of planets, moves in an orbit around the center of the Milky Way galaxy at approximately 140 miles per second. It takes the sun about 230 million years to complete one orbit. In the 4.6 billion years since it was formed, how many times has the sun circled the center of the galaxy?

We would divide the 4.6 billion years by the 230 million years to determine the number of revolutions.

$$\frac{4.6 \text{ billion}}{230 \text{ million}} = \frac{4.6 \times 10^9}{230 \times 10^6}$$ 1 billion = 10^9
   1 million = 10^6

$$= \frac{4.6 \times 10^9}{2.3 \times 10^2 \times 10^6}$$ $230 = 2.3 \times 10^2$

$$= \frac{4.6 \times 10^9}{2.3 \times 10^8}$$ $10^2 \times 10^6 = 10^8$

$$= 2.0 \times 10^{9-8}$$ Divide

$$= 2.0 \times 10^1$$

$$= 20$$ Standard form

Therefore the sun has circled the center of the galaxy 20 times.

☞ **Quick check**   Perform the indicated operations using scientific notation. Leave the answer in standard form.

$$\frac{(102,000,000)(0.00105)}{(1,190)(0.012)}$$

Any scientific calculator will accept numbers in scientific notation. In fact this is the only way to enter a very large or small number. Calculators usually have a key marked $\boxed{\text{EE}}$, for Enter Exponent, or $\boxed{\text{EXP}}$, for exponent. For example, to enter 3.5×10^{-19} into a calculator a typical key sequence would be 3.5 $\boxed{\text{EE}}$ 19 $\boxed{+/-}$.

To perform a calculation such as $(3.8 \times 10^8)(-2.5 \times 10^{-12})$ one would enter a sequence like

Scientific calculator: 3.8 $\boxed{\text{EE}}$ 8 $\boxed{\times}$ 2.5 $\boxed{+/-}$ $\boxed{\text{EE}}$ 12 $\boxed{+/-}$ $\boxed{=}$ or
Graphing calculator: 3.8 $\boxed{\text{EE}}$ 8 $\boxed{\times}$ $\boxed{(-)}$ 2.5 $\boxed{\text{EE}}$ $\boxed{(-)}$ 12 $\boxed{\text{ENTER}}$
The result is -9.5×10^{-4}.

● **Example 3–5 D**
Calculators and scientific notation

Calculate $(3,500,000,000,000,000)(51,000,000,000)$. Leave the result in scientific notation.

$$(3,500,000,000,000,000)(51,000,000,000)$$
$$= (3.5 \times 10^{15})(5.1 \times 10^{10})$$
$$= (3.5)(5.1)(10^{15} \times 10^{10})$$
$$= 17.85 \times 10^{25}$$
$$= 1.785 \times 10^{26}$$

Calculator steps: 3.5 $\boxed{\text{EE}}$ 15 $\boxed{\times}$ 5.1 $\boxed{\text{EE}}$ 10 $\boxed{=}$ (or $\boxed{\text{ENTER}}$)

The following correspondences between scientific notation and terms used to describe large and small values are worth noting.

Trillion	10^{12}	3.2 trillion = 3.2×10^{12}	U.S. national debt in dollars in 1990
Billion	10^9	5 billion = 5×10^9	Population of earth in 1990
Million	10^6	31.5 million = 31.5×10^6	Approximate number of seconds in one year

Millionth 10^{-6} 100 microns = 100 millionths of a meter = 100×10^{-6} meter

Approximate width of a human hair

Billionth 10^{-9} 1 nanosecond = 1 billionth of a second = 10^{-9} seconds

Time it takes light to travel about 1 foot

Trillionth 10^{-12} 12 picofarads = 12 trillionths of a farad = 12×10^{-12} farads

Typical capacity of small capacitors in electronics

MASTERY POINTS

Can you
- Express a number in scientific notation?
- Convert a number from scientific notation to standard form?
- Do computations using scientific notation?

Exercise 3–5

Directions
Express the following numbers in scientific notation. See example 3–5 A.

Quick Check

Examples 4,380   0.00592

Solutions   = 4.38×10^3 Three places to the left,   = 5.92×10^{-3} Three places to the right,
exponent is 3   exponent is −3

Directions   Express the following numbers in scientific notation. See example 3–5 A.

1. 255
2. 65,000,000
3. 12,345
4. 14,800
5. 155,000
6. 14.36
7. 855.076
8. 1,570.7
9. 1,007,600

10. 6,000,736
11. 0.00012
12. 0.0863
13. 0.0000081
14. 0.0000147
15. 0.0007
16. 0.12079
17. 0.000000000094
18. 456

19. 4,500
20. 0.00087
21. 5,850,000
22. 0.0567
23. 45.78
24. 34,000,000
25. 0.00000002985

Directions
Convert the following numbers in scientific notation to their standard form. See example 3–5 B.

Quick Check

Examples 9.98×10^{-4} 5.63×10^4

Solutions   = 0.000998   Exponent is −4, move 4   = 56,300   Exponent is 4, move 4
places to the left   places to the right

Directions   Convert the following numbers in scientific notation to their standard form. See example 3–5 B.

26. 2.07×10^3
27. 4.99×10^7
28. 5.061×10^5
29. 7.23×10^0
30. 1.073×10^4

31. 4.2×10^{-3}
32. 7.611×10^{-7}
33. 1.47×10^{-6}
34. 5.0×10^{-2}
35. 7.89×10^{-4}

36. 2.3×10^5
37. 4.82×10^{-9}
38. 2.61×10^2
39. 4.92×10^{-6}
40. 9.3×10^8

Directions Perform the indicated operations using scientific notation. Leave the answer in both scientific notation and standard form. See examples 3–5 C and D.	☞ *Quick Check*
	Example $\dfrac{(102{,}000{,}000)(0.00105)}{(1{,}190)(0.012)}$
	Solution $= \dfrac{(1.02 \times 10^8)(1.05 \times 10^{-3})}{(1.19 \times 10^3)(1.2 \times 10^{-2})}$ Scientific notation
	$= \dfrac{(1.02)(1.05)10^8 \cdot 10^{-3}}{(1.19)(1.2)10^3 \cdot 10^{-2}}$ Commutative and associative properties
	$= \dfrac{(1.02)(1.05)}{(1.19)(1.2)} \times 10^4$ Properties of exponents
	$= 0.75 \times 10^4$ Multiplication and division
	$= 7.5 \times 10^3$ Scientific notation
	$= 7{,}500$ Standard form

Directions Perform the indicated operations using scientific notation. Leave the answer in both scientific notation and standard form. See examples 3–5 C and D.

41. $(6{,}370{,}000) \cdot (19{,}200{,}000)$

42. $(18{,}700{,}000) \cdot (52{,}600{,}000)$

43. $(9.41 \times 10^{12}) \cdot (3.86 \times 10^{-14})$

44. $(4.49 \times 10^{-18}) \cdot (5.89 \times 10^{27})$

45. $(0.00341) \cdot (0.0000519)$

46. $(0.00827) \cdot (0.0196)$

47. $(5.93 \times 10^{-4}) \cdot (8.17 \times 10^{11})$

48. $(177{,}000) \div (0.15)$

49. $(1.344 \times 10^{-8}) \div (9.6 \times 10^{-12})$

50. $(1.036 \times 10^{15}) \div (3.7 \times 10^{-4})$

51. $\dfrac{(92{,}000{,}000) \cdot (0.0036)}{(0.018) \cdot (4{,}000)}$

52. $\dfrac{(39{,}600) \cdot (0.00264)}{(0.00000132) \cdot (66{,}000{,}000)}$

53. $(31{,}000{,}000{,}000{,}000) \div (5{,}300{,}000{,}000{,}000{,}000)$

54. $(5{,}000{,}000{,}000{,}000)(0.000\,000\,000\,25)$

55. $(39{,}100{,}000{,}000)^2$

56. $(0.000\,001\,21)^2$

57. $\sqrt{4{,}000{,}000{,}000{,}000{,}000{,}000{,}000}$

58. $\sqrt{0.000\,000\,000\,225}$

59. It is estimated that there are 5 billion people on earth. The average human body manufactures approximately 2 trillion red blood cells each day. Approximate the number of red blood cells that are manufactured on earth each day.

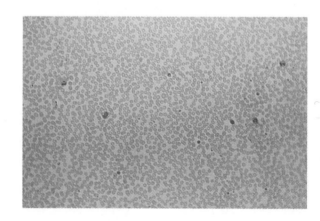

60. Show that there are about 31.5 million seconds in 1 year.

61. A light year is the distance light travels in 1 year. Light travels at about 300 million meters per second. Use this fact and the value in exercise 60 to compute an approximation to a light year.

62. There are about 3.3×10^{22} molecules of water in 1 cubic centimeter of water. There are one million cubic centimeters in a cubic meter. It is estimated that the world's oceans, lakes, and rivers contain about 1.7×10^{18} cubic meters. Use this to estimate how many molecules of water there are in the world's oceans, lakes, and rivers.

63. The amount of solar radiation reaching the surface of the earth is about 3.9 million exajoules a year. An exajoule is one billion joules of energy. The combustion of a ton of oil releases about 45.5 joules of energy. (a) Compute how many tons of oil would have to be burned to equal the total amount of solar radiation reaching the surface of the earth in one hour. (b) The annual consumption of global energy is about 350 exajoules. How many tons of oil would have to be burned to yield this amount of energy?

64. The U.S. national debt is about $3.2 trillion. (a) There are about 300 million people in the United States. How much money is the national debt per person in the United States? (b) There are about 5 billion people on earth. How much money is the U.S. national debt per person on earth?

⊛ Explore to Learn

It is theoretically possible that, with every breath you take, you breathe in at least one molecule of air that was also breathed in by George Washington. Use scientific notation and the properties of exponents to show why this is true, considering these facts: There are about 1.7×10^{21} liters of air. There are about 2.7×10^{22} molecules of air per liter. One lungfull of air is about 1 liter.

Review Exercises

1. Given $x = 2$, $y = -3$ and $z = -1$, evaluate the expression $\dfrac{4x - y}{2y + z}$. See section 2–1.

2. A piece of lumber 16 feet long is to be divided into two pieces so that one piece is 1 foot longer than twice the length of the other piece. Find the lengths of the two pieces of lumber. See section 2–7.

Directions   Simplify the following expressions. Assume all denominators are nonzero. Express answers with positive exponents only. See section 3–4.

3. $(5x^{-2}y^3)^0$

4. $\dfrac{3xy^2}{3^{-2}xy^{-1}}$

5. $(-2x^3y^2)^3$

3–6   The Quotient of Two Polynomials

In section 3–3, we observed the process of dividing a monomial by a monomial. We shall first review this process before we deal with other types of polynomial division. Recall the quotient property for division of expressions that have like bases, $a^m \div a^n = a^{m-n}$, $a \neq 0$.

● **Example 3–6 A**

Dividing a monomial by a monomial

Find the indicated quotients. Assume all variables are nonzero.

1. $x^7 \div x^4 = x^{7-4}$ Subtract exponents when dividing
    $= x^3$

2. $\dfrac{2^2 a^5}{2a^3} = 2^{2-1} a^{5-3}$ Divide like bases by subtracting exponents

    $= 2^1 a^2$ Perform indicated subtractions
    $= 2a^2$ $2^1 = 2$

☛ ***Quick check***   Find the quotient. $\dfrac{5^3 x^4}{5x}$

Division of a Polynomial by a Monomial

Consider the indicated division.

$$\frac{3x^3 - 9x^2 + 15x}{3x}$$

To perform this division, we use a principle of fractions.

$$\frac{a}{c} + \frac{b}{c} = \frac{a+b}{c}, c \neq 0$$

By reversing this equation, the principle can be used to divide a polynomial by a monomial.

Division of a Polynomial by a Monomial

$$\frac{a+b}{c} = \frac{a}{c} + \frac{b}{c}, c \neq 0$$

Concept

To divide a polynomial by a monomial, divide each term of the polynomial by the monomial.

● **Example 3–6 B**

Dividing a polynomial by a monomial

Find the indicated quotients. Assume all denominators are nonzero.

1. $\dfrac{8a^4 + 4a^2 - 12a}{4a} = \dfrac{8a^4}{4a} + \dfrac{4a^2}{4a} - \dfrac{12a}{4a}$ Divide each term of numerator by the monomial denominator

    $= 2a^3 + a - 3$ Simplify each term by reducing

2. $\dfrac{5a^7 + 15a^5 - 10a}{5a^2} = \dfrac{5a^7}{5a^2} + \dfrac{15a^5}{5a^2} - \dfrac{10a}{5a^2}$ Divide each term of numerator by the monomial denominator

    $= a^5 + 3a^3 - \dfrac{2}{a}$ Simplify each term by reducing

Recall that we can check our division by

(quotient)(divisor) = dividend

In example 1,

$$(2a^3 + a - 3)(4a) = 2a^3 \cdot 4a + a \cdot 4a - 3 \cdot 4a \qquad \text{Distributive property}$$
$$= 8a^4 + 4a^2 - 12a \qquad \text{Dividend}$$

☛ *Quick check*   Find the quotient: $\dfrac{16x^5 + 20x^3 - 4x^2}{4x^2}$

Division of a Polynomial by a Polynomial

Consider a quotient in which the divisor is not a monomial. For example,

$$\frac{y^2 - y - 2}{y - 2} \quad \xleftarrow{} \quad \text{Dividend}$$
$$\xleftarrow{} \quad \text{Divisor}$$

which involves the division of a trinomial by a binomial. We handle this just like long division with whole numbers. Set it up in the form

$$y - 2 \overline{)\, y^2 - y - 2}$$

Note. The divisor and dividend must be arranged in descending powers of one variable with zeros inserted to hold the position of any missing term.

The following table demonstrates writing a polynomial in descending powers of the variable and inserting zeros to hold the position of missing terms.

Dividend	Dividend arranged in descending powers
$x^3 + 2x + 3x^4 + 4x^2 - 1$	$3x^4 + x^3 + 4x^2 + 2x - 1$
$x^3 + x - 9$	$x^3 + 0x^2 + x - 9$
$x^4 - 1$	$x^4 + 0x^3 + 0x^2 + 0x - 1$

The method for dividing polynomials is similar to the long division used in dividing whole numbers. To demonstrate this, we divide 972 by 36 step-by-step as we divide $(y^2 - y - 2)$ by $(y - 2)$.

$$36 \overline{)\, 972} \qquad\qquad y - 2 \overline{)\, y^2 - y - 2}$$

Step 1   Divide 36 into 97, which goes 2 times. Place 2 over 7 in the dividend.

$$\begin{array}{r} 2 \\ 36 \overline{)\, 972} \end{array}$$

Divide y into y^2, which goes y times. Place y over y in the dividend.

$$\begin{array}{r} y \\ y - 2 \overline{)\, y^2 - y - 2} \end{array}$$

Step 2   Multiply 2 times 36, place 72 below 97 in the dividend.

$$\begin{array}{r} 2 \\ 36 \overline{)\, 972} \\ \underline{72} \end{array}$$

Multiply y times $(y - 2)$, place $y^2 - 2y$ below $y^2 - y$ in the dividend.

$$\begin{array}{r} y \\ y - 2 \overline{)\, y^2 - y - 2} \\ \underline{y^2 - 2y} \end{array}$$

Step 3 Subtract 72 from 97. The difference is 25.

$$
\begin{array}{r}
2 \\
36\overline{)972} \\
-72 \\
\hline
25
\end{array}
$$

Subtract $y^2 - 2y$ from $y^2 - y$.

$(y^2 - y) - (y^2 - 2y) =$
$y^2 - y - y^2 + 2y = y$

$$
\begin{array}{r}
y \\
y-2\overline{)y^2 - y - 2} \\
-(y^2 - 2y) \\
\hline
y
\end{array}
$$

Change the sign of each term and add

Step 4 Bring down the next digit of the dividend, 2.

$$
\begin{array}{r}
2 \\
36\overline{)972} \\
72 \\
\hline
252
\end{array}
$$

Bring down the next term of the dividend, −2.

$$
\begin{array}{r}
y \\
y-2\overline{)y^2 - y - 2} \\
y^2 - 2y \\
\hline
y - 2
\end{array}
$$

Step 5 Divide 36 into 252, which goes 7 times. Place 7 over 2 in the dividend.

$$
\begin{array}{r}
27 \\
36\overline{)972} \\
72 \\
\hline
252
\end{array}
$$

Divide y into y, which goes 1 time. Place 1 over 2 in the dividend with a plus sign between y and 1.

$$
\begin{array}{r}
y + 1 \\
y-2\overline{)y^2 - y - 2} \\
y^2 - 2y \\
\hline
y - 2
\end{array}
$$

Step 6 Multiply 7 times 36, which is 252. Place this product below 252 at the bottom.

$$
\begin{array}{r}
27 \\
36\overline{)972} \\
72 \\
\hline
252 \\
-252 \\
\hline
0
\end{array}
$$

Multiply 1 times $(y - 2)$, which is $y - 2$. Place this below $y - 2$ at the bottom.

$$
\begin{array}{r}
y + 1 \\
y-2\overline{)y^2 - y - 2} \\
y^2 - 2y \\
\hline
y - 2 \\
-(y - 2) \\
\hline
0
\end{array}
$$

Change signs and add

Step 7 Subtract $252 - 252 = 0$. There is no remainder.

$972 \div 36 = 27$

Subtract $(y - 2) - (y - 2) = 0$. There is no remainder.

$(y^2 - y - 2) \div (y - 2)$
$= y + 1$

Step 8   Check your division by multiplying the quotient by the divisor to see if you get the original dividend.

$$27 \cdot 36 = 972 \qquad\qquad (y + 1)(y - 2) = y^2 - y - 2$$

Note A common error is committed when we subtract polynomials as we did in step 3. Remember, to subtract two polynomials, *change the sign of each term in the second polynomial and then add.*

$$\begin{aligned} y^2 - y &\rightarrow \quad y^2 - y \\ (-)y^2 - 2y &\rightarrow \underline{-y^2 + 2y} \\ & \qquad\quad\ \ 0 + y = y \end{aligned}$$

The large majority of errors in this type of problem occur when polynomials are subtracted.

Observe that we always divide by only the first term of the divisor (y in the example above; we never actually divide by $y - 2$). The steps for dividing a polynomial by a polynomial may be summarized as follows.

To Divide a Polynomial by a Polynomial

1. Arrange the terms of the dividend and the divisor in descending powers of the same variable.
2. Insert zeros for missing terms.
3. Repeat the following four steps as necessary.
   Divide the first term into the first term; write the result in the quotient.
   Multiply each term of the divisor by this result.
   Subtract (change signs and add).
   Bring down the next term(s).

● **Example 3–6 C**

Dividing a polynomial by a polynomial

Perform the indicated divisions and check the answer. Assume that all denominators are nonzero.

1. $\dfrac{x^2 + 3x - 4}{x + 4}$

$$\begin{array}{r} x - 1 \\ x + 4 \overline{)\, x^2 + 3x - 4} \\ \underline{-(x^2 + 4x)} \\ -x - 4 \\ \underline{-(-x - 4)} \\ 0 \end{array}$$

$x(x + 4) = x^2 + 4x$

Subtract to get $-x$ and bring down -4

$-1(x + 4) = -x - 4$

Subtract to get 0

Therefore, $\dfrac{x^2 + 3x - 4}{x + 4} = x - 1$

Check: $(x - 1)(x + 4) = x^2 + 4x - x - 4 = x^2 + 3x - 4$

If we still have a remainder after "bringing down" all of the terms of the dividend, handle it as follows:

2. $\dfrac{a^2 + 5a + 6}{a - 2}$

$$\begin{array}{r} a + 7 \\ a - 2 \overline{)\, a^2 + 5a + 6} \\ \underline{-(a^2 - 2a)} \\ 7a + 6 \\ \underline{-(7a - 14)} \\ 20 \end{array}$$

$a(a - 2) = a^2 - 2a$

Subtract to get $7a$. Bring down 6

$7(a - 2) = 7a - 14$

$(7a + 6) - (7a - 14) =$
$7a + 6 - 7a + 14 = 20$

Hence, $\dfrac{a^2 + 5a + 6}{a - 2} = a + 7 + \dfrac{20}{a - 2}$, where the remainder 20 is placed over the divisor $a - 2$.

To check our answer, we add the remainder of 20 to the product of $(a + 7)$ and $(a - 2)$.

$$(a + 7)(a - 2) + (20) = a^2 - 2a + 7a - 14 + (20)$$
$$= a^2 + 5a + 6$$

3. $\dfrac{x^3 - x + 2}{x - 3}$

Note that there is no term in the dividend that contains x^2. The division will be easier to perform if the term $0x^2$ is inserted as a placeholder so that all powers of the variable x are present in descending order. Thus, we have

$$\dfrac{x^3 + 0x^2 - x + 2}{x - 3}$$

and the value of the dividend has not been changed since we have added $0x^2$, which is 0. Therefore, to perform the division, we get

$$
\begin{array}{r}
x^2 + 3x + 8 \\
x - 3 \overline{\smash{\big)}\, x^3 + 0x^2 - x + 2} \\
\underline{-(x^3 - 3x^2)} \\
3x^2 - x \\
\underline{-(3x^2 - 9x)} \\
8x + 2 \\
\underline{-(8x - 24)} \\
26
\end{array}
$$

$$\dfrac{x^3 - x + 2}{x - 3} = x^2 + 3x + 8 + \dfrac{26}{x - 3}$$

Check: $(x^2 + 3x + 8)(x - 3) + (26)$
$$= x^3 - 3x^2 + 3x^2 - 9x + 8x - 24 + (26)$$
$$= x^3 - x + 2$$

Note When performing division, always count the number of terms in the denominator (divisor). Use long division only when the denominator (divisor) has 2 or more terms.

🚩 *Quick check*   Perform the indicated division and check the answers. Assume that the denominator is nonzero. $\dfrac{6x^2 - 7x - 3}{2x - 3}$ ●

MASTERY POINTS

Can you
• Divide a monomial by a monomial?
• Divide a polynomial by a monomial?
• Divide a polynomial by a polynomial?
• Check the answer?

Exercise 3–6	☞ *Quick Check*
Directions Perform the indicated divisions and check the answers. Assume that all denominators are nonzero. See example 3–6 A.	**Example** $\dfrac{5^3 x^4}{5x}$ **Solution** $= 5^{3-1} x^{4-1}$ Divide like bases by subtracting exponents $\quad\quad = 5^2 x^3$ Perform subtractions $\quad\quad = 25x^3$ $5^2 = 25$

Directions   Perform the indicated divisions and check the answers. Assume that all denominators are nonzero. See example 3–6 A.

1. $\dfrac{8x^3}{2x}$

2. $\dfrac{-15x^5}{3x^2}$

3. $\dfrac{-65x^4 y^2 z}{13xy}$

4. $\dfrac{-28a^3 b}{-7ab}$

5. $\dfrac{3(a-b)^2}{a-b}$

6. $\dfrac{5(x+y)^3}{x+y}$

7. $\dfrac{6a^2(b-c)^2}{3a(b-c)}$

8. $\dfrac{-10x^3(y-z)^3}{2x^2(y-z)}$

9. $\dfrac{12a^3 b^2 c(x+y)^3}{-3abc(x+y)^2}$

Directions	☞ *Quick Check*
Directions Perform the indicated divisions. Assume that all denominators are nonzero. See example 3–6 B.	**Example** $\dfrac{16x^5 + 20x^3 - 4x^2}{4x^2}$ **Solution** $= \dfrac{16x^5}{4x^2} + \dfrac{20x^3}{4x^2} - \dfrac{4x^2}{4x^2}$ Divide the denominator into each term of the numerator $\quad\quad = 4x^3 + 5x - 1$ Divide constants and apply properties of exponents

Directions   Perform the indicated divisions. Assume that all denominators are nonzero. See example 3–6 B.

10. $\dfrac{6x - 9}{3}$

11. $\dfrac{24a^2 - 12a}{-6}$

12. $\dfrac{bx^2 - bx}{bx}$

13. $\dfrac{a^3 - 3a^2 + 2a}{a}$

14. $\dfrac{12x^3 - 8x^2 + 3x}{4x}$

15. $\dfrac{15a^3 - 9a^2 + 12a - 6}{3a}$

16. $\dfrac{13a - a^2 b^2 + a^2 b}{a^2 b}$

17. $\dfrac{x^2 y - xy^2 - 2xy^3}{-xy}$

18. $\dfrac{14a^2 b^3 - 21a^2 b^2 - 28ab}{7ab}$

19. $\dfrac{30x^3 y^4 + 21x^2 y^2 - 18x^2 y^4}{3x^2 y^2}$

20. $\dfrac{-21m^2 n^5 + 35m^3 n^2 - 14m^2 n^2}{-7m^2 n^2}$

21. $\dfrac{a(b-1) - c(b-1)}{b-1}$

22. $\dfrac{a(x-y) - b(x-y)}{x-y}$

Directions	⌐ *Quick Check*
Perform the indicated divisions. Assume that all denominators are nonzero. See example 3–6 C.	**Example** $\dfrac{6x^2 - 7x - 3}{2x - 3}$

Solution

$$\begin{array}{r} 3x + 1 \\ 2x-3{\overline{\smash{\big)}\,6x^2 - 7x - 3}} \\ -(6x^2 - 9x) \qquad \scriptstyle 3x(2x-3)=6x^2-9x \\ 2x - 3 \qquad \scriptstyle \text{Subtract to get } 2x \text{ and bring down } -3 \\ -(2x - 3) \qquad \scriptstyle 1(2x-3)=2x-3 \\ 0 \qquad \scriptstyle (2x-3)-(2x-3)=2x-3-2x+3=0 \end{array}$$

$$\frac{6x^2 - 7x - 3}{2x - 3} = 3x + 1$$

The check is left to the student.

Directions Perform the indicated divisions. Assume that all denominators are nonzero. See example 3–6 C.

23. $\dfrac{x^2 + 7x + 12}{x + 3}$

24. $\dfrac{a^2 + 7a + 10}{a + 5}$

25. $\dfrac{a^2 - 7a + 12}{a - 4}$

26. $\dfrac{x^2 - 7x + 10}{x - 5}$

27. $\dfrac{x^2 - 4x - 21}{x - 7}$

28. $\dfrac{x^2 - 6x - 16}{x - 8}$

29. $\dfrac{a^2 + 7a + 10}{a - 2}$

30. $\dfrac{x^2 + 8x + 15}{x + 5}$

31. $\dfrac{a^2 + 5a + 10}{a + 3}$

32. $\dfrac{x^2 - x - 72}{x + 8}$

33. $(a^2 + 6a + 10) \div (a + 3)$

34. $(4a^2 + 1 + 4a) \div (2a + 1)$

35. $\dfrac{2x^2 - 7x - 13}{x + 5}$

36. $\dfrac{2a^2 + a - 13}{a + 3}$

37. $\dfrac{6x^2 - 19x + 18}{2x - 3}$

38. $\dfrac{6x^2 - 23x + 25}{3x - 4}$

39. $(9a^2 - 24a + 12) \div (3a - 4)$

40. $(27a^3 - 1) \div (3a - 1)$

41. $(x^3 - 8) \div (x - 2)$

42. $(x^4 - 14) \div (x - 2)$

43. $\dfrac{x^3 + 4x^2 + 7x + 6}{x + 2}$

44. $\dfrac{2a^3 - 3a^2 - 13a + 12}{a - 5}$

45. $\dfrac{b^3 + 6b^2 + 7b - 8}{b - 1}$

46. $\dfrac{6x^4 - x^3 - 2x^2 - 7x - 19}{2x - 3}$

47. $(15a^2 + 28a - 32) \div (5a - 4)$

48. $(x^4 - 2x^3 + 4x^2 - x + 3) \div (x^2 - x + 4)$

49. $(x^4 + 3x^3 - 6x^2 + 3x - 8) \div (x^2 + 3x - 5)$

50. $(y^4 + 2y^3 - 4y + 2) \div (y^2 - y + 1)$

51. $(y^4 + 2y - 3) \div (y^2 + 2y - 5)$

52. A jet aircraft is flying at 400 miles per hour, which is also about 600 feet per second. It launches a missile that will accelerate and eventually fly into space to place a satellite in orbit. This missile accelerates at 20 feet per second per second. Under these conditions, the distance traveled by the aircraft in t seconds is $600t$. The distance traveled by the missile after launch is $600t + 10t^2$. Write an expression that describes the ratio of the distance traveled by the missile to the distance traveled by the aircraft. Transform this expression using division.

53. What polynomial when divided by $-2x + 5$ yields the quotient $3x^3 - 2x + 6$?

54. What polynomial when divided by $3x - 2$ yields the quotient $2x^2 + 3x - 5$?

🌀 Explore to Learn

Algebraic long division will show that
$$\frac{x^3 - 3x^2 - x + 5}{x - 1} = x^2 - 2x - 3 + \frac{2}{x - 1}.$$
Suppose that the quotient $\dfrac{x^3 - 3x^2 - x + 5}{x - 1}$ has to be evaluated thousands of times, over and over. This could easily happen in a computer program that is being applied to an engineering application. Suppose that an error of 1% would be acceptable in the result (also typical of applied situations). The polynomial $x^2 - 2x - 3$ could be evaluated much faster than the quotient. Could this be used instead?

a. As x gets larger and larger the value of $\dfrac{2}{x - 1}$ gets smaller and smaller. If we think of this as the error between $\dfrac{x^3 - 3x^2 - x + 5}{x - 1}$ and $x^2 - 2x - 3$, there is a point where the error is less than 1%. Assume that $x > 1$. Find the smallest integer value of x so that the percent of error is less than 1%. This requires calculating $\dfrac{\text{error}}{\text{value}} \cdot 100$ for larger and larger values of x until the result is less than 1. Thus, evaluate

$$\frac{\dfrac{2}{x - 1}}{\dfrac{x^3 - 3x^2 - x + 5}{x - 1}} \cdot 100$$ for larger and larger values of

x until the result is less than 1. For all values of x greater than or equal to this value the computer program can just calculate $x^2 - 2x - 3$.

b. Most computers take longer to multiply/divide than add/subtract. Assume that a certain microcomputer takes 16 machine cycles to multiply or divide and 2 machine cycles to add or subtract. Suppose also that expressions like x^3 are calculated as $x \cdot x \cdot x$. Calculate how many machine cycles are required to compute the quotient $\dfrac{x^3 - 3x^2 - x + 5}{x - 1}$ and how many cycles are required to compute $x^2 - 2x - 3$. What is the time savings as a percent?

Review Exercises

Directions　Perform the indicated addition or subtraction. See section 2–2.

1. $3x + 2x$

2. $6a^2 - 3a^2$

3. $5ab + 7ab$

Directions　Perform the indicated multiplication and simplify. See section 3–2.

4. $x^2(x + 2)$

5. $3a(2a - 5)$

6. $x^2y(3x + 2y - 7)$

Chapter 3 Lead-in Problem

There are approximately 3.3×10^{22} molecules of water in one cubic centimeter (cm³) of water. There are one million cubic centimeters in one cubic meter (m³). It is estimated that the five Great Lakes contain about 22.7 trillion cubic meters of water. Use this information to estimate how many molecules of water there are in the Great Lakes.

Solution

Number of molecules of water in 1 cm³	1,000,000 cm³ = 1m³	22.7 trillion m³ is 2.27×10^{13}	Number of molecules of water in the Great Lakes

$$3.3 \times 10^{22} \cdot 1 \times 10^6 \cdot 2.27 \times 10^{13} = 7.491 \times 10^{41}$$

Chapter 3 Summary

• Glossary

exponential form (page 185)   a base with an exponent
FOIL (page 194)   an acronym indicating the multiplication to be performed when multiplying two binomials.

scientific notation (page 210)   a number written as $a \times 10^n$, where $1 \leq a < 10$, and n is an integer.

• New Symbols and Notation

a^n a to the nth power

a^{-n} a to the negative nth power

a^0 a to the zero power

$a \times 10^n$ scientific notation

• Properties and Definitions

Definition of exponents (page 186)

$a^n = \underbrace{a \cdot a \cdot a \cdots a}_{n \text{ factors of } a}$, where n is a positive integer.

Product property of exponents (page 187)
$$a^m \cdot a^n = a^{m+n}$$

Group of factors to a power property of exponents (page 188)
$$(ab)^n = a^n b^n$$

Power to a power property of exponents (page 189)
$$(a^m)^n = a^{m \cdot n}$$

The square of a binomial (page 196)
In general, for real numbers a and b,
$$(a + b)^2 = a^2 + 2ab + b^2$$
and
$$(a - b)^2 = a^2 - 2ab + b^2$$

The difference of two squares (page 196)
For real numbers a and b,
$$(a + b)(a - b) = a^2 - b^2$$

Fraction to a power property of exponents (page 200)
$$\left(\frac{a}{b}\right)^n = \frac{a^n}{b^n}, b \neq 0$$

Quotient property of exponents (page 201)
$$a^m \div a^n = \frac{a^m}{a^n} = a^{m-n}, a \neq 0$$

Definition of negative exponents (page 202)
$$a^{-n} = \frac{1}{a^n}, a \neq 0$$

Definition of zero as an exponent (page 203)
$$a^0 = 1, a \neq 0$$

Division of a polynomial by a monomial (page 216)
$$\frac{a + b}{c} = \frac{a}{c} + \frac{b}{c}, c \neq 0$$

• Procedures

[3–2]

Multiplying two polynomials (page 194)
When we are multiplying two polynomials, we multiply each term in the first polynomial by each term in the second polynomial. We then combine like terms.

The square of a binomial (page 195)

1. The first term of the product is the square of the first term of the binomial.

2. The second term of the product is two times the product of the two terms of the binomial.

3. The third term of the product is the square of the second term of the binomial.

[3–5]

To write a number in scientific notation (page 210)

1. Move the decimal point in the original number to the right of the first nonzero digit.

2. Count the number of places, n, the decimal point has been moved. This number is the exponent of 10.

3. If

   a. the decimal point is moved to the left, n is positive.

   b. the decimal point is moved to the right, n is negative.

   c. the decimal point already follows the first nonzero digit, n is zero.

To convert a number in scientific notation to its standard form (page 211)

When the power of 10, n, is

1. positive, the decimal point is moved to the right n places.

2. negative, the decimal point is moved to the left $|n|$ places.

3. zero, the decimal point is not moved.

[3–6]

To divide a polynomial by a polynomial (page 219)

1. Arrange the terms of the dividend and the divisor in descending powers of the same variable.

2. Insert zeros for missing terms.

3. Repeat the following four steps as necessary.

   Divide the first term into the first term; write the result in the quotient.

   Multiply each term of the divisor by this result.

   Subtract (change signs and add).

   Bring down the next term(s).

☼ Chapter 3 Error Analysis

1. Exponents
   Example: $y^3 = 3 \cdot y$
   Correct answer: $y^3 = y \cdot y \cdot y$
   What error was made? *(see page 186)*

2. Multiplication of like bases
   Example: $4^2 \cdot 4^3 = 16^5$
   Correct answer: $4^2 \cdot 4^3 = 4^5$
   What error was made? *(see page 187)*

3. Power of a power
   Example: $(a^3)^2 = a^5$
   Correct answer: a^6
   What error was made? *(see page 189)*

4. Multiplying unlike bases
   Example: $x^3 \cdot y = (xy)^4$
   Correct answer: $x^3 \cdot y = x^3 y$
   What error was made? *(see page 187)*

5. Product of a monomial and a multinomial
   Example: $-x(x^2 - 2x + 1) = x^3 - 2x^2 + x$
   Correct answer: $-x^3 + 2x^2 - x$
   What error was made? *(see page 193)*

6. Squaring a binomial
   Example: $(4x + y)^2 = (4x)^2 + (y)^2 = 16x^2 + y^2$
   Correct answer: $16x^2 + 8xy + y^2$
   What error was made? *(see page 196)*

7. Dividing like bases
   Example: $\dfrac{x^3}{x} = x^4$
   Correct answer: x^2
   What error was made? *(see page 201)*

8. Negative exponents
   Example: $\dfrac{1}{a^{-3}} = -a^3$
   Correct answer: $\dfrac{1}{a^{-3}} = a^3$
   What error was made? *(see page 202)*

9. Zero exponent
Example: $(x + y)^0 = x^0 + y^0 = 1 + 1 = 2$
Correct answer: $(x + y)^0 = 1$
What error was made? *(see page 203)*

10. Evaluate absolute value
Example: $-|-4| = 4$
Correct answer: $-|-4| = -4$
What error was made? *(see page 40)*

11. Dividing a polynomial by a monomial
Example: $\dfrac{y - y^2}{y} = \dfrac{\overset{1}{\cancel{y}} - y^2}{\underset{1}{\cancel{y}}} = 1 - y^2$
Correct answer: $1 - y$
What error was made? *(see page 216)*

12. Dividing a polynomial by a polynomial
Example: $\dfrac{x^2 - 3x - 4}{x + 1}$

$$
\begin{array}{r}
x - 2 \\
x + 1 \overline{)\, x^2 - 3x - 4} \\
\underline{x^2 + x} \\
-2x - 4 \\
\underline{-2x - 2} \\
-6
\end{array}
= x - 2 - \dfrac{6}{x + 1}
$$

Correct answer: $x - 4$
What error was made? *(see page 219)*

13. Negative exponents
Example: $\dfrac{1}{2y^{-2}} = 2y^2$
Correct answer: $\dfrac{y^2}{2}$
What error was made? *(see page 202)*

14. Scientific notation
Example: $(4 \times 10^3)(2 \times 10^4) = 8 \times 10^{12}$
Correct answer: 8×10^7
What error was made? *(see page 211)*

Chapter 3 Critical Thinking

1. Given the number 52^2, determine a method by which you can square 52 mentally.

2. Find two monomials
 a. whose sum is $6x^5$
 b. whose product is $6x^5$
 c. whose difference is $6x^5$
 d. whose quotient is $6x^5$

3. Our current notation for exponents developed only about 300 years ago. Other notations could have developed. In fact, this is almost required in a computer language like Pascal, which has no exponents. In Pascal one could create what is called a Pascal expression for exponents. In the notation Pascal requires, 5^3 might look like POWER(5, 3), and x^{-2} might be POWER(X, −2). Since Pascal represents multiplication with ∗, an expression like $3x^2 - 2x^3y^5$ would be written 3∗POWER(X, 2)−2∗POWER(X, 3)∗POWER(Y, 5). Translate all the *Properties and definitions of exponents* listed in the chapter 3 summary into this notation. For example, $a^0 = 1$ translates into POWER(A, 0) = 1. *Note:* Pascal uses / to indicate division; there are no fractions. Thus, $\dfrac{a}{b}$ is A/B in Pascal. Addition and subtraction are the conventional + and −.

4. A person once saved the life of a certain very rich monarch. The monarch, wishing to reward the individual, asked how to repay this person, who told the monarch "Take a chess board; put 1 cent on the first square, 2 cents on the second, 4 cents on the third, and so on, doubling this amount each time. I wish to be paid the amount which you put on the last (64th) square." This sounded good to the monarch, who agreed. Estimate the amount paid to the individual. If this person shared this money equally with every person on earth, how much would each person get? (Assume that there are 5 billion people.)

5. The weight of the earth is about 1.32×10^{25} pounds. The average radius of the earth is about 3,960 miles. The volume of a sphere (the earth is approximately a sphere) is $\frac{4}{3}\pi r^3$, where r is the radius of the sphere. Compute the average density of the earth in pounds per cubic inch as follows:

   a. Convert the radius of the earth to inches. Remember there are 5,280 feet in one mile, and 12 inches in one foot.

   b. Compute the volume of the earth in cubic inches using the result of part a.

   c. Divide the result of part b into the weight of the earth.

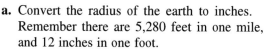
Chapter 3 Review

[3–1]

Directions   Simplify by using the properties of exponents.

1. $a^5 \cdot a^7$

2. $a \cdot a^4 \cdot a^9$

3. $4^3 \cdot 4^2$

4. $(xy)^4$

5. $(a^3)^5$

6. $(5ab^3)(4a^3b^2)$

7. $(3x^2y^3)(2xy^4)$

8. $(-5x^2)(3x^3)$

9. $(2a^2b)(3ab^4)$

10. $(5x^2y)(2x^3y^4)$

11. $(-3a^2b^3)(2a^4b^7)$

12. Write an expression for 4 times the cube of x.

13. Write an expression for the square of x, increased by 6.

14. Write an expression for the product of x squared and y cubed.

15. Write an expression for 10 more than the square of a number.

[3–2]

Directions   Perform the indicated multiplication and simplify.

16. $5x(3x - 2y)$

17. $-3a^2b(2a^2 - 3ab + 4b^2)$

18. $(y - 5)^2$

19. $(x + 3)(x - 4)$

20. $(x + 5)^2$

21. $(a - 7)(a + 7)$

22. $(5x - y)(3x + 2y)$

23. $(x - 2y)(x^2 + 3xy + y^2)$

24. $(0.2x + 0.3)(0.6x - 0.4)$

25. $(2a + b)^2$

[3–3]

Directions   Simplify and leave the answers with only positive exponents. Assume that all variables are nonzero.

26. $\dfrac{b^5}{b^7}$

27. $5a^{-2}$

28. $a^{-5} \cdot a^9$

29. $\dfrac{x^3 x^2}{x^8}$

30. $3b^0$

31. $(a^{-3})^2$

32. $x^5 x^{-2} x^0$

33. $\left(\dfrac{a}{b}\right)^5$

34. $\left(\dfrac{2yz}{x}\right)^2$

35. $\dfrac{2a^2 b^4}{2^3 a^5 b}$

36. $\dfrac{a^5 b^{-2}}{a^{-4} b}$

[3–4]

Directions   Simplify and leave the answers with only positive exponents. Assume that all variables are nonzero.

37. $(2a^2 b^3)^3$

38. $(3^3 x^4 y^5)^4$

39. $(2xy^{-3})^{-2}$

40. $\dfrac{2x^{-1} y^0 z^3}{4x^{-2} y^{-3}}$

41. $\dfrac{8a^{-5} b^{-4} c^0}{4a^{-7} b^2 c^{-3}}$

42. $(x^5 y^4)^4 (2x^2 y^3)^3$

43. $(2a^2 b)^3 (3a^4)^2$

44. $\left(\dfrac{a^3 b^0}{c^4}\right)^3$

45. $\left(\dfrac{3a^3 b^2}{c^5}\right)^2$

46. $\left(\dfrac{2xy^4}{z^6}\right)^5$

[3–5]

Directions   Express the following numbers in scientific notation.

47. 1,840

48. 0.00157

49. 107,000,000

50. 849,000,000,000

51. 37.5

52. 0.00543

Directions   Express the following numbers in standard form.

53. 5.04×10^5

54. 6.39×10^{-3}

55. 5.96×10^2

56. 8.86×10^{-3}

57. 7.35×10^{-7}

58. 8.12×10^8

Directions Perform the indicated operations using scientific notation. Leave the answer in scientific notation.

59. $(456,000,000) \cdot (0.000587)$

60. $(0.0000183) \cdot (0.000846)$

61. $(756,000) \div (105,000,000)$

62. $(0.00525) \div (42,000)$

63. The average human body manufactures approximately 2×10^{12} red blood cells each day. In 72 years there are approximately 2.6×10^4 days. Approximately how many red blood cells does the average human body produce in 72 years?

64. One gallon of oil releases 1.4×10^5 BTUs of heat energy when burned. If the Grutz Crude Oil Refinery has 6.3×10^7 gallons in its storage tanks, what is the total amount of energy, in BTUs, contained in the tanks?

Directions Find the indicated quotients. Assume that all denominators are nonzero.

65. $\dfrac{8a^3(a + b)^2}{4a^2(a + b)}$

66. $\dfrac{2a^2 - 3a + 5a^3}{a}$

67. $\dfrac{5x^2y - 3xy^4 + x^2y^2}{xy}$

68. $\dfrac{8a^3b + 12a^2b^2 - 24a^3b^7}{4a^2b}$

69. $\dfrac{8a^2 - 2a - 3}{2a - 1}$

70. $\dfrac{3a^2 - 17a + 11}{a - 5}$

71. $\dfrac{x^2 - 49}{x + 7}$

72. $\dfrac{20x^3 - 19x^2 - 13x + 12}{4x - 3}$

Chapter 3 Test

Directions Perform the indicated operations. Leave all answers with only positive exponents. Assume that the variables are nonzero.

1. $2^3 \cdot 2^2$

2. a^2aa^5

3. $2a(3a - 2b)$

4. $5x^0$

5. $a^3a^0a^{-1}$

6. $(y - 7)^2$

7. $\dfrac{x^4x^3}{x^5}$

8. $(3x^2y^4z)^3$

9. $(3x - 2)(3x + 2)$

10. $\left(\dfrac{3a}{b}\right)^3$

11. $\dfrac{3^2a^4b}{3a^2b^7}$

12. $(-2x^2y)(3xy^3)$

13. $(x + 6)^2$

14. $(2x^2y^{-2})^{-2}$

15. $\left(\dfrac{2a^2}{b^3}\right)^3$

16. $(2a + 3b)(a - 4b)$

17. $\dfrac{4x^{-2}y^2}{8x^{-1}y^{-3}z^0}$

18. $\dfrac{5.98 \times 10^{27}}{9.2 \times 10^{22}}$

Directions   Find the indicated quotient. Assume that all denominators are nonzero.

19. $\dfrac{9x^2(y-z)^2}{3x(y-z)}$

20. $\dfrac{16a^5 - 12a^4 + 2a^2}{2a^2}$

21. $(x^3 + 3x^2 - x + 12) \div (x + 4)$

22. $(4a^3 - 3a - 2) \div (2a - 1)$

23. $(3x^3 - 10x^2 + 3x + 11) \div (x - 2)$

24. Write an expression for five times the square of y.

25. A 12,000-gallon swimming pool holds approximately 4.5×10^7 cubic centimeters of water. If each cubic centimeter of water has approximately 3.3×10^{22} water

molecules, about how many molecules of water would this swimming pool hold?

Chapter 3 Cumulative Test

Directions   Perform the indicated operations, if possible, and simplify. Assume that the variables in 14 and 15 and the denominator in 16 are nonzero.

1. $\dfrac{(8)}{(-4)}$

2. $\dfrac{(-9)}{0}$

3. $(-2)(4)(0)(-4)$

4. $48 - 24 \div 8 - 3 - 2^2$

5. $(3x^2y - 4xy + 2xy^2) - (2x^2y - 4xy + 3x^2y^2)$

6. $(3x - y)^2$

7. $(2a^2b^5)^2$

8. -5^2

9. $10 - 10 \div 10 \cdot 10 - 10 + 10$

10. $(3x - 2y)(3x + 2y)$

11. $(3x^2y)(2xy^4)$

12. $2[5(7 - 4) - 6 + 4]$

13. $(x + 1)(x^2 - x - 1)$

14. $x^{-3}x^5x^0$

15. $\dfrac{a^{-5}}{a^{-9}}$

16. Divide $(x^2 - 8x + 13) \div (x - 2)$

Directions   Find the solution set.

17. $3x - 4 = x + 10$

18. $2(x - 4) + 7 = 8x - 11$

19. $\dfrac{2}{3}x + 4 = \dfrac{5}{6}$

Directions   Find the solution.

20. $8 - 3x < 9$

21. $6x + 5 - 4 > 2$

22. $-9 \le 2x + 7 \le 5$

23. If a number is decreased by 17 and that result is then divided by 5, the answer is 16. Find the number.

24. A rectangular picture that is 9 inches long and 5 inches wide is to be enlarged so that the enlargement will be 24 inches wide. What should be the length of the enlargement?

25. Brenda invested part of $30,000 at 8% and the rest at 7%. If her income for 1 year from the 8% investment was $675 more than that from the 7% investment, how much was invested at each rate?

4

Factoring and Solution of Quadratic Equations by Factoring

The formula $s = vt - 16t^2$ gives the height s in feet that an object will travel in t seconds if it is propelled directly upward with an initial velocity of v feet per second. If an object is thrown upward at 96 feet per second, how long will it take the object to reach a height of 144 feet?

4–1   Factors; the Greatest Common Factor

Greatest Common Factor

To find the solution set of certain equations that are not linear, we will need to study a technique called **factoring a polynomial**. Factoring polynomials will also be useful in dealing with algebraic fractions since, as we have seen with arithmetic fractions, we must have the numerator and the denominator of the fractions in a factored form to reduce or to find the least common denominator.

The ability to factor is also related to the security of a newly developed cryptography (coded messages) system that uses large numbers that are the product of two primes as a defense against being decoded quickly.

The first type of factoring that we will do involves finding the **greatest common factor** (GCF) of all the terms in the polynomial. Recall the statement of the distributive property.

$$a(b + c) = ab + ac$$

$a(b + c)$ is called the **factored form** of $ab + ac$. We now use the distributive property to write $3x + 6$ in the factored form. First we notice that $3x + 6$ can be written as

$$3 \cdot x + 3 \cdot 2 \qquad \text{3 is a common factor in both terms}$$

Then thinking $ab + ac = a(b + c)$, we can apply the distributive property to get

$$3x + 6 = 3 \cdot x + 3 \cdot 2 = 3(x + 2)$$

$3(x + 2)$ is the factored form of $3x + 6$.

This type of factoring, as its name implies, involves looking for numbers or variables that are **common factors** in *all* of the original terms. In our example, the number 3 was common to all of the original terms, and we were able to factor it out.

————————————————————————————————— *Multiplying*

Polynomial *(terms)*	Distributive Property *(determine the GCF)*	Factored Form *(factors)*
$3x + 6$	$\mathbf{3} \cdot x + \mathbf{3} \cdot 2$	$3(x + 2)$
$10x^2 + 15y$	$\mathbf{5} \cdot 2x^2 + \mathbf{5} \cdot 3y$	$5(2x^2 + 3y)$
$12a - 42b$	$\mathbf{6} \cdot 2a - \mathbf{6} \cdot 7b$	$6(2a - 7b)$
$18xy + 12xz$	$\mathbf{6x} \cdot 3y + \mathbf{6x} \cdot 2z$	$6x(3y + 2z)$

Factoring ———————————————————————————————→

Factoring the Greatest Common Factor

1. Factor each term such that it is the product of primes[1] and variables in exponential form.
2. Write down all the factors from step 1 that are common to **every** term and raise each of them to the lowest power that they are raised to in any of the terms. **The product of these factors is the greatest common factor**, GCF.
3. Determine the polynomial within the parentheses by dividing each term of the polynomial being factored by the GCF.
4. The answer is the product of steps 2 and 3.

● **Example 4–1 A**

Factoring out the greatest common factor

Write in completely factored form.

1. $7a^3 + 14a$

$= 7a(\quad + \quad)$ ← Polynomial factor will have as many terms as the original expression

GCF is $7a$

$= 7a(a^2 + 2)$ Completely factored form

$$\frac{14a}{7a}$$

$$\frac{7a^3}{7a}$$

If we want to check the answer, we apply the distributive property and perform the multiplication as follows:

$7a(a^2 + 2) = 7a \cdot a^2 + 7a \cdot 2$ Distributive property

$= 7a^3 + 14a$ Carry out the multiplication

[1]Primes and prime factorization are covered in section 1–2.

2. $9x^5 + 6x^3 - 18x^2$

$\quad = 3^2x^5 + 2 \cdot 3x^3 - 2 \cdot 3^2x^2$ Factor each term

$\quad = 3x^2(\quad + \quad - \quad)$ Determine the GCF

$\quad = 3x^2(3x^3 + 2x - 6)$ Completely factored form

3. $3x^3y^2 + 15x^2y^4 + 3xy^2$

$\quad = 3 \cdot x^3 \cdot y^2 + 3 \cdot 5 \cdot x^2 \cdot y^4 + 3 \cdot x \cdot y^2$ Factor each term

$\quad = 3xy^2(\quad + \quad + \quad)$ Determine the GCF

$\quad = 3xy^2(x^2 + 5xy^2 + 1)$ Completely factored form

<u>Note</u> In example 3, the last term in the factored form is 1. This situation occurs when a term and the GCF are the same, that is, whenever we are able to factor all of the numbers and variables out of a given term. For example, $\dfrac{3xy^2}{3xy^2} = 1$. The number of terms inside the parentheses must be equal to the number of terms in the original polynomial.

📏 **Quick check** Write $9x + 6y + 3$ and $9x^3y^2 - 18x^2y^3$ in completely factored form. ●

Remember that when an expression is within a grouping symbol, we treat the quantity as just one factor. Therefore if we have a quantity common to all of the terms, we can factor it out of the polynomial.

● **Example 4–1 B**

Factoring out a common quantity

Write in completely factored form.

1. $x(a - 2b) + y(a - 2b)$

The quantity $(a - 2b)$ is common to both terms. We then factor the common quantity out of each term and place the remaining factors from each term in the second parentheses.

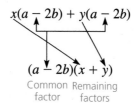

$$x(a - 2b) + y(a - 2b)$$

$$(a - 2b)(x + y)$$

Common   Remaining
factor   factors

2. $x(a + b) + (a + b)$

$\quad = (a + b)(\quad + \quad)$ Determine the GCF

$\quad = (a + b)(x + 1)$ Completely factored form

3. $3x^2(2a - b) - 9x(2a - b)$

$\quad = 3x^2(2a - b) - 3^2x(2a - b)$ Factor each term

$\quad = 3x(2a - b)(\quad - \quad)$ Determine the GCF

$\quad = 3x(2a - b)(x - 3)$ Completely factored form

📏 **Quick check** Write in completely factored form. $x(y + 5) - z(y + 5)$. ●

Four-term Polynomials

Consider $ax + ay + bx + by$. We observe that this is a *four-term polynomial* and we will *try* to factor it by grouping.

$$ax + ay + bx + by = (ax + ay) + (bx + by)$$

There is a common factor of a in the first two terms and a common factor of b in the last two terms.

$$(ax + ay) + (bx + by) = a(x + y) + b(x + y)$$

The quantity $(x + y)$ is common to both terms. Factoring it out, we have

$$a(x + y) + b(x + y) = (x + y)(a + b)$$

Therefore we have factored the polynomial by grouping.

Factoring a Four-term Polynomial by Grouping

1. Arrange the four terms so that the first two terms have a common factor and the last two terms have a common factor.
2. For each pair of terms determine the GCF and factor it out.
3. If step 2 produces a binomial factor common to both terms, factor it out.
4. If step 2 does not produce a binomial factor common to both terms, try grouping the terms of the original polynomial in different possible ways.
5. If step 4 does not produce a binomial factor common to both terms, the polynomial will not factor by this procedure.

● **Example 4–1 C**

Factoring by grouping

Write in completely factored form.

1. $ax + 2ay + bx + 2by$
$= (ax + 2ay) + (bx + 2by)$ Group in pairs
$= a(x + 2y) + b(x + 2y)$ Factor out the GCF
$= (x + 2y)(a + b)$ Factor out the common binomial

2. $3ac + 6ad - 2bc - 4bd$
$= (3ac + 6ad) - (2bc + 4bd)$ Group in pairs
$= 3a(c + 2d) - 2b(c + 2d)$ Factor out the GCF
$= (c + 2d)(3a - 2b)$ Factor out the common binomial

3. $6ax + by + 3ay + 2bx$
$= 6ax + 3ay + 2bx + by$ Rearrange the terms
$= (6ax + 3ay) + (2bx + by)$ Group in pairs
$= 3a(2x + y) + b(2x + y)$ Factor out the GCF
$= (2x + y)(3a + b)$ Factor out the common binomial

<u>Note</u> As in example 3, sometimes the terms must be rearranged so that the pairs will have a common factor.

☞ **Quick check**   Write in completely factored form. $3ax + 6bx + 2ay + 4by$. ●

It is important to remember to look for the greatest common factor first *when we attempt to determine the completely factored form of any polynomial. If we fail to do this, the answer may not be in a completely factored form or we may not see how to factor the polynomial by an appropriate procedure.*

MASTERY POINTS

Can you
- Factor out the greatest common factor?
- Factor a four-term polynomial by grouping?

Exercise 4–1

Directions
Write in completely factored form. See example 4–1 A.

🏴 **Quick Check**

Examples $9x + 6y + 3$ $\qquad\qquad\qquad\qquad 9x^3y^2 - 18x^2y^3$

Solutions $= 3 \cdot 3x + 3 \cdot 2y + 3 \cdot 1$ Factor each term $= 9x^2y^2 \cdot x - 9x^2y^2 \cdot 2y$ Factor each term

$\qquad\qquad = 3(\quad + \quad + \quad)$ Determine the GCF $= 9x^2y^2(\quad - \quad)$ Determine the GCF

$\qquad\qquad = 3(3x + 2y + 1)$ Completely factored form $= 9x^2y^2(x - 2y)$ Completely factored form

---

Directions   Write in completely factored form. See example 4–1 A.

1. $2y + 6$
2. $3a - 12$
3. $4x^2 + 8y$
4. $8y^2 + 10x^2$
5. $3x^2y + 15z$
6. $5r^2 + 10rs - 20s$
7. $7a - 14b + 21c$
8. $8x - 12y + 16z$
9. $15xy - 18z + 3x^2$
10. $18ab - 27a + 3ac$
11. $42xy - 21y^2 + 7$
12. $15a^2 - 27b^2 + 12ab$
13. $8x - 10y + 12z - 18w$
14. $15L^2 - 21W^2 + 36H$
15. $20a^2b - 60ab + 45ab^2$

16. $4x^2 + 8x$
17. $3x^2y + 6xy$
18. $8x^3 + 4x^2$
19. $2R^4 - 6R^2$
20. $3x^2 - 3xy + 3x$
21. $2x^3 - x^2 + x$
22. $24a^2 + 12a - 6a^3$
23. $15ab + 18ab^2 - 3a^2b$
24. $2x^4 - 6x^2 + 8x$
25. $xy^2 + xyz + xy$
26. $3R^2S - 6RS^2 + 12RS$
27. $2L^3 - 18L + 2L^2$
28. $V^2 + V^3 - V^4 + 2V$
29. $5p^2 + 10p + 15p^3$
30. $16x^3y - 3x^2y^2 + 24x^2y^3$

Directions
Write in completely factored form. See example 4–1 B.

🏴 **Quick Check**

Example $x(y + 5) - z(y + 5)$

Solution $= (y + 5)(\quad - \quad)$ Determine the GCF
$\qquad\quad = (y + 5)(x - z)$ Completely factored form

Directions   Write in completely factored form. See example 4–1 B.

31. $x(a + b) + y(a + b)$
32. $3a(x - y) + b(x - y)$

33. $15x(2a + b) + 10y(2a + b)$
34. $21R(L + 2N) - 35S(L + 2N)$

35. $3x(a + 4b) + 6y(a + 4b)$

36. $4RS(2P + q) - 8RT(2P + q)$

37. $8a(b + 6) - (b + 6)$

38. $x(x + 2) + 3(x + 2)$

39. $a(a - 3) + 5(a - 3)$

40. $3x(2x - 1) - 4(2x - 1)$

41. $2x(4x - 3) - 3(4x - 3)$

Directions	Quick Check
Write in completely factored form. See example 4–1 C.	**Example** $3ax + 6bx + 2ay + 4by$ **Solution** $= (3ax + 6bx) + (2ay + 4by)$ Group in pairs $= 3x(a + 2b) + 2y(a + 2b)$ Factor out the GCF $= (a + 2b)(3x + 2y)$ Factor out the common binomial

Directions Write in completely factored form. See example 4–1 C.

42. $rt + ru + st + su$

43. $ac + ad + bc + bd$

44. $5ax - 3by + 15bx - ay$

45. $x^2 + 3x + 4x + 12$

46. $x^2 + 2x + 4x + 8$

47. $x^2 + 3x + 3x + 9$

48. $x^2 + 4x + 4x + 16$

49. $a^2 + 3a + 5a + 15$

50. $a^2 + 6a + a + 6$

51. $x^2 + 7x + x + 7$

52. $y^2 + 3y + y + 3$

53. $6ax - 2by + 3bx - 4ay$

54. $2ax^2 - bx^2 + 6a - 3b$

55. $4ax + 2ay - 2bx - by$

56. $ac + 3ad - 4bc - 12bd$

57. $20x^2 + 5xz - 12xy - 3yz$

58. $a^2x + 3a^2y - 3x - 9y$

59. $4ax + 12bx - 3ay - 9by$

60. $ac + ad - 2bc - 2bd$

61. $2ac + 6bc - ay - 3by$

62. $2ac + bc - 4ay - 2by$

63. $2ac + 3bc + 8ay + 12by$

64. $5ac - 3by + 15bc - ay$

65. $6ax + by + 2ay + 3bx$

66. $2ax - ad + 4bx - 2bd$

67. $3ax - 2bd - 6ad + bx$

68. $6ax + 3bd - 2ad - 9bx$

69. $2a^3 + 15 + 10a^2 + 3a$

70. $3a^3 - 6a^2 + 5a - 10$

71. $8a^3 - 4a^2 + 6a - 3$

Directions Write in completely factored form. See example 4–1 A.

72. The area of the surface of a cylinder is determined by $A = 2\pi rh + 2\pi r^2$. Factor the right side. (π is the Greek letter pi.)

73. The total surface area of a right circular cone is given by $A = \pi rs + \pi r^2$. Factor the right side.

74. The equation for the distance traveled by a rocket fired vertically upward into the air is given by $S = 560t - 16t^2$, where the rocket is S feet from the ground after t seconds. Factor the right side.

75. In engineering, the equation for deflection of a beam is given by

$$Y = \frac{2wx^4}{48EI} - \frac{3\ell wx^3}{48EI} - \frac{\ell^3 wx}{48EI}$$

Factor the right side.

Horner's method

Suppose you needed to evaluate the polynomial $x^2 - 3x - 4$ for many values of x. A method called Horner's method can help. Write $x^2 - 3x - 4$ as $x(x - 3) - 4$ by factoring the first two terms. Now suppose x is 5. Then evaluating $x(x - 3) - 4$ for $x = 5$ means calculate

$$5(5 - 3) - 4 \qquad \text{Subtract 3 from 5}$$
$$5(2) - 4 \qquad \text{Multiply the result by 5}$$
$$10 - 4 \qquad \text{Subtract 4}$$
$$6$$

To evaluate $x(x - 3) - 4$ for any value of x consists of the same simple steps: (1) start with the value of x; (2) subtract 3; (3) multiply by the value; (4) subtract 4. This is much easier than computing $5^2 - 3(5) - 4$ (try it).

Even more work is saved for something like $2x^3 - 5x^2 + 3x - 6$; it is factored as

$$x(2x^2 - 5x + 3) - 6$$
$$x[x(2x - 5) + 3] - 6$$

which corresponds to the following steps for evaluating for a given value; (1) start with the value of x; (2) multiply by 2; (3) subtract 5; (4) multiply by the value; (5) add 3; (6) multiply by the value; (7) subtract 6.

Directions   Use Horner's method to evaluate the expression in each problem for $x = 2, 5, -3, \frac{1}{2}$.

76. $x^2 + 4x - 9$

77. $2x^2 - 3x - 1$

78. $x^3 - 3x^2 + 2x - 1$

79. $2x^3 - 2x^2 + 3x - 5$

⚙ Explore to Learn

Most computers take much longer to multiply than to add or subtract. This means that an expression like $x[x(2x - 5) + 3] - 6$ can be much more efficient than the equivalent expression $2x^3 - 5x^2 + 3x - 6$. Explain why by counting the number of operations necessary to compute the expression in each form.

Review Exercises

Directions   Perform the indicated multiplication and simplify. See section 3–2.

1. $(x + 6)(x + 3)$

2. $(a + 8)(a - 3)$

3. $(x + 6)(x + 2)$

4. $(a - 6)(a + 4)$

5. $(a - 3)(a + 2)$

6. $(b - 4)(b - 2)$

7. $(x - 5)(x - 2)$

8. $(x - 4)(x - 6)$

9. $2(a + 5)(a - 2)$

10. $3(a - 8)(a + 2)$

11. The length of a rectangle is 9 feet more than its width. The perimeter of the rectangle is 90 feet. Find the dimensions. See section 2–7.

4–2   Factoring Trinomials of the Form $x^2 + bx + c$

Determining When a Trinomial Will Factor

In section 3–2, we learned how to multiply two binomials as follows:

Factors   Terms
$$(x + 2)(x + 6) = x^2 + 6x + 2x + 12 = x^2 + 8x + 12$$
Multiplying ————————————→

In this section, we are going to reverse the procedure and factor the trinomial.

Terms   Factors
$$x^2 + 8x + 12 = (x + 2)(x + 6)$$
Factoring ————————→

The following group of trinomials will enable us to see how a trinomial factors.

1.
$$12 = 2 \cdot 6 \qquad \text{Product}$$
$$x^2 + 8x + 12 = (x + 2)(x + 6)$$
$$8 = 2 + 6 \qquad \text{Sum}$$

2.
$$12 = (-2) \cdot (-6) \qquad \text{Product}$$
$$x^2 - 8x + 12 = (x - 2)(x - 6)$$
$$-8 = (-2) + (-6) \qquad \text{Sum}$$

3.
$$-12 = (-2) \cdot 6 \qquad \text{Product}$$
$$x^2 + 4x - 12 = (x - 2)(x + 6)$$
$$4 = (-2) + 6 \qquad \text{Sum}$$

4.
$$-12 = 2 \cdot (-6) \qquad \text{Product}$$
$$x^2 - 4x - 12 = (x + 2)(x - 6)$$
$$-4 = 2 + (-6) \qquad \text{Sum}$$

In general,

$$(x + m)(x + n) = x^2 + (m + n)x + m \cdot n$$

The trinomial $x^2 + bx + c$ will factor with integer coefficients only if there are two integers, which we will call m and n, such that $m + n = b$ and $m \cdot n = c$.

Sum   Product
$m + n$ $m \cdot n$
$$x^2 + bx + c = (x + m)(x + n)$$

Factoring a Trinomial of the Form $x^2 + bx + c$

1. Factor out the GCF. If there is a common factor, make sure to include it as part of the final factorization.
2. Determine if the trinomial is factorable by finding m and n such that $m + n = b$ and $m \cdot n = c$. If m and n do not exist, we conclude that the trinomial will not factor.
3. Using the m and n values from step 2, write the trinomial in factored form.

The Signs ($+$ or $-$) for m and n

1. If c is positive, then m and n have the same sign as b.
2. If c is negative, then m and n have different signs and the one with the greater absolute value has the same sign as b.

● **Example 4–2 A**

Factoring $x^2 + bx + c$

Factor completely each trinomial.

1. $a^2 + 11a + 18$ $m + n = 11$ and $m \cdot n = 18$
   Since $b = 11$ and $c = 18$ are both positive, then m and n are both positive.

List the factorizations of 18	Sum of the factors of 18
$1 \cdot 18$	$1 + 18 = 19$
$2 \cdot 9$	$2 + 9 = 11$ ◄——— Correct sum
$3 \cdot 6$	$3 + 6 = 9$

   The m and n values are 2 and 9. The factorization is

   $$a^2 + 11a + 18 = (a + 2)(a + 9)$$

   The answer can be checked by performing the indicated multiplication.

   $$(a + 2)(a + 9) = a^2 + 9a + 2a + 18 = a^2 + 11a + 18$$

   Note The commutative property allows us to write the factors in any order. That is, $(a + 2)(a + 9) = (a + 9)(a + 2)$.

2. $b^2 - 2b - 15$ $m + n = -2$ and $m \cdot n = -15$
   Since $b = -2$ and $c = -15$ are both negative, then m and n have different signs and the one with the greater absolute value is negative.

Factorizations of -15, where the negative sign goes with the factor with the greater absolute value	Sum of the factors of -15
$1 \cdot (-15)$	$1 + (-15) = -14$
$3 \cdot (-5)$	$3 + (-5) = -2$ ◄——— Correct sum

   The m and n values are 3 and -5. The factorization is

   $$b^2 - 2b - 15 = (b + 3)(b - 5)$$

3. $5x - 24 + x^2$

It is easier to identify b and c if we write the trinomial in descending powers of the variable, which is called **standard form**.

$$x^2 + 5x - 24 \qquad m + n = 5 \quad \text{and} \quad m \cdot n = -24$$

Since $b = 5$ is positive and $c = -24$ is negative, m and n have different signs and the one with the greater absolute value is positive.

Factorizations of -24, where the positive factor is the one with the greater absolute value

Factorizations of -24	Sum of the factors of -24
$(-1) \cdot 24$	$(-1) + 24 = 23$
$(-2) \cdot 12$	$(-2) + 12 = 10$
$(-3) \cdot 8$	$(-3) + 8 = 5$ ⟵ Correct sum
$(-4) \cdot 6$	$(-4) + 6 = 2$

The m and n values are -3 and 8. The factorization is

$$x^2 + 5x - 24 = (x - 3)(x + 8)$$

4. $c^2 - 9c + 14 \qquad m + n = -9$ and $m \cdot n = 14$

Since $b = -9$ is negative and $c = 14$ is positive, m and n are both negative.

List the factorizations of 14	Sum of the factors of 14
$(-1)(-14)$	$(-1) + (-14) = -15$
$(-2)(-7)$	$(-2) + (-7) = -9$ ⟵ Correct sum

The m and n values are -2 and -7. The factorization is

$$c^2 - 9c + 14 = (c - 2)(c - 7)$$

5. $x^2 + 5x + 12 \qquad m + n = 5$ and $m \cdot n = 12$

Since $b = 5$ and $c = 12$ are both positive, m and n are both positive.

Factorizations of 12	Sum of the factors of 12
$1 \cdot 12$	$1 + 12 = 13$
$2 \cdot 6$	$2 + 6 = 8$
$3 \cdot 4$	$3 + 4 = 7$

No sum equals 5

Since none of the factorizations of 12 add to 5, there is no pair of integers (m and n) and the trinomial will not factor using integer coefficients. We call this a **prime polynomial**.

6. $x^4 - 4x^3 - 21x^2 = x^2(x^2 - 4x - 21)$ The GCF is x^2

To complete the factorization, we see if the trinomial $x^2 - 4x - 21$ will factor. We need to find m and n that add to -4 and multiply to -21. The values are 3 and -7. The completely factored form is

$$x^4 - 4x^3 - 21x^2 = x^2(x + 3)(x - 7)$$

Note A common error when the polynomial has a common factor is to factor it out but to forget to include it as one of the factors in the completely factored form.

7. $x^2y^2 + 9xy + 20$

Rewriting the polynomial as $(xy)^2 + 9xy + 20$, we want to find values for m and n that add to 9 and multiply to 20. The numbers are 4 and 5. The factorization is

$$x^2y^2 + 9xy + 20 = (xy + 4)(xy + 5)$$

8. $x^2 - 5ax + 6a^2 \qquad m + n = -5a \qquad$ and $\qquad m \cdot n = 6a^2$

We need to find m and n that add to $-5a$ and multiply to $6a^2$. The values are $-2a$ and $-3a$. The factorization is

$$x^2 - 5ax + 6a^2 = (x - 2a)(x - 3a)$$

☛ *Quick check*   Factor completely $z^2 + 8z - 20$

MASTERY POINTS

Can you
• Determine two integers whose product is one number and whose sum is another number?
• Recognize when the trinomial $x^2 + bx + c$ will factor and when it will not?
• Factor trinomials of the form $x^2 + bx + c$?
• Always remember to look for the greatest common factor before applying any of the factoring rules?

Exercise 4–2	☛ *Quick Check*	
Directions	*Examples* 4, −4	−27, −6
Find two integers such that their product is the first number and their sum is the second number. See procedure box on page 239.	*Solutions* Since $(-2)(-2) = 4$ and $(-2) + (-2) = -4$, then −2 and −2 are the integers.	Since $(-9)(3) = -27$ and $(-9) + (3) = -6$, then −9 and 3 are the integers.

Directions   Find two integers such that their product is the first number and their sum is the second number. See procedure box on page 239.

1. −16, 0	**7.** −12, −1	**13.** 35, 12
2. 25, 10	**8.** 48, 16	**14.** −15, 2
3. 20, −9	**9.** 0, −7	**15.** −18, 3
4. −11, 10	**10.** −36, −9	**16.** −30, −1
5. −30, 1	**11.** −8, 7	**17.** −12, 1
6. −72, −21	**12.** −9, 0	

Directions
Factor completely each trinomial. See example 4–2 A.

☞ *Quick Check*

Example $z^2 + 8z - 20$

Solution Since $b = 8$ is positive and $c = -20$ is negative, m and n have different signs and the one with the greater absolute value is positive.

Factorizations of -20, where the positive factor is the one with the greater absolute value	Sum of the factors of -20
$(-1) \cdot 20$	$(-1) + 20 = 19$
$(-2) \cdot 10$	$(-2) + 10 = 8$ ⟵ Correct sum
$(-4) \cdot 5$	$(-4) + 5 = 1$

The m and n values are -2 and 10. The factorization is

$$z^2 + 8z - 20 = (z - 2)(z + 10)$$

Directions   Factor completely each trinomial. If a trinomial will not factor, so state. See example 4–2 A.

18. $x^2 + 5x + 4$

19. $x^2 + 7x + 6$

20. $x^2 + 5x + 6$

21. $x^2 + 7x + 10$

22. $a^2 + 12a + 20$

23. $x^2 + 15x + 14$

24. $b^2 + 22b + 40$

25. $b^2 + 14b + 40$

26. $x^2 + 13x + 36$

27. $a^2 + 9a + 18$

28. $c^2 + 9c + 20$

29. $x^2 + 11x - 12$

30. $x^2 + 13x + 12$

31. $y^2 + 13y - 30$

32. $a^2 + 9a + 14$

33. $x^2 - 14x + 24$

34. $b^2 - 10b + 21$

35. $a^2 + 5a - 24$

36. $y^2 + 9y - 36$

37. $x^2 + 8x + 12$

38. $c^2 + 8c + 15$

39. $a^2 - 2a - 24$

40. $z^2 - 5z - 36$

41. $2x^2 + 6x - 20$

42. $2a^2 + 26a + 24$

43. $3x^2 - 18x - 48$

44. $a^2 - 9a + 4$

45. $x^2 + 5x + 7$

46. $x^2 - 4x + 6$

47. $y^2 + 17y + 30$

48. $b^2 + 13b + 40$

49. $4x^2 - 4x - 24$

50. $5y^2 + 5y - 30$

51. $5a^2 - 15a - 50$

52. $x^2y^2 - 4xy - 21$

53. $x^2y^2 - 3xy - 18$

54. $x^2y^2 - xy - 30$

55. $x^2y^2 + 13xy + 12$

56. $4a^2b^2 - 32ab + 28$

57. $3x^2y^2 - 3xy - 36$

58. $3x^2y^2 + 21xy + 36$

59. $x^2 + 3xy + 2y^2$

60. $a^2 - ab - 2b^2$

61. $a^2 - 2ab - 3b^2$

62. $a^2 - 7ab + 10b^2$

63. $a^2 - ab - 6b^2$

64. $x^2 + 2xy - 8y^2$

65. $x^2 - 2xy - 15y^2$

66. $a^2 + 7ab + 12b^2$

⬥ Explore to Learn

Find all integer values of b for which $x^2 + bx - 120$ factors.

Review Exercises

Directions   Factor completely. See section 4–1.

1. $ax^2 + bx^2 + cx^2$

2. $3x^3 + 12x^2 - 6x$

3. $3x(2x + 1) + 5(2x + 1)$

4. $2x(3x - 2) + 3(3x - 2)$

5. $4x(5x + 1) + (5x + 1)$

6. $6x(2x + 3) - (2x + 3)$

7. $x(3x - 5) - 2(3x - 5)$

8. $7x(x - 9) - 3(x - 9)$

9. Robert had $37,000, part of which he invested at 8% interest and the rest at 6%. If his total income was $2,600 from the two investments, how much did he invest at each rate? See section 2–7.

4–3   Factoring Trinomials of the Form $ax^2 + bx + c$

How to Factor Trinomials

In this section, we are going to factor trinomials of the form $ax^2 + bx + c$. This is called the **standard form** of a trinomial, where we have a single variable and the terms of the polynomial are arranged in descending powers of that variable. The a, b, and c in our standard form represent integer constants, and a is called the **leading coefficient**. For example,

$$2x^2 + 9x + 9$$

is a trinomial in standard form, where $a = 2$, $b = 9$, and $c = 9$.

Consider the product

$$(2x + 3)(x + 3)$$

By multiplying these two quantities together, we get a trinomial.

$$(2x + 3)(x + 3) = 2x^2 + 6x + 3x + 9$$
$$= 2x^2 + 9x + 9$$

To completely factor the trinomial $2x^2 + 9x + 9$ entails reversing this procedure to get

$$(2x + 3)(x + 3)$$

The trinomial will factor with integer coefficients if we can find a pair of integers (m and n) whose sum is equal to b, and whose product is equal to $a \cdot c$. In the trinomial $2x^2 + 9x + 9$, b is equal to 9, and $a \cdot c$ is $2 \cdot 9 = 18$. Therefore we want $m + n = 9$ and $m \cdot n = 18$. The values for m and n are 3 and 6.

In section 4–2 when the m and n values were found, we could immediately write the trinomial in factored form. That procedure only works when the leading coefficient is 1. When the leading coefficient is not 1, we will use the following procedure.

If we observe the multiplication process in our example, we see that m and n appear as the coefficients of the middle terms that are to be combined for our final answer.

$$(2x + 3)(x + 3) = 2x^2 + 6x + 3x + 9$$
$$= 2x^2 + 9x + 9$$

This is precisely what we do with the m and n values. We replace the coefficient of the middle term in the trinomial with these values. In our example, m and n are 6 and 3 and we replace the 9 with them.

$$2x^2 + 9x + 9 = 2x^2 + \overbrace{6x + 3x}^{9x} + 9$$

Our next step is to group the first two terms and the last two terms.

$$(2x^2 + 6x) + (3x + 9)$$

Now we factor out what is common in each pair. We see that the first two terms contain the common factor $2x$ and the last two terms contain the common factor 3.

$$2x(x + 3) + 3(x + 3)$$

When we reach this point, what is inside the parentheses in each term will be the same. Since the quantity $(x + 3)$ is common to both terms, we can factor it out.

$$2x\underbrace{(x + 3)}_{} + 3\underbrace{(x + 3)}_{}$$

Common to both terms

Having factored out what is common, what is left in each term is placed in the second parentheses.

$$2x(x + 3) + 3(x + 3)$$
$$(x + 3)(2x + 3)$$

Common   Remaining
factors   factors

The trinomial is factored.

A summary of the steps follows:

Factoring a Trinomial of the Form $ax^2 + bx + c$

1. Determine if the trinomial $ax^2 + bx + c$ is factorable by finding m and n such that $m \cdot n = a \cdot c$ and $m + n = b$. If m and n do not exist, we conclude that the trinomial will not factor.
2. Replace the middle term, bx, by the sum of mx and nx.
3. Place parentheses around the first and second terms and around the third and fourth terms. Factor out what is common to each pair.
4. Factor out the common quantity of each term and place the remaining factors from each term in the second parentheses.

Note Steps 3 and 4 are the procedure that we used to factor by grouping. There are two important differences between factoring by grouping and factoring $ax^2 + bx + c$. First, when we replace the middle terms, *bx*, by the sum of *mx* and *nx*, the four terms are automatically arranged in a correct order. Second, if we are able to find values for *m* and *n*, the trinomial will factor.

We determine the signs (+ or −) for *m* and *n* in a fashion similar to that of section 4–2.

The Signs (+ or −) for *m* and *n*

1. If $a \cdot c$ is positive, then *m* and *n* have the same sign as *b*.
2. If $a \cdot c$ is negative, then *m* and *n* have different signs and the one with the greater absolute value has the same sign as *b*.

● **Example 4–3 A**

Factoring $ax^2 + bx + c$ by grouping

Factor completely the following trinomials. If a trinomial will not factor, so state.

1. $6x^2 + 13x + 6$

> **Step 1** $m \cdot n = 6 \cdot 6 = 36$ and $m + n = 13$
>
> We determine by inspection that *m* and *n* are 9 and 4.

$$13x$$

Step 2 $= 6x^2 + 9x + 4x + 6$	Replace *bx* with *mx* and *nx*
Step 3 $= (6x^2 + 9x) + (4x + 6)$	Group the first two terms and the last two terms
$= 3x(2x + 3) + 2(2x + 3)$	Factor out what is common to each pair
Step 4 $= (2x + 3)(3x + 2)$	Factor out the common quantity

Note The order in which we place *m* and *n* into the polynomial will not change the answer.

(Alternate) $13x$

Step 2 $= 6x^2 + 4x + 9x + 6$	Replace *bx* with *nx* and *mx*
Step 3 $= (6x^2 + 4x) + (9x + 6)$	Group the first two terms and the last two terms
$= 2x(3x + 2) + 3(3x + 2)$	Factor out what is common to each pair
Step 4 $= (3x + 2)(2x + 3)$	Factor out the common quantity

We see that the outcome in step 4 is the same regardless of the order of *m* and *n* in the polynomial.

Note The order in which the two factors are written in the answer does not matter. That is, $(2x + 3)(3x + 2) = (3x + 2)(2x + 3)$

2. $3x^2 + 5x + 2$

> **Step 1** $m \cdot n = 3 \cdot 2 = 6$ and $m + n = 5$
>
> *m* and *n* are 2 and 3.

$$5x$$

Step 2 $= 3x^2 + 2x + 3x + 2$	Replace *bx* with *mx* and *nx*
Step 3 $= (3x^2 + 2x) + (3x + 2)$	Group the first two terms and the last two terms
$= x(3x + 2) + 1(3x + 2)$	Factor out what is common to each pair

We observe in the last two terms that the greatest common factor is only 1 or −1. We factor out 1 so that we have the same quantity inside the parentheses.

Step 4 $= (3x + 2)(x + 1)$ Factor out the common quantity

3. $4x^2 − 11x + 6$

Step 1 $m \cdot n = 4 \cdot 6 = 24$ and $m + n = −11$
        m and n are $−3$ and $−8$.

$$\overbrace{}^{-11x}$$

Step 2 $= 4x^2 − 3x − 8x + 6$ Replace bx with mx and nx
Step 3 $= (4x^2 − 3x) + (−8x + 6)$ Group the first two terms and the last two terms
        $= x(4x − 3) − 2(4x − 3)$ Factor out what is common to each pair

We have 2 or −2 as the greatest common factor in the last two terms. We factor out −2 so that we will have the same quantity inside the parentheses.

Step 4 $= (4x − 3)(x − 2)$ Factor out the common quantity

Note If the third term in step 2 is preceded by a minus sign, we will usually factor out the negative factor.

4. $6x^2 − 9x − 4$
$m \cdot n = 6 \cdot (−4) = −24$ and $m + n = −9$
Our m and n values are not obvious by inspection.

Note If you cannot determine the m and n values by inspection, then you should use the following systematic procedure to list all the possible factorizations of $a \cdot c$. This way you will either find m and n or verify that the trinomial will not factor using integers.

1. Take the natural numbers $1, 2, 3, 4, \cdots$ and divide them into the $a \cdot c$ product. Those that divide into the product evenly we write as a factorization using the correct m and n signs.

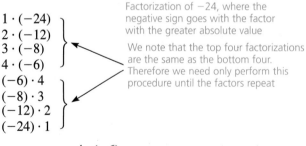

$1 \cdot (−24)$
$2 \cdot (−12)$
$3 \cdot (−8)$
$4 \cdot (−6)$
$(−6) \cdot 4$
$(−8) \cdot 3$
$(−12) \cdot 2$
$(−24) \cdot 1$

Factorization of −24, where the negative sign goes with the factor with the greater absolute value

We note that the top four factorizations are the same as the bottom four. Therefore we need only perform this procedure until the factors repeat

$4 \cdot (−6)$

Factors repeat

$(−6) \cdot 4$

2. Find the sum of the factorizations of $a \cdot c$. If there is a sum equal to b, the trinomial will factor. If there is no sum equal to b, then the trinomial will not factor with integer coefficients.

Factorizations of -24	Sum of the factors of -24	
$1 \cdot (-24)$	$1 + (-24) = -23$	
$2 \cdot (-12)$	$2 + (-12) = -10$	
$3 \cdot (-8)$	$3 + (-8) = -5$	$\longleftarrow$ Passed -9
$4 \cdot (-6)$	$4 + (-6) = -2$	
	No sum equals -9	

Since none of the factorizations of -24 add to -9, there is no pair of integers (m and n) and the trinomial will not factor.

Note Regardless of the signs of *m* and *n*, the column of values of the sum of the factors will either be increasing or decreasing. Therefore, once the desired value has been passed, the process can be stopped and the trinomial will not factor.

5. $24x^2 - 39x - 18$

Before we attempt to apply any factoring rule, recall that we must always factor out what is common to each term. Therefore we have

$$24x^2 - 39x - 18 = 3(8x^2 - 13x - 6) \qquad \text{Common factor of 3}$$

Now we are ready to factor the trinomial $8x^2 - 13x - 6$.

Step 1 $m \cdot n = 8(-6) = -48$ and $m + n = -13$
    m and n are 3 and -16.

$$\overbrace{}^{-13x}$$

Step 2	$= 3(8x^2 + 3x - 16x - 6)$	Replace *bx* with *mx* and *nx*
Step 3	$= 3[(8x^2 + 3x) + (-16x - 6)]$	Group the first two terms and the last two terms
	$= 3[x(8x + 3) - 2(8x + 3)]$	Factor out what is common to each pair
Step 4	$= 3(8x + 3)(x - 2)$	Factor out the common quantity

Note In example 5, we factored out 3 that was common to all the original terms. A common error is to forget to include it as one of the factors in the answer.

☞ ***Quick check***   Factor completely $6x^2 + 23x + 15$ and $12x^2 + 12x - 9$ ●

Factoring by Inspection—an Alternative Approach

In the beginning of this section, we studied a systematic procedure for determining if a trinomial will factor and how to factor it. In many instances, we can determine how the trinomial will factor by inspecting the trinomial rather than by applying this procedure.

Factoring by inspection is accomplished as follows: Factor $7x + 2x^2 + 3$.

Step 1 Write the trinomial in standard form.

$$2x^2 + 7x + 3 \qquad \text{Arrange terms in descending powers of } x$$

Step 2 Determine the possible combinations of first-degree factors of the first term.

$$(2x \quad)(x \quad) \qquad \text{The only factorization of } 2x^2 \text{ is } 2x \cdot x$$

Step 3 Combine with the factors of step 2 all the possible factors of the third term.

$(2x \quad 3)(x \quad 1)$ The only factorization of 3 is $3 \cdot 1$
$(2x \quad 1)(x \quad 3)$

Step 4 Determine the possible symbol ($+$ or $-$) between the terms in each binomial.

$(2x + 3)(x + 1)$
$(2x + 1)(x + 3)$

The rules of real numbers given in chapter 1 provide the answer to step 4.

1. If the third term is preceded by a $+$ sign and the middle term is preceded by a $+$ sign, then the symbols will be

$(\quad + \quad)(\quad + \quad)$

2. If the third term is preceded by a $+$ sign and the middle term is preceded by a $-$ sign, then the symbols will be

$(\quad - \quad)(\quad - \quad)$

3. If the third term is preceded by a $-$ sign, then the symbols will be

$(\quad + \quad)(\quad - \quad)$

or

$(\quad - \quad)(\quad + \quad)$

<u>Note</u> It is assumed that the first term is preceded by a $+$ sign or no sign. If it is preceded by a $-$ sign, these rules could still be used if (-1) is first factored out of all the terms.

Step 5 Determine which factors, if any, yield the correct middle term.

$(+3x) + (+2x) = +5x$

$(+x) + (+6x) = +7x$ Correct middle term

The second set of factors gives us the correct middle term. Therefore

$(2x + 1)(x + 3)$ is the factorization of $2x^2 + 7x + 3$.

● **Example 4–3 B**

Factoring $ax^2 + bx + c$ by inspection

Factor completely the following trinomials by inspection.

1. $6x^2 + 13x + 6$

Step 1 $6x^2 + 13x + 6$ Standard form

Step 2 $(6x \quad)(x \quad)$ $6x^2 = 6x \cdot x$ or $3x \cdot 2x$
        $(3x \quad)(2x \quad)$

Step 3 $(6x \quad 1)(x \quad 6)$
$(6x \quad 6)(x \quad 1)$
$(6x \quad 2)(x \quad 3)$
$(6x \quad 3)(x \quad 2)$
$(3x \quad 6)(2x \quad 1)$
$(3x \quad 1)(2x \quad 6)$
$(3x \quad 3)(2x \quad 2)$
$(3x \quad 2)(2x \quad 3)$

These six combinations can be eliminated since if the trinomial does not have a common factor, the binomials cannot have a common factor

Step 4 $(6x + 1)(x + 6)$

$(3x + 2)(2x + 3)$

Using the rules of signed numbers, determine the possible signs between the terms

Step 5 $(6x + 1)(x + 6)$

$+1x$

$+36x$

$(+1x) + (+36x) = +37x$

$(3x + 2)(2x + 3)$

$+4x$

$+9x$

$(+4x) + (+9x) = +13x$ Correct middle term

Hence, $(3x + 2)(2x + 3)$ is the factorization of $6x^2 + 13x + 6$.

Note As was pointed out in step 3, if the trinomial that we are trying to factor does not have a common factor, then neither of its binomial factors can have a common factor. Using this fact, we were able to eliminate six possible combinations. This will often help eliminate some of the possible combinations.

2. $4x^2 - 5x + 1$

Step 1 $4x^2 - 5x + 1$ Standard form

Step 2 $(4x \quad)(x \quad)$ $4x^2 = 2x \cdot 2x$ or $4x \cdot x$
$(2x \quad)(2x \quad)$

Step 3 $(4x \quad 1)(x \quad 1)$ The only factorization of 1 is $1 \cdot 1$
$(2x \quad 1)(2x \quad 1)$

Step 4 $(4x - 1)(x - 1)$ Determine the possible signs between the terms
$(2x - 1)(2x - 1)$

Step 5 $(4x - 1)(x - 1)$

$-x$

$-4x$

$(-x) + (-4x) = -5x$ Correct middle term

$(2x - 1)(2x - 1)$

$-2x$

$-2x$

$(-2x) + (-2x) = -4x$

The first set of factors gives us the correct factorization.

$4x^2 - 5x + 1 = (4x - 1)(x - 1)$

3. $13x - 5 + 6x^2$

Step 1 $6x^2 + 13x - 5$ Write in standard form

Step 2 $(6x\ )(x\ )$ $6x^2 = 6x \cdot x$ or $2x \cdot 3x$
$(2x\ )(3x\ )$

Step 3 $(6x\quad 5)(x\quad 1)$
$(6x\quad 1)(x\quad 5)$ The only factorization of 5 is $5 \cdot 1$
$(2x\quad 5)(3x\quad 1)$
$(2x\quad 1)(3x\quad 5)$

Step 4 $(6x + 5)(x - 1)$ or $(6x - 5)(x + 1)$ Determine the possible signs
$(6x + 1)(x - 5)$ or $(6x - 1)(x + 5)$ between the terms
$(2x + 5)(3x - 1)$ or $(2x - 5)(3x + 1)$
$(2x + 1)(3x - 5)$ or $(2x - 1)(3x + 5)$

Step 5 $(6x + 5)(x - 1)$ or $(6x - 5)(x + 1)$

$+5x$ $-5x$

$-6x$ $+6x$

$(+5x) + (-6x) = -x$ or $(-5x) + (+6x) = +x$

<u>Note</u> Switching the signs in both parentheses only changes the sign of the middle term.

$(6x + 1)(x - 5)$ or $(6x - 1)(x + 5)$

$+x$ $-x$

$-30x$ $+30x$

$(+x) + (-30x) = -29x$ or $(-x) + (+30x) = +29x$

$(2x + 5)(3x - 1)$ or $(2x - 5)(3x + 1)$

$+15x$ $-15x$

$-2x$ $+2x$

$(+15x) + (-2x) = +13x$ or $(-15x) + (+2x) = -13x$

Correct middle term

$(2x + 1)(3x - 5)$ or $(2x - 1)(3x + 5)$

$+3x$ $-3x$

$-10x$ $+10x$

$(+3x) + (-10x) = -7x$ or $(-3x) + (+10x) = +7x$

The factorization of $6x^2 + 13x - 5$ is $(2x + 5)(3x - 1)$.

MASTERY POINTS

Can you
- Determine two integers whose product is one number and whose sum is another number?
- Recognize when the trinomial $ax^2 + bx + c$ will factor and when it will not?
- Factor trinomials of the form $ax^2 + bx + c$?
- Always remember to look for the greatest common factor before applying any of the factoring rules?

Exercise 4–3

Directions

Factor completely each trinomial. If a trinomial will not factor, so state. See examples 4–3 A and B.

☞ *Quick Check*

Example $6x^2 + 23x + 15$

Solution $m \cdot n = 6 \cdot 15 = 90$ Determine m and n
$m + n = 23$
m and n are 5 and 18.

$$= 6x^2 + \overbrace{5x + 18x}^{23x} + 15$$ Replace bx with mx and nx
$= (6x^2 + 5x) + (18x + 15)$ Group first two terms and last two terms
$= x(6x + 5) + 3(6x + 5)$ Factor out what is common to each pair
$= (6x + 5)(x + 3)$ Factor out the common quantity

Example $12x^2 + 12x - 9$

Solution $= 3(4x^2 + 4x - 3)$ Common factor of 3
$m \cdot n = 4(-3) = -12$
$m + n = 4$
m and n are -2 and 6.   Determine m and n

$$= 3[4x^2 \overbrace{- 2x + 6x}^{4x} - 3]$$ Replace bx with mx and nx
$= 3[2x(2x - 1) + 3(2x - 1)]$ Group first two terms and last two terms, factor out what is common to each pair
$= 3(2x - 1)(2x + 3)$ Factor out the common quantity

Directions   Factor completely each trinomial. If a trinomial will not factor, so state. See examples 4–3 A and B.

1. $3x^2 + 10x + 3$
2. $4x^2 + 9x + 2$
3. $5x^2 + 11x + 2$
4. $2x^2 + 7x + 6$
5. $3x^2 + 10x + 8$
6. $4x^2 + 15x + 9$
7. $5x^2 + 13x + 6$
8. $9x^2 + 9x + 2$
9. $8x^2 + 18x + 9$
10. $10x^2 + 19x + 6$
11. $2x^2 + x - 6$
12. $3x^2 + 7x - 6$
13. $2x^2 + 3x + 1$
14. $4x^2 - 5x + 1$
15. $2R^2 - 7R + 6$
16. $R^2 - 4R + 6$

17. $5x^2 - 7x - 6$

18. $2x^2 - x - 1$

19. $9x^2 - 6x + 1$

20. $8x^2 - 17x + 2$

21. $5x^2 + 4x + 6$

22. $2x^2 - 11x + 12$

23. $6x^2 + 13x + 6$

24. $2r^2 + 13r + 18$

25. $4x^2 + 20x + 21$

26. $7R^2 + 20R - 3$

27. $4x^2 - 2x + 5$

28. $4x^2 - 4x - 3$

29. $9y^2 - 21y - 8$

30. $6x^2 - 23x - 4$

31. $10x^2 + 7x - 6$

32. $10x^2 + 9x + 2$

33. $2x^2 - 9x + 10$

34. $7x^2 - 3x + 6$

35. $4x^2 + 14x + 12$

36. $5R^2 - 9R - 2$

37. $4x^2 + 10x + 6$

38. $6x^2 - 17x + 12$

39. $6x^2 + 7x - 3$

40. $3x^2 + 12x + 12$

41. $2x^2 + 6x - 20$

42. $3a^2 + 8a - 4$

43. $6x^2 + 5x - 6$

44. $3x^2 - 19x + 20$

45. $4x^2 + 12x + 9$

46. $9z^2 - 30z + 25$

47. $7x^2 - 36x + 5$

48. $3x^2 + 2x + 4$

49. $15P^2 + 2P - 1$

50. $12x^2 + 13x - 4$

51. $2x^3 - 6x^2 - 20x$

52. $4x^2 + 10x + 4$

53. $2a^3 + 15a^2 + 7a$

54. $9x^2 + 27x + 8$

55. $8x^2 - 14x - 15$

56. $8x^2 - 18x + 9$

57. When an object is thrown vertically into the air, the height S of the object at any instant in time t is given by $S = -16t^2 + 32t$. Factor the right side.

58. Multiplication and factoring of expressions sometimes have a geometric interpretation. For example, since the area of a rectangle is the product of its two dimensions, the square shown represents $(x + y)^2 = x^2 + 2xy + y^2$. Construct a rectangle that would represent $2x^2 + 7xy + 3y^2 = (2x + y)(x + 3y)$.

	x	y
x	x^2	xy
y	xy	y^2

59. (See exercise 58.) Construct a square that represents $(x - y)^2 = x^2 - 2xy + y^2$.

60. How does the middle term change in $(x + 5)(x - 4)$ and $(x - 5)(x + 4)$? In general, how does the middle term change in $(x + a)(x - b)$ and $(x - a)(x + b)$ if $a > b$, $a > 0$, and $b > 0$?

61. How does the middle term change in $(x + 5)(x + 4)$ and $(x - 5)(x - 4)$? In general, how does the middle term change in $(x + a)(x + b)$ and $(x - a)(x - b)$ if $a > 0$ and $b > 0$?

⚙ Explore to Learn

Find all integer values of b for which $2x^2 + bx - 120$ factors.

Review Exercises

Directions   Use the special product rules to carry out the indicated multiplication. See section 3-2.

1. $(x - y)(x + y)$

2. $(3a - 2b)(3a + 2b)$

3. $(x - y)^2$

4. $(5a + 4b)(5a - 4b)$

5. $(2a + b)^2$

6. $(4x - y)^2$

7. $(x^2 + 1)(x^2 - 1)$

8. $(a^2 - 4)(a^2 + 4)$

9. Nine times a number is decreased by 4, leaving 122. What is the number? See section 2-7.

10. One half of a number minus one third of the number is 8. Find the number. See section 2-7.

4-4   Factoring the Difference of Two Squares and Perfect Square Trinomials

In section 3-2, we saw that the product of $(a + b)(a - b)$ was $a^2 - b^2$. We refer to the indicated product $(a + b)(a - b)$ as the *product* of the *sum* and *difference* of the same two terms. Notice that in one factor we *add* the terms and in the other we find the *difference* between these same terms. The product will *always* be the *difference of the squares* of the two terms. To factor the **difference of two squares**, we must be able to recognize **perfect squares**.

● **Example 4-4 A**

Perfect squares

Write the following as a quantity squared, if possible.

1. $25a^2 = 5a \cdot 5a$ 25 is $5 \cdot 5$ and a^2 is $a \cdot a$
  $= (5a)^2$ It is now rewritten as a square

2. $9y^4 = 3y^2 \cdot 3y^2$ 9 is $3 \cdot 3$ and y^4 could be written as $y^2 \cdot y^2$
  $= (3y^2)^2$ It is now rewritten as a square

☞ **Quick check**   Write 64 and $9x^4$ each as a quantity squared.   ●

Factoring the Difference of Two Squares

1. Identify that we have a perfect square *minus* another perfect square.
2. Rewrite the binomial as a first term squared minus a second term squared.

$$(\text{first term})^2 - (\text{second term})^2$$

3. Factor the binomial into the first term plus the second term times the first term minus the second term.

$$(\text{first term} + \text{second term})(\text{first term} - \text{second term})$$

● **Example 4–4 B**

Factoring the difference of two squares

Write the following in completely factored form.

	Step 1		Step 2		Step 3
	Identify		Rewrite		Factor
$a^2 - b^2$	$=$	$(a)^2 - (b)^2$	$=$	$(a+b)(a-b)$

1. $x^2 - 9 = (x)^2 - (3)^2 = (x+3)(x-3)$

2. $4a^2 - b^2 = (2a)^2 - (b)^2 = (2a+b)(2a-b)$

3. $4p^2 - 25v^2 = (2p)^2 - (5v)^2 = (2p+5v)(2p-5v)$

4. $r^4 - 49 = (r^2)^2 - (7)^2 = (r^2+7)(r^2-7)$

⚐ *Quick check*   Write $t^2 - 64$ and $4a^2 - b^2c^2$ in completely factored form.   ●

Our first step in any factoring problem is to look for any common factors. Often an expression that does not appear to be factorable becomes so by taking out the greatest common factor. When we have applied a factoring rule to a problem, we must inspect all parts of our answer to make sure that nothing will factor further.

● **Example 4–4 C**

Factoring using more than one procedure

Write the following in completely factored form.

1. $2x^2 - 18y^2 = 2(x^2 - 9y^2)$ Factor out what is common, 2
   $= 2[(x)^2 - (3y)^2]$ Identify and rewrite as squares
   $= 2(x+3y)(x-3y)$ Factor and inspect the factors

2. $5a^4 - 45a^2b^2 = 5a^2(a^2 - 9b^2)$ Common factor of $5a^2$
   $= 5a^2[(a)^2 - (3b)^2]$ Identify and rewrite
   $= 5a^2(a+3b)(a-3b)$ Factor and inspect the factors

3. $a^4 - 16 = (a^2)^2 - (4)^2$ Identify and rewrite
   $= (a^2+4)(a^2-4)$ Factor and inspect the factors
   $= (a^2+4)[(a)^2 - (2)^2]$ Identify and rewrite
   $= (a^2+4)(a+2)(a-2)$ Factor and inspect

<u>Note</u> In example 3, $a^2 + 4$ is called the *sum of two squares.* This will *not* factor using integers.   ●

Perfect Square Trinomials

In section 3–2, two of the special products that we studied were the squares of a binomial. We will now restate those special products.

$$a^2 + 2ab + b^2 = (a+b)^2$$

and

$$a^2 - 2ab + b^2 = (a-b)^2$$

The right side of each of these equations is called the square of a binomial, and the left side is called a **perfect square trinomial**. Perfect square trinomials can always be factored by our factoring procedure. However if we observe that the first and last terms of a trinomial are positive and perfect squares, we should see if the trinomial will factor as the square of a binomial. To factor a trinomial as the square of a binomial, the following three conditions need to be met.

Necessary Conditions for a Perfect Square Trinomial

1. The first term must have a positive coefficient and be a perfect square, a^2.
2. The last term must have a positive coefficient and be a perfect square, b^2.
3. The middle term must be twice the product of the bases of the first and last terms, $2ab$ or $-2ab$.

We observe that

$$9x^2 + 12x + 4$$
$$= (3x)^2 + 2(3x)(2) + (2)^2$$

Condition 1   Condition 3   Condition 2

Therefore it is a perfect square trinomial and factors into

$$(3x + 2)^2$$

● **Example 4–4 D**

Factoring perfect square trinomials

The following examples show the factoring of some other perfect square trinomials.

		Condition 1		Condition 3		Condition 2		Square of a binomial
1. $4x^2 + 20x + 25$	$=$	$(2x)^2$	$+$	$2(2x)(5)$	$+$	$(5)^2$	$=$	$(2x + 5)^2$
2. $9x^2 - 6x + 1$	$=$	$(3x)^2$	$-$	$2(3x)(1)$	$+$	$(1)^2$	$=$	$(3x - 1)^2$
3. $16x^2 + 24x + 9$	$=$	$(4x)^2$	$+$	$2(4x)(3)$	$+$	$(3)^2$	$=$	$(4x + 3)^2$
4. $9y^2 - 30y + 25$	$=$	$(3y)^2$	$-$	$2(3y)(5)$	$+$	$(5)^2$	$=$	$(3y - 5)^2$

MASTERY POINTS

Can you
- Identify and rewrite a perfect square?
- Factor the difference of two squares?
- Remember that $a^2 + b^2$ will not factor using integers?
- Factor out any common factors before applying other factoring rules?
- Inspect all factors to make sure the problem is completely factored?
- Factor perfect square trinomials?

Exercise 4–4

Directions

Write the following as a quantity squared, if possible. See example 4–4 A.

☞ *Quick Check*

Examples   64 $9x^4$

Solutions $= 8 \cdot 8$ Identify $= 3x^2 \cdot 3x^2$ Identify
    $= (8)^2$ Rewrite $= (3x^2)^2$ Rewrite

Directions   Write the following as a quantity squared, if possible. See example 4–4 A.

1. 36

2. 25

3. c^2

4. e^2

5. $16x^2$

6. $49b^2$

7. $4z^4$

8. $25b^2$

Directions Write in completely factored form. See examples 4–4 B, C, and D.	☞ *Quick Check*
	Examples $t^2 - 64$ Identify $4a^2 - b^2c^2$ Identify
	Solutions $= (t)^2 - (8)^2$ Rewrite $= (2a)^2 - (bc)^2$ Rewrite
	$= (t + 8)(t - 8)$ Factor $= (2a + bc)(2a - bc)$ Factor

Directions   Write in completely factored form. See examples 4–4 B, C, and D.

9. $x^2 - 1$

10. $x^2 - 25$

11. $a^2 - 4$

12. $r^2 - s^2$

13. $9 - E^2$

14. $49 - R^2$

15. $1 - k^2$

16. $4y^2 - 9$

17. $9b^2 - 16$

18. $x^2 - 16z^2$

19. $b^2 - 36c^2$

20. $16x^2 - y^2$

21. $4a^2 - 25b^2$

22. $16a^2 - b^2$

23. $25p^2 - 81$

24. $r^2 - 4s^2$

25. $8x^2 - 32y^2$

26. $3a^2 - 27b^2$

27. $5r^2 - 125s^2$

28. $20 - 5b^2$

29. $50 - 2x^2$

30. $x^2y^2 - 4z^2$

31. $r^2s^2 - 25t^2$

32. $a^4 - 25$

33. $x^4 - 9$

34. $x^4 - 1$

35. $r^4 - 81$

36. $16t^4 - 1$

37. $49x^2 - 64y^4$

38. $125p^2 - 20v^2$

39. $98x^2y^2 - 50p^2c^2$

40. $a^2 + 10a + 25$

41. $x^2 + 4x + 4$

42. $y^2 - 2y + 1$

43. $a^2 + 16a + 64$

44. $x^2 - 20x + 100$

45. $x^2 - 18x + 81$

46. $b^2 - 16b + 64$

47. $c^2 - 14c + 49$

48. $b^2 + 8b + 16$

49. $a^2 + 6a + 9$

50. $x^2 - 12x + 36$

51. $y^2 - 6y + 9$

52. $a^2 + 6ab + 9b^2$

53. $4a^2 - 12ab + 9b^2$

54. $x^2 - 16xy + 64y^2$

55. $9c^2 - 12cd + 4d^2$

56. $9a^2 - 30ab + 25b^2$

57. $4x^2 + 12xy + 9y^2$

58. $9a^2 + 12ab + 4b^2$

59. $4x^2 + 20xy + 25y^2$

60. $25a^2 - 20ab + 4b^2$

61. In engineering, the equation of transverse shearing stress in a rectangular beam is given by

$$T = \frac{V}{8I}(h^2 - 4v_1^2)$$

Factor the right side.

62. The area of a circle is $A = \pi r^2$, where r is the radius. The area of an annular ring with radius r_1 and outer radius r_2 would therefore be $\pi r_2^2 - \pi r_1^2$. Factor this expression.

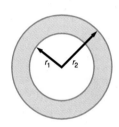

63. A freely falling body near the surface of the earth falls a distance $S = \frac{1}{2}gt^2$ in time t.

The difference of two such distances measured for two different times would therefore be

$$\frac{1}{2}gt_2^2 - \frac{1}{2}gt_1^2.$$

Factor this expression.

⊛ Explore to Learn

In this section we discuss a method for factoring perfect square trinomials. This is certainly an efficient way to factor these expressions, but can any of the other factoring methods already covered be used to factor these perfect square trinomials? Explain and give an example.

Review Exercises

Directions   Factor completely. See sections 4–1, 4–2, 4–3, and 4–4.

1. $x^2 + 8x + 12$
2. $49a^2 - 81$
3. $3ax + bx - 12ay - 4by$
4. $2x^3 + 14x^2 + 24x$
5. $10a^2 + 21a + 9$
6. $4a^2 - 20a + 25$

7. $x^2y^2 + 8xy + 15$
8. $x^2 + 4xy + 4y^2$
9. The width of a rectangle is 3 feet less than its length. The perimeter of the rectangle is 110 feet. Find the dimension. See section 2–7.

4–5   Other Types of Factoring

The Difference of Two Cubes

In section 4–4, we factored expressions that involved the difference of two squares. To factor these types of expressions, we identified the two terms as perfect squares and applied the procedure. In this section, we will factor the *sum and difference of two cubes* in a similar fashion.

Consider the indicated product of $(a - b)(a^2 + ab + b^2)$. If we carry out the multiplication, we have

$$(a - b)(a^2 + ab + b^2) = a^3 + a^2b + ab^2 - a^2b - ab^2 - b^3$$
$$= a^3 - b^3$$

Therefore $(a - b)(a^2 + ab + b^2) = a^3 - b^3$ and $(a - b)(a^2 + ab + b^2)$ is the factored form of $a^3 - b^3$.

To use this factoring technique, we must be able to recognize **perfect cubes**.

● **Example 4–5 A**

Perfect cubes

Write the following as a quantity cubed, if possible.

1. $8a^3 = 2a \cdot 2a \cdot 2a$ 8 is $2 \cdot 2 \cdot 2$ and a^3 is $a \cdot a \cdot a$
   $\quad = (2a)^3$ It is now rewritten as a cube

2. $64x^6 = 4x^2 \cdot 4x^2 \cdot 4x^2$ 64 is $4 \cdot 4 \cdot 4$ and x^6 is $x^2 \cdot x^2 \cdot x^2$
   $\quad = (4x^2)^3$ It is now rewritten as a cube

☞ *Quick check*   Write the following a a quantity cubed, if possible. $27y^3$ ●

Factoring the Difference of Two Cubes

1. Identify that we have a perfect cube minus another perfect cube.
2. Rewrite the problem as a first term cubed minus a second term cubed.

$$(\text{1st term})^3 - (\text{2nd term})^3$$

3. Factor the expression into the first term minus the second term, times the first term squared plus the first term times the second term plus the second term squared.

$$(\text{1st term} - \text{2nd term})[(\text{1st term})^2 + (\text{1st term} \cdot \text{2nd term}) + (\text{2nd term})^2]$$

● **Example 4–5 B**

Factor the difference of two cubes

Factor completely.

1. $x^3 - 27$ We rewrite x^3 as a cube and 27 as a cube.

$$x^3 - 27 = (x)^3 - (3)^3$$

The first term is x and the second term is 3. Then we write the procedure for factoring the difference of two cubes.

$$(\quad - \quad)[(\quad)^2 + (\quad)(\quad) + (\quad)^2]$$

$$\underset{\text{1st}}{\uparrow} \ \underset{\text{2nd}}{\uparrow} \ \underset{\text{1st}}{\uparrow} \quad \underset{\text{1st}}{\uparrow} \ \underset{\text{2nd}}{\uparrow} \quad \underset{\text{2nd}}{\uparrow}$$

Now substitute x where the first term is in the procedure and 3 where the second term is.

$$(x - 3)[(x)^2 + (x)(3) + (3)^2]$$

$$\underset{\text{1st}}{\uparrow} \ \underset{\text{2nd}}{\uparrow} \ \underset{\text{1st}}{\uparrow} \quad \underset{\text{1st}}{\uparrow} \underset{\text{2nd}}{\uparrow} \quad \underset{\text{2nd}}{\uparrow}$$

Finally we simplify.

$$(x - 3)(x^2 + 3x + 9)$$

Therefore $x^3 - 27 = (x - 3)(x^2 + 3x + 9)$.

2. $8x^3 - y^3 = (2x)^3 - (y)^3$ First term is $2x$ and second term is y

Then $(\quad - \quad)[(\quad)^2 + (\quad)(\quad) + (\quad)^2]$ Factoring procedure ready for substitution

$= (2x - y)[(2x)^2 + (2x)(y) + (y)^2]$ The first term is $2x$, the second term is y

$= (2x - y)(4x^2 + 2xy + y^2)$ Simplify within the second group

3. $a^{15} - 64b^3 = (a^5)^3 - (4b)^3$ Rewrite as cubes

Then $(\quad - \quad)[(\quad)^2 + (\quad)(\quad) + (\quad)^2]$ Factoring procedure ready for substitution

$= (a^5 - 4b)[(a^5)^2 + (a^5)(4b) + (4b)^2]$ The first term is a^5, the second term is $4b$.

$= (a^5 - 4b)(a^{10} + 4a^5b + 16b^2)$ Simplify within the second group

Note   In example 3, we observe that a number raised to a power that is a multiple of 3 can be written as a cube by dividing the exponent by 3. The quotient is the exponent of the number inside the parentheses and the 3 is the exponent outside the parentheses. For example, $y^{12} = (y^4)^3$ or $z^{24} = (z^8)^3$.

☞ *Quick check*   Factor completely $16R^3 - 54$

The Sum of Two Cubes

If we carry out the indicated multiplication in $(a + b)(a^2 - ab + b^2)$, we have

$$(a + b)(a^2 - ab + b^2) = a^3 - a^2b + ab^2 + a^2b - ab^2 + b^3$$
$$= a^3 + b^3$$

Therefore $(a + b)(a^2 - ab + b^2) = a^3 + b^3$ and $(a + b)(a^2 - ab + b^2)$ is the factored form of $a^3 + b^3$.

Factoring the Sum of the Two Cubes

1. Identify that we have a perfect cube plus another perfect cube.
2. Rewrite the problem as a first term cubed plus a second term cubed.

   (1st term)^3 + (2nd term)^3

3. Factor the expression into the first term plus the second term, times the first term squared minus the first term times the second term plus the second term squared.

   (1st term + 2nd term)[(1st term)^2 – (1st term)(2nd term) + (2nd term)^2]

Note In the factorization of both the sum and the difference of two cubes, the middle term of the trinomial factor has the opposite sign of the second term of the binomial factor. That is

$$(a - b)(a^2 + ab + b^2) \text{ and } (a + b)(a^2 - ab + b^2)$$

Opposite signs

● **Example 4–5 C**

Factoring the sum of two cubes

Factor completely.

1. $a^3 + 8 = (a)^3 + (2)^3$

We now write the procedure for the sum of two cubes.

$$(\quad + \quad)[(\quad)^2 - (\quad)(\quad) + (\quad)^2]$$

1st 2nd 1st   1st 2nd   2nd

Substituting,

$$(a + 2)[(a)^2 - (a)(2) + (2)^2]$$

1st 2nd 1st   1st 2nd   2nd

simplifying, $(a + 2)(a^2 - 2a + 4)$
therefore $a^3 + 8 = (a + 2)(a^2 - 2a + 4)$

2. $x^3 + 125 = (x)^3 + (5)^3$ Rewrite as cubes
   Then $(\quad + \quad)[(\quad)^2 - (\quad)(\quad) + (\quad)^2]$ Factoring procedure ready for substitution

$$= (x + 5)[(x)^2 - (x)(5) + (5)^2]$$ The first term is x, the second term is 5

$$= (x + 5)(x^2 - 5x + 25)$$ Simplify within the second group

3. $8a^3 + b^{21} = (2a)^3 + (b^7)^3$ Rewrite as cubes

Then $(\quad + \quad)[(\quad)^2 - (\quad)(\quad) + (\quad)^2]$ Factoring procedure ready for substitution

$= (2a + b^7)[(2a)^2 - (2a)(b^7) + (b^7)^2]$ The first term is 2*a*, the second term is b^7

$= (2a + b^7)(4a^2 - 2ab^7 + b^{14})$ Simplify within the second group

☞ **Quick check**   Factor completely $27x^3 + y^3$

MASTERY POINTS

Can you
- Identify and rewrite a perfect cube?
- Factor the sum and difference of two cubes?

Exercise 4–5	☞ *Quick Check*
Directions	**Example** $27y^3$
Write the following as a quantity cubed, if possible. See example 4–5 A.	**Solution** $= 3y \cdot 3y \cdot 3y$ 27 is $3 \cdot 3 \cdot 3$ and y^3 is $y \cdot y \cdot y$
	$= (3y)^3$ It is now rewritten as a cube

Directions   Write the following as a quantity cubed, if possible. See example 4–5 A.

1. 64	**4.** 1	**7.** a^6	**9.** $8b^{15}$
2. 8	**5.** $27x^3$	**8.** x^9	**10.** $64c^{21}$
3. 125	**6.** $64a^3$		

Directions
Factor completely. If an expression will not factor, so state. See examples 4–5 B and C.

☞ *Quick Check*

Example $16R^3 - 54$

Solution $= 2[8R^3 - 27]$ Common factor of 2

$= 2[(2R)^3 - (3)^3]$ Rewrite as cubes

$2(\quad - \quad)[(\quad)^2 + (\quad)(\quad) + (\quad)^2]$ Factoring procedure ready for substitution

$= 2(2R - 3)[(2R)^2 + (2R)(3) + (3)^2]$ The first term is 2*R*, the second term is 3

$= 2(2R - 3)(4R^2 + 6R + 9)$ Simplify within the second group

Example $27x^3 + y^3$

Solution $= (3x)^3 + (y)^3$ Rewrite as cubes

$(\quad + \quad)[(\quad)^2 - (\quad)(\quad) + (\quad)^2]$ Factoring procedure ready for substitution

$= (3x + y)[(3x)^2 - (3x)(y) + (y)^2]$ The first term is 3*x*, the second term is *y*

$= (3x + y)(9x^2 - 3xy + y^2)$ Simplify within the second group

Directions   Factor completely. If an expression will not factor, so state. See examples 4–5 B and C.

11. $r^3 + s^3$

12. $L^3 + 8$

13. $8x^3 + y^3$

14. $27r^3 + 8$

15. $h^3 - k^3$

16. $p^3 - q^3$

17. $a^3 - 8$

18. $b^3 + 64$

19. $x^3 - 8y^3$

20. $27a^3 - b^3$

21. $64x^3 - y^3$

22. $r^3 - 27$

23. $27x^3 - 8y^3$

24. $64a^3 - 8$

25. $8a^3 + 27b^3$

26. $64s^3 + 1$

27. $2a^3 + 16$

28. $3x^3 + 81$

29. $2x^3 - 16$

30. $81a^3 - 3b^3$

31. $x^5 + 27x^2y^3$

32. $16a^3 + 2b^3$

33. $x^6 + y^3$

34. $x^3 + y^9$

35. $a^9 - b^3$

36. $a^6 - 8$

37. $x^{12} - 27$

38. $x^{15} + 64$

39. $8a^2b^3 - a^5$

40. $2x^3 - 54y^3$

41. $54r^3 + 2s^3$

42. $b^5 + 64b^2c^3$

43. $x^3y^3 - z^3$

44. $x^3y^9 - 1$

45. $a^{15}b^6 - 8c^9$

46. $x^{18}y^9 - 27z^3$

47. $a^3b^3 + 8$

48. $x^3y^6 + z^3$

49. $x^9y^{12} + z^{15}$

50. $a^{12}b^{15} + c^{24}$

✵ Explore to Learn

ax^my^n is a perfect cube, a, m, and n are integers, and $50 < a < 100$. What can you say about a, m, and n?

Review Exercises

Directions   Write in completely factored form. See sections 4–1, 4–2, 4–3, and 4–4.

1. $a^2 - 7a + 10$

2. $6ax + 2bx - 3ay - by$

3. $x^2 + 4xy + 4y^2$

4. $6a^2 - ab - b^2$

5. $5a^3 - 40a^2 + 75a$

6. $6x^2 + x - 12$

7. Alanzo has \$16,875, part of which he invests at 10% interest and the rest at 8%. If his income from each investment is the same, how much did he invest at each rate? See section 2–7.

4–6   Factoring: A General Strategy

In this section, we will review the different methods of factoring that we have studied in the previous sections. The following outline gives a general strategy for factoring polynomials.

I. Factor out all common factors.
   Examples
   1. $5a^3 - 25a^2 = 5a^2(a - 5)$
   2. $c(a - 2b) + 2d(a - 2b) = (a - 2b)(c + 2d)$

II. Count the number of terms.
   A. Two terms: Check to see if the polynomial is the difference of two squares, the difference of two cubes, or the sum of two cubes.
      Examples
      1. $a^2 - 16b^2 = (a - 4b)(a + 4b)$ Difference of two squares
      2. $8a^3 - b^3 = (2a - b)(4a^2 + 2ab + b^2)$ Difference of two cubes
      3. $m^3 + 64n^3 = (m + 4n)(m^2 - 4mn + 16n^2)$ Sum of two cubes
   B. Three terms: Check to see if the polynomial is a perfect square trinomial. If it is not, use one of the general methods for factoring a trinomial.
      Examples
      1. $a^2 + 6a + 9 = (a + 3)^2$ Perfect square trinomial
      2. $a^2 + 5a - 14 = (a + 7)(a - 2)$ General trinomial, leading coefficient of 1
      3. $6a^2 + 7a - 20 = (2a + 5)(3a - 4)$ General trinomial, leading coefficient other than 1
   C. Four terms: Check to see if we can factor by grouping.
      Examples
      1. $ac + 3a - 2bc - 6b = (a - 2b)(c + 3)$
      2. $a^3 + 2a^2 - 3a - 6 = (a^2 - 3)(a + 2)$

III. Check to see if any of the factors we have written can be factored further. Any common factors that were missed in part I can still be factored out here.
   Examples
   1. $c^4 - 11c^2 + 28 = (c^2 - 4)(c^2 - 7)$
      $\qquad\qquad\qquad\quad = (c - 2)(c + 2)(c^2 - 7)$ Difference of two squares
   2. $4a^2 - 36b^2 = (2a - 6b)(2a + 6b)$
      $\qquad\qquad\quad = 2(a - 3b)2(a + 3b)$ Overlooked common factor
      $\qquad\qquad\quad = 4(a - 3b)(a + 3b)$

The following examples illustrate our strategy for factoring polynomials.

● **Example 4–6 A**

Factoring polynomials

Completely factor the following polynomials.

1. $3x^3 - 3xy^2$
   I. First we look for any common factors.
   $$3x^3 - 3xy^2 = 3x(x^2 - y^2) \qquad \text{Common factor of } 3x$$
   II. The factor $x^2 - y^2$ has two terms and is the difference of two squares.
   $$x^2 - y^2 = (x - y)(x + y) \qquad \text{Factoring the binomial}$$
   III. After checking to see if any of the factors will factor further, we conclude that $3x(x - y)(x + y)$ is the completely factored form. Therefore
   $$3x^3 - 3xy^2 = 3x(x - y)(x + y)$$

2. $3ax + bx + 6ay + 2by$

    I. There is no common factor (other than 1 or -1).

    II. The polynomial has four terms and we factor it by grouping.

$$(3ax + bx) + (6ay + 2by)$$ Group in pairs
$$= x(3a + b) + 2y(3a + b)$$ Factor out what is common to each pair
$$= (3a + b)(x + 2y)$$ Factor out the common quantity

    III. None of the factors will factor further.

$$3ax + bx + 6ay + 2by = (3a + b)(x + 2y)$$

3. $3a^2 - 2a - 8$

    I. There is no common factor (other than 1 or -1).

    II. The polynomial has three terms and the coefficient of a^2 is not 1. Therefore we must find m and n and factor the trinomial. $m + n = -2$ and $m \cdot n = -24$, the values for m and n are -6 and 4.

$$= 3a^2 - 6a + 4a - 8$$ Replace $-2a$ with $-6a + 4a$
$$= (3a^2 - 6a) + (4a - 8)$$ Group the first two terms and the last two terms
$$= 3a(a - 2) + 4(a - 2)$$ Factor out what is common to each pair
$$= (a - 2)(3a + 4)$$ Factor out the common quantity

    III. None of the factors will factor further.

$$3a^2 - 2a - 8 = (a - 2)(3a + 4)$$

⌐ ***Quick check***   Factor completely $4x^2 - 36y^2$

MASTERY POINTS

Can you
- Factor out the greatest common factor?
- Factor the difference of two squares?
- Factor the sum and difference of two cubes?
- Factor trinomials?
- Factor perfect square trinomials?
- Factor a four-term polynomial?
- Use the general strategy for factoring polynomials?

Exercise 4–6

Directions

Completely factor the following polynomials. If a polynomial will not factor, so state. See the outline of the general strategy for factoring polynomials and example 4–6 A.

⌐ ***Quick Check***

Example $4x^2 - 36y^2$

Solution   I. $4x^2 - 36y^2 = 4(x^2 - 9y^2)$ Common factor of 4

           II. $= 4(x - 3y)(x + 3y)$ Factoring the difference of two squares

           III. $4x^2 - 36y^2 = 4(x - 3y)(x + 3y)$ Completely factored form

Directions   Completely factor the following polynomials. If a polynomial will not factor, so state. See the outline of the general strategy for factoring polynomials and example 4–6 A.

1. $n^2 - 49$

2. $a^2 + 6a + 5$

3. $7b^2 + 36b + 5$

4. $2x^2 + 15x + 18$

5. $x^2y^2 + 2xy - 8$

6. $y^2 + 11y + 10$

7. $36 - y^2$

8. $25a^2(3b + c) + 5a(3b + c)$

9. $10a^2 - 20ab + 10b^2$

10. $a^2b^2 - 5ab - 14$

11. $4a^2 - 16b^2$

12. $12x^3y^2 - 18x^2y^2 + 16xy^4$

13. $3ax + 6ay - bx - 2by$

14. $5x^2 + 18x - 60$

15. $6x^2 + 7x - 5$

16. $9x^5y - 6x^3y^3 + 3x^2y^2$

17. $6am + 4bm - 3an - 2bn$

18. $5x^2 - 32x - 21$

19. $7b^2 + 16b - 15$

20. $16b^2 + 8b + 1$

21. $4x^2 + 4x + 1$

22. $a^3 + a^2b + ab^2 + b^3$

23. $3ax - 6bx + 4ay - 8by$

24. $4a^2 - 8ab + 4b^2$

25. $x^2 - 8x + 16$

26. $5x^2 - 27x + 10$

27. $a^2 - 8a - 20$

28. $b^2 - 9b - 36$

29. $x^2y + 3x^2 + 2y + 6$

30. $3a^2 + 13a + 4$

31. $4x^2 + 17x - 15$

32. $5y^2 + 16y + 12$

33. $6x^2 - 24xy - 48y^2$

34. $4ab(x + 3y) - 8a^2b^2(x + 3y)$

35. $3x^2y(m - 4n) + 15xy^2(m - 4n)$

36. $4x^2 - 20xy + 25y^2$

37. $9a^2 - 30ab + 25b^2$

38. $80y^4 - 5y$

39. $3a^5 - 48a$

40. $3a^5b - 18a^3b^3 + 27ab^5$

41. $3a^3b^3 + 6a^2b^4 + 3ab^5$

42. $3b^2 + 8b - 91$

43. $3b^2 - 32b - 91$

44. $b^4 - 81$

45. $3ax + 6bx + 2ay + 4by$

46. $12ax + 4bx - 3ay - by$

47. $6x^2 + 11x - 2$

48. $6x^2 - 17x - 3$

49. $3x^4 - 48x^2$

50. $3x^3 + 3x^2 - 18x$

51. $y^3 + 27z^3$

52. $8b^3 - c^3$

53. $x^3 - y^9$

54. $a^3b^3 + 64$

55. $a^3 - 125$

56. $2a^3 - 54b^3$

57. $x^{27} + 27$

58. $x^3y^3 + z^3$

59. $a^{15} - b^{15}$

60. $x^9 + y^9$

✸ Explore to Learn

Factor $x^6 - 1$, first as a difference of two squares, then as a difference of two cubes. Draw a conclusion about the expression $x^4 + x^2 + 1$.

Review Exercises

Directions   Find the solution set of the following equations. See section 2–5.

1. $2x + 6 = 0$ **3.** $6x + 4 = 0$ **5.** $3x = 0$

2. $4x - 12 = 0$ **4.** $4x - 1 = 0$

Directions   Write an algebraic expression for each of the following. See section 2–1.

6. 7 more than a number

7. A number decreased by 11

8. 6 times the sum of x^2 and x

9. The sum of x^2 and $2x$, divided by 8

Directions   Solve the following word problems. See section 2–7.

10. Three times a number is increased by 12 and the results is 51. Find the number.

11. One number is two more than five times another number. If their sum is 38, find the numbers.

12. The sum of three consecutive even integers is 72. Find the integers.

4–7   Solving Quadratic Equations by Factoring

The Standard Form of a Quadratic Equation

In this section, we will find the solutions to an equation that contains the second, but no higher, power of that variable. Such an equation is a **second-degree equation**, also called a **quadratic equation**.

Quadratic Equation

A quadratic equation can be written in the form
$$ax^2 + bx + c = 0 \qquad a \neq 0$$
where a, b, and c are real numbers. This is called the **standard form** of a quadratic equation.

Note   It is necessary that $a \neq 0$. If $a = 0$ and $b \neq 0$, then $0 \cdot x^2 + bx + c$ becomes $bx + c = 0$, which is a linear equation.

Solution of Quadratic Equations in Factored Form

Some quadratic equations can be solved by factoring and applying the **zero product property**.

Zero Product Property

Given real numbers p and q, if $pq = 0$, then $p = 0$ or $q = 0$.

Concept

If the product of two factors is zero, then at least one of the factors is zero.

● **Example 4–7 A**

Solving quadratic equations in factored form

Find the solution set of the following quadratic equations.

1. $(x + 5)(x - 4) = 0$

Since the product of $(x + 5)$ and $(x - 4)$ is 0, by the zero product property, one or both factors must equal zero. Using this fact, we set each of the factors equal to 0 and solve the resulting equations.

$(x + 5)(x - 4) = 0$	The equation is in factored form
$x + 5 = 0$ or $x - 4 = 0$	Set each factor equal to 0 and solve
$x = -5$ $x = 4$	The solutions

To check the solutions, we substitute the values into the original equation.

Check for $x = -5$	Check for $x = 4$	
$(-5 + 5)(-5 - 4) = 0$	$(4 + 5)(4 - 4) = 0$	Substitute the solution for x
$0 \cdot (-9) = 0$	$9 \cdot 0 = 0$	Order of operations
$0 = 0$	$0 = 0$	True, both solutions check

To express the answer as a solution set, we would write the solutions in any order, separated by a comma and enclosed within a pair of braces.

The solution set is $\{-5, 4\}$.

Note In future examples, we will not always show a check of the solutions, but checking your solutions is always an important part of the problem.

2. $(x - 3)(3x + 1) = 0$ The equation is in factored form

$x - 3 = 0$ or $3x + 1 = 0$	Set each factor equal to 0 and solve
$x = 3$ $\qquad 3x = -1$	Add 3, subtract 1
$x = -\dfrac{1}{3}$	Divide by 3
$x = 3$ or $\qquad x = -\dfrac{1}{3}$	The solutions

The solution is $\left\{ -\dfrac{1}{3}, 3 \right\}$.

☞ *Quick check*   Find the solution set of $(2x + 3)(x + 1) = 0$ ●

Solving Quadratic Equations by Factoring

In general, to find the solution set of a quadratic equation by factoring, we use the following procedure:

Solving a Quadratic Equation by Factoring

1. Write the quadratic equation in standard form with the squared term positive.
2. Completely factor the side of the equation that is not 0.
3. Set each of the factors containing the variable equal to 0 and solve the resulting equations.

● **Example 4–7 B**

Solving quadratic equations

Find the solution set of the following quadratic equations.

1. $x^2 + 5x = -6$

$x^2 + 5x + 6 = 0$ Write the equation in standard form.

$(x + 2)(x + 3) = 0$ Factor $x^2 + 5x + 6 = (x + 2)(x + 3)$

$x + 2 = 0$ or $x + 3 = 0$ Set each factor equal to 0

$x = -2$ $x = -3$ Solve each equation, giving the solutions

The solution set is $\{-3, -2\}$.

2. $x^2 = 2x$

$x^2 - 2x = 0$ Write the equation in standard form

$x(x - 2) = 0$ Factor completely

$x = 0$ or $x - 2 = 0$ Set each factor equal to 0

$x = 0$ $x = 2$ Solve each equation, giving the solutions

The solution set is $\{0, 2\}$.

3. $x^2 = 16$

$x^2 - 16 = 0$ Write the equation in standard form.

$(x - 4)(x + 4) = 0$ Factor $x^2 - 16 = (x - 4)(x + 4)$

$x - 4 = 0$ or $x + 4 = 0$ Set each factor equal to 0

$x = 4$ $x = -4$ Solve each equation, giving the solutions

The solution set is $\{-4, 4\}$.

4. $4x^2 = 20x - 25$

$4x^2 - 20x + 25 = 0$ Write the equation in standard form

$(2x - 5)^2 = 0$ $4x^2 - 20x + 25 = (2x - 5)^2$

$2x - 5 = 0$ Set the repeated factor equal to 0

$x = \dfrac{5}{2}$ Solve the equation for x, giving the solution

The solution set is $\left\{\dfrac{5}{2}\right\}$.

<u>Note</u> In example 4, we have *two* factors, $(2x - 5)$ and $(2x - 5)$, but since they are the same, the equation has only *one* distinct solution.

☞ *Quick check*   Find the solution set of $4y^2 = 9$.   ●

Applications of Quadratic Equations

Many word problems require the use of a quadratic equation for their solution. It is important that we check our solutions to see that they satisfy the problem and are physically possible.

● **Example 4–7 C**

Problem solving

1. The product of two consecutive even integers is 168. Find the integers.

Let x represent the lesser even integer. Then $x + 2$ is the next consecutive even integer.

<u>Note</u> Consecutive even or odd integers are given by $x, x + 2, x + 4, \cdots$.

two consecutive even integers   is   168

product of

$$x \quad \cdot \quad (x + 2) \quad = \quad 168$$

$x^2 + 2x = 168$	Original equation
$x^2 + 2x - 168 = 0$	Write in standard form
$(x + 14)(x - 12) = 0$	Factor the left side
$x + 14 = 0 \quad \text{or} \quad x - 12 = 0$	Set each factor equal to zero
$x = -14 \qquad\qquad x = 12$	Solve each equation

When $x = -14$, then $x + 2 = -14 + 2 = -12$.
When $x = 12$, then $x + 2 = 12 + 2 = 14$.
Check: Since $(-14)(-12) = 168$ and $(12)(14) = 168$, and both solutions are consecutive even integers, the conditions of the problem are met. Therefore, the two integers are -14 and -12 or 12 and 14.

2. The area of a rectangle is $A = \ell w$, where ℓ is the length and w is the width of the rectangle. The length of a rectangle is 2 inches more than three times the width. If $A = 33$ square inches, find the length and width of the rectangle.

Let w represent the width of the rectangle. Then the length ℓ is $3w + 2$

area of a rectangle is   length   times   width

$$A \quad = \quad \ell \quad \cdot \quad w$$
$$33 \quad = \quad (3w + 2) \quad \cdot \quad w$$

$w(3w + 2) = 33$	Original equation
$3w^2 + 2w = 33$	Distribute multiplication
$3w^2 + 2w - 33 = 0$	Write in standard form
$(3w + 11)(w - 3) = 0$	Factor the left side
$3w + 11 = 0 \quad \text{or} \quad w - 3 = 0$	Set each factor equal to zero
$3w = -11 \qquad\qquad w = 3$	Solve each equation
$w = -\dfrac{11}{3} \qquad\qquad w = 3$	Solutions of the equation

The solution set of the equation we wrote is $\left\{-\dfrac{11}{3}, 3\right\}$. Since the width cannot be negative, $w = -\dfrac{11}{3}$ is not a solution of the problem, even though it *is* a solution of the equation. So $w = 3$ is the only physical solution and

$$\ell = 3w + 2 = 3(3) + 2 = 9 + 2 = 11$$

Check: Since 11 is two more than 3 times 3 and $(3)(11) = 33$, the conditions of the problem are met.
The rectangle is 3 inches wide and 11 inches long.

3. Current in a circuit flows according to the equation $i = 16 - 16t^2$, where i is the current in amperes and t is the time in seconds. Find the time t when $i = 0$ amperes (no current).

Replacing i by 0, we have the equation

$0 = 16 - 16t^2$	Substitute 0 for i
$0 = 16(1 - t^2)$	Factor the common factor 16
$0 = 16(1 - t)(1 + t)$	Factor $1 - t^2 = (1 - t)(1 + t)$
$1 - t = 0$ or $1 + t = 0$	Set each factor containing the variable equal to 0
$t = 1$ $t = -1$	Solutions of the equation

Since time cannot be negative, $t = 1$ second.

MASTERY POINTS

Can you
- Find the solution set of a quadratic equation in factored form whose product is equal to zero?
- Find the solution set of a quadratic equation by factoring?
- Set up a quadratic equation for a word problem and solve the equation?
- Substitute values into a formula and solve the resulting quadratic equation?

Exercise 4-7

Directions
Find the solution set of the following quadratic equations. See example 4-7 A.

⚑ *Quick Check*

Example $(2x + 3)(x + 1) = 0$ The equation is in factored form

Solution Set each factor equal to 0 and solve the equations.

$2x + 3 = 0$	or $x + 1 = 0$	The factors are set equal to 0
$2x = -3$		Solve each equation
$x = -\dfrac{3}{2}$	$x = -1$	The solutions

The solution set is $\left\{ -\dfrac{3}{2}, -1 \right\}$

Directions   Find the solution set of the following quadratic equations. See example 4-7 A.

1. $(x + 5)(x - 5) = 0$

2. $(x - 1)(x + 1) = 0$

3. $x(x + 6) = 0$

4. $x(x - 8) - 0$

5. $3a(a - 7) = 0$

6. $5p(p + 9) = 0$

7. $(3x - 9)(2x + 3) = 0$

8. $(8x - 4)(5x + 10) = 0$

9. $(2x + 1)(3x - 2) = 0$

10. $(4y - 3)(5y + 2) = 0$

11. $(5x - 1)(5x + 1) = 0$

12. $(4a - 1)(4a + 1) = 0$

13. $(8 - y)(7 - y) = 0$

14. $(1 - x)(3 - x) = 0$

15. $(4 - 3u)(8 - 5u) = 0$

16. $(7 + 3y)(2 - 3y) = 0$

17. $(5 - x)(5 + x) = 0$

18. $(8 - x)(8 + x) = 0$

Directions Find the solution set of the following quadratic equations by factoring. Check the solutions. See example 4–7 B.	***Quick Check*** **Example** $4y^2 = 9$

Solution

$$4y^2 - 9 = 0$$ Write the equation in standard form.

$$(2y + 3)(2y - 3) = 0$$ Factor $4y^2 - 9 = (2y + 3)(2y - 3)$

$$2y + 3 = 0 \quad \text{or} \quad 2y - 3 = 0$$ Set each factor equal to 0

$$2y = -3 \qquad\qquad 2y = 3$$ Solve each equation

$$y = -\frac{3}{2} \qquad\qquad y = \frac{3}{2}$$ The solutions

Check by replacing y with $-\dfrac{3}{2}$ and $\dfrac{3}{2}$ in the original equation.

The solution set is $\left\{ -\dfrac{3}{2}, \dfrac{3}{2} \right\}$.

Directions   Find the solution set of the following quadratic equations by factoring. Check the solutions. See example 4–7 B.

19. $x^2 + 4x = 0$

20. $y^2 - 4y = 0$

21. $3a^2 - 5a = 0$

22. $4x^2 + 7x = 0$

23. $10a^2 = -15a$

24. $4y^2 = -6y$

25. $a^2 = 25$

26. $y^2 = 49$

27. $5y^2 - 45 = 0$

28. $8x^2 - 18 = 0$

29. $y^2 + 6y - 16 = 0$

30. $x^2 - 3x - 4 = 0$

31. $a^2 + 14a + 49 = 0$

32. $x^2 - 16x + 64 = 0$

33. $b^2 + 5b - 14 = 0$

34. $x^2 + x - 42 = 0$

35. $x^2 + 3x + 2 = 0$

36. $y^2 = -11y - 10$

37. $a^2 - 11a = 12$

38. $x^2 - 14x = 15$

39. $y^2 - 32 = 4y$

40. $x^2 = 27 - 6x$

41. $2x^2 - 7x - 9 = 0$

42. $2y^2 - y - 3 = 0$

43. $6a^2 - 5a + 1 = 0$

44. $3p^2 + 10p - 8 = 0$

45. $6x^2 + x - 12 = 0$

46. $-6x = -3x^2 - 3$

47. $6p^2 - 7p = 20$

48. $9y^2 + 20 = -27y$

49. $3x^2 + 6x = -3$

50. $12y^2 - 15 = 8y$

51. $3x^2 - 4x - 28 = x$

52. $6z^2 + z = -10z - 3$

53. $2b^2 + 2b = 9 - b$

54. $3a - 3 = 15a^2 + 2a - 5$

Directions   Solve the following word problems by setting up a quadratic equation. See example 4–7 C.

55. The product of two consecutive odd integers is 143. Find the integers.

56. The product of two consecutive integers is 132. Find the integers.

57. The product of two consecutive even integers is 288. Find the integers.

58. The product of two consecutive integers is 306. Find the integers.

59. One integer is six more than a second integer. The product of the two integers is 91. Find the integers.

60. One integer is eight less than a second integer. The product of the two integers is 153. Find the integers.

61. The product of two consecutive even integers is four more than two times their sum. Find the integers.

62. The product of two consecutive odd integers is five more than six times the lesser integer. Find the integers.

63. The sum of two integers is -13 and their product is 36. Find the integers.

64. The sum of two integers is -3 and their product is -70. Find the integers.

65. One number is one more than three times the other. Their product is 14. Find the numbers.

66. One number is two more than the other and their product is -1. Find the numbers.

67. The length of a rectangle is 2 meters less than twice the width. If the area is 24 square meters, what are the dimensions of the rectangle?

68. The area of a rectangle is 21 square feet. What are its dimensions if the length is 5 feet less than four times the width?

69. The area of a rectangle is numerically equal to twice the length. If the length is 3 feet more than the width, what are the dimensions of the rectangle?

70. The length of a rectangle is three less than twice the width. If the area is numerically five times the length, find the dimensions of the rectangle.

71. The height of a page of a book is 3 inches more than the width. If the area of the page is six less than ten times the width, find the width of the page.

72. The height of a page of a book is 4 inches more than the width. If there is a margin of 1 inch all around the printed matter of the page and the area of the printed matter is 32 square inches, what is the height of the page?

73. An object with initial velocity v undergoes an acceleration a for time t. The displacement s of the object for this time is given by the equation

$$s = vt + \frac{1}{2}at^2$$

a. Find t when $s = 8$, $v = 2$, $a = 2$
b. Find t when $s = 6$, $v = 3$, $a = 6$

74. The current in a circuit flows according to the equation $i = 12 - 12t^2$, where i is the current in amperes and t is the time in seconds. Find t when i is (a) 0 amperes, (b) 9 amperes.

75. The power output of a generator armature is given by $P_0 = E_g I - r_g I^2$.
a. Find I when $P_0 = 120$, $E_g = 22$, $r_g = 1$
b. Find I when $P_0 = 120$, $E_g = 16$, $r_g = \frac{1}{2}$

76. The output power P of a 100-volt electric generator is defined by

$$P = 100I - 5I^2$$

where I is in amperes. Find I when (a) $P = 480$, (b) $P = 375$

77. A ball rolls down a slope and travels a distance $d = 6t + \frac{t^2}{2}$ feet in t seconds. Find t when (a) $d = 14$ feet, (b) $d = 32$ feet.

78. Because of gravity, an object falls a distance s feet according to the formula $s = 16t^2$, where t seconds is the time it falls. How long will it take the object to fall (a) 256 feet; (b) 49 feet; (c) $2\frac{1}{4}$ feet; (d) 1,024 feet?

Directions   The formula $s = v_0t - 16t^2$ gives the height s in feet that an object will travel in t seconds if it is propelled directly upward with an *initial* velocity of v_0 feet per second. (Use this formula in exercises 79 through 82.)

79. If an object is thrown upward at 96 feet per second, how long will it take the object to reach a height of 80 feet?

80. How long will it take before the object of exercise 79 hits the ground? (Hint: $s = 0$ when this happens.)

81. A projectile is fired upward with an initial velocity of 144 feet per second. How long will it take before the projectile strikes an object 288 feet directly overhead?

82. How long will it take before the projectile in exercise 81 falls back to the ground?

83. The formula for the area A of a trapezoid is $A = \frac{1}{2}h(b_1 + b_2)$, where h is the altitude (height) of the trapezoid and the parallel bases are b_1 and b_2 (see diagram). If the area of the trapezoid is 63 square inches, base b_1 is 10 inches long, and the altitude h is 1 inch less than the length of b_2, find the length of h and b_2.

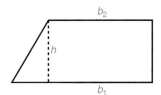

84. If the area of a trapezoid is 21 square feet, base b_2 is 5 feet long, and base b_1 is 6 feet longer than the altitude h, find the altitude of the trapezoid.

85. The altitude and base b_2 of a trapezoid have the same length. If the area of the trapezoid is 24 square meters and base b_1 is twice as long as base b_2, find the dimensions of the trapezoid.

86. One base of a trapezoid is three times the length of the other base. If the altitude is twice as long as the shorter base and the area is 36 square centimeters, find the dimensions of the trapezoid.

87. The volume of a box (rectangular solid) is given by $V = \ell wh$, where ℓ is the length, w is the width, and h is the height of the box (see diagram). If the box is 4 feet tall, the length is 1 foot longer than the width, and the volume is 224 cubic feet, find the length and the width of the box.

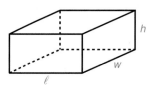

88. A box is 9 inches long and has a volume of 162 cubic inches. If the width of the box is twice the height, find the width and the height of the box.

89. A storage room is three times as long as it is wide. If the room contains 756 cubic feet of space and has a ceiling 7 feet high, find the length and width of the room.

90. Compare a linear equation ($ax + b = 0, a \neq 0$) with a quadratic equation ($ax^2 + bx + c = 0, a \neq 0$). Include in your discussion the degree of each equation and the number of solutions.

91. A cardboard box has a volume of 108 cubic inches. If the length of the box is 1 inch more than two times its width and the box has a height of 3 inches, find the length and the width of the box.

✦ Explore to Learn

1. a. What's so special about the zero product property? For example, what does the statement $ab = 0$ tell us that $ab = 1$ does not?

   b. The problem $2ab = 0$ is on a test. A student writes the solution $2 = 0$ or $a = 0$ or $b = 0$. Is this student wrong?

2. In this section the formula $s = v_o t - 16t^2$ is given. It gives the height s of an object t seconds after it had velocity v_0. Suppose the velocity of a fireworks rocket v_0 is 304 feet per second. Make a table of values with two columns, t and s. Start with $t = 0$ and fill in values for s for every second. It should become apparent when to stop. How high does the rocket go?

Review Exercises

Directions   Find the solution set. See sections 2–6 and 4–7.

1. $3x - 2 = 0$

2. $4a - 5 = 0$

3. $x^2 - 9 = 0$

4. $x^2 + 6x + 8 = 0$

Directions   Evaluate the following expressions if $a = 4$, $b = -3$, $c = -6$, and $d = 5$. See section 1–9.

5. $ab - cd$

6. $b^2 - c^2$

7. $\dfrac{c - 2b}{d - a}$

8. $\dfrac{c + d}{3a + 4b}$

9. Peter has invested $5,000 at an 8% rate. How much more must he invest at 10% to make the total income for one year from both sources a 9% rate? See section 2–7.

Chapter 4 Lead-in Problem

The formula $s = vt - 16t^2$ gives the height s in feet that an object will travel in t seconds if it is propelled directly upward with an initial velocity of v feet per second. If an object is thrown upward at 96 feet per second, how long will it take the object to reach a height of 144 feet?

Solution

$s = vt - 16t^2$	Original equation
$(144) = (96)t - 16t^2$	Substitute 144 for s and 96 for v
$16t^2 - 96t + 144 = 0$	Standard form
$16(t^2 - 6t + 9) = 0$	Common factor
$16(t - 3)(t - 3) = 0$	Factor the trinomial
$t - 3 = 0$	Set the repeated factor equal to zero
$t = 3$	Solve

The object will reach a height of 144 feet in 3 seconds.

Chapter 4 Summary

• Glossary

Common factors (page 232)   are numbers or variables that appear as a factor in all of the original terms

factor (page 60)   the numbers or variables in an indicated multiplication

factored form (page 231)   when a polynomial is written as the product of two or more polynomials. That is, $a(b + c)$ is the factored form of $ab + ac$.

factoring a polynomial (page 231)   finding the factored form of a polynomial

greatest common factor (GCF) (page 231)   the greatest common factor of a polynomial is the greatest factor that divides every term of the polynomial.

leading coefficient (page 243)   the coefficient of the first term of a polynomial written in standard form.

perfect square trinomial (page 254)   a trinomial that can be factored as the square of a binomial.

prime polynomial (page 240)   a polynomial that cannot be factored using integer coefficients.

standard form (page 243)   when a polynomial is written in descending powers of the variable.

• Properties and Definitions

Quadratic equation (page 265)
A quadratic equation can be written in the form

$$ax^2 + bx + c = 0, \qquad a \neq 0$$

where a, b, and c are real numbers. This is called the standard form of a quadratic equation.

Zero product property (page 265)
Given real numbers p and q, if $pq = 0$, then $p = 0$ or $q = 0$.

• Procedures

[4–1]

Factoring the greatest common factor (page 232)

1. Factor each term such that it is the product of primes and variables in exponential form.

2. Write down all the factors from step 1 that are common to every term and raise each of them to the lowest power that they are raised to in any of the terms. The product of these factors is the greatest common factor, GCF.

3. Determine the polynomial within the parentheses by dividing each term of the polynomial being factored by the GCF.

4. The answer is the product of steps 2 and 3.

Factoring a four-term polynomial by grouping (page 234)

1. Arrange the four terms so that the first two terms have a common factor and the last two terms have a common factor.

2. For each pair of terms determine the GCF and factor it out.

3. If step 2 produces a binomial factor common to both terms, factor it out.

4. If step 2 does not produce a binomial factor common to both terms, try grouping the terms of the original polynomial in different possible ways.

5. If step 4 does not produce a binomial factor common to both terms, the polynomial will not factor by this procedure.

[4–2]

Factoring a trinomial of the form $x^2 + bx + c$ (page 239)

1. Factor out the GCF. If there is a common factor, make sure to include it as part of the final factorization.

2. Determine if the trinomial is factorable by finding m and n such that $m + n = b$ and $m \cdot n = c$. If m and n do not exist, we conclude that the trinomial will not factor.

3. Using the m and n values from step 2, write the trinomial in factored form.

The signs (+ or −) for m and n (page 239)

1. If c is positive, then m and n have the same sign as b.

2. If c is negative, then m and n have different signs and the one with the greater absolute value has the same sign as b.

[4–3]

Factoring a trinomial of the form $ax^2 + bx + c$ (page 244)

1. Determine if the trinomial $ax^2 + bx + c$ is factorable by finding m and n such that $m \cdot n = a \cdot c$ and $m + n = b$. If m and n do not exist, we conclude that the trinomial will not factor.

2. Replace the middle term, bx, by the sum of mx and nx.

3. Place parentheses around the first and second terms and around the third and fourth terms. Factor out what is common to each pair.

4. Factor out the common quantity of each term and place the remaining factors from each term in the second parentheses.

The signs (+ or −) for m and n (page 245)

1. If $a \cdot c$ is positive, then m and n have the same sign as b.

2. If $a \cdot c$ is negative, then m and n have different signs and the one with the greater absolute value has the same sign as b.

[4–4]

Factoring the difference of two squares (page 253)

1. Identify that we have a perfect square *minus* another perfect square.

2. Rewrite the problem as a first term squared minus a second term squared.

$$(\text{first term})^2 - (\text{second term})^2$$

3. Factor the problem into the first term plus the second term times the first term minus the second term.

$$(\text{first term} + \text{second term})(\text{first term} - \text{second term})$$

Necessary conditions for a perfect square trinomial (page 255)

1. The first term must have a positive coefficient and be a perfect square, a^2.

2. The last term must have a positive coefficient and be a perfect square, b^2.

3. The middle term must be twice the product of the bases of the first and last terms, $2ab$ or $-2ab$.

[4–5]

Factoring the difference of two cubes (page 258)

1. Identify that we have a perfect cube minus another perfect cube.

2. Rewrite the problem as a first term cubed minus a second term cubed.

$$(\text{1st term})^3 - (\text{2nd term})^3$$

3. Factor the expression into the first term minus the second term, times the first term squared plus the first term times the second term plus the second term squared.

Factoring the sum of two cubes (page 259)

1. Identify that we have a perfect cube plus another perfect cube.

2. Rewrite the problem as a first term cubed plus a second term cubed.

$$(\text{1st term})^3 + (\text{2nd term})^3$$

3. Factor the expression into the first term plus the second term, times the first term squared minus the first term times the second term plus the second term squared.

[4–6]

Factoring: A general strategy (page 262)

I. Factor out all common factors.

II. Count the number of terms.

   A. Two terms: Check to see if the polynomial is the difference of two squares, the difference of two cubes, or the sum of two cubes.

   B. Three terms: Check to see if the polynomial is a perfect square trinomial. If it is not, use one of the general methods for factoring a trinomial.

   C. Four terms: Check to see if we can factor by grouping.

III. Check to see if any of the factors we have written can be factored further. Any common factors that were missed in part I can still be factored out here.

[4–7]

Solving a quadratic equation by factoring (page 266)

1. Write the quadratic equation in standard form with the squared term positive.

2. Completely factor the side of the equation that is not 0.

3. Set each of the factors containing the variable equal to 0 and solve the resulting equations.

☼ Chapter 4 Error Analysis

1. Factoring a common factor
   Example: $3x^2 - 6x - 3 = 3(x^2 - 2x)$
   Correct answer: $3(x^2 - 2x - 1)$
   What error was made? (*see page 233*)

2. Completely factoring a trinomial
   Example: $3x^2 - 6x + 3 = 3(x^2 - 2x + 1)$
   $= (x - 1)^2$
   Correct answer: $3(x - 1)^2$
   What error was made? (*see page 255*)

3. Factoring the sum of two squares
   Example: $4x^2 + 9y^2 = (2x + 3y)^2$
   Correct answer: $4x^2 + 9y^2$ is not factorable.
   What error was made? (*see page 254*)

4. Factoring the difference of two squares
   Example: $16x^2 - 4y^2 = (4x)^2 - (2y)^2$
   $= (4x + 2y)(4x - 2y)$
   Correct answer: $4(2x + y)(2x - y)$
   What error was made? (*see page 254*)

5. Factoring the sum of two cubes
   Example: $x^3 + y^3 = (x + y)(x^2 - 2xy + y^2)$
   Correct answer: $x^3 + y^3 = (x + y)(x^2 - xy + y^2)$
   What error was made? (*see page 259*)

6. Order between real numbers
   Example: $-15 > 4$
   Correct answer: $-15 < 4$
   What error was made? (*see page 39*)

7. Combining using grouping symbols
   Example: $6 - (10 + 3) = 6 - 10 + 3 = -1$
   Correct answer: -7
   What error was made? (*see page 56*)

8. Multiplication of real numbers
   Example: $5\dfrac{3}{4} = 5 \cdot \dfrac{3}{4} = \dfrac{15}{4}$
   Correct answer: $\dfrac{23}{4}$
   What error was made? (*see page 11*)

9. Division by zero
   Example: $\dfrac{5}{0} = 0$
   Correct answer: $\dfrac{5}{0}$ is undefined.
   What error was made? (*see page 65*)

10. Exponents
    Example: $-2^2 = 4$
    Correct answer: -4
    What error was made? (*see page 70*)

☐ Chapter 4 Critical Thinking

1. If n is an integer, for what values of n will $4n^2 + 10n + 4$ represent an even number?

2. In July 1992 the United States had 2 billion used automobile tires stockpiled. Another 280 million more are stockpiled each year. A power plant in Connecticut burns 11 million tires a year and produces enough electricity to serve 15,000 people.
   a. If no one else used the two billion tires, how long could the Connecticut plant burn the existing tires before running out?
   b. The United States has about 300 million people. If plants like the one in Connecticut were used to serve everyone in the United States, how long would the two billion tires last?

c. How many people could be served on an ongoing basis with the 280 million tires stockpiled each year?

Chapter 4 Review

[4–1]

Directions   Write in completely factored form.

1. $x^2y + xyz + xy^2z$

2. $a^3b + a^3b^2$

3. $3R^3 - R^2 + 5R^4$

4. $4y^2 + 8y + 12y^3$

5. $x^4 + 3x^3 + 9x^2$

6. $16R^3S^2 - 12R^4S^3 + 24R^2S^2$

7. $10a^4b^3 + 15a^2b^2 - 20a^3b^2$

8. $2(a + b) + x(a + b)$

9. $y(x - 3z) + 4(x - 3z)$

10. $a(3R + 1) + b(3R + 1)$

11. $2a(x - 3y) - 3b(x - 3y)$

12. $6ax - 3ay - 2bx + by$

13. $4ax + 6by + 8ay + 3bx$

14. $ax + 3bx - 4a - 12b$

15. $ax^2 - 2bx^2 + 4a - 8b$

[4–2]

Directions   Write in completely factored form.

16. $x^2 - 9x + 14$

17. $2a^3 - 8a^2 - 10a$

18. $a^2 + 14a + 24$

19. $x^2 - 4x - 32$

20. $a^2 - 16a - 36$

21. $3x^2 - 9x - 30$

22. $x^3 - x^2 - 6x$

23. $x^3 - 4x^2 - 21x$

24. $a^2b^2 + ab - 6$

25. $a^2b^2 + 10ab + 24$

26. $a^2b^2 - 9ab + 18$

27. $a^2b^2 - 8ab - 20$

[4–3]

Directions   Write in completely factored form.

28. $4x^2 + 4x + 1$

29. $9r^2 - 36r + 36$

30. $4x^2 - 5x + 1$

31. $9a^2 + 9a - 10$

32. $8a^2 - 2a - 3$

33. $24x^2 + 22x + 3$

34. $8a^2 - 18a + 9$

35. $2a^2 + 15a + 18$

[4–4]

Directions   Write in completely factored form.

36. $4a^2 - 9$

37. $36b^2 - c^2$

38. $25 - a^2$

39. $16x^2 - 4y^2$

40. $9x^2 - y^4$

41. $x^4 - 16$

42. $y^4 - 81$

43. $b^2 + 12b + 36$

44. $c^2 - 10c + 25$

45. $4x^2 - 12x + 9$

46. $9x^2 - 12x + 4$

[4–5]

Directions   Write in completely factored form.

47. $R^3 + 8S^3$

48. $16x^3 - 54$

49. $27a^3 + 125b^3$

50. $x^3y^3 - 1$

51. $2x^9 + 250$

52. $64x^{12} - y^{15}$

53. $a^3b^6 + c^9$

[4–6]

Directions   Write in completely factored form.

54. $12x^4 - 3x^3$

55. $a^2 - 3a - 10$

56. $4a^2 - 21a + 5$

57. $9y^2 - 4$

58. $6ax + 9bx - 4a - 6b$

59. $b^2 - b - 20$

60. $9x^2 + 21x + 10$

61. $a^2 + 14a + 49$

62. $12x^5 - 3x^3$

63. $c^3 + 9c^2 + 20c$

64. $16a^2 - 8a + 1$

65. $b^4 - 1$

[4–7]

Directions   Determine the solution set.

66. $(x - 1)(x + 3) = 0$

67. $3x(x - 8) = 0$

68. $(5x + 1)(3x - 7) = 0$

69. $(7x - 1)(5x - 8) = 0$

70. $(4 - 3x)(9 - x) = 0$

71. $(5x - 4)(5x + 4) = 0$

Directions   Find the solution set of the following quadratic equations by factoring.

72. $4x^2 - 9x = 0$

73. $y^2 = 1$

74. $2a^2 = 128a$

75. $3x^2 - 75 = 0$

76. $x^2 - x - 30 = 0$

77. $2y = y^2 + 1$

78. $4a^2 + 13a + 3 = 0$

79. $5x^2 - 4 = 8x$

80. $8y^2 + y - 6 = -y$

81. $4a - 12 = 2a - 2a^2$

Directions   Solve the following word problems.

82. If the product of two consecutive integers is ten less than the square of the greater integer, determine the integers.

83. The length of a rectangular flower garden is 5 feet more than its width. If the area of the garden is 104 square feet, find the dimensions of the flower garden.

84. A farmer has some cattle to sell to a slaughterhouse. When the manager of the slaughterhouse asked how many cattle he had to sell, the farmer replied, "If you triple the square of the number of cattle you get 1,200." How many cattle did the farmer have to sell?

85. The height h in feet of a projectile launched vertically upward from the top of a 96-foot tall tower when time $t = 0$ is given by $h = 96 + 80t - 16t^2$. How long will it take the projectile to strike the ground?

Chapter 4 Test

Directions   Completely factor the following polynomials. If a polynomial will not factor, so state.

1. $4x(3a + b) - (3a + b)$

2. $x^2 - 9x + 14$

3. $6y^2 + 9y - 4y - 6$

4. $4x^2 + 5x - 6$

5. $a^2 + 8a + 12$

6. $9x^5 - 12x^3 + 6x^2$

7. $z^2 - 49$

8. $x^3 + 27y^3$

9. $10a^2 + 23a + 12$

10. $6ax - 2ay - 3bx + by$

11. $b^2 - 11b + 30$

12. $10a^2 - 20ab + 10b^2$

13. $3x^2 + x + 5$

14. $6a^2 + 19a + 3$

15. $a^2b^2 - ab - 12$

16. $4a^2 - 9b^2$

17. $4y^2 - 20y + 25$

18. $2ac - 2bc + 3bd - 3ad$

19. $8a^3 - b^3$

20. $x^2 - 3xy - 4y^2$

21. $d^4 - 81$

22. $81x^3 - 3y^3$

23. $8x^2 - 19x + 6$

24. $a^5 + 27a^2b^3$

Directions Find the solution set of the following equations.

25. $a^2 + 6 = 7a$

26. $(2x - 5)(3x + 7) = 0$

27. $4y^2 = 2y$

28. $x^2 - 18x + 81 = 0$

29. The area of the rectangle is 84 square meters. If the length is 5 meters longer than the width, what are the dimensions of the rectangle?

30. The product of two consecutive even integers is 168. Find the integers.

Chapter 4 Cumulative Test

Directions Perform the indicated operations and simplify. Express answers with positive exponents. Assume that no denominator is equal to zero.

1. $40 - 2 \cdot 8 \div 4 - 6 + 3$

2. $(2a^2b)^3$

3. $(a + 2b)^2$

4. $a^3 \cdot a^2 \cdot a$

5. $3[6 - 4(5 - 2) + 10]$

6. $(x^2y^3)(x^3y)$

7. $(3x - y) - (4y - 3x) - (x + 2y)$

8. $(3x - 2y)(3x + 2y)$

9. $\left(\dfrac{2a^2}{b}\right)^3$

10. $x^3 \cdot x^{-5}$

11. $3x - (x - y) - (2x + 3y)$

12. $(3x^{-3}y^2)^{-2}$

13. $(2x^2 - 5x - 15) \div (x - 4)$

Directions Solve the inequalities and find the solution set for the equations.

14. $3(3x - 2) = 4x + 3$

15. $\dfrac{3}{4}x + 2 < \dfrac{1}{2}x + 5$

16. $2(2x - 1) = 6(x - 2) - 4$

17. $2x + 7 \geq 13$

18. $x^2 - 9 = 0$

19. $x^2 - 7x + 10 = 0$

20. $3(2x + 1) < 4x + 8$

21. $-2 \leq 3x - 8 \leq 6$

Directions   Solve for the specified variable.

22. $3x - y = x + 5y$ for x

23. $2x + 5 - 3y = 5x + 3 - 8y$ for x

Directions   Write in completely factored form.

24. $2ab - 4a^2b^2 - 8a^3b^5$

27. $4a^2 - 4a - 15$

25. $4a^2 + 12a + 9$

28. $x^2 + 9x + 18$

26. $25c^2 - 9d^2$

Directions   Set up an equation and solve for the unknown(s).

29. One number is eleven more than twice a second number. If their sum is 53, what are the numbers?

30. The product of two consecutive positive odd integers is 195. Find the integers.

31. Terry has $15,000. He invests part of this money at 8% and the rest at 6%, and his income for one year from these investments totals $1,100. How much was invested at each rate?

32. The area of a rectangle is six more than ten times the width. If the length is 5 meters longer than the width, what are the dimensions of the rectangle?

5

Rational Expressions and Equations

Marc owns $\frac{5}{8}$ share of a print shop and his uncle owns $\frac{1}{4}$ share of the shop. In a given year, they had a combined earnings of $140,000 from the print shop. How much did the shop earn that year?

5–1   Simplifying Rational Expressions

A Rational Expression

In chapter 1, we defined a rational number. We extend that definition to involve the quotient of two polynomials and define a **rational expression**.

Rational Expressions

A **rational expression** is an expression of the form

$$\frac{P}{Q}$$

where P and Q are polynomials, with $Q \neq 0$.

Concept

A rational expression is an expression that can be written as the quotient of two polynomials with the denominator not zero.

For example,

$$\frac{3}{4}, \quad \frac{2x}{x+1}, \quad \frac{x^2 - 1}{x^2 - x - 6}, \quad \text{and} \quad \frac{x^2 + x}{5}$$

are all rational expressions.

Just as a rational number has a numerator and a denominator, so does a rational expression. In the rational expression

$$\frac{x^2 - 1}{x^2 - x - 6}$$

the polynomial on the top, $x^2 - 1$, is called the **numerator** and the polynomial on the bottom, $x^2 - x - 6$, is called the **denominator**.

<u>Note</u> Any polynomial is a rational expression since the denominator can be considered to be 1. For example, $x^2 - 4 = \dfrac{x^2 - 4}{1}$.

Domain of a Rational Expression

Domain of a Rational Expression

The set of all replacement values of the variable for which a rational expression is defined determines the **domain** of the rational expression.

When determining the domain of a rational expression, we are most concerned with those replacement values of the variable that will make the denominator zero, because the rational expression becomes undefined when the denominator is zero. The values that make the denominator zero are called the **restrictions on the variable** and are excluded from the domain. The following procedure can be used to find the domain of a rational expression.

Finding the Domain of a Rational Expression

1. Factor the denominator into a product of prime polynomials, if possible.
2. Set each factor of the denominator containing the variable equal to zero (using the zero product property).
3. Solve the resulting equations. **The solutions are the restrictions placed on the variable.**
4. The domain is all real numbers excluding any restricted values found in step 3.

● **Example 5–1 A**

Determining the domain of a rational expression

Determine the domain of each of the following rational expressions.

1. $\dfrac{a - 3}{a - 4}$

$$\begin{aligned} a - 4 &= 0 && \text{Set denominator equal to 0} \\ a &= 4 && \text{Solve equation for } a \end{aligned}$$

Our solution tells us that when $a = 4$, the denominator is zero. This value must be excluded from the domain. The domain is all real numbers except 4, that is $a \neq 4$.

<u>Note</u> We look *only* at the denominator. The value(s) of the variable for which the numerator is zero is of no concern to us in determining the domain.

2. $\dfrac{x-3}{x^2-x}$

$$\dfrac{x-3}{x^2-x} = \dfrac{x-3}{x(x-1)} \qquad \text{Factor the denominator}$$

$x = 0 \quad$ or $\quad x - 1 = 0 \qquad$ Set each factor in the denominator equal to 0

$x = 0 \qquad\qquad\quad x = 1 \qquad$ Solve each equation for x

The restrictions are $x \neq 0$ and $x \neq 1$. Domain is all real numbers except 0 and 1.

Note In example 2, a common error is to forget to place restrictions on the factor _x_.

3. $\dfrac{x+3}{x^2+4}$

$x^2 + 4$ does not factor. If there is a restriction, it will occur when x^2 is -4. Since x^2 is never negative, the sum $x^2 + 4$ can never be zero and there are no restrictions on the variable. Thus, the domain is all real numbers.

> **Quick check**   Determine the domain: $\dfrac{x+3}{x^2+x-6}$

The Fundamental Principle of Rational Expressions

One of the most important procedures we can use when we work with rational expressions is the simplification of the rational expression. To do this, we use a principle called the **fundamental principle of rational expressions**.

Fundamental Principle of Rational Expressions

If _P_ is any polynomial and _Q_ and _R_ are nonzero polynomials, then

$$\dfrac{PR}{QR} = \dfrac{P}{Q} \qquad \text{and} \qquad \dfrac{P}{Q} = \dfrac{PR}{QR}$$

Concept

To change the appearance of a rational expression without changing its value, we may multiply or divide both the numerator and the denominator by the same nonzero polynomial.

This property is based on 1 being the identity element for multiplication. That is,

$$\dfrac{PR}{QR} = \dfrac{P}{Q} \cdot \dfrac{R}{R} = \dfrac{P}{Q} \cdot 1 = \dfrac{P}{Q}$$

This property permits us to **reduce rational expressions to lowest terms**. A rational expression is _completely reduced_ if the greatest factor common to both the numerator and the denominator is 1 or -1. We can see that the key to reducing rational expressions is _finding_ and dividing out _factors_ that are common to both the numerator and the denominator.

> ## To Reduce a Rational Expression to Lowest Terms
>
> 1. Write the numerator and the denominator in factored form.
> 2. Divide the numerator and the denominator by all common factors.

● **Example 5–1 B**

Reducing a rational expression

Simplify the following rational expressions by reducing to lowest terms. Assume that all denominators are nonzero.

1. $\dfrac{14x^3}{10x^2} = \dfrac{2x^2 \cdot 7x}{2x^2 \cdot 5}$ Factor the numerator and the denominator

$\qquad = \dfrac{7x}{5}$ Divide by the common factor $2x^2$

2. $\dfrac{5a - 15}{4a - 12} = \dfrac{5(a - 3)}{4(a - 3)}$ Factor numerator and denominator

$\qquad = \dfrac{5}{4}$ Divide numerator and denominator by common factor $(a - 3)$

3. $\dfrac{y - 7}{y^2 - 49} = \dfrac{y - 7}{(y + 7)(y - 7)}$ Factor denominator

$\qquad = \dfrac{1}{y + 7}$ Divide numerator and denominator by common factor $(y - 7)$

<u>Note</u> Any restrictions on the variables in the original rational expression remain with the simplified form. In example 2, the expression simplifies to $\dfrac{5}{4}$ provided that $a \neq 3$ and example 3 simplifies to $\dfrac{1}{y + 7}$ provided that $y \neq -7$ and $y \neq 7$.

☛ *Quick check*   Reduce $\dfrac{25z^5}{15z^3}$ and $\dfrac{a^2 - 36}{a^2 - a - 30}$ to lowest terms.   ●

Reducing $\dfrac{a - b}{b - a}$

Consider the rational expression $\dfrac{x - 5}{5 - x}$, which does not appear to be reducible by a common factor. However,

$$5 - x = -x + 5 \qquad \text{Commute the terms}$$
$$= -1(x - 5) \qquad \text{Factor out } -1$$

Thus, $5 - x = -(x - 5)$. That is, $5 - x$ is equal to the **opposite** of $x - 5$.

$$\dfrac{x - 5}{5 - x} = \dfrac{x - 5}{-1(x - 5)}$$
$$= \dfrac{1}{-1} \qquad \text{Reduce by common factor } (x - 5)$$
$$= -1$$

Reducing by Opposite Factors

In general, for all real numbers a and b, $a \neq b$,

$$\frac{a - b}{b - a} = -1$$

Note Whenever there are factors, one in the numerator and the other in the denominator, that are opposites, we divide them out and leave -1 as a factor in the numerator.

● **Example 5–1 C**

Reducing a rational expression involving opposites

Simplify the following rational expressions by reducing to lowest terms. Assume that no denominator equals zero.

1. $\dfrac{6 - 3x}{x - 2} = \dfrac{3(2 - x)}{x - 2}$ Factor the numerator

$\phantom{\dfrac{6 - 3x}{x - 2}} = 3(-1)$ $\dfrac{2 - x}{x - 2} = -1$

$\phantom{\dfrac{6 - 3x}{x - 2}} = -3$ Multiply by -1

Note $\dfrac{a - b}{a + b}$ will not reduce since $a + b$ and $a - b$ are not opposites.

2. $\dfrac{4 - x}{x^2 - 16}$

$ = \dfrac{4 - x}{(x - 4)(x + 4)}$ Completely factor the denominator, $4 - x$ and $x - 4$ are opposite factors

$ = \dfrac{-1}{x + 4}$ Divide out the opposite factors and leave -1 as a factor in the numerator

☛ **Quick check**   Reduce $\dfrac{16 - y^2}{y - 4}$ to lowest terms.

MASTERY POINTS

Can you
- Determine the restrictions on the variable in a rational expression?
- Determine the domain of a rational expression?
- Reduce a rational expression to lowest terms using the fundamental principle of rational expressions?
- Recognize factors $a - b$ and $b - a$ and use $\dfrac{a - b}{b - a} = -1$?

Exercise 5–1

Directions
Determine the domain of the given rational expression. See example 5–1 A.

☞ *Quick Check*

Example $\dfrac{x + 3}{x^2 + x - 6}$

Solution $\dfrac{x + 3}{(x + 3)(x - 2)}$ Factor the denominator

$x + 3 = 0$ or $x - 2 = 0$ Set each factor equal to zero

$x = -3$ $x = 2$ Solve for x

The restrictions are $x \neq -3$ and $x \neq 2$ and the domain is the set of all real numbers except -3 and 2.

Directions   Determine the domain of the given rational expression. See example 5–1 A.

1. $\dfrac{5}{4x}$

2. $\dfrac{8}{x - 2}$

3. $\dfrac{10}{x - 5}$

4. $\dfrac{x}{x + 7}$

5. $\dfrac{3x^2}{x + 3}$

6. $\dfrac{x + 1}{2x - 1}$

7. $\dfrac{a + 9}{4a - 3}$

8. $\dfrac{p - 3}{5 - 2p}$

9. $\dfrac{y + 4}{8 - 3y}$

10. $\dfrac{x + 7}{x^2 + 3x - 18}$

11. $\dfrac{8b + 1}{b^2 - 7b + 6}$

12. $\dfrac{5s^2 + 7}{2s^2 - s - 3}$

13. $\dfrac{8z}{3z^2 + 2z - 8}$

14. $\dfrac{4}{x^2 - 4}$

15. $\dfrac{5x}{9x^2 - 4}$

16. $\dfrac{a - 2}{4a^2 - 16}$

17. $\dfrac{b + 3}{5b^2 - 45}$

18. $\dfrac{7x}{3x - 15}$

19. $\dfrac{5b - 1}{9b - 21}$

20. $\dfrac{16x}{8x^2 - 18}$

21. $\dfrac{17q}{q^2 + 16}$

22. $\dfrac{23}{x^2 + 1}$

23. $\dfrac{5x - 3}{x^2 + 4}$

24. $\dfrac{2x^2}{x^2 + 9}$

Directions
Simplify the following rational expressions by reducing to lowest terms. Assume that no denominator equals zero. See example 5–1 B.

📬 *Quick Check*

Example $\dfrac{25z^5}{15z^3}$

Solution $= \dfrac{5z^3 \cdot 5z^2}{5z^3 \cdot 3}$ Factor the numerator and the denominator

$= \dfrac{5z^2}{3}$ Divide by the common factor of $5z^3$

Example $\dfrac{a^2 - 36}{a^2 - a - 30}$

Solution $= \dfrac{(a + 6)(a - 6)}{(a + 5)(a - 6)}$ Factor numerator and denominator

$= \dfrac{a + 6}{a + 5}$ Divide numerator and denominator by $(a - 6)$

Directions Simplify the following rational expressions by reducing to lowest terms. Assume that no denominator equals zero. See example 5–1B.

25. $\dfrac{54}{72}$

26. $\dfrac{75}{145}$

27. $\dfrac{6x}{15}$

28. $\dfrac{8a}{10}$

29. $\dfrac{16x^2}{12x}$

30. $\dfrac{15b^3}{20b}$

31. $\dfrac{-8x^2}{6x^4}$

32. $\dfrac{3a^6}{-9a^3}$

33. $\dfrac{16a^2b}{20ab^2}$

34. $\dfrac{15a^2x^3}{35ax^2}$

35. $\dfrac{20ab^2c^3}{-4ab^2c^3}$

36. $\dfrac{-72x^4y^3z^2}{9x^4y^3z^2}$

37. $\dfrac{10(x + 5)}{8(x + 5)}$

38. $\dfrac{24(x - 3)}{15(x - 3)}$

39. $\dfrac{6(x - 2)}{(x + 3)(x - 2)}$

40. $\dfrac{-8(x + 1)}{4(x + 1)(x - 6)}$

41. $\dfrac{a + b}{a^2 - b^2}$

42. $\dfrac{x^2 - y^2}{x - y}$

43. $\dfrac{3m - 6}{5m - 10}$

44. $\dfrac{8b + 12}{10b + 15}$

45. $\dfrac{3x - 3}{6x + 6}$

46. $\dfrac{6y - 6}{8y^2 - 8}$

47. $\dfrac{x^2 - 9}{x^2 + 6x + 9}$

48. $\dfrac{a^2 - 10a + 25}{a^2 - 25}$

49. $\dfrac{x^2 - 3x - 10}{x^2 - x - 6}$

50. $\dfrac{y^2 - y - 42}{y^2 + 12y + 36}$

51. $\dfrac{2y^2 - 3y - 9}{4y^2 - 13y + 3}$

52. $\dfrac{4m^2 - 15m - 4}{8m^2 - 18m - 5}$

53. $\dfrac{x - 3}{x^3 - 27}$

54. $\dfrac{x + 2}{x^3 + 8}$

55. $\dfrac{a^2 - b^2}{a^3 + b^3}$

56. $\dfrac{x^3 + y^3}{x^2 - y^2}$

Directions Simplify by reducing to lowest terms. Assume that no denominator is equal to zero. See example 5–1 C.	☞ **Quick Check**

Example $\dfrac{16 - y^2}{y - 4}$

Solution $= \dfrac{(4 - y)(4 + y)}{(y - 4)}$ Factor numerator

$= \dfrac{4 - y}{y - 4} \cdot \dfrac{4 + y}{1}$

$= -1 \cdot (4 + y)$ $\dfrac{4 - y}{y - 4} = -1$

   Multiply the numerator by -1

$= -4 - y$ or $-y - 4$ Alternative forms of answer

Directions Simplify by reducing to lowest terms. Assume that no denominator is equal to zero. See example 5–1 C.

57. $\dfrac{4x - 4y}{y - x}$

58. $\dfrac{8b - 8a}{a - b}$

59. $\dfrac{2x - 8}{12 - 3x}$

60. $\dfrac{12a - 8b}{10b - 15a}$

61. $\dfrac{2y^2 - 2x^2}{x - y}$

62. $\dfrac{3p - 3q}{6q^2 - 6p^2}$

63. $\dfrac{(x - y)^2}{y^2 - x^2}$

64. $\dfrac{a - b}{b^2 - a^2}$

65. $\dfrac{n^2 - m^2}{(m + n)^2}$

66. $\dfrac{p^2 - q^2}{q^2 - p^2}$

67. $\dfrac{4x - 4y}{y^2 - x^2}$

68. $\dfrac{4 - y}{2y^2 - 7y - 4}$

69. In your own words, explain what the domain of a rational expression means.

70. Explain in your own words why $\dfrac{1}{x - 2}$ and $\dfrac{1}{2 - x}$ are opposites.

71. Explain in your own words why $\dfrac{1}{x + 2}$ and $\dfrac{1}{2 + x}$ are not opposites.

✺ **Explore to Learn**

1. Graphing calculators, and some others as well, allow the entry of an expression for convenience. This can then be used to evaluate the expression for different values of the variable. The following shows how to enter the rational expression $\dfrac{16 - x^2}{2x^2 - 7x - 4}$ on a Texas Instruments TI-81/82 and Casio 7700 GE graphing calculator, and evaluate it at the value $x = 3$, then -4.

TI $\boxed{Y =}$ $\boxed{(}$ $\boxed{(}$ 16 $\boxed{-}$ $\boxed{X \mid T}$ $\boxed{x^2}$ $\boxed{)}$ $\boxed{\div}$ $\boxed{(}$ 2 $\boxed{X \mid T}$ $\boxed{x^2}$ $\boxed{-}$ 7 $\boxed{X \mid T}$ $\boxed{-}$ 4 $\boxed{)}$ $\boxed{ENTER}$ $\boxed{2nd}$ $\boxed{QUIT}$ 3 $\boxed{STO\triangleright}$
$\boxed{X \mid T}$ $\boxed{ENTER}$ $\boxed{2nd}$ $\boxed{Y\text{-}VARS}$ { $\boxed{1\text{:Function}}$ (TI-82 only)} $\boxed{1\text{:}Y1}$ $\boxed{ENTER}$

   Note: On the TI-82 the $\boxed{X \mid T}$ key looks like $\boxed{X, T, \theta}$.

Casio $\boxed{AC}$ $\boxed{(}$ $\boxed{(}$ 16 $\boxed{-}$ $\boxed{X, \theta, T}$ $\boxed{x^2}$ $\boxed{)}$ $\boxed{\div}$ $\boxed{(}$ 2 $\boxed{X, \theta, T}$ $\boxed{x^2}$ $\boxed{-}$ 7 $\boxed{X, \theta, T}$ $\boxed{-}$ 4 $\boxed{)}$ $\boxed{SHIFT}$ $\boxed{[F] \; MEM}$ $\boxed{F1\text{:}STO}$ 1
$\boxed{EXE}$ $\boxed{EXIT}$ 3 $\boxed{\rightarrow}$ $\boxed{X, \theta, T}$ $\boxed{EXE}$ $\boxed{SHIFT}$ $\boxed{[F] \; MEM}$ $\boxed{F2\text{:}RCL}$ 1 $\boxed{EXE}$

When $x = 3$, the answer is -1. Then to evaluate this same expression at the value $x = -4$ do the following.

TI $\boxed{(-)}$ 4 $\boxed{\text{STO}\triangleright}$ $\boxed{\text{X}|\text{T}}$ $\boxed{\text{ENTER}}$ $\boxed{\text{2nd}}$ $\boxed{\text{Y-VARS}}$ { $\boxed{\text{1:FUNCTION}}$ (TI-82 only)} $\boxed{\text{1:Y1}}$ $\boxed{\text{ENTER}}$

Casio $\boxed{(-)}$ 4 $\boxed{\rightarrow}$ $\boxed{\text{X, }\theta\text{, T}}$ $\boxed{\text{EXE}}$ $\boxed{\text{SHIFT}}$ $\boxed{\text{F}\,\text{MEM}}$ $\boxed{\text{F2:RCL}}$ 1 $\boxed{\text{EXE}}$

When $x = -4$, the answer is 0.

2. The fundamental principle of rational expressions allows us to change rational expressions to a simpler form when there is a common factor in the numerator and denominator. The example $\dfrac{16 - x^2}{2x^2 - 7x - 4}$ (above) can be transformed into $-\dfrac{x + 4}{2x + 1}$.

It is important to understand that these two expressions represent exactly the same value for any value of x except where either denominator is not defined. To see if this is true for the expressions above, compute the value of each expression for the values of x shown. Use the result of part 1 above if you have access to a graphing calculator.

x	-10	-4	$-\dfrac{1}{2}$	0	3	4	10
$\dfrac{16 - x^2}{2x^2 - 7x - 4}$							
$-\dfrac{x + 4}{2x + 1}$							

Review Exercises

Directions   Solve the following equations. See sections 2–3 and 2–6.

1. $6y + 5 = y - 4$ (Find the solution set.)

2. $P = 2\ell + 2w$ for w

3. The product of two consecutive even integers is 168. Find the integers. See section 4–7.

4. Solve the inequality $3x - 2 \leq 2(x + 1)$. See section 2–9.

Directions   Completely factor each expression. See sections 4–2 and 4–4.

5. $16x^2 - y^2$

6. $x^2 - 16x - 17$

7. $5x^2 - 5x - 10$

8. Tammy had \$6,000. She invested part of her money at $7\dfrac{1}{2}\%$ interest and the rest at 9%. If her income from the two investments was \$511.50, how much did she invest at each rate? See section 2–7.

5–2   Multiplication and Division of Rational Expressions

Multiplication of Rational Expressions

Recall that to multiply two real number fractions we multiply the numerators and multiply the denominators.

Multiplication Property of Fractions

If a, b, c, and d are real numbers, then

$$\frac{a}{b} \cdot \frac{c}{d} = \frac{a \cdot c}{b \cdot d} \ (b, d \neq 0)$$

<u>Note</u>   Any possible reduction should be performed *before* the multiplications takes place.

● **Example 5–2 A**
Multiplying fractions

Multiply the fractions $\frac{3}{7}$ and $\frac{14}{27}$ and simplify the product.

$$\frac{3}{7} \cdot \frac{14}{27} = \frac{3 \cdot 14}{7 \cdot 27}$$

Multiply the numerators
Multiply the denominators

$$= \frac{3 \cdot 2 \cdot 7}{7 \cdot 3 \cdot 3 \cdot 3}$$

Factor the numerator and the denominator

$$= \frac{2 \cdot (3 \cdot 7)}{3 \cdot 3 \cdot (3 \cdot 7)}$$

Group the common factors (3 · 7)

$$= \frac{2}{3 \cdot 3}$$

Divide numerator and denominator by (3 · 7)

$$= \frac{2}{9}$$

Multiply the remaining factors   ●

This same procedure is followed when we multiply two rational expressions.

Multiplication Property of Rational Expressions

Given rational expressions $\frac{P}{Q}$ and $\frac{R}{S}$, then

$$\frac{P}{Q} \cdot \frac{R}{S} = \frac{P \cdot R}{Q \cdot S} \ (Q, S \neq 0)$$

Since we want the resulting product to be stated in lowest terms, we apply the *fundamental principle of rational expressions* and divide both the numerator and the denominator by their common factors. That is, we *reduce* by dividing out the common factors.

Multiplication of Rational Expressions

1. State the numerators and denominators as indicated products. (Do not multiply.)
2. Factor the numerator and the denominator.
3. Divide the numerators and the denominators by the factors that are common.
4. Multiply the remaining factors in the numerator and place this product over the product of the remaining factors in the denominator.

● **Example 5–2 B**

Multiplying rational expressions

Perform the indicated multiplication and simplify your answer. Assume that no denominator equals zero.

1. $\dfrac{4}{9x} \cdot \dfrac{3x^2}{2}$

$= \dfrac{4 \cdot 3x^2}{9x \cdot 2}$ Multiply numerators
Multiply denominators

$= \dfrac{2 \cdot 2 \cdot 3 \cdot x \cdot x}{3 \cdot 3 \cdot 2 \cdot x}$ Factor numerator and denominator

$= \dfrac{2 \cdot x \cdot (2 \cdot 3 \cdot x)}{3 \cdot (2 \cdot 3 \cdot x)}$ Identify common factors

$= \dfrac{2x}{3}$ Divide numerator and denominator by $(2 \cdot 3 \cdot x)$

2. $\dfrac{x+1}{3-x} \cdot \dfrac{(x-3)^2}{x-2}$

$= \dfrac{(x+1) \cdot (x-3)^2}{(3-x) \cdot (x-2)}$ Multiply numerators
Multiply denominators

$= \dfrac{x-3}{3-x} \cdot \dfrac{(x+1)(x-3)}{x-2}$ Identify opposites

$= \dfrac{-1 \cdot (x+1)(x-3)}{x-2}$ Divide opposites leaving -1 as a factor in the numerator

$= \dfrac{-1(x^2 - 2x - 3)}{x-2}$ Multiply as indicated

$= \dfrac{-x^2 + 2x + 3}{x-2}$

3. $\dfrac{x^2 - 8x + 16}{x^2 + 3x - 10} \cdot \dfrac{x^2 - 4}{x^2 - 5x + 4}$

$= \dfrac{(x^2 - 8x + 16)(x^2 - 4)}{(x^2 + 3x - 10)(x^2 - 5x + 4)}$ Multiply numerators and denominators

$= \dfrac{(x-4)(x-4)\,(x-2)(x+2)}{(x+5)(x-2)\,(x-4)(x-1)}$ Factor numerator and denominator

$= \dfrac{(x-4)(x+2)}{(x+5)(x-1)}$ Reduce by common factors $(x-4)$ and $(x-2)$

$= \dfrac{x^2 - 2x - 8}{x^2 + 4x - 5}$ Multiply remaining factors

☞ ***Quick check***   Multiply $\dfrac{12}{5y} \cdot \dfrac{15y^2}{4}$

Division of Rational Expressions

Recall that to divide two fractions $\frac{a}{b}$ and $\frac{c}{d}$, we multiply $\frac{a}{b}$ by the *reciprocal* of $\frac{c}{d}$, which is $\frac{d}{c}$.

Division Property of Fractions

If a, b, c, and d are real numbers, then

$$\frac{a}{b} \div \frac{c}{d} = \frac{a}{b} \cdot \frac{d}{c} = \frac{a \cdot d}{b \cdot c} \quad (b, c, d \neq 0)$$

● **Example 5–2 C**

Dividing fractions

Find the indicated quotient. Reduce your answer to lowest terms.

$$\frac{18}{25} \div \frac{9}{5}$$

$$= \frac{18}{25} \cdot \frac{5}{9} \qquad \text{Multiply by the reciprocal of } \frac{9}{5}$$

$$= \frac{2 \cdot 3 \cdot 3 \cdot 5}{5 \cdot 5 \cdot 3 \cdot 3} \qquad \text{Factor in numerator and denominator}$$

$$= \frac{2 \cdot (3 \cdot 3 \cdot 5)}{5 \cdot (3 \cdot 3 \cdot 5)} \qquad \text{Group the common factors } (3 \cdot 3 \cdot 5)$$

$$= \frac{2}{5} \qquad \text{Reduce by the common factors } (3 \cdot 3 \cdot 5)$$

Division of rational expressions is done in the same way.

Division Property of Rational Expressions

If $\frac{P}{Q}$ and $\frac{R}{S}$ are rational expressions, then

$$\frac{P}{Q} \div \frac{R}{S} = \frac{P}{Q} \cdot \frac{S}{R} = \frac{P \cdot S}{Q \cdot R} \quad (Q, R, S \neq 0)$$

Notice that once the operation of division has been changed to multiplication, we proceed exactly as we did with the multiplication of rational expressions.

Division of Rational Expressions

1. Multiply the first rational expression by the reciprocal of the second.
2. Proceed as in the multiplication of rational expressions.

● **Example 5–2 D**

Dividing rational expressions

Find the indicated quotients. Express the answer in reduced form.

1. $\dfrac{3ab}{5} \div \dfrac{9abc}{10}$

$= \dfrac{3ab}{5} \cdot \dfrac{10}{9abc}$ Multiply by the reciprocal of $\dfrac{9abc}{10}$

$= \dfrac{3ab \cdot 2 \cdot 5}{5 \cdot 3 \cdot 3 \cdot abc}$ Factor numerator and denominator

$= \dfrac{2 \cdot (3 \cdot 5 \cdot ab)}{3 \cdot c \cdot (3 \cdot 5 \cdot ab)}$ Identify common factors ($3 \cdot 5 \cdot ab$)

$= \dfrac{2}{3c}$ Divide numerator and denominator by common factors ($3 \cdot 5 \cdot ab$)

2. $\dfrac{x^2 - 4}{5} \div \dfrac{x - 2}{15}$

$= \dfrac{x^2 - 4}{5} \cdot \dfrac{15}{x - 2}$ Multiply by the reciprocal of $\dfrac{x-2}{15}$

$= \dfrac{(x - 2)(x + 2) \cdot 3 \cdot 5}{5 \cdot (x - 2)}$ Factor numerator and denominator

$= \dfrac{3 \cdot (x + 2) \cdot 5(x - 2)}{5(x - 2)}$ Identify common factors $5(x - 2)$

$= \dfrac{3(x + 2)}{1} = 3x + 6$ Reduce to lowest terms by dividing numerator and denominator by $5(x - 2)$

3. $\dfrac{4x + 2}{x - 1} \div \dfrac{2x - 1}{1 - x}$

$= \dfrac{4x + 2}{x - 1} \cdot \dfrac{1 - x}{2x - 1}$ Multiply by the reciprocal of $\dfrac{2x-1}{1-x}$

$= \dfrac{2(2x + 1)(1 - x)}{(x - 1)(2x - 1)}$ Factor the numerator

$= \dfrac{-1 \cdot 2(2x + 1)}{2x - 1}$ Divide opposites leaving -1 as a factor in the numerator

Note $2x + 1$ and $2x - 1$ are not opposites.

$= \dfrac{-4x - 2}{2x - 1}$ Multiply as indicated

4. $\dfrac{x^2 - 9}{2x + 1} \div \dfrac{x - 3}{2x^2 + 7x + 3}$

$= \dfrac{x^2 - 9}{2x + 1} \cdot \dfrac{2x^2 + 7x + 3}{x - 3}$ Multiply by the reciprocal of $\dfrac{x - 3}{2x^2 + 7x + 3}$

$= \dfrac{(x - 3)(x + 3) \cdot (2x + 1)(x + 1)}{(2x + 1)(x - 3)}$ Factor numerators

$= (x + 3)(x + 3)$ Reduce by $(2x + 1)(x - 3)$
$= x^2 + 6x + 9$ Multiply as indicated

⚑ *Quick check*   Divide $\dfrac{x^2 - 4}{4x - 1} \div \dfrac{x - 2}{4x^2 + 3x - 1}$

MASTERY POINTS

Can you
- Multiply rational expressions?
- Divide rational expressions?

Exercise 5–2

Directions
Find the indicated product or quotient. Write your answer in simplest form. Assume all denominators are nonzero. See examples 5–2 A–D.

📌 *Quick Check*

Example $\dfrac{12}{5y} \cdot \dfrac{15y^2}{4}$

Solution $= \dfrac{12 \cdot 15y^2}{5y \cdot 4}$ Multiply numerator and denominator

$= \dfrac{2 \cdot 2 \cdot 3 \cdot 3 \cdot 5 \cdot y \cdot y}{5 \cdot y \cdot 2 \cdot 2}$ Factor numerator and denominator

$= \dfrac{3 \cdot 3 \cdot y(5 \cdot y \cdot 2 \cdot 2)}{1(5 \cdot y \cdot 2 \cdot 2)}$ Identify common factors $(5 \cdot y \cdot 2 \cdot 2)$

$= \dfrac{9y}{1}$ Divide by common factors $(5 \cdot y \cdot 2 \cdot 2)$

$= 9y$

Example $\dfrac{x^2 - 4}{4x - 1} \div \dfrac{x - 2}{4x^2 + 3x - 1}$

Solution $= \dfrac{x^2 - 4}{4x - 1} \cdot \dfrac{4x^2 + 3x - 1}{x - 2}$ Multiply the reciprocal of $\dfrac{x - 2}{4x^2 + 3x - 1}$

$= \dfrac{(x + 2)(x - 2) \cdot (4x - 1)(x + 1)}{(4x - 1)(x - 2)}$ Factor numerator and denominator

$= (x + 2)(x + 1)$ Reduce by $(4x - 1)(x - 2)$

$= x^2 + 3x + 2$ Multiply as indicated

Directions   Find the indicated product or quotient. Write your answer in simplest form. Assume all denominators are nonzero. See examples 5–2 A–D.

1. $\dfrac{24}{35} \cdot \dfrac{7}{8}$

2. $\dfrac{3}{8} \cdot \dfrac{5}{9}$

3. $\dfrac{7}{10} \div \dfrac{21}{25}$

4. $\dfrac{56}{39} \div \dfrac{8}{13}$

5. $\dfrac{4a}{5} \cdot \dfrac{5}{2}$

6. $\dfrac{16b}{7a} \cdot \dfrac{5}{4}$

7. $\dfrac{14}{3a} \div \dfrac{7}{15a}$

8. $\dfrac{6x}{5y} \div \dfrac{21x}{15y}$

9. $\dfrac{5}{6} \cdot \dfrac{3x}{10y}$

10. $\dfrac{7a}{12b} \cdot \dfrac{9b}{28}$

11. $\dfrac{9x^2}{8} \cdot \dfrac{4}{6x}$

12. $\dfrac{36p^2}{7q} \cdot \dfrac{14q^2}{28p^3}$

13. $\dfrac{24a}{35x} \div 6a$

14. $\dfrac{14y}{23x} \div 7y$

15. $6a \div \dfrac{24a}{35x}$

16. $7y \div \dfrac{14y}{23}$

17. $\dfrac{21ab}{16c} \cdot \dfrac{8c^2}{3ab^2}$

18. $\dfrac{18x^2y^2}{5ab} \cdot \dfrac{25a^2b}{12xy}$

19. $\dfrac{5x^2}{9y^3} \div \dfrac{20x}{6y}$

20. $\dfrac{28m}{15n} \div \dfrac{7m^2}{3n^3}$

21. $\dfrac{24abc}{7xyz^2} \cdot \dfrac{14x^2yz}{9a^2}$

22. $\dfrac{80x^2yz^3}{11mn^2} \cdot \dfrac{33mn^2}{25xyz}$

23. $\dfrac{3ab}{8x^2} \div \dfrac{15b^3}{16x}$

24. $\dfrac{20mn^3}{9x^2} \div \dfrac{4mn}{3xy^2}$

25. $\dfrac{x+y}{3} \cdot \dfrac{12}{(x+y)^2}$

26. $\dfrac{5(a-b)}{8} \cdot \dfrac{12}{10(a-b)}$

27. $\dfrac{9-p}{7} \div \dfrac{4(p-9)}{21}$

28. $\dfrac{4x-2}{15} \div \dfrac{1-2x}{27}$

29. $\dfrac{3b-6}{4b+8} \cdot \dfrac{5b+10}{2-b}$

30. $\dfrac{8y+16}{3-y} \cdot \dfrac{4y-12}{3y+6}$

31. $\dfrac{4a+12}{a-5} \div (a+3)$

32. $\dfrac{9-3z}{2z+8} \div (6-2z)$

33. $(x^2-4x+4) \cdot \dfrac{18}{x^2-4}$

34. $\dfrac{21}{a^2-9} \cdot (a^2+a-12)$

35. $\dfrac{x^2-4}{25y} \cdot \dfrac{24y^2}{x+2}$

36. $\dfrac{16a^2}{b^2-9} \cdot \dfrac{b-3}{12a^2}$

37. $\dfrac{r^2-16}{r+1} \div \dfrac{r+4}{r^2-1}$

38. $\dfrac{p^2+2p+1}{4p-1} \div \dfrac{p^2-1}{16p^2-1}$

39. $\dfrac{9-x^2}{x+y} \cdot \dfrac{4x+4y}{x-3}$

40. $\dfrac{b^2-a^2}{2a+4b} \cdot \dfrac{a+b}{a-b}$

41. $\dfrac{a^2-5a+6}{a^2-9a+20} \cdot \dfrac{a^2-5a+4}{a^2-3a+2}$

42. $\dfrac{a^2-5a-14}{a^2-9a-36} \cdot \dfrac{a^2+10a+21}{a^2+4a-77}$

43. $\dfrac{x^2-2x-3}{x^2+3x-4} \div \dfrac{x^2-x-6}{x^2+x-12}$

44. $\dfrac{y^2+3y+2}{y^2+5y+4} \div \dfrac{y^2+5y+6}{y^2+10y+24}$

45. $\dfrac{2x^2-15x+7}{x^2-9x+8} \cdot \dfrac{x^2-2x+1}{x^2-49}$

46. $\dfrac{4x^2-4}{3x^2-13x-10} \cdot \dfrac{x^2-6x+5}{4x+4}$

47. $\dfrac{6r^2-r-7}{12r^2+16r-35} \div \dfrac{r^2-r-2}{2r^2+r-10}$

48. $\dfrac{4x^2-9}{x^2-9x+18} \div \dfrac{2x^2-5x-12}{x^2-10x+24}$

49. $(3x^2-2x-8) \div \dfrac{x^2-4}{x+2}$

50. $(8a^2-16a) \div \dfrac{a^3-16a}{a-4}$

51. $\dfrac{3x-4}{2x+1} \div (6x^2-5x-4)$

52. $\dfrac{m^2-3m-10}{m^2-4} \div (2m^2-9m-5)$

53. $\dfrac{10}{a^3-27} \cdot \dfrac{a^2+3a-18}{15}$

54. $\dfrac{x^3-8}{16} \cdot \dfrac{24}{x^2+2x-8}$

55. $\dfrac{z^2-5z-14}{z-4} \div \dfrac{5z^3+40}{z^2-z-12}$

56. $\dfrac{3b^3+3}{b-2} \div \dfrac{b^2+2b+1}{b^2+6b-16}$

57. $\dfrac{y^2+8y+16}{y+4} \cdot \dfrac{y^2-25}{y^2+9y+20} \cdot \dfrac{y^2+5y}{y^2-5y}$

58. $\dfrac{6m^2-7m+2}{6m^2+5m+1} \cdot \dfrac{2m^2+m}{4m^2-1} \cdot \dfrac{12m^2-5m-3}{12m^2-17m+6}$

59. The figure shows an aircraft that has launched an object toward the ground. The rectangle describes the area a fixed camera would have to cover to view the descent of the object. The ratio of the height to width of the picture is called its aspect ratio. If the picture's upper left corner is 120 feet to the right and 48 feet below the aircraft, and if the object is accelerating horizontally at 11 feet per second per second to the right, then the ratio of length to width (aspect ratio) is $\dfrac{x^2 + 22x + 120}{x^2 + 16x + 48}$. Under the conditions where the object accelerates horizontally at 10 feet per second per second, and where the picture's upper left corner is 96 feet to the right of, and 64 feet below the aircraft at launch, the

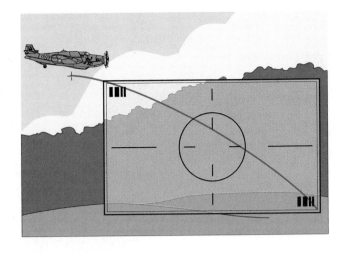

aspect ratio is $\dfrac{x^2 + 20x + 96}{x^2 + 16x + 64}$. Find and simplify an expression that describes the ratio of these two aspect ratios by dividing the first expression by the second expression and simplifying the result.

60. Explain in your own words why Q, R, and S cannot be zero in the division property of rational expressions.
$$\frac{P}{Q} \div \frac{R}{S} = \frac{P}{Q} \cdot \frac{S}{R} = \frac{P \cdot S}{Q \cdot R}$$

❂ Explore to Learn

Make up a problem of the form $\dfrac{a_1 x^2 + b_1 x + c_1}{a_2 x^2 + b_2 x + c_2} \div \dfrac{a_3 x^2 + b_3 x + c_3}{a_4 x^2 + b_4 x + c_4}$ in which the answer is $\dfrac{x + 6}{x - 3}$ and none of the constants a_i, b_i, and c_i are zero.

Review Exercises

Directions Add or subtract the following. See section 1–2.

1. $\dfrac{3}{4} + \dfrac{5}{6}$

2. $\dfrac{7}{8} - \dfrac{5}{12}$

Directions Completely factor the following. See sections 4–2 and 4–3.

3. $2x^2 - 50$

4. $x^2 + 9x - 22$

5. $x^2 + 8x + 16$

Directions Solve the following proportions. See section 2–8.

6. $\dfrac{3}{x} = \dfrac{5}{8}$

7. $\dfrac{5}{9} = \dfrac{y}{27}$

8. Write the number 0.0000789 in scientific notation. See section 3–5.

9. One third of a number is 12 less than one half of the number. Find the number. See section 2–7.

10. One number is seven times another. If their difference is 28, what are the numbers? See section 2–7.

5–3 Addition and Subtraction of Rational Expressions—I

Recall that to add or subtract fractions having the same denominator, we add, or subtract, the numerators and place this sum, or difference, over the same denominator.

Addition and Subtraction Properties for Fractions

If a, b, and c are real numbers, $b \neq 0$, then

$$\frac{a}{b} + \frac{c}{b} = \frac{a+c}{b} \quad \text{and} \quad \frac{a}{b} - \frac{c}{b} = \frac{a-c}{b}$$

● **Example 5–3 A**

Adding and subtracting fractions with a common denominator

Add or subtract as indicated.

1. $\dfrac{3}{11} + \dfrac{4}{11} = \dfrac{3+4}{11}$ Add numerators

$\qquad = \dfrac{7}{11}$ $3 + 4 = 7$

2. $\dfrac{3}{8} - \dfrac{1}{8} = \dfrac{3-1}{8}$ Subtract numerators

$\qquad = \dfrac{2}{8}$ $3 - 1 = 2$

$\qquad = \dfrac{1}{4}$ Reduce

We use the following similar procedure to add or subtract rational expressions.

Addition and Subtraction Properties for Rational Expressions

If $\dfrac{P}{R}$ and $\dfrac{Q}{R}$ are rational expressions, $R \neq 0$, then

$$\frac{P}{R} + \frac{Q}{R} = \frac{P+Q}{R} \quad \text{and} \quad \frac{P}{R} - \frac{Q}{R} = \frac{P-Q}{R}$$

Addition and Subtraction of Rational Expressions with a Common Denominator

1. Add or subtract the numerators.
2. Place the sum or difference over the common denominator.
3. Reduce the resulting rational expression to lowest terms.

● **Example 5–3 B**

Adding and subtracting rational expressions with a common denominator

Find the indicated sum or difference. Assume all denominators are nonzero.

1. $\dfrac{15y}{3y+5} - \dfrac{9y}{3y+5} = \dfrac{15y-9y}{3y+5}$ Subtract numerators and place over $3y+5$

$\phantom{\dfrac{15y}{3y+5} - \dfrac{9y}{3y+5}} = \dfrac{6y}{3y+5}$ $15y - 9y = 6y$

2. $\dfrac{2x-1}{x^2+5x+6} - \dfrac{4-x}{x^2+5x+6}$

$= \dfrac{(2x-1)-(4-x)}{x^2+5x+6}$ Place numerators in parentheses and subtract

$= \dfrac{2x-1-4+x}{x^2+5x+6}$ Remove parentheses and subtract

$= \dfrac{3x-5}{x^2+5x+6}$ Combine like terms

Note Notice that when we subtracted $4 - x$ from $2x - 1$, we placed parentheses around each numerator. This step is *important* to avoid the common mistake of failing to change signs in the second numerator when subtraction is involved.

3. $\dfrac{2x-1}{x^2+5x+6} + \dfrac{4-x}{x^2+5x+6}$

$= \dfrac{(2x-1)+(4-x)}{x^2+5x+6}$ Place numerators in parentheses and add

$= \dfrac{x+3}{x^2+5x+6}$ Remove parentheses and combine like terms

$= \dfrac{x+3}{(x+3)(x+2)}$ Factor denominator

$= \dfrac{1}{x+2}$ Reduce by common factor $x + 3$

Note In the last step, always look for possible common factors to be sure that the answer is reduced to lowest terms.

☞ *Quick check* $\dfrac{4m-5}{m^2-9} - \dfrac{3m-8}{m^2-9}$ ●

When one denominator is the opposite of the other, as in the indicated sum

$$\frac{2x}{3} + \frac{5}{-3},$$

where 3 and -3 are opposites, we first multiply one of the expressions by $\dfrac{-1}{-1}$ to obtain equivalent expressions with the same denominator.

● **Example 5–3 C**

Adding and subtracting rational expressions with denominators that are opposites

Find the indicated sum or difference. Assume all denominators are not zero.

1. $\dfrac{2x}{3} + \dfrac{5}{-3} = \dfrac{2x}{3} + \dfrac{-1}{-1} \cdot \dfrac{5}{-3}$ Multiply $\dfrac{5}{-3}$ by $\dfrac{-1}{-1}$

$= \dfrac{2x}{3} + \dfrac{-1(5)}{-1(-3)}$ Multiply numerators and denominators

$= \dfrac{2x}{3} + \dfrac{-5}{3}$ Common denominator of 3

$= \dfrac{2x + (-5)}{3}$ Add numerators and place over 3

$= \dfrac{2x - 5}{3}$ Definition of subtraction

2. $\dfrac{5y - 1}{y - 4} + \dfrac{2y + 3}{4 - y}$

$= \dfrac{5y - 1}{y - 4} + \dfrac{-1}{-1} \cdot \dfrac{2y + 3}{4 - y}$ Multiply $\dfrac{2y + 3}{4 - y}$ by $\dfrac{-1}{-1}$

$= \dfrac{5y - 1}{y - 4} + \dfrac{-1(2y + 3)}{-1(4 - y)}$

$= \dfrac{5y - 1}{y - 4} + \dfrac{-2y - 3}{y - 4}$ Same denominator: $-1(4 - y) = y - 4$

$= \dfrac{(5y - 1) + (-2y - 3)}{y - 4}$ Add numerators in parentheses

$= \dfrac{5y - 1 - 2y - 3}{y - 4}$ Remove parentheses

$= \dfrac{3y - 4}{y - 4}$ Combine like terms

<u>*Note*</u> We could have multiplied $\dfrac{5y - 1}{y - 4}$ by $\dfrac{-1}{-1}$. The resulting denominator would then have been $4 - y$ and the numerator would have been $4 - 3y$, and $\dfrac{1 - 5y}{4 - y} + \dfrac{2y + 3}{4 - y}$ would add to $\dfrac{4 - 3y}{4 - y}$. We can show that this answer is equivalent to $\dfrac{3y - 4}{y - 4}$ by multiplying either answer by $\dfrac{-1}{-1}$, which will produce the other answer.

3. $\dfrac{2x + 1}{x - 5} - \dfrac{x - 4}{5 - x}$

$= \dfrac{2x + 1}{x - 5} - \dfrac{-1}{-1} \cdot \dfrac{x - 4}{5 - x}$ Multiply $\dfrac{x - 4}{5 - x}$ by $\dfrac{-1}{-1}$

$= \dfrac{2x + 1}{x - 5} - \dfrac{-1(x - 4)}{-1(5 - x)}$

$= \dfrac{2x + 1}{x - 5} - \dfrac{4 - x}{x - 5}$ Same denominator: $-1(x - 4) = 4 - x$ $-1(5 - x) = x - 5$

$= \dfrac{(2x + 1) - (4 - x)}{x - 5}$ Place "( )" around numerators

$$= \frac{2x + 1 - 4 + x}{x - 5} \qquad \text{Definition of subtraction}$$

$$= \frac{3x - 3}{x - 5} \qquad \text{Combine like terms}$$

☞ *Quick check* $\dfrac{x + 7}{x - 1} + \dfrac{3x + 1}{1 - x}$ ●

The Least Common Denominator (LCD)

If the fractions to be added or subtracted do not have the same denominator, we must change at least one of the fractions to an equivalent fraction so the fractions do have a common denominator. There are many such numbers we could use as a common denominator. However, the most convenient denominator to use is the smallest (least) number that is exactly divisible by each of the denominators—called the **least common denominator**, denoted by LCD. For example, the least common denominator (LCD) of the two fractions

$$\frac{5}{6} \quad \text{and} \quad \frac{2}{9}$$

is 18, since 18 is the smallest (least) number that is exactly divisible by both 6 and 9.

Finding the Least Common Denominator

1. Completely factor each denominator. Write each factorization using exponential notation.
2. Choose the greatest power of each different factor. The product of these factors is the LCD.

Note The LCD of two or more rational expressions is also called the least common multiple (LCM) of their denominators.

● **Example 5–3 D**
Finding the LCD

Find the LCD of rational expressions having the given denominators.

1. $16a$ and $8a^3$

$$16a = 2 \cdot 2 \cdot 2 \cdot 2 \cdot a = 2^4 \cdot a$$
$$8a^3 = 2 \cdot 2 \cdot 2 \cdot a \cdot a \cdot a = 2^3 \cdot a^3$$

Since the different factors are 2 and a, the greatest power of 2 is 2^4, and the greatest power of a is a^3, the LCD is $2^4 \cdot a^3 = 16a^3$.

2. $50x^3y^2$ and $20x^2y^3$

$$50x^3y^2 = 2 \cdot 5 \cdot 5 \cdot x^3 \cdot y^2 = 2 \cdot 5^2 \cdot x^3 \cdot y^2$$
$$20x^2y^3 = 2 \cdot 2 \cdot 5 \cdot x^2 \cdot y^3 = 2^2 \cdot 5 \cdot x^2 \cdot y^3$$

The different factors are 2, 5, x, and y. Since the greatest power of 2 is 2^2, of 5 is 5^2, of x is x^3, and of y is y^3, the LCD is $2^2 \cdot 5^2 \cdot x^3 \cdot y^3 = 100x^3y^3$.

3. $x^2 + x - 12$ and $x^2 + 2x - 8$

$$x^2 + x - 12 = (x + 4)(x - 3)$$
$$x^2 + 2x - 8 = (x + 4)(x - 2)$$

The different factors are $x + 4$, $x - 3$, and $x - 2$.
Each factor is to the first power, so the LCD is $(x + 4)(x - 3)(x - 2)$.

☞ **Quick check** Find the LCD of the rational expressions having the following denominators. $3x - 6$, $x^2 - 4x + 4$, $x^2 - 2x$. ●

MASTERY POINTS

Can you
- Add and subtract rational expressions having a common denominator?
- Add and subtract rational expressions having denominators that are opposites?
- Find the least common denominator (LCD) of a set of rational expressions?

Exercise 5–3

Directions
Combine the given rational expressions and reduce the answer to lowest terms. Assume all denominators are nonzero. See example 5–3 B.

☞ **Quick Check**

Example $\dfrac{4m - 5}{m^2 - 9} - \dfrac{3m - 8}{m^2 - 9}$

Solution $= \dfrac{(4m - 5) - (3m - 8)}{m^2 - 9}$ Place numerators in parentheses

$= \dfrac{4m - 5 - 3m + 8}{m^2 - 9}$ Remove parentheses and change signs where necessary

$= \dfrac{m + 3}{(m + 3)(m - 3)}$ Combine like terms in the numerator and factor the denominator

$= \dfrac{1}{m - 3}$ Reduce to lowest terms

Directions   Combine the given rational expressions and reduce the answer to lowest terms. Assume all denominators are nonzero. See example 5–3 B.

1. $\dfrac{5}{x} + \dfrac{3}{x}$

2. $\dfrac{8}{y^2} + \dfrac{10}{y^2}$

3. $\dfrac{9}{p} - \dfrac{2}{p}$

4. $\dfrac{18}{m^2} - \dfrac{5}{m^2}$

5. $\dfrac{5x}{x + 2} + \dfrac{9x}{x + 2}$

6. $\dfrac{8y}{y - 1} + \dfrac{-3y}{y - 1}$

7. $\dfrac{x - 1}{2x} - \dfrac{x + 3}{2x}$

8. $\dfrac{3y - 2}{y^2} - \dfrac{4y - 1}{y^2}$

9. $\dfrac{3x + 5}{x^2 - 1} - \dfrac{2x + 3}{x^2 - 1}$

10. $\dfrac{b^2 + 2}{b + 3} - \dfrac{b^2 + 2b - 3}{b + 3}$

11. $\dfrac{3a - 8}{a^2 - 16} + \dfrac{4 - 2a}{a^2 - 16}$

12. $\dfrac{x - 3}{x^2 - 1} + \dfrac{4}{x^2 - 1}$

13. $\dfrac{2y - 5}{y^2 - 9} - \dfrac{y - 2}{y^2 - 9}$

14. $\dfrac{5x - 3}{x^2 - 4} - \dfrac{4x - 5}{x^2 - 4}$

15. $\dfrac{a + 1}{a^2 + 6a + 9} + \dfrac{2}{a^2 + 6a + 9}$

16. $\dfrac{x - 3}{x^2 + 8x + 16} + \dfrac{7}{x^2 + 8x + 16}$

17. $\dfrac{4x - 3}{x^2 - 10x + 25} - \dfrac{3x + 2}{x^2 - 10x + 25}$

18. $\dfrac{2y - 1}{y^2 + 2y + 1} - \dfrac{y - 2}{y^2 + 2y + 1}$

19. $\dfrac{3a - 1}{a^2 + 5a + 6} - \dfrac{2a - 3}{a^2 + 5a + 6}$

20. $\dfrac{4a - 1}{a^2 + 7a + 6} - \dfrac{3a - 2}{a^2 + 7a + 6}$

21. $\dfrac{2x - 5}{x^2 - 2x - 8} + \dfrac{1 - x}{x^2 - 2x - 8}$

22. $\dfrac{x - 9}{x^2 - 4x - 12} + \dfrac{3}{x^2 - 4x - 12}$

Directions	☞ **Quick Check**
See example 5–3 C.	**Example** $\dfrac{x + 7}{x - 1} + \dfrac{3x + 1}{1 - x}$

Solution

$= \dfrac{x + 7}{x - 1} + \dfrac{-1}{-1} \cdot \dfrac{3x + 1}{1 - x}$ Multiply $\dfrac{3x + 1}{1 - x}$ by $\dfrac{-1}{-1}$

$= \dfrac{x + 7}{x - 1} + \dfrac{-1(3x + 1)}{-1(1 - x)}$

$= \dfrac{x + 7}{x - 1} + \dfrac{-3x - 1}{x - 1}$ $-1(1 - x) = x - 1$

$= \dfrac{x + 7 - 3x - 1}{x - 1}$ Add numerators

$= \dfrac{-2x + 6}{x - 1}$ Combine like terms

Directions See example 5–3 C.

23. $\dfrac{5}{7} + \dfrac{6}{-7}$

24. $\dfrac{9}{10} - \dfrac{3}{-10}$

25. $\dfrac{4}{z} - \dfrac{5}{-z}$

26. $\dfrac{6}{y} + \dfrac{9}{-y}$

27. $\dfrac{5}{x - 2} + \dfrac{12}{2 - x}$

28. $\dfrac{1}{x - 7} - \dfrac{5}{7 - x}$

29. $\dfrac{5y}{y - 6} - \dfrac{4y}{6 - y}$

30. $\dfrac{4z}{z - 3} + \dfrac{z}{3 - z}$

31. $\dfrac{x + 1}{x - 5} + \dfrac{2x - 3}{5 - x}$

32. $\dfrac{4y + 3}{y - 9} - \dfrac{2y - 7}{9 - y}$

33. $\dfrac{2y - 5}{2y - 3} - \dfrac{y + 7}{3 - 2y}$

34. $\dfrac{z + 5}{4z - 3} + \dfrac{4z - 1}{3 - 4z}$

35. $\dfrac{2x + 5}{5 - 2x} + \dfrac{x + 9}{2x - 5}$

36. $\dfrac{5 - y}{6 - 5y} - \dfrac{9y + 1}{5y - 6}$

Directions	☞ **Quick Check**
Find the least common denominator (LCD) of rational expressions having the following denominators. See example 5–3 D.	**Example** $3x - 6$, $x^2 - 4x + 4$, and $x^2 - 2x$

Solution

$3x - 6 = 3(x - 2)$

$x^2 - 4x + 4 = (x - 2)^2$ Factor each denominator

$x^2 - 2x = x(x - 2)$

LCD is $3x(x - 2)^2$

Directions   Find the least common denominator (LCD) of rational expressions having the following denominators. See example 5–3 D.

37. $6x$ and $9x$

38. $8a$ and $12a$

39. $16x^2$ and $24x$

40. $6b^2$ and $14b$

41. $28y^2$ and $35y^3$

42. $9z^3$ and $7z^4$

43. $32a^2$ and $64a^4$

44. $4x^2$, $3x$, and $8x^3$

45. $10a^2$, $12a^3$, and $9a$

46. $4x - 2$ and $2x - 1$

47. $x - 4$ and $3x - 12$

48. $6x - 12$ and $9x - 18$

49. $18y^3$ and $9y - 36$

50. $32z^2$ and $16z - 32$

51. $a^2 + a$ and $a^2 - 1$

52. $(z - 1)^2$ and $z^2 - 1$

53. $8a + 16$ and $a^2 + 3a + 2$

54. $9p - 18$ and $p^2 - 7p + 10$

55. $a^2 - 5a + 6$ and $a^2 - 4$

56. $y^2 - y - 12$ and $y^2 + 6y + 9$

57. $x^2 - 1$ and $x^2 + 2x + 1$

58. $x^2 - 8x + 16$ and $x^2 - 2x - 8$

59. $a^2 + 10a + 25$ and $a^2 - 25$

60. $y^2 - 2y - 3$ and $y^2 - 6y + 9$

61. $a^2 - 9$ and $a^2 - 5a + 6$

62. $p^2 - 9$, $p^2 + p - 6$, and $p^2 - 4p + 4$

63. $a^2 - 4a + 4$, $a^2 + 4a + 4$, and $a^2 - 4$

64. $x^2 - 2x + 1$, $x^2 - 1$, and $x^2 + 2x + 1$

65. $a^2 + a - 6$, $a^2 + 7a + 12$, and $a^2 + 2a - 8$

66. $x^2 - x - 6$, $x^2 + x - 2$, and $x^2 - 4x + 3$

67. In your own words describe the least common denominator.

68. In your own words describe the relationship between the least common denominator of a set of rational expressions and the least common multiple of a set of denominators.

✺ Explore to Learn

This section stated that $4 - y = -1(y - 4)$. What does this mean? Substitute some values of y into both sides of this equation and evaluate to suggest this meaning.

Review Exercises

1. The statement $4(y + 3) = 4(3 + y)$ demonstrates what property of real numbers? See section 1–8.

Directions   Completely factor the following expressions. See sections 4–3 and 4–4.

2. $5y^2 - 20$

3. $x^2 + 20x + 100$

4. $3y^2 - y - 4$

Directions   Find the solution set of the following equations. See sections 2–5 and 4–7.

5. $4(x + 3) = 5(4 - 3x)$

6. $x^2 - 2x = 15$

7. Jeremy has $6,000 invested at 6%; how much must he invest at 10% to realize a net return of 9%? See section 2–7.

5–4   Addition and Subtraction of Rational Expressions—II

Now that we can find the least common denominator (LCD) of a set of rational expressions, let us review the process for changing a fraction (or rational expression) to an equivalent fraction with a new denominator.

● **Example 5–4 A**

Forming equivalent fractions

1. Change $\dfrac{7}{15}$ to an equivalent fraction having denominator 60.

We want $\dfrac{7}{15} = \dfrac{?}{60}$.

Since $60 = 15 \cdot 4$ (the factor 4 is found by $60 \div 15 = 4$), we multiply the given fraction by $\dfrac{4}{4}\left(\dfrac{4}{4} = 1\right)$.

$$
\begin{aligned}
\frac{7}{15} &= \frac{7}{15} \cdot \frac{4}{4} \\
&= \frac{7 \cdot 4}{15 \cdot 4} && \text{Multiply numerators} \\
&&& \text{Multiply denominators} \\
&= \frac{28}{60}
\end{aligned}
$$

Thus $\dfrac{7}{15} = \dfrac{28}{60}$

2. Change $\dfrac{x+1}{x-4}$ to an equivalent rational expression having denominator $x^2 - 2x - 8$.

We want $\dfrac{x+1}{x-4} = \dfrac{?}{x^2 - 2x - 8}$.

Since $x^2 - 2x - 8 = (x-4)(x+2)$, we multiply the given rational expression by $\dfrac{x+2}{x+2}$.

$$
\begin{aligned}
\frac{x+1}{x-4} &= \frac{x+1}{x-4} \cdot \frac{x+2}{x+2} \\
&= \frac{(x+1)(x+2)}{(x-4)(x+2)} && \text{Multiply numerators} \\
&&& \text{Multiply denominators} \\
&= \frac{x^2 + 3x + 2}{x^2 - 2x - 8} && \text{Perform indicated operations}
\end{aligned}
$$

Thus $\dfrac{x+1}{x-4} = \dfrac{x^2 + 3x + 2}{x^2 - 2x - 8}$

Once equivalent rational expressions are obtained with the LCD as the denominator, we add or subtract as previously learned. Use the following steps to add or subtract rational expressions having different denominators.

> ## Addition and Subtraction of Rational Expressions Having Different Denominators
>
> 1. Find the LCD of the rational expressions.
> 2. Write each rational expression as an equivalent rational expression with the LCD as the denominator.
> 3. Perform the indicated addition or subtraction as before.
> 4. Reduce the results to lowest terms.

● **Example 5–4 B**

Adding rational expressions

Add the following rational expressions. Assume the denominators are not equal to zero. Reduce all answers to lowest terms.

1. $\dfrac{8x}{15} + \dfrac{x}{6}$

   Step 1 Find the LCD.

   $$15 = 3 \cdot 5$$
   $$6 \ = 2 \cdot 3$$

   The LCD is $2 \cdot 3 \cdot 5 = 30$.

   Step 2 Rewrite each fraction as an equivalent fraction with the LCD.

   $$= \frac{8x(2)}{15(2)} + \frac{x(5)}{6(5)} = \frac{16x}{30} + \frac{5x}{30}$$

   Step 3 Add the numerators and place the sum over the LCD.

   $$= \frac{16x + 5x}{30} = \frac{21x}{30}$$

   Step 4 Reduce to lowest terms.

   $$= \frac{3 \cdot 7x}{3 \cdot 10} = \frac{7x}{10}$$

2. $\dfrac{3a + 2}{a^2 - 4} + \dfrac{-2}{a - 2}$

   Step 1 Find the LCD.

   $$a^2 - 4 = (a + 2)(a - 2)$$

   $a - 2$ will not factor

   The LCD is $(a + 2)(a - 2)$

   Step 2 Rewrite with the LCD.

   $$= \frac{3a + 2}{(a + 2)(a - 2)} + \frac{-2(a + 2)}{(a - 2)(a + 2)}$$
   $$= \frac{3a + 2}{(a + 2)(a - 2)} + \frac{-2a - 4}{(a + 2)(a - 2)}$$

   Step 3 Add numerators.

   $$= \frac{(3a + 2) + (-2a - 4)}{(a + 2)(a - 2)}$$

<u>*Note*</u> When the numerators have two or more terms as in example 2, we place the numerators in parentheses when we add the numerators. This is a good practice to avoid a *most common* mistake when subtracting.

$$= \frac{3a + 2 - 2a - 4}{(a + 2)(a - 2)} = \frac{a - 2}{(a + 2)(a - 2)}$$

Step 4 Reduce if possible.

$$= \frac{1}{a + 2}$$

3. $\dfrac{3y + 2}{y^2 - 16} + \dfrac{y - 4}{3y + 12}$

$\left. \begin{array}{l} y^2 - 16 = (y + 4)(y - 4) \\ 3y + 12 = 3(y + 4) \end{array} \right\}$ LCD is $3(y + 4)(y - 4)$ Find the LCD

$$= \frac{3(3y + 2)}{3(y + 4)(y - 4)} + \frac{(y - 4)(y - 4)}{3(y + 4)(y - 4)}$$

Multiply $\dfrac{3y + 2}{(y + 4)(y - 4)}$

by $\dfrac{3}{3}$ and $\dfrac{y - 4}{3(y + 4)}$

by $\dfrac{y - 4}{y - 4}$

$$= \frac{9y + 6}{3(y + 4)(y - 4)} + \frac{y^2 - 8y + 16}{3(y + 4)(y - 4)}$$ Multiply in the numerators

$$= \frac{(9y + 6) + (y^2 - 8y + 16)}{3(y + 4)(y - 4)}$$ Add numerators

$$= \frac{9y + 6 + y^2 - 8y + 16}{3(y + 4)(y - 4)}$$ Remove parentheses

$$= \frac{y^2 + y + 22}{3(y + 4)(y - 4)}$$ Combine like terms in the numerator

Since $y^2 + y + 22$ will not factor, the fraction will not reduce. It is convenient to leave the denominator in factored form, since it is easier to identify common factors and reduce to lowest terms.

☛ *Quick check*   Add the rational expressions $\dfrac{20}{3y} + \dfrac{25}{12y^2}$.

● **Example 5–4 C**

Subtracting rational expressions

Subtract the following rational expressions. Assume the denominators are not equal to zero. Reduce to lowest terms.

1. $\dfrac{5y}{2y - 1} - \dfrac{y + 1}{y + 2}$

The LCD of the rational expressions is $(2y - 1)(y + 2)$.

$$= \frac{5y(y + 2)}{(2y - 1)(y + 2)} - \frac{(y + 1)(2y - 1)}{(2y - 1)(y + 2)}$$ Multiply numerators and denominators

$$= \frac{(5y^2 + 10y) - (2y^2 + y - 1)}{(2y - 1)(y + 2)}$$ Subtract numerators

$$= \frac{5y^2 + 10y - 2y^2 - y + 1}{(2y - 1)(y + 2)}$$ Remove parentheses and change signs when subtracting

$$= \frac{3y^2 + 9y + 1}{(2y - 1)(y + 2)}$$ Combine like terms

   Don't forget the parentheses

Note The numerator $3y^2 + 9y + 1$ cannot be factored so we are unable to reduce. We should always check this!

2. $\dfrac{5x - 4}{x^2 - 2x + 1} - \dfrac{3x}{x^2 + 4x - 5}$

$\left.\begin{array}{l} x^2 - 2x + 1 = (x - 1)^2 \\ x^2 + 4x - 5 = (x + 5)(x - 1) \end{array}\right\}$ LCD is $(x - 1)^2(x + 5)$ Find the LCD

$= \dfrac{(5x - 4)(x + 5)}{(x - 1)^2(x + 5)} - \dfrac{3x(x - 1)}{(x + 5)(x - 1)(x - 1)}$ Multiply $\dfrac{5x - 4}{(x - 1)^2}$

by $\dfrac{x + 5}{x + 5}$ and

$\dfrac{3x}{(x + 5)(x - 1)}$

by $\dfrac{x - 1}{x - 1}$

$= \dfrac{(5x - 4)(x + 5)}{(x - 1)^2(x + 5)} - \dfrac{3x(x - 1)}{(x - 1)^2(x + 5)}$ Multiply numerators and denominators

$= \dfrac{(5x^2 + 21x - 20) - (3x^2 - 3x)}{(x - 1)^2(x + 5)}$ Subtract numerators

$= \dfrac{5x^2 + 21x - 20 - 3x^2 + 3x}{(x - 1)^2(x + 5)}$ Remove parentheses and change signs

$= \dfrac{2x^2 + 24x - 20}{(x - 1)^2(x + 5)}$ Combine like terms

Don't forget the parentheses

🏳 ***Quick check*** Subtract and reduce. $\dfrac{y + 1}{y^2 - y - 12} - \dfrac{3y + 2}{y^2 - 9y + 20}$ ⬤

Problem Solving

⬤ **Example 5–4 D**

Problem solving

Set up a rational expression for the following word statements.

1. If Lamar can mow his lawn in h hours, what part of the lawn can he mow in 1 hour?

If it takes Lamar 2 hours to mow his lawn, then in 1 hour he could mow $\dfrac{1}{2}$ of the lawn. If it takes him 3 hours to mow his lawn, then in 1 hour he could mow $\dfrac{1}{3}$ of the lawn. By looking at a specific time such as 2 hours or 3 hours to mow the lawn, we see that if it takes him h hours to mow the lawn, then in 1 hour he could mow $\dfrac{1}{h}$ of the lawn.

2. If the area (A) of a rectangle is n square inches, what is the expression for the length, ℓ, if the width, w, is 12 inches? ($A = \ell \cdot w$)

Using $A = \ell \cdot w$, we have $\ell = \dfrac{A}{w}$ and so the length

$\ell = \dfrac{n}{12}$ inches.   Replace A with n and w with 12

🏳 ***Quick check*** If the area of a rectangle is p square yards, write an expression for the width, w, if the length is 5 yards. ⬤

MASTERY POINTS

Can you
- Add and subtract rational expressions having different denominators?
- Can you write a rational expression for a word statement?

Exercise 5–4

Directions
Perform the indicated addition and reduce the answer to lowest terms. Assume all denominators are not equal to zero. See example 5–4 B.

⚐ *Quick Check*

Example $\dfrac{20}{3y} + \dfrac{25}{12y^2}$

Solution $\left.\begin{array}{l} 3y = 3 \cdot y \\ 12y^2 = 2 \cdot 2 \cdot 3 \cdot y \cdot y \end{array}\right\}$ LCD is $2 \cdot 2 \cdot 3 \cdot y \cdot y = 12y^2$

$\dfrac{20}{3y} + \dfrac{25}{12y^2} = \dfrac{20}{3y} \cdot \dfrac{4y}{4y} + \dfrac{25}{12y^2}$ Multiply numerator and denominator of $\dfrac{20}{3y}$ by $4y$

$= \dfrac{80y}{12y^2} + \dfrac{25}{12y^2}$ Perform indicated operations

$= \dfrac{80y + 25}{12y^2}$ Add numerators and place over common denominator

---

Directions Perform the indicated addition and reduce the answer to lowest terms. Assume all denominators are not equal to zero. See example 5–4 B.

1. $\dfrac{x}{6} + \dfrac{3}{4}$

2. $\dfrac{3z}{10} + \dfrac{2z}{15}$

3. $\dfrac{2x - 1}{16} + \dfrac{x + 2}{24}$

4. $\dfrac{4}{3x} + \dfrac{5}{2x}$

5. $\dfrac{3a + 1}{a} + \dfrac{2a - 3}{3a}$

6. $\dfrac{4}{x - 1} + \dfrac{5}{x + 3}$

7. $\dfrac{8}{y + 4} + \dfrac{7}{y - 5}$

8. $\dfrac{x}{x + 2} + \dfrac{3x}{4x - 1}$

9. $\dfrac{15}{5y - 10} + \dfrac{14}{2y - 4}$

10. $\dfrac{21}{6x + 12} + \dfrac{15}{2x + 4}$

11. $\dfrac{12}{x^2 - 4} + \dfrac{7}{4x - 8}$

12. $\dfrac{16}{2y + 6} + \dfrac{5}{y^2 - 9}$

13. $5 + \dfrac{4x}{x + 8}$

14. $9 + \dfrac{y + 9}{y - 1}$

15. $\dfrac{x}{x - 1} + \dfrac{3x}{x^2 - 1}$

16. $\dfrac{2y}{y^2 - 16} + \dfrac{5y}{2y - 8}$

17. $\dfrac{4}{x^2 - x - 6} + \dfrac{5}{x^2 - 9}$

18. $\dfrac{6}{x^2 - 4x - 12} + \dfrac{5}{x^2 - 36}$

19. $\dfrac{2y}{y^2 - 6y + 9} + \dfrac{5y}{y^2 - 2y - 3}$

20. $\dfrac{4z}{z^2 + z - 20} + \dfrac{z}{z^2 - 8z + 16}$

21. $\dfrac{y-2}{y^2-3y-10}+\dfrac{y+1}{y^2-y-6}$

22. $\dfrac{2x+1}{x^2+6x+5}+\dfrac{4x-3}{x^2-x-30}$

Directions
Perform the indicated subtraction and reduce to lowest terms. Assume all denominators are not zero. See example 5-4 C.

☞ **Quick Check**

Example $\dfrac{y+1}{y^2-y-12}-\dfrac{3y+2}{y^2-9y+20}$

Solution $\left.\begin{array}{l} y^2-y-12=(y-4)(y+3) \\ y^2-9y+20=(y-4)(y-5) \end{array}\right\}$ LCD is $(y-4)(y+3)(y-5)$.

$$=\dfrac{(y+1)(y-5)}{(y-4)(y+3)(y-5)}-\dfrac{(3y+2)(y+3)}{(y-4)(y-5)(y+3)}$$

Multiply numerator and denominator of
$\dfrac{y+1}{(y-4)(y+3)}$ by
$y-5$ and of
$\dfrac{3y+2}{(y-4)(y-5)}$ by
$y+3$

$$=\dfrac{(y^2-4y-5)-(3y^2+11y+6)}{(y-4)(y+3)(y-5)}$$

Multiply in each numerator, place parentheses around each product, and subtract

$$=\dfrac{y^2-4y-5-3y^2-11y-6}{(y-4)(y+3)(y-5)}$$

Remove parentheses, change signs, and subtract

$$=\dfrac{-2y^2-15y-11}{(y-4)(y+3)(y-5)}$$

Combine like terms in numerator

Don't forget the parentheses

Note The numerator will not factor. The denominator is left in factored form since it is easier to identify common factors, if there had been any.

Directions   Perform the indicated subtraction and reduce to lowest terms. Assume all denominators are not zero. See example 5-4 C.

23. $\dfrac{y}{9}-\dfrac{5}{6}$

24. $\dfrac{5y}{12}-\dfrac{y}{8}$

25. $\dfrac{9}{14y}-\dfrac{1}{21y}$

26. $\dfrac{7}{12z}-\dfrac{10}{9z}$

27. $\dfrac{5a+3}{12}-\dfrac{a-4}{10}$

28. $\dfrac{2x+9}{8}-\dfrac{x-7}{20}$

29. $\dfrac{2x+5}{6x}-\dfrac{x-5}{9x}$

30. $\dfrac{4y-1}{3y}-\dfrac{2y-3}{15y}$

31. $\dfrac{7}{2x-3}-\dfrac{6}{x-5}$

32. $\dfrac{7}{4x-6}-\dfrac{12}{3x+9}$

33. $\dfrac{12}{3y+6}-\dfrac{11}{7y+14}$

34. $\dfrac{14}{5x-15}-\dfrac{8}{2x-6}$

35. $9 - \dfrac{6}{x + 8}$

36. $12 - \dfrac{7}{z - 12}$

37. $\dfrac{2x}{3x + 1} - 9$

38. $\dfrac{4y}{5y - 4} - 10$

39. $\dfrac{-3}{a^2 - 5a + 6} - \dfrac{3}{a^2 - 4}$

40. $\dfrac{8}{x^2 - 25} - \dfrac{7}{x^2 + 3x - 10}$

41. $\dfrac{20}{y^2 - 2y - 24} - \dfrac{8}{y^2 + y - 12}$

42. $\dfrac{2p}{p^2 - 9p + 20} - \dfrac{5p - 2}{p - 5}$

43. $\dfrac{2a - 3}{a^2 - 5a + 6} - \dfrac{3a}{a - 2}$

Directions   Add and subtract as indicated.

44. $\dfrac{13}{12b} - \dfrac{2}{9b} + \dfrac{5}{4b}$

45. $\dfrac{5}{6z} + \dfrac{4}{8z} - \dfrac{1}{4z}$

46. $\dfrac{3x}{8} - \dfrac{2x}{5} + \dfrac{7x}{10}$

47. $\dfrac{4a}{5} + \dfrac{7a}{15} - \dfrac{a}{9}$

48. $\dfrac{a - 1}{5} - \dfrac{2a + 3}{15} + \dfrac{3a - 1}{25}$

49. $\dfrac{5b + 1}{6} + \dfrac{3b - 2}{9} - \dfrac{b + 1}{12}$

50. $\dfrac{4a}{a^2 + 2a - 15} + \dfrac{3a}{2a^2 + 11a + 5} - \dfrac{5a}{2a^2 - 5a - 3}$

51. $\dfrac{5}{z^2 - 4} - \dfrac{z}{z^2 - 1} + \dfrac{4}{z^2 + z - 2}$

Directions   Solve the following problems.

52. For a lens maker to manufacture lenses that will refract light by exactly the right amounts, the following formula is used:

$$\frac{1}{f} = (u - 1)\left(\frac{1}{R_1} + \frac{1}{R_2}\right)$$

Add the expressions containing R_1 and R_2.

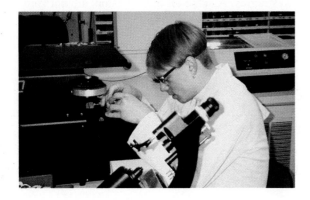

53. Three individuals A, B, and C can complete a given job in a, b, and c hours, respectively. Working together they can complete in one hour $\dfrac{1}{a} + \dfrac{1}{b} + \dfrac{1}{c}$ of the job. By combining, obtain a single expression for what they can do together in one hour.

54. In electricity, the total resistance of any parallel circuit may be given by

$$\frac{1}{R_t} = \frac{I_1}{E_1} + \frac{I_2}{E_2} + \frac{I_3}{E_3}$$

Combine the expression in the right side.

Directions	
See example 5–4 D.	☞ *Quick Check*
	Example If the area of a rectangle is p square yards, write an expression for the width, w, if the length is 5 yards.
	Solution Using $A = \ell \cdot w$, we are given that $A = p$ square yards and $\ell = 5$ yards. Substituting, we obtain
	$$p = 5 \cdot w$$
	and solving for w, we divide each side by 5. Thus $w = \dfrac{p}{5}$.

Directions   See example 5–4 D.

55. If one printer can print x pages per minute and another can print $x + 2$ pages per minute then the combined rate for these printers is $\dfrac{1}{x} + \dfrac{1}{x + 2}$. Combine the rational expression.

56. If x is the speed of a boat in still water (in knots), and $x + 3$ is the speed of the boat in a 3-knot current, then the difference in times it will take to cover one knot under these two different conditions is $\dfrac{1}{x} - \dfrac{1}{x + 3}$. Combine the rational expression.

57. If two investments have a rate of return of r_1 and r_2, with $r_1 > r_2$, then the difference in time it would take each investment to produce an amount of interest I on a principal P is $\dfrac{I}{Pr_2} - \dfrac{I}{Pr_1}$. Combine this expression.

58. In an electrical circuit 100 volts is applied to a resistive load. Originally the current flowing was I amperes, but later it increased to $I + 40$ amperes. Under these conditions, the change in resistance in the circuit can be described by $\dfrac{100}{I + 40} - \dfrac{100}{I}$. Combine this expression into one term.

59. A faucet, when fully open, can fill the sink in m minutes. What part of the sink can it fill in 3 minutes?

60. An inlet pipe to a swimming pool can fill the pool in h hours. What part of the pool can it fill in 9 hours?

61. An outlet pipe can drain a swimming pool in 36 hours. What part of the pool can it drain in h hours?

62. Jane can paint her house in h hours. What part of the house can she paint in 1 hour?

63. The product of two numbers is 48. If one of the numbers is m, what is the other number?

64. The area of a rectangle is 54 square centimeters. What is the length ℓ if the width is w centimeters?

65. The area of a rectangle is A square feet. What is the width if the length is 23 feet?

66. The area of a triangle is 21 square yards. If the triangle has a base length b, what is the altitude h of the triangle? $\left(A = \frac{1}{2}bh\right)$

67. The area of a triangle is A square rods. If the altitude h is 9 rods, what is the length of the base b? (See problem 66.)

Directions   Solve the following problems.

72. $\frac{2x-1}{x^2-9} + m = \frac{x}{x-3}$; what is the value of m?

73. $b + \frac{x}{x^2-x-6} = \frac{10}{x-3}$; what is the value of b?

68. John drives 25 miles in h hours. At what speed, r, did he travel? [*Hint:* Use distance traveled $(d) =$ rate $(r) \times$ time (t).]

69. Mabel travels d miles at a rate of 55 miles per hour. Write an expression for the time t that she traveled. (See problem 68.)

70. What is the reciprocal of the natural number n?

71. What is the reciprocal of the fraction $\frac{a}{b}$?

74. $\frac{5x}{x+3} \div y = \frac{x}{x-1}$; what is the value of y?

75. $\frac{5x}{x+3} \cdot y = \frac{x}{x-1}$; what is the value of y?

⚛ Explore to Learn

A computerized test generator is being programmed to generate problems for an algebra book. It generates the answers at the same time. One of the ways it will generate fraction problems is using the model $\frac{x+1}{x+3} + \frac{x}{x-1} - \frac{x-2}{(x-1)(x+3)}$. For example, if $x = 10$ is used this generates the problem $\frac{11}{13} + \frac{10}{9} - \frac{8}{117}$.

a. Combine the three rational expressions into one.

b. Then use this one expression to compute the answer to the arithmetic problem.

c. Create two more problems and their answers with this problem generator.

d. For what value(s) of x will this generator only give two fractions to be combined?

e. What values of x must be avoided so that the problems do not become undefined?

Review Exercises

Directions   Completely factor the following expressions. See sections 4–2, 4–3, and 4–4.

1. $x^2 - 14x + 49$

2. $2x^2 - 11x + 5$

3. $4x^2 - 16$

Directions   Multiply the following expressions. See section 3–2.

4. $(x+9)(x-9)$

5. $(4x+3)^2$

6. $(2x+1)(x-8)$

Directions   Find the LCD of the rational expressions having the following denominators. See section 5–3.

7. $16, 12, 6$

8. $4x, 2x^2, 6$

9. $x^2 - 9; (x-3)^2; x+3$

Directions   Add or subtract the following. See section 5–4.

10. $\frac{5}{x} + \frac{3}{2x}$

11. $\frac{5}{x-2} - \frac{3}{x-1}$

12. One side of a triangle is three times as long as the second side and the third side is 1 more than two times the second side. If the perimeter is 37 meters, find the lengths of the sides. See section 2–7.

5–5   Complex Fractions

A Complex Fraction

A **complex fraction** is a fraction (**complex rational expression**) whose numerator, denominator, or both contain fractions (or rational expressions). The expressions

$$\frac{\frac{3}{4}}{\frac{5}{6}}, \qquad \frac{\frac{1}{3}+2}{1-\frac{1}{2}}, \qquad \frac{\frac{1}{m^2}-\frac{1}{n^2}}{m+n}, \qquad \frac{x-y}{\frac{1}{x}-\frac{1}{y}}, \qquad \text{and} \qquad \frac{\frac{x-1}{x-2}}{\frac{x+3}{x}}$$

are examples of complex fractions. Given a complex fraction, we simplify the fraction by *eliminating the fractions within the numerator and/or the denominator* to obtain a simple fraction.

We name the parts of a complex fraction as shown in the following examples.

$$\frac{\dfrac{3}{x}+\dfrac{4}{y}}{\dfrac{1}{x}-\dfrac{2}{y}}$$

The numerator of the complex fraction

Fraction bar

The denominator of the complex fraction

To simplify a complex fraction, we use one of the following methods.

> ### Simplifying a Complex Fraction
>
> **Method 1** Find the LCD of all fractions within the complex fraction, then multiply the numerator and the denominator of the complex fraction by this LCD. Reduce the resulting fraction to lowest terms.
>
> **Method 2** Form a single fraction in the numerator and in the denominator. Multiply the fraction in the numerator by the reciprocal of the fraction in the denominator and simplify.

● **Example 5–5 A**

Simplifying complex fractions by method 1

Simplify each complex fraction using method 1.

1. $\dfrac{\dfrac{3}{4}}{\dfrac{5}{6}} = \dfrac{\dfrac{3}{4}\cdot 12}{\dfrac{5}{6}\cdot 12}$ The LCD of all the fractions within the complex fraction is 12

Multiply the numerator and the denominator of the complex fraction by the LCD

$\qquad = \dfrac{3\cdot 3}{5\cdot 2}$ Divide 4 into 12

Divide 6 into 12

$\qquad = \dfrac{9}{10}$ Multiply in numerator and denominator

2. $\dfrac{\dfrac{3}{a}-1}{1+\dfrac{4}{b}}$

The LCD of all the fractions within the complex fraction is ab. We multiply the numerator, $\dfrac{3}{a} - 1$, and the denominator, $1 + \dfrac{4}{b}$, by ab. Remember to multiply the LCD times *every term* in the numerator and the denominator.

$$= \frac{\left(\dfrac{3}{a} - 1\right) \cdot ab}{\left(1 + \dfrac{4}{b}\right) \cdot ab}$$

Multiply the numerator and the denominator of the complex fraction by ab

$$= \frac{\dfrac{3}{a} \cdot ab - 1 \cdot ab}{1 \cdot ab + \dfrac{4}{b} \cdot ab}$$

Multiply each term of the numerator and the denominator of the complex fraction by ab

$$= \frac{3b - ab}{ab + 4a}$$

Perform indicated multiplications

$$= \frac{b(3 - a)}{a(b + 4)}$$

Factor the numerator and the denominator. No common factors, the fraction will not reduce

3. $\dfrac{\dfrac{4}{x} - \dfrac{3}{y}}{\dfrac{5}{x} + \dfrac{7}{y}}$

The LCD of all the fractions within the complex fraction is xy. We multiply the numerator, $\dfrac{4}{x} - \dfrac{3}{y}$, and the denominator, $\dfrac{5}{x} + \dfrac{7}{y}$, by xy and get

$$= \frac{\left(\dfrac{4}{x} - \dfrac{3}{y}\right) \cdot xy}{\left(\dfrac{5}{x} + \dfrac{7}{y}\right) \cdot xy} = \frac{\dfrac{4}{x} \cdot xy - \dfrac{3}{y} \cdot xy}{\dfrac{5}{x} \cdot xy + \dfrac{7}{y} \cdot xy}$$

Multiply each term of the numerator and the denominator of the complex fraction by xy

$$= \frac{4y - 3x}{5y + 7x}$$

Perform indicated multiplications

4. $\dfrac{\dfrac{1}{a} + \dfrac{1}{b}}{\dfrac{1}{a^2} - \dfrac{1}{b^2}}$

$$= \frac{\left(\dfrac{1}{a} + \dfrac{1}{b}\right)a^2b^2}{\left(\dfrac{1}{a^2} - \dfrac{1}{b^2}\right)a^2b^2}$$

a^2b^2 is the LCD of all the fractions within the complex fraction

$$= \frac{\dfrac{1}{a} \cdot a^2b^2 + \dfrac{1}{b} \cdot a^2b^2}{\dfrac{1}{a^2} \cdot a^2b^2 - \dfrac{1}{b^2} \cdot a^2b^2}$$

Distribute the multiplication

$$= \frac{ab^2 + a^2b}{b^2 - a^2}$$

Perform the indicated multiplication

$$= \frac{ab(b + a)}{(b - a)(b + a)}$$

Factor and look for common factors

$$= \frac{ab}{b - a}$$

Divide out the common factor of $(b + a)$

➤ *Quick check*   Simplify $\dfrac{\dfrac{3}{4} + \dfrac{2}{3}}{\dfrac{5}{6} - \dfrac{1}{12}}$.

● **Example 5–5 B**

Simplifying complex fractions by method 2

Simplify each complex fraction using method 2.

1. $\dfrac{\dfrac{3}{4}}{\dfrac{5}{6}} = \dfrac{3}{4} \div \dfrac{5}{6} = \dfrac{3}{4} \cdot \dfrac{6}{5}$ Multiply by the reciprocal of $\dfrac{5}{6}$

$\qquad\qquad = \dfrac{3 \cdot 6}{4 \cdot 5}$ Multiply

$\qquad\qquad = \dfrac{9}{10}$ Reduce to lowest terms

2. $\dfrac{\dfrac{3}{a} - 1}{1 + \dfrac{4}{b}}$

We first express the numerator, $\dfrac{3}{a} - 1$, and the denominator, $1 + \dfrac{4}{b}$, as single fractions. Thus,

$$\dfrac{3}{a} - 1 = \dfrac{3}{a} - \dfrac{a}{a} = \dfrac{3 - a}{a}$$ Subtract the terms in the numerator of the complex fraction

and

$$1 + \dfrac{4}{b} = \dfrac{b}{b} + \dfrac{4}{b} = \dfrac{b + 4}{b}$$ Add the terms in the denominator of the complex fraction

$$\dfrac{\dfrac{3}{a} - 1}{1 + \dfrac{4}{b}} = \dfrac{\dfrac{3 - a}{a}}{\dfrac{b + 4}{b}} = \dfrac{3 - a}{a} \cdot \dfrac{b}{b + 4}$$ Multiply by the reciprocal of the denominator

$$\qquad\qquad = \dfrac{b(3 - a)}{a(b + 4)}$$

<u>Note</u>   Always form a single fraction in the numerator and in the denominator *before* we invert and multiply. That is,

$$\dfrac{\dfrac{3}{a} - 1}{1 + \dfrac{4}{b}} \text{ does not equal } \left(\dfrac{3}{a} - 1\right) \cdot \left(1 + \dfrac{b}{4}\right)$$

3. $\dfrac{\dfrac{a^2 - x^2}{x}}{\dfrac{a + x}{x^2}}$

$$= \frac{a^2 - x^2}{x} \cdot \frac{x^2}{a + x}$$

Multiply by the reciprocal of the denominator

$$= \frac{(a - x)(a + x)}{x} \cdot \frac{x^2}{a + x}$$

Factor and look for common factors

$$= (a - x) \cdot x$$

Reduce by the common factor $x(a + x)$

$$= ax - x^2$$

Multiply as indicated

In examples 5–5 A numbers 1 and 2 and 5–5 B numbers 1 and 2 you can see that either method could be used to simplify the complex fraction. You should choose one method and use it until you are proficient with it. Eventually you may wish to know both methods, so that you can apply whichever one is best suited to the problem.

Quick check   Simplify $\dfrac{\dfrac{3}{4} + \dfrac{2}{3}}{\dfrac{5}{6} - \dfrac{1}{12}}$

MASTERY POINTS

Can you
- Simplify complex fractions?

Exercise 5–5

Directions
Simplify each complex fraction. See example 5–5 A and B.

Quick Check

Example $\dfrac{\dfrac{3}{4} + \dfrac{2}{3}}{\dfrac{5}{6} - \dfrac{1}{12}}$

Solution   **Method 1**

$$\frac{\dfrac{3}{4} + \dfrac{2}{3}}{\dfrac{5}{6} - \dfrac{1}{12}} = \frac{\left(\dfrac{3}{4} + \dfrac{2}{3}\right)12}{\left(\dfrac{5}{6} - \dfrac{1}{12}\right)12} = \frac{\dfrac{3}{4} \cdot 12 + \dfrac{2}{3} \cdot 12}{\dfrac{5}{6} \cdot 12 - \dfrac{1}{12} \cdot 12}$$

Multiply numerator and denominator by 12

$$= \frac{9 + 8}{10 - 1}$$

Multiply as indicated

$$= \frac{17}{9}$$

Combine in the numerator and the denominator

Method 2

$$\frac{\dfrac{3}{4} + \dfrac{2}{3}}{\dfrac{5}{6} - \dfrac{1}{12}} = \frac{\dfrac{9}{12} + \dfrac{8}{12}}{\dfrac{10}{12} - \dfrac{1}{12}} = \frac{\dfrac{17}{12}}{\dfrac{9}{12}}$$

Add in the numerator

Subtract in the denominator

$$= \frac{17}{12} \cdot \frac{12}{9}$$

Multiply by the reciprocal of $\dfrac{9}{12}$

$$= \frac{17}{9}$$

Directions   Simplify each complex fraction. See example 5–5 A and B.

1. $\dfrac{\dfrac{2}{3}}{\dfrac{4}{5}}$

2. $\dfrac{\dfrac{7}{8}}{\dfrac{5}{6}}$

3. $\dfrac{\dfrac{4}{3}}{\dfrac{8}{9}}$

4. $\dfrac{\dfrac{9}{10}}{\dfrac{7}{6}}$

5. $\dfrac{1 + \dfrac{3}{5}}{2 - \dfrac{1}{5}}$

6. $\dfrac{5 - \dfrac{3}{4}}{1 + \dfrac{5}{8}}$

7. $\dfrac{7}{2 + \dfrac{4}{5}}$

8. $\dfrac{10}{4 - \dfrac{11}{12}}$

9. $\dfrac{4 + \dfrac{3}{5}}{6}$

10. $\dfrac{10 - \dfrac{7}{8}}{3}$

11. $\dfrac{\dfrac{3}{4} + \dfrac{5}{8}}{\dfrac{1}{2} - \dfrac{1}{4}}$

12. $\dfrac{\dfrac{6}{7} - \dfrac{5}{14}}{\dfrac{3}{14} - \dfrac{5}{7}}$

13. $\dfrac{x + \dfrac{1}{4}}{x - \dfrac{3}{4}}$

14. $\dfrac{y - \dfrac{5}{6}}{y + \dfrac{1}{2}}$

15. $\dfrac{\dfrac{1}{a} + 3}{\dfrac{2}{a} - 4}$

16. $\dfrac{5 - \dfrac{3}{b}}{6 + \dfrac{5}{b}}$

17. $\dfrac{\dfrac{3}{a^2} + 4}{5 - \dfrac{3}{a}}$

18. $\dfrac{\dfrac{5}{x} - 5}{6 + \dfrac{4}{x^3}}$

19. $\dfrac{a - \dfrac{3}{b}}{a + \dfrac{4}{b}}$

20. $\dfrac{x + \dfrac{4}{y}}{x - \dfrac{5}{y}}$

21. $\dfrac{\dfrac{1}{x} + \dfrac{1}{y}}{\dfrac{1}{x} - \dfrac{1}{y}}$

22. $\dfrac{\dfrac{3}{x^2} - \dfrac{4}{y}}{\dfrac{5}{x} + \dfrac{2}{y^2}}$

23. $\dfrac{x + y}{\dfrac{1}{x} + \dfrac{1}{y}}$

24. $\dfrac{a - b}{\dfrac{2}{a} + \dfrac{3}{b}}$

25. $\dfrac{\dfrac{4}{a^2} - \dfrac{5}{b}}{a - b}$

26. $\dfrac{\dfrac{1}{a} - \dfrac{1}{b}}{\dfrac{1}{a^2} - \dfrac{1}{b^2}}$

27. $\dfrac{\dfrac{1}{a} - a}{\dfrac{1 - a^2}{a}}$

28. $\dfrac{\dfrac{3}{x} - x}{\dfrac{3 - x^2}{x}}$

29. $\dfrac{\dfrac{4}{x} - x}{\dfrac{4 - x^2}{x^2}}$

30. $\dfrac{x - \dfrac{16}{x}}{\dfrac{x^2 - 16}{x^2}}$

31. $\dfrac{\dfrac{9}{x} - x}{9 - x^2}$

32. $\dfrac{a - \dfrac{1}{a}}{a^2 - 1}$

33. $\dfrac{a + \dfrac{1}{a}}{a^2 + 1}$

34. $\dfrac{x + y}{\dfrac{x + y}{x^2 y^2}}$

35. $\dfrac{x + \dfrac{1}{x}}{\dfrac{x^2 + 1}{4}}$

36. $\dfrac{a + \dfrac{4}{a}}{\dfrac{a^2 + 4}{5}}$

37. $\dfrac{\dfrac{1}{x + y} - \dfrac{1}{x - y}}{\dfrac{1}{x + y} + \dfrac{1}{x - y}}$

38. $\dfrac{\dfrac{a}{a + b} - \dfrac{b}{a - b}}{\dfrac{b}{a + b} + \dfrac{a}{a - b}}$

39. $\dfrac{\dfrac{x + y}{x - y} + \dfrac{x - y}{x + y}}{x^2 - y^2}$

40. $\dfrac{\dfrac{b}{a + b} - \dfrac{a}{a - b}}{a^2 - b^2}$

41. $\dfrac{\dfrac{2}{ab} + \dfrac{3}{ab^2}}{a^2 b^2}$

42. $\dfrac{\dfrac{5}{x^2 y^2} - \dfrac{4}{xy}}{xy}$

43. $\dfrac{\dfrac{2}{a+3}+\dfrac{1}{a-2}}{\dfrac{3}{a-2}-\dfrac{4}{a+3}}$

44. $\dfrac{\dfrac{7}{x-4}-\dfrac{5}{x+3}}{\dfrac{5}{x+3}+\dfrac{9}{x-4}}$

45. $\dfrac{\dfrac{5}{x^2-x-12}}{\dfrac{4}{x+3}-\dfrac{5}{x-4}}$

46. $\dfrac{\dfrac{7}{b-7}+\dfrac{8}{b-5}}{\dfrac{6}{b^2-12b+35}}$

47. $\dfrac{3-\dfrac{4}{a+4}}{\dfrac{5}{a+4}-\dfrac{6}{a-1}}$

48. $\dfrac{\dfrac{6}{x-5}+7}{\dfrac{8}{x-5}-\dfrac{9}{x+3}}$

49. $\dfrac{\dfrac{y^2-y-6}{y^2+2y+1}}{\dfrac{y^2-2y-8}{y^2+7y+6}}$

50. $\dfrac{\dfrac{a^2+3a-10}{a^2-5a-14}}{\dfrac{a^2+6a+5}{a^2-8a+7}}$

Directions   Solve the following problems.

51. A refrigeration coefficient-of-performance formula for the ideal refrigerator is given by

$$cp = \dfrac{1}{\dfrac{T_2}{T_1}-1}$$

Simplify the right side.

52. In electronics, a formula for coupled inductance in parallel with fields aiding is given by

$$L_t = \dfrac{1}{\dfrac{1}{L_1+M}+\dfrac{1}{L_2+M}}$$

Simplify the right side.

53. In electronics, a formula for self-inductance of circuits in parallel is given by

$$L_t = \dfrac{1}{\dfrac{1}{L_1}+\dfrac{1}{L_2}+\dfrac{1}{L_3}}$$

Simplify the right side.

54. Coupled inductance with circuits connected in parallel with opposing fields is given by

$$L_t = \dfrac{1}{\dfrac{1}{L_1-M}+\dfrac{1}{L_2-M}}$$

Simplify the right side.

✸ Explore to Learn

1. a. $\dfrac{\dfrac{x}{x-3}}{m}$ simplifies to $x+1$. What is the value of m?

b. $\dfrac{\dfrac{x}{x-3}}{\dfrac{x+4}{m}}$ simplifies to x. What is the value of m?

2. Simplify the following. When you apply the definition of a negative exponent the following will become complex fractions.

a. $\dfrac{x^{-1}-y^{-1}}{x^{-2}-y^{-2}}$

b. $\dfrac{x^{-1}+y^{-1}}{x^{-2}-y^{-2}}$

c. $(a^{-1}+b^{-1})^{-1}$

Review Exercises

Directions   Reduce the following expressions to lowest terms. See section 5–1.

1. $\dfrac{36}{42}$

2. $\dfrac{x^2 - 5x - 14}{x^2 + 4x + 4}$

3. Subtract $(3x^3 - 2x^2 + x - 12) - (x^3 - 5x^2 + 9)$. See section 2–2.

Directions   Find the solution set for the following equations. See sections 2–5 and 4–7.

4. $4x + 3x = 21$

5. $8y - 4 = 5y - 10$

6. $x^2 + 2x - 3 = 0$

Directions   Determine the domain of the following rational expressions. See section 5–1.

7. $\dfrac{3}{x + 7}$

8. $\dfrac{y - 4}{y^2 - 4}$

9. Perform the indicated operations, if possible. See section 1–7.

a. $\dfrac{4}{0}$ b. $\dfrac{0}{-3}$

10. The length of a rectangle is 1 inch less than three times the width. Find the dimensions if the perimeter is 70 inches. See section 2–7.

5–6   Rational Equations

A Rational Equation

An algebraic equation that contains *at least one* rational expression is called a **rational equation**.

The basic operations for solving equations that you learned in chapter 2 will apply to rational equations once the denominators in the equation are eliminated. We accomplish this by using the multiplication property of equality. The multiplier is the least common denominator of all the rational expressions of the equation.

Solving a Rational Equation

1. Find the LCD of all the rational expressions.
2. Eliminate the denominators by multiplying each term of both sides of the equation by the LCD.
3. Use the four steps from section 2–5 to solve the resulting equation.

Note We eliminate all the denominators *only* when solving **rational equations**. This is *never done* when adding or subtracting **rational expressions**. To illustrate, given the equation $\dfrac{6}{x} - \dfrac{4}{x^2} = 0$, we multiply each term by x^2, whereas given the subtraction problem $\dfrac{6}{x} - \dfrac{4}{x^2}$, we do *not* multiply each term by x^2, we change the denominators to x^2 (the LCD) and perform the subtraction.

● **Example 5–6 A**

Solving an equation involving rational expressions

Find the solution set of each of the following rational equations.

1. $\dfrac{x-3}{4} = \dfrac{x}{8}$

The LCD of the rational expressions is 8.

$$8 \cdot \frac{(x-3)}{4} = 8 \cdot \frac{x}{8} \qquad \text{Multiply each side by 8}$$

$$2(x-3) = x \qquad\qquad \text{Reduce in each side}$$
$$2x - 6 = x \qquad\qquad \text{Multiply as indicated}$$
$$x - 6 = 0 \qquad\qquad \text{Subtract } x \text{ from each side}$$
$$x = 6 \qquad\qquad \text{Add 6 to each side}$$

The solution set is $\{6\}$. If we wish to check our work, we substitute 6 for x in the original equation.

$$\frac{(6)-3}{4} = \frac{(6)}{8} \qquad \text{Replace } x \text{ with 6 in the original equation}$$

$$\frac{3}{4} = \frac{3}{4} \qquad \text{(True)}$$

2. $\dfrac{t}{4} - \dfrac{t-4}{5} = \dfrac{7}{10}$

The LCD of the rational expressions is 20.

$$20 \cdot \frac{t}{4} - 20 \cdot \frac{t-4}{5} = 20 \cdot \frac{7}{10} \qquad \text{Multiply each term by the LCD 20}$$

$$5t - 4(t-4) = 2 \cdot 7 \qquad\qquad \text{Reduce each term}$$
$$5t - 4t + 16 = 14 \qquad\qquad \text{Multiply as indicated}$$
$$t + 16 = 14 \qquad\qquad \text{Combine like terms}$$
$$t = -2 \qquad\qquad \text{Subtract 16 from each side}$$

The solution set is $\{-2\}$. Check your answer by replacing t with -2 in the original equation.

<u>Note</u> A common error that is made when multiplying $-4(t-4)$ is to get $-4t - 16$. Do not forget that you are using the distributive property to multiply -4 times each term in the group $(t-4)$. The correct result is $-4t + 16$.

3. $\dfrac{5}{3a} + \dfrac{4}{9} = \dfrac{5}{12a}$

We determine that the LCD is $36a$.

$$36a \cdot \frac{5}{3a} + 36a \cdot \frac{4}{9} = 36a \cdot \frac{5}{12a} \qquad \text{Multiply each term by the LCD } 36a$$

$$12 \cdot 5 + 4a \cdot 4 = 3 \cdot 5 \qquad\qquad \text{Reduce each term}$$
$$60 + 16a = 15 \qquad\qquad \text{Multiply as indicated}$$
$$16a = -45 \qquad\qquad \text{Subtract 60 from each side}$$
$$a = -\frac{45}{16} \qquad\qquad \text{Divide each side by 16}$$

The solution set is $\left\{-\dfrac{45}{16}\right\}$. Check your solution.

<u>Note</u> The domain of the variable is every real number *except* 0 since two of the terms are undefined when *a* = 0. This is an important observation to make as shown in example 4.

4. $\dfrac{3}{y} = \dfrac{4}{y^2 - 2y} - \dfrac{2}{y - 2}$

Factor the denominator $y^2 - 2y$ to get $y(y - 2)$. We determine the LCD to be $y(y - 2)$.

$$y(y - 2) \cdot \dfrac{3}{y} = y(y - 2) \cdot \dfrac{4}{y(y - 2)} \qquad \text{Multiply each term by the LCD } y(y - 2),\ (y \neq 0, 2)$$

$$- y(y - 2) \cdot \dfrac{2}{y - 2}$$

$$(y - 2) \cdot 3 = 4 - y \cdot 2 \qquad \text{Reduce each term}$$

$$3y - 6 = 4 - 2y \qquad \text{Multiply as indicated}$$

$$3y + 2y = 4 + 6 \qquad \text{Add 6 and } 2y \text{ to each side}$$

$$5y = 10 \qquad \text{Combine like terms}$$

$$y = 2 \qquad \text{Divide each side by 5}$$

Two *is not in the domain* of the variable y, since $y = 2$ makes the denominator $y - 2 = 0$. So 2 cannot be a solution of the equation. Therefore, the solution set is $\varnothing$.

We conclude there is no solution for the equation of example 4. The number 2 in that example is called an **extraneous solution of a rational equation**. An extraneous solution can occur whenever the variable appears in the denominator of one or more of the terms of the equation. Thus, you should *always* check your solution(s) of a rational equation when a variable appears in the denominator.

☛ *Quick check*   Find the solution set of $\dfrac{6}{4z} + \dfrac{7}{8} = \dfrac{9}{16z}$

Rational Equations in More Than One Variable

In scientific fields, equations and formulas involving rational expressions and *more than one variable* are common. It is often desirable to solve such equations for one variable in terms of the other variables in the equation. The procedures for finding such solutions are identical to those used in solving the preceding equations.

Solving a Rational Equation in More Than One Variable

1. Remove the fractions by multiplying each side of the equation by the LCD of all the rational expressions in the equation.
2. Collect all terms containing the variable you are solving for on one side of the equation and all other terms on the other side.
3. Factor out the variable you are solving for if it appears in more than one term.
4. Divide each side by the coefficient of the variable for which you are solving.

● **Example 5–6 B**

Solving for a specified variable

Solve each rational equation for the indicated variable. Assume all denominators are nonzero.

1. Solve $\dfrac{a}{3} + \dfrac{3x}{2} = c$ for x.

The LCD of the rational expressions is 6.

$$6 \cdot \dfrac{a}{3} + 6 \cdot \dfrac{3x}{2} = 6 \cdot c \qquad \text{Multiply each term by 6}$$
$$2 \cdot a + 3 \cdot 3x = 6c \qquad \text{Reduce where possible}$$
$$2a + 9x = 6c \qquad \text{Multiply in left side}$$
$$2a + 9x - 2a = 6c - 2a \qquad \text{Subtract } 2a \text{ from each side}$$
$$9x = 6c - 2a \qquad \text{Combine like terms}$$
$$\dfrac{9x}{9} = \dfrac{6c - 2a}{9} \qquad \text{Divide each side by the coefficient 9}$$
$$x = \dfrac{6c - 2a}{9}$$

2. Solve $\dfrac{1}{a} = \dfrac{1}{b} + \dfrac{1}{c}$ for c.

The LCD of the rational expressions is abc.

$$abc \cdot \dfrac{1}{a} = abc \cdot \dfrac{1}{b} + abc \cdot \dfrac{1}{c} \qquad \text{Multiply each term by } abc$$
$$bc = ac + ab \qquad \text{Reduce where possible}$$

<u>Note</u> Since we are solving for c, we must get all the terms containing c on one side of the equation.

$$bc - ac = ac + ab - ac \qquad \text{Subtract } ac \text{ from each side}$$
$$bc - ac = ab$$
$$(b - a)c = ab \qquad \text{Factor } c \text{ from each term on the left side}$$
$$\dfrac{(b - a)c}{(b - a)} = \dfrac{ab}{(b - a)} \qquad \text{The coefficient of } c \text{ is } (b - a), \text{ divide each side by it}$$
$$c = \dfrac{ab}{b - a} \qquad \text{Reduce on left side}$$

☛ **Quick check** Solve $\dfrac{5}{a} - \dfrac{3}{b} = 3$ for a.

● **Example 5–6 C**

Problem solving

The coefficient of linear expansion, k, of a solid when heated is given by $k = \dfrac{L_t - L_0}{L_0 t}$, where L_0 is the length at $0°$ C, L_t is the length at $t°$ C, and t is any given temperature in Celsius. Solve for L_t.

$$k = \dfrac{L_t - L_0}{L_0 t} \qquad \text{Original equation}$$
$$k \cdot L_0 t = \dfrac{L_t - L_0}{L_0 t} \cdot L_0 t \qquad \text{Multiply both sides by } L_0 t$$
$$k L_0 t = L_t - L_0 \qquad \text{Simplify}$$
$$k L_0 t + L_0 = L_t \qquad \text{Add } L_0 \text{ to both sides}$$
$$L_t = k L_0 t + L_0 \qquad \text{Symmetric property}$$

The equation is solved for L_t in terms of the other variables. In this form, we are able to easily determine the length L_t of the solid at $t°$ C.   ●

MASTERY POINTS

Can you
• Solve equations containing rational expressions?
• Solve rational equations for one variable in terms of the other variables?
• Determine when a solution is extraneous?

Exercise 5–6

Directions
Find the solution set of each rational equation. See example 5–6 A.

👈 *Quick Check*

Example $\dfrac{6}{4z} + \dfrac{7}{8} = \dfrac{9}{16z}$

Solution   The LCD of the rational expressions is $16z$.

$$16z \cdot \frac{6}{4z} + 16z \cdot \frac{7}{8} = 16z \cdot \frac{9}{16z} \qquad \text{Multiply each term by } 16z, \ (z \neq 0)$$

$$4 \cdot 6 + 2z \cdot 7 = 9 \qquad\qquad \text{Reduce in each term}$$

$$24 + 14z = 9 \qquad\qquad\quad \text{Multiply in each term}$$

$$14z = -15 \qquad\qquad\quad\ \text{Subtract 24 from each side}$$

$$z = -\frac{15}{14} \qquad\qquad\quad\ \text{Divide each side by 14}$$

The solution set is $\left\{ -\dfrac{15}{14} \right\}$.

Directions   Find the solution set of each rational equation. See example 5–6 A.

1. $\dfrac{y}{4} = \dfrac{2}{3}$

2. $\dfrac{4x}{5} - \dfrac{2}{3} = 4$

3. $\dfrac{p}{6} = \dfrac{7}{9}$

4. $\dfrac{a}{3} + \dfrac{5}{2} = 6$

5. $\dfrac{z}{8} + 3 = \dfrac{1}{4}$

6. $\dfrac{3R}{4} - 5 = \dfrac{5}{6}$

7. $\dfrac{3a}{6} + \dfrac{2a}{5} = 1$

8. $\dfrac{5x}{8} - \dfrac{x}{12} = 3$

9. $\dfrac{2x+1}{7} - \dfrac{2x-3}{14} = 1$

10. $\dfrac{3a+1}{9} + \dfrac{1}{12} = \dfrac{2a-1}{3}$

11. $\dfrac{3}{2x} = \dfrac{4}{5} + \dfrac{2}{x}$

12. $\dfrac{4}{b} - \dfrac{7}{3b} = \dfrac{2}{5}$

13. $\dfrac{2}{3R} + \dfrac{3}{2R} + \dfrac{1}{R} = 4$

14. $\dfrac{3}{w} - \dfrac{6}{5w} + \dfrac{1}{2w} = 5$

15. $\dfrac{4}{6y} + 5 = \dfrac{1}{9y} + 2$

16. $\dfrac{16}{5a} - 1 = 5 + \dfrac{3}{4a}$

17. $\dfrac{5}{3b} - \dfrac{1}{2} = \dfrac{7}{6b}$

18. $3 - \dfrac{5}{9x} = \dfrac{4}{6x}$

19. $\dfrac{3p+2}{7p} - 3 = \dfrac{p}{14p}$

20. $\dfrac{5-x}{8x} = \dfrac{2x+5}{6x}$

21. $\dfrac{a-4}{3a} = \dfrac{2a-1}{4a}$

22. $\dfrac{R+2}{10R} + \dfrac{4R-1}{4R} = 2$

23. $\dfrac{4}{x-4} = \dfrac{5}{x+4}$

24. $\dfrac{9}{3-x} = \dfrac{8}{2x+1}$

25. $\dfrac{10}{3z + 1} = \dfrac{3}{5}$

26. $\dfrac{3}{2a + 3} = \dfrac{1}{7}$

27. $\dfrac{y}{y - 2} - 2 = \dfrac{2}{y - 2}$

28. $\dfrac{1}{x + 5} = \dfrac{3}{x - 5} - \dfrac{10}{x^2 - 25}$

29. $\dfrac{b - 1}{b^2 - 4} = \dfrac{6}{b + 2}$

30. $\dfrac{5}{R + 3} = \dfrac{4R + 3}{R^2 - 9}$

31. $\dfrac{5}{a^2 - 25} + \dfrac{3}{a - 5} = \dfrac{4}{a + 5}$

32. $\dfrac{6}{2p - 4} + \dfrac{4}{p - 2} = \dfrac{1}{p^2 - 4}$

33. $\dfrac{5}{x^2 + x - 6} = \dfrac{2}{x^2 + 3x - 10}$

34. $\dfrac{8}{a^2 - 6a + 8} = \dfrac{1}{a^2 - 16}$

35. $\dfrac{1}{a - 3} + \dfrac{2}{a + 4} = \dfrac{6}{a^2 + a - 12}$

36. $\dfrac{\dfrac{5}{2x - 1} - \dfrac{7}{3x + 2}}{\dfrac{9}{6x^2 + x - 2}} =$

37. $3x^2 + 4x + \dfrac{4}{3} = 0$

38. $3x^2 + \dfrac{11}{2}x + \dfrac{3}{2} = 0$

39. $b^2 + \dfrac{3}{2}b = \dfrac{9}{2}$

40. $\dfrac{2}{3}x^2 + x = \dfrac{20}{3}$

41. $\dfrac{3}{4}z^2 = 2 - \dfrac{5}{2}z$

42. $x^2 - \dfrac{5}{6}x = \dfrac{2}{3}$

43. $x^2 = \dfrac{2}{3} - \dfrac{x}{3}$

44. $x^2 + \dfrac{4}{3}x = -\dfrac{4}{9}$

Directions
Solve the following equations for the indicated letter. Assume all denominators are nonzero. See example 5–6 B.

> **Quick Check**
>
> **Example** $\dfrac{5}{a} - \dfrac{3}{b} = 3$ for a
>
> **Solution** $ab \cdot \dfrac{5}{a} - ab \cdot \dfrac{3}{b} = ab \cdot 3$ Multiply each term by ab
>
> $\qquad\qquad 5b - 3a = 3ab$ Reduce each term
>
> $\qquad\qquad\quad 5b = 3ab + 3a$ Add $3a$ to each side
>
> So $\qquad\qquad 5b = a(3b + 3)$ Factor a in the right side
>
> $\qquad\qquad\quad\; a = \dfrac{5b}{3b + 3}$ Divide each side by $(3b + 3)$

Directions Solve the following equations for the indicated letter. Assume all denominators are nonzero. See example 5–6 B.

45. $\dfrac{2}{x} + \dfrac{1}{y} = 3$ for x

46. $\dfrac{5}{I} - 6 = \dfrac{8}{E}$ for I

47. $\dfrac{1}{c} = \dfrac{1}{c_1} + \dfrac{1}{c_2}$ for c_1

48. $\dfrac{1}{R} = \dfrac{1}{R_1} + \dfrac{1}{R_2} + \dfrac{1}{R_3}$ for R

49. $\dfrac{1}{8} = \dfrac{1}{a} + \dfrac{1}{b}$ for a

50. $\dfrac{1}{x} - \dfrac{1}{y} - \dfrac{1}{z} = 6$ for y

51. $\dfrac{3}{a} - \dfrac{4}{b} = \dfrac{5}{ab}$ for a

52. $\dfrac{5}{m} + 4 = \dfrac{6a}{2m} + 3b$ for m

53. $\dfrac{x + 4}{2} + \dfrac{y - 3}{5} = \dfrac{2}{10}$ for x

Directions   Solve the following problems. See example 5–6 C.

54. The principal amount of money P in a savings account paying interest rate r, over a given period of time t, which pays interest I is given by $P = \dfrac{I}{rt}$. Solve this equation for r.

55. The pressure per square inch of steam or water in a pipe, p, is given by $p = \dfrac{P}{LD}$, where

P = the total pressure on a diametral plane,
L = the length of the pipe in inches, and
D = the diameter of the pipe in inches.
Solve for D.

56. The safe internal unit pressure, p, in a given pipe of given thickness is given by $p = \dfrac{2st}{D}$, where

s = the unit tensile stress,
t = the thickness of the pipe, and
D = the diameter of the pipe. Solve for s.

57. The pitch diameter of a gear, P, is given by $P = \dfrac{D \cdot N}{N + 2}$, where D = the outside diameter of the gear and N = the number of teeth in the gear. Solve for N.

58. The rule governing the speeds of two gears, one the driver gear and the other the driven gear, is given by $\dfrac{T_A}{T_B} = \dfrac{R_B}{R_A}$, where T_A = the number of teeth in the driven gear, T_B = the number of teeth in the driver gear, T_B = the revolutions per minute of the driven gear, and R_A = the revolutions per minute of the driver gear. Solve for T_B.

59. Given F is the force on the large piston and f is the force on the small piston of a hydraulic press, then $\dfrac{F}{f} = \dfrac{A}{a}$, where A is the area of the large piston and a is the area of the small piston. Solve for f.

60. The kinetic energy of a body, E, is computed by $E = \dfrac{Wv^2}{2g}$, where v is the velocity expressed in feet per second, W is the weight in pounds, and g is acceleration due to gravity. Solve for W.

61. Charles' law is in regard to the relationship between pressure P, volume V, and the absolute temperature T of a gas as it expands when heated. Charles' law is represented by $\dfrac{P_1 V_1}{T_1} = \dfrac{P_2 V_2}{T_2}$, where P_1, V_1, and T_1 are the initial conditions of the gas and P_2, V_2, and T_2 are the final conditions. Solve for T_2.

62. The coefficient of linear expansion, k, of a solid when heated is given by $k = \dfrac{L_t - L_0}{L_0 t}$, where L_0 is the length at $0°$ C, L_t is the length at $t°$ C, and t is any given temperature in Celsius. Solve for L_0.

63. The coefficient of linear expansion α (alpha) when metal is heated is given by $\alpha = \dfrac{L_2 - L_1}{L_1(t_2 - t_1)}$. Solve for L_1.

64. Carnot's ideal efficiency of any heat engine operating between the temperature limits T_1 and T_2 is given by $E = \dfrac{T_1 - T_2}{T_1}$. Solve for T_1.

65. A formula for resistors in parallel is given by $R = \dfrac{R_1 R_2}{R_1 + R_2}$. Solve for R_1.

66. In electronics, a formula for the unknown resistance R_x in a battery is given by $R_x = R_m \left(\dfrac{E_1}{E_2} - 1 \right)$. Solve for E_2.

67. In your own words, explain what it means when two equations are equivalent.

68. In your own words, what is the solution set?

69. In your own words, explain how to check your solution.

70. In your own words, explain what an extraneous solution is.

71. Simplify $\frac{x}{3} + \frac{2x}{7} + 4$ and solve $\frac{x}{3} + \frac{2x}{7} = 4$. Are the results the same? Why or why not? Include in your explanation the effect of the use of the LCD in each procedure.

⊛ Explore to Learn

In most computer languages / indicates division. To indicate $\frac{x}{x-1}$, most computer languages would write X/(X−1). These languages obey the order of operations as given in chapter 1, and blank spaces have no meaning, just as with algebraic expressions. Show what mathematical expression is being computed for each of the following programming language expressions.

a. 2X/X+3

b. 5X+1/2X−1

c. (3X−1)(X+3)/(X+1)(X−2)

d. 5X/(X−1)/(X+1)

e. (X+1)/(X+2)/(X+3)/(X+4)

Review Exercises

Directions Evaluate each expression for the given values. See sections 1–9.

1. $5x - 3y + z$ when $x = 1$, $y = -2$, and $z = 3$

2. $\frac{3a - b}{2a + b}$ when $a = 4$ and $b = -5$

3. $\frac{y_1 - y_2}{x_1 - x_2}$ when $x_1 = 3$, $x_2 = 1$, $y_1 = 3$, and $y_2 = -5$

Directions Solve each equation for y. See section 2–6.

4. $5x + y = 4$

5. $2x + 3y = 6$

6. $x - 4y = 8$

Directions Simplify the following expressions. Assume all denominators are nonzero. Answer with positive exponents only. See section 3–4.

7. $(3y^{-1})(y^2 x^{-2})$

8. $\frac{x^{-3} y^3}{x^2 y^{-1}}$

9. The width of a rectangle is 3 meters less than the length. If the perimeter of the rectangle is 126 meters, find the dimensions of the rectangle. See section 2–7.

5–7 Rational Expression Applications

Rational expressions occurring in rational equations have many applications in the physical and scientific world. We now wish to discuss some of the more common applications.

● **Example 5–7 A**
  Problem solving

Choose a variable, set up an appropriate equation, and solve the following problems.

1. A water holding tank is fed by two pipes. If it takes the smaller pipe 12 hours to fill the tank and the larger pipe 9 hours to fill the tank, how long would it take to fill the tank if both pipes are open? (This is called a *work* problem.)

   Let x represent the number of hours it takes the two pipes to fill the tank. Now,

   a. the smaller pipe can fill $\dfrac{1}{12}$ of the tank in 1 hour,

   b. the larger pipe can fill $\dfrac{1}{9}$ of the tank in 1 hour,

   c. the two pipes together can fill $\dfrac{1}{x}$ of the tank in 1 hour.

   The amount of the tank filled by the smaller pipe in 1 hour plus the amount of the tank filled by the larger pipe in 1 hour must be equal to the amount of the tank filled by the two pipes together in 1 hour. Thus, the equation is

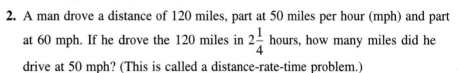

   $$\underset{\substack{\text{Amount} \\ \text{by smaller} \\ \text{pipe}}}{\longrightarrow}\; \frac{1}{12} + \overset{\substack{\text{Amount by larger pipe} \\ \downarrow}}{\frac{1}{9}} = \frac{1}{x} \longleftarrow \text{Amount together}$$

   Multiply each term by the LCD, which is $36x$.

   $$36x \cdot \frac{1}{12} + 36x \cdot \frac{1}{9} = 36x \cdot \frac{1}{x}$$

   $$3x + 4x = 36 \qquad \text{Reduce in each term}$$
   $$7x = 36 \qquad \text{Combine on left side}$$
   $$x = \frac{36}{7} \qquad \text{Divide each side by 7}$$

   Therefore, together the two pipes can fill the tank in $\dfrac{36}{7}$ hours, or $5\dfrac{1}{7}$ hours, that is, approximately 5 hours and 9 minutes.

2. A man drove a distance of 120 miles, part at 50 miles per hour (mph) and part at 60 mph. If he drove the 120 miles in $2\dfrac{1}{4}$ hours, how many miles did he drive at 50 mph? (This is called a distance-rate-time problem.)

   <u>Note</u>   In a distance (d)-rate (r)-time (t) problem, we use $d = rt$, or the alternate forms of the equation solved for t or r. They are $t = \dfrac{d}{r}$, or $r = \dfrac{d}{t}$.

   Let x represent the distance he drove at 50 mph. Then $120 - x$ represents the distance he drove at 60 mph.

   The time traveled at 50 mph plus the time traveled at 60 mph equals the total time traveled, $2\dfrac{1}{4}$ hr $\left(\text{or } \dfrac{9}{4} \text{ hr}\right)$. We use the following table for distance-rate-time problems.

distance (*d*)	rate (*r*)	time (*t*)
x	50	$\dfrac{x}{50}$
$120 - x$	60	$\dfrac{120 - x}{60}$

$$t = \frac{d}{r}$$

The equation is then

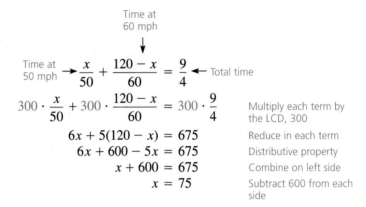

Time at 60 mph

Time at 50 mph → $\dfrac{x}{50} + \dfrac{120 - x}{60} = \dfrac{9}{4}$ ← Total time

$$300 \cdot \frac{x}{50} + 300 \cdot \frac{120 - x}{60} = 300 \cdot \frac{9}{4} \qquad \text{Multiply each term by the LCD, 300}$$

$$6x + 5(120 - x) = 675 \qquad \text{Reduce in each term}$$

$$6x + 600 - 5x = 675 \qquad \text{Distributive property}$$

$$x + 600 = 675 \qquad \text{Combine on left side}$$

$$x = 75 \qquad \text{Subtract 600 from each side}$$

Thus, the man drove 75 miles at 50 mph.

Note He drove $120 - x = 120 - 75 = 45$ miles at 60 mph.

3. The denominator of a fraction is 3 more than the numerator. If 4 is added to the numerator and the denominator, the resulting fraction is $\dfrac{3}{4}$. Find the original fraction.

Let *x* represent the numerator of the original fraction. Then $x + 3$ represents the denominator of the original fraction. The equation we get is

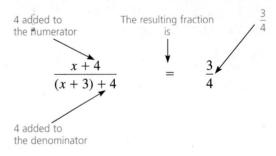

4 added to the numerator

The resulting fraction is

$\dfrac{3}{4}$

$$\frac{x + 4}{(x + 3) + 4} = \frac{3}{4}$$

4 added to the denominator

Then

$$\frac{x + 4}{x + 7} = \frac{3}{4}$$

The LCD is $4(x + 7)$.

$$4(x + 7) \cdot \frac{x + 4}{x + 7} = 4(x + 7) \cdot \frac{3}{4} (x \neq -7)$$

Multiply each term by $4(x + 7)$

$$4(x + 4) = (x + 7) \cdot 3$$

Reduce each term

$$4x + 16 = 3x + 21$$

Multiply as indicated

$$x = 5$$

Subtract $3x$ and 16 from each side

and $x + 3 = 8$

The original fraction is $\frac{5}{8}$.

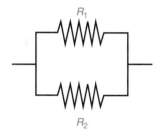

4. In an electrical circuit, when two resistors are connected in parallel, the total resistance R of the circuit in ohms is given by

$$\frac{1}{R} = \frac{1}{R_1} + \frac{1}{R_2}$$

where R_1 and R_2 are the resistances of the two resistors in ohms and the circuit is connected in parallel as shown in the diagram. Find the total resistance R of an electrical circuit having two resistors connected in parallel if their resistances are 4 ohms and 6 ohms.

We want R if $R_1 = 4$ ohms and $R_2 = 6$ ohms.

$$\frac{1}{R} = \frac{1}{4} + \frac{1}{6}$$

Multiply both sides by the LCD, which is $12R$.

$$12R \cdot \frac{1}{R} = 12R \cdot \frac{1}{4} + 12R \cdot \frac{1}{6}$$

$$12 = 3R + 2R$$

$$12 = 5R$$

$$\frac{12}{5} = R$$

Therefore the total resistance is $\frac{12}{5}$, or $2\frac{2}{5}$, ohms.

▶ **Quick check** If the same number is added to the numerator and the denominator of the fraction $\frac{3}{4}$, the resulting fraction is $\frac{5}{6}$. What is the number?

MASTERY POINTS

Can you
- Solve work problems?
- Solve distance-rate-time problems?
- Solve number problems?
- Solve for resistance in electric circuits problems?

Exercise 5–7

Directions

Choose a variable, set up an equation, and solve the following problems.

▶ *Quick Check*

Example   If the same number is added to the numerator and the denominator of the fraction $\frac{3}{4}$, the resulting fraction is $\frac{5}{6}$.

What is the number?

Solution   Let x represent the number to be added to the numerator and the denominator. The equation is then

$$\frac{3 + x}{4 + x} = \frac{5}{6}$$

$$6(4 + x) \cdot \frac{3 + x}{4 + x} = 6(4 + x) \cdot \frac{5}{6} \qquad \text{Multiply each side by the LCD } 6(4 + x)$$

$$6(3 + x) = (4 + x) \cdot 5 \qquad \text{Reduce in each side}$$

$$18 + 6x = 20 + 5x \qquad \text{Multiply as indicated}$$

$$x = 2 \qquad \text{Solve for } x$$

The number to be added is 2. $\left(Check : \frac{3 + 2}{4 + 2} = \frac{5}{6} \right)$

Directions   Solve the following problems. See example 5–7 A number 1.

1. Jim can mow his parents' lawn in 50 minutes. His younger brother Kenny can mow the lawn in 70 minutes. How long would it take for the boys to mow the lawn if they mow together?

2. In a factory, worker A can do a certain job in 6 hours while worker B can do the same job in 5 hours. How long would it take workers A and B to do the same job working together?

3. Three different sized pipes feed water into a swimming pool. If the pipes can fill the same pool individually in 6 hours, 8 hours, and 9 hours, respectively, how long would it take to fill the same pool if all three pipes were open?

4. During "clean-up week" in Dexter, Jane, Jack, and Ruth work to clean up a vacant lot. If the people can do the job individually in 2 hours, 3 hours, and 4 hours, respectively, how long will it take them, working together, to clean the lot?

5. It takes two painters working together 4 hours to paint the exterior of a house. If one painter could do the job in 6 hours working alone, how long would it take the other to paint the house alone?

6. A water tank has two drain pipes. If the larger pipe could empty the tank in 45 minutes and the two pipes together could drain the tank in 30 minutes, how long would it take for the smaller pipe to drain the tank?

7. Three combines—A, B, and C—working together can harvest a field of oats in $1\frac{1}{2}$ hours. If A could do the job alone in 5 hours and B could do it alone in 6 hours, how long would it take for combine C to do the same job alone?

8. In a pizzeria, three people are at work making pizzas. The three could make 50 pizzas in $1\frac{1}{2}$ hours working together. If two of the people could make all the pizzas in 5 hours and 6 hours, respectively, working alone, how long would it take the third person to make the 50 pizzas working alone?

9. A tank has one inlet pipe and one outlet pipe. If the inlet pipe can fill the tank in $3\frac{1}{3}$ hours and the outlet pipe can empty the tank in 5 hours, how long would it take to fill the tank if both pipes are left open? (*Hint*: Subtract the drainage.)

10. A sink drain, when left open, can empty a sink full of water in 4 minutes. If the cold water and hot water faucets can, when fully open, fill the sink in $2\frac{1}{2}$ minutes and 3 minutes, respectively, how long would it take to fill the sink if all three are open simultaneously?

Directions   Solve the following problems. See example 5–7 A number 2.

11. Rafi drove 320 miles in $5\frac{1}{2}$ hours. If he drove part of the trip averaging 55 mph and the rest averaging 60 mph, how many miles did he drive at each speed?

12. A. J. Foyt, when driving the Indianapolis 500 Race, averaged 198 mph over part of the race. Due to an accident on the track, he averaged 160 mph over the rest. If the race took 2 hours and 40 minutes to run, how many miles did he drive at an average of 198 mph? (Round the answer to the nearest integer.)

13. Car *A* travels 120 miles in the same time that car *B* travels 150 miles. If car *B* averages 10 mph faster than car *A*, how fast is each car traveling?

14. A freight train travels 260 kilometers in the same time a passenger train travels 320 kilometers. If the passenger train averages 15 kilometers per hour faster than the freight train, what was the average speed of the freight train?

15. Sheila can row a boat 2 mph in still water. How fast is the current of a river if she takes the same length of time to row 4 miles upstream as she does to row 10 miles downstream? (*Hint*: Subtract current upstream and add downstream.)

16. An airplane flew 1,000 miles with the wind in the same length of time it took to fly 850 miles against the wind. If the wind was blowing at 25 miles per hour, what was the average air speed of the plane?

17. On a trip from Detroit to Columbus, Ohio, Mrs. Smith drove at an average speed of 60 mph. Returning, her average speed was 55 mph. If it took her $\frac{1}{3}$ hour longer on the return trip, how far is it from Detroit to Columbus?

18. A jet plane flew at an average speed of 240 miles per hour going from city *A* to city *B* and averaged 300 miles per hour on the return flight. Its return flight took 1 hour and 40 minutes less time. How far is it from city *A* to city *B*? (Disregard any wind.)

Directions Solve the following problems. See example 5–7 A number 3.

19. The numerator of a given fraction is 4 less than the denominator. If 5 is added to both the numerator and the denominator, the resulting fraction is $\frac{5}{7}$. What is the original fraction?

20. The denominator of a fraction exceeds the numerator by 7. If 3 is added to the numerator and 1 is subtracted from the denominator, the resulting fraction is $\frac{4}{5}$. Find the original fraction.

21. One number is four times another number. The sum of their reciprocals is $\frac{5}{12}$. What are the numbers?

22. One number is four times another number. The sum of their reciprocals is $\frac{1}{4}$. What are the numbers?

23. If $\frac{1}{2}$ is added to three times the reciprocal of a number, the result is 1. Find the number.

24. If $\frac{1}{2}$ is subtracted from four times the reciprocal of a number, the result is 0. Find the number.

25. When a certain number is added to the numerator and subtracted from the denominator of the fraction $\frac{2}{5}$, the result is 6. What is the number?

26. When a certain number is added to the numerator and subtracted from the denominator of the fraction $\frac{5}{9}$, the result is $\frac{4}{7}$. What is the number?

27. A prescription for an illness calls for a child's dosage to be $\frac{3}{5}$ of the number of pills that an adult takes. Together they use 24 pills. How many pills are taken by each?

28. An apprentice electrician receives $\frac{3}{8}$ of the hourly wage of a journeyman electrician. Together their hourly wage is $29.70. Find the hourly wage of each.

29. Pat and Mike earn a total of $65.50 per week delivering papers. If Pat's earnings are $\frac{5}{7}$ of Mike's, how much does Mike earn each day?

Directions Use the formula $\frac{1}{R} = \frac{1}{R_1} + \frac{1}{R_2}$ to solve problems 30–33. See example 5–7 A number 4.

30. Two resistors in an electric circuit have resistances of 6 ohms and 8 ohms and are connected in parallel. Find the total resistance of the circuit.

31. Two resistors of an electric circuit are connected in parallel. If one has a resistance of 5 ohms and the other has a resistance of 12 ohms, what is the total resistance in the circuit?

32. The total resistance in a parallel wiring circuit is 12 ohms. If the resistance in one branch is 30 ohms, what is the resistance in the other branch?

33. The resistance in one branch of a two-resistor parallel wiring circuit is 10 ohms. If the total resistance in the circuit is 6 ohms, what is the resistance in the other branch?

34. Three resistors connected in parallel have resistances of 4 ohms, 6 ohms, and 10 ohms. What is the total resistance in the electric circuit? $\left(\textit{Hint}: \text{Use } \frac{1}{R} = \frac{1}{R_1} + \frac{1}{R_2} + \frac{1}{R_3}.\right)$

35. A three-resistor parallel wiring circuit has a total resistance of 10 ohms. If two of the branches of the circuit have resistances of 20 ohms and 30 ohms, what is the resistance in the third branch? (See hint in exercise 34.)

✸ Explore to Learn

An enormous tank in an oil tanker (ship) can be filled by two pipes (pipe 1 and pipe 2) in 14 hours 20 minutes. Because of a stuck valve, the tank was once filled by pipe 1 alone. It took 35 hours 30 minutes that time. Today pipe 1's valve was stuck, so pipe 2 was used to begin filling the tank. After 6 hours pipe 1's valve was fixed. At this point the captain asked the engineer in charge to estimate how much more time it would take to fill the tank. What is this time, to the nearest minute?

Review Exercises

Directions   Find the solution set of the following equation. See section 2–5.

1. $2y + 3 = 4y - 1$

Directions   Solve for y. See section 2–6.

2. $2y - 3x = 6$

Directions   Factor completely the following expressions. See sections 4–2, 4–3, and 4–4.

3. $8y^2 - 32$ **4.** $x^2 + 20x + 100$ **5.** $3y^2 - 4y - 4$

Directions   Combine the following rational expressions. See section 5–4.

6. $\dfrac{3x}{x - 1} + \dfrac{2x}{x + 3}$

7. $\dfrac{4y}{2y + 1} - \dfrac{3y}{y - 5}$

 8. Susan had $19,000, part of which she invested at 9% interest and the rest at 7%. If her income from the 7% investment was $40 more than that from the 9% investment, how much did she invest at each rate? See section 2–7.

9. Donald made two investments totaling $17,500. One investment made him a 13% profit, but on the other investment he took a 9% loss. If his net loss was $475, how much was each investment? See section 2–7.

Chapter 5 Lead-in Problem

Marc owns $\frac{5}{8}$ share of a print shop and his uncle owns $\frac{1}{4}$ share of the shop. In a given year, they had a combined earnings of $140,000 from the print shop. How much did the shop earn that year?

	Share	Shop earnings	Income
Marc	$\frac{5}{8}$	x	$\frac{5}{8}x$
Uncle	$\frac{1}{4}$	x	$\frac{1}{4}x$

Solution

Let x represent the total earnings of the print shop that year.

$$\text{Then } \frac{5}{8}x + \frac{1}{4}x = 140{,}000 \qquad \text{Together they earned \$140,000}$$

$$8 \cdot \frac{5}{8}x + 8 \cdot \frac{1}{4}x = 8 \cdot 140{,}000 \qquad \text{Multiply each term by the LCD, which is 8}$$

$$5x + 2x = 1{,}120{,}000 \qquad \text{Perform indicated multiplications}$$

$$7x = 1{,}120{,}000 \qquad \text{Combine like terms}$$

$$x = 160{,}000 \qquad \text{Divide each side by 7}$$

The shop earned $160,000 that year.

Chapter 5 Summary

• Glossary

complex fraction (complex rational expression)
(page 313) a fraction (rational expression) whose
numerator, denominator, or both contain fractions
(or rational expressions).

domain of a rational expression (page 282) the set
of all replacement values of the variable for which
a rational expression is defined.

extraneous solution of a rational equation (page 321)
a solution that is not in the domain of the variable.

least common denominator (LCD) (page 300) the
smallest (least) number or expression that is
exactly divisible by each of the denominators.

rational equation (page 319) an algebraic equation
that contains at least one rational expression.

rational expression (page 281) an expression that can
be written as the quotient of two polynomials with
the denominator not zero.

reduced to lowest terms (page 283) when the
greatest factor common to both the numerator
and the denominator is 1 or -1.

restrictions on the variable (page 282) replacement
values for the variable that make the denominator
zero.

• Properties and Definitions

Rational expressions (page 281)
A rational expression is an expression of the form

$$\frac{P}{Q}$$

where P and Q are polynomials, with $Q \neq 0$.

Domain of a rational expression (page 282)
The set of all replacement values of the variable for
which a rational expression is defined determines the
domain of the rational expression.

Fundamental principle of rational expressions
(page 283)
If P is any polynomial and Q and R are nonzero
polynomials, then

$$\frac{PR}{QR} = \frac{P}{Q} \quad \text{and} \quad \frac{P}{Q} = \frac{PR}{QR}$$

Reducing by opposite factors (page 285)
In general, for all real numbers a and b, $a \neq b$,

$$\frac{a - b}{b - a} = -1$$

Multiplication property of fractions (page 290)
If a, b, c, and d are real numbers, then

$$\frac{a}{b} \cdot \frac{c}{d} = \frac{a \cdot c}{b \cdot d} \quad (b, d \neq 0)$$

Multiplication property of rational expressions
(page 290)

Given rational expressions $\dfrac{P}{Q}$ and $\dfrac{R}{S}$, then

$$\frac{P}{Q} \cdot \frac{R}{S} = \frac{P \cdot R}{Q \cdot S} \quad (Q, S \neq 0)$$

Division property of fractions (page 292)
If a, b, c, and d are real numbers, then

$$\frac{a}{b} \div \frac{c}{d} = \frac{a}{b} \cdot \frac{d}{c} = \frac{a \cdot d}{b \cdot c} \quad (b, c, d \neq 0)$$

Division property of rational expressions (page 292)

If $\dfrac{P}{Q}$ and $\dfrac{R}{S}$ are rational expressions, then

$$\frac{P}{Q} \div \frac{R}{S} = \frac{P}{Q} \cdot \frac{S}{R} = \frac{P \cdot S}{Q \cdot R} \quad (Q, R, S \neq 0)$$

Addition and subtraction properties for fractions
(page 297)
If a, b, and c are real numbers, $b \neq 0$, then

$$\frac{a}{b} + \frac{c}{b} = \frac{a + c}{b} \quad \text{and} \quad \frac{a}{b} - \frac{c}{b} = \frac{a - c}{b}$$

**Addition and subtraction properties for rational
expressions** (page 297)

If $\dfrac{P}{R}$ and $\dfrac{Q}{R}$ are rational expressions, $R \neq 0$, then

$$\frac{P}{R} + \frac{Q}{R} = \frac{P + Q}{R} \quad \text{and} \quad \frac{P}{R} - \frac{Q}{R} = \frac{P - Q}{R}$$

A rational equation (page 319)
An algebraic equation that contains at least one rational
expression is called a rational equation.

• **Procedures**

[5–1]

Finding the domain of a rational expression (page 282)

1. Factor the denominator into a product of prime polynomials, if possible.

2. Set each factor of the denominator containing the variable equal to zero (using the zero product property).

3. Solve the resulting equations. The solutions are the restrictions placed on the variable.

4. The domain is all real numbers excluding any restricted values found in step 3.

To reduce a rational expression to lowest terms (page 284)

1. Write the numerator and the denominator in factored form.

2. Divide the numerator and the denominator by all common factors.

[5–2]

Multiplication of rational expressions (page 291)

1. State the numerators and denominators as indicated products. (Do not multiply.)

2. Factor the numerator and the denominator.

3. Divide the numerators and the denominators by the factors that are common.

4. Multiply the remaining factors in the numerator and place this product over the product of the remaining factors in the denominator.

Division of rational expressions (page 292)

1. Multiply the first rational expression by the reciprocal of the second.

2. Proceed as in the multiplication of rational expressions.

[5–3]

Addition and subtraction of rational expressions with a common denominator (page 297)

1. Add or subtract the numerators.

2. Place the sum or difference over the common denominator.

3. Reduce the resulting rational expression to lowest terms.

Finding the least common denominator (page 300)

1. Completely factor each denominator. Write each factorization using exponential notation.

2. Choose the greatest power of each different factor. The product of these factors is the LCD.

[5–4]

Addition and subtraction of rational expressions having different denominators (page 305)

1. Find the LCD of the rational expressions.

2. Write each rational expression as an equivalent rational expression with the LCD as the denominator.

3. Perform the indicated addition or subtraction as before.

4. Reduce the results to lowest terms.

[5–5]

Simplifying a complex fraction (page 313)

Method 1 Find the LCD of all fractions within the complex fraction, then multiply the numerator and the denominator of the complex fraction by this LCD. Reduce the resulting fraction to lowest terms.

Method 2 Form a single fraction in the numerator and in the denominator. Multiply the fraction in the numerator by the reciprocal of the fraction in the denominator and simplify.

[5–6]

Solving a rational equation (page 319)

1. Find the LCD of all the rational expressions.

2. Eliminate the denominators by multiplying each term of both sides of the equation by the LCD.

3. Use the four steps from section 2–5 to solve the resulting equation.

Solving a rational equation in more than one variable (page 321)

1. Remove the fractions by multiplying each side of the equation by the LCD of all the rational expressions in the equation.

2. Collect all terms containing the variable you are solving for on one side of the equation and all other terms on the other side.

3. Factor out the variable you are solving for if it appears in more than one term.

4. Divide each side by the coefficient of the variable for which you are solving.

☀ Chapter 5 Error Analysis

1. Finding the domain of a rational expression

Example: $\dfrac{3x+1}{x^2+2x} = \dfrac{3x+1}{x(x+2)}$

Domain is all real numbers except -2.

Correct answer: Domain is all real numbers except 0 and -2.

What error was made? (*see page 283*)

2. Subtracting rational expressions

Example: $\dfrac{3x-1}{x-2} - \dfrac{x+2}{x-2}$

$$\dfrac{3x-1}{x-2} - \dfrac{x+2}{x-2} = \dfrac{3x-1-x+2}{x-2} = \dfrac{2x+1}{x-2}$$

Correct answer: $\dfrac{2x-3}{x-2}$

What error was made? (*see page 298*)

3. Solving rational equations

Example: Find the solution set of $\dfrac{y}{3} - \dfrac{y+1}{4} = \dfrac{2}{12}$

$$12 \cdot \dfrac{y}{3} - 12 \cdot \dfrac{y+1}{4} = 12 \cdot \dfrac{2}{12}$$
$$4y - 3(y+1) = 2$$
$$4y - 3y + 3 = 2$$
$$y = -1 \quad \{-1\}$$

Correct answer: $\{5\}$

What error was made? (*see page 320*)

4. Adding and subtracting rational expressions

Example: $\dfrac{3}{x-5} + \dfrac{5}{5-x}$

$$\dfrac{3}{x-5} + \dfrac{5}{5-x} = \dfrac{3x+5}{x-5} = \dfrac{8}{x-5}$$

Correct answer: $\dfrac{-2}{x-5}$ or $\dfrac{2}{5-x}$

What error was made? (*see page 299*)

5. Squaring a binomial

Example: $(x-6)^2 = (x)^2 - (6)^2 = x^2 - 36$

Correct answer: $x^2 - 12x + 36$

What error was made? (*see page 196*)

6. Reducing to lowest terms by "cancelling"

Example: $\dfrac{x-3}{x^2-9} = \dfrac{\overset{1}{\cancel{x}} - \overset{1}{\cancel{3}}}{\underset{x}{\cancel{x^2}} - \underset{3}{\cancel{9}}} = \dfrac{1-1}{x-3} =$

$$\dfrac{0}{x-3} = 0$$

Correct answer: $\dfrac{1}{x+3}$

What error was made? (*see page 284*)

7. Adding and subtracting rational expressions

Example: $\dfrac{3}{x^2} + \dfrac{2}{x^2}$

$$\dfrac{3}{x^2} + \dfrac{2}{x^2} = x^2 \cdot \dfrac{3}{x^2} + x^2 \cdot \dfrac{2}{x^2} = 3 + 2 = 5$$

Correct answer: $\dfrac{5}{x^2}$

What error was made? (*see page 319*)

8. Order of operations

Example: $3^2 + 12 \cdot 2 - 8 \div 2 = 17$

Correct answer: 29

What error was made? (*see page 71*)

9. Combining polynomials

Example: $(4a^2b - 2ab^2) - (a^2b + ab) = 3a^2b - ab^2$

Correct answer: $3a^2b - 3ab^2$

What error was made? (*see page 106*

10. Solving linear inequalities

Example: If $4 < -2x \le 6$, then $-2 < x \le -3$.

Correct answer: If $4 < -2x \le 6$, then $-2 > x \ge -3$.

What error was made? (*see page 166*)

11. Properties of exponents

Example: $(3x)^3 = 3x^3$

Correct answer: $27x^3$

What error was made? (*see page 188*)

12. Division using zero

Example: $\dfrac{0}{-3}$ is undefined

Correct answer: $\dfrac{0}{-3} = 0$

What error was made? (*see page 65*)

Chapter 5 Critical Thinking

1. If m and n represent two integers where $n > m$, how many integers are there from m to n?

2. Which integers can be written as the sum of three consecutive integers?
Example: $6 = 1 + 2 + 3$

3. Part 7 of Error Analysis shows a common mistake that certainly must be avoided. The mistake of multiplying the expression by the least common denominator is tempting, however, because it eliminates denominators and thereby simplifies the work. Can you think of a way to combine expressions under addition and subtraction that allows this "mistake" and arrives at the correct result?

4. A person is going climb a mountain along the single trail on the mountain. This person leaves the starting point at the base of the mountain at about 8 A.M. and arrives at the top at about 6 P.M. The climber stops for lunch and occasional rests, and does not walk at the same rate for the whole climb. The climber spends the night at the top, and at about 8 A.M. the next day, this person starts the descent. Taking more breaks and a lunch break as well, and not walking at the same rate all the time, the hiker reaches the starting point of the day before at about 6 P.M. The question posed is this: Is there any point along the path at which the hiker arrived at exactly the same time on both trips?

Chapter 5 Review

[5–1]

Directions   Determine the domain of the given rational expression.

1. $\dfrac{x + 1}{x}$

2. $\dfrac{y - 3}{y + 7}$

3. $\dfrac{3x + 1}{x - 9}$

4. $\dfrac{2z - 5}{3z + 2}$

5. $\dfrac{x}{5x - 3}$

6. $\dfrac{x^2 - x + 4}{x^2 + x - 12}$

7. $\dfrac{x^2 + 3x + 2}{x^2 - 1}$

Directions   Reduce the following rational expressions to lowest terms. Assume all denominators are nonzero.

8. $\dfrac{18ab^2}{6a^2b}$

9. $\dfrac{45x^2yz^3}{30xy^3z^2}$

10. $\dfrac{x^2 - 49}{x^2 + 14x + 49}$

11. $\dfrac{x^2 - 3x - 18}{x^2 + x - 42}$

12. $\dfrac{18a - 6b}{15a - 5b}$

13. $\dfrac{x^2 - y^2}{y - x}$

14. $\dfrac{3p^2 - 8p + 4}{5p^2 - 9p - 2}$

15. $\dfrac{2R^2 - 32}{6R^2 + 22R - 8}$

16. $\dfrac{20 - 9n + n^2}{8 + 2n - n^2}$

[5–2]

Directions   Perform the indicated multiplication and reduce the product to lowest terms. Assume all denominators are nonzero.

17. $\dfrac{15}{8a} \cdot \dfrac{12a}{5}$

18. $\dfrac{24b}{7a} \cdot \dfrac{21a^2}{8b^2}$

19. $\dfrac{x - 3y}{2x + y} \cdot \dfrac{2x - y}{x - 3y}$

20. $\dfrac{5x - 10}{x + 3} \cdot \dfrac{3x + 9}{15}$

21. $\dfrac{m^2 - n^2}{14x} \cdot \dfrac{35x^2}{5m + 5n}$

22. $\dfrac{y}{y^2 - 1} \cdot \dfrac{y + 1}{y^2 - y}$

23. $\dfrac{b^2 - 36}{b^2 + 12b + 36} \cdot \dfrac{2b + 12}{b - 6}$

24. $\dfrac{5 - x}{6a - 3} \cdot \dfrac{2a - 1}{x^2 - 25}$

25. $\dfrac{x^2 + x - 2}{x^2 - 4x - 12} \cdot \dfrac{x^2 - 7x + 6}{x^2 - 1}$

Directions   Find the indicated quotients and state the answer reduced to lowest terms. Assume all denominators are nonzero.

26. $\dfrac{14a}{9} \div \dfrac{7}{3}$

27. $\dfrac{10}{9b} \div \dfrac{35}{12b^2}$

28. $\dfrac{24ab}{7} \div \dfrac{16a^2b^2}{21}$

29. $\dfrac{2x - 1}{3x + 4} \div \dfrac{3x}{7x}$

30. $\dfrac{x + 6}{x^2 - 4} \div \dfrac{(x + 6)^2}{x + 2}$

31. $\dfrac{4a - 8}{2a + 1} \div \dfrac{6a - 12}{10a + 5}$

32. $\dfrac{x^2 + 16x + 64}{x^2 + 9x + 8} \div \dfrac{x^2 - 64}{x + 1}$

33. $\dfrac{1 - b}{(b + 3)^2} \div \dfrac{b^2 - 2b + 1}{b^2 - 9}$

34. $\dfrac{9a^2 + 15a + 6}{a^2 + 3a - 4} \div \dfrac{36a^2 - 16}{3a^2 + 10a - 8}$

[5–3]

Directions   Find the least common denominator of rational expressions having the given denominators. Assume all denominators are nonzero.

35. $9x^2$ and $12x$

36. $14ab$ and $21a^2b^2$

37. $x^2 - 9$ and $2x + 6$

38. $y^2 - 2y - 15$ and $y^2 - 25$

39. $4z - 8$, $z^2 - 4$, and $z + 1$

40. $x^2 + x$, $x^2 + 2x + 1$, and $3x^2 - 2x - 5$

Directions   Find the indicated sum or difference. Assume all denominators are nonzero.

41. $\dfrac{4}{x} + \dfrac{1}{-x}$

42. $\dfrac{y}{y - 2} - \dfrac{3y}{2 - y}$

43. $\dfrac{3x - 2}{3x - 1} + \dfrac{x + 6}{3x - 1}$

44. $\dfrac{x + 7}{2x - 5} - \dfrac{2x - 6}{2x - 5}$

45. $\dfrac{2x + 1}{x^2 - 9} - \dfrac{x - 2}{x^2 - 9}$

46. $\dfrac{x + 1}{x^2 + 4x + 4} - \dfrac{1}{x^2 + 4x + 4}$

[5–4]

Directions   Add or subtract as indicated and reduce the answer to lowest terms. Assume all denominators are nonzero.

47. $\dfrac{4x}{15} + \dfrac{8x}{21}$

48. $\dfrac{25}{16a} - \dfrac{13}{12a}$

49. $\dfrac{5}{3x + 1} + \dfrac{9}{4x - 3}$

50. $\dfrac{4x}{x + 4} - \dfrac{7x}{x^2 - 16}$

51. $\dfrac{4}{ab^2} + \dfrac{12}{5a^2b} - \dfrac{3}{4ab}$

52. $4a + \dfrac{6}{a+1}$

53. $\dfrac{7}{x^2+1} - 10$

54. $\dfrac{x+1}{4x} - \dfrac{3x-5}{8x^2} + \dfrac{5x+4}{12x}$

55. $\dfrac{4y}{y^2-7y-18} + \dfrac{9y}{y^2-4}$

56. $\dfrac{5a}{x+3} + \dfrac{2}{x^2-9} - \dfrac{7a}{x-3}$

57. An outlet pipe drains a water tank in 20 hours. What part of the tank can it drain in h hours?

58. The area of a rectangle is 84 square meters. What is the width w if the length is ℓ meters?

[5–5]

Directions   Simplify the given complex fractions and reduce to lowest terms. Assume all denominators are nonzero.

59. $\dfrac{\dfrac{4}{7}}{\dfrac{9}{4}}$

60. $\dfrac{\dfrac{4}{5}+1}{2-\dfrac{1}{5}}$

61. $\dfrac{\dfrac{1}{x}-\dfrac{1}{y}}{\dfrac{1}{x}+\dfrac{1}{y}}$

62. $\dfrac{\dfrac{4}{x^2}+\dfrac{3}{x}}{\dfrac{2}{x^2}-\dfrac{5}{x}}$

63. $\dfrac{\dfrac{a}{b}-\dfrac{b}{a}}{\dfrac{a}{b}+\dfrac{b}{a}}$

64. $\dfrac{\dfrac{1}{x}-\dfrac{1}{y}}{\dfrac{1}{xy}}$

65. $\dfrac{\dfrac{1}{x}-\dfrac{1}{y}}{\dfrac{1}{x^2}-\dfrac{1}{y^2}}$

[5–6]

Directions   Find the solution set of the following rational equations. Assume all denominators are nonzero.

66. $\dfrac{x}{8} - 3 = \dfrac{2x}{12} + 1$

67. $\dfrac{4a+1}{4} + \dfrac{5}{6} = 3a - \dfrac{5}{8}$

68. $\dfrac{12}{4a} - \dfrac{5}{6a} = 4$

69. $\dfrac{7}{a+1} - \dfrac{1}{a^2-1} = \dfrac{2}{a+1}$

70. $\dfrac{1}{y+3} - \dfrac{6}{y} = \dfrac{7}{y} + \dfrac{2}{y+3}$

71. $\dfrac{7}{2c^2} + \dfrac{5}{6c^2} = \dfrac{7}{12c^3}$

72. $\dfrac{x}{x+4} + 4 = \dfrac{-4}{x+4}$

73. $\dfrac{2}{3}y^2 = \dfrac{3}{2}$

74. $\dfrac{x^2}{2} - \dfrac{15}{2} = -x$

75. $\dfrac{a^2}{3} = \dfrac{5a}{2} - 3$

Directions   Solve for the indicated variable. Assume all denominators are nonzero.

76. $\dfrac{a}{x} + \dfrac{b}{x} = 3$ for x

77. $\dfrac{y}{x+1} = \dfrac{y^2}{x-3}$ for x

78. $\dfrac{3}{4-y} = \dfrac{a}{b}$ for y

79. $\dfrac{4}{y-c} - \dfrac{5}{y+b} = 0$ for y

Directions   Solve the following problems. Assume all denominators are nonzero.

80. An equation for tensile and compressional stresses is given by $y = \dfrac{F\ell}{Ae}$. Solve for ℓ.

81. The efficiency of a screw jack is calculated by the formula $E = \dfrac{Wp}{2 + LF}$. Solve for L.

82. A formula for the theoretical mechanical advantage, M, of a chain fall is given by $M = \dfrac{2R}{R - r}$. Solve for R.

83. The total reaction force F of air against a plane at the bottom of a vertical loop in centripetal force is given by $F = \dfrac{mv^2}{r} + m \cdot g$. Solve for m.

[5–7]

84. With different equipment, one painter can paint a house three times faster than a second painter. Working together they can do it in 4 hours. How long would it take each of them to paint the house working alone?

85. A man can row his boat at a rate of 4 miles per hour in still water. It takes him as long to row 20 miles downstream as it takes him to row 8 miles upstream. What is the rate of the current?

86. The sum of three times a number and twice its reciprocal is 5. Find the number.

Chapter 5 Test

Directions   Determine the domain of the given rational expression.

1. $\dfrac{a + 6}{a - 4}$

2. $\dfrac{x - 5}{x^2 + 2x}$

Directions   Find the least common denominator (LCD) of the rational expressions having the following denominators.

3. $4a^2$, $6a^3$, and $8a$

4. $x^2 - 4$, $x^2 - 4x + 4$

Directions   Perform the indicated operations and reduce to lowest terms. Assume all denominators are nonzero.

5. $\dfrac{x^2 + 7x + 10}{x^2 - 4}$

6. $\dfrac{14a^2b^3}{10xy^4} \div \dfrac{21ab^3}{6x^2y^2}$

7. $\dfrac{x - 4}{x^2 - 9} + \dfrac{7}{x^2 - 9}$

8. $\dfrac{x + 2}{9x} + \dfrac{x - 1}{3x^2}$

9. $\dfrac{\dfrac{1}{4} + \dfrac{3}{8}}{\dfrac{3}{4} - \dfrac{1}{2}}$

10. $\dfrac{3a + 1}{2a - 8} - \dfrac{a - 2}{a - 4}$

11. $\dfrac{2a - 1}{a^2 + 5a + 4} - \dfrac{a - 2}{a^2 + 5a + 4}$

12. $\dfrac{\dfrac{1}{2a} + \dfrac{1}{b}}{\dfrac{1}{2b} - \dfrac{1}{a}}$

13. $\dfrac{y}{2y + 2} + \dfrac{3}{y^2 - 1}$

14. $\dfrac{2x + 6}{x^2 + 7x + 10} \cdot \dfrac{x^2 + 10x + 25}{4x + 12}$

15. $\dfrac{3x - 6}{8 - 4x}$

16. $\dfrac{y^2 - 5y + 6}{y^2 + y - 6} \div \dfrac{y^2 - 4y + 3}{y^2 + 5y + 4}$

17. $\dfrac{a - \dfrac{9}{a}}{\dfrac{a^2 - 9}{a^2}}$

18. $\dfrac{a^2 - a - 12}{a^2 + 10a + 21} \cdot \dfrac{a^2 + 9a + 14}{a^2 - 5a + 4}$

19. Tara can wax her car in h hours. What part of the car can she wax in 1 hour?

Directions   Find the solution set for each rational equation.

20. $\dfrac{5}{a^2 - 9} = \dfrac{-2}{a^2 + 5a + 6}$

21. $x^2 = \dfrac{5}{6}x + \dfrac{2}{3}$

22. Solve $\dfrac{6}{x} = 4 + \dfrac{7}{y}$ for y

23. One number is twice another number. The sum of their reciprocals is $\dfrac{15}{8}$. Find the numbers.

24. Tanya, Bill, and Neysa work in a bakery and can mix 12 loaves of bread individually in 36 minutes, 40 minutes, and 30 minutes, respectively. If they worked together, in how many minutes could they mix the 12 loaves?

Chapter 5 Cumulative Test

Directions   Simplify the following expressions with only positive exponents.

1. $x^8 \cdot x^3 \cdot x^0$

2. $\dfrac{y^{-5}}{y^4}$

3. $(-5x^2y^{-3})^2$

4. $(9x^3 - 3x^2 + 4x - 8) - (-x^3 + x^2 - 7x + 1)$

Directions   Solve the inequalities and find the solution set of the equations.

5. $-5(2x + 5) = -3x + 1$

6. $\dfrac{1}{3}x - \dfrac{3}{4} = 6$

7. $2y^2 - y = 6$

8. $3x - 4 \le 5x + 6$

Directions   Completely factor each expression.

9. $2x^2 - 18$

10. $6x^5 - 36x^3 + 9x^2$

11. $4x^2 + 16x + 15$

12. $3 - 12x^6$

Directions Perform the indicated operations.

13. $(7x - 6)^2$

14. $(2x - 3)(3x^2 - 5x + 11)$

15. $(3 - 2x^2)(5 - 4x^2)$

16. Find x when $\dfrac{x}{24} = \dfrac{5}{7}$.

17. Power (in foot-pounds per minute) is given by the ratio of work, w, to time, t. Find the power exerted by pushing an 80-pound load 180 feet in 6 minutes. (*Hint:* $w = $ weight $\cdot$ length.)

Directions Perform the indicated operations and reduce the answers to lowest terms. Assume all denominators are nonzero.

18. $\dfrac{a^2 - 36}{a^2 - a - 42}$

19. $\dfrac{y^2 - y - 20}{y^2 - 25}$

20. $\dfrac{y^2 - 4}{y + 3} \cdot \dfrac{y^2 - 9}{y + 2}$

21. $\dfrac{2}{x^2 y} + \dfrac{4}{xy^2}$

22. $\dfrac{6}{x - y} - \dfrac{3}{x + y}$

23. $\dfrac{7}{x^2 - 49} + \dfrac{6}{x^2 - 5x - 14}$

Directions Simplify the following complex fractions. Assume all denominators are nonzero.

24. $\dfrac{3 - \dfrac{1}{y}}{4 + \dfrac{5}{y}}$

25. $\dfrac{\dfrac{1}{x} + \dfrac{1}{y}}{\dfrac{3}{y} - \dfrac{4}{x}}$

Directions Find the solution set of the following equations.

26. $\dfrac{4}{x} = \dfrac{6}{2x - 1}$

27. $x^2 - 2 = \dfrac{17x}{3}$

28. Solve the equation $\dfrac{1}{f} = \dfrac{1}{p} + \dfrac{1}{q}$ for q.

Directions Choose a variable, set an appropriate equation, and solve the following problems.

29. Tina can build a fence in 6 days and Ashley can build the same fence in 4 days. How long would it take them to build the fence working together?

30. The denominator of a fraction is one more than two times the numerator. If 2 is added to the numerator and 2 is subtracted from the denominator, the resulting fraction is $\dfrac{6}{7}$. Find the original fraction.

CHAPTER

6

Linear Equations in Two Variables

Tickets to a football game at Chelsea University cost $8 for students and $17 for nonstudents. The total receipts for a game were $165,575. Write an equation using **x** for the number of students and **y** for the number of nonstudents who purchased tickets to the game.

6–1 Ordered Pairs and the Rectangular Coordinate System

Linear Equations in Two Variables

In chapter 2, we developed methods for solving linear equations (first-degree equations) in one variable. All such equations could be stated in the form

$$ax + b = 0$$

where a and b are real number constants and $a \neq 0$. In this chapter, we expand our work to linear equations in *two variables*.

Definition

A **linear equation in two variables** x and y is any equation that can be written in the form

$$ax + by = c$$

where a, b, and c are real numbers and a and b are not both zero.

Our primary concern with all equations is finding their **solution(s)**. These are the replacement values for the variable(s) that satisfy the equation. In an equation in two variables, x and y, any *pair* of values for x and y that satisfies the equation is a solution of the equation.

343

Ordered Pairs of Numbers

The pairs of values for x and y may be written as a pair of numbers. We separate them by a comma and place them inside the parentheses. The value of x is *always* given first. That is, the pair of numbers is written (x, y).

Pairs of numbers written in this special *order* (with the x value always first) are called **ordered pairs of numbers**. The *first number* of the ordered pair (the value of x) is called the *first component* of the ordered pair. The *second number* (the value of y) is called the *second component* of the ordered pair.

● **Example 6–1 A**

Determining if an ordered pair is a solution

Given the linear equation $3x + 2y = 6$, determine if the given ordered pair is a solution of the equation.

1. $(2, 0)$, which represents $x = 2$ and $y = 0$

$$3x + 2y = 6$$
$$3(2) + 2(0) = 6 \qquad \text{Replace } x \text{ with 2 and } y \text{ with 0 in the equation}$$
$$6 + 0 = 6 \qquad \text{Multiply as indicated}$$
$$6 = 6 \qquad \text{(True)}$$

We see that the given values satisfy the equation, so $x = 2$ and $y = 0$ is a solution. That is, $(2, 0)$ is a solution.

2. $(3, 1)$, which represents $x = 3$ and $y = 1$

$$3x + 2y = 6$$
$$3(3) + 2(1) = 6 \qquad \text{Replace } x \text{ with 3 and } y \text{ with 1 in the equation}$$
$$9 + 2 = 6$$
$$11 = 6 \qquad \text{(False)}$$

We conclude that $x = 3$ and $y = 1$ *do not* satisfy the equation and hence *do not form a solution*. That is, $(3, 1)$ is not a solution.

☞ *Quick check* Given the linear equation $3x + y = 2$, determine if $(1, -1)$ and $(2, 0)$ are solutions of the equation. ●

Finding Ordered Pair Solutions

To determine ordered pairs that are solutions of an equation in two variables, we use the following procedure:

Finding Ordered Pair Solutions

1. Choose a value for one of the variables.
2. Replace the variable with this known value and solve the resulting equation for the other variable.

● **Example 6–1 B**

Finding the other component of an ordered pair

Using the given value of one of the variables, find the value of the other variable. Write the ordered pair solution of the equation, $y = 2x + 1$.

1. Let $x = 3$ 　　　　　　　　　Select any value for x

$$y = 2x + 1$$
$$y = 2(3) + 1$$ 　　　Replace x with 3 in the equation
$$y = 6 + 1$$ 　　　Multiply as indicated
$$y = 7$$ 　　　Add on the right side

The ordered pair (3,7) is a solution.

2. Let $x = -\dfrac{1}{2}$ 　　　　　Select any other value for x

$$y = 2x + 1$$
$$y = 2\left(-\dfrac{1}{2}\right) + 1$$ 　　Replace x with $-\dfrac{1}{2}$ in the equation
$$y = -1 + 1$$ 　　　Solve for y
$$y = 0$$

The ordered pair $\left(-\dfrac{1}{2}, 0\right)$ is a solution.

3. Let $y = 5$ 　　　　　　　　This time choose any value for y

$$y = 2x + 1$$
$$(5) = 2x + 1$$ 　　　Replace y with 5 in the equation
$$4 = 2x$$ 　　　Solve for x
$$2 = x$$

The ordered pair (2,5) is a solution.

<u>Note</u> We can choose *infinitely many* values for *x* and get a corresponding value of *y* for each one. That means there are infinitely many solutions.

Many times a **table of values** is used to show ordered pairs of an equation. The table could be set up either horizontally or vertically.

● **Example 6–1 C**

Finding ordered pairs using a table of values

Complete the table of values for the given equation.

1. $3x - 2y = 4$

x	2		$\dfrac{1}{3}$
y		7	

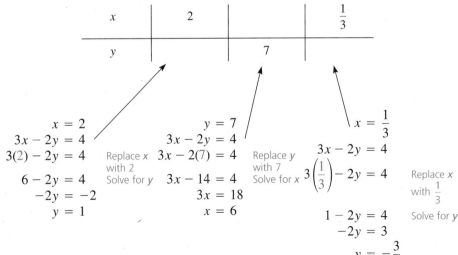

$$x = 2$$
$$3x - 2y = 4$$
$$3(2) - 2y = 4$$ 　Replace x with 2
$$6 - 2y = 4$$ 　Solve for y
$$-2y = -2$$
$$y = 1$$

$$y = 7$$
$$3x - 2y = 4$$
$$3x - 2(7) = 4$$ 　Replace y with 7
$$3x - 14 = 4$$ 　Solve for x
$$3x = 18$$
$$x = 6$$

$$x = \dfrac{1}{3}$$
$$3x - 2y = 4$$
$$3\left(\dfrac{1}{3}\right) - 2y = 4$$ 　Replace x with $\dfrac{1}{3}$
$$1 - 2y = 4$$ 　Solve for y
$$-2y = 3$$
$$y = -\dfrac{3}{2}$$

The completed table is as follows:

x	2	6	$\frac{1}{3}$
y	1	7	$-\frac{3}{2}$

The ordered pairs are $(2,1)$, $(6,7)$, and $\left(\frac{1}{3}, -\frac{3}{2}\right)$.

2. $y = 6$

x	-2	0	5
y			

In the equation $y = 6$, we think of the equation as $0x + y = 6$, which shows that no matter what value we choose for x, y is 6. For example, if

$$x = -2$$
$$0(-2) + y = 6$$
$$y = 6$$

The completed table is as follows:

x	-2	0	5
y	6	6	6

The ordered pairs are $(-2,6)$, $(0,6)$, and $(5,6)$.

Note No matter what *value* we choose for *x*, *y* will *always* be 6, so *every* ordered pair will have a 6 as the second component.

3. $x = -3$

x			
y	-1	$\frac{1}{2}$	4

We think of this equation as $x + 0y = -3$, which shows that no matter what value we choose for y, x is -3. For example, if $y = \frac{1}{2}$,

$$x + 0\left(\frac{1}{2}\right) = -3$$
$$x = -3$$

The completed table is as follows:

x	-3	-3	-3
y	-1	$\frac{1}{2}$	4

The ordered pairs are $(-3,-1)$, $\left(-3, \frac{1}{2}\right)$, and $(-3,4)$.

Note No matter what *value* we choose for *y*, *x* will *always* be −3, so every ordered pair will have a −3 as the first component.

From examples 2 and 3, we can see that solutions of linear equations in two variables that can be written in the form

$$x = a \quad \text{or} \quad y = b$$

where *a* and *b* are constants, have very special characteristics. Given

1. $x = a$, the *first component* in every ordered pair is always the number *a*. The equation could be written $x + 0y = a$.

2. $y = b$, the *second component* in every ordered pair is always the number *b*. The equation could be written $0x + y = b$.

Quick check   In the following equations, find the value of the other variable for each given value.
a. $2x + y = 1$; $x = 2$, $x = -3$
b. $2y - 3x = 1$; $y = -4$, $y = 0$

The Rectangular Coordinate Plane

In chapter 1, we associated the set of real numbers with points on a straight line and called this the number line. Now we associate the solutions of a linear equation *in two variables* with *points* on a flat surface, called a *plane*.

We take two number lines, one horizontal and the other vertical, on the plane, and call them *axes*. These two number lines are represented in figure 6–1. The horizontal line (*x*-axis) is associated with values of *x*, and the vertical line (*y*-axis) is associated with values of *y*. Together, the *x*- and *y*-axes form the **rectangular coordinate plane** (or Cartesian coordinate plane).[1]

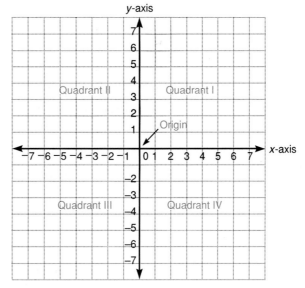

Figure 6–1

---

[1]The invention of this graphing method (relating the algebraic concept of an ordered number pair with the geometric concept of a point in the plane) is attributed to French mathematician and philosopher René Descartes (1596–1650). This combination of algebra and geometry has become known as *analytic geometry* or *coordinate geometry*.

The point at which the two axes intersect is their common zero point and is called the **origin**. The origin corresponds to the ordered pair (0,0). The two axes separate the coordinate plane into four regions called **quadrants**. These quadrants are named as shown in figure 6–1. Points that lie on the *x*-axis or *y*-axis *do not* lie in any of the quadrants.

On the *x*-axis, numbers to the right of the origin are positive and those to the left of the origin are negative. On the *y*-axis, the positive numbers are above the origin, and the negative numbers are below the origin. For each point on the *x*-axis, *y* = 0, and for each point on the *y*-axis, *x* = 0.

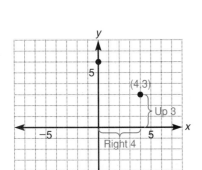

Figure 6–2

Note We have chosen to make each unit on the axes equal to 1. Other choices are possible and, in fact, may be necessary in some instances.

Each ordered pair (*x*, *y*) corresponds to *exactly* one point, called the *graph* of the ordered pair. To find the location of such a point is called *plotting* the point. We can plot any ordered pair, (*x*, *y*), on the coordinate plane if we consider the ordered pair as two instructions to direct us from the origin to the proper location of the point. To plot the point that corresponds to the ordered pair (4,3), we start at the origin. Since the *x*-value, also called the **abscissa** of the point, is 4, we move four units *to the right* (the positive direction) along the *x*-axis. From this position, the *y*-value, also called the **ordinate** of the point, instructs us to move *up* (the positive direction) three units parallel to the *y*-axis. The abscissa and ordinate are usually called the **coordinates of a point**. We have plotted the point that is the graph of the ordered pair (4,3) as shown in figure 6–2.

Similarly, to plot the graph of (−3,−5), we start at the origin. Since *x* = −3, we move three units *to the left* (the negative direction). Next, *y* = −5 instructs us to move *down* (the negative direction) five units parallel to the *y*-axis. We have plotted the graph of the ordered pair (−3,−5) as shown in figure 6–3.

Points are always named by capital letters. The notation *A*(*x*, *y*) indicates that the *name* of the point is *A* and the *coordinates* of the point are (*x*, *y*).

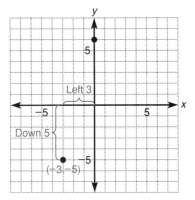

Figure 6–3

● **Example 6–1 D**
Graphing a point

Plot the following points.

a. $A(2,4)$

b. $B(-1,-3)$

c. $C(-4,3)$

d. $D(5,0)$

e. $E(2,-3)$

f. $F(0,4)$

g. $G(-3,0)$

h. $H\left(2,\dfrac{3}{2}\right)$

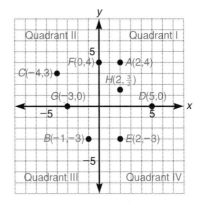

Note Whenever either one of the coordinates of the point is zero, the point is located on an axis. If the **x** component is zero, the point is on the **y**-axis. If the **y** component is zero, the point is on the **x**-axis. If both components are zero, the point is the origin.

We can see in the diagram that a point is in

1. quadrant I when x and y are both positive,

2. quadrant II when x is negative and y is positive,

3. quadrant III when x and y are both negative,

4. quadrant IV when x is positive and y is negative.

Graphs of Solutions—Linear Equations in Two Variables

Now consider the graphs of some of the ordered pairs that are solutions of the linear equation in two variables $y = 2x + 1$. Suppose we let $x = -3$, $x = 0$, $x = 2$, and $x = 3$. Then

when $x = -3$, $y = 2(-3) + 1 = -6 + 1 = -5$ Replace x with -3
when $x = 0$, $y = 2(0) + 1 = 0 + 1 = 1$ Replace x with 0
when $x = 2$, $y = 2(2) + 1 = 4 + 1 = 5$ and   Replace x with 2
when $x = 3$, $y = 2(3) + 1 = 6 + 1 = 7$ Replace x with 3

The ordered pairs $(-3, -5)$, $(0, 1)$, $(2, 5)$, and $(3, 7)$ are solutions of the linear equation $y = 2x + 1$.

We can also express the ordered pairs in a vertical table as follows:

x	y
-3	-5
0	1
2	5
3	7

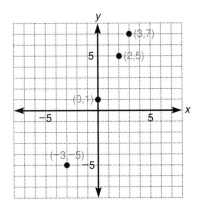

Figure 6–4

We now plot these points in figure 6–4.

MASTERY POINTS

Can you
- Determine whether or not an ordered pair is a solution of a given equation?
- Find the value of one variable, given the value of the other variable?
- Plot ordered pairs in the rectangular coordinate plane?
- Plot ordered pair solutions of linear equations?

Exercise 6–1

Directions
Determine which ordered pairs are solutions of the given equation. See example 6–1 A.

🏳 ***Quick Check***

Example $3x + y = 2$; $(1, -1)$, $(2, 0)$.

Solution   a. $3x + y = 2$
$3(1) + (-1) = 2$ Replace x with 1 and y with -1
$3 + (-1) = 2$
$2 = 2$ (True)

Therefore $(1, -1)$ is a solution.

b. $3x + y = 2$
$3(2) + (0) = 2$ Replace x with 2 and y with 0
$6 + 0 = 2$
$6 = 2$ (False)

Therefore $(2, 0)$ is *not* a solution.

Directions   Determine which ordered pairs are solutions of the given equation. See example 6–1 A.

1. $y = 3x - 1$; (1,2), (−1,−4), (2,3)

2. $y = 2x + 4$; (−1,3), (0,4), (2,8)

3. $x + 2y = 3$; (1,2), (−1,2), (3,0)

4. $3x - y = 4$; (1,−1), $\left(\frac{1}{3},2\right)$, (0,−4)

5. $3y - 4x = 2$; (1,2), (−2,1), $\left(\frac{1}{2},1\right)$

6. $5x - 2y = 6$; (2,2), (0,−3), (4,−2)

7. $3x = 2y$; (2,3), (3,2), (0,0)

8. $3y = -4x$; (4,−3), (−4,3), (8,−6)

9. $x = -4$; (−4,1), (4,2), (−4,−4)

10. $y = 3$; (2,3), (−5,2), $\left(\frac{3}{4},3\right)$

11. $x + 5 = 0$; (3,−5), (−5,3), (−5,8)

12. $y - 2 = 0$; (−2,2), $\left(\frac{2}{3},-2\right)$, (5,2)

Directions	**☞ Quick Check**
Find the value for y corresponding to the given values for x in each equation. Express the answer as an ordered pair. See example 6–1 B and C.	**Example** $2x + y = 1$; $x = 2$, $x = -3$.

Solution   a. Let $x = 2$, then

$$2x + y = 1$$
$$2(2) + y = 1 \quad \text{Replace } x \text{ with 2}$$
$$4 + y = 1 \quad \text{Solve for } y$$
$$y = -3$$

The ordered pair is $(2,-3)$.

b. Let $x = -3$, then

$$2x + y = 1$$
$$2(-3) + y = 1 \quad \text{Replace } x \text{ with } -3$$
$$-6 + y = 1 \quad \text{Solve for } y$$
$$y = 7$$

The ordered pair is $(-3,7)$.

Directions   Find the value for y corresponding to the given values for x in each equation. Express the answer as an ordered pair. See example 6–1 B and C.

13. $y = 3x + 2$; $x = 1$, $x = -2$, $x = 0$

14. $y = 4x - 3$; $x = -1$, $x = 2$, $x = 0$

15. $y = 3 - 2x$; $x = 3$, $x = -4$, $x = 0$

16. $y = -5 - x$; $x = \frac{1}{5}$, $x = 5$, $x = 0$

17. $3x + y = 4$; $x = 3$, $x = -2$, $x = 0$

18. $y - 4x = 1$; $x = 1$, $x = \frac{5}{4}$, $x = 0$

19. $x - 5y = 3$; $x = -2$, $x = 3$, $x = 0$

20. $x + 4y = 0$; $x = -4$, $x = 8$, $x = 0$

21. $5x + 2y = -3$; $x = 1$, $x = -1$, $x = 0$

22. $2x - 3y = 1$; $x = \frac{1}{4}$, $x = -4$, $x = 0$

23. $y = 5$; $x = 1$, $x = -6$, $x = 0$

24. $y = -4$; $x = -5$, $x = -4$, $x = 0$

25. $y + 1 = 0$; $x = 7$, $x = -\frac{3}{5}$, $x = 0$

26. $y - 4 = 0$; $x = 1$, $x = 2$, $x = 0$

<table>
<tr><td>

Directions

Find the value for x corresponding to the given value for y in each equation. Express the answer as an ordered pair. See example 6–1 B and C.

</td><td>

📌 *Quick Check*

Example $2y - 3x = 1$; $y = -4$, $y = 0$.
Solution   a. Let $y = -4$, then   b. Let $y = 0$, then

a.
$$2y - 3x = 1$$
$$2(-4) - 3x = 1 \quad \text{Replace } y \text{ with } -4$$
$$-8 - 3x = 1 \quad \text{Solve for } x$$
$$-3x = 9$$
$$x = -3$$

The ordered pair is $(-3, -4)$.

b.
$$2y - 3x = 1$$
$$2(0) - 3x = 1 \quad \text{Replace } y \text{ with } 0$$
$$0 - 3x = 1 \quad \text{Solve for } x$$
$$-3x = 1$$
$$x = -\frac{1}{3}$$

The ordered pair is $\left(-\frac{1}{3}, 0\right)$.

</td></tr>
</table>

Directions   Find the value for x corresponding to the given value for y in each equation. Express the answer as an ordered pair. See example 6–1 B and C.

27. $x = 2y + 3$; $y = 1$, $y = -2$, $y = 0$

28. $x = -3y + 1$; $y = -1$, $y = 2$, $y = 0$

29. $3y - 2x = 0$; $y = 2$, $y = -4$, $y = 0$

30. $2x + y = 3$; $y = -3$, $y = 5$, $y = 0$

31. $x = 1$; $y = -2$, $y = 7$, $y = 0$

32. $x + 7 = 0$; $y = -1$, $y = 3$, $y = 0$

Directions   Solve the following problems.

33. The total cost c in dollars of producing x units of a certain commodity is given by the equation $c = 2x + 20$. Find the cost of producing (a) 75; (b) 300; (c) 1,000 units of the commodity. Write the answers as ordered pairs (x, c).

34. In exercise 33, find the number of units produced when the total cost is (a) \$430; (b) \$700; (c) \$1,400. Write the answers as ordered pairs (x, c).

35. The cholesterol level in the blood, y, is related to the dosage of a new anticholesterol drug in grams, x, by the equation $y = 240 - 2x$. Find the cholesterol level in the blood when the dosage is (a) 2 grams, (b) 12 grams, (c) 0 grams. Write the answers as ordered pairs (x, y).

36. In exercise 35, determine the number of grams in the dosage when the cholesterol level in the blood is (a) 200, (b) 0, (c) 210. Write the answers as ordered pairs (x, y).

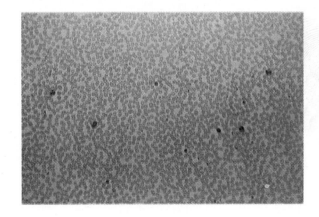

37. Suppose the equation $y = 3x + 20$ represents the estimated number of minutes y it takes to do a student's mathematics assignment consisting of x problems. This comes from estimating 20 minutes to look over the lecture and text examples, and 3 minutes per problem, including checking each answer. Find the number of minutes required for an assignment with the number of problems given. (a) 10, (b) 15, (c) 8. Write the answers as ordered pairs (x, y).

38. Referring to exercise 37, how many problems were in an assignment (to the nearest whole number) if it took the given number of minutes to do the assignment? (a) 32 minutes (b) 53 minutes (c) 1 hour (d) $1\frac{3}{4}$ hours. Write the answers as ordered pairs (x, y).

39. If the distance, y (in miles), that Mary Jane travels in x hours of driving is given by the equation $y = 55x$, how far does Mary Jane travel in (a) 3 hours, (b) 8 hours, (c) 5 hours and 12 minutes $\left(5\frac{1}{5} \text{ hours}\right)$? Write the answers as ordered pairs (x, y).

40. In exercise 39, how many hours does it take Mary Jane to drive (a) 110 miles, (b) 385 miles, (c) 500 miles? Write the answers as ordered pairs (x, y).

Directions Plot the following ordered pairs on a rectangular coordinate plane. See example 6–1 D.

41. $(2, 4)$

42. $(5, 1)$

43. $(-1, 3)$

44. $(-4, 4)$

45. $(-6, -1)$

46. $(-2, -3)$

47. $(0, 4)$

48. $(0, 2)$

49. $(5, 0)$

50. $(-4, 0)$

51. $(-7, 0)$

52. $(0, 0)$

53. $\left(\frac{1}{2}, 3\right)$

54. $\left(\frac{2}{3}, -2\right)$

55. $\left(\frac{3}{2}, 0\right)$

Directions Determine the coordinates, (x, y), of the given points in the diagram.

56. A

57. B

58. C

59. D

60. E

61. F

62. G

63. H

64. I

65. J

66. K

Directions State the quadrant in which each point lies.	**Example** $(-1,3)$ lies in the second quadrant because the x-component is negative and the y-component is positive.

Directions   State the quadrant in which each point lies.

67. $(-2,-5)$

68. $(4,-1)$

69. $(5,3)$

70. $(-7,-9)$

71. $\left(\dfrac{1}{2},-4\right)$

72. $\left(-\dfrac{2}{3},7\right)$

73. $\left(-\dfrac{5}{2},-\dfrac{3}{4}\right)$

74. $\left(\dfrac{7}{8},-\dfrac{7}{8}\right)$

75. $(5,-14)$

76. $(-3,-3)$

Directions   Solve the following problems.

77. In what quadrant does a point (x,y) lie if
(a) $x>0$ and $y<0$, (b) $x<0$ and $y>0$,
(c) $x<0$ and $y<0$, (d) $x>0$ and $y>0$?

78. What is the value of x for any point on the y-axis?

79. What is the value of y for any point on the x-axis?

80. Using the diagram for exercises 56–66, what is the *abscissa* of each of the points A, C, E, G, I, and K?

81. Using the diagram for exercises 56–66, what is the *ordinate* of each of the points B, D, F, H, and J?

82. Plot five points whose abscissa is -3. Connect the plotted points. Describe the resulting figure.

83. Using the same axes as in exercise 82, plot five points whose ordinate is 4. Connect the plotted points. Describe the resulting figure.

84. Choose five ordered pairs whose first and second coordinates are the same. Plot these points and connect the points. What kind of figure do you get? In what quadrants does the figure lie?

85. Choose five ordered pairs whose first component is the opposite of the second component. Plot the points and connect them. What kind of figure do you get? In what quadrants does the figure lie?

Directions
Plot the points in the following exercises using the indicated scale.

<u>Note</u> Exercises 86 to 90 require the use of different scales on the two axes. This is sometimes necessary in certain problems, especially in applications in the sciences and other fields.

Example   1 square = 5 units (on *x*-axis)
1 square = 10 units (on *y*-axis)
$A(0,-10), B(5,20), C(-15,20), D(-25,-40), E(20,0)$

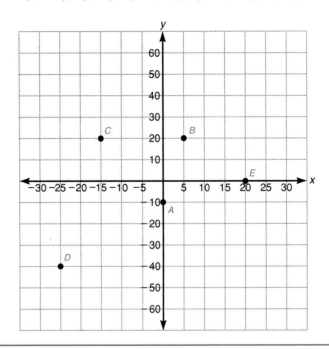

Directions   Plot the points in the following exercises using the indicated scale.

86. 1 square = 10 units
$A(20,20), B(-30,40), C(0,-50), D(40,0), E(50,-70)$

87. 1 square = 1 unit (on *x*-axis)
1 square = 5 units (on *y*-axis)
$A(4,10), B(-6,35), C(5,5), D(0,-45), E(-3,-50)$

88. 1 square = 5 units (on *x*-axis)
1 square = 1 unit (on *y*-axis)
$A(-20,3), B(-5,5), C(40,-6), D(25,0), E(-20,-4)$

89. 1 square = 10 units
$A(-35,20), B(35,-40), C(25,100), D(0,0), E(-40,90)$

90. 1 square = 25 units (on *x*-axis)
1 square = 100 units (on *y*-axis)
$A(75,-600), B(200,-200), C(0,-100), D(-175,500),$
$E(-225,0)$

Directions
Find the coordinates of the given points.

Example The following set of points represents the percent grades Bruce received on ten algebra quizzes.

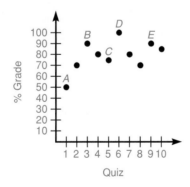

Find the coordinates of points A, B, C, D, and E and interpret the coordinates.

Solution $A(1,50)$: Grade was 50 percent on first quiz
$B(3,90)$: Grade was 90 percent on third quiz
$C(5,75)$: Grade was 75 percent on fifth quiz
$D(6,100)$: Grade was 100 percent on sixth quiz
$E(9,90)$: Grade was 90 percent on ninth quiz

Directions   Find the coordinates of the given points.

91. The following sets of points represent the average temperature in Detroit during a seven-day period last summer. Find the approximate coordinates of points *A*, *B*, *C*, and *D* and interpret the coordinates.

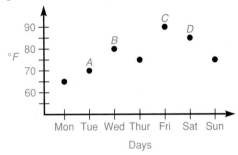

92. The following sets of points represent the approximate number of housing starts in the United States for the years 1973 to 1979. Find the coordinates of points *A*, *B*, *C*, and *D* and interpret the coordinates.

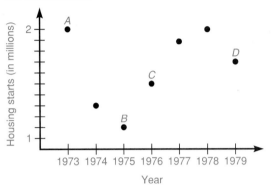

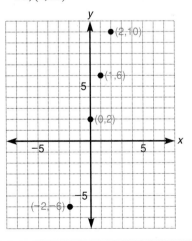

Directions
Find the missing component in each ordered pair using the given equation. Then plot the ordered pairs using a separate coordinate system for each problem. (See the diagram.)

Example $y = 4x + 2$; $(0,\ )$, $(-2,\ )$, $(1,\ )$, $(2,\ )$

Solution When $x = 0$, $y = 4(0) + 2 = 0 + 2 = 2$; $(0,2)$
$x = -2$, $y = 4(-2) + 2 = -8 + 2 = -6$; $(-2,-6)$
$x = 1$, $y = 4(1) + 2 = 4 + 2 = 6$; $(1,6)$
$x = 2$, $y = 4(2) + 2 = 8 + 2 = 10$; $(2,10)$

or in table form

x	0	-2	1	2
y	2	-6	6	10

Directions   Find the missing component in each ordered pair using the given equation. Then plot the ordered pairs using a separate coordinate system for each problem. (See the diagram.)

93. $y = x + 4$; $(0, \ )$, $(-4, \ )$, $\left(\dfrac{5}{2}, \ \right)$, $(2, \ )$

97. $y = 2x$; $(0, \ )$, $\left(\dfrac{1}{2}, \ \right)$, $(2, \ )$, $(-1, \ )$

94. $y = x - 3$; $(0, \ )$, $(3, \ )$, $(1, \ )$, $\left(-\dfrac{1}{2}, \ \right)$

98. $y = 3x$; $(0, \ )$, $(3, \ )$, $(-3, \ )$, $\left(\dfrac{1}{3}, \ \right)$

95. $y = 2x + 1$; $(0, \ )$, $\left(-\dfrac{1}{2}, \ \right)$, $(2, \ )$, $(-1, \ )$

99. $y = -x + 3$; $(0, \ )$, $(-3, \ )$, $(3, \ )$, $\left(\dfrac{5}{2}, \ \right)$

96. $y = 3x - 4$; $(0, \ )$, $(-1, \ )$, $\left(-\dfrac{1}{3}, \ \right)$, $(2, \ )$

100. $y = -x + 1$; $(0, \ )$, $(1, \ )$, $(-2, \ )$, $(-4, \ )$

101. $y = -2x + 3$; $(0, \quad), (3, \quad), (-2, \quad), \left(\dfrac{3}{2}, \quad\right)$

103. $y = 4 - x$; $(0, \quad), (-3, \quad), (2, \quad), (3, \quad)$

102. $3x + y = 1$; $(0, \quad), \left(\dfrac{1}{3}, \quad\right), (-2, \quad), (1, \quad)$

104. $2y - 3x = 2$; $(0, \quad), \left(\dfrac{2}{3}, \quad\right), (-2, \quad), (2, \quad)$

✦ Explore to Learn

In chapter 1 we learned that the decimal representation of rational numbers eventually terminates or repeats. The decimal representation of irrational numbers never repeats or terminates. Some examples of rational numbers are

$$\frac{3}{8} = 0.37500000000\ldots$$

$$\frac{1}{3} = 0.333\ldots$$

$$\frac{2}{7} = 0.285714285714285714\ldots$$

Some examples of irrational numbers and their decimal forms are

$$\sqrt{2} \approx 1.414213562373095049$$

$$\pi \approx 3.141592653589793238$$

We can actually graph the decimal form of a number. For example, for $\dfrac{3}{8}$ we could graph the ordered pairs $(1,3), (2,7), (3,5), (4,0), (5,0)$, etc. Each ordered pair is the digit's position (1st, 2nd, 3rd, etc.) and the digit itself. (a) Graph the first fifteen digits of each of the numbers above. (b) What will be true of the graphs of the rational numbers that will not be true of the graphs of the irrational numbers?

Review Exercises

Directions Perform the indicated operations. See sections 5–2 and 5–3.

1. $\dfrac{x^2 - 4}{3x} \cdot \dfrac{6x^2}{x - 2}$

2. $\dfrac{2x - 1}{x + 3} \div \dfrac{4x^2 - 1}{x^2 + 6x + 9}$

3. $\dfrac{4}{x - 2} - \dfrac{3}{x + 1}$

4. Find $-|-8|$. See section 1–4.

5. Multiply $x^2 \cdot x \cdot x^0$. See section 3–3.

6. Multiply $(-2)(-3)(4)(0)$. See section 1–7.

Directions   Solve the following problems. See sections 2–8 and 5–7.

7. Stock *A* sells for $19 per share and stock *B* sells for $46 per share. A woman wishes to buy seven times as many shares of stock *A* as stock *B*. If she invests $8,055, how many shares of each stock does she buy?

8. Tim and Sara can mow the lawn in 3 hours working together. If Sara could do the job in 5 hours working alone, how long would it take Tim to do the job working alone?

6–2   Graphing Linear Equations in Two Variables

Graphing a Linear Equation

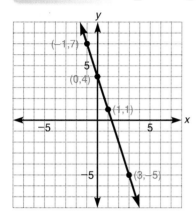

Figure 6–5

In section 6–1, we learned that there are infinitely many ordered pairs that will satisfy an equation in two variables. That is, given the linear equation $3x + y = 4$, we can find many ordered pairs (as many as we wish) that are solutions of the equation. To list all of these solutions is impossible. However, these solutions can be represented geometrically by a graph of the ordered pairs that are solutions.

In section 6–1, we plotted the graphs of several of the ordered pairs that satisfied the given equations. To illustrate again, consider the equation $3x + y = 4$. In figure 6–5, we plot the ordered pairs

$$(0,4), (1,1), (-1,7), \text{ and } (3,-5)$$

that are solutions of the equation.

Connecting the points, we find they all lie on the same straight line. We have drawn arrowheads in each direction at each end of the line to indicate that the line goes on indefinitely in each direction. *Any point whose coordinates satisfy the equation $3x + y = 4$ will lie on this line*, and *the coordinates of any point on this line will satisfy the equation*. We have a graphical representation of a portion of the solutions of the equation. The straight line in figure 6–5 is called the *graph of the linear equation* $3x + y = 4$.

Straight Line

In general, the graph of *any* linear equation in two variables is a *straight line*. A geometric fact we now use is that *through any two given points in the plane we can draw one and only one straight line*. Thus, since we know the graph of a linear equation in two variables is a straight line, we can determine the graph of the equation using only two points. However, it is a good idea to find a third point as a check on our work. (Remember, the word *line* appears in the name *line*ar equation.)

● **Example 6–2 A**

Graphing a line using three points

Graph the equation $y = 2x + 4$.

We choose *three* arbitrary values of *x* and find corresponding values for *y* by substituting the value of *x* and solving for *y*.

x	*y*
0	4
−2	0
2	8

Let $x = 0$, then $y = 2(0) + 4 = 0 + 4 = 4$
$x = -2$, then $y = 2(-2) + 4 = -4 + 4 = 0$
$x = 2$, then $y = 2(2) + 4 = 4 + 4 = 8$

Now we graph the points $(0,4)$, $(-2,0)$, and $(2,8)$ and draw the straight line through these points.

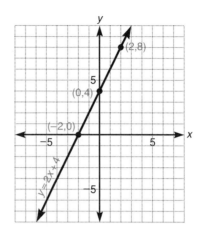

The x- and y-Intercepts

Notice the graph of $y = 2x + 4$ crosses the y-axis at $(0,4)$ and the x-axis at $(-2,0)$. The points $(0,4)$ and $(-2,0)$ are called the **y-intercept** and the **x-intercept**, respectively. Since we need only two points to sketch the graph of a linear equation in two variables, in many cases we use the x- and y-intercepts. Observe from the example that when the line crosses the x-axis, the value of y is zero. When the line crosses the y-axis, the value of x is zero.

Finding x- and y-Intercepts

1. To find the x-intercept, we let $y = 0$ and find the corresponding value of x. This is the point $(x,0)$.
2. To find the y-intercept, we let $x = 0$ and find the corresponding value for y. This is the point $(0,y)$.

● **Example 6–2 B**

Graphing a line using x- and y-intercepts

Graph the following linear equations using x- and y-intercepts.

1. $y = -2x + 3$

Let $x = 0$ to find the y-intercept; let $y = 0$ to find the x-intercept.

$y = -2x + 3$
$y = -2(0) + 3$ Replace x with 0
$y = 0 + 3$ Multiply as indicated
$y = 3$

The point $(0,3)$ is the y-intercept.

$y = -2x + 3$
$(0) = -2x + 3$ Replace y with 0
$2x = 3$ Add $2x$
$x = \dfrac{3}{2}$

The point $\left(\dfrac{3}{2},0\right)$ is the x-intercept.

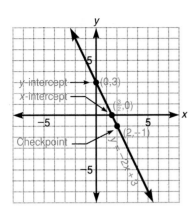

To find another point, choose $x = 2$.

$$y = -2x + 3$$
$$y = -2(2) + 3 \quad \text{Replace } x \text{ with 2}$$
$$y = -4 + 3 \quad \text{Multiply as indicated}$$
$$y = -1$$

The third point is $(2, -1)$.

We now plot the three points $(0, 3)$, $\left(\dfrac{3}{2}, 0\right)$, and $(2, -1)$ and draw the straight line through these points.

2. $3y - 2x = 9$

Let $x = 0$ and $y = 0$ to find the y- and x-intercepts. Let x be some other value to obtain a third point. Here we select $x = 3$. This is shown in the table of values.

Fill in the table as was done in section 6–1. This results in the second table shown. Plot the points shown in the table, $(0, 3)$, $\left(-4\dfrac{1}{2}, 0\right)$, and $(3, 5)$, and draw the straight line that passes through these points.

x	0		3
y		0	

x	0	$-4\dfrac{1}{2}$	3
y	3	0	5

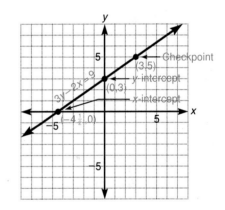

In the previous examples, the x- and y-intercepts were different points. For some equations, the x- and y-intercepts are the same point, as demonstrated in example 3.

3. $y = 3x$

If we let $x = 0$, then $y = 3(0) = 0$, giving the ordered pair $(0, 0)$. When $y = 0$, then $0 = 3x$ and $x = 0$, giving the same point $(0, 0)$. We must choose two additional values for x or y. Let $x = 1$ and $x = -1$.

x	$y = 3x$	Ordered pair (x, y)	
0	$3(0) = 0$	$(0, 0)$	x- and y-intercepts
1	$3(1) = 3$	$(1, 3)$	Arbitrary second point
-1	$3(-1) = -3$	$(-1, -3)$	Checkpoint

Plot the points $(0,0)$, $(1,3)$, and $(-1,-3)$ and draw the straight line through them.

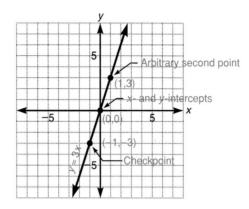

Note Any linear equation that can be written in the form

$$y = kx \quad \text{or} \quad x = ky$$

where k is a real number, will pass through the origin $(0,0)$.

4. $y = -3$

Recall that this equation could be written $0x + y = -3$ and that for *any* value of x we might choose, y is *always* equal to -3. Therefore, we choose any three values for x and obtain $y = -3$.

x	y
-2	-3
0	-3
2	-3

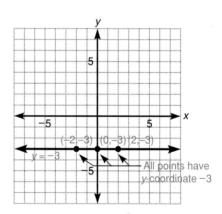

Note The graph has a *y*-intercept, -3, but no *x*-intercept. The graph is a horizontal straight line (parallel to the *x*-axis) passing through the point $(0,-3)$. In fact, the graph of any equation of the form $y = b$ will be a horizontal line passing through the point with coordinates $(0,b)$, the *y*-intercept.

5. $x = 4$

Recall that this equation can be written $x + 0y = 4$ and that the value of x will be 4 for *any* value of y.

x	y
4	-2
4	0
4	2

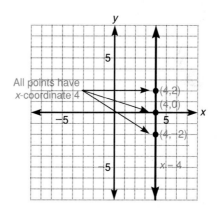

Note   The graph has an *x*-intercept, 4, but no *y*-intercept. The graph is a **vertical** straight line (parallel to the *y*-axis) passing through the point with coordinates (4,0). In fact, the graph of any equation of the form **x** = **a** will be a vertical line passing through the point with coordinates (**a**,0), the *x*-intercept.

Appendix A shows how to graph nonvertical straight lines with a graphing calculator.

Quick check   a. Find the *x*- and *y*-intercepts for the graph of $5x + 3y = 15$.
b. Graph the linear equation $2y - 3x = 12$ using the *x*- and *y*-intercepts.

We now summarize the different forms that linear equations might take.

1. $ax + by = c$ Graph by finding the   *Example:* $2x + y = 4$
                          x-intercept (let $y =$
                          0), the *y*-intercept
                          (let $x = 0$), and a
                          third checkpoint by
                          choosing any value for
                          x or *y* not yet used.

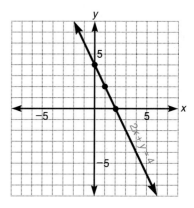

2. $y = kx$ or
   $x = ky$

   Graph goes through the origin $(0,0)$. Find two other points by choosing values for x or y other than 0.

   Examples: $y = 2x$ and $x = -3y$

   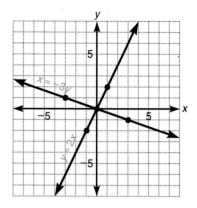

3. $x = a$

   Graph is a vertical line through $(a,0)$.

   Example: $x = 1$

   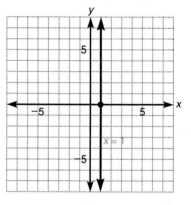

4. $y = b$

   Graph is a horizontal line through $(0,b)$.

   Example: $y = -6$

   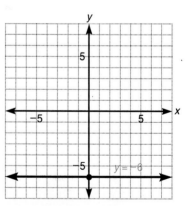

MASTERY POINTS

Can you
- Graph linear equations using ordered pairs?
- Find the *x*- and *y*-intercepts of a linear equation?
- Graph linear equations using the *x*- and *y*-intercepts?
- Graph the equations $x = a$ and $y = b$, where *a* and *b* are constants?

Exercise 6–2

Directions
Find the *x*- and *y*-intercepts. Write ordered pairs representing the points where the line crosses the axes. See example 6–2 B.

⚑ *Quick Check*

Example $5x + 3y = 15$

Solution   Let $x = 0$, then $5(0) + 3y = 15$ Replace *x* with 0
$$0 + 3y = 15$$
$$3y = 15$$
$$y = 5$$ *y*-intercept
Let $y = 0$, then $5x + 3(0) = 15$ Replace *y* with 0
$$5x + 0 = 15$$
$$5x = 15$$
$$x = 3$$ *x*-intercept

The line crosses the *y*-axis at (0,5) and the *x*-axis at (3,0).

Directions   Find the *x*- and *y*-intercepts. Write ordered pairs representing the points where the line crosses the axes. See example 6–2 B.

1. $y = 2x + 4$
2. $y = 5x - 10$
3. $y = 3x + 1$
4. $y = 2x - 3$
5. $2x + 3y = 6$
6. $x + 4y = 12$
7. $2x + 5y - 11 = 0$
8. $x - y = 4$
9. $y = 5x$
10. $y = -2x$

11. $3x - 2y = 0$
12. $y - 4x = 0$
13. $(1.2)x + (2.4)y = 4.8$
14. $(0.3)x - (0.4)y = 0.7$
15. $y = \frac{1}{2}x - \frac{3}{2}$
16. $y = \frac{2}{3}x - \frac{1}{3}$

Directions
Graph the given linear equations using the *x*- and *y*-intercepts. See example 6–2 B.

⚑ *Quick Check*

Example $2y - 3x = 12$

Solution

When $x = 0$
$$2y - 3x = 12$$
$$2y - 3(0) = 12$$ Replace *x* with 0
$$2y - 0 = 12$$
$$2y = 12$$
$$y = 6$$ *y*-intercept

The point (0,6) is the *y*-intercept.

When $y = 0$
$$2y - 3x = 12$$
$$2(0) - 3x = 12$$ Replace *y* with 0
$$0 - 3x = 12$$
$$-3x = 12$$
$$x = -4$$ *x*-intercept

The point (−4,0) is the *x*-intercept.

When $x = -2$
$$2y - 3x = 12$$
$$2y - 3(-2) = 12$$ Replace *x* with −2
$$2y - (-6) = 12$$
$$2y + 6 = 12$$
$$2y = 6$$
$$y = 3$$

The checkpoint is (−2,3).

continued

We now plot the points (0,6), (−4,0), and (−2,3) and draw the line through them.

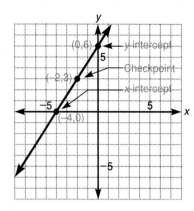

Directions   Graph the given linear equations using the *x*- and *y*-intercepts. See example 6–2 B.

17. $y = 3x + 6$ **20.** $y = 2x + 4$ **23.** $y = x$

18. $y = x + 5$ **21.** $y = 2x − 8$ **24.** $y = −3x$

19. $y = x − 2$

22. $y = 3x + 3$

25. $y = −2x$

26. $x + y = 0$

27. $y - 2x = 0$

28. $x - 3y = 0$

29. $4x - 3y = 12$

30. $3x - 2y = 6$

31. $2y + 5x = 10$

32. $2x + 2y = 3$

33. $5x - 6y = 30$

34. $3y - 5x = 15$

35. $y = 6$

36. $y = -2$

37. $x = 5$

38. $x = -2$

39. $x = 0$

40. $y = 0$

Directions
Solve the following linear equations for y in terms of x. Write the equations in the form $y = mx + b$, where m and b are rational numbers. Identify m and b. See section 2–6 for solving a literal equation for a specified variable.

Example $3y - 2x = 4$

Solution $3y = 2x + 4$ Add $2x$ to each side

$$y = \frac{2}{3}x + \frac{4}{3}$$ Divide each side by 3

$$m = \frac{2}{3} \text{ and } b = \frac{4}{3}$$

Directions   Solve the following linear equations for y in terms of x. Write the equations in the form $y = mx + b$, where m and b are rational numbers. Identify m and b. See section 2–6 for solving a literal equation for a specified variable.

41. $y - 2x + 7 = 0$

42. $y - 3x - 4 = 0$

43. $3y - 4x = 9$

44. $5y - 2x = 7$

45. $7x + 3y = 10$

46. $5x + 3y = 4$

47. $x - 5y + 7 = 0$

48. $x - 3y + 9 = 0$

49. $5y - 8x + 14 = 0$

50. $7x + 5y - 11 = 0$

51. Graph the equation $y = -2x + b$ for (a) $b = 5$, (b) $b = 0$, and (c) $b = -3$ all on the same coordinate system.

52. Graph the equation $y = mx + 1$ for (a) $m = 1$, (b) $m = \frac{1}{2}$, and (c) $m = -2$ all on the same coordinate system.

Directions
Write an equation for each of the statements in exercises 53 to 56 and graph the equation.

Example The value of y is 6 more than twice the value of x.

Solution The equation is $y = 2x + 6$.

x	y	
0	6	y-intercept
-3	0	x-intercept
-2	2	Checkpoint

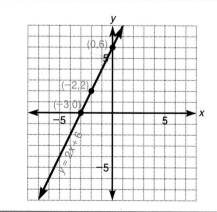

Directions   Write an equation for each of the statements in exercises 53 to 56 and graph the equation.

53. The value of y is 3 less than two times the value of x.

54. The value of x is 4 more than the value of y.

55. Two times x taken away from three times y is 6.

56. Five times x less the product of 2 and y gives 20.

57. The figure shows a balance in a physics lab with weights of 10, 20, and 25 grams hung as shown. The 20-gram weight is hung 5 centimeters from the balance point, and the other distances are x and y as shown. It has been observed that the net moment acting on one side of a beam like this is the product of the weight and distance from the point of balance. For the balance beam to be level, the net moment on each side must be equal. Under these conditions, for the scale to be in balance, the following statement must be true: $5(20) + 10x = 25y$. Create a graph that shows all possible combinations of x and y for which the beam will be in balance. Note that we presume that $x \geq 0$ and $y \geq 0$.

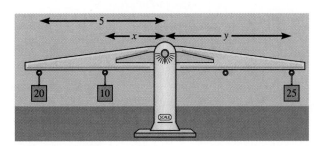

Explore to Learn

A Diophantine equation, named after Diophantus of Alexandria, Egypt (about A.D. 250), is an equation that accepts only integer solutions. Consider the linear equation in two variables $2x + 3y = 8$.

    a. Graph the equation on graph paper.

    b. Does the equation seem to have any integer solutions (ordered pairs (x, y) where both x and y are integers)?

    c. Devise an algebraic method for finding an unlimited number of integer solutions to the equation.

Review Exercises

Directions   Reduce the following expressions to lowest terms. See sections 1–2 and 5–1.

1. $\dfrac{18}{15}$

2. $\dfrac{x-3}{x^2 + x - 12}$

3. $\dfrac{x^2 - 8x + 16}{x^2 + x - 20}$

4. Given $\dfrac{a-b}{c-d}$, evaluate the expression when $a = 2$, $b = 1$, $c = -3$, and $d = 2$. See section 1–9.

5. Given $y = mx + b$, find y when $m = -3$ and $b = 6$. See section 1–9.

6. If the product of a number and three times that number is 48, find the number. See section 4–8.

7. If 2 quarts of water are added to 10 quarts of punch that is 15% strawberry, what percent of the new punch is strawberry? See section 2–7.

8. A number is decreased by 7 and twice this result is 52. What is the number? See section 2–7.

6–3   The Slope of a Line

The Slope

Consider the portions of the two roadways denoted by R_1 and R_2 (read "R sub-one" and "R sub-two") shown in figure 6–6.

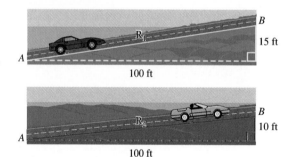

Figure 6–6

We would say roadway R_1 is "steeper" than roadway R_2. In moving from point A to point B on each roadway, a horizontal change in position of 100 feet, the vertical change in position is

$$15 \text{ feet on roadway } R_1$$

and

$$10 \text{ feet on roadway } R_2$$

If we measure this "steepness" by the ratio

$$\frac{\text{vertical change}}{\text{horizontal change}}$$

the roadway

$$R_1 \text{ has "steepness"} = \frac{15 \text{ ft}}{100 \text{ ft}} = \frac{3}{20}$$

and

$$R_2 \text{ has ``steepness''} = \frac{10 \text{ ft}}{100 \text{ ft}} = \frac{1}{10}$$

Note $\frac{3}{20}$ is greater than $\frac{1}{10}$, so R_1 is "steeper" than R_2.

When applying this concept to any straight line, "steepness" is called the **slope** of the line. The slope of any line L is given by $\dfrac{\text{vertical change}}{\text{horizontal change}}$. Observe that the *slope is a ratio*. (See figure 6–7.)

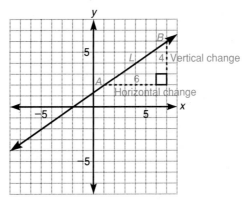

Figure 6–7

In figure 6–7, line L has from point A to point B

$$\text{a vertical change} = 4 \text{ units}$$

and

$$\text{a horizontal change} = 6 \text{ units}$$

Thus, the

$$\text{slope of } L = \frac{\text{vertical change}}{\text{horizontal change}}$$
$$= \frac{4}{6} = \frac{2}{3}$$

Definition of the Slope of a Nonvertical Line

The slope m of the nonvertical line through the points $P_1(x_1, y_1)$ and $P_2(x_2, y_2)$ is given by

$$m = \frac{y_2 - y_1}{x_2 - x_1} \text{ if } x_1 \ne x_2$$

Concept

The slope of a nonvertical line is determined by dividing the change in y-values by the change in x-values of any two points on the line. See figure 6–8.

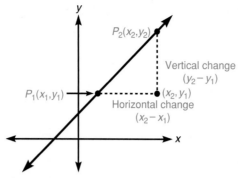

Figure 6–8

<u>Note</u> The vertical change is sometimes called the *rise* and the horizontal change is called the *run*. Thus, slope *m* can be defined

$$m = \frac{\text{rise}}{\text{run}}$$

● **Example 6–3 A**

Finding the slope of a line given two points

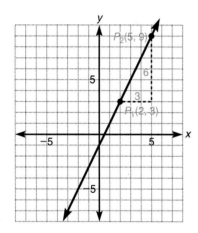

1. Find the slope of the line through the points $P_1(2,3)$ and $P_2(5,9)$ and graph the line.

$$\text{slope } m = \frac{y_2 - y_1}{x_2 - x_1}$$

$$= \frac{(9) - (3)}{(5) - (2)} \quad \text{Replace } y_2 \text{ with 9, } y_1 \text{ with 3,}$$
$$\phantom{= \frac{(9) - (3)}{(5) - (2)}} \quad x_2 \text{ with 5, and } x_1 \text{ with 2}$$

$$= \frac{6}{3} = \frac{2}{1}$$

$$= 2$$

This means that for every unit moved to the right, there is a rise of 2 units.

<u>Note</u> It makes no difference which point you label (x_1, y_1) or (x_2, y_2). If (5,9) was labeled (x_1, y_1) and (2,3) was labeled (x_2, y_2) the answer is the same

$$m = \frac{3 - 9}{2 - 5} = \frac{-6}{-3} = 2$$

2. Find the slope of the line through points $P_1(-3,2)$ and $P_2(5,-4)$ and graph the line.

$$\text{slope } m = \frac{y_2 - y_1}{x_2 - x_1}$$

$$= \frac{(-4) - (2)}{(5) - (-3)}$$

$$= \frac{-6}{8}$$

$$= \frac{-3}{4}$$

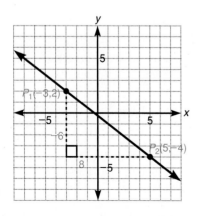

This means for every move of 4 units to the right, there is a "negative rise" (fall) of 3 units.

3. Find the slope of the horizontal line through $P_1(-3,4)$ and $P_2(2,4)$ and graph the line.

$$\text{slope } m = \frac{y_2 - y_1}{x_2 - x_1}$$

$$= \frac{(4) - (4)}{(2) - (-3)}$$

$$= \frac{0}{5}$$

$$= 0$$

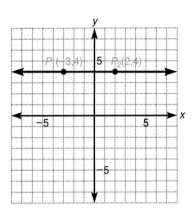

The slope $m = 0$ and the line is a horizontal line. Recall that the equation of any horizontal line is of the form $y = b$ or $y - b = 0$, where b is the y-intercept. In this case, the equation of the horizontal line is $y = 4$. Therefore we can generalize as follows:

Slope of a Horizontal Line

The slope m of any horizontal line is $m = 0$.

4. Find the slope of the vertical line through $P_1(-4,1)$ and $P_2(-4,-3)$ and graph the line.

$$\text{slope } m = \frac{y_2 - y_1}{x_2 - x_1}$$

$$= \frac{(-3) - (1)}{(-4) - (-4)}$$

$$= \frac{-4}{0} \quad \text{(Undefined)}$$

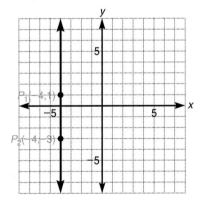

The line is a vertical line and the slope is undefined. Recall that the equation of any vertical line is of the form $x = a$, where a is the x-intercept. In this case, the equation of the vertical line is $x = -4$.

Slope of a Vertical Line

Any vertical line *has undefined slope.*

☞ *Quick check* Find the slope of the line passing through the points $(-2,1)$ and $(3,4)$. Draw the graph of the line. ●

It is important for us to realize that the *slope of a nonvertical line is the same no matter what two points on the line we use to compute the slope.* Consider figure 6–9.

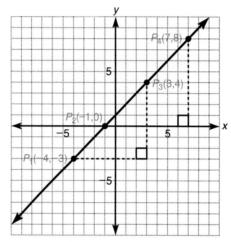

Figure 6–9

Using P_1 and P_3, $m = \dfrac{(4) - (-3)}{(3) - (-4)} = \dfrac{7}{7} = 1$ and

using P_2 and P_4, $m = \dfrac{(8) - (0)}{(7) - (-1)} = \dfrac{8}{8} = 1$

We obtain the same slope, $m = 1$

The Slope of a Line Through Two Plotted Points

Suppose we are given the graph of a line and we wish to determine its slope. (See figure 6–10.) We inspect the graph and look for two points on the line that apparently meet the grid lines at points of intersection. In figure 6–10, we can see that the graph goes through the points $(-1,4)$ and $(3,-2)$. Then the slope m is

$$m = \frac{(-2) - (4)}{(3) - (-1)} = \frac{-6}{4} = \frac{-3}{2}$$

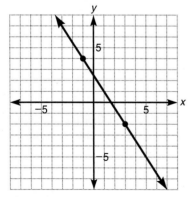

Figure 6–10

● **Example 6–3 B**

*Finding the slope of a line
from the graph of the line*

Find the slope of the line whose graph is plotted.

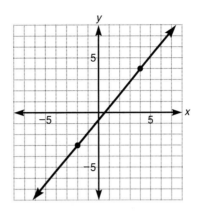

The line passes through points whose coordinates are (4,4) and (−2,−3).
Therefore

$$m = \frac{(4) - (-3)}{(4) - (-2)} = \frac{4+3}{4+2} = \frac{7}{6}$$

☞ *Quick check*   Find the slope of the line whose graph is plotted.

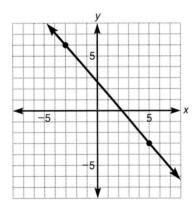

MASTERY POINTS

Can you
● Find the slope of a line given two points on the line?
● Remember the slope of a horizontal line and of a vertical line?
● Find the slope of a line given the graph of the line?

Exercise 6–3

Directions

Find the slope of the line passing through each of the following pairs of points. Draw the graph of the line. See example 6–3 A.

☛ *Quick Check*

Example $(-2,1)$ and $(3,4)$

Solution $m = \dfrac{(4) - (1)}{(3) - (-2)}$

$\qquad\quad = \dfrac{3}{5}$

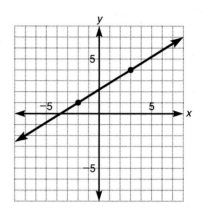

Directions   Find the slope of the line passing through each of the following pairs of points. Draw the graph of the line. See example 6–3 A.

1. $(5,2), (3,3)$

4. $(-2,3), (5,5)$

7. $(6,-7), (6,1)$

2. $(4,3), (2,2)$

5. $(-4,3), (2,3)$

8. $(-1,4), (-1,-3)$

3. $(-4,-1), (2,3)$

6. $(4,-4), (2,-4)$

Directions   Find the slope of the line passing through each of the following pairs of points. See example 6–3 A.

9. $(-3,1)$, $(0,3)$

10. $(4,0)$, $(-2,8)$

11. $(-3,9)$, $(-3,-5)$

12. $(5,0)$, $(-3,0)$

13. $(5,6)$, $(3,4)$

14. $(0,0)$, $(4,3)$

15. $(0,7)$, $(0,-8)$

16. $(4,-8)$, $(0,-2)$

17. $(-8,-3)$, $(-1,-2)$

18. $(-6,-5)$, $(-4,-3)$

19. $(-10,4)$, $(2,-5)$

20. $(7,-3)$, $(8,-7)$

21. $(5,7)$, $(-6,-3)$

22. $(9,4)$, $(-1,-1)$

Directions
Find the slope of each of the following lines. See example 6–3 B.

☛ *Quick Check*

Example

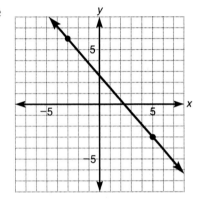

Solution   We note that the line passes through the points $(-3,6)$ and $(5,-3)$.

$$\text{Then slope } m = \frac{(6) - (-3)}{(-3) - (5)}$$

$$= \frac{9}{-8}$$

$$= -\frac{9}{8}$$

Directions   Find the slope of each of the following lines. See example 6–3 B.

23.

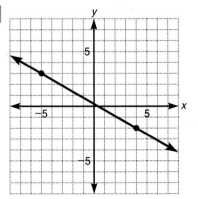

24.

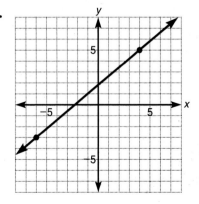

25.

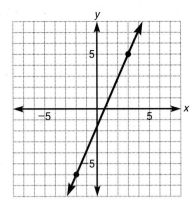

27.

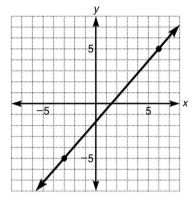

26.

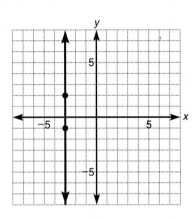

28.

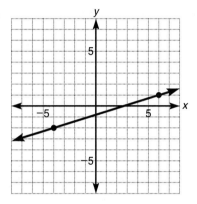

Directions Solve the following problems.	**Example**	A child is flying a kite. The kite is 35 feet above the ground, and the distance from the child to a point directly below the kite is 40 feet. If the string from the child to the kite forms a straight line, what is the slope of the string?
	Solution	Using $m = \dfrac{\text{vertical change}}{\text{horizontal change}}$, the vertical change is 35 feet and the horizontal change is 40 feet, then $$m = \frac{35\text{ ft}}{40\text{ ft}} = \frac{7}{8}$$

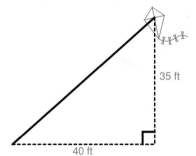

Directions   Solve the following problems.

29. The roof of a home rises vertically a distance of 8 feet through a horizontal distance of 12 feet. Find the pitch (slope) of the roof.

30. The roof of a factory rises vertically 6 feet through a horizontal run of 27 feet. What is the pitch of the roof?

31. A ladder leaning against the side of a building touches the building at a point 12 meters from the ground. If the foot of the ladder is 18 meters from the base of the building, what is the slope of the ladder?

32. A guy wire is attached to a telephone pole. If the wire is attached to the ground at a point 15 feet from the base of the pole and to the pole at a point 10 feet up on the pole, what is the slope of the wire?

33. A company's profits (P) are related to the number of items produced (x) by a linear equation. If profits rise by $1,000 for every 250 items produced, what is the slope of the graph of the equation?

34. A company's profits (P) are related to increases in a worker's average pay (x) by a linear equation. If the company's profits drop by $1,500 per month for every increase of $450 per year in the worker's average pay, what is the slope of the graph of the equation?

35. The diagram shows a linear representation of a jogger's heartbeat in beats per minute as his speed is increased (in feet per second). What is the slope of the line?

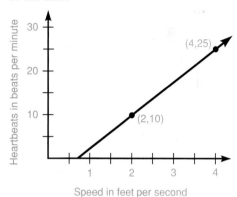

36. The diagram shows a linear representation of the bacteria count in a culture as related to the hours it exists. What is the slope of the line?

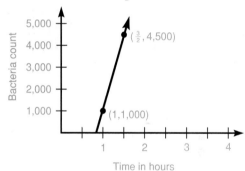

✺ Explore to Learn

Consider the line $2y + 3x = 6$. Graph this line carefully. Looking at the graph, find another point which is on the line. Take the (x, y) coordinate for this point and substitute the values into the equation. What happens? Take the (x, y) coordinate of some point which is not on the line and substitute the values into the equation. What happens? Make up a statement related to what happened about all the points in the plane and the line you graphed.

Review Exercises

1. Simplify the expression $[3 - 4(5 - 2)]$. See section 1–8.

2. Evaluate $4^2 - 2^3 - 3 \cdot 5 - 2 \cdot 3$. See section 1–8.

Directions   Simplify the following. Use only positive exponents. Assume all variables are nonzero. See section 3–4.

3. $(-2)^{-3}$

4. $\dfrac{a^{-2}b^3}{a^3 b^{-2}}$

5. $x^{-2} \cdot x^3 \cdot x^0$

6. When four times a number is increased by 12, the result is 64. What is the number? See section 2–7.

7. When $4\frac{1}{2}$ cans of water are added to a 12-ounce can of frozen lemonade concentrate, the resulting mixture is 10% lemon. What is the percent of pure lemon in the concentrate? See section 2–7.

Directions   Solve the following equations for y. See section 2–6.

8. $3x + y = -1$

9. $3x - 3y = 6$

6–4   The Equation of a Line

The Standard Form

The **standard form** of the equation of a straight line is stated here.

Standard Form of the Equation of a Line
$ax + by = c$
where a, b, and c are real numbers, $a \geq 0$, a and b not both zero.

To illustrate, the equations

$$2x + 3y = 6, \ 4x - 2y = 0, \ \text{and} \ x - 6y = -2$$

are written in standard form. In most situations, equations will be written in this form.

The Point-slope Form

Consider again the definition for the slope of a line

$$m = \frac{y_2 - y_1}{x_2 - x_1} \qquad (x_1 \neq x_2)$$

where (x_1, y_1) and (x_2, y_2) are coordinates of two *known* points on the line. Suppose we replace the known point (x_2, y_2) with *any other arbitrary point* (x, y) on the line. Then the slope is given by

$$m = \frac{y - y_1}{x - x_1} \qquad (x_1 \neq x)$$

$$m(x - x_1) = y - y_1 \qquad \text{Multiply both sides of the equation by } x - x_1$$

$$y - y_1 = m(x - x_1) \qquad \text{Symmetric property}$$

We call this the **point-slope form** of the equation of a line, where (x_1, y_1) is a known point on the line, m is the slope, and (x, y) is *any* other unknown point on the line.

Point-slope Form of the Equation of a Line

$$y - y_1 = m(x - x_1)$$

where (x_1, y_1) is a known point on the line and m is the slope of the line.

We can use this form to find the equation of the line if we know the slope of the line and the coordinates of at least one point on the line.

● **Example 6–4 A**

Finding the equation of a line using the point-slope form

1. Find the standard form of the equation of a line having slope $m = 2$ and passing through the point $(4, -3)$.

Using the point-slope form, we know $m = 2$ and $(x_1, y_1) = (4, -3)$.

$$y - y_1 = m(x - x_1)$$
$$y - (-3) = 2(x - 4) \qquad \text{Replace } y_1 \text{ with } -3, x_1 \text{ with 4, and } m \text{ with 2}$$
$$y + 3 = 2x - 8 \qquad \text{Multiply and subtract as indicated}$$
$$y = 2x - 11 \qquad \text{Add } -3 \text{ to both sides}$$
$$-2x + y = -11 \qquad \text{Add } -2x \text{ to both sides}$$
$$2x - y = 11 \qquad \text{Multiply each side by } -1$$

Note We wrote our final answer in *standard form, ax + by = c* where $a > 0$.

2. Find the standard form of the equation of the line passing through the points $(-3, 2)$ and $(5, 1)$.

We use the two points to find the slope and then choose either one of the two points to find the equation.

$$m = \frac{y_2 - y_1}{x_2 - x_1}$$
$$m = \frac{(1) - (2)}{(5) - (-3)} \qquad \text{Replace } y_2 \text{ with 1, } y_1 \text{ with 2,} \atop x_2 \text{ with 5, and } x_1 \text{ with } -3$$
$$m = \frac{-1}{8} = -\frac{1}{8}$$

Choosing *either* of the points $(-3, 2)$ or $(5, 1)$ together with the slope, we use the point-slope form.

$$y - y_1 = m(x - x_1)$$
$$y - 2 = -\frac{1}{8}[x - (-3)] \qquad \text{Use the point } (-3, 2). \text{ Replace } y_1 \text{ with 2, } x_1$$
$$\text{with } -3, \text{ and } m \text{ with } -\frac{1}{8}$$

$$y - 2 = -\frac{1}{8}(x + 3) \qquad x - (-3) = x + 3$$
$$8(y - 2) = -1(x + 3) \qquad \text{Multiply each side by 8}$$
$$8y - 16 = -x - 3 \qquad \text{Perform indicated multiplications}$$
$$8y = -x + 13 \qquad \text{Add 16 to both sides}$$
$$x + 8y = 13 \qquad \text{Add } x \text{ to each side}$$

Note If we had used the point (5,1), then

$$y - y_1 = m(x - x_1)$$
$$y - 1 = -\frac{1}{8}(x - 5)$$
$$8(y - 1) = -1(x - 5)$$
$$8y - 8 = -x + 5$$
$$8y = -x + 13$$
$$x + 8y = 13 \qquad \text{(Produces the same equation.)}$$

☛ *Quick check*   a. Find the equation of the line passing through $(-1,3)$ and having slope $m = -2$. Write the equation in standard form.
b. Find the equation of the line passing through the points $(-1,2)$ and $(3,4)$. Write the equation in standard form.

The Slope-intercept Form

Suppose a given line L, having slope m, passes through the point $(0,b)$, the y-intercept of the line. Using the point-slope form of the equation of a line,

$$y - b = m(x - 0) \qquad \text{Replace } y_1 \text{ with } b \text{ and } x_1 \text{ with } 0$$

Then $y - b = mx \qquad x - 0 = x$

and $y = mx + b \qquad \text{Add } b \text{ to each side}$

Slope ⟶   ⟵ y-intercept

We call $y = mx + b$ the **slope-intercept** form of the equation of a line.

Slope-intercept Form of the Equation of a Line

When a linear equation is stated in the form $y = mx + b$, m is the slope of the line and b is the y-intercept.

Note When we say that the y-intercept is b, this refers to the point $(0,b)$. When we say that the x-intercept is a, this refers to the point $(a,0)$.

We can use the slope-intercept form of the equation of a line to find the slope and the y-intercept of a line and to graph a linear equation in two variables.

● **Example 6–4 B**

Finding the slope and y-intercept from the equation of the line

1. Find the slope and the y-intercept of the line given by the equation $3y - 5x = 9$.

$$3y - 5x = 9 \qquad \text{Solve for } y$$
$$3y = 5x + 9 \qquad \text{Add } 5x \text{ to each side}$$
$$y = \frac{5x + 9}{3} \qquad \text{Divide each side by 3}$$
$$y = \frac{5}{3}x + 3 \qquad \text{Write in slope-intercept form by dividing each of the terms in the numerator by 3}$$

Then slope $m = \frac{5}{3}$ and y-intercept $b = 3$. [The y-intercept is the point $(0,3)$.]

☛ *Quick check*   Write the equation $3x + 4y = -12$ in slope-intercept form and determine the slope m and the y-intercept.

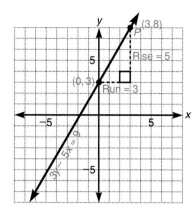

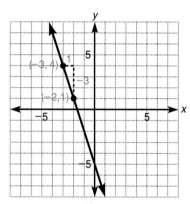

2. Graph the equation $3y - 5x = 9$ using slope and y-intercept. From example 1, we determined the slope $m = \dfrac{5}{3}$ and the y-intercept $b = 3$. Then the point $(0,3)$ is on the graph and we use the slope $\dfrac{5}{3}$ to find another point. To do this, we plot the y-intercept $(0,3)$. Using the slope $\dfrac{5}{3}$, from the point $(0,3)$ move 3 units to the *right* (the run, denominator) and from this point move 5 units *up* (the rise, numerator) to find a second point P $(3,8)$. Draw the line through this point P and the y-intercept to obtain the graph.

3. Graph the line through $(-3,4)$ having slope $m = -3$. First, plot the point $(-3,4)$. Write the slope

$$m = -3 = \frac{-3}{1} = \frac{\text{negative rise (fall)}}{\text{run}}$$

From the point $(-3,4)$, move 1 unit to the right and then 3 units *down* (because of the negative sign) to find a second point P $(-2,1)$. Draw the line through point P and the given point $(-3,4)$.

▶ ***Quick check***   a. Graph the equation $3x + y = 2$ using the slope m and the y-intercept b.

b. Graph the line through $(-3,2)$ with slope $m = \dfrac{1}{3}$.   ●

Given the slope of a line and its y-intercept, it is then possible to determine the equation of the line.

● **Example 6-4 C**

Finding the equation of a line from given information

1. Find the equation of the line having slope $m = \dfrac{2}{3}$ and passing through the point $(0,-4)$. Leave your answer in slope-intercept form.

Since the point $(0,-4)$ is on the y-axis, it is the y-intercept, so $b = -4$.

$$y = mx + b$$
$$y = \left(\frac{2}{3}\right)x + (-4) \qquad \text{Replace } m \text{ with } \tfrac{2}{3} \text{ and } b \text{ with } -4$$
$$y = \frac{2}{3}x - 4$$

2. Find the equation of the line with slope $m = 0$ and having y-intercept 3.

$$y = mx + b$$
$$y = (0)x + (3) \qquad \text{Replace } m \text{ with } 0 \text{ and } b \text{ with } 3$$
$$y = 0 + 3$$
$$y = 3$$

<u>Note</u> The equation is of the form $y = b$ so the graph is a horizontal line.

3. Find the equation of the line with undefined slope and passing through $(3,-4)$.

Since m is undefined, the line is vertical and the equation is of the form $x = a$. Thus, $x = 3$ is the equation.   ●

Parallel and Perpendicular Lines

Given two distinct straight lines in a plane, they will either be **parallel lines** (never meet no matter how far they are extended) or intersect in one point.

For two nonvertical lines to be parallel, they must *have the same slope and different y-intercepts.*

Slopes of Parallel Lines

Two distinct nonvertical lines having slopes m_1 and m_2 are parallel if and only if $m_1 = m_2$.

Note All vertical lines (whose slopes are undefined) are parallel to one another.

● **Example 6–4 D**
Parallel lines

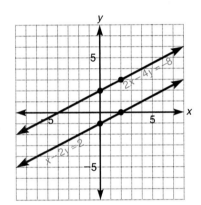

Show that distinct lines $x - 2y = 2$ and $2x - 4y = -8$ are parallel lines and graph the lines.

We solve each equation for y to write them in slope-intercept form $y = mx + b$.

$$
\begin{aligned}
x - 2y &= 2 & 2x - 4y &= -8 \\
-2y &= -x + 2 & -4y &= -2x - 8 \\
y &= \frac{-x + 2}{-2} & y &= \frac{-2x - 8}{-4} \\
y &= \frac{1}{2}x - 1 & y &= \frac{1}{2}x + 2 \\
m_1 &= \frac{1}{2} & m_2 &= \frac{1}{2}
\end{aligned}
$$

Since $m_1 = \dfrac{1}{2} = m_2$, the two lines are parallel.

Two lines that intersect in a single point can be *perpendicular.* **Perpendicular lines** make *right angles* with one another. Two nonvertical lines having slopes m_1 and m_2, respectively, are perpendicular if and only if the *product* of their slopes is -1.

Slopes of Perpendicular Lines

Two nonvertical lines having slopes m_1 and m_2 are perpendicular if and only if $m_1 m_2 = -1$.

To illustrate, lines having slopes $\dfrac{3}{4}$ and $-\dfrac{4}{3}$ are perpendicular since

$$\left(\frac{3}{4}\right)\left(-\frac{4}{3}\right) = -1$$

Note The slopes $\dfrac{3}{4}$ and $-\dfrac{4}{3}$ are **negative reciprocals** of each other. From this, we can conclude that the slopes of perpendicular lines will be negative reciprocals.

● **Example 6–4 E**
Perpendicular lines

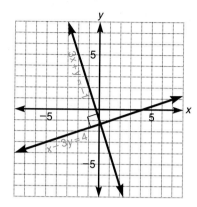

Show that the lines $x - 3y = 4$ and $3x + y = -1$ are perpendicular lines and graph the lines.

We solve for y to write each equation in slope-intercept form $y = mx + b$.

$$x - 3y = 4 \qquad\qquad 3x + y = -1$$
$$-3y = -x + 4 \qquad\qquad y = -3x - 1$$
$$y = \frac{1}{3}x - \frac{4}{3} \qquad\qquad m_2 = -3$$
$$m_1 = \frac{1}{3}$$

Since $m_1 m_2 = \dfrac{1}{3} \cdot (-3) = -1$, the two lines are perpendicular.

Note $\dfrac{1}{3}$ and -3 are *negative reciprocals* of each other. The box where the lines intersect indicates perpendicular lines.

☞ **Quick check** Find the slopes of the lines $3x + y = 4$ and $2x - y = 1$ and determine if the lines are parallel, perpendicular, or neither. ●

All vertical lines are perpendicular to all horizontal lines even though the product of their slopes does not exist. For example, the lines $x = -2$ and $y = 4$ are perpendicular.

MASTERY POINTS

Can you
• Write the equation of a line in standard form?
• Find the equation of a line knowing the slope and a point or two points on the line?
• Find the slope and y-intercept of a line knowing the equation of the line?
• Graph a linear equation in two variables using the slope and y-intercept?
• Graph a linear equation in two variables using the slope and a point of the line?
• Find the equation of a line given the slope and the y-intercept?
• Determine whether two lines are parallel or perpendicular?

Exercise 6–4

Directions
Find the equation of the line passing through the given point and having the given slope. Write the equation in standard form $ax + by = c$, where a, b, and c are integers, $a \geq 0$. See example 6–4 A–1, C.

☞ *Quick Check*

Example $(-1, 3); m = -2$

Solution Use the *point-slope form* of the equation of a line.

$$y - y_1 = m(x - x_1)$$
$$y - (3) = (-2)[x - (-1)] \qquad \text{Replace } y_1 \text{ with 3, } x_1 \text{ with } -1, \text{ and } m \text{ with } -2$$
$$y - 3 = -2(x + 1) \qquad \text{Definition of subtraction}$$
$$y - 3 = -2x - 2$$
$$y = -2x + 1 \text{ or } 2x + y = 1 \qquad \text{Written in standard form } ax + by = c$$

Directions   Find the equation of the line passing through the given point and having the given slope. Write the equation in standard form $ax + by = c$, where a, b, and c are integers, $a \geq 0$. See example 6–4 A–1, C.

1. (1,3); $m = 4$

2. (−3,1); $m = 2$

3. (0,−1); $m = -2$

4. (0,4); $m = -3$

5. (2,7); $m = \dfrac{3}{5}$

6. (−1,5); $m = \dfrac{4}{7}$

7. (5,0); $m = -\dfrac{7}{6}$

8. (−3,0); $m = -\dfrac{5}{8}$

9. (−3,−7); $m = \dfrac{5}{4}$

10. (−9,−2); $m = \dfrac{3}{2}$

11. (2,−3); $m = 0$

12. (−7,3); $m = 0$

13. (1,−8); slope is undefined.

14. (0,4); slope is undefined.

Directions
Find the equation of the line passing through each pair of given points. Write the equation in standard form $ax + by = c$, where a, b, and c are integers, $a \geq 0$. See example 6–4 A–2.

📌 **Quick Check**

Example   (−1,2) and (3,4)

Solution   Find the slope using

$$m = \frac{y_2 - y_1}{x_2 - x_1}$$

$$= \frac{(4) - (2)}{(3) - (-1)} \qquad \text{Replace } y_2 \text{ with 4, } y_1 \text{ with 2, } x_2 \text{ with 3, and } x_1 \text{ with } -1$$

$$= \frac{2}{4} \qquad \text{Subtract as indicated}$$

$$= \frac{1}{2}$$

Use the point-slope form of the equation of a line and *one of the given points*, (3,4).

$$y - y_1 = m(x - x_1)$$

$$y - (4) = \left(\frac{1}{2}\right)[x - (3)] \qquad \text{Replace } y_1 \text{ with 4, } x_1 \text{ with 3, and } m \text{ with } \frac{1}{2}$$

$$2y - 8 = x - 3 \qquad \text{Multiply each side by 2 to clear denominator}$$

$$2y = x + 5 \qquad \text{Add 8 to each side}$$

$$-x + 2y = 5 \qquad \text{Subtract } x \text{ from each side}$$

$$x - 2y = -5 \qquad \text{Multiply each side by } -1$$

Directions   Find the equation of the line passing through each pair of given points. Write the equation in standard form $ax + by = c$, where a, b, and c are integers, $a \geq 0$. See example 6–4 A–2.

15. (2,1) and (6,3)

16. (3,2) and (4,5)

17. (−3,2) and (5,−1)

18. (−6,2) and (4,−3)

19. (0,4) and (−2,2)

20. (1,7) and (0,−3)

21. (−6,0) and (1,−1)

22. (5,6) and (5,0)

23. (0,8) and (−3,0)

24. (4,0) and (0,−1)

25. (0,0) and (−5,8)

26. (7,−1) and (0,0)

Directions
Write the following equations in slope-intercept form and determine the slope m and y-intercept. See example 6–4 B–1.

🏴 **Quick Check**

Example $3x + 4y = -12$

Solution Solve for y.

$$3x + 4y = -12$$
$$4y = -3x - 12 \qquad \text{Add } -3x \text{ to each side}$$
$$y = \frac{-3x - 12}{4} \qquad \text{Divide each side by 4}$$
$$y = -\frac{3}{4}x - 3 \qquad \text{Divide } -3 \text{ and } -12 \text{ by 4}$$

Then slope $m = -\dfrac{3}{4}$ and y-intercept $b = -3$, which is the point $(0, -3)$.

Directions   Write the following equations in slope-intercept form and determine the slope m and y-intercept. See example 6–4 B–1.

27. $x + y = 2$

28. $y - x = 3$

29. $3x + y = -2$

30. $y - 4x = 5$

31. $2x + 5y = 10$

32. $3x + 2y = 8$

33. $7x - 2y = -4$

34. $9x - 3y = 3$

35. $2y - 9x = -6$

36. $4y - 5x = -12$

37. $8x - 9y = 1$

38. $-7x + 4y = -5$

Directions
Use the slope m and the y-intercept b to graph the following equations. See example 6–4 B–2.

🏴 **Quick Check**

Example $3x + y = 2$

Solution Write the equation in slope-intercept form $y = mx + b$.
$y = -3x + 2$ so $m = -3$ and $b = 2$

Since $-3 = \dfrac{-3}{1}$, from the point $(0,2)$, move 1 unit to the right and 3 units down to get a second point $P(1, -1)$. Draw the straight line through the two points.

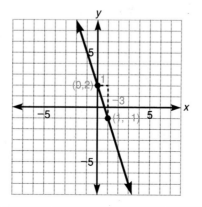

Directions   Use the slope m and the y-intercept b to graph the following equations. See example 6–4 B–2.

39. $y = 2x - 4$

43. $y = \dfrac{2}{3}x - 1$

47. $5x - 2y = 6$

40. $y = 3x + 1$

44. $y = -\dfrac{4}{3}x + 3$

48. $3x - 4y = 8$

41. $y = -5x + 2$

45. $2x + y = -3$

49. $4x + 3y = -9$

42. $y = -2x - 3$

46. $3x + y = -2$

50. $3x + 8y = -16$

Directions
Graph the line through the given point having the given slope m. See example 6–4 B–3.

⚐ *Quick Check*

Example $(-3,2)$; $m = \dfrac{1}{3}$

Solution   Plot the point $(-3,2)$. Then move 3 units to the right and 1 unit up to find a second point, $(0,3)$. Draw the straight line through the two points.

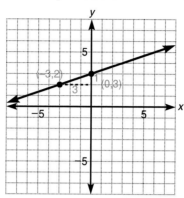

Directions   Graph the line through the given point having the given slope m. See example 6–4 B–3.

51. $(4,3)$; $m = \dfrac{3}{4}$

54. $(2,-1)$; $m = -\dfrac{3}{2}$

57. $(0,3)$; $m = -1$

52. $(-2,-2)$; $m = \dfrac{2}{3}$

55. $(-4,-5)$; $m = 4$

58. $(1,5)$; $m = -3$

56. $(2,-3)$; $m = 2$

53. $(-1,2)$; $m = -\dfrac{1}{3}$

Directions Graph the line through the given point having the given slope m. (*Hint:* For exercises 59 to 62, recall the types of lines that have slope $m = 0$ or undefined slope.)

59. (5,6); $m = 0$

61. $(-1,-1)$; slope is undefined

60. $(-3,1)$; $m = 0$

62. (4,7); slope is undefined

Directions Find the equation of the line (in slope-intercept form) having the following characteristics. See example 6–4 C.

63. Having slope $m = 4$ and y-intercept $b = -1$

66. Having slope $m = \dfrac{3}{4}$ and passing through (0,5)

64. Having slope $m = -8$ and y-intercept $b = 6$

67. Having slope $m = 0$ and passing through $(0,-1)$

65. Having slope $m = -\dfrac{5}{3}$ and passing through (0,2)

68. Having slope $m = 0$ and y-intercept 4

Directions Find the slope of each line and determine if each pair of lines is parallel, perpendicular, or neither. See examples 6–4 D and E.	☛ *Quick Check*
	Example $3x + y = 4$ $2x - y = 1$ **Solution** Write each equation in slope-intercept form. $\begin{aligned} 3x + y &= 4 \\ y &= -3x + 4 \\ \text{so } m_1 &= -3 \end{aligned}$ $\qquad \begin{aligned} 2x - y &= 1 \\ y &= 2x - 1 \\ \text{so } m_2 &= 2 \end{aligned}$ Since $m_1 \neq m_2$, the lines are *not* parallel; and since $m_1 m_2 = -3 \cdot 2 = -6$, the lines are not perpendicular. Therefore the lines are neither.

Directions Find the slope of each line and determine if each pair of lines is parallel, perpendicular, or neither. See examples 6–4 D and E.

69. $x + y = 4$
$\quad\; x + y = -7$

71. $x + y = 5$
$\quad\; -x + y = -1$

70. $-x + y = 9$
$\quad\; -x + y = -2$

72. $-x + y = 8$
$\quad\;\; x + y = 3$

73. $3x - y = 2$
$6x + 2y = -5$

74. $x + 2y = 5$
$-6x + 3y = -1$

75. $5x + y = 1$
$10x + 3y = 2$

76. $8x - 9y = 1$
$16x + 7y = 3$

77. $4x - y = 5$
$12x - 3y = 1$

78. $4x + 6y = -3$
$6x + 9y = 4$

79. $y = 5$
$x = 2$

80. $x - 3y = -6$
$3x - 9y = -18$

Directions
Find the equation for the given line using the slope, one of the points, and the point-slope form $y - y_1 = m(x - x_1)$, where point (x_1, y_1) is a known point on the line. Write the equation in standard form.

Example The line passes through points $(-4, 3)$ and $(3, -4)$.

Solution Then $m = \dfrac{(3) - (-4)}{(-4) - (3)}$

$= \dfrac{7}{-7} = -1$

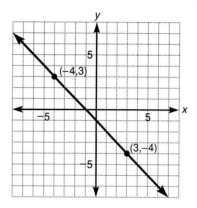

Use $y - y_1 = m(x - x_1)$ and the point $(-4, 3)$.

$$y - (3) = (-1)[x - (-4)]$$
$$y - 3 = -1(x + 4)$$
$$y - 3 = -x - 4$$
$$y = -x - 1$$
$$x + y = -1$$

Directions Find the equation for the given line using the slope, one of the points, and the point-slope form $y - y_1 = m(x - x_1)$, where point (x_1, y_1) is a known point on the line. Write the equation in standard form.

81.

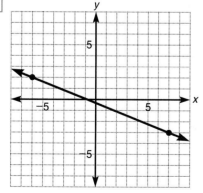

82.

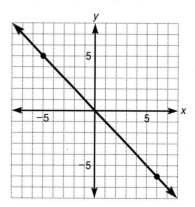

83.

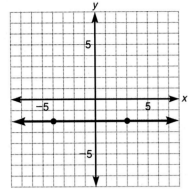

85.

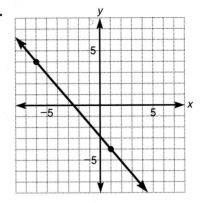

84.

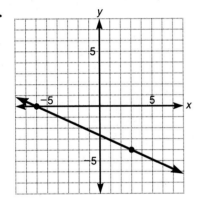

86.

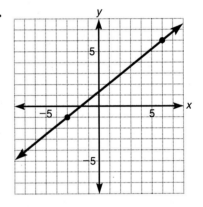

 ## Explore to Learn

The given data represents the world's record for the 1-mile run in the years shown. Some of the times have been converted to seconds. Finish converting all the times to seconds, then plot the data using (year, time in seconds) ordered pairs. Use the first and last data points (for 1875 and 1993) to create the equation of the straight line that models this data. Graph the line. Does it seem to come close to the data? Use the equation of the line to predict when someone will run a 3:42 mile.

Year	Time	Time (seconds)
1875	4:24.5	264.5
1895	4:17	257
1915	4:12.6	
1923	4:10.4	
1934	4:06.8	
1945	4:01.4	
1954	3:59.4	
1965	3:53.6	
1975	3:49.4	
1979	3:49.0	
1980	3:48.8	
1981	3:48.40	
1985	3:46.31	
1993	3:44.39	224.39

Review Exercises

1. From the numbers 0, -3, $\frac{3}{4}$, 9, $-\frac{7}{8}$, $\frac{9}{3}$, and $\frac{0}{4}$, choose the integers. See section 1–4.

Directions   Find the solution set of the following equations. See section 2–5.

2. $4y + 1 = 2y - 5$

3. $3(2y - 1) = -1(y + 2)$

4. $\dfrac{3x}{5} - \dfrac{x}{2} = 1$

Directions   Factor completely the following expressions. See sections 4–2, 4–3, and 4–4. If the expression does not factor, so state.

5. $3x^2 + 4x - 4$

6. $x^2 + 5x + 7$

7. $8y^2 - 32x^2$

8. A 100-foot extension cord is to be cut into two pieces so that one piece is 17 feet longer than the other. What is the length of each piece? See section 2–7.

9. A grocer wishes to mix candy worth \$4 per pound with 50 pounds of candy worth \$1 per pound. How many pounds of the \$4 candy must be used to make a mixture worth \$3 per pound? See section 2–7.

6–5   Graphing Linear Inequalities in Two Variables

In section 6–2, we graphed linear equations in two variables such as $2x + y = 6$. In this section, we consider the graphs of *linear inequalities* in two variables such as

$$2x + y < 6 \qquad \text{and} \qquad 2x + y \geq 6$$

where the "is equal to" sign has been replaced by one of the four inequality symbols

$$<, \leq, >, \text{ or } \geq$$

The graph of the equation $2x + y = 6$ is a straight line in the plane. This line divides the plane into two regions called *half-planes* and serves as the *boundary line* for each half-plane. See figure 6–11.

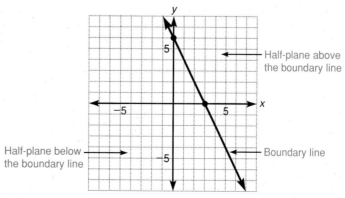

Figure 6–11

Graph of a Linear Inequality

The graph of any linear inequality in two variables is a half-plane.

The inequality $2x + y \geq 6$ is read "$2x + y$ is greater than or equal to 6." So,

$$2x + y > 6 \quad \text{or} \quad 2x + y = 6$$

The graph *will consist of the boundary line* and the proper half-plane. The inequality $2x + y > 6$ is read "$2x + y$ is greater than 6" so the boundary line *is not* a part of the graph of the inequality. To indicate this, the boundary line is drawn as a dashed line rather than a solid line.

1. For $2x + y > 6$, the boundary line $2x + y = 6$ is a *dashed* line to show that the points on the line are *not* included. See figure 6–12(a).

2. For $2x + y \geq 6$, the boundary line $2x + y = 6$ is a *solid* line, to show that the points on the line are included. See figure 6–12(b).

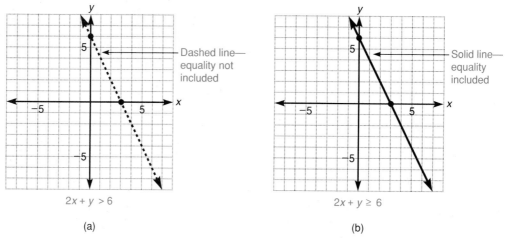

(a)   (b)

Figure 6–12

To determine the proper half-plane to shade, we choose a *test point* in one of the half-planes [usually the origin, (0,0)] and substitute into the inequality.

1. If the test point satisfies the inequality, shade the half-plane containing that point.

2. If the test point *does not satisfy* the inequality, shade the half-plane that does not contain the point.

<u>Note</u> We will use the origin (0,0) as the test point in all cases except when the boundary line passes through the origin.

To illustrate, consider the inequality $2x + y > 6$ and the test point (0,0).

$$2x + y > 6$$
$$2(0) + (0) > 6 \qquad \text{Replace } x \text{ with 0 and } y \text{ with 0}$$
$$0 + 0 > 6$$
$$0 > 6 \qquad \text{(False)}$$

Shade the half-plane that *does not* contain the origin (0,0). See figure 6–13(a). If the inequality was $2x + y \geq 6$, the graph would be as shown in figure 6–13(b).

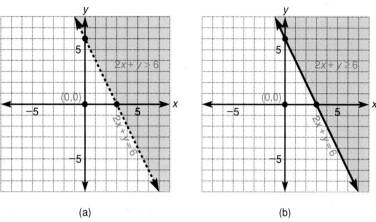

(a)   (b)

Figure 6–13

Graphing a Linear Inequality in Two Variables

1. Replace the inequality symbol with the equality symbol.
2. Graph the resulting equation for the boundary line which will be a
   a. *solid line* if the inequality symbol is $\leq$ or $\geq$,
   b. *dashed line* if the inequality symbol is $<$ or $>$.
3. Choose some test point that is *not* on the line [usually the origin (0,0)] and substitute the coordinates of the point into the inequality.
4. If the resulting statement is
   a. *true,* shade the half-plane containing the test point,
   b. *false,* shade the half-plane that *does not* contain the test point.

● **Example 6–5 A**

Graphing a linear inequality

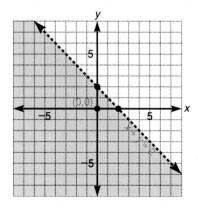

Graph the following linear inequalities.

1. $x + y < 2$
   a. Replace $<$ with $=$ to get the equation $x + y = 2$.
   b. Graph $x + y = 2$ as a *dashed* line since we have $<$.
   c. Since the boundary line does not go through the origin, choose test point (0,0).

$$x + y < 2$$
$$(0) + (0) < 2 \qquad \text{Replace } x \text{ with 0 and } y \text{ with 0}$$
$$0 < 2 \qquad \text{(True)}$$

   d. Shade the half-plane containing the origin (0,0).

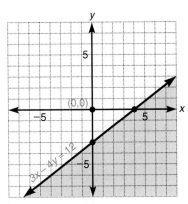

2. $3x - 4y \geq 12$
   a. Replace $\geq$ with $=$ to get the equation $3x - 4y = 12$.
   b. Graph $3x - 4y = 12$ as a *solid* line since we have $\geq$.
   c. Choose test point $(0,0)$.

$$3x - 4y \geq 12$$
$$3(0) - 4(0) \geq 12$$
$$0 \geq 12 \quad \text{(False)}$$

   d. Shade the half-plane that *does not* contain the origin $(0,0)$.

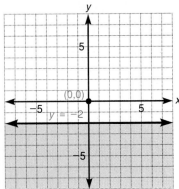

3. $y \leq -2$
   a. Replace $\leq$ with $=$ to get the equation $y = -2$.
   b. Graph $y = -2$ as a *solid* horizontal line since we have $\leq$.
   c. Choose the test point $(0,0)$.

$$y \leq -2$$
$$0 \leq -2 \quad \text{(False)}$$

   d. Shade the half-plane that *does not* contain the origin.

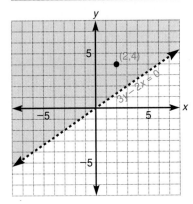

4. $3y - 2x > 0$
   a. Replace $>$ with $=$ to get the equation $3y - 2x = 0$.
   b. Graph $3y - 2x = 0$ as a *dashed* line since we have the inequality $>$.
   c. Since the graph goes through the origin, choose another test point not on the line, say $(2,4)$.

$$3y - 2x > 0$$
$$3(4) - 2(2) > 0 \quad \text{Replace } x \text{ with 2 and } y \text{ with 4}$$
$$12 - 4 > 0$$
$$8 > 0 \quad \text{(True)}$$

   d. Shade the half-plane containing the point $(2,4)$.

☞ *Quick check* Graph the linear inequality. $2y + x < 5$

The Casios have a graph mode for inequalities. The inequality must be solved for y. For example 4 above, to graph the solution to the inequality $3y - 2x > 0$, first solve for y to get $y > \frac{2}{3}x$, set $\boxed{\text{RANGE} -9, 9, 1, -6, 6, 1}$ then proceed:

Casio fx 7700/9700GE $\boxed{\text{MENU}}$ 6   Graph mode
$\boxed{\text{F3:TYPE}}$ $\boxed{\text{F4:INEQ}}$ Graph inequalities, not equalities
2 $\boxed{a^b/c}$ 3 $\boxed{\text{X,}\theta\text{,T}}$ $\boxed{\text{F1:STO}}$ $\boxed{\text{F3:Y}>}$ Enter the inequality
$\boxed{\text{F6:DRAW}}$ Draw the graph

Exercise 6–5

Directions

Complete the graph of the inequality by shading the correct half-plane. The correct boundary line has been drawn. See example 6–5 A.

Examples a. $x \geq -3$

Solutions Since (0,0) makes $x \geq -3$ true, $(0) \geq -3$, shade to the right of $x = -3$.

b. $4x - y < 8$

Since (0,0) makes $4x - y < 8$ true, $4(0) - (0) < 8$, shade to the left of $4x - y = 8$.

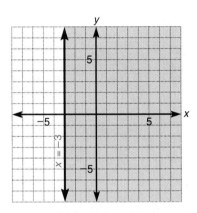

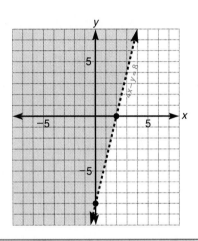

Directions   Complete the graph of the inequality by shading the correct half-plane. The correct boundary line has been drawn. See example 6–5 A.

1. $x < 4$

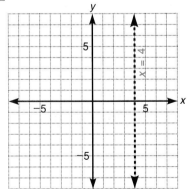

2. $y \geq -5$

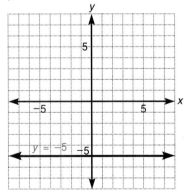

3. $x + y \geq 4$

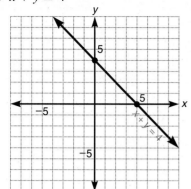

5. $3x - y \leq 4$

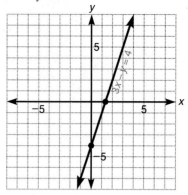

4. $x - y < 1$

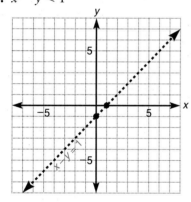

6. $2x - 4y > 8$

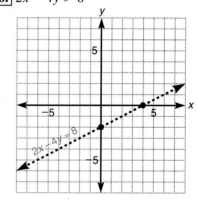

7. $3x + 5y \geq 0$

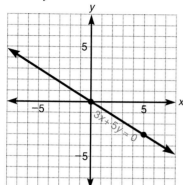

9. $2y \leq 3x$

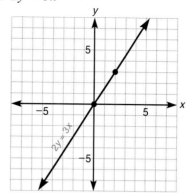

8. $y - 2x < 0$

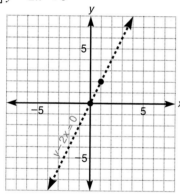

10. $-4x \geq y$

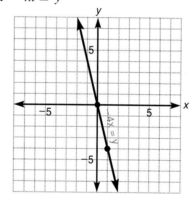

Directions
Graph the given linear inequality. See example 6–5 A.

🏴 *Quick Check*

Example $2y + x < 5$

Solution Graph the equation $2y + x = 5$ making the line *dashed*. Substitute test point $(0,0)$ into the inequality.

$$2y + x < 5$$
$$2(0) + (0) < 5 \qquad \text{Replace } x \text{ by 0 and } y \text{ by 0}$$
$$0 + 0 < 5$$
$$0 < 5 \qquad \text{(True)}$$

Shade the half-plane containing $(0,0)$.

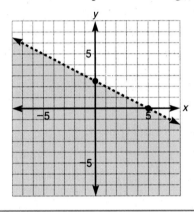

Directions   Graph the given linear inequality. See example 6–5 A.

11. $x + y \le -2$

14. $x - y \ge -5$

17. $y \ge 4x + 1$

12. $x + y > 3$

15. $x + 3y > 1$

18. $y < 2x - 3$

13. $x - y < 6$

16. $x + 2y \le -2$

19. $3x - 5y < -15$

20. $2x - 7y > 14$

21. $5x - 4y \leq 20$

22. $5x + 2y \leq -10$

23. $x > 3$

24. $x > -4$

25. $x \leq -2$

26. $x < 5$

27. $x + 4 > 0$

28. $x + 5 \leq -2$

29. $y + 3 \leq 5$

30. $y - 2 > 1$

31. $y > x - 2$

32. $3x \geq y$

33. $x < -y$

34. $y \leq 4x$

35. $x \geq 3y$

36. $x + y \geq 0$

37. $8x - 3y \geq 0$

38. $x - y < 0$

39. $3x + 2y < 0$

40. Write an inequality describing all points in the plane that lie (a) below the x-axis, (b) to the right of the y-axis.

41. Write inequalities describing all points that lie in (a) quadrant I, (b) quadrant II, (c) quadrant III, (d) quadrant IV.

⚜ Explore to Learn

Suppose a computer had to compute the value of $\dfrac{5x + 2}{x}$ several thousand times per second for positive values of x. Division shows that this is equivalent to $5 + \dfrac{2}{x}$. Suppose further that an error of 1% can be tolerated in the calculation. The value of $\dfrac{2}{x}$ gets smaller and smaller as x gets larger and larger, which means that for some value of x the difference between $5 + \dfrac{2}{x}$ and 5 should be less than 1%. In this case, the computer

does not need to do the computation; it knows the result is within 1% of 5. By trial and error find the smallest value of x, to the nearest 0.5, such that the computer does not have to actually do the computation. Note that this is equivalent to solving the inequality

$$\dfrac{\dfrac{5x + 2}{x} - 5}{\dfrac{5x + 2}{x}} \le 0.01.$$ This inequality does not

correspond to a straight line. Formal methods for solving such an inequality are studied in more advanced courses.

Review Exercises

Directions Find the domain of the following rational expressions. See section 5–1.

1. $\dfrac{2x + 1}{x - 3}$

2. $\dfrac{y - 4}{y^2 + 3y - 10}$

3. $\dfrac{4 - x}{x^2 + 1}$

Directions Find the standard form of the equation of the straight line having the following characteristics. See section 6–4.

4. Through points $(-2, 3)$ and $(4, 1)$

5. Having slope $\dfrac{3}{4}$ and y-intercept -2

6. Vertical line through $(-3, 6)$

7. Horizontal line through $(8, -3)$

8. One number is 33 more than another. The smaller number is one fourth of the larger number. Find the numbers. See section 2–8.

9. Mario can do a job in 6 hours. Working together, Mario and Teresa can do the job in $3\dfrac{1}{2}$ hours. How long would it take Teresa to do the job working alone? See section 5–7.

10. How many gallons of 2% milk must be mixed with skim milk to make 20 gallons of $\dfrac{1}{2}$% milk? See section 2–7.

6–6   Functions Defined by Linear Equations in Two Variables

Functions

In section 6–1, we studied equations in two variables such as

$$3x + 2y = 4 \quad \text{or} \quad y = 4 - 3x$$

which related values for x and y that satisfied the given equation. We showed this relationship by ordered pairs (x, y), where, in *most* cases, the value of y was dependent on a chosen value for x.

In all phases of mathematics, from the most elementary to the most sophisticated, the idea of a function is a cornerstone for each mathematical development. Phrases such as "price is a function of cost" or "price is determined by the cost to manufacture," in the business world, and "velocity is a function of time" or "velocity depends on time," of a falling object in science, are common. What is being said in each case is that "the price of an item will change when the cost of producing the item changes" and "the velocity of a falling object will change as the falling time changes."

In mathematics, a function is a special kind of relation, so we first define a **relation**.

Relation

A **relation** is any set of ordered pairs.

Thus, the set of ordered pairs $\{(1, 2), (-3, 6), (0, -5), (3, 4)\}$ is a relation. Every relation has a **domain** and a **range**.

Domain and Range of a Relation

1. The **domain** of a relation is the set of all first components of the ordered pairs.
2. The **range** of a relation is the set of all second components of the ordered pairs.

Thus, in the relation $\{(1, 2), (-3, 6), (0, -5), (3, 4)\}$,

$$\text{the domain } = \{1, -3, 0, 3\} \text{ and the range } = \{2, 6, -5, 4\}$$

The correspondence between the elements of the domain and the elements of the range is shown in figure 6–14.

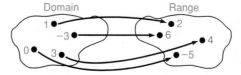

Figure 6–14

Recall, we initially said that a function is a special relation—a set of ordered pairs with a special characteristic.

Function

A **function** is a relation in which no two distinct ordered pairs have the same first component.

From this definition, each element of the domain is assigned to one and only one element of the range.

Note Since a function is a relation, the **domain of a function** is the set of all first components of the ordered pairs, and the **range of a function** is the set of all second components of the ordered pairs.

● **Example 6–6 A**

Determining whether a relation is a function

Determine whether the following relations are functions. If the relation is a function, state the domain and range.

1. $\{(1, 2), (-3, 6), (0, -5), (3, 4)\}$
   No two distinct ordered pairs have the same first component, so the relation is a function. The domain is $\{-3, 0, 1, 3\}$, and the range is $\{-5, 2, 4, 6\}$.

2. $\{(-3, 5), (2, 4), (2, -6), (-7, 0)\}$
   Since the ordered pairs $(2, 4)$ and $(2, -6)$ have the same first component, this *is not* a function.

3. $\{(1, 3), (-2, 3), (0, 3), (4, 3)\}$
   Even though the second component of every ordered pair is the same number, this does not violate the definition. This relation is a function. The domain is $\{-2, 0, 1, 4\}$, and the range is $\{3\}$.

☞ *Quick check*   Determine whether the following relation is a function. If it is a function, state the domain and range.
$\{(-2, -1), (-1, 0), (0, 1), (1, 2)\}$ ●

Vertical Line Test for a Function

Many useful functions have infinitely many ordered pairs and must be defined by an equation that tells how to get the second component when we are given the first component.

   Usually we use x and y to denote the variables in a mathematical function. By our definition, **for y to be a function of x, the two variables must be related so that for each value assigned to x, there is a unique (one and only one) corresponding value assigned to y.** This is a very important concept for us to remember. When we use an equation to determine the outcome in a given situation, we want our equation to be of the type that gives us only one result.

   Recall that two or more points having the same first component lie on the same vertical line. Also we have learned that a function cannot have two or more ordered pairs having the same first component (the abscissa of the point). These facts lead us to a visual test for determining if a particular graph does or does not represent a function.

Vertical Line Test for a Function

If every vertical line drawn in the plane intersects the graph of a relation in *at most one point,* the relation is a function.

● **Example 6–6 B**

Using the vertical line test

Determine by the vertical line test if the following graphs represent functions.

1.

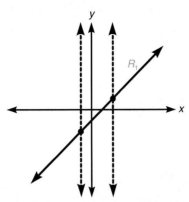

2.

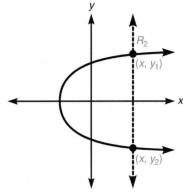

Since any vertical line drawn in the plane will intersect the graph of relation R_1 in *at most one point*, R_1 is a function.

Many lines can be drawn in the plane that will intersect the graph of relation R_2 *in two points*, so R_2 *is not* a function.

⚐ **Quick check**   Determine by the vertical line test if the graph represents a function.

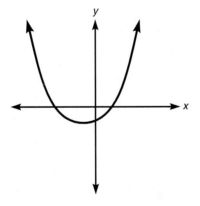

Functions Defined by Formulas: $f(x)$ Notation

It is impractical to specify all functions as sets of ordered pairs, or as tables. Most functions have too many ordered pairs for this, and some have an unlimited number of ordered pairs. There are many notations for functions; one of the most popular in mathematics is called "f of x" notation. In this notation[2] x represents any value in the domain, and $f(x)$ (read f of x) represents the corresponding range element. The lowercase letters g and h are also commonly used to denote a function. For example, if f is the function $\{(-1, 6), (0, 8), (1, 3), (2, 5)\}$, then $f(-1)$ is 6, $f(0)$ is 8, $f(1)$ is 3, and $f(2)$ is 5.

An expression is used to describe the range values of a function whenever possible. For example, if we needed a function that would double a number and add 1 to the result then we will describe this by writing $f(x) = 2x + 1$. For

[2] $f(x)$ notation was first used in a standard way by Leonhard Euler in 1734.

the domain elements -3, 5, and 0, we could compute the corresponding range elements as follows:

$$f(-3) = 2(-3) + 1 = -5 \qquad \text{Replace } x \text{ in } f(x) = 2x + 1 \text{ by } -3$$
$$f(5) = 2(5) + 1 = 11 \qquad \text{Replace } x \text{ in } f(x) = 2x + 1 \text{ by } 5$$
$$f(0) = 2(0) + 1 = 1 \qquad \text{Replace } x \text{ in } f(x) = 2x + 1 \text{ by } 0$$

Linear Functions

Throughout this chapter we have been graphing lines. Straight-line graphs (except for vertical lines) are graphs of **linear functions**.

Definition of a Linear Function

A linear function is a function that can be written in the form

$$f(x) = mx + b$$

where m and b are real numbers, and m is the slope and the point $(0, b)$ is the y-intercept.

Since a function is a set of ordered pairs, it can be graphed. Example 6–6 C illustrates the ideas above.

● **Example 6–6 C**

Graphing a linear function

Given the linear equation $3x - y = 2$,

a. Rewrite the equation as a linear function.

b. Give the set of ordered pairs that correspond to the given set of domain elements $\{-2, -1, 0, 1, 2\}$.

c. List the ordered pairs from part (b) using a table.

d. Graph the function.

a. To be able to write an equation in x and y as a function, we must solve the equation for y in terms of x and replace y with $f(x)$.

$$
\begin{array}{ll}
3x - y = 2 & \text{Original equation} \\
3x = y + 2 & \text{Add } y \text{ to both sides} \\
3x - 2 = y & \text{Subtract 2 from both sides} \\
y = 3x - 2 & \text{Symmetric property} \\
f(x) = 3x - 2 & \text{Replace } y \text{ with } f(x)
\end{array}
$$

b.

x	$f(x)$	Ordered pair
$x = -2$	$f(-2) = 3(-2) - 2 = -8$	$(-2, -8)$
$x = -1$	$f(-1) = 3(-1) - 2 = -5$	$(-1, -5)$
$x = 0$	$f(0) = 3 \cdot 0 - 2 = -2$	$(0, -2)$
$x = 1$	$f(1) = 3 \cdot 1 - 2 = 1$	$(1, 1)$
$x = 2$	$f(2) = 3 \cdot 2 - 2 = 4$	$(2, 4)$

c.

x	-2	-1	0	1	2
y	-8	-5	-2	1	4

Note In the table we used y in place of $f(x)$. We generally replace $f(x)$ with y when we are going to graph a function.

We will now plot the values from the table that are the ordered pairs from part (b), and draw the straight line through these points.

d. The graph is shown.

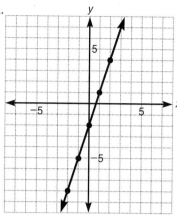

<u>Note</u> In the function of example 6–6 C, we selected five elements from the domain of the function. Unless we specify a domain, it is understood that the domain is all real numbers for which the function is defined.

☛ *Quick check*   Given the following linear equation (a) rewrite the equation as a linear function, (b) give the set of ordered pairs that correspond to the given set of domain elements, (c) list the ordered pairs from part (b) using a table, and (d) graph the function. $y - 3x = 2; \{-3, -1, 0, 5\}$ ●

A Word on Notation

In mathematics we will describe a function by writing $f(x)$, or $h(x)$, or $g(x)$, or something similar. In a programming language such as BASIC we might use a statement $A = ABS(x)$. In this case, ABS is the name of the function that computes $|x|$.

In science and engineering the "f of x" notation is often not used, just implied. For example, a scientist might write $F = \frac{9}{5}C + 32$ is the function that changes a temperature given in $°C$ (degrees Celsius) to $°F$ (degrees Fahrenheit). A mathematician would prefer $f(x) = \frac{9}{5}x + 32$, but this abstract statement is not as meaningful as the first statement.

When discussing arbitrary functions involving the variables x and y, we often see the statement

$$y = f(x)$$

This states that "y is a function of x."

Graphs of Linear Functions

A function is a set of ordered pairs (x, y). The x values may be substituted into an expression defining the function to obtain the corresponding y value. That is, each y in an ordered pair (x, y) is the value of $f(x)$. As a practical matter it is easier to graph any function by replacing the symbol $f(x)$ by y. Example 6–6 D illustrates this.

● **Example 6–6 D**

Graphing a linear function

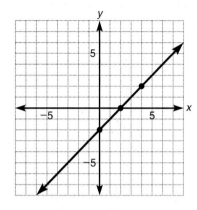

Graph the linear function $f(x) = x - 2$.

Replace $f(x)$ by y to obtain $y = x - 2$. This is a straight line and can be graphed by the methods illustrated earlier in this chapter to obtain the graph shown. One way is to compute the x- and y-intercepts and an arbitrary third point, and plot them.

$$
\begin{array}{llll}
x = 0 : & y = (0) - 2 = -2 & (0, -2) & \text{y-intercept} \\
y = 0 : & 0 = x - 2 & & \\
 & 2 = x & (2, 0) & \text{x-intercept}
\end{array}
$$

To find a third point, choose $x = 4$.

$$y = (4) - 2 = 2$$

The arbitrary third point is $(4, 2)$.

☞ *Quick check*   Graph the linear function $f(x) = -x + 3$.   ●

MASTERY POINTS

Can you
- Determine whether or not a given set of ordered pairs represents a function?
- Determine the domain and range of a function?
- Understand what $f(x)$ means and find particular values of $f(x)$?
- Use the vertical line test to determine if a graph represents a function?
- If given a formula and a given set of domain elements,
    Rewrite the linear equation as a linear function?
    Give the set of ordered pairs that correspond to the given set of domain elements?
    Use a table to list ordered pairs?
    Graph the function?
- Graph a linear function?

Exercise 6–6

Directions
Determine whether the following relations are functions. If a relation is a function, state the domain and range. See example 6–6 A.

☞ *Quick Check*

Example $\{(-2, -1), (-1, 0), (0, 1), (1, 2)\}$

Solution   Yes (no two distinct ordered pairs have the same first component)
The domain is $\{-2, -1, 0, 1\}$, the range is $\{-1, 0, 1, 2\}$.

Directions   Determine whether the following relations are functions. If a relation is a function, state the domain and range. See example 6–6 A.

1. $\{(2,4), (6,8), (1,3), (5,7)\}$

2. $\{(5,4), (3,2), (2,3), (4,5)\}$

3. $\left\{\left(\frac{1}{2},\frac{1}{4}\right), (0,2), (3,0)\right\}$

4. $\{(0.7,1.7), (2.1,0.1), (-1.1,3.1)\}$

5. $\{(-1,3), (0,5), (-1,4), (1,6)\}$

6. $\{(3,5), (2,-4), (-4,2), (3,-5)\}$

7. $\{(-2,3), (-1,3), (0,3), (1,3)\}$

8. $\left\{\left(1,\frac{1}{2}\right), \left(3,\frac{1}{4}\right), (2,0), \left(1,-\frac{1}{4}\right)\right\}$

9. $\{(-2,1), (-1,3), (0,5), (-2,7), (2,9)\}$

10. $\{(4,2), (3,4), (4,6), (2,8), (1,10)\}$

Directions	⌐ *Quick Check*
Use the vertical line test to identify which of the following graphs represent functions. See example 6–6 B.	**Example** 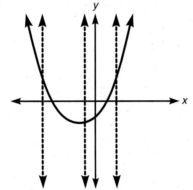

Solution Yes
(Since any vertical line drawn will intersect the graph in at most one point, the graph is a function.)

Directions Use the vertical line test to identify which of the following graphs represent functions. See example 6–6 B.

11.

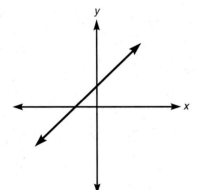

12.

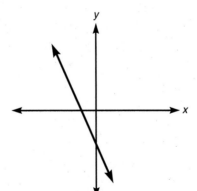

13.

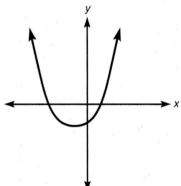

14.

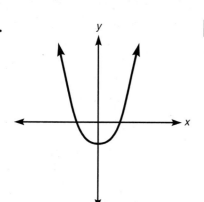

17.

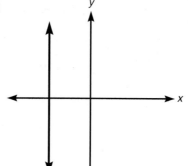

20.

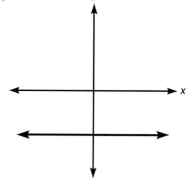

15.

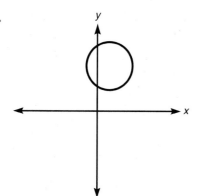

18.

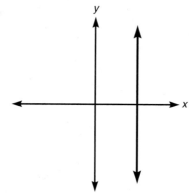

21.

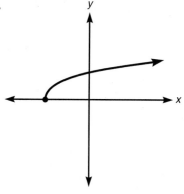

16.

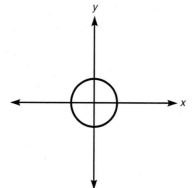

19.

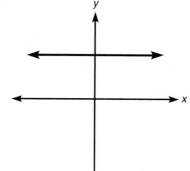

22.

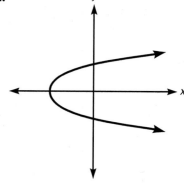

Directions
Given the following linear equations
(a) rewrite the equation as a linear function, (b) give the set of ordered pairs that correspond to the given set of domain elements, (c) list the ordered pairs from part (b) using a table, and (d) graph the function. See example 6–6 C.

☞ *Quick Check*

Example $y - 3x = 2;\ \{-3, -1, 0, 5\}$

Solution **a.** $y - 3x = 2$ Original equation
$y = 3x + 2$ Add $3x$ to both sides
$f(x) = 3x + 2$ Replace y with $f(x)$

b. Compute $f(x)$ for each of the values in the given set of domain elements and give the corresponding ordered pairs.

$f(-3) = 3(-3) + 2 = -7$
$f(-1) = 3(-1) + 2 = -1$
$f(0) = 3(0) + 2 = 2$
$f(5) = 3(5) + 2 = 17$

c. $\{(-3, -7), (-1, -1), (0, 2), (5, 17)\}$

x	-3	-1	0	5
y	-7	-1	2	17

d.

Directions Given the following linear equations (a) rewrite the equation as a linear function, (b) give the set of ordered pairs that correspond to the given set of domain elements, (c) list the ordered pairs from part (b) using a table, and (d) graph the function. See example 6–6 C.

23. $y - x = 2,\ \{-3, -2, -1, 0\}$

24. $x - y = 1, \{-1, 0, 1, 2\}$ **27.** $4x + 2y = 8, \{0, 1, 2, 3\}$ **30.** $2y - 3x = 5, \{0, 1, 2, 3\}$

25. $2x - y = 2, \{-1, 0, 1, 2\}$ **28.** $6x - 3y = 12, \{0, 1, 2, 3\}$

31. $x - y = 0, \{-1, 0, 1, 2\}$

26. $2x + y = 2, \{-1, 0, 1, 2\}$ **29.** $2x + 3y = 4, \{0, 1, 2, 3\}$

32. $x + y = 0, \{-1, 0, 1, 2\}$

Directions
Graph the following linear functions. See example 6–6 D.

☛ *Quick Check*

Example $f(x) = -x + 3$

Solution

$y = -x + 3$		Replace $f(x)$ with y
$y = -(0) + 3 = 3$	$(0,3)$	y-intercept
$(0) = -x + 3$		
$x = 3$	$(3,0)$	x-intercept
$y = -(1) + 3 = 2$	$(1,2)$	Arbitrary third point

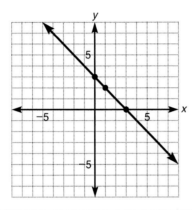

Directions   Graph the following linear functions. See example 6–6 D.

33. $f(x) = x + 1$

34. $f(x) = x - 2$

35. $f(x) = 2x - 1$

36. $f(x) = 3x + 2$

37. $f(x) = -3x + 2$

38. $f(x) = -2x - 3$

39. $f(x) = \dfrac{1}{2}x + 4$

41. $f(x) = \dfrac{2}{3}x - 2$

40. $f(x) = \dfrac{1}{3}x - 3$

42. $f(x) = -\dfrac{3}{4}x + 3$

Directions   Solve the following problems.

43. The temperature in degrees Fahrenheit (F) can be expressed as a function of degrees Celsius (C) by $F = f(C)$. If $f(C) = \dfrac{9}{5}C + 32$, find $f(100)$, $f(0)$, $f(25)$, $f(-10)$.

44. The temperature in degrees Celsius (C) can be expressed as a function of degrees Fahrenheit (F) by $C = g(F)$. If $g(F) = \dfrac{5}{9}(F - 32)$, find $g(32)$, $g(212)$, $g(-4)$, $g(59)$.

45. The cost c in cents of sending a letter by first-class mail is a function of the weight w in ounces of the letter defined by $c = 23w + 9$. Then $c = h(w) = 23w + 9$. Find $h(2)$, $h(3)$, $h(4)$.

46. The perimeter P of a square can be expressed as a function of its sides s. If $P = 4s$, then $P = f(s) = 4s$. Find $f(4)$, $f\left(\dfrac{3}{4}\right)$, $f(4.7)$, $f\left(\dfrac{5}{2}\right)$. What is the domain of f?

47. The resistance R in an electric circuit can be expressed as a function of the voltage E. If $R = 0.07E$, then $R = g(E) = 0.07E$. Find $g(2)$, $g(3)$, $g(1.2)$.

48. When two objects are pressed together, the frictional force F_f between their surfaces is related to the perpendicular (or normal) force N holding the surfaces together by $F_f = \mu N$, where μ is a constant coefficient of friction. For wood on wood, the coefficient of static friction is $\mu = 0.5$. Using h, define F_f as a function of N and find $h(100)$, $h(50)$, $h(25)$.

49. Using exercise 48, for a rubber tire on dry concrete, the coefficient of sliding friction is $\mu = 0.7$. Define a function g and find $g(5)$, $g(10)$, $g(3)$.

50. The circumference C of a circle may be expressed as a function of its radius r. Given $C = 2\pi r$, where 2π is a constant, express C as a function of r using f and find $f(7)$, $f(0.8)$, and $f\left(\dfrac{5}{2}\right)$. State the domain of function f.

51. The temperature F (in degrees Fahrenheit) can be approximated by the formula

$$F = \frac{n}{4} + 40$$

where n is the number of times a cricket chirps in 1 minute. Express F as a function of n using h to name the function and find $h(50)$, $h(12)$, $h(120)$.

52. Simple interest, I, on a loan for one year at 12% interest is given by $I = (0.12)P$, where P is the amount borrowed. Using f to name the function, express I as a function of P and find $f(1,000)$, $f(5,000)$, $f(12,350)$.

53. The gross wages W of an hourly worker are determined by $W =$ hourly rate $(r) \times$ hours worked (h) or $W = rh$. Define W as a function of h when $r = \$10.50$. Calling this function g, find $g(40)$, $g(48)$, $g(28)$.

54. A company's income statement, I, for one quarter is given by $I = R - E$, where R is the revenue in sales and gains and E is the expenses incurred. Define $I = h(R, E) = R - E$ and find $h(1,000; 350)$, $h(875; 490)$, $h(2,515; 1,031)$. (*Hint:* $h(1,000; 350)$ means that $R = \$1,000$ and $E = 350$.)

55. The selling price, S, of some merchandise is given by $S = C + M$, where C is the cost price of the merchandise and M is the markup (expenses incurred in operations). Using f, define S as a function of C and M; that is $S = f(C, M) = C + M$. Find $f(25, 7)$, $f(32, 11)$, $f(146, 27)$.

✵ Explore to Learn

This is related to the explore to learn problem in section 6–4. In this section and others we have used two data points to establish a linear equation or function. In practice, we often have many more than two points. In this case, we try to find a linear equation that comes closest to all the points. This process is called linear regression. Although we will not investigate this topic in depth, the following should give a flavor for this subject. The data in the table represents the world's record for the 1-mile run in the year shown, with time in both hours and minutes and seconds. Plot the data, using years and seconds. Use the data points for 1875 and 1993 to create the equation of the straight line that models this data. Graph the line on as large a piece of graph paper as practical. Does it seem to come close to the data? It turns out that the best equation for predicting this data is $y = 922.4 - 0.35x$. Graph this line also. Does it seem to be "closer" to the data, on average?

Year	Time	Time (seconds)
1875	4:24.5	264.5
1895	4:17	257
1915	4:12.6	252.6
1923	4:10.4	250.4
1934	4:06.8	246.8
1945	4:01.4	241.4
1954	3:59.4	239.4
1965	3:53.6	233.6
1975	3:49.4	229.4
1979	3:49.0	229
1980	3:48.8	228.8
1981	3:48.40	228.4
1985	3:46.31	226.31
1993	3:44.39	224.39

Review Exercises

Directions   Solve the following equations for *y*. See section 2–6.

1. $2x + y = 7$
2. $3x - 4y = 8$
3. Given $x = 4y - 2$ and $3y - 2x = 1$, replace *x* with $4y - 2$ in the second equation and solve for *y*. See sections 1–9 and 2–6.

4. Graph the equations $3x - 2y = 6$ and $x + y = 2$ on the same set of axes. Determine the coordinates of the point of intersection. See section 6–2.

Directions   Find the value of the unknown variable in each of the following equations. See section 6–1.

5. $x + 3y = 4$ when $y = -2$
6. $4y - 5x = 3$ when $x = 1$
7. The sum of the number of teeth on two gears is 74 and their difference is 22. How may teeth are on each gear? See section 2–7.

8. The sum of two resistances in a series is 30 ohms and their difference is 14 ohms. How many ohms are in each resistor? See section 2–7.

9. A 12-foot board is cut into two pieces so that one piece is 4 feet longer than the other. How long is each piece? See section 2–7.

6–7   Representation of Data

Bar Graphs

Graphing is an important tool in business. A graph allows a person to be able to visualize data that has been accumulated or to see the correspondence between interrelated quantities. A graph is a means of summarizing data so a reader can quickly understand a relationship.

Data can be represented in many different ways. In most instances how we present data can be left up to our individual creativity. In this section we examine several common graphs, but you should remember that there are many variations possible for the representation of data.

The first graphing technique we examine is called the **bar graph**. The bar graph is made up of several parallel rectangles called **bars**. These bars are of lengths proportional to the quantities that they represent. The bars are equally spaced and their widths are all equal. They can be constructed either vertically or horizontally. To construct a bar graph, we choose a scale that will accommodate our data and then determine the length of each bar. Consider the following.

● **Example 6–7 A**
Constructing a bar graph

The four tallest buildings in the United States are the Sears Tower (1,454 ft), the World Trade Center (1,377 ft), the Empire State Building (1,250 ft), and the AMOCO Building (1,136 ft). Construct a bar graph of these data.

We want the graph to show clearly the differences in the heights of these buildings. If we want a 5-in. graph, we must choose a scale that will allow us to draw the Sears Tower as approximately 5 in. high, and then the other buildings will be of lesser heights. We therefore choose a scale of 300 ft = 1 in. (figure 6–15).

Building	Height (ft)	Length of Bar (in.)
Sears Tower	1,454	$\frac{1,454}{300} \approx 4.8$
World Trade Center	1,3″7	$\frac{1,377}{300} \approx 4.6$
Empire State Building	1,250	$\frac{1,250}{300} \approx 4.2$
AMOCO	1,136	$\frac{1,136}{300} \approx 3.8$

Source: Information Please Almanac.

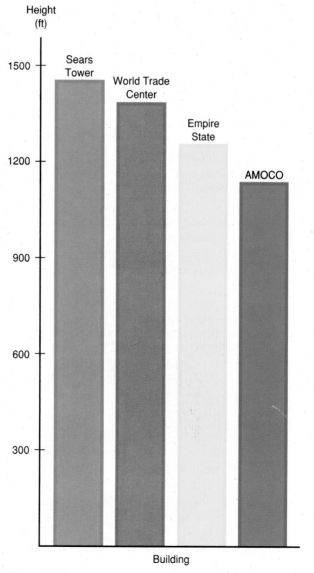

Figure 6–15

To read a bar graph, you would locate the particular category or item and approximate the value from the given scale. If you were reading an article about buildings that contained this bar graph, and you wanted to estimate the height of the tallest building, you would go to the top of the bar representing the Sears Tower and then go to the right and read off of the vertical axis. An approximate answer might be 1,450 feet.

Circle Graphs

The **circle graph** is useful when we want to show the relationship between parts and the whole quantity they comprise. To construct a circle graph, we first express the data in terms of a percent of the whole quantity. The portion that each part will occupy in the graph will be its percent of 360°. (The whole circle has 360°.) Observe the following example.

● **Example 6–7 B**

Constructing a circle graph

The following table gives the results of the 1792 presidential election. The circle graph is shown in figure 6–16.

Candidate	Electoral votes	Percentage of electoral votes	Number of degrees of the circle
George Washington	132	49	176
John Adams	77	28.5	103
George Clinton	50	18.5	67
Thomas Jefferson, Aaron Burr, and votes not cast	11	4	14
Totals	270	100*	360*

*Because of rounding, the total percentage may not be exactly 100% and the total for the number of degrees in the circle may not be exactly 360°.
Source: Information Please Almanac.

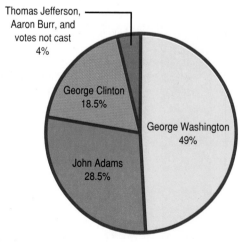

Figure 6–16

To read a circle graph, you would locate the particular category or item within the circle. If the percentage was not given, you would approximate what percentage of the circle represented the category or item and multiply that (in decimal form) by the total in the problem. If you were reading a book on American history that contained this circle graph, and you wanted to estimate the number of electoral votes that George Washington received, you would estimate the portion of the circle graph representing the electoral votes that Washington received (it is shown to be 49%, an estimate might have been 50%). You would then multiply the percent in decimal form, 0.49, by the total number of electoral votes, 270, to find the approximate number, $(0.49)(270) = 132.3$, of electoral votes that Washington received.

Broken-line Graphs

The last type of graph we examine is the **broken-line graph**. The broken-line graph is used to show changing conditions. We choose an appropriate scale for each quantity and then plot the different points. These points are then joined by straight line segments. In many instances the changes will be over a certain interval of time, and if time is one of the quantities, it is generally placed horizontally; the other quantity is then placed vertically. The following example shows a broken-line graph for housing starts over a 6-year period.

● **Example 6–7 C**

Constructing a broken-line graph

The following table gives the number of housing starts in the United States for the years 1988–1993.

Year	Number of units
1988	1,480,000
1989	1,376,000
1990	1,193,000
1991	1,014,000
1992	1,200,000
1993	1,288,000

Source: Construction Reports,
Housing starts.

To construct the broken-line graph, we place the number of housing starts on the vertical axis, and the year along the horizontal axis. We connect the points with line segments.

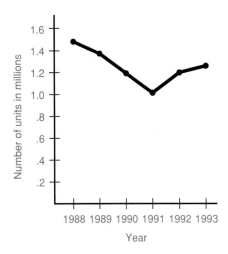

<u>Note</u> The scales that we use on the horizontal and vertical axes need not be the same.

To read a broken-line graph, you would locate the particular time element along the horizontal axis and approximate on the vertical axis the corresponding value. If you were reading an article about housing starts that contained this broken-line graph, and you wanted to estimate the number of housing starts in 1990, you would locate 1990 on the horizontal axis, find the point above that corresponds to it, and go to the left to approximate the value on the vertical scale. An approximation might be 1.2 on the vertical axis, which would correspond to 1,200,000 housing starts.

🏴 **Quick check**   Construct a bar graph to represent the given data. Choose a scale you feel is most appropriate for the data. (There are many different possible scales.)

The five tallest mountains in the United States are: Mt. McKinley—20,320 ft; Mt. St. Elias—18,008 ft; Mt. Foraker—17,400 ft; Mt. Bona—16,421 ft; and Mt. Blackburn—16,390 ft.

MASTERY POINTS

Can you
- Construct a bar graph?
- Read a bar graph?
- Construct a circle graph?
- Read a circle graph?
- Construct a broken-line graph?
- Read a broken-line graph?

Exercise 6–7

Directions   Seven of the busiest airports in the world are represented in the following bar graph. The vertical axis represents the number of passengers in millions using these airports each year. Use this bar graph to answer exercises 1–4. See example 6–7 A.

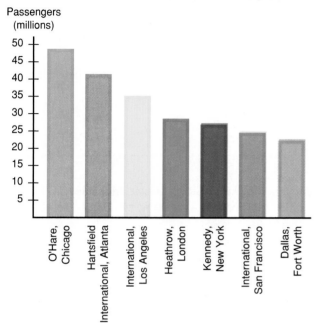

Passengers (millions)

1. What was the busiest airport and approximately how many passengers used that airport?

2. What is the third busiest airport and approximately how many passengers used that airport?

3. Approximately how many more passengers used Hartsfield International than Heathrow airport?

4. Approximately how many more passengers used Los Angeles International than San Francisco International airport?

Source: Airport Operators Council International.

Directions   The following circle graph shows the approximate disbursement of the estimated 133 billion barrels of oil reserves in the Western hemisphere. Use this circle graph to answer exercises 5–8. See example 6–7 B.

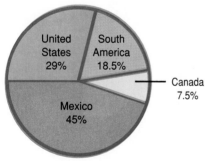

Oil Reserves – Western Hemisphere

Source: Energy Information Administration.

5. Which area has the greatest oil reserve and approximately how many barrels of oil is it?

6. Which area has the least oil reserves and approximately how many barrels of oil is it?

7. Approximately how many more barrels of oil does Mexico have in reserve than the United States?

8. Approximately how many more barrels of oil does South America have in reserve than Canada?

Directions The following broken-line graph shows the temperatures recorded in Detroit on March 27 from noon until midnight. Use this broken-line graph to answer exercises 9–12. See example 6–7 C.

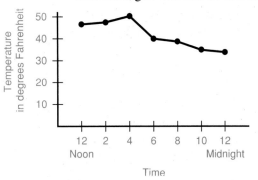

Source: Detroit Metropolitan Airport.

9. What was the warmest temperature recorded in that time period?

10. What was the coldest temperature recorded in that time period?

11. What was the change in temperature between 4 P.M. and 6 P.M.?

12. What was the change in temperature between noon and midnight?

Directions
Construct a bar graph to represent the given data. Choose a scale you feel is most appropriate for the data. (There are many different possible scales.) See example 6–7 A.

🏳 *Quick Check*

Example The five tallest mountains in the United States are Mt. McKinley—20,320 ft; Mt. St. Elias—18,008 ft; Mt. Foraker—17,400 ft; Mt. Bona—16,421 ft; and Mt. Blackburn—16,390 ft.
Source: Department of the Interior.

Solution Since the tallest mountain is approximately 20,000 feet, we might choose the scale to go up to 20,000 feet and mark the vertical scale in thousands of feet. Since we do not need to start at zero on the vertical scale, we might decide to have it begin at 9,000 feet. We then make bars of equal widths, whose heights represent the height of the mountain.

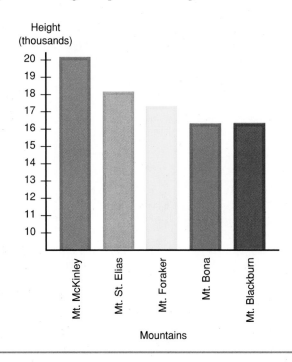

Directions   Construct a bar graph to represent the given data. Choose a scale you feel is most appropriate for the data. (There are many different possible scales.) See example 6–7 A.

13. The six largest U.S. businesses are

Company	Sales in millions of dollars
General Motors	133,600
Ford Motor Company	108,500
Exxon	97,800
IBM	62,700
General Electric	60,800
Mobil Oil	56,600

Source: Fortune 500.

14. The Consumer Price Index for all urban consumers, with 1982–1984 as a base of 100 are 1960—29.6, 1970—38.8, 1980—82.4, 1990—130.7, and 1993—144.5.

Source: Monthly Labor Review.

15. The estimated maximum speeds in miles per hour of the six fastest animals are cheetah—70, pronghorn antelope—61, wildebeest—50, lion—50, Thomson's gazelle—50, and quarterhorse—47.

Source: Natural History Magazine.

16. The estimated number of women in the working population over the last 60 years are 1930—10,752,000; 1940—12,845,000; 1950—18,410,000; 1960—23,270,000; 1970—31,560,000; 1980—44,010,000; and 1990—56,554,000.

Source: Department of Labor.

17. The following is a list of advertising expenditures in billions of dollars relative to the different media:

Media	Amount (in billions of dollars)
Radio	3.4
Television	10.4
Magazines	2.8
Newspapers	14.5
Direct mail	6.5
Billboards	0.6

Source: National Advertising Council.

18. The following is a projection of the number of people that will be in each age group in the year 2000:

Age	Numbers (in millions)
Under 19	100.7
20–34	55.4
35–44	41.4
45–54	35.9
55–64	23.0
65–74	17.1
75 & Over	13.6

Source: U.S. Bureau of the Census.

Directions   Construct a circle graph to represent the given information. See example 6–7 B.

19. The blue-collar work force has approximately 28,065,000 members. The following is a listing of the major categories:

Occupational group	Number
Craft and kindred workers	11,850,000
Operators	9,240,000
Transport equipment operators	3,100,000
Nonfarm laborers	3,875,000

Source: Department of Labor.

20. The approximate number of engineers in the United States is 1,073,000. This group is composed of the following areas:

Type of engineer	Number
Aeronautical and Astronautical	64,000
Civil	164,000
Electrical and Electronic	366,000
Industrial	238,000
Mechanical	241,000

Source: Department of Labor.

21. Of the 7,120,224 life insurance policy dividends paid in the United States last year, 1,823,472 were taken in cash, 2,691,792 were used to purchase additional paid-up life insurance, and 2,604,960 were left with the insurance companies to earn interest.

Source: American Council of Life Insurance.

22. In the 1994 Winter Olympic games, 61 gold medals were awarded in the following fashion: Russia—11, Norway—10, Germany—9, Italy—7, United States—6, and others—18.

Source: Information Please Almanac.

23. In the United States last year, there were 71,000 accidental deaths in the following categories:

Category	Number
Motor vehicle	40,300
Falls	12,400
Drowning	4,300
Fires	4,000
Firearms	1,400
Others	8,600

Source: National Safety Council.

24. It is estimated that last year, the total vehicle fuel consumption was 126,087,094,000,000 gal. The breakdown of the use of fuel was as follows:

Vehicle	Fuel consumed
Passenger vehicles	84,794,300,000,000
Buses	1,021,514,000,000
Single unit trucks	27,780,000,000,000
Tractor-trailer trucks	12,491,280,000,000

Source: Department of Commerce.

Directions   Construct a broken-line graph to represent the following information. See example 6–7 C.

25. The following is a list of the number of mobile homes shipped during the years 1988—1993: 1988—218,000; 1989—198,000; 1990—188,000; 1991—171,000; 1992—211,000; 1993—254,000.
Source: Manufactured Housing Institute.

26. The annual railroad carloadings for a 30-year period were as follows:

Year	Total
1960	27,800,000
1965	28,300,000
1970	27,000,000
1975	22,900,000
1980	22,200,000
1985	19,500,000
1990	21,400,000

Source: Association of American Railroads.

27. The following is a list of the number of home runs that Babe Ruth hit during the years that he was with the New York Yankees: 1920—54, 1921—59, 1922—35, 1923—41, 1924—46, 1925—25, 1926—47, 1927—60, 1928—54, 1929—46, 1930—49, 1931—46, 1932—41, 1933—34, 1934—22.
Source: The Complete Handbook of Baseball.

28. The following table gives the amount of material necessary to make a cubical container with a certain length of an edge.

Length of edge (m)	Material (m²)
1	6
2	24
3	54
4	96
5	150
6	216

29. The following is the median age at first marriage for males in the United States from 1900 to 1990: 1900—25.6, 1910—25.1, 1920—24.6, 1930—24.3, 1940—24.3, 1950—22.8, 1960—22.8, 1970—23.2, 1980—24.7, and 1990—26.1.
Source: U.S. Bureau of the Census.

30. The following are the number of motor vehicle deaths in selected years from 1975 to 1992: 1975—45,853; 1980—53,172; 1985—45,600; 1990—46,300; 1991—43,500; 1992—40,300.
Source: National Safety Council.

⚜ Explore to Learn

The following table gives the results of the Ogema mayoral election.

	Votes	Percent of total
Terry Ricter	618	47
Ruben Blumberg	347	26
Cornelius Vetten	291	22
Chester Dvojack	32	2
Ralph Trinkle	26	2
Simon Willard	12	1
Total	1326	100

Since the top three vote getters account for 95% of the votes, what would be a way to present the data in a circle graph so that you avoided several very small portions of the circle representing the bottom three vote getters?

Review Exercises

Directions   Solve for the variable whose value is not given. See section 6–1.

1. $3x - 2y = 6$, when $y = -3$.

2. $x - 5y = 8$, when $x = 3$

Directions   Solve the following equations for y. See section 2–6.

3. $\dfrac{4x}{3} + \dfrac{2y}{3} = 4$

4. $\dfrac{5x}{3} - \dfrac{5y}{4} = \dfrac{10}{3}$

Directions   Solve the following word problems.

5. The product of a positive integer and the next consecutive integer is 156. What is the integer? See section 3–9.

6. Your roommate padlocked the TV cabinet with a lock with a three-number combination, then left town. The roommate also left this note. "You shouldn't watch TV unless you can do some math. On the combination, the second number is 1 less than twice the first. The third is the sum of the first two. If you add up all three numbers you get 82. Find the numbers and enjoy." See section 2–7.

7. Someone is buying a lottery ticket in a store and you hear them say "I play the number 31 because it's the sum of the ages of my two children. One is 5 years older than the other." How old are this person's children?

Chapter 6 Lead-in Problem

Tickets to a football game at Chelsea University cost $8 for students and $17 for nonstudents. The total receipts for a game were $165,575. Write an equation using x for the number of students and y for the number of nonstudents who purchased tickets to the game.

Solution x is the number of students purchasing tickets
y is the number of nonstudents purchasing tickets

	Cost per ticket	Number	Receipts
Students	8	x	$8x$
Nonstudent	17	y	$17y$

We add the individual receipts from the students and the nonstudents to obtain the total receipts, $165,575.

$$8x + 17y = 165{,}575$$

Chapter 6 Summary

• Glossary

abscissa (page 348)   the first component of an ordered pair (the x value).

coordinates of a point (page 348)   the values in an ordered pair (the abscissa and the ordinate).

domain (page 403)   the domain of a relation is the set of all first components of the ordered pairs.

function (page 404)   a relation in which no two distinct ordered pairs have the same first component.

ordered pairs of numbers (page 344)   a pair of numbers written in a special order [(x, y) where the x-coordinate is first and the y-coordinate is second].

ordinate (page 348)   the second component of an ordered pair (the y value).

origin (page 348)   the point at which the x-axis and the y-axis intersect, $(0, 0)$.

parallel lines (page 384)   two distinct straight lines in a plane that never intersect.

perpendicular lines (page 384)   lines that make right angles with one another.

quadrant (page 348)   the four regions into which the rectangular coordinate system divides the plane.

range (page 403)   the range of a relation is the set of all second components of the ordered pairs.

rectangular coordinate plane (Cartesian coordinate plane) (page 347)   the x-axis and the y-axis intersecting at right angles and used when drawing graphs.

relation (page 403)   any set of ordered pairs.

slope (page 371)   the ratio of vertical change to horizontal change.

x-intercept (page 360)   the value of x at which the graph crosses the x-axis. The point is $(x, 0)$.

y-intercept (page 360)   the value of y at which the graph crosses the y-axis. The point is $(y, 0)$.

• New Symbols and Notation

$f(x)$ functional notation read f of x or f at x

• Properties and Definitions

Definition (page 343)
A linear equation in two variables x and y is any equation that can be written in the form

$$ax + by = c$$

where a, b, and c are real numbers and a and b are not both zero.

Definition of the slope of a nonvertical line (page 371)
The slope m of the nonvertical line through the points $P_1(x_1, y_1)$ and $P_2(x_2, y_2)$ is given by

$$m = \frac{y_2 - y_1}{x_2 - x_1} \quad (x_1 \neq x_2)$$

Slope of a horizontal line (page 373)

The slope m of any horizontal line is $m = 0$.

Slope of a vertical line (page 374)

Any vertical line has undefined slope.

Standard form of the equation of a line (page 380)

$$ax + by = c$$

where a, b, and c are real numbers, $a \geq 0$, a and b not both zero.

Point-slope form of the equation of a line (page 381)

$$y - y_1 = m(x - x_1)$$

where (x_1, y_1) is a known point on the line and m is the slope of the line.

Slope-intercept form of the equation of a line (page 382)

When a linear equation is stated in the form $y = mx + b$, m is the slope of the line and b is the y-intercept.

Slopes of parallel lines (page 384)

Two distinct nonvertical lines having slopes m_1 and m_2 are parallel if and only if $m_1 = m_2$.

Slopes of perpendicular lines (page 384)

Two nonvertical lines having slopes m_1 and m_2 are perpendicular if and only if $m_1 m_2 = -1$.

• Procedures

[6–1]

Finding ordered pair solutions (page 344)

1. Choose a value for one of the variables.

2. Replace the variable with this known value and solve the resulting equation for the other variable.

[6–2]

Finding x- and y-intercepts (page 360)

1. To find the x-intercept, we let $y = 0$ and find the corresponding value of x. This is the point $(x, 0)$.

2. To find the y-intercept, we let $x = 0$ and find the corresponding value for y. This is the point $(0, y)$.

Graph of a linear inequality (page 394)

The graph of any linear inequality in two variables is a half-plane.

Relation (page 403)

A relation is any set of ordered pairs.

Domain and range of a relation (page 403)

1. The domain of a relation is the set of all first components of the ordered pairs.

2. The range of a relation is the set of all second components of the ordered pairs.

Function (page 404)

A function is a relation in which no two distinct ordered pairs have the same first component.

Vertical line test for a function (page 404)

If every vertical line drawn in the plane intersects the graph of a relation in *at most one point*, the relation is a function.

Definition of a linear function (page 406)

A linear function is a function that can be written in the form

$$f(x) = mx + b$$

where m and b are real numbers, and m is the slope and the point $(0, b)$ is the y-intercept.

[6–5]

Graphing a linear inequality in two variables (page 395)

1. Replace the inequality symbol with the equality symbol.

2. Graph the resulting equation for the boundary line which will be a

   a. solid line if the inequality symbol is $\leq$ or $\geq$,
   b. dashed line if the inequality symbol is $<$ or $>$.

3. Choose some test point that is not on the line [usually the origin $(0, 0)$] and substitute the coordinates of the point into the inequality.

4. If the resulting statement is

   a. true, shade the half-plane containing the test point,
   b. false, shade the half-plane that does not contain the test point.

Chapter 6 Error Analysis

1. Ordered pair solutions of linear equations in two variables
   Example: The ordered pair (1,5) is a solution of $x = 5$.
   Correct answer: The ordered pair (5,1) is a solution of $x = 5$.
   What error was made? *(see page 363)*

2. Coordinates of points on axes
   Example: The point (0,4) lies on the *x*-axis.
   Correct answer: The point (4,0) lies on the *x*-axis.
   What error was made? *(see page 360)*

3. Ordered pairs on the graph of an equation
   Example: The ordered pair $(-1,2)$ lies on the graph of $2x - y = 3$.
   Correct answer: $(-1,2)$ is not a solution, the point does not lie on the graph.
   What error was made? *(see page 344)*

4. Finding the slope of a line
   Example: The slope of the line through $(-2,3)$ and $(2,1)$ is
   $$m = \frac{3-1}{2-(-2)} = \frac{2}{2+2} = \frac{2}{4} = \frac{1}{2}.$$
   Correct answer: $-\dfrac{1}{2}$
   What error was made? *(see page 371)*

5. Zero exponents
   Example: $(a^2 b^3 c^0)^2 = a^4 b^6 c^2$
   Correct answer: $a^4 b^6$
   What error was made? *(see page 203)*

6. Negative exponents
   Example: $(-3)^{-3} = \dfrac{1}{3^3} = \dfrac{1}{27}$
   Correct answer: $-\dfrac{1}{27}$
   What error was made? *(see page 203)*

7. Multiplying like bases
   Example: $3^2 \cdot 3^3 = 9^5$
   Correct answer: 3^5
   What error was made? *(see page 187)*

8. Exponents
   Exponents: $-(5)^2 = 25$
   Correct answer: -25
   What error was made? *(see page 70)*

9. Graphing linear inequalities
   Example: The graph of $-2 \le x < 4$ is

   Correct answer:

   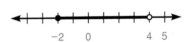

   What error was made? *(see page 162)*

10. Power to a power
    Example: $(-5x^2)^3 = -5x^5$
    Correct answer: $-125x^6$
    What error was made? *(see page 189)*

Chapter 6 Critical Thinking

1. The owner of a store has made up a work schedule for Bill and Ted so that at least one of them will be at the store each day (they never will both be off at the same time). Starting today Bill has every fourth day off and starting tomorrow Ted will have every sixth day off. Will the owner's plan work and explain why or why not.

2. A company makes resistors for electronic components. These resistors are guaranteed to be able to survive a voltage that will cause them to heat up to double their rated value. The company produced 5,000 of these resistors in its latest production run. It selected 200 at random and caused them to heat up to double their rated value. Of the 200, 184 survived. On this basis, how many of the remaining resistors would be expected to survive if each one were tested?

Chapter 6 Review

[6–1]

Directions   Find the value of y corresponding to the given values for x. Express the answer as an ordered pair.

1. $y = 3x + 4$; $x = -1$, $x = 0$, $x = 4$

2. $2x - 3y = -1$; $x = -2$, $x = 0$, $x = 1$

3. $y + 3 = 0$; $x = -7$, $x = 0$, $x = 5$

4. $5x + y = 0$; $x = -3$, $x = 0$, $x = 3$

Directions   Find the value of x corresponding to the given values of y. Express the answer as an ordered pair.

5. $x = -3y + 1$; $y = -2$, $y = 0$, $y = 5$

6. $4x + 2y = 7$; $y = -1$, $y = 0$, $y = 3$

7. $x - 1 = 0$; $y = -8$, $y = 0$, $y = 2$

8. $3x - 2y = 0$; $y = -3$, $y = 0$, $y = 1$

Directions   Plot the following ordered pairs on a rectangular coordinate system.

9. $(1,5)$

10. $(4,-4)$

11. $(-1,-6)$

12. $(0,-4)$

13. $\left(2, \dfrac{1}{2}\right)$

14. $\left(5, -\dfrac{2}{3}\right)$

Directions   Determine to the nearest integer the coordinates, (x, y), of the given points.

15. A

16. B

17. C

18. D

19. E

20. F

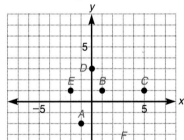

Directions Find the missing component in each ordered pair using the given equation. Then plot the ordered pairs using a separate coordinate system for each problem.

21. $y = 3x + 4$; $(2, \ )$, $(0, \ )$, $\left(\dfrac{4}{3}, \ \right)$, $\left(-\dfrac{2}{3}, \ \right)$

23. $y = x^2 + 2$; $(1, \ )$, $(-1, \ )$, $(0, \ )$, $(2, \ )$

24. $y = x^2 - 16$; $(4, \ )$, $(3, \ )$, $(0, \ )$, $(-4, \ )$

22. $y = -2x + 5$; $(-2, \ )$, $\left(\dfrac{5}{2}, \ \right)$, $(0, \ )$, $(3, \ )$

[6–2]

Directions Find the x- and y-intercepts. State the answer as an ordered pair.

25. $y = 3x + 5$

26. $y = -4x$

27. $y + 2 = 0$

28. $x - 6 = 0$

29. $2x - 7y = 4$

30. $4y - x = 0$

Directions Graph the following linear equations using the x- and y-intercepts, where possible.

31. $y = -x + 7$

33. $y = -x$

34. $y = 7$

32. $y = \dfrac{1}{2}x - 1$

35. $x = -4$

37. $5x - 3y = 15$

36. $2x - 3y = -6$

[6–3]

Directions   Find the slope of the line passing through the given pairs of points.

38. $(-4, 3)$ and $(1, 0)$

40. $(1, -2)$ and $(5, -2)$

39. $(-5, 3)$ and $(-5, -1)$

41. $(-4, -4)$ and $(1, 1)$

[6–4]

Directions   Express the following equations in *slope-intercept* form $y = mx + b$ and determine the slope m and y-intercept b.

42. $3x - 4y = 8$

44. $8x - 3y + 1 = 0$

43. $4x + 3y = 2$

45. $y - 5x = -10$

Directions   Find the slope of the following lines and find the equation of the line using the point-slope form of the equation of a line.

46.

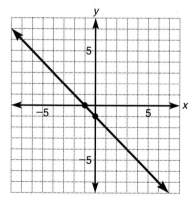

47.

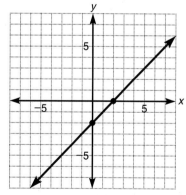

48. A roadway rises vertically 150 feet through a horizontal distance of 1,050 feet. What is the slope of the roadway?

Directions   Find the equation of the lines having the given conditions. Write the answer in standard form.

49. Passing through $(4, 5)$ having slope $m = -4$

50. Having slope $m = \dfrac{1}{3}$ and y-intercept 4

51. Passing through the points $(5, 2)$ and $(-3, 1)$

52. Passing through the points $(0, 4)$ and $(-3, 0)$

Directions   Graph the following equations using the slope m and y-intercept b.

53. $y = 3x + 4$

55. $4y - 3x = 8$

54. $2x - y = 5$

56. $3y + 4x = 9$

[6–5]

Directions   Graph the following inequalities.

57. $x \le 2y + 1$

60. $y \le 5$

58. $y > 3x - 2$

61. $x < -1$

59. $2x - y \ge 4$

62. $2y - 5x > 0$

63. $3x + 2y \geq 9$

64. $5x - 2y < 10$

[6–6]

Directions   Determine whether the following relations are functions. If a relation is a function, state the domain and range.

65. $\{(5,4), (4,4), (3,4), (2,4)\}$

66. $\{(2,3), (3,2), (4,5), (5,4)\}$

Directions   Use the vertical line test to identify which of the following graphs represent functions.

67.

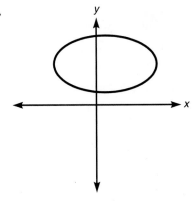

68.

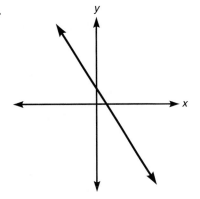

Directions   Given the following linear equations (a) rewrite the equation as a linear function, (b) give the set of ordered pairs that correspond to the given set of domain elements, (c) list the ordered pairs from part (b) using a table, and (d) graph the function.

69. $3x - y = 6, \{-2, -1, 0, 1\}$

70. $4x - 2y = 4, \{-1, 0, 1, 2\}$

Directions Graph the following linear functions.

71. $f(x) = 3x$

72. $f(x) = -x + 2$

Directions Solve the following problem.

73. The perimeter P of a regular hexagon (six-sided figure whose sides are the same length and whose angles have the same measure) is given by $P = 6s$, where s is the length of a side. Express P as a function of s using f and find $f(3)$ and $f(14)$. What is the domain of f?

[6–7]

Directions The following data represents the temperatures in degrees Celsius recorded in Chicago on July 11.

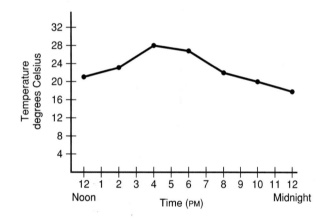

74. What was the coldest temperature recorded in that time period?

75. What was the change in temperature from noon to midnight?

Directions The following circle graph represents the percentage of working adults ages 18–24 without a high school education in various earnings groups.

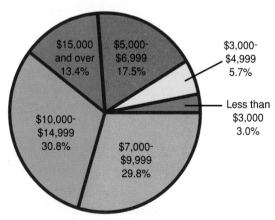

Earnings groups

3,070,000 adults working without a highschool education

76. To the nearest ten thousand approximately how many adults ages 18–24 working without a high school education earn less than $7,000 per year?

77. To the nearest ten thousand approximately how many adults ages 18–24 working without a high school education earn $15,000 and over per year?

78. To the nearest ten thousand approximately how many adults ages 18–24 working without a high school education earn from $7,000 to $14,999 per year?

79. Construct a bar graph to represent the following data. The following is the percent of unemployed workers in selected years since 1929.

Year	1929	1932	1941	1945	1950	1960	1970	1975	1980	1985	1990
Percent unemployed	3.2	23.6	9.9	1.9	5.3	5.5	4.9	8.5	7.8	7.2	5.5

Source: U.S. Department of Labor.

Chapter 6 Test

1. Find the value for y corresponding to the given value of x in the equation $y = \frac{1}{2}x - 1$; $x = -2$, $x = 0$, $x = 2$

2. Find the x- and y-intercepts for the equation $x + 3y = -6$. Write ordered pairs representing the points where the line crosses the axes.

3. Find the slope of the line passing through $(-1,0)$ and $(3,5)$.

4. Determine whether the following relation is a function. If it is a function, state the domain and range. $\{(-1,3), (2,1), (4,3), (0,0)\}$.

Directions   Find the slope of each line and determine if each pair of lines is parallel, perpendicular, or neither.

5. $y = -\frac{1}{2}x + 5$
$y = 2x + 3$

6. $2x - 3y = 4$
$3x + 2y = -4$

Directions   Write the following equations in slope-intercept form and determine the slope m and the y-intercept b.

7. $4x - 2y = 8$

8. $3x + 4y = 12$

Directions   Find the equation of the line (in slope-intercept form) using the following information.

9. Having slope $m = 2$ and y-intercept $b = -2$.

10. Through the points $(2,3)$ and $(-3,-1)$.

11. Having slope $m = -1$ and passing through the point $(0,4)$.

12. If the distance, y (in kilometers), that Ryan travels in x hours of driving is given by the equation $y = 88x$, how far does Ryan travel in (a) 2 hours, (b) $4\frac{1}{2}$ hours, (c) 6 hours? Write the answers as ordered pairs (x,y).

13. A guy wire is attached to a radio antenna. If the wire is attached to the ground at a point 40 feet from the base of the antenna and to the antenna at a point 180 feet up on the antenna, what is the slope of the wire?

Directions   Use the vertical line test to identify which of the following graphs represent functions.

14.

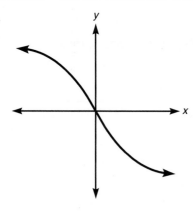

15.

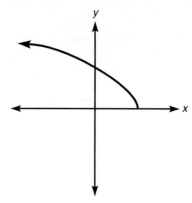

Directions   Graph the following.

16. $y = -3x + 2$

20. $f(x) = -3x + 6$

17. $2x - 4y = 8$

21. $x - 2y \geq -1$

18. $f(x) = -\frac{4}{3}x + 4$

22. $x + 2y = 0$

19. $x + y < 1$

23. $y = 3$

Directions The following circle graph represents the 238 recipients of the Medal of Honor during the Vietnam War and to which branch of the armed services they belonged.

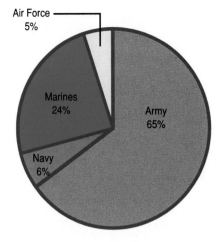

Air Force
5%

Marines
24%

Army
65%

Navy
6%

Source: The Congressional Medal of Honor Society, New York, N.Y.

24. Which branch of the armed services had the greatest number of Medal of Honor recipients, and approximately how many recipients were there?

25. Which branch of the armed services had the second greatest number of Medal of Honor recipients, and approximately how many recipients were there?

Chapter 6 Cumulative Test

1. Evaluate the expression
$xy + (-xy)^2 - x + y$ when $x = 3$ and $y = -2$.

Directions Find the solution set of the following equations.

2. $6(x + 2) = 9x + 4$

3. $\dfrac{3}{4} = \dfrac{5}{x} - \dfrac{2}{3}$

4. $-4(y + 1) - 8 = -10 - (-3 + y)$

5. $3 - \dfrac{4}{x^2} = \dfrac{4}{x}$

Directions Solve the inequality.

6. $-7 \le 5 - 3x < 8$

Directions Perform the indicated operations.

7. $(4y - 3x)^2$

8. $\dfrac{2x}{x^2 - x - 42} + \dfrac{3x}{x^2 - 49}$

9. $\dfrac{a + 1}{a + 5} - \dfrac{a - 3}{a^2 + 3a - 10}$

10. $\dfrac{2y - 6}{y + 8} \cdot \dfrac{3y + 24}{y^2 - 9}$

Directions Simplify the following. Express the answer with positive exponents. Assume that no variable is equal to zero.

11. $\dfrac{x^{-6}}{x^3}$

12. $2^{-3} \cdot 2^{-4}$

13. $(-5a^{-3})^3$

14. Divide. $(3x^3 - 2x^2 + 4x - 5) \div (x + 2)$

Directions   Graph the following equations using the x- and y-intercepts if possible.

15. $2x - y = 7$

17. Graph the linear inequality.
$$3x - 5y > 15$$

16. $x - 7 = 0$

18. Find the slope of the line through the points $(2,3)$ and $(-1,-4)$.

Directions   Find the equation of the line in standard form.

19. Passing through the points $(1,4)$ and $(-3,1)$

20. Having slope $-\dfrac{3}{5}$ and passing through the point $(4,3)$

Directions   Determine if the following lines are parallel, perpendicular, or neither.

21. $2x - 3y = 4$ and $4x - 6y = -1$

22. $x + 2y = 6$ and $3x - 6y = 1$

Directions   Solve the following word problems.

23. Three fourths of what number is 102?

24. The product of a whole number times the next consecutive even whole number is 168. What is the whole number?

25. A right triangle has a hypotenuse of 13 inches. If one of the legs is 7 inches longer than the other, find the lengths of the two legs. (*Hint:* Use $c^2 = a^2 + b^2$, where c is the hypotenuse and a and b are the legs.)

7

Systems of Linear Equations

A company has $100,000 to invest. The company has two places it can put the money. One choice is to invest in a low-risk stock portfolio that pays 4%. A second choice is to invest in a new company. This company is expected to earn 12% on its capital. Unfortunately any new company must be considered a high risk. The board of directors of the company is demanding at least a 6% return on investment, and directs the company to divide up the $100,000 into the two choices, but to only put just enough of the money into the high-risk venture so that the net return is 6% (if all goes well). How much money should be put into each investment choice to meet these conditions?

7–1 Solving Systems of Two Linear Equations by Graphing

In chapter 6 we investigated situations in which one linear equation was involved. This involves some relationship involving two quantities, typically represented by x and y. Recall that a graphical interpretation of such a situation is a straight line. It turns out that many situations in the real world are best described by two linear equations, such as the problem described at the beginning of this chapter. In fact, the theory of what we study in this chapter can be expanded to include many equations involving many more variables than just two. Large companies and the government use this theory to solve problems that involve literally thousands of variables.

In this chapter we will restrict ourselves to situations that can be handled by two linear equations in two variables—in other words, situations whose graphical interpretation is two straight lines. It is surprising how many practical, realistic situations fall into this category.

Two or more linear equations involving the same variables form a **system of linear equations**. Here are examples of systems of linear equations.

$$
\begin{array}{llll}
\text{I} & \text{II} & \text{III} & \text{IV} \\
4x - y = 1 & 3x - 7y = 5 & x = 3y + 1 & x = 4 \\
x + y = 4 & 5x + 2y = 11 & x + 3y = 4 & 3x - 2y = 5
\end{array}
$$

To solve a system of two linear equations, we must find the ordered pair(s) of real numbers (x, y) that satisfy both equations at the same time.

Solutions of Systems of Linear Equations

A **solution** of a system of linear equations is an ordered pair of numbers that is a solution to each equation in the system.

● **Example 7–1 A**

Determining when an ordered pair is a solution

Determine whether the given ordered pair is a solution of the system of linear equations.

$$
\begin{array}{l}
4x - y = 1 \\
x + y = 4
\end{array} \quad (1, 3)
$$

$$
\begin{array}{ll}
4x - y = 1 & \\
4(1) - (3) = 1 & \text{Replace } x \text{ with 1} \\
& \text{and } y \text{ with 3} \\
4 - 3 = 1 & \\
1 = 1 \quad \text{(True)} &
\end{array}
\qquad
\begin{array}{l}
x + y = 4 \\
(1) + (3) = 4 \\
4 = 4 \quad \text{(True)}
\end{array}
$$

Both resulting statements are true. Therefore, the ordered pair $(1, 3)$ satisfies both equations and is a solution.

⌐ *Quick check*   Determine if the ordered pair $(1, 1)$ is a solution of the system.
$$
\begin{array}{l}
3x - y = 2 \\
4x + 3y = 9
\end{array}
$$

Graphical Solutions

The first method for solving these systems that we shall examine is by graphing. Using the x- and y-intercepts and a checkpoint (as in chapter 6) we graph each equation. Then we look for the coordinates of any point where the graphs intersect. This point of intersection will give a solution of the system.

● **Example 7–1 B**

Solution by graphing

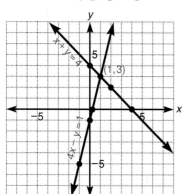

Find the solution of each system by graphing.

1. $4x - y = 1$
  $\ x + y = 4$

$$4x - y = 1 \qquad\qquad\qquad x + y = 4$$

x	y	
0	-1	y-intercept; $x = 0$
$\dfrac{1}{4}$	0	x-intercept; $y = 0$
-1	-5	Checkpoint

x	y
0	4
4	0
2	2

It appears that the point of intersection of the two lines is the point (1,3). We must check this as the possible solution.

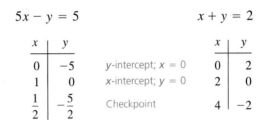

$$4x - y = 1$$
$$4(1) - (3) = 1$$ Replace x with 1 and y with 3
$$4 - 3 = 1$$
$$1 = 1$$ (True)

$$x + y = 4$$
$$(1) + (3) = 4$$
$$4 = 4$$ (True)

The solution of the system is the ordered pair (1,3).

2. $5x - y = 5$
$\quad x + y = 2$

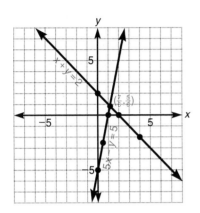

$$5x - y = 5$$

x	y	
0	-5	y-intercept; $x = 0$
1	0	x-intercept; $y = 0$
$\dfrac{1}{2}$	$-\dfrac{5}{2}$	Checkpoint

$$x + y = 2$$

x	y
0	2
2	0
4	-2

An estimate of the point of intersection is $(1,1)$. A check shows this is not a correct solution.

$$5x - y = 5$$
$$5(1) - (1) = 5$$ Replace x with 1 and y with 1
$$4 = 5$$ (False)

$$x + y = 2$$
$$(1) + (1) = 2$$
$$2 = 2$$ (True)

We will see in section 7–2 that the actual point of intersection is $\left(\dfrac{7}{6}, \dfrac{5}{6}\right)$.

 Quick check   Estimate the solution of the system by graphing.
$$3x - y = 6$$
$$2y + x = 4$$

It should be obvious that the graphical method of finding the solution may not be exact because the drawing of the graph depends on measuring devices.

A graphing calculator can be used to find an approximation to the point of intersection of two lines. Example B of appendix A, Introduction to Graphing Calculators, shows how. The only restriction is that the equations must be solved for y in order to be entered into the graphing calculator.

Inconsistent and Dependent Systems of Equations

When the graph of a system of linear equations consists of intersecting lines and thus has one ordered pair for a solution, the system is said to be a **consistent and independent system of linear equations**. Sometimes the graphs of the two linear equations in the system do not intersect at all or are one and the same line.

● **Example 7–1 C**

Inconsistent and dependent systems

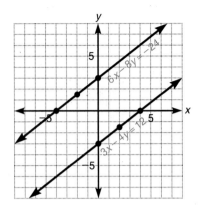

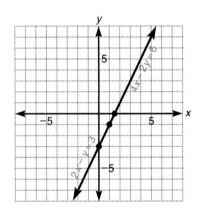

1. $3x - 4y = 12$
$6x - 8y = -24$

$3x - 4y = 12$				$6x - 8y = -24$	
x	y			x	y
0	-3	y-intercept; $x = 0$		0	3
4	0	x-intercept; $y = 0$		-4	0
2	$-\dfrac{3}{2}$	Checkpoint		-2	$\dfrac{3}{2}$

The lines appear to be parallel (they are) and will not intersect. When this condition exists, the system is said to be an **inconsistent system** and there is no solution (no point of intersection). We showed in section 6–4 how to determine if two lines are parallel by comparing their slopes. We will show in section 7–2 another way to determine whether or not the lines in a system of linear equations are parallel.

2. $4x - 2y = 6$
$2x - y = 3$

$4x - 2y = 6$				$2x - y = 3$	
x	y			x	y
0	-3	y-intercept; $x = 0$		0	-3
$\dfrac{3}{2}$	0	x-intercept; $y = 0$		$\dfrac{3}{2}$	0
1	-1	Checkpoint		1	-1

We observe that the two intercepts and checkpoint $(1, -1)$, are the same for both lines. Therefore, the two lines coincide and the graph is a single line. Such a system is called a **dependent system** of equations. In a dependent system of equations, the solutions are all of the ordered pairs that are on the line. Thus a dependent system of equations has an infinite number of solutions.   ●

To Solve a System of Two Linear Equations by Graphing

1. Graph each equation on the same set of axes.
2. Their graphs may intersect in
   a. one point—one solution (the system is *consistent* and *independent*).
   b. no point—the lines are parallel and there is no solution (the system is *inconsistent*).
   c. all points—the lines are one and the same (the system is *dependent*).
3. If the graphs intersect in one point, estimate the coordinates of the point and check this estimate in each equation.

MASTERY POINTS

Can you
• Determine if an ordered pair is a solution of a system of linear equations?
• Solve a system of two linear equations by graphing?
• Recognize an inconsistent and a dependent system of linear equations?

Exercise 7–1

Directions
Determine if the given ordered pair is a solution to the system of equations. See example 7–1 A.

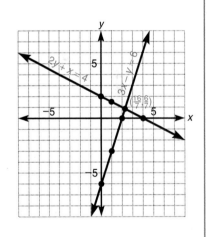

🏴 *Quick Check*

Example Determine if the ordered pair $(1,1)$ is a solution of the system.
$$3x - y = 2 \qquad 4x + 3y = 9$$

Solution Substitute 1 for x and 1 for y in each equation.

$3(1) - (1) = 2$ *Replace x with 1 and y with 1* $4(1) + 3(1) = 9$
$\qquad 3 - 1 = 2$ $4 + 3 = 9$
$\qquad\qquad 2 = 2$ (True) $7 = 9$ (False)

Since $(1,1)$ satisfies only one equation, it is *not* a solution of the system of equations.

Directions Determine if the given ordered pair is a solution to the system of equations. See example 7–1 A.

1. $x + y = 2$ $(3,-1)$
$\quad 3x - y = 10$

2. $x + 2y = 4$ $(-2,3)$
$\quad x + 4y = 10$

3. $x - y = 1$ $(3,2)$
$\quad x + y = 5$

4. $3x - y = -10$ $(-1,7)$
$\quad 3x + y = 4$

5. $2x + y = 2$ $(3,-4)$
$\quad 6x - y = 22$

6. $x - 3y = -1$ $(-1,0)$
$\quad 3x + y = -2$

7. $3x + 4y = 14$ $(1,4)$
$\quad 2x + y = 6$

8. $3x + y = 12$ $(2,6)$
$\quad 2x + 3y = 15$

9. $x = 6$ $(6,9)$
$\quad 2x - y = 3$

10. $y = 4x - 3$ $(-1,-1)$
$\quad y = -1$

11. $3x - 2y = 4$ $(10,2)$
$\quad y = x + 3$

12. $y = 2 - 5x$ $(-1,5)$
$\quad x - y = 6$

Directions
Estimate the solution of each system by graphing. All answers in the back of the book will be given exactly. If the system is inconsistent or dependent, so state. See examples 7–1 B and C.

🏴 *Quick Check*

Example $3x - y = 6 \qquad 2y + x = 4$

Solution $3x - y = 6 \qquad\qquad 2y + x = 4$

x	y	
0	-6	*y*-intercept
2	0	*x*-intercept
1	-3	Checkpoint

x	y
0	2
4	0
1	$\dfrac{3}{2}$

We estimate the solution to be $\left(\dfrac{9}{4}, \dfrac{3}{4}\right)$.

<u>Note</u> The actual solution is $\left(\dfrac{16}{7}, \dfrac{6}{7}\right)$.

Directions Estimate the solution of each system by graphing. All answers in the back of the book will be given exactly. If the system is inconsistent or dependent, so state. See examples 7–1 B and C.

13. $x - y = 1$
$\quad x + y = 5$

14. $x + 2y = 4$
$\quad x + 4y = 10$

15. $3x - y = 10$
$\quad x + y = 2$

16. $x - y = 3$
$2x + 3y = 11$

21. $2x + y = 2$
$-6x - 3y = 6$

26. $2x - y = 3$
$6x - 3y = 9$

17. $2x - y = 3$
$y + 2x = -7$

22. $x = 4$
$3x + 2y = 4$

27. $5x + y = -5$
$-10x - 2y = 10$

18. $y = 1$
$3x - 2y = -9$

23. $3x + y = -3$
$x = -1$

28. $\dfrac{3}{2}x - \dfrac{1}{2}y = -5$
$x + \dfrac{1}{3}y = \dfrac{4}{3}$

19. $2x + 3y = 13$
$3x - 2y = 0$

24. $y + 2x = 6$
$4y + 3x = 4$

29. $\dfrac{1}{3}x + \dfrac{1}{6}y = 1$
$\dfrac{1}{4}x + \dfrac{1}{3}y = 1$

20. $3x + y = 10$
$6x + 2y = 5$

25. $3x + y = 3$
$6x + 2y = 3$

30. $6x - 5y = 14$
     $y = 8$

31. Explain what an inconsistent system of linear equations is.

32. Explain what a dependent system of linear equations is.

33. Explain what the solution to a consistent and independent system of two linear equations represents.

Explore to Learn

1. A teacher wants to create a linear equation $Ax + By = C$ to which the ordered pair $(-2,5)$ is a solution. The teacher also does not want either A or B to be zero. Make up such a problem.

2. The same teacher wants the same type of problem, with the same solution, but wants to make sure that the x- and y-intercepts are integers. Make up such a problem.

3. Make up a system of two linear equations to which the solution is the ordered pair $(-2,5)$.

Review Exercises

1. Find the solution set of the equation $\dfrac{3x}{2} - \dfrac{x}{4} = 1$. See section 5–6.

2. Divide as indicated. $(3x^3 - x^2 + 1) \div (x + 3)$. See section 3–6.

Directions   Find the solution of the following inequalities. See section 2–9.

3. $4y - 3 \leq 2y + 7$

4. $-5 < 2x + 1 \leq 7$

Directions   Perform the indicated operations. See sections 3–2 and 3–3.

5. $x^{-3} \cdot x^5 \cdot x^2$

6. $x^3(x^2 - 2x + 1)$

7. $(2x - y)^2$

8. Two gears have a total of 59 teeth. One gear has 15 fewer teeth than the other. How many teeth are on each gear? See section 2–7.

9. A woman wishes to mix 20 pounds of tea costing $1.80 per pound with tea costing $2.40 per pound. How many pounds of the $2.40 tea must she add to obtain tea worth $2.00 per pound? See section 2–8.

7–2   Solving Systems of Two Linear Equations by Elimination

In section 7–1, we found the solution of a system of consistent and independent linear equations by graphing the equations and by observing the coordinates of the point of intersection of the two lines. This technique will often produce only an approximate solution.

An algebraic method for obtaining an *exact* solution to a system involves the operation of *addition*, together with the following property of real numbers.

Addition Property of Equality

Given a, b, c, and d are real numbers, if $a = b$ and $c = d$, then

$$a + c = b + d$$

Concept

Adding equal quantities to each side of an equation will result in equal sums.

This property of real numbers is used to *eliminate* one of the variables in the system. The resulting single equation has only one unknown that we solve to obtain one of the components of the solution. We call this solution by elimination. We illustrate the use of this property in the following example.

● **Example 7–2 A**

Solution by elimination and checking the solution

Find the solution set of the following system of linear equations by elimination.

$$x = 4 - 3y$$
$$2x - 3y = 2$$

To solve the system by elimination, like terms must be in a column and the equations must be in standard form.

$$\begin{array}{ll} x + 3y = 4 & \text{Add } 3y \text{ to each side of the first equation} \\ \underline{2x - 3y = 2} & \\ 3x \quad\;\; = 6 & \text{Add left and right sides} \\ \quad\; x = 2 & \text{Divide each side by 3} \end{array}$$

Replace x with 2 in either equation and solve for y.

$$\begin{array}{ll} x + 3y = 4 & \text{Choose either equation } (x + 3y = 4) \\ (2) + 3y = 4 & \text{Replace } x \text{ with 2} \\ 3y = 2 & \text{Subtract 2 from each side} \\ y = \dfrac{2}{3} & \text{Divide each side by 3} \end{array}$$

The solution of the system is the ordered pair $\left(2, \dfrac{2}{3}\right)$. These are the coordinates of the point of intersection of the two lines. The solution set is $\left\{\left(2, \dfrac{2}{3}\right)\right\}$. Check the solution in *both original* equations. ●

In example 7–2 A, we observe that one of the variables was easily eliminated by addition because in the original equations that variable had coefficients that were opposites. When this condition is not present in the equation, we must first use the following procedure to obtain the necessary opposite coefficients.

> ## To Solve a System of Two Linear Equations by Elimination
>
> 1. Write the system of linear equations so that each equation is in standard form $ax + by = c$.
> 2. If necessary multiply both sides of one or both equations by a constant that will make the coefficients of one of the variables opposites of each other.
> 3. Add the corresponding sides of the two equations, and solve this new equation for the remaining variable.
> 4. Substitute the value found in step 3 into either of the original equations and solve this equation for the other variable. The solution is the ordered pair of numbers obtained in step 3 and this step.
> 5. If the system has a solution found in step 4, check the solution in both equations.
>
> **Note**
>
> If, when adding in step 3,
>
> a. both variables are eliminated and a *false statement* is obtained, there is *no solution* and the system is *inconsistent* (parallel lines).
> b. both variables are eliminated and a *true statement* such as $0 = 0$ is obtained, there are an infinite number of solutions and the system is *dependent* (same line).

● **Example 7–2 B**

Solution by elimination

Find the solution set of each system of linear equations by elimination.

1. $7x + 5y = -9$ Multiply each side by 2→ $14x + 10y = -18$
$3x - 2y = -8$ Multiply each side by 5→ $\underline{15x - 10y = -40}$
　　　　　Add the equations $29x \quad\quad = -58$
　　　　　Solve for x $x = -2$

Using $7x + 5y = -9$, we substitute and solve for y.

$$7(-2) + 5y = -9 \quad \text{Replace } x \text{ with } -2 \text{ in } 7x + 5y = -9$$
$$-14 + 5y = -9 \quad \text{Solve for } y$$
$$5y = 5$$
$$y = 1$$

Solution: $(-2, 1)$. The solution set is $\{(-2, 1)\}$. Check the solution.

Note In the previous example, both equations had to be multiplied by a constant to form opposite coefficients of *y*. We could have eliminated *x* by multiplying the first equation by 3 and the second equation by −7, forming opposite coefficients of 21 and −21. It is an arbitrary choice to decide which variable to eliminate, but for ease in solving the system, it is best to eliminate the variable whose opposite coefficients will be least in value.

2. $3x = 5y - 1$
$2x - 3y = 8$

We must write the first equation in standard form, $ax + by = c$, so that a variable can be eliminated when we add the corresponding sides of the equations once the coefficients of one of the variables are opposites of each other.

$$3x - 5y = -1 \quad \text{Subtract } 5y \text{ from each side}$$

The system of linear equations is now

$$3x - 5y = -1 \quad (1)$$
$$2x - 3y = 8 \quad (2)$$

One way to eliminate one variable in this case (and there are other ways) is to multiply new equation (1) by 2 and equation (2) by -3.

$$3x - 5y = -1 \qquad \text{Multiply by 2} \rightarrow \qquad 6x - 10y = -2$$
$$2x - 3y = 8 \qquad \text{Multiply by } -3 \rightarrow \quad \underline{-6x + 9y = -24}$$

Coefficients of x are opposites of each other

$$\text{Add} \qquad \qquad -y = -26$$
$$y = 26$$

Substitute 26 for y in equation (2) and solve for x.

$$2x - 3(26) = 8 \qquad \text{Replace } y \text{ with } 26$$
$$2x - 78 = 8 \qquad \text{Solve for } x$$
$$2x = 86$$
$$x = 43$$

The solution set is $\{(43,26)\}$.

Note We could have multiplied equation (1) by 3 and equation (2) by -5, or equation (1) by -3 and equation (2) by 5. In either case, we would eliminate y through adding.

3. $\dfrac{1}{2}x + \dfrac{2}{3}y = \dfrac{1}{6}$

$\dfrac{3}{4}x - \dfrac{1}{4}y = \dfrac{1}{2}$

The first step in this problem will be to clear the fractions.

$$\dfrac{1}{2}x + \dfrac{2}{3}y = \dfrac{1}{6} \qquad \begin{array}{l}\text{Multiply each side by}\\ \text{the LCD, 6}\end{array} \qquad 6\left(\dfrac{1}{2}x + \dfrac{2}{3}y\right) = 6 \cdot \dfrac{1}{6}$$
$$3x + 4y = 1$$
$$\dfrac{3}{4}x - \dfrac{1}{4}y = \dfrac{1}{2} \qquad \begin{array}{l}\text{Multiply each side by}\\ \text{the LCD, 4}\end{array} \qquad 4\left(\dfrac{3}{4}x - \dfrac{1}{4}y\right) = 4 \cdot \dfrac{1}{2}$$
$$3x - y = 2$$

We now solve the system.

$$3x + 4y = 1 \qquad \qquad \qquad \qquad \quad 3x + 4y = 1$$
$$3x - y = 2 \qquad \text{Multiply each side by } (-1) \rightarrow \quad \underline{-3x + y = -2}$$
$$\text{Add the equations} \qquad \qquad \qquad 5y = -1$$
$$\text{Solve for } y \qquad \qquad \qquad \qquad y = -\dfrac{1}{5}$$

$$3x - \left(-\dfrac{1}{5}\right) = 2 \qquad \text{Replace } y \text{ with } -\dfrac{1}{5} \text{ in } 3x - y = 2$$

$$3x + \dfrac{1}{5} = 2 \qquad \text{Definition of subtraction}$$

$$15x + 1 = 10 \qquad \text{Multiply each term by the LCD, 5}$$
$$15x = 9 \qquad \text{Subtract 1 from each side}$$
$$x = \dfrac{3}{5} \qquad \text{Divide each term by 15 and reduce}$$

Solution: $\left(\dfrac{3}{5}, -\dfrac{1}{5}\right)$. The solution set is $\left\{\left(\dfrac{3}{5}, -\dfrac{1}{5}\right)\right\}$.

4. $4x - 2y = 6$

$2x - y = 3$ Multiply each side by (-2) →

$$\begin{array}{r} 4x - 2y = 6 \\ -4x + 2y = -6 \\ \hline 0x + 0y = 0 \\ 0 = 0 \end{array}$$ (True)

Add the equations

At this point, both variables have been eliminated, and we have formed the true statement $0 = 0$. This indicates that any ordered pair that satisfies either equation is a solution and there are an infinite number of solutions. The system is **dependent** and the solution set is all ordered pairs satisfying equation $2x - y = 3$ or $4x - 2y = 6$. Each equation represents the same line.

5. $3x - 4y = 12$ Multiply each side by (-2) → $-6x + 8y = -24$

$6x - 8y = -24$

$$\begin{array}{r} 6x - 8y = -24 \\ \hline 0x + 0y = -48 \\ 0 = -48 \end{array}$$ (False)

Add the equations

The resulting statement, $0 = -48$, is false, which indicates there is no ordered pair that can satisfy this system. The system is **inconsistent** and the solution set is $\varnothing$ (the null set). The equations represent parallel lines.

☞ *Quick check*   Find the solution set of the system of linear equations by elimination.

$$5x - 2y = 0$$
$$y - 3x = 4$$

MASTERY POINTS

Can you
• Find the solution set for a system of two linear equations by elimination?
• Recognize a dependent or an inconsistent system of linear equations while solving by elimination?

Exercise 7–2

Directions
Find the solution set of the following systems of linear equations by elimination. If a system is inconsistent or dependent, so state. See examples 7–2 A and B.

☞ *Quick Check*

Example $5x - 2y = 0$
$y - 3x = 4$

Solution   Rewrite the second equation to get the system.

$$\begin{array}{l} 5x - 2y = 0 \\ -3x + y = 4 \end{array}$$ Multiply each term by 2 →

$$\begin{array}{r} 5x - 2y = 0 \\ -6x + 2y = 8 \\ \hline -x = 8 \\ x = -8 \end{array}$$

Add equations

Multiply by -1

$5(-8) - 2y = 0$ Replace x with -8 in $5x - 2y = 0$

$-40 - 2y = 0$ Solve for y

$-2y = 40$

$y = -20$

Solution: $(-8, -20)$. The solution set is $\{(-8, -20)\}$.

Directions   Find the solution set of the following systems of linear equations by elimination. If a system is inconsistent or dependent, so state. See examples 7–2 A and B.

1. $x - y = 1$
$x + y = 5$

2. $x + y = 2$
$3x - y = 10$

3. $3x - y = 10$
$6x + y = 5$

4. $2x + y = 2$
$-6x - y = 4$

5. $x - 2y = 4$
$-x + 2y = 4$

6. $-x + y = 6$
$x + 4y = 4$

7. $-3x + y = -3$
$3x - 3y = -1$

8. $2x + y = 1$
$-2x + y = 1$

9. $x + 2y = 4$
$x + 4y = 10$

10. $x - y = 3$
$2x + 3y = 11$

11. $3x = 2y - 9$
$2x + y = 1$

12. $5x = y + 4$
$x + 3y = 2$

13. $4x = 11 - y$
$2x + 3y = 4$

14. $x = 2y + 1$
$3x - 6y = 3$

15. $8x - 4y = 12$
$2x - y = 3$

16. $-2x + 3y = 6$
$4x + y = 1$

17. $10x + y = 5$
$2x + 5y = -23$

18. $10x - 5y = 7$
$2x - y = 4$

19. $-6x + 3y = 9$
$2x - y = -3$

20. $3x - 2y = 6$
$4x - 3y = 7$

21. $5x + 2y = 3$
$3x - 3y = 4$

22. $4x + 7y = 11$
$8x - 3y = -4$

23. $5x - 2y = 4$
$3x + 3y = 5$

24. $5x - 3y = 34$
$4x + 5y = 5$

25. $3x - 2y = 19$
$5x + 3y = 19$

26. $4x + 5y = -2$
$3x - 2y = -36$

27. $\dfrac{1}{2}x + \dfrac{1}{3}y = 1$
$\dfrac{2}{3}x - \dfrac{1}{4}y = \dfrac{1}{12}$

28. $\dfrac{2}{3}x - \dfrac{1}{4}y = 4$
$\dfrac{1}{3}x + y = 2$

29. $\dfrac{1}{2}x + \dfrac{2}{5}y = \dfrac{7}{10}$
$\dfrac{3}{2}x + \dfrac{6}{5}y = \dfrac{3}{10}$

30. $\dfrac{6}{7}x - \dfrac{3}{5}y = \dfrac{9}{10}$
$\dfrac{2}{7}x - \dfrac{1}{5}y = \dfrac{1}{2}$

31. $\dfrac{9}{2}x + \dfrac{1}{2}y = -10$
$\dfrac{1}{3}x + \dfrac{1}{4}y = \dfrac{3}{4}$

32. $\dfrac{7}{2}x - y = -\dfrac{17}{2}$
$\dfrac{2}{3}x + \dfrac{1}{3}y = 1$

33. $(0.4)x - (0.7)y = 0.7$
$(0.6)x - (0.5)y = 2.7$

34. $x + (0.4)y = 3.4$
$(0.6)x - (1.4)y = 0.4$

35. Let x represent the width of a rectangle and y represent the length of the rectangle. Write an equation stating that the length is five times the width.

36. The perimeter of the rectangle in exercise 35 is 60 inches. If the perimeter of a rectangle equals twice the width plus twice the length, write an equation for the perimeter of the rectangle.

37. Let x represent the speed of automobile A and y represent the speed of automobile B. Write an equation stating that automobile B travels 20 mph faster than automobile A.

38. If the two automobiles in exercise 37 travel in opposite directions, write an equation that states the two automobiles are 500 miles apart after 3 hours. (*Note:* distance = speed times the time traveled.)

39. Let x represent the amount of money Jane invests at 8% and y represent the amount she invests at 7%. Write an equation stating that Jane invests a total of $14,000.

40. In exercise 39, Jane receives a total of $1,000 from her investments. If income equals the amount invested times the rate of interest, write an equation relating Jane's total income to her two investments.

41. Given the system of linear equations

$$15x + 3y = 5$$
$$7x - 2y = 8$$

Which variable would be the easiest to eliminate and why?

⚜ Explore to Learn

Consider the system

$$3x - 5y = 9$$
$$3x + 7y = 2$$

The coordinates of the ordered pair solution are fractions and the arithmetic necessary to find the other component once you have found one of the components is tedious.

An alternative to substituting a fraction into one of the original equations is to eliminate one of the variables, which will give one of the components, and then eliminate the other variable, which will give the other component. Try this approach on this system of equations.

Review Exercises

1. Given $2x + y = 3$, solve for x when $y = x + 2$. See sections 1–9 and 2–6.

Directions   Find the equation in standard form of each line having the given properties. See section 6–4.

2. Through $(-1, 6)$, slope $\dfrac{2}{3}$

3. Through the points $(0, 5)$ and $(-7, 1)$

Directions   Factor completely the following expressions. See sections 4–3 and 4–4.

4. $16x^2 - 4y^2$

5. $9x^2 - 12x + 4$

6. $5y^2 - 6y - 8$

7. Find the solution set of the equation $4(3x - 2) + 2x = 6$. See section 2–5.

8. The sum of two currents is 45 amperes. If the greater current is 15 amperes more than the lesser current, find their values. See section 2–7.

9. The outlet pipe to a swimming pool can empty the pool in 15 hours, while the inlet pipe can fill the pool in 10 hours. A maintenance person leaves both pipes open but discovers the error after 2 hours, and closes the outlet pipe. How much longer will it take to fill the pool from that point? See section 5–7.

7–3   Solving Systems of Two Linear Equations by Substitution

A basic property of equality is that if two expressions are equal (represent the same quantity), one expression may replace the other expression in any equation to form an equivalent equation. We use this property to solve systems of equations by a third method, called the **substitution** method. This method is useful when

1. one of the equations is solved for one variable in terms of the other; for example, $y = 3x - 4$, or
2. one of the equations can *easily* be solved for one variable in terms of the other (usually when the coefficient of the variable is 1 or -1), for example, $x - 2y = 3$.

To illustrate, consider the system of linear equations

$$3x + 2y = 9$$
$$y = x + 2$$

Since the second equation is already solved for y in terms of x, we *substitute* $x + 2$ for y in the first equation.

$$3x + 2y = 9$$
$$3x + 2(x + 2) = 9 \qquad \text{Replace } y \text{ with } x + 2$$
$$3x + 2x + 4 = 9 \qquad \text{Solve for } x$$
$$5x + 4 = 9$$
$$5x = 5$$
$$x = 1$$

Now substitute 1 for x in the second equation.

$$y = x + 2$$
$$y = (1) + 2 \qquad \text{Replace } x \text{ with } 1$$
$$y = 3$$

The solution is $(1,3)$. The solution set is $\{(1,3)\}$.

Note We can check our solution by substituting 1 for *x* and 3 for *y* in *both* original equations.

$$
\begin{array}{ll}
3x + 2y = 9 & y = x + 2 \\
3(1) + 2(3) = 9 & (3) = (1) + 2 \\
3 + 6 = 9 & 3 = 3 \\
9 = 9 &
\end{array}
$$

● **Example 7–3 A**
Solution by substitution

Find the solution set of the following systems of linear equations by substitution.

1. $3x + 2y = 14$
   $y = 2x - 7$

$$3x + 2y = 14$$
$$3x + 2(2x - 7) = 14 \qquad \text{Replace } y \text{ with } 2x - 7$$
$$3x + 4x - 14 = 14 \qquad \text{Multiply as indicated}$$
$$7x - 14 = 14 \qquad \text{Combine like terms}$$
$$7x = 28 \qquad \text{Add 14 to each side}$$
$$x = 4 \qquad \text{Divide each side by 7}$$

Using the equation

$$y = 2x - 7$$
$$y = 2(4) - 7 \qquad \text{Replace } x \text{ with } 4$$
$$y = 8 - 7$$
$$y = 1$$

The solution of the given system is $(4,1)$. Check the solution. The solution set is $\{(4,1)\}$.

2. $2x - 4y = 7$
$x = 3y - 2$

$$2x - 4y = 7$$
$$2(3y - 2) - 4y = 7 \quad \text{Replace } x \text{ with } 3y - 2$$
$$6y - 4 - 4y = 7 \quad \text{Multiply as indicated}$$
$$2y = 11 \quad \text{Combine like terms; add 4 to each side}$$
$$y = \frac{11}{2} \quad \text{Divide each side by 2}$$

Using the equation $x = 3y - 2$,

$$x = 3\left(\frac{11}{2}\right) - 2 \quad \text{Replace } y \text{ with } \frac{11}{2}$$
$$x = \frac{33}{2} - 2$$
$$x = \frac{29}{2}$$

The system has the solution $\left(\frac{29}{2}, \frac{11}{2}\right)$. Check the solution. The solution set is $\left\{\left(\frac{29}{2}, \frac{11}{2}\right)\right\}$.

3. $3x + y = 12$
$2x + 5y = 8$
To use substitution, we solve the first equation for y in terms of x.

$$3x + y = 12$$
$$y = 12 - 3x \quad \text{Subtract } 3x \text{ from each side}$$

Using the equation $2x + 5y = 8$,

$$2x + 5(12 - 3x) = 8 \quad \text{Replace } y \text{ with } 12 - 3x$$
$$2x + 60 - 15x = 8 \quad \text{Solve for } x$$
$$60 - 13x = 8 \quad \text{Combine like terms}$$
$$-13x = -52 \quad \text{Subtract 60 from each side}$$
$$x = 4$$

Using the equation $y = 12 - 3x$,

$$y = 12 - 3x$$
$$y = 12 - 3(4) \quad \text{Replace } x \text{ with 4}$$
$$y = 12 - 12 \quad \text{Multiply in right side}$$
$$y = 0$$

The solution is the ordered pair $(4,0)$. Check the solution. The solution set is $\{(4,0)\}$.

4. $8x + 2y = 6$
$y = -4x + 3$
Using the equation $8x + 2y = 6$,

$$8x + 2(-4x + 3) = 6 \quad \text{Replace } y \text{ with } -4x + 3$$
$$8x - 8x + 6 = 6$$
$$6 = 6 \quad \text{(True)}$$

Since $6 = 6$ is a true statement, the system is dependent and each equation represents the same line. The solution set is all ordered pairs satisfying the equation $8x + 2y = 6$, which is the same as $y = -4x + 3$.

5. $-4x + 2y = 7$
$y = 2x + 9$
Since we have $y = 2x + 9$, we substitute $2x + 9$ for y in the equation $-4x + 2y = 7$.

$$-4x + 2y = 7$$
$$-4x + 2(2x + 9) = 7 \qquad \text{Replace } y \text{ with } 2x + 9$$
$$-4x + 4x + 18 = 7$$
$$18 = 7 \qquad \text{(False)}$$

Since $18 = 7$ is a *false* statement, there is no solution, and the system is inconsistent. The solution set is $\varnothing$ (the null set). Lines are parallel.

Quick check   Find the solution set of the system of linear equations using substitution.
$3x - y = 7$
$4x - 5y = 2$

To Solve a System of Two Linear Equations by Substitution

1. Solve one of the equations for one of the variables in terms of the other (if this is not already done).
2. Substitute the expression obtained in step 1 into the other equation and solve.
3. Substitute the value obtained in step 2 into either equation and solve for the other variable.
4. The solution is the ordered pair obtained from steps 2 and 3.
5. Check the solution in both equations.

Note

If step 2 results in

   a. the variables being eliminated and a true statement is obtained, the system is dependent and there are an infinite number of solutions (same line).
   b. the variables being eliminated and a false statement is obtained, the system is inconsistent and there is no solution (parallel lines).

As a final note, when the method of solution for a system of linear equations is not specified, the following are good criteria to select a method.

1. The process of elimination is generally the easiest.
2. Substitution is most useful when the coefficient of one of the variables is 1 or -1 or an equation is already solved for one variable in terms of the other variable.
3. Graphical solutions are approximations and can be used when exact answers are not necessary or to verify the results that we find algebraically.

MASTERY POINTS

Can you
- Find the solution set for a system of two linear equations by substitution?
- Choose an appropriate method for finding the solution set for a system of two linear equations?

Exercise 7–3

Directions

Find the solution set of the following systems of linear equations by *substitution*. If a system is inconsistent or dependent, so state. See example 7–3 A.

☞ *Quick Check*

Example $3x - y = 7$
$4x - 5y = 2$

Solution   Solve $3x - y = 7$ for y. Then $y = 3x - 7$.

$$4x - 5y = 2$$
$$4x - 5(3x - 7) = 2 \qquad \text{Replace } y \text{ with } 3x - 7$$
$$4x - 15x + 35 = 2 \qquad \text{Solve for } x$$
$$-11x + 35 = 2$$
$$-11x = -33$$
$$x = 3$$

Using the equation

$$3x - y = 7$$
$$3(3) - y = 7 \qquad \text{Replace } x \text{ with } 3$$
$$9 - y = 7$$
$$-y = -2$$
$$y = 2$$

The solution is $(3,2)$ and the solution set is $\{(3,2)\}$.

Directions   Find the solution set of the following systems of linear equations by *substitution*. If a system is inconsistent or dependent, so state. See example 7–3 A.

1. $3x - y = 10$
   $y = -x + 2$

2. $2x + 3y = 9$
   $x = y - 3$

3. $-2x + 5y = 17$
   $y = 2x + 5$

4. $5x + y = 10$
   $x = 2 - 3y$

5. $2y - x = 3$
   $x = 4y - 1$

6. $3x + 4y = -1$
   $y = 2 - 3x$

7. $4y - x = 3$
   $x = 2y + 1$

8. $3y + 4x = -4$
   $x = 4 - y$

9. $5x - 2y = 3$
   $y = x - 4$

10. $3x + y = 10$
    $2x + y = 5$

11. $2x + y = 13$
    $3x + y = 17$

12. $3y + x = -10$
    $3y - x = 4$

13. $2x - y = 2$
    $6x - y = 22$

14. $x - y = 3$
    $2x + 3y = 11$

15. $2x + y = 3$
    $3x - y = 4$

16. $x + 5y = 7$
    $2x + 3y = 5$

17. $5x - 3y = 4$
    $x + 2y = -2$

18. $x + y = 4$
    $3x - 5y = 7$

19. $x - y = 3$
    $2x - 2y = 11$

20. $x + y = 5$
    $3x + 3y = 3$

21. $3x - y = 7$
    $6x - 2y = 14$

22. $-x + 3y = -3$
    $9y - 3x = -9$

23. $3x + y = -3$
$x - 3y = -1$

24. $2x + y = 1$
$-3x + y = 1$

25. $2x + y = -3$
$5x - 4y = 1$

26. $5x + 2y = 11$
$7x - y = 4$

27. $2x + 3y = 2$
$6x - y = 5$

28. $x - 2y = 4$
$3x + 2y = 4$

29. $2x + y = 6$
$3x + 4y = 4$

30. $-x - 4y = 3$
$2x + 8y = -6$

31. $x + 4y = -3$
$-2x - 8y = 6$

32. $x - y = 7$
$3x + 3y = 4$

33. $5x - 3y = 11$
$x + y = 5$

34. $-x - 2y = 6$
$6x + 4y = 5$

35. $-3x - 2y = 6$
$6x + 4y = 5$

36. $4x + 3y = 1$
$y = -4$

37. $3x - 5y = -6$
$x = 5$

38. $-5x + 2y = 11$
$x = -3$

39. $3x - 7y = 14$
$y = 2$

40. $y = 2x + 1$
$y = -x + 3$

41. $y = 5x + 3$
$y = 2x - 4$

42. $y = 3x - 4$
$y = 3x + 5$

Directions   Find the solution set of each system of linear equations by either elimination or substitution. Try to choose the most suitable method.

43. $3x - 2y = -1$
$2x + 2y = 1$

44. $x - y = 2$
$-5x + 2y = -2$

45. $x - 3 = 0$
$3x + 2y = 1$

46. $y + 2x = 4$
$x = -6y + 1$

47. $\dfrac{1}{2}x - y = 3$

$2x + \dfrac{1}{3}y = 1$

48. $\dfrac{2x}{3} - \dfrac{y}{2} = 1$

$3x + \dfrac{y}{4} = 2$

49. $\dfrac{y}{3} - \dfrac{x}{4} = 3$

$\dfrac{x}{2} + \dfrac{y}{5} = -2$

50. $4x - 3y = -10$
$y + 2 = 0$

51. $4x - y = 8$
$5x + 3y = -4$

52. $7x - 2y = 3$
$4x + 3y = -2$

53. Let x represent one current in an electrical circuit and y represent a second current. If the first current has twice as many amperes as the second, write an equation stating this.

54. If the sum of the two currents in exercise 53 is 56 amperes, write an equation stating this.

55. A clothier has two kinds of suits. Let x represent the number of suits selling for one price and y represent the number of suits selling for another price. If he has a total of 80 suits, write an equation stating this.

56. In exercise 55, if the cost of the first kind of suit (x) is \$190 per suit and the cost per suit of the second kind (y) is \$250, write an equation stating that the total income from the suits was \$8,400.

57. Given the system of linear equations

$$y = 2x - 5$$
$$y = 3x - 7$$

Which method, elimination or substitution, would you use and why?

✸ Explore to Learn

Consider the system

$$Ax + By = C$$
$$y = ax + b$$

Use substitution to solve for x and y in terms of the variables A, B, C, a, and b.

Review Exercises

1. A 42-foot piece of wood is cut into two pieces. If one piece is 6 feet less than twice the length of the other piece, how long are the pieces of wood? See section 2–7.

2. The length of a rectangle is 1 yard longer than twice its width. If the area of the rectangle is 36 square yards, find the length and width of the rectangle. See section 4–7. (*Hint:* Area = length · width.)

3. Solve the equation $A = \dfrac{1}{2}h(b + c)$ for b. See section 2–6.

4. Find the slope of the line through the points $(-4, 4)$ and $(5, 6)$. See section 6–3.

5. Find the solution set for the rational equation $\dfrac{4}{x} - \dfrac{3}{2x} = 4$. See section 5–6.

6. Solve the inequality $-4 \le 1 - 5x < 6$. See section 2–9.

7–4   Applications of Systems of Two Linear Equations

In chapter 2, we learned how to take a verbal statement and translate it into an algebraic equation in one unknown. Many practical problems that can be solved using single equations in one unknown can more easily be solved by translating into *two* equations involving *two* unknowns.

While there is no standard procedure *for solving applied problems*, the following guidelines should be useful.

To Solve a Word Problem Using Systems of Two Linear Equations

Read **Analyze-Visualize**	1. Read the problem carefully, usually several times. Determine what information is given, and what information you are asked to find. If possible, draw a picture, make a diagram, or construct a table to help visualize the information.
Choose two variables	2. Choose two variables to represent the unknown quantities. Write a sentence stating exactly what each variable represents. Label any picture, diagram, or chart using the chosen variables.
Write two equations	3. Use the unknowns from step 2 to translate seperate statements from the word problem into two linear equations. There may be some underlying relationship or formula that you need to know to write the equation. If it is a geometry problem, the formulas inside the back cover may be useful. Other commonly used formulas are interest equals principal times rate times time ($I = prt$) or distance equals rate times time ($d = rt$). If there is no applicable formula, then the words in the problem give the information necessary to write the equation.

continued

Solve the system of linear equations   4. Solve the system of linear equations by one of the methods we have studied.

Answer the question(s)   5. Answer the question or questions in the problem.

Check your results   6. Check your results in the original statement of the problem. Make sure your answer makes sense. Remember that in a geometry problem the dimensions of the figure cannot be negative. If you were determining how fast a car was traveling, 1,000 miles per hour would not make sense.

● **Example 7–4 A**

Problem solving

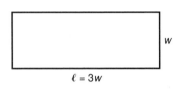

$\ell = 3w$

1. The length of a rectangle is 3 times as long as its width. The perimeter of the rectangle is 40 inches. What are the dimensions of the rectangle?

Note We need the prior knowledge that the perimeter, *P*, of a rectangle is given by the formula $P = 2\ell + 2w$, where ℓ is the length and w is the width of the rectangle.

Let ℓ represent the length of the rectangle and w represent the width of the rectangle.

length	is	3 times the width
ℓ	$=$	$3w$

We use the formula for the perimeter, $P = 2\ell + 2w$, and since the perimeter is 40 inches, we have $40 = 2\ell + 2w$. Therefore, we have the system of linear equations

$$\ell = 3w$$
$$2\ell + 2w = 40$$

Using substitution and the equation $2\ell + 2w = 40$,

$$2(3w) + 2w = 40 \qquad \text{Replace } \ell \text{ with } 3w$$
$$8w = 40 \qquad \text{Solve for } w$$
$$w = 5$$

Substituting 5 for w in the equation $\ell = 3w$, we get

$$\ell = 3(5) = 15$$

The rectangle is 15 inches long and 5 inches wide.
Check:
a. The perimeter is 40 inches.

$$2(15) + 2(5) = 40$$
$$30 + 10 = 40$$
$$40 = 40 \qquad \text{(True)}$$

b. The length is 3 times the width.

$$15 = 3(5)$$
$$15 = 15 \qquad \text{(True)}$$

2. A woman has $10,000, part of which she invests at 11% and the rest at 8% annual simple interest. If her total income for 1 year from the two investments is $980, how much does she invest at each rate?

Note The prior knowledge needed for this problem is that Simple interest = Principal · Rate · Time. Time in this example is 1 year. When using this formula, time should be stated in years.

Let x represent amount invested at 11% and y be the amount invested at 8%. Then, $(0.11)x =$ interest from the 11% investment for one year
$(0.08)y =$ interest from the 8% investment for one year

11% investment plus   8% investment   is $980

$(0.11)x \quad + \quad (0.08)y \quad = \quad 980$

From "A woman has $10,000...,"

$$x + y = 10,000$$

The system of linear equations is

$$x + y = 10,000$$
$$(0.11)x + (0.08)y = 980$$

Multiply the second equation by 100 to clear the decimal fractions.

$$100(0.11)x + 100(0.08)y = 100(980)$$
$$11x + 8y = 98,000$$

We then have the system

$x + y = 10,000$ Multiply each term by $(-8) \rightarrow$

$11x + 8y = 98,000$

$$-8x - 8y = -80,000$$
$$\underline{11x + 8y = \quad 98,000}$$
$$3x \quad\quad = \quad 18,000 \quad \text{Add}$$
$$x = \quad 6,000 \quad \text{Divide by 3}$$

Using $x + y = 10,000$,

$$(6,000) + y = 10,000 \qquad \text{Replace } x \text{ with 6,000}$$
$$y = 4,000$$

The woman invested $6,000 at 11% and $4,000 at 8% interest.
Check:
a. The total yearly income is $980.

$$(0.11)(6,000) + (0.08)(4,000) = 980$$
$$660 + 320 = 980$$
$$980 = 980 \qquad \text{(True)}$$

b. The woman has $10,000.

$$6,000 + 4,000 = 10,000$$
$$10,000 = 10,000 \qquad \text{(True)}$$

3. The sum of two lengths of wire is 11 meters. If four times the first length is added to the second length, the result is 25 meters. Find the two lengths of wire.

Let x represent the first length of wire and y be the second length of wire.

the sum of the two lengths   is   11 meters

$x + y \qquad\qquad = \qquad 11$

	four times the first length	added to	the second length	is	25 meters
	$4x$	$+$	y	$=$	25

Therefore, we have the system of linear equations

$$x + y = 11 \quad \text{Multiply each term by } (-1)$$
$$4x + y = 25$$

$$-x - y = -11$$
$$\underline{4x + y = 25}$$
$$3x = 14 \quad \text{Add}$$
$$x = \frac{14}{3} = 4\frac{2}{3}$$

Using the equation $x + y = 11$, substitute $\dfrac{14}{3}$ for x and solve for y.

$$\left(\frac{14}{3}\right) + y = 11 \qquad\qquad \text{Replace } x \text{ with } \frac{14}{3}$$

$$y = 11 - \frac{14}{3} \qquad\qquad \text{Subtract } \frac{14}{3} \text{ from each side}$$

$$y = \frac{33}{3} - \frac{14}{3} = \frac{19}{3} \qquad \text{Change to common denominator 3 and subtract}$$

$$y = 6\frac{1}{3}$$

Thus the two lengths are $4\dfrac{2}{3}$ and $6\dfrac{1}{3}$ meters.

Check:

a. The sum of the two lengths is 11.

$$\frac{14}{3} + \frac{19}{3} = 11$$
$$\frac{33}{3} = 11$$
$$11 = 11 \qquad \text{(True)}$$

b. Four times the first length added to the second length is 25.

$$4\left(\frac{14}{3}\right) + \frac{19}{3} = 25$$
$$\frac{56}{3} + \frac{19}{3} = 25$$
$$\frac{75}{3} = 25$$
$$25 = 25 \qquad \text{(True)}$$

4. What quantities of silver alloy that are 65% silver and 45% silver must be fused together to get 100 grams of 50% silver?

Note   65% silver alloy means the metal is 65% silver and 35% of some other metal(s).

Let x represent the number of grams of 65% silver alloy, and y be the number of grams of 45% silver alloy. We can use the following table to summarize the given information.

Alloy	Amount of alloy	Percent of silver	Amount of silver in alloy
First	x	65%	$0.65x$
Second	y	45%	$0.45y$
Final	100	50%	$0.50(100) = 50$

$$x + y = 100 \qquad 0.65x + 0.45y = 50$$

We must now solve the system of linear equations.

$$x + y = 100$$
$$0.65x + 0.45y = 50$$

Multiply the second equation by 100 to clear decimals.

$$\begin{aligned} -65x - 65y &= -6{,}500 \\ \underline{65x + 45y &= 5{,}000} \\ -20y &= -1{,}500 \\ y &= 75 \end{aligned}$$

Multiply $x + y = 100$ by -65

Add equations

Using the equation,

$$\begin{aligned} x + y &= 100 \\ x + (75) &= 100 \\ x &= 25 \end{aligned}$$

Replace y with 75

Then 25 grams of 65% silver alloy must be fused with 75 grams of 45% silver alloy to get 100 grams of 50% silver alloy.

Check:

1. $25 + 75 = 100$ — Total of 100 grams
2. $0.65(25) + 0.45(75) = 50$ — Sum of alloys yields 50 grams
   $16.25 + 33.75 = 50$
   $50 = 50$

5. Two automobiles start at the same place and go in opposite directions. After traveling for 3 hours, they are 351 miles apart. If one automobile is averaging 13 miles per hour faster than the other, what is the average speed of each automobile in miles per hour?

Use the formula that relates distance, rate, and time, $d = rt$. Time $t = 3$ since each automobile travels 3 hours.

Let x represent the average speed of the slower automobile and y be the average speed of the faster automobile.

We can use the following table to organize the information.

	r	t	d
Slower automobile	x	3	$3x$
Faster automobile	y	3	$3y$

$$3x + 3y = 351$$

From "If one automobile is averaging 13 miles per hour faster than the other," we get the equation

$$y = x + 13$$

Thus we solve the system of linear equations

$$y = x + 13$$
$$3x + 3y = 351$$

Substitute $x + 13$ for y in the second equation.

$$3x + 3(x + 13) = 351 \qquad \text{Replace } y \text{ with } x + 13$$
$$3x + 3x + 39 = 351$$
$$6x + 39 = 351$$
$$6x = 312$$
$$x = 52$$

Replace x by 52 in the equation.

$$y = x + 13$$
$$y = (52) + 13 = 65 \qquad \text{Replace } x \text{ with } 52$$

The faster automobile is traveling at 65 mph and the slower automobile is traveling at 52 mph.

Check:

$$3x + 3y = 351 \qquad\qquad y = x + 13$$
$$3(52) + 3(65) = 351 \qquad (65) = (52) + 13$$
$$156 + 195 = 351 \qquad\qquad 65 = 65$$
$$351 = 351$$

Quick check Karen and her mother are in cities 400 miles apart. They meet at a location on a line between the two cities. If Karen drives 20 miles per hour faster than her mother, and they meet after 4 hours, how fast was each person driving?

MASTERY POINTS

Can you
• Set up and solve a system of two linear equations in two variables given a word problem?

Exercise 7–4

Directions
Solve the following problems by setting up a system of two linear equations in two unknowns. See example 7–4 A number 1.

☛ *Quick Check*

Example Karen and her mother are in cities 400 miles apart. They meet at a location on a line between the two cities. If Karen drives 20 miles per hour faster than her mother, and they meet after 4 hours, how fast was each person driving?

Solution Let x represent the speed of Karen's automobile and y be the speed of her mother's automobile.

We use the formula distance (d) = rate (r) · time (t). Time t for each automobile is 4 since each drives 4 hours. The following chart compiles this information using $r \cdot t = d$.

	rate (r)	time (t)	distance (d)
Karen	x	4	$4x$
Mother	y	4	$4y$

Then, since they are 400 miles apart,

$$\text{total distance} = 400 \text{ miles}$$
$$4x + 4y = 400$$
$$x + y = 100 \qquad \text{\small Divide each term by 4}$$

Since Karen drives 20 miles per hour faster than her mother,

$$x = y + 20$$

We solve the system of linear equations.

$$x + y = 100$$
$$x = y + 20$$

Using substitution,

$$(y + 20) + y = 100 \qquad \text{\small Replace } x \text{ with } y + 20 \text{ in } x + y = 100$$
$$2y + 20 = 100$$
$$2y = 80$$
$$y = 40$$

Since $x = y + 20$
$$= (40) + 20 \qquad \text{\small Replace } y \text{ with } 40$$
$$= 60$$

Karen was driving 60 miles per hour and her mother was driving 40 miles per hour.

Directions   Solve the following problems by setting up a system of two linear equations in two unknowns. See example 7–4 A number 1.

1. The perimeter of a rectangle is 36 meters and the length is 2 meters more than three times the width. Find the dimensions.

2. The perimeter of a room is 40 feet, and twice the length increased by three times the width is 48 feet. What are the dimensions?

3. The perimeter of a rectangle is 100 meters. The rectangle is 22 meters longer than it is wide. Find the dimensions.

4. A rectangular field has a length that is 21 yards more than twice the width. The perimeter is 620 yards. What are the dimensions of the field?

5. The length of a room is 2 feet less than twice its width. If its perimeter is 62 feet, what are the dimensions?

Directions Solve the following problems by setting up a system of two linear equations in two unknowns. See example 7–4 A number 2.

6. Phil has $20,000, part of which he invests at 8% interest and the rest at 6%. If his total income for one year from the two investments was $1,460, how much did he invest at each rate?

7. Yoko has $18,000. She invests part of her money at $3\frac{1}{2}$% interest and the rest at 5%. If her yearly income from the two investments was $862.50, how much did she invest at each rate?

8. Rich has $21,000, part of which he invests at 6% interest and the rest at 4%. If his yearly income from each investment was the same, what did he invest at each rate?

9. Rico has $31,600, part of which he invests at 6% interest and the rest at 5%. If his yearly income from the 5% investment was $106 more than that from the 6% investment, how much was invested at each rate?

10. Kimberly made two investments totaling $13,900. On one investment she made an 8% profit, but on the other investment she took a 3% loss. If her yearly net gain was $496, how much was in each investment?

11. Anne made two investments totaling $13,600. One investment made her a 7% profit, but on the other investment she took a 5% loss. If her net yearly loss was $152, how much was in each investment?

Directions Solve the following problems by setting up a system of two linear equations in two unknowns. See example 7–4 A number 3.

12. A piece of pipe is 19 feet long. The pipe must be cut so that one piece is 5 feet longer than the other piece. What are the lengths of the two pieces of pipe?

13. A 12-foot board is cut into two pieces so that one piece is 4 feet longer than the other. How long is each piece?

14. A 24-foot rope is cut into two pieces so that one piece is twice as long as the other. How long is each piece?

15. The sum of two currents is 80 amperes. If the greater current is 24 amperes more than the lesser current, find their values.

16. The resistance of one resistor exceeds that of another resistor by 25 ohms and their sum is 67 ohms. How many ohms are in each resistor?

17. A clothing store sells suits at $125 and $185 each. The store owners observe that they sold 40 suits for a total of $5,720. How many suits of each type did they sell?

18. Three times the number of bolts in a piece of machinery is 3 less than twice the number of spot welds, while seven times the number of bolts is 5 more than four times the number of spot welds. How many bolts and spot welds are there?

19. When two batteries are connected in series, we add the voltages together to get a total voltage of 27 V. When the batteries are connected in opposition, the resulting voltage, 5 V, is their difference. Find the voltages of the two batteries.

20. When two batteries are connected in series, the total voltage is 53 V. When the batteries are connected in opposition, the total voltage is 7 V. Find the voltage of the two batteries.

21. Three times the number of teeth on a first gear is 1 more than the number of teeth on a second gear. Also, five times the number of teeth on the first gear is 4 less than twice the number of teeth on the second gear. Find the number of teeth on each gear.

22. The specific gravity of an object is defined as its weight in air divided by the difference between its weight in air and its weight when submerged in water. If an object has a specific gravity of 3 and the sum of its weight in air and in water is 30 pounds, find its weight in air and its weight in water.

23. The perimeter of a rectangular flower garden is 82 feet. If the length of the rectangle is 5 feet longer than twice the width, what are the dimensions of the rectangle?

24. If twice the length of a rectangular floor is increased by three times the width, the sum is 48 feet. The perimeter of the room is 40 feet. What are the dimensions of the floor?

25. The distance around a rectangular flower garden is 64 feet. If the length is three times the width, what are the dimensions?

26. The perimeter of a rectangular plot of ground is 30 meters. Three times the length minus four times the width is 3 meters. Find the length and the width of the plot.

27. A hardware supply company sells two types of doorknobs. The chromium-plated knob sells at $8 per knob, and the solid brass knob sells for $11.50 per knob. The company sold 420 doorknobs for $3,622.50. How many of each type were sold?

28. A road construction crew consists of cat operators working at $90 per day and laborers working at $50 per day. The total payroll per day is $1,600. If there are 3 laborers doing odd jobs and 4 laborers are assigned to work with each cat operator, how many laborers and operators are there in the crew?

Directions   Solve the following problems by setting up a system of two linear equations in two unknowns. See example 7–4 A number 4.

29. An auto mechanic has two bottles of battery acid. One contains a 10% solution and the other a 4% solution. How many cubic centimeters (cc) of each solution must be used to make 120 cubic centimeters of a solution that is 6% acid?

30. A metallurgist wishes to form 2,000 kilograms (kg) of an alloy that is 80% copper. This alloy is to be obtained by fusing some alloy that is 68% copper and some that is 83% copper. How many kilograms of each alloy must be used?

31. If a jeweler wishes to form 12 ounces of 75% gold from sources that are 60% and 80% gold, how much of each substance must be mixed together to produce this?

32. A chemist wishes to make 1,000 liters of a 3.5% acid solution by mixing a 2.5% solution with a 4% solution. How many liters of each solution are necessary?

33. How many grams of silver that is 60% pure must be mixed together with silver that is 35% pure to obtain a mixture of 90 grams of silver that is 45% pure?

34. A pharmacist is able to fill 200 3-grain and 2-grain capsules using 500 grains of a certain drug. How many capsules of each kind does he fill?

35. A solution that is 38% silver nitrate is to be mixed with a solution that is 3% silver nitrate to obtain 100 centiliters (cl) of solution that is 5% silver nitrate. How many centiliters of each solution should be used in the mixture?

36. A druggist has two solutions, one 60% hydrogen peroxide and the other 30% hydrogen peroxide. How many liters of each should be mixed to obtain 30 liters of a solution that is 40% hydrogen peroxide?

Directions   Solve the following problems by setting up a system of two linear equations in two unknowns. See example 7–4 A number 5.

37. Two cars are 100 miles apart. If they drive toward each other they will meet in 1 hour. If they drive in the same direction they will meet in 2 hours. Find their speeds.

38. A cyclist and a pedestrian are 20 miles apart. If they travel toward each other, they will meet in 75 minutes, but if they travel in the same direction, the cyclist will overtake the pedestrian in 150 minutes. What are their speeds in miles per hour?

39. Jane and Jim leave from a drugstore at the same time, walking in opposite directions. After 1 hour they are 9,680 yards apart. If Jim walked at a rate of 0.5 mph faster than Jane, how fast was each walking in miles per hour? (*Hint:* 1 mile = 1,760 yards.)

40. A boat can travel 24 miles downstream in 2 hours and 16 miles upstream in the same amount of time. What is the speed of the boat in still water and what is the speed of the current? (*Hint:* Add the speed of the current when going downstream and subtract the speed of the current when going upstream.)

41. An airplane can fly at 460 miles per hour with the wind and 322 miles per hour against the wind. What is the speed of the wind and what is the speed of the airplane in still air?

42. If a boat takes $1\frac{1}{2}$ hours to go 12 miles downstream and 6 hours to return, what is the speed of the current and of the boat in still water?

Directions
Solve the following problems by setting up a system of two linear equations in two unknowns.

Example   A line whose equation is of the form $y = mx + b$ passes through a point (x, y) if and only if the coordinates of the point satisfy the equation. Find the values of m and b so that the line passes through the points $(-1, -7)$ and $(2, 5)$. Write the equation of the line.

Solution   The ordered pairs $(-1, -7)$ and $(2, 5)$ are solutions of the equation. We then substitute these values for x and y into $y = mx + b$ to obtain two linear equations in the variables m and b.

1. When $(x, y) = (-1, -7)$

$$y = mx + b$$
$$(-7) = m(-1) + b \quad \text{Replace } y \text{ with } -7 \text{ and } x \text{ with } -1$$
$$-7 = -m + b$$
$$-m + b = -7$$
$$m - b = 7 \quad \text{Write in standard form}$$

2. When $(x, y) = (2, 5)$

$$y = mx + b$$
$$(5) = m(2) + b \quad \text{Replace } y \text{ with } 5 \text{ and } x \text{ with } 2$$
$$5 = 2m + b$$
$$2m + b = 5 \quad \text{Write in standard form}$$

Then we have the system of linear equations.

$$
\begin{aligned}
m - b &= 7 \\
2m + b &= 5 \\
\hline
3m &= 12 \quad \text{Add the equations} \\
m &= 4
\end{aligned}
$$

Substitute 4 for m in $2m + b = 5$.

$$2(4) + b = 5 \quad \text{Replace } m \text{ with } 4$$
$$8 + b = 5$$
$$b = -3$$

Replace b with -3 and m with 4 in $y = mx + b$ and the equation of the line passing through the two points is $y = 4x - 3$.

Directions   Solve the following problems by setting up a system of two linear equations in two unknowns.

43. The line with equation $y = mx + b$ passes through a point (x, y) if and only if the coordinates of the point satisfy the equation. Find the values of m and b so that the line will contain the points $(1, -1)$ and $(2, 2)$. Write the equation of the line.

44. Find the values of m and b for the line that passes through the points $(2, 1)$ and $(-1, 7)$. Write the equation of the line. (Refer to exercise 43.)

45. Find the values of m and b for the line that passes through the points $(-2, 3)$ and $(3, -7)$. Write the equation of the line. (Refer to exercise 43.)

46. Find the values of m and b for the line that passes through the points $(-1, 7)$ and $(2, -2)$. Write the equation of the line. (Refer to exercise 43.)

47. Find the values of m and b and write the equation of the line that passes through $(0, -4)$ and $(5, 0)$. (Refer to exercise 43.)

48. Find the values of m and b and write the equation of the line that passes through $(0, 0)$ and $(-4, 5)$. (Refer to exercise 43.)

✦ Explore to Learn

The x-intercepts of the two lines are $(2, 0)$ and $(5, 0)$ and their respective y-intercepts are $(0, -1)$ and $\left(0, 2\dfrac{1}{2}\right)$ as shown in the figure. Find the exact coordinates of point P.

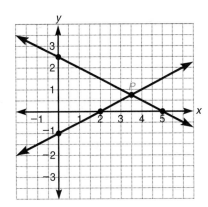

Review Exercises

Directions   Perform the indicated operations. Express your answer with positive exponents. Assume all variables are nonzero. See sections 3–3 and 3–4.

1. $(-4x^2 y^3)^2$

2. $\dfrac{x^{-2} y^3}{xy^2}$

3. $x^{-2} x^5$

Directions   See sections 2–2 and 3–2.

4. $(2x^2 - x + 3) - (x^2 + 4x - 1)$

5. $4x^2(3x^2 + 2x - 3)$

6. $(3y + 2)(3y - 2)$

7. $(4x - 5y)^2$

8. Find the solution set of the equation $5(x - 2) = -3(2 - 3x)$. See section 2–5.

9. Find the solution set of the quadratic equation $x^2 - 3x - 10 = 0$. See section 4–7.

7–5 Solving Systems of Linear Inequalities by Graphing

In section 6–5, we learned how to graph a linear inequality in two variables. To review this method, let us consider the inequality $3x + y > 4$. We first replace the inequality symbol, $>$, with the equal symbol, $=$, and graph the resulting equation $3x + y = 4$. Since we have the symbol $>$, the coordinates of the points on the line do not satisfy the inequality and the line should be dashed. We then choose a point on either side of the line, usually $(0,0)$, and substitute its coordinates into the inequality.

$$3x + y > 4$$
$$3(0) + (0) > 4$$
$$0 > 4 \quad \text{(False)}$$

We then shade the half-plane that does *not* contain $(0,0)$ to represent the graph of the inequality. (See figure 7–1.)

In this section, we will consider the graphical method of obtaining the solution of systems of linear inequalities such as

$$2x + y > 3$$
$$x - y > 1$$

A system of linear inequalities has as its solution all ordered pairs of real numbers that are solutions of the inequalities of the system.

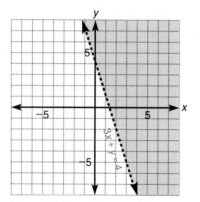

Figure 7–1

● **Example 7–5 A**

Graphical solution of a system of inequalities

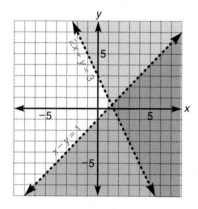

Graph the solution of the following systems of linear inequalities.

1. $2x + y > 3$
   $x - y > 1$
   First, graph each inequality on the same axes, shading each graph. (See the diagram.) The solution of the system is the overlap of the regions of the two graphs—the double-shaded region.

Note The dashed lines that show the ordered pairs of numbers for the points on the lines *are not solutions*.

2. $2x + 3y \le 4$
   $x - 4y \le 1$
   Graph $2x + 3y \le 4$ and $x - 4y \le 1$ on the same axes. The graph of the solution of the system is the overlapping (double-shaded) region of the diagram.

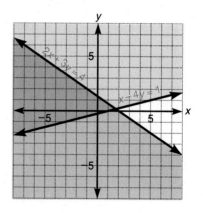

Note The symbol, ≤, involved in each inequality indicates that the points of that portion of the line bordering the double-shaded region are included as solutions. Therefore when graphing the equations, make the lines **solid**.

The solutions are in the shaded region and *include both* half-lines that form the boundary of this region.

3. $\dfrac{1}{2}x - \dfrac{2}{3}y > 1$

  $y \leq 3$

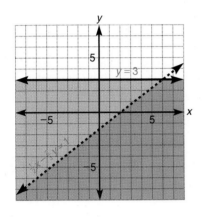

Graph the inequalities $\dfrac{1}{2}x - \dfrac{2}{3}y > 1$ and $y \leq 3$ on the same axes. First, clear

the denominators in the first inequality by multiplying by the LCD, 6.

$$6 \cdot \dfrac{1}{2}x - 6 \cdot \dfrac{2}{3}y > 6 \cdot 1$$

$$3x - 4y > 6$$

Recall, the graph of $y = 3$ is a horizontal line passing through the point (0,3). The graph of the solutions is the double shading in the diagram, together with that portion of the line $y = 3$ that borders the double-shaded region.

Note The point of intersection (6,3), of the dashed line and the solid line is *not* in the solution.

 Quick check Graph the solution of the system of linear inequalities.

  $2x + y \geq 1$

  $2y < x + 2$

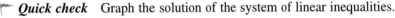

The Casio fx-7700 GE and 9700 GE can graph systems of linear inequalities in two variables. To do number 2 of example 7–5 A on the Casio, proceed as follows.

Solve each inequality for y: $y \leq -\dfrac{2}{3}x + \dfrac{4}{3}$ and $y \geq \dfrac{1}{4}x - \dfrac{1}{4}$.

Use GRAPH mode: MENU 6.

Set the range to RANGE −7.5,7.5,1,−5,5,1. Use the Range and EXIT keys to do this.

Set the equation type to inequality: F3:Type F4:INEQ.

Enter the first inequality into Y1: (−) 2 $a^b/_c$ 3 x, θ, T + 4 $a^b/_c$ 3 F1:STO F6:Y≤.

Select Y2 using the down cursor key ▼.

Enter the second inequality into Y2: 1 $a^b/_c$ 4 x, θ, T − 1 $a^b/_c$ 4 F1:STO F5:Y≥.

Now graph: F6:Draw.

The trace feature F1:Trace can be used to find approximate values of the points where the edges of the graphs cross. The right/left cursor keys trace along one line. Use the up/down cursor keys to jump between the different inequalities.

MASTERY POINTS

Can you
- Solve a system of two linear inequalities by graphing?

Exercise 7–5

Directions
Graph the solution of the following systems of linear inequalities. See example 7–5 A.

▶ *Quick Check*

Example $2x + y \geq 1$
$2y < x + 2$

Solution   Graph $2x + y \geq 1$ and $2y < x + 2$ on the same axes. A solid line is used in the graph of the first inequality and a dashed line is used in the graph of the second inequality. The graph is the double-shaded portion of the diagram, together with that portion of the line $2x + y = 1$ bordering the double-shaded region.

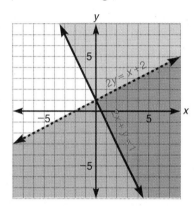

Directions   Graph the solution of the following systems of linear inequalities. See example 7–5 A.

1. $x + y < 2$
   $x - y < 4$

3. $2x + y > 1$
   $x - 2y > 2$

5. $-4x + y < 5$
   $2x - y < 0$

2. $x + y < 3$
   $x - y > 1$

4. $x + 3y > 4$
   $2x - y < 0$

6. $5x - y < 5$
   $2x + y \geq 1$

7. $2x + 3y > 5$
$3x - 2y < 4$

12. $x - 4y \geq 3$
$3y \leq 2x$

17. $2x - 3y \leq 0$
$4x - 6y \geq -12$

8. $4x + 3y > 12$
$5x - 3y > 15$

13. $y \leq 2x - 7$
$2y + x \geq 5$

18. $-4x + 10y \leq 5$
$-2x + 5y > -1$

9. $x + 2y < 4$
$2x - 3y > 6$

14. $4x + 5y \geq 20$
$y \geq x + 1$

19. $\frac{1}{3}x - y \geq 2$

$\frac{1}{3}x + 2y \leq 4$

10. $5x - 2y > 15$
$2x - 3y < 6$

15. $5y + 2x \geq 3$
$x \leq 2y - 1$

20. $\frac{3}{4}x - \frac{2}{3}y \leq 1$

$2x + \frac{1}{2}y \geq 3$

11. $3x - 2y \geq 1$
$2x \geq 5y$

16. $x + y \geq 5$
$x \geq 3y + 2$

21. $\frac{1}{5}x \geq 3y + 1$

$x - \frac{2}{3}y \leq 2$

24. $x - 4y \geq 6$
$x > -2$

28. $x \leq 5$
$y > -3$

25. $5x - 2y \leq 0$
$y > -1$

29. $x \geq 0$
$y \geq 0$

22. $\frac{3}{2}y - 4x \geq 2$

$\frac{1}{3}y > 1 - x$

26. $2y - x > 1$
$y \leq 3$

30. $x \leq 0$
$y \geq 0$

23. $3x - 5y > 2$
$x \leq 3$

27. $x \geq -2$
$y < 4$

Note Exercises 29 and 30 define quadrants I and II, respectively, together with a portion of each axis.

31. State systems of linear inequalities that define (a) quadrant III, (b) quadrant IV.

32. In your own words, describe the steps for finding the solution of a system of two linear inequalities.

✸ Explore to Learn

Describe a system of linear inequalities for which the graph could be the solution. The x- and y-intercepts of the dashed lines are the integer values shown.

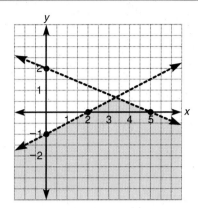

Review Exercises

Directions Perform the indicated operations. In exercises 5–7 assume that no variables equal zero. See sections 2–2, 3–2, and 3–4.

1. $(3x^2 - 2x + 1) - (x^2 - 5x - 6)$

2. $(4x + 1)(x - 3)$

3. $(5y - 2)^2$

4. $(7z + 1)(7z - 1)$

5. $a^3 a^2 a^{-1}$

6. $\dfrac{x^2 y^3}{xy}$

7. $(-y^3)^3$

Directions Completely factor the following expressions. See sections 4–1 and 4–3.

8. $5x^2 - 3x$

9. $4x^2 + 8x - 5$

10. $9y^2 - 30y + 25$

Chapter 7 Lead-in Problem

A company has $100,000 to invest. The company has two places it can put the money. One choice is to invest in a low-risk stock portfolio that pays 4%. A second choice is to invest in a new company. This company is expected to earn 12% on its capital. Unfortunately any new company must be considered a high risk. The board of directors of the company is demanding at least a 6% return on investment, and directs the company to divide up the $100,000 into the two choices, but to only put just enough of the money into the high-risk venture so that the net return is 6% (if all goes well). How much money should be put into each investment choice to meet these conditions?

Solution

Let x be the amount invested at 4% and y be the amount invested at 12%. Then $x + y = 100,000$. We want a 6% return on the $100,000, which means a return of $6,000. This comes from 4% of x and 12% of y, so that

$$4\% \text{ of } x + 12\% \text{ of } y = 6,000$$
$$0.04x + 0.12y = 6,000$$
$$4x + 12y = 600,000$$
$$x + 3y = 150,000$$

Solve the system $x + y = 100,000$
$\qquad\qquad\qquad\quad x + 3y = 150,000$

Multiply the first equation by -1:

$$\begin{aligned} -x - y &= -100,000 \\ x + 3y &= 150,000 \qquad \text{Add} \\ 2y &= 50,000 \\ y &= 25,000 \\ x + y &= 100,000 \\ x + 25,000 &= 100,000 \qquad \text{Substitute 25,000 for } y \\ x &= 75,000 \end{aligned}$$

Thus invest $75,000 at 4% and $25,000 at 12%.

Chapter 7 Summary

• Glossary

consistent and independent system of linear equations (page 443) when the graph of a system of linear equations consists of intersecting lines and thus has one ordered pair for a solution.

dependent system (page 444) when the graphs of the equations of a system are the same.

inconsistent system (page 444) when the system has no solution.

system of linear equations (page 441) two or more linear equations involving the same variables.

• Properties and Definitions

Solutions of systems of two linear equations (page 442)

A **solution** of a system of two linear equations is an ordered pair of numbers that is a solution to each equation in the system.

Addition property of equality (page 448)

Given a, b, c, and d are real numbers, if $a = b$ and $c = d$, then

$$a + c = b + d$$

• Procedures

[7–1]

To solve a system of two linear equations by graphing (page 444)

1. Graph each equation on the same set of axes.

2. Their graphs may intersect in

   a. one point—one solution (the system is *consistent* and *independent*).

   b. no point—the lines are parallel and there is no solution (the system is *inconsistent*).

   c. all points—the lines are one and the same (the system is *dependent*).

3. If the graphs intersect in one point, estimate the coordinates of the point and check this estimate in each equation.

[7–2]

To solve a system of two linear equations by elimination (page 449)

1. Write the system of linear equations so that each equation is in standard form, $ax + by = c$.

2. If necessary multiply both sides of one or both equations by a constant that will make the coefficients of one of the variables opposites of each other.

3. Add the corresponding sides of the two equations, and solve this new equation for the remaining variable.

4. Substitute the value found in step 3 into either of the original equations and solve this equation for the other variable. The solution is the ordered pair of numbers obtained in step 3 and this step.

5. If the system has a solution found in step 4, check the solution in both equations.

[7–3]

To solve a system of two linear equations by substitution (page 456)

1. Solve one of the equations for one of the variables in terms of the other (if this is not already done).

2. Substitute the expression obtained in step 1 into the other equation and solve.

3. Substitute the value obtained in step 2 into either equation and solve for the other variable.

4. The solution is the ordered pair obtained from steps 2 and 3.

5. Check the solution in both equations.

[7–4]

To solve a word problem using systems of two linear equations (page 459)

Read Analyze-Visualize	1. Read the problem carefully, usually several times. Determine what information is given, and what information you are asked to find. If possible, draw a picture, make a diagram, or construct a table to help visualize the information.

Choose a variable 2. Choose a variable to represent one of the unknowns and write a sentence stating exactly what that variable represents. Use this variable to write algebraic expressions that represent any other unknowns in the problem. Label any picture, diagram, or chart using the chosen variable.

Write an equation 3. Use the unknowns from step 2 to translate the word problem into an equation. There may be some underlying relationship or formula that you need to know to write the equation. If it is a geometry problem, the formulas inside the front cover may be useful. Other commonly used formulas are: interest equals principal times rate times time ($I = prt$) or distance equals rate times time ($d = rt$). If there is no applicable formula, then the words in the problem give the information necessary to write the equation.

Solve the equation 4. Solve the equation.

Answer the question(s) 5. Answer the question or questions in the problem. If there is more than one unknown in the problem, substitute the solution of the equation into the algebraic expressions that represent the other unknowns to determine the other unknown values in the problem.

Check your results 6. Check your results in the original statement of the problem. Make sure your answer makes sense. Remember that in a geometry problem the dimensions of the figure cannot be negative. If you were determining how fast a car was traveling, 1,000 miles per hour would not make sense.

☀ Chapter 7 Error Analysis

1. Solving systems of linear equations
Example: Solve the system.

$$2x - 3y = -1$$
$$x + 4y = -2 \quad \text{Multiply by 2} \rightarrow$$

$$2x - 3y = -1$$
$$2x + 8y = -4$$
$$\text{Add} \quad 4x + 5y = -5$$

Correct answer:

$$2x - 3y = -1$$
$$x + 4y = -2 \quad \text{Multiply by } -2 \rightarrow$$

$$2x - 3y = -1$$
$$-2x - 8y = 4$$
$$\text{Add} \quad -11y = 3$$
$$y = -\frac{3}{11}$$

What error was made? (*see page 449*)

2. Solutions of systems of linear equations
Example: (2,3) is the solution of the system
$$4x - 3y = -1$$
$$2x + y = 6$$
Check: $4(2) - 3(3) = -1$
$$8 - 9 = -1$$
$$-1 = -1$$
(2,3) is the solution.
Correct answer: (2,3) is not a solution of $2x + y = 6$ so it is not a solution of the system.
What error was made? (*see page 442*)

3. Subtracting real numbers
Example: $(-10) - (-5) = -15$
Correct answer: -5
What error was made? (*see page 55*)

4. Division of real numbers
Example: $\dfrac{-16}{-4} = -4$

Correct answer: $\dfrac{-16}{-4} = 4$

What error was made? (*see page 64*)

5. Product of a monomial and a polynomial
Example: $-4y(y^2 + 3y - 4) = 4y^3 + 12y^2 - 16y$
Correct answer: $-4y^3 - 12y^2 + 16y$
What error was made? (*see page 193*)

6. Dividing unlike bases
Example: $\dfrac{x^3}{y} = x^2$

Correct answer: $\dfrac{x^3}{y} = \dfrac{x^3}{y}$

What error was made? (*see page 201*)

7. Zero exponent
   Example: $4y^0 = 1$
   Correct answer: 4
   What error was made? (*see page 203*)

8. Scientific notation
   Example: $4.21 \times 10^{-2} = 421$
   Correct answer: 0.0421
   What error was made? (*see page 211*)

9. Factoring the sum of two squares
   Example: $25a^2 + 36b^2 = (5a + 6b)(5a - 6b)$
   Correct answer: $25a^2 + 36b^2$ cannot be factored.
   What error was made? (*see page 254*)

10. Completely factoring the difference of two squares
    Example: $x^4 - y^4 = (x^2 + y^2)(x^2 - y^2)$
    Correct answer: $(x^2 + y^2)(x + y)(x - y)$
    What error was made? (*see page 254*)

Chapter 7 Critical Thinking

1. Pick any integer from 20 to 29 and subtract the sum of its digits from it. The answer is always 18. Why is this true?

2. M85 automobile fuel is 85% methanol and 15% gasoline. It is used to lower emissions. M85 has about 60% of the energy of gasoline, so that an automobile would only go about 60% as far as on a gallon of gasoline. Suppose gasoline costs $1.20 per gallon and M85 costs $0.80 per gallon. Which is more cost effective? State the answer as a percent of the cost of gasoline.

3. Starting out with $32 you bet six times. Each time you bet one half of what you have. Suppose you win three times and lose three times. Would you come out ahead, behind, or even?

Chapter 7 Review

[7–1]

Directions Find the solution of the following systems of linear equations by graphing. All answers in the back of the book will be given exactly. If the system is inconsistent or dependent, so state.

1. $x + y = 4$
   $x - y = 2$

2. $3x - y = 2$
   $x + y = 6$

3. $x - 2y = 0$
   $2x - y = 6$

4. $x + y = 4$
   $-x + y = 8$

5. $x - \dfrac{1}{3}y = 1$
   $2x - \dfrac{2}{3}y = -2$

6. $x - y = 4$
   $-3x + 3y = -12$

[7–2]

Directions   Find the solution set of the following systems of linear equations by elimination. If a system is inconsistent or dependent, so state.

7. $x - 2y = 3$
  $x + y = 3$

8. $x - y = 5$
  $3x + 2y = 25$

9. $2x + 3y = 1$
  $-x + y = 2$

10. $5x + y = 4$
  $5x - 3y = 8$

11. $x + 3y = 6$
  $2x = 3y - 6$

12. $\dfrac{1}{2}x + \dfrac{1}{4}y = \dfrac{7}{4}$
  $\dfrac{2}{3}x - \dfrac{1}{3}y = \dfrac{13}{3}$

13. $(0.3)x - (0.2)y = 1.1$
  $(0.3)x + (0.2)y = 1.9$

14. $x - 2y = 5$
  $-3x + 6y = 4$

15. $2x - y = 3$
  $3x + y = 7$

16. $3x - y = 2$
  $9x - 3y = 6$

[7–3]

Directions   Find the solution set of the following systems of linear equations by substitution. If a system is inconsistent or dependent, so state.

17. $x - y = 2$
  $2x - 3y = 1$

18. $2x + y = 7$
  $2x - y = 13$

19. $3x - 2y = 4$
  $6x - 4y = 5$

20. $2x - 3y = 3$
  $3x + y = 7$

21. $x + y = 4$
  $x - y = 2$

22. $x - y = 5$
  $\dfrac{1}{2}x + \dfrac{1}{3}y = \dfrac{25}{6}$

23. $12x - 9y = 15$
  $4x - 3y = 5$

24. $5x + 3y = 4$
  $2x - y = 5$

[7–4]

Directions   Solve the following problems by setting up a system of two linear equations in two unknowns.

25. The perimeter of a rectangle is 64 feet. Find the dimensions if three times the width is 4 less than the length.

26. Bruce made two investments totaling $18,000. On one investment he made a 14% profit, but on the other investment he took a 23% loss. If his net yearly loss was $70, how much was in each investment?

27. Two trains start from points A and B, 534 miles apart, and travel toward each other. Train A travels 25 miles per hour faster than train B. They meet in 6 hours. How fast was each train traveling?

28. A couple wishes to invest $8,200, part at $5\dfrac{1}{4}\%$ interest and part at 10% interest. They earn $677.50 in interest at the end of the year. How much was invested at each rate?

29. An airplane travels 1,200 miles in 5 hours against a headwind. If the plane can travel 1,800 miles in the same time with a tailwind, find the speed of the wind and the speed of the plane in still air.

30. Find the values of the slope, m, and the y-intercept, b, and write the equation of the line passing through the points $(3,2)$ and $(-3,2)$.

[7–5]
Directions   Graph the solution of the following systems of linear inequalities.

31. $x + y \geq 5$
$\quad 2x - y < 3$

33. $4y - 3x \geq 2$
$\quad y \leq 1 - 2x$

35. $x < 4y + 3$
$\quad y \leq 4$

32. $4x - 3y < 1$
$\quad 2x + 5y \leq 2$

34. $2y \leq 4 - x$
$\quad 5y - x > 2$

36. $x > 1$
$\quad 5y - x \leq 0$

Chapter 7 Test

Directions   In exercises 1 and 2, estimate the solution of each system by graphing. All answers in the back of the book will be given exactly.

1. $x - y = -1$
$\quad 3x - y = 3$

2. $x - y = -2$
$\quad 5x - y = 2$

Directions   In exercises 3–10 find the solution set of each system of linear equations. Use the elimination method for exercise 3. Use the substitution method for exercise 4. Use either elimination or substitution for exercises 5–10. If the system is inconsistent or dependent, so state.

3. $2x + 3y = 2$
$\quad 3x - 2y = 3$

4. $x + 2y = -3$
$\quad x + y = 4$

5. $3x - y = 4$
$\quad 2x + y = 1$

6. $3x - 2y = 6$
$\quad 5x + y = 10$

7. $\dfrac{2}{3}x + \dfrac{3}{5}y = \dfrac{14}{3}$

$\quad x - \dfrac{2}{5}y = -6$

8. $2x - y = 0$
$\quad x + y = 1$

9. $x - 2y = 4$
$\quad 3x - 6y = 9$

10. $x + 10y = -7$
$\quad -2x + 5y = 4$

Directions In exercises 11 and 12 graph the solution of the following systems of linear inequalities.

11. $2x - 3y \le 12$
    $x + 2y \ge 4$

12. $2x - y > 5$
    $x + y < 6$

13. Peter invests $20,000 in two accounts, at 5% and 7% interest. The yearly interest from the 7% account is $440 more than the 5% account. How much did he invest at each rate?

14. A boat travels 36 miles in 4 hours upstream. In the same amount of time the boat can travel 48 miles downstream. Find the rate of the current and the rate of the boat in still water.

15. Find the values of the slope, m, and the y-intercept, b, and write the equation of the line passing through points $(4, -1)$ and $(-4, 1)$.

16. The length of a rectangle is 5 inches longer than the width. The perimeter is 36 inches. Find the two dimensions.

17. The length of a certain rectangle is 3 meters longer than the width. The perimeter is 120 meters. Find the two dimensions.

Chapter 7 Cumulative Test

1. Evaluate the expression $\dfrac{ab - bc}{a}$, when $a = \dfrac{1}{2}$, $b = \dfrac{2}{3}$, and $c = \dfrac{3}{4}$.

Directions Perform the indicated operations and simplify the results. Assume all denominators are nonzero.

2. $\dfrac{4x}{5y} \div \dfrac{24x^2}{15y^3}$

3. $\dfrac{5}{x^2 - 9x} - \dfrac{3}{x}$

4. $\dfrac{36x^3y^2 - 56xy^3}{4xy}$

5. $\dfrac{a^2 - 8a}{a^2 - a - 56} \cdot \dfrac{a^2 - 49}{7a}$

6. $(5x - 3y)^2$

7. $(8y^2 + 10y - 42) \div (4y - 7)$

Directions Find the solution set of the following equations.

8. $\dfrac{4a - 1}{4} = \dfrac{5a + 2}{7}$

9. $x^2 \doteq -9x$

10. $\dfrac{3}{4x} - 2 = \dfrac{5}{3x} + 1$

11. $7x^2 + 4x = 3$

Directions Find the solution set for the following systems of linear equations by the appropriate method. If a system is inconsistent or dependent, so state.

12. $3y - x = 2$
$8y + x = 20$

13. $4x - y = 6$
$3x + 2y = -1$

14. $5x - 2y = 4$
$x = -3y + 1$

15. $x - 2y = -6$
$3x - 6y = 1$

16. $8x - 3y = 2$
$y - 1 = 0$

17. $y + 3x = 8$
$6x + 3y = 24$

18. Graph the solution of the system of linear inequalities.
$3x + 2y \le 6$
$x - 5y > 5$

19. The sum of two numbers is 79. Their difference is 5. Find the two numbers.

20. The sum of the squares of two consecutive integers is 61. What are the integers?

21. The square of a number is equal to five times that number. Find the number.

22. The sum of a number and twice its reciprocal is 3. Find the number.

23. A tank can be filled in 4 hours by pipe A and in 5 hours by pipe B. How long will it take to fill the tank if both pipes are open?

24. The length of a rectangle is 1 inch less than twice its width. If the perimeter is 40 inches, what are the dimensions of the rectangle?

25. Mary and Karen attend universities 345 miles apart. If Karen travels 15 miles per hour faster than Mary and they meet at a location on a line between the schools in three hours, how fast was each woman traveling?

8

Roots and Radicals

Roger has a ladder that will extend to a length of 21 feet. For the ladder to be safe to climb on, it must be placed 7 feet away from the house. The roof is 20 feet above the ground. Will the ladder be able to reach the roof safely? If not, how far up will the ladder reach? (Leave your answer rounded to one decimal place.)

8–1   Finding Roots

Square Root

In chapter 4, quadratic equations were solved by factoring, but many quadratic equations such as

$$x^2 - 7 = 0 \quad \text{or} \quad x^2 + 3x + 1 = 0$$

will not factor over the set of rational numbers. We need to be able to solve equations that involve a squared variable. Therefore, we want a process that is the inverse of squaring a number.

In chapter 1, we discussed how to square a number. Consider the following examples:

$$\text{If } x = 3, \quad \text{then } x^2 = (3)^2 = (3)(3) = 9$$
$$\text{If } x = -3, \text{ then } x^2 = (-3)^2 = (-3)(-3) = 9$$

Reversing the process, we ask:

$$\text{If } x^2 = 9, \text{ then what number is } x \text{ equal to?}$$

This inverse operation is called *finding the square root of a number*.

Definition of a Square Root

For every pair of real numbers a and b, if $a^2 = b$, then a is called a square root of b.

Concept

A square root of a number is one of two equal factors of the number.

From this discussion and the definition of square root, we can see that the answer to the question we asked,

If $x^2 = 9$, then what number is x equal to?

is 3 or -3 since $(3)^2 = 9$ and $(-3)^2 = 9$. To distinguish between the two square roots, **we define the principal square root of a positive number to be positive**. The $\sqrt{}$ symbol[1] denotes the principal square root. Thus, if $x = \sqrt{9}$, then $x = 3$ and we say 3 is the principal square root of 9.

$$\sqrt{9} = 3 \text{ (principal square root)}$$

The parts of the principal square root are

Radical sign   Radicand

The entire expression is called a *radical* and is read "the principal square root of a."

● **Example 8–1 A**

Finding the principal square root

Find the principal square root.

1. $\sqrt{16} = 4$, since $4 \cdot 4 = 4^2 = 16$.

2. $\sqrt{49} = 7$, since $7 \cdot 7 = 7^2 = 49$.

3. $\sqrt{25} = 5$, since $5 \cdot 5 = 5^2 = 25$.

4. $\sqrt{36} = 6$, since $6 \cdot 6 = 6^2 = 36$.

5. $\sqrt{0} = 0$, since $0 \cdot 0 = 0^2 = 0$.

Note In the examples, 0, 16, 49, 25, and 36 are called **perfect-square integers** because their square roots are integers. ●

Whenever we wish to express the negative value of the square root of a number, we use the symbol $-\sqrt{}$. For example, $-\sqrt{9}$ would indicate that we want the negative square root value. That is, $-\sqrt{9} = -3$.

● **Example 8–1 B**

Finding the negative square root

Find the indicated square root.

1. $-\sqrt{4} = -2$

2. $-\sqrt{49} = -7$

---

[1]The symbol $\sqrt{}$, sometimes called the radix symbol, was introduced by Rudolff in 1526.

3. $-\sqrt{25} = -5$

4. $-\sqrt{36} = -6$

▷ *Quick check*   Find the square root. $-\sqrt{16}$ ●

Our first examples of finding the square root of a number have dealt only with perfect-square integers. We shall now try to find the $\sqrt{2}$. We could use 1.414 from a calculator but when we square 1.414, we do not get 2 as an answer. We can show that no matter how many decimal places the answer is carried to, when the result is squared, it will be close to, but not equal, 2. The $\sqrt{2}$ is called an **irrational number** because it has the property that it cannot be expressed as a terminating or a repeating decimal number. If an integer is not a perfect square its square root is irrational. Another number that is irrational is π, which is used in geometric formulas involving circles.

Whenever you work with irrational numbers in a problem, you may have to approximate the number to as many decimal places as are needed in the problem by using a calculator.

● **Example 8–1 C**

Finding square roots with a calculator

Compute the following irrational numbers to three decimal places with a calculator.

1. $\sqrt{3} \approx 1.732$ ≈ means "is approximately equal to"
Typical calculator keystrokes   3 $\boxed{\sqrt{x}}$
Typical graphing calculator keystrokes $\boxed{2\text{nd}}$ $\boxed{\sqrt{}}$ 3 $\boxed{\text{ENTER}}$

2. $-\sqrt{41} \approx -6.403$
Typical calculator keystrokes   41 $\boxed{\sqrt{x}}$ $\boxed{+/-}$
Typical graphing calculator keystrokes $\boxed{(-)}$ $\boxed{2\text{nd}}$ $\boxed{\sqrt{}}$ 41 $\boxed{\text{ENTER}}$

3. $5\sqrt{6} \approx 12.247$
Typical calculator keystrokes   5 $\boxed{\times}$ 6 $\boxed{\sqrt{x}}$ $\boxed{=}$
Typical graphing calculator keystrokes   5 $\boxed{2\text{nd}}$ $\boxed{\sqrt{}}$ 6 $\boxed{\text{ENTER}}$

▷ *Quick check*   Find the decimal approximation to three decimal places by using a calculator. $\sqrt{48}$ and $-\sqrt{56}$ ●

Not all real numbers have a rational or an irrational square root. Consider the following:

$$\sqrt{-4} = \text{ what?}$$

We know that all real numbers are either positive, negative, or zero. If we square a real number, the product is never negative. Hence, there is no real number that when squared produces a negative answer. *The square root of a negative number does not exist in the set of real numbers.*

● **Example 8–1 D**

Problem solving

Suppose an automotive engineer wishes to determine the diameter of the cylinder bore (D) required to produce H horsepower from N cylinders of an engine that is turning 1,000 rpm. The engineer will use the formula

$$D = \sqrt{\frac{H}{(0.4)N}}$$

What would be the bore diameter (in inches) required to produce 40 horsepower at 1,000 rpm from a 4-cylinder engine?

We substitute 40 for H and 4 for N.

$$D = \sqrt{\frac{(40)}{(0.4)(4)}} = \sqrt{\frac{40}{1.6}} = \sqrt{25} = 5$$

Therefore, we need a bore diameter of 5 inches.

Pythagorean Theorem

The following is an important property of right triangles called the **Pythagorean theorem**.[2]

In a right triangle, the square of the length of the hypotenuse (the side opposite the right angle) is equal to the sum of the squares of the lengths of the two legs (the sides that form the right angle). If c is the length of the hypotenuse and a and b are the lengths of the legs, this property can be stated as:

$$c^2 = a^2 + b^2$$

● **Example 8–1 E**

Finding the length of a hypotenuse

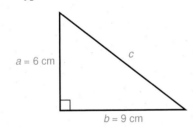

Find the length of the hypotenuse of a right triangle whose legs are 6 centimeters and 9 centimeters to the nearest tenth of a centimeter.

We want to find c when $a = 6$ cm and $b = 9$ cm.
By the Pythagorean theorem,

$c^2 = a^2 + b^2$	Pythagorean theorem
$c^2 = (6)^2 + (9)^2$	Substitute
$c^2 = 36 + 81$	Square values
$c^2 = 117$	Add
$c = \sqrt{117} \approx 10.8$	Rounded to the nearest tenth

We choose the positive square root since c represents the length of the hypotenuse of the right triangle.

☞ *Quick check*   Find the length of the second leg of a right triangle whose hypotenuse has length 13 inches and whose first leg is 5 inches long.   ●

*n*th Roots

The concept of square root can be extended to find cube roots (third root of a number), fourth roots, fifth roots, and so on. A cube root is one of three equal factors of a number. The symbol that is used to express the cube root is $\sqrt[3]{}$. The 3 is called the **index** of the radical expression. The index denotes what root we are looking for. The fourth root would be indicated by $\sqrt[4]{}$. General notation for the *n*th root would be $\sqrt[n]{}$, where n is a natural number greater than 1.

---

[2]Named for the Greek mathematician Pythagoras of Samos (ca. 580–500 B.C.), but known to the Mesopotamians 4,000 years ago.

Definition of an *n*th Root

The *n*th root of number *a*, denoted by

$$\sqrt[n]{a}$$

is one of *n* equal factors such that $\sqrt[n]{a} = b$ and

$$\overbrace{b \cdot b \cdot b \cdots b}^{n \text{ factors}} = b^n = a$$

where n *is a natural number greater than 1.*

The parts of the *n*th root are

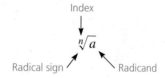

Note If there is no index associated with a radical symbol, it is understood to be 2. The symbol $\sqrt[n]{}$ to denote the *n*th root of *a* was used in 1891 by G. Peano.

● **Example 8–1 F**

Finding an nth root

Find the indicated root.

1. $\sqrt[4]{16} = 2$, since $2 \cdot 2 \cdot 2 \cdot 2 = 2^4 = 16$.

2. $\sqrt[3]{-27} = -3$, since $(-3)(-3)(-3) = (-3)^3 = -27$.

3. $\sqrt[3]{-125} = -5$, since $(-5)(-5)(-5) = (-5)^3 = -125$.

4. $\sqrt{-16}$ Does not exist in the set of real numbers.

☞ *Quick check*   Find the cube root. $\sqrt[3]{-64}$

MASTERY POINTS

Can you
- Find the principal square root of a perfect-square integer?
- Find the negative square root of a number?
- Find the decimal approximation with a calculator for a root that is an irrational number?
- Use the Pythagorean theorem?
- Find an *n*th root?

Exercise 8–1	☞ *Quick Check*
Directions Find the indicated square root. See examples 8–1 A and B.	**Example** $-\sqrt{16}$ **Solution** $= -4$ Since $4 \cdot 4 = 16$, and we want the negative value of the square root

Directions Find the indicated square root. See examples 8–1 A and B.

1. $\sqrt{100}$
2. $\sqrt{9}$
3. $\sqrt{4}$

4. $\sqrt{64}$
5. $-\sqrt{144}$
6. $-\sqrt{81}$

7. $-\sqrt{121}$
8. $-\sqrt{1}$
9. $\sqrt{121}$

10. $\sqrt{81}$
11. $-\sqrt{9}$
12. $-\sqrt{64}$

Directions Find the decimal approximation to three decimal places of the indicated square root by using a calculator. See example 8–1 C.	☞ *Quick Check*
	Examples $\sqrt{48}$ $-\sqrt{56}$ **Solutions** ≈ 6.928 ≈ -7.483

Directions Find the decimal approximation to three decimal places of the indicated square root by using a calculator. See example 8–1 C.

13. $\sqrt{18}$
14. $\sqrt{24}$

15. $\sqrt{41}$
16. $\sqrt{47}$

17. $-\sqrt{52}$
18. $-\sqrt{10}$

Directions Compute the following values exactly when appropriate or to three decimal places when the value is irrational. See examples 8–1 A, B, and C.

19. $6\sqrt{25}$
20. $-3\sqrt{4}$
21. $-\dfrac{1}{2}\sqrt{144}$
22. $\dfrac{1}{3}\sqrt{9}$
23. $4\sqrt{19}$

24. $-2\sqrt{40}$
25. $-10\sqrt{60}$
26. $\dfrac{3}{4}\sqrt{2}$
27. $\dfrac{2}{3}\sqrt{3}$

28. $-\dfrac{4}{5}\sqrt{10}$
29. $-\dfrac{3}{8}\sqrt{5}$
30. $\dfrac{1}{3}\sqrt{24}$

Directions Solve the following problems. See example 8–1 D.

31. The current I (amperes) in a circuit is found by the formula $I = \sqrt{\dfrac{\text{watts}}{\text{ohms}}}$. What is the current of a circuit that has 3 ohms resistance and uses 1,728 watts?

32. What is the current of a circuit that has 2 ohms resistance and uses 450 watts? (Refer to exercise 31.)

33. The slant height, S, of a right circular cone is found by the formula $S = \sqrt{r^2 + h^2}$, where r is the radius of the base and h is the height of the cone. What is the slant height of a right circular cone whose base radius is 5 units and whose height is 12 units?

34. What is the slant height of a right circular cone whose base radius is 3 units and whose height is 4 units? (Refer to exercise 33.)

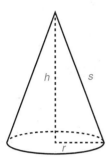

Directions	▶ *Quick Check*
In the following right triangles, find the length of the unknown side. See example 8–1 E. Round the answer to the nearest tenth when necessary.	**Example** Find the length of the second leg of a right triangle whose hypotenuse has length 13 inches and whose first leg is 5 inches long. **Solution** We want to find b given that $c = 13$ in. and $a = 5$ in. $c^2 = a^2 + b^2$ Pythagorean theorem $(13)^2 = (5)^2 + b^2$ Substitute $169 = 25 + b^2$ Square values $144 = b^2$ Subtract 25 $12 = b$ Square root We want the positive square root since b is the side of a right triangle. Therefore, the length of the second leg is 12 in.

Directions   In the following right triangles, find the length of the unknown side. See example 8–1 E. Round the answer to the nearest tenth when necessary.

35. $a = 3$ m, $b = 4$ m

36. $a = 8$ ft, $c = 10$ ft

37. $a = 12$ in., $b = 5$ in.

38. $a = 15$ cm, $b = 8$ cm

39. $a = 6$ yd, $c = 10$ yd

40. $b = 16$ m, $c = 20$ m

41. $a = 12$ mm, $b = 16$ mm

42. $a = 10$ in., $b = 24$ in.

43. $a = 6$ m, $b = 7$ m

44. $a = 8$ in., $b = 2$ in.

45. $a = 5$ ft, $c = 10$ ft

46. $a = 4$ m, $c = 13$ m

47. $b = 6$ cm, $c = 11$ cm

48. Find the width of a rectangle whose diagonal is 13 cm and whose length is 12 cm.

49. How long of a length of wire is needed to reach the top of a 15-foot telephone pole if the wire is to be attached to a point on the ground 8 feet from the base of the pole?

50. Find the diagonal of a rectangle whose length is 8 meters and whose width is 6 meters.

51. A 22-foot ladder is placed against the wall of a house. If the bottom of the ladder is 9 feet from the house, how far from the ground is the top of the ladder? Round the answer to the tenth of a foot.

52. Find the width of a rectangle whose length is 10 inches and whose diagonal is 14 inches. Round the answer to the tenth of an inch.

Directions Find the indicated root. See example 8–1 F.	☛ *Quick Check* **Example** $\sqrt[3]{-64}$ **Solution** $= -4$ Since $(-4)^3 = -64$

Directions   Find the indicated root. See example 8–1 F.

53. $\sqrt[3]{8}$ **58.** $\sqrt[3]{-64}$ **63.** $\sqrt[5]{32}$ **68.** $\sqrt[14]{1}$

54. $\sqrt[3]{27}$ **59.** $\sqrt[4]{81}$ **64.** $\sqrt[6]{64}$ **69.** $\sqrt[9]{-1}$

55. $\sqrt[3]{125}$ **60.** $\sqrt[4]{16}$ **65.** $-\sqrt[5]{243}$ **70.** $\sqrt[15]{-1}$

56. $\sqrt[3]{64}$ **61.** $-\sqrt[4]{81}$ **66.** $\sqrt[5]{-32}$

57. $\sqrt[3]{-8}$ **62.** $-\sqrt[4]{625}$ **67.** $\sqrt[10]{1}$

Directions   Solve the following problems. See example 8–1 D.

71. The formula for finding the length of an edge, e, of a cube when the volume, v, is known is $e = \sqrt[3]{v}$. What is the length of the edge of a cube whose volume is 729 cubic units?

72. What is the length of the edge of a cube whose volume is 216 cubic units? (Refer to exercise 71.)

73. Explain why $\sqrt{-16}$ has no real number solution.

⊛ Explore to Learn

As with square roots, most nth roots are irrational. Although calculators can generally be used to find approximate values, we do not study this until section 8–6 because we need the notation that is introduced there. However, to explore this topic now consider how one might find $\sqrt[3]{27}$ on a calculator. (We know it is 3.)

Consider that we want a solution to the equation $x^3 = 27$. This equation can be transformed into $x^2 = \dfrac{27}{x}$, or $x = \sqrt{\dfrac{27}{x}}$. This is called an iterative formula. To use it, make a guess about the value

of $\sqrt[3]{27}$. Say you choose 4. Then calculate $\sqrt{\dfrac{27}{4}}$. This gives a new value for x. Use this new value to replace x in $\sqrt{\dfrac{27}{x}}$. This gives a third value for x. Each time this process is done, we get a value closer to the cube root of 27. Try this, starting with 4, and count how many times it takes to get the result 3 correct to three decimal places. (This means that when the result is rounded, it is between 2.999 and 3.000.)

Review Exercises

Directions   Write each number as a product of prime factors. See section 1–2.

1. 9   **3.** 8   **5.** 50   **7.** 64

2. 12   **4.** 40   **6.** 81   **8.** 16

9. Four times the number of teeth on a first gear is 6 less than the number of teeth on a second gear. If the total number of teeth on both gears is 66, how many teeth are on each gear? See section 2–7.

10. A chemist wishes to make 120 milliliters (ml) of a 75% acid solution by mixing a 60% solution with an 80% solution. How many milliliters of each solution are necessary? See section 2–7.

8–2   Product Property for Radicals

Multiplying Square Roots

In this section, we are going to develop properties for simplifying radicals. Consider the example

$$\sqrt{4} \cdot \sqrt{25} = 2 \cdot 5 = 10$$

We also observe that

$$\sqrt{4 \cdot 25} = \sqrt{100} = 10$$

From our example, we can conclude that

$$\sqrt{4} \cdot \sqrt{25} = \sqrt{4 \cdot 25}$$

We now can generalize the product property for square roots.

Product Property for Square Roots

For all nonnegative real numbers a and b,

$$\sqrt{a} \cdot \sqrt{b} = \sqrt{ab}$$

Concept

The product of the square roots of two numbers is equal to the square root of the product of the two numbers.

● **Example 8–2 A**
Multiplying square roots

Perform the indicated operations. Assume that all variables represent nonnegative real numbers.

1. $\sqrt{3}\sqrt{5} = \sqrt{3 \cdot 5} = \sqrt{15}$
2. $\sqrt{6}\sqrt{7} = \sqrt{6 \cdot 7} = \sqrt{42}$
3. $\sqrt{3}\sqrt{a} = \sqrt{3a}$
4. $\sqrt{x}\sqrt{y}\sqrt{z} = \sqrt{xyz}$

Simplifying Principal Square Roots

An important use of the product property is in simplifying radicals. Consider the following example:

Since 12 can be factored into $4 \cdot 3$, by the product property, we can write

$$\sqrt{12} = \sqrt{4 \cdot 3} = \sqrt{4}\sqrt{3} = 2\sqrt{3}$$

We are able to simplify the radical because the radicand contains a perfect-square integer factor, 4. In our example, $2\sqrt{3}$ is called the *simplified form* of $\sqrt{12}$.

Simplifying the Principal Square Root

1. If the number under the $\sqrt{\ }$ symbol (radicand) is a perfect square, write the corresponding square root.
2. If the number under the $\sqrt{\ }$ symbol has any perfect-square factors, write the corresponding square root as a coefficient of the radical. ($\sqrt{12} = \sqrt{4 \cdot 3} = 2\sqrt{3}$)
3. The square root is in simplest form when the radicand has no perfect-square integer or variable factors other than 1.

The following property is used when changing radicals involving variables in the radicand to simplest form.

$\sqrt{a^2}$ Property

If a is any nonnegative real number, then

$$\sqrt{a^2} = a$$

● **Example 8–2 B**

Simplifying square roots

Simplify the following expressions. Assume that all variables represent nonnegative real numbers.

1. $\sqrt{50} = \sqrt{25 \cdot 2}$ Factor having a perfect-square integer

$\qquad = \sqrt{25}\sqrt{2}$ Product property

$\qquad = 5\sqrt{2}$ $\sqrt{25} = 5$

2. $\sqrt{28} = \sqrt{4 \cdot 7} = \sqrt{4}\sqrt{7} = 2\sqrt{7}$

3. $\sqrt{9a} = \sqrt{9 \cdot a} = \sqrt{9}\sqrt{a} = 3\sqrt{a}$

4. $\sqrt{a^3} = \sqrt{a^2 \cdot a} = \sqrt{a^2}\sqrt{a} = a\sqrt{a}$

5. $\sqrt{a^3 b^4} = \sqrt{a^2 \cdot a \cdot b^2 \cdot b^2} = \sqrt{a^2}\sqrt{a}\sqrt{b^2}\sqrt{b^2} = a\sqrt{a}\,bb = ab^2\sqrt{a}$

6. $\sqrt{a^2 + b^2}$ will not simplify because we are not able to factor the radicand. *The radicand must always be in a factored form before we can simplify.*

Note $\sqrt{a^2 + b^2} \neq \sqrt{a^2} + \sqrt{b^2}$. For example, $\sqrt{3^2 + 4^2} = \sqrt{9 + 16} \neq \sqrt{9} + \sqrt{16}$, since $\sqrt{9 + 16} = \sqrt{25} = 5$, whereas $\sqrt{9} + \sqrt{16} = 3 + 4 = 7$.

7. $\sqrt{2}\sqrt{6} = \sqrt{2 \cdot 6} = \sqrt{12} = \sqrt{4 \cdot 3} = \sqrt{4} \cdot \sqrt{3} = 2\sqrt{3}$

8. $\sqrt{10}\sqrt{5} = \sqrt{10 \cdot 5} = \sqrt{50} = \sqrt{25 \cdot 2} = \sqrt{25}\sqrt{2} = 5\sqrt{2}$

9. $\sqrt{14x}\sqrt{2x} = \sqrt{14x \cdot 2x} = \sqrt{28x^2} = \sqrt{4 \cdot 7 \cdot x^2} = \sqrt{4}\sqrt{7}\sqrt{x^2}$
$\qquad = 2\sqrt{7}x = 2x\sqrt{7}$

☛ *Quick check*   Simplify. $\sqrt{8a^4}$ and $\sqrt{3a}\sqrt{6a}$. Assume a represents nonnegative real numbers.

● **Example 8–2 C**
Problem solving

The formula for approximating the velocity V in miles per hour of a car based on the length of its skid marks S (in feet) on wet pavement is given by $V = 2\sqrt{3S}$. If the skid marks are 100 feet long, what is the velocity? Round the answer to the nearest mile per hour.

$$V = 2\sqrt{3S}$$ Original formula
$$V = 2\sqrt{3(100)}$$ Substitute
$$V = 2\sqrt{300}$$ Multiply
$$V = 2 \cdot 17.320508$$ Square root
$$V \approx 35$$ Multiply and round

Therefore, the car was traveling at approximately 35 miles per hour.   ●

Multiplying *n*th Roots

In section 8–1, we observed that even roots of negative numbers do not exist in the set of real numbers. Therefore, **we will consider all variables to represent nonnegative real numbers whenever the index of the radical is even**.

The product property for square roots can be extended to radicals with any index.

Product Property for Radicals

$$\sqrt[n]{a}\,\sqrt[n]{b} = \sqrt[n]{ab}$$

Concept

When we multiply two radicals *having the same index*, we multiply the radicands and put the product under a radical symbol with the common index.

● **Example 8–2 D**
Multiplying nth roots

Perform the indicated operations.

1. $\sqrt[3]{3}\,\sqrt[3]{2} = \sqrt[3]{3 \cdot 2} = \sqrt[3]{6}$

2. $\sqrt[5]{7}\,\sqrt[5]{9} = \sqrt[5]{7 \cdot 9} = \sqrt[5]{63}$

3. $\sqrt[3]{3}\,\sqrt[4]{5}$ These radicals cannot be multiplied together in this form since they do not have the same index.   ●

Simplifying *n*th Roots

We simplify *n*th roots, where n is greater than 2, as we did square roots. As long as the radicand can be factored so that one or more factors is a

    1. perfect cube when the index is 3,
    2. perfect fourth-power when the index is 4,
    3. perfect fifth-power when the index is 5, and so on,

the radical can be simplified. To do this, we use the following property.

$\sqrt[n]{a^n}$ Property

If a is any nonnegative real number, then

$$\sqrt[n]{a^n} = a$$

● **Example 8–2 E**
Simplifying nth roots

Simplify the following radical expressions. Assume that all variables represent nonnegative real numbers.

1. $\sqrt[3]{81} = \sqrt[3]{27 \cdot 3}$ Factor having a perfect-cube integer

$= \sqrt[3]{27}\sqrt[3]{3}$ Product property

$= 3\sqrt[3]{3}$ $\sqrt[3]{27} = 3$

2. $\sqrt[4]{32} = \sqrt[4]{16 \cdot 2} = \sqrt[4]{16}\sqrt[4]{2} = 2\sqrt[4]{2}$

3. $\sqrt[3]{x^5} = \sqrt[3]{x^3 \cdot x^2} = \sqrt[3]{x^3}\sqrt[3]{x^2} = x\sqrt[3]{x^2}$

4. $\sqrt[5]{y^{10}} = \sqrt[5]{y^5 \cdot y^5} = \sqrt[5]{y^5}\sqrt[5]{y^5} = y \cdot y = y^2$

Note In example 4, the exponent 10 is evenly divisible by the index 5, and the radical is eliminated. When the exponent of a factor is evenly divisible by the index, that factor will no longer remain under the radical symbol.

5. $\sqrt[4]{x^7y^4} = \sqrt[4]{x^4 \cdot x^3 \cdot y^4} = \sqrt[4]{x^4}\sqrt[4]{x^3}\sqrt[4]{y^4} = x\sqrt[4]{x^3}y = xy\sqrt[4]{x^3}$

6. $\sqrt[3]{a^7b^2} = \sqrt[3]{a^3a^3ab^2} = \sqrt[3]{a^3}\sqrt[3]{a^3}\sqrt[3]{ab^2} = a \cdot a\sqrt[3]{ab^2} = a^2\sqrt[3]{ab^2}$

Note No simplification relative to *b* is possible because the exponent of *b* is less than the value of the index.

7. $\sqrt[3]{54x^3y^5} = \sqrt[3]{27 \cdot 2 \cdot x^3 \cdot y^3 \cdot y^2} = \sqrt[3]{27}\sqrt[3]{x^3}\sqrt[3]{y^3}\sqrt[3]{2y^2} = 3xy\sqrt[3]{2y^2}$

☛ *Quick check*   Simplify. $\sqrt[3]{8a^4}$. Assume *a* represents nonnegative real numbers.

Observe from the preceding examples that **we can simplify a radical if the radicand has a factor(s) whose exponent is equal to or greater than the index**.

● **Example 8–2 F**
Multiplying and simplifying nth roots

Perform the indicated multiplication and simplify. Assume that all variables represent nonnegative real numbers.

1. $\sqrt[3]{a^2}\sqrt[3]{a} = \sqrt[3]{a^2 \cdot a}$ Product property

$= \sqrt[3]{a^3}$ Multiply like bases

$= a$ Perfect cube

2. $\sqrt[4]{8a^2}\sqrt[4]{4a^3} = \sqrt[4]{8a^2 \cdot 4a^3} = \sqrt[4]{32a^5} = \sqrt[4]{16 \cdot 2 \cdot a^4 \cdot a}$

$= \sqrt[4]{16}\sqrt[4]{a^4}\sqrt[4]{2a} = 2a\sqrt[4]{2a}$

☛ *Quick check*   Simplify. $\sqrt[3]{9a^2}\sqrt[3]{3a^2}$. Assume *a* represents nonnegative real numbers.

Note A very common error in problems involving radicals is to forget to carry along the correct index for the radical symbol.

MASTERY POINTS

Can you
- Multiply and simplify square roots?
- Multiply and simplify *n*th roots?

Exercise 8–2

Directions

Perform any indicated operations and simplify. Assume that all variables represent nonnegative real numbers. See examples 8–2 A and B.

☛ *Quick Check*

Examples $\sqrt{8a^4}$ $\sqrt{3a}\sqrt{6a}$

Solutions $= \sqrt{4 \cdot 2a^2a^2}$ Factor having perfect squares $= \sqrt{3a \cdot 6a}$ Product property

$= \sqrt{4}\sqrt{a^2}\sqrt{a^2}\sqrt{2}$ Product property $= \sqrt{18a^2}$ Multiply

$= 2aa\sqrt{2}$ $\sqrt{4} = 2, \sqrt{a^2} = a$ $= \sqrt{9 \cdot 2 \cdot a^2}$ Factor having perfect squares

$= 2a^2\sqrt{2}$ Multiply $= \sqrt{9}\sqrt{2}\sqrt{a^2}$ Product property

$= 3\sqrt{2a}$ $\sqrt{9} = 3, \sqrt{a^2} = a$

$= 3a\sqrt{2}$ Multiply

Directions   Perform any indicated operations and simplify. Assume that all variables represent nonnegative real numbers. See examples 8–2 A and B.

1. $\sqrt{16}$

2. $\sqrt{63}$

3. $\sqrt{20}$

4. $\sqrt{75}$

5. $\sqrt{45}$

6. $\sqrt{48}$

7. $\sqrt{32}$

8. $\sqrt{27}$

9. $\sqrt{80}$

10. $\sqrt{54}$

11. $\sqrt{98}$

12. $\sqrt{96}$

13. $\sqrt{a^7}$

14. $\sqrt{a^5}$

15. $\sqrt{4a^2b^3}$

16. $\sqrt{9ab^4c^3}$

17. $\sqrt{27a^3b^5}$

18. $\sqrt{24x^5yz^3}$

19. $\sqrt{6}\sqrt{3}$

20. $\sqrt{27}\sqrt{6}$

21. $\sqrt{15}\sqrt{15}$

22. $\sqrt{11}\sqrt{11}$

23. $\sqrt{6}\sqrt{10}$

24. $\sqrt{18}\sqrt{24}$

25. $\sqrt{25}\sqrt{15}$

26. $\sqrt{20}\sqrt{20}$

27. $\sqrt{5}\sqrt{15}$

28. $\sqrt{2a}\sqrt{3a}$

29. $\sqrt{5x}\sqrt{15x}$

30. $\sqrt{6x}\sqrt{14xy}$

31. $\sqrt{2a}\sqrt{24b^2}$

Directions   Solve the following problems. See example 8–2 C.

32. A square-shaped television picture tube has a surface area of 121 square inches. What is the length of the side of the tube? (*Hint:* Area of a square is found by squaring the length of a side. $A = s^2$.)

33. A room in the shape of a square is 169 square feet. What is the length of a side? (See exercise 32.)

34. The formula for approximating the velocity V in miles per hour of a car based on the length of its skid marks S (in feet) on wet pavement is given by $V = 2\sqrt{3S}$. If the skid marks are 75 feet long, what was the velocity?

35. The formula for approximating the velocity V in miles per hour of a car based on the length of its skid marks S (in feet) on dry pavement is given by $V = 2\sqrt{6S}$. If the skid marks are 24 feet long, what was the velocity?

Directions	☛ *Quick Check*		
Perform any indicated operations and simplify. Assume that all variables represent nonnegative real numbers. See examples 8–2 D, E, and F.	**Examples** $\sqrt[3]{8a^4}$		$\sqrt[3]{9a^2}\sqrt[3]{3a^2}$

Solutions
$$= \sqrt[3]{8 \cdot a^3 \cdot a} \quad \text{Factor having perfect cubes}$$
$$= \sqrt[3]{8}\sqrt[3]{a^3}\sqrt[3]{a} \quad \text{Product property}$$
$$= 2a\sqrt[3]{a} \quad \sqrt[3]{8} = 2, \sqrt[3]{a^3} = a$$

$$= \sqrt[3]{9a^2 \cdot 3a^2} \quad \text{Product property}$$
$$= \sqrt[3]{27a^4} \quad \text{Multiply radicands}$$
$$= \sqrt[3]{27 \cdot a^3 \cdot a} \quad \text{Factor having perfect cubes}$$
$$= \sqrt[3]{27}\sqrt[3]{a^3}\sqrt[3]{a} \quad \text{Product property}$$
$$= 3a\sqrt[3]{a} \quad \sqrt[3]{27} = 3, \sqrt[3]{a^3} = a$$

Directions Perform any indicated operations and simplify. Assume that all variables represent nonnegative real numbers. See examples 8–2 D, E, and F.

36. $\sqrt[3]{48}$

37. $\sqrt[5]{64}$

38. $\sqrt[4]{32}$

39. $\sqrt[3]{24}$

40. $\sqrt[5]{a^7}$

41. $\sqrt[3]{b^8}$

42. $\sqrt[3]{x^9}$

43. $\sqrt[5]{y^{15}}$

44. $\sqrt[3]{a^{12}}$

45. $\sqrt[3]{4a^2b^3}$

46. $\sqrt[3]{8r^2s^8}$

47. $\sqrt[3]{16a^4b^5}$

48. $\sqrt[5]{64x^{10}y^{14}}$

49. $\sqrt[3]{81a^5b^{11}}$

50. $\sqrt[3]{a^2}\sqrt[3]{a}$

51. $\sqrt[3]{b^2}\sqrt[3]{b^2}$

52. $\sqrt[5]{b^4}\sqrt[5]{b^3}$

53. $\sqrt[5]{a}\sqrt[5]{a^4}$

54. $\sqrt[3]{5a^2b}\sqrt[3]{75a^2b^2}$

55. $\sqrt[3]{3ab^2}\sqrt[3]{18a^2b^2}$

56. $\sqrt[4]{8a^3b}\sqrt[4]{4a^2b^2}$

57. $\sqrt[4]{27a^2b^3}\sqrt[4]{9ab}$

58. $\sqrt[3]{25x^5y^7}\sqrt[3]{15xy^3}$

59. $\sqrt[3]{16a^{11}b^4}\sqrt[3]{12a^4b^6}$

60. $\sqrt[4]{8xy}\sqrt[4]{4x^3y^3}$

61. The moment of inertia for a rectangle is given by the formula $I = \dfrac{bh^3}{12}$. If we know the values of I and b, we can solve for h as follows: $h = \sqrt[3]{\dfrac{12I}{b}}$. Find h if $I = 2$ in.^{4} and $b = 3$ in.

62. Use exercise 61 to find h if $I = 27$ in.^{4} and $b = 4$ in.

63. The moment of inertia for a circle is given by the formula $I = \dfrac{\pi r^4}{4}$. If we know the value of I, we can solve for r as follows: $r = \sqrt[4]{\dfrac{4I}{\pi}}$. Find r if $I = 12.56$ in.^{4} and we use 3.14 for π.

64. Use exercise 63 to find r if $I = 63.585$ in.^{4}

65. Explain why in the product property of square roots both a and b must be nonnegative real numbers. ($\sqrt{a}\sqrt{b} = \sqrt{ab}$)

Explore to Learn

1. The product property for radicals, $\sqrt[n]{a}\sqrt[n]{b} = \sqrt[n]{ab}$, requires that a and b be nonnegative when n is even. Give a numeric example in which one side of this property is true but the other is not that illustrates why this restriction must be made.

2. Give two examples that illustrate why a and b do not have to be nonnegative in the product property for radicals when n is odd. In one example let a be positive and b be negative; in the second example let both a and b be negative values.

Review Exercises

Directions Reduce the following fractions and rational expressions to lowest terms. Assume that no variable is equal to zero. See sections 1–2, 3–3, and 5–1.

1. $\dfrac{49}{56}$

2. $\dfrac{16x^3y^2}{-4xy}$

3. $\dfrac{2y^2 - 50}{y^2 - 4y - 5}$

Directions Perform the indicated operations and simplify. Assume that all variables represent nonnegative real numbers. See sections 8–1 and 8–2.

4. $\sqrt{5}\sqrt{5}$

5. $\sqrt{3}\sqrt{3}$

6. $\sqrt{2}\sqrt{8}$

7. $\sqrt{12}\sqrt{3}$

8. $\sqrt{x}\sqrt{x}$

9. How many liters of a 6% salt solution must a druggist mix with 40 liters of a 12% salt solution to obtain a 10% salt solution? See section 2–7.

10. A woman can row her boat 25 miles downstream in the same time she can row 10 miles upstream. How fast can she row her boat in still water if the stream had a current of 4 miles per hour? See section 5–7.

8–3   Quotient Property for Radicals

The Square Root of a Fraction

The following example will help us develop a property for division involving radicals.

$$\sqrt{\dfrac{4}{9}} = \sqrt{\left(\dfrac{2}{3}\right)^2} = \dfrac{2}{3}$$

We also observe that

$$\frac{\sqrt{4}}{\sqrt{9}} = \frac{2}{3}$$

From our example, we can conclude that

$$\sqrt{\frac{4}{9}} = \frac{\sqrt{4}}{\sqrt{9}}$$

We can now generalize this idea.

Quotient Property for Square Roots

For any nonnegative real numbers a and b, where $b \neq 0$,

$$\sqrt{\frac{a}{b}} = \frac{\sqrt{a}}{\sqrt{b}}$$

Concept

The square root of a fraction can be written as the square root of the numerator divided by the square root of the denominator.

● **Example 8–3 A**

Simplifying a square root with a fraction

Simplify the following expressions. Assume that all variables represent positive real numbers.

1. $\sqrt{\dfrac{16}{25}} = \dfrac{\sqrt{16}}{\sqrt{25}}$ Rewrite as the square root of the numerator over the square root of the denominator and simplify

$$= \frac{4}{5}$$

2. $\sqrt{\dfrac{36}{49}} = \dfrac{\sqrt{36}}{\sqrt{49}} = \dfrac{6}{7}$

3. $\sqrt{\dfrac{81}{100}} = \dfrac{\sqrt{81}}{\sqrt{100}} = \dfrac{9}{10}$

4. $\sqrt{\dfrac{x^4}{64}} = \dfrac{\sqrt{x^4}}{\sqrt{64}} = \dfrac{x^2}{8}$

5. $\sqrt{\dfrac{x^3}{y^4}} = \dfrac{\sqrt{x^3}}{\sqrt{y^4}} = \dfrac{x\sqrt{x}}{y^2}$

☛ *Quick check*   Simplify. $\sqrt{\dfrac{16}{49}}$

Rationalizing the Denominator

When simplifying and evaluating radical expressions containing a radical in the denominator, it is common practice to eliminate the radical in the denominator. For example,

$$\sqrt{\frac{4}{5}} = \frac{\sqrt{4}}{\sqrt{5}} = \frac{2}{\sqrt{5}}$$

Since $\sqrt{5} \cdot \sqrt{5} = 5$, we can eliminate the radical $\sqrt{5}$ in the denominator by multiplying the numerator and the denominator of the fraction by $\sqrt{5}$.

$$\frac{2}{\sqrt{5}} \cdot \frac{\sqrt{5}}{\sqrt{5}} = \frac{2\sqrt{5}}{\sqrt{25}} = \frac{2\sqrt{5}}{5}$$

The process of changing the denominator from a radical to a rational number is called **rationalizing the denominator.**

To Rationalize a Square Root Denominator

1. Multiply the numerator and the denominator by the square root that is in the denominator. The radicand in the denominator will be a perfect-square integer.
2. Simplify the radical expressions in the numerator and the denominator.
3. Reduce the resulting fraction if possible.

● **Example 8–3 B**

Rationalizing a square root denominator

Simplify the following expressions. Leave no radicals in the denominator. Assume that all variables represent positive real numbers.

1. $\dfrac{5}{\sqrt{7}} = \dfrac{5}{\sqrt{7}} \cdot \dfrac{\sqrt{7}}{\sqrt{7}}$ Multiply numerator and denominator by $\sqrt{7}$

$\qquad = \dfrac{5\sqrt{7}}{\sqrt{49}}$ Multiply in numerator and denominator

$\qquad = \dfrac{5\sqrt{7}}{7}$ $\sqrt{49} = 7$

2. $\dfrac{4}{\sqrt{6}} = \dfrac{4}{\sqrt{6}} \cdot \dfrac{\sqrt{6}}{\sqrt{6}} = \dfrac{4\sqrt{6}}{\sqrt{36}} = \dfrac{4\sqrt{6}}{6} = \dfrac{2\sqrt{6}}{3}$

Note In example 2, we were able to reduce the fraction as a final step. Always check to see that the answer is in reduced form.

3. $\dfrac{a}{\sqrt{a}} = \dfrac{a}{\sqrt{a}} \cdot \dfrac{\sqrt{a}}{\sqrt{a}} = \dfrac{a\sqrt{a}}{\sqrt{a^2}} = \dfrac{a\sqrt{a}}{a} = \sqrt{a}$

4. $\sqrt{\dfrac{a^3}{b}} = \dfrac{\sqrt{a^3}}{\sqrt{b}} = \dfrac{a\sqrt{a}}{\sqrt{b}} \cdot \dfrac{\sqrt{b}}{\sqrt{b}} = \dfrac{a\sqrt{ab}}{\sqrt{b^2}} = \dfrac{a\sqrt{ab}}{b}$

☛ *Quick check*   Simplify. $\dfrac{6}{\sqrt{6}}$ and $\sqrt{\dfrac{a^2}{b}}$. Assume that all variables represent positive real numbers.   ●

The following is a summary of the conditions necessary for a radical expression to be in **simplest form**, also called **standard form**.

1. The radicand contains no factors that can be written with an exponent greater than or equal to the index. ($\sqrt{a^3}$ violates this.)

2. The radicand contains no fractions. $\left(\sqrt{\dfrac{a}{b}} \text{ violates this.}\right)$

3. No radicals appear in the denominator. $\left(\dfrac{1}{\sqrt{a}} \text{ violates this.}\right)$

The *n*th Root of a Fraction

Our quotient property for square roots can be extended to radicals with any index. As with the product property for radicals, the quotient property requires that a and b represent nonnegative values when n is even.

Quotient Property for Radicals

$$\sqrt[n]{\dfrac{a}{b}} = \dfrac{\sqrt[n]{a}}{\sqrt[n]{b}} \qquad (b \neq 0)$$

Concept

The *n*th root of a fraction can be written as the *n*th root of the numerator divided by the *n*th root of the denominator.

● **Example 8–3 C**

Simplifying the nth root of a fraction

Simplify the following expressions. Assume that all variables represent positive real numbers.

1. $\sqrt[3]{\dfrac{8}{27}} = \dfrac{\sqrt[3]{8}}{\sqrt[3]{27}}$ Rewrite as the cube root of the numerator over the cube root of the denominator and simplify

     $= \dfrac{2}{3}$

2. $\sqrt[5]{\dfrac{32}{a^5}} = \dfrac{\sqrt[5]{32}}{\sqrt[5]{a^5}} = \dfrac{2}{a}$

3. $\sqrt[3]{\dfrac{x^5}{y^6}} = \dfrac{\sqrt[3]{x^5}}{\sqrt[3]{y^6}} = \dfrac{x\sqrt[3]{x^2}}{y^2}$

☛ **Quick check**   Simplify. $\sqrt[3]{\dfrac{1}{27}}$ and $\sqrt[3]{\dfrac{8a^5}{b^3}}$. Assume that all variables represent positive real numbers.

Rationalizing the Denominator (*n*th Root)

The following example will help us develop a procedure for rationalizing a denominator that has a single term.

$$\sqrt[3]{\dfrac{1}{a}} = \dfrac{\sqrt[3]{1}}{\sqrt[3]{a}} = \dfrac{1}{\sqrt[3]{a}}$$

At this point, a radical still remains in the denominator. We must now determine what we can do to the fraction to remove the radical from the denominator.

Observations:

1. We can multiply the numerator and the denominator by the same number and form equivalent fractions.
2. If we multiply by a radical, the indices must be the same to carry out the multiplication.
3. To bring a factor out from under the radical symbol and not leave any of the factor behind, the index must divide evenly into the exponent.

With these observations in mind, we rationalize the fraction as follows:

$$= \frac{1}{\sqrt[3]{a}} \cdot \frac{\sqrt[3]{}}{\sqrt[3]{}} \qquad \text{Indices are the same}$$

$$= \frac{1}{\sqrt[3]{a}} \cdot \frac{\sqrt[3]{a^2}}{\sqrt[3]{a^2}} \qquad \text{Multiply numerator and denominator by the same number}$$

$$= \frac{\sqrt[3]{a^2}}{\sqrt[3]{a^3}} \qquad \text{The sum of the exponents of } a \text{ in the denominator is equal to the index}$$

$$= \frac{\sqrt[3]{a^2}}{a} \qquad \text{The index divides evenly into the exponent, the radical is eliminated}$$

To Rationalize an *n*th Root Denominator

1. Multiply the numerator and the denominator by a radical with the same index as the radical that we wish to eliminate from the denominator.
2. The exponent of each factor under the radical must be such that when we add it to the original exponent of the factor under the radical in the denominator, the sum will be equal to or divisible by the index of the radical.
3. Carry out the multiplication and reduce the fraction if possible.

● **Example 8–3 D**

Rationalizing an nth root denominator

Simplify the following expressions. Leave no radicals in the denominator. Assume that all variables represent positive real numbers.

1. $\dfrac{1}{\sqrt[3]{7}} = \dfrac{1}{\sqrt[3]{7}} \cdot \dfrac{\sqrt[3]{7^2}}{\sqrt[3]{7^2}}$ Multiply numerator and denominator by $\sqrt[3]{7^2}$ to get $\sqrt[3]{7^3}$ in the denominator

$\qquad = \dfrac{\sqrt[3]{7^2}}{\sqrt[3]{7^3}}$ Multiply in numerator and denominator ($\sqrt[3]{7}\,\sqrt[3]{7^2} = \sqrt[3]{7^3}$)

$\qquad = \dfrac{\sqrt[3]{7^2}}{7}$ $\sqrt[3]{7^3} = 7$

$\qquad = \dfrac{\sqrt[3]{49}}{7}$ $7^2 = 49$

2. $\dfrac{a}{\sqrt[5]{b^2}} = \dfrac{a}{\sqrt[5]{b^2}} \cdot \dfrac{\sqrt[5]{b^3}}{\sqrt[5]{b^3}} = \dfrac{a\,\sqrt[5]{b^3}}{\sqrt[5]{b^5}} = \dfrac{a\,\sqrt[5]{b^3}}{b}$

3. $\dfrac{x}{\sqrt[4]{x}} = \dfrac{x}{\sqrt[4]{x}} \cdot \dfrac{\sqrt[4]{x^3}}{\sqrt[4]{x^3}} = \dfrac{x\sqrt[4]{x^3}}{\sqrt[4]{x^4}} = \dfrac{x\sqrt[4]{x^3}}{x} = \sqrt[4]{x^3}$

4. $\dfrac{1}{\sqrt[5]{a^2 b}} = \dfrac{1}{\sqrt[5]{a^2 b}} \cdot \dfrac{\sqrt[5]{a^3 b^4}}{\sqrt[5]{a^3 b^4}} = \dfrac{\sqrt[5]{a^3 b^4}}{\sqrt[5]{a^5 b^5}} = \dfrac{\sqrt[5]{a^3 b^4}}{ab}$

☞ *Quick check* Simplify. $\dfrac{1}{\sqrt[5]{b^2}}$. Assume that b represents positive real numbers.

MASTERY POINTS

Can you
- Simplify radicals containing fractions?
- Rationalize a square root denominator?
- Rationalize an nth root denominator?

Exercise 8–3

Directions
Simplify the following expressions. Leave no radicals in the denominator. Assume that all variables represent positive real numbers. See examples 8–3 A and B.

☞ *Quick Check*

Examples $\sqrt{\dfrac{16}{49}}$ $\qquad$ $\sqrt{\dfrac{a^2}{b}}$ $\qquad$ $\dfrac{6}{\sqrt{6}}$

Solutions $= \dfrac{\sqrt{16}}{\sqrt{49}}$ Rewrite as the square root of the numerator over the square root of the denominator and simplify

$= \dfrac{4}{7}$

$= \dfrac{\sqrt{a^2}}{\sqrt{b}}$

$= \dfrac{a}{\sqrt{b}} \cdot \dfrac{\sqrt{b}}{\sqrt{b}}$

$= \dfrac{a\sqrt{b}}{b}$

$= \dfrac{6}{\sqrt{6}} \cdot \dfrac{\sqrt{6}}{\sqrt{6}}$ Multiply the numerator and the denominator by the square root in the denominator and simplify

$= \dfrac{6\sqrt{6}}{6}$

$= \sqrt{6}$

Directions Simplify the following expressions. Leave no radicals in the denominator. Assume that all variables represent positive real numbers. See examples 8–3 A and B.

1. $\sqrt{\dfrac{9}{25}}$

2. $\sqrt{\dfrac{25}{36}}$

3. $\sqrt{\dfrac{25}{49}}$

4. $\sqrt{\dfrac{81}{100}}$

5. $\sqrt{\dfrac{3}{4}}$

6. $\sqrt{\dfrac{5}{9}}$

7. $\sqrt{\dfrac{64}{a^2}}$

8. $\sqrt{\dfrac{y^4}{16}}$

9. $\sqrt{\dfrac{1}{2}}$

10. $\sqrt{\dfrac{1}{3}}$

11. $\sqrt{\dfrac{4}{7}}$

12. $\sqrt{\dfrac{9}{11}}$

13. $\sqrt{\dfrac{1}{15}}$

14. $\sqrt{\dfrac{1}{14}}$

15. $\sqrt{\dfrac{4}{75}}$

16. $\sqrt{\dfrac{5}{12}}$

17. $\dfrac{2}{\sqrt{2}}$

18. $\dfrac{6}{\sqrt{3}}$

19. $\dfrac{10}{\sqrt{8}}$

20. $\dfrac{15}{\sqrt{27}}$

21. $\dfrac{10}{5\sqrt{2}}$

22. $\dfrac{6}{3\sqrt{3}}$

23. $\dfrac{9}{\sqrt{3}}$

24. $\dfrac{6}{\sqrt{2}}$

25. $\dfrac{5}{\sqrt{10}}$

26. $\dfrac{3}{\sqrt{6}}$

27. $\dfrac{15}{\sqrt{10}}$

28. $\dfrac{12}{\sqrt{15}}$

29. $\dfrac{8}{\sqrt{8}}$

30. $\dfrac{14}{\sqrt{12}}$

31. $\sqrt{\dfrac{x^2}{y}}$

32. $\sqrt{\dfrac{1}{a}}$

33. $\sqrt{\dfrac{1}{x}}$

34. $\sqrt{\dfrac{a^2}{b^3}}$

35. $\dfrac{\sqrt{a^5}}{\sqrt{a}}$

36. Find the width of a rectangle whose diagonal is 17 feet and length is 8 feet.

37. Find the diagonal of a rectangle whose length is 5 meters and whose width is 4 meters.

38. A 13-foot ladder is placed against the wall of a house. If the bottom of the ladder is 5 feet from the house, how far from the ground is the top of the ladder?

39. At a height of h feet above the sea or level ground, the distance d in miles that a person can see an object is found by using the equation $d = \sqrt{\dfrac{3h}{2}}$.

How far can someone see who is in a tower 96 feet above the ground?

Directions
Simplify the following expressions. Leave no radicals in the denominator. Assume that all variables represent positive real numbers. See examples 8-3 C and D.

☛ *Quick Check*

Examples $\sqrt[3]{\dfrac{1}{27}}$ $\sqrt[3]{\dfrac{8a^5}{b^3}}$ $\dfrac{1}{\sqrt[5]{b^2}}$

Solutions $= \dfrac{\sqrt[3]{1}}{\sqrt[3]{27}}$ $= \dfrac{\sqrt[3]{8a^5}}{\sqrt[3]{b^3}}$ Rewrite as the cube root of the numerator over the cube root of the denominator and simplify $= \dfrac{1}{\sqrt[5]{b^2}} \cdot \dfrac{\sqrt[5]{b^3}}{\sqrt[5]{b^3}}$ Multiply the numerator and the denominator by $\sqrt[5]{b^3}$

$\qquad = \dfrac{1}{3}$ $= \dfrac{\sqrt[3]{2^3a^5}}{b}$

$\qquad\qquad = \dfrac{2a\sqrt[3]{a^2}}{b}$ $= \dfrac{\sqrt[5]{b^3}}{\sqrt[5]{b^5}}$ Perfect 5th root

$\qquad\qquad\qquad = \dfrac{\sqrt[5]{b^3}}{b}$ $\sqrt[5]{b^5} = b$

Directions   Simplify the following expressions. Leave no radicals in the denominator. Assume that all variables represent positive real numbers. See examples 8-3 C and D.

40. $\sqrt[3]{\dfrac{8}{27}}$

41. $\sqrt[3]{\dfrac{1}{8}}$

42. $\sqrt[4]{\dfrac{16}{81}}$

43. $\sqrt[3]{\dfrac{27}{125}}$

44. $\sqrt[3]{\dfrac{a^2}{b^2}}$

45. $\sqrt[3]{\dfrac{3a^6}{b^3}}$

46. $\sqrt[3]{\dfrac{x}{y^{12}}}$

47. $\sqrt[5]{\dfrac{a^4}{b^{10}}}$

48. $\sqrt[5]{\dfrac{32x^4}{y^5}}$

49. $\sqrt[4]{\dfrac{a^4b^9}{c^{11}}}$

50. $\sqrt[4]{\dfrac{a^9b^{13}}{c^8}}$

51. $\sqrt[5]{\dfrac{x^3y^2}{z^{15}}}$

52. $\sqrt[3]{\dfrac{8}{9}}$

56. $\sqrt[4]{\dfrac{3}{4}}$

60. $\dfrac{xy}{\sqrt[5]{y^3}}$

64. $\dfrac{a}{\sqrt[5]{a^2b^4}}$

53. $\sqrt[3]{\dfrac{4}{25}}$

57. $\sqrt[3]{\dfrac{x^3}{y^2}}$

61. $\sqrt[3]{\dfrac{a^3}{b^2c}}$

65. $\dfrac{ab}{\sqrt[3]{ab^2}}$

54. $\sqrt[3]{\dfrac{27}{16}}$

58. $\sqrt[3]{\dfrac{x^6}{y}}$

62. $\sqrt[3]{\dfrac{8}{xy^2}}$

66. $\dfrac{xy^2}{\sqrt[5]{x^4y}}$

55. $\sqrt[4]{\dfrac{16}{125}}$

59. $\dfrac{ab}{\sqrt[3]{a^2}}$

63. $\sqrt[3]{\dfrac{a^2}{b^2c}}$

67. If we wish to construct a sphere of specific volume, we can find the length of radius necessary by the formula $r = \sqrt[3]{\dfrac{3V}{4\pi}}$. Find the radius necessary for a sphere to have a volume of 113.04 cubic units. (Use 3.14 for π.)

68. Use exercise 67 to find r if $V = 904.32$ cubic units. (Use 3.14 for π.)

69. Explain why in the quotient property of square roots both a and b must be nonnegative real numbers.
$$\left(\sqrt{\dfrac{a}{b}} = \dfrac{\sqrt{a}}{\sqrt{b}}\right)$$

✵ Explore to Learn

Section 8–1 discussed when $\sqrt{a}$ is an integer. Use this information and the quotient property for radicals to state the necessary constraints on a and b if a and b are integers and $\sqrt{\dfrac{a}{b}}$ is a rational number (fraction with both numerator and denominator integers).

Review Exercises

Directions   Combine like terms. See section 2–2.

1. $4x + 2x$

2. $9y - 5y$

3. $5ab + 3ab$

4. $xy + 4xy$

Directions   Multiply the following. See section 3–2.

5. $(x + 3)(x - 3)$

6. $(x + y)(x - y)$

7. The sum of two voltages is 89 and their difference is 32. Find the two voltages. See section 2–7.

8. Jack can paint the barn twice as fast as Gary can. Working together they can paint the barn in 3 days. How long would it take Jack to paint the barn alone? See section 5–7.

8–4   Sums and Differences of Radicals

Like Radicals

We have learned that in addition and subtraction of algebraic expressions, we can only combine like terms. This same idea applies when we are dealing with radicals. **We can add or subtract only like radicals.** Like radicals are radicals having the same index and the same radicand. For example, $3\sqrt{5x}$ and $-2\sqrt{5x}$ are like radicals, but $5\sqrt{7x}$ and $7\sqrt{5x}$ are not, because the radicands are not the same.

Addition and Subtraction Involving Square Roots

Addition and subtraction of radicals follow the same procedure as addition and subtraction of algebraic expressions. That is, *once we have determined that we have like radicals, the operations of addition and subtraction are performed only with the numerical coefficients.*

● **Example 8–4 A**

Addition and subtraction of like square roots

Perform the indicated operations and simplify. Assume that all variables represent nonnegative real numbers.

1. $5\sqrt{2} + 3\sqrt{2} = (5+3)\sqrt{2} = 8\sqrt{2}$ Apply the distributive property

2. $12\sqrt{3} - \sqrt{3} = (12-1)\sqrt{3} = 11\sqrt{3}$

3. $2\sqrt{a} + 3\sqrt{a} = 5\sqrt{a}$

☛ *Quick check*   Simplify. $3\sqrt{6} + 2\sqrt{6} - \sqrt{6}$ ●

Consider the example

$$\sqrt{27} + 4\sqrt{3}$$

It appears that the indicated addition cannot be performed since we do not have like radicals. However we should have observed that the $\sqrt{27}$ can be simplified as

$$\sqrt{27} = \sqrt{9 \cdot 3} = 3\sqrt{3}$$

Our problem then becomes

$$\sqrt{27} + 4\sqrt{3} = 3\sqrt{3} + 4\sqrt{3} = 7\sqrt{3}$$

and we are able to add the like radicals. Therefore, *whenever we are working with radicals, we must be certain that all radicals are in simplest form.*

To Combine Like Radicals

1. Perform any simplification within the terms.
2. Use the distributive property to combine terms that have like radicals.

● **Example 8–4 B**

Simplifying and combining square roots

Perform the indicated operations. Assume that all variables represent nonnegative real numbers.

1. $\sqrt{45} + \sqrt{20} = \sqrt{9 \cdot 5} + \sqrt{4 \cdot 5}$ Factor $45 = 9 \cdot 5$ and $20 = 4 \cdot 5$

$\qquad\qquad\qquad = 3\sqrt{5} + 2\sqrt{5}$ $\sqrt{4} = 2;\ \sqrt{9} = 3$

$\qquad\qquad\qquad = 5\sqrt{5}$ Add coefficients

2. $\sqrt{32} + 5\sqrt{8} = \sqrt{16 \cdot 2} + 5\sqrt{4 \cdot 2}$ Factor $32 = 16 \cdot 2;\ 8 = 4 \cdot 2$

$\qquad\qquad\qquad = 4\sqrt{2} + 5 \cdot 2\sqrt{2}$ $\sqrt{16} = 4$ and $\sqrt{4} = 2$

$\qquad\qquad\qquad = 4\sqrt{2} + 10\sqrt{2}$ Multiply $5 \cdot 2 = 10$

$\qquad\qquad\qquad = 14\sqrt{2}$ Add coefficients

3. $3\sqrt{3a} - \sqrt{12a} + 5\sqrt{48a}$

$\qquad = 3\sqrt{3a} - \sqrt{4 \cdot 3a} + 5\sqrt{16 \cdot 3a}$ Factor $12 = 4 \cdot 3;\ 48 = 16 \cdot 3$

$\qquad = 3\sqrt{3a} - 2\sqrt{3a} + 5 \cdot 4\sqrt{3a}$ $\sqrt{4} = 2;\ \sqrt{16} = 4$

$\qquad = 3\sqrt{3a} - 2\sqrt{3a} + 20\sqrt{3a}$ $5 \cdot 4 = 20$

$\qquad = 21\sqrt{3a}$ Combine coefficients

☞ ***Quick check*** Simplify. $5\sqrt{2} + \sqrt{18}$ ●

Addition and Subtraction Involving *n*th Roots

Addition and subtraction of radicals other than square roots follow the same procedure as addition and subtraction of expressions containing square roots. That is, *once we have determined that we have like radicals, the operations of addition and subtraction are performed only with the numerical coefficients.*

● **Example 8–4 C**

Simplifying and combining nth roots

Perform the indicated operations and simplify. Assume that all variables represent nonnegative real numbers.

1. $\sqrt[3]{5} + 6\sqrt[3]{5} = (1 + 6)\sqrt[3]{5} = 7\sqrt[3]{5}$ Combine coefficients

2. $4\sqrt[3]{81} - \sqrt[3]{24} = 4\sqrt[3]{27 \cdot 3} - \sqrt[3]{8 \cdot 3}$ Factor $81 = 27 \cdot 3;\ 24 = 8 \cdot 3$

$\qquad\qquad\qquad = 4 \cdot 3\sqrt[3]{3} - 2\sqrt[3]{3}$ $\sqrt[3]{27} = 3;\ \sqrt[3]{8} = 2$

$\qquad\qquad\qquad = 12\sqrt[3]{3} - 2\sqrt[3]{3}$ $4 \cdot 3 = 12$

$\qquad\qquad\qquad = 10\sqrt[3]{3}$ Subtract coefficients

3. $\sqrt[3]{16x^2y} + \sqrt[3]{54x^2y} = \sqrt[3]{8 \cdot 2x^2y} + \sqrt[3]{27 \cdot 2x^2y}$ Factor $16 = 8 \cdot 2;\ 54 = 27 \cdot 2$

$\qquad\qquad\qquad = 2\sqrt[3]{2x^2y} + 3\sqrt[3]{2x^2y}$ $\sqrt[3]{8} = 2;\ \sqrt[3]{27} = 3$

$\qquad\qquad\qquad = 5\sqrt[3]{2x^2y}$ Add coefficients   ●

MASTERY POINTS

Can you
- Identify like radicals?
- Add and subtract like radicals?

Exercise 8–4

Directions

Perform the indicated operations and simplify. Assume that all variables represent nonnegative real numbers. See examples 8–4 A and B.

☞ *Quick Check*

Examples $3\sqrt{6} + 2\sqrt{6} - \sqrt{6}$

Solutions $= (3 + 2 - 1)\sqrt{6}$ Distributive property

$= 4\sqrt{6}$ Combine coefficients

$5\sqrt{2} + \sqrt{18}$

$= 5\sqrt{2} + \sqrt{9 \cdot 2}$ Factor $18 = 9 \cdot 2$

$= 5\sqrt{2} + 3\sqrt{2}$ $\sqrt{9} = 3$

$= (5 + 3)\sqrt{2}$ Distributive property

$= 8\sqrt{2}$ Add coefficients

Directions   Perform the indicated operations and simplify. Assume that all variables represent nonnegative real numbers. See examples 8–4 A and B.

1. $5\sqrt{3} + 4\sqrt{3}$

2. $8\sqrt{7} - 2\sqrt{7}$

3. $6\sqrt{5} + 4\sqrt{5}$

4. $9\sqrt{6} - 6\sqrt{6}$

5. $2\sqrt{3} + 7\sqrt{3} - 3\sqrt{3}$

6. $5\sqrt{5} - 4\sqrt{5} + 6\sqrt{5}$

7. $\sqrt{7} + 5\sqrt{7} - 3\sqrt{7}$

8. $2\sqrt{10} + 11\sqrt{10} - 9\sqrt{10}$

9. $\sqrt{a} + 2\sqrt{a}$

10. $3\sqrt{x} + 4\sqrt{x}$

11. $5\sqrt{a} - 4\sqrt{a} + 7\sqrt{a}$

12. $6\sqrt{y} - \sqrt{y} + 4\sqrt{y}$

13. $5\sqrt{xy} + 2\sqrt{xy}$

14. $3\sqrt{x} + 2\sqrt{y} - \sqrt{x}$

15. $5\sqrt{a} + 2\sqrt{ab} + 3\sqrt{a}$

16. $\sqrt{ab} + 2\sqrt{ab} + 3\sqrt{a}$

17. $5\sqrt{xy} - \sqrt{xy} + 3\sqrt{y}$

18. $\sqrt{20} + 3\sqrt{5}$

19. $\sqrt{8} + 5\sqrt{2}$

20. $\sqrt{12} + \sqrt{75}$

21. $\sqrt{48} - \sqrt{27}$

22. $2\sqrt{3} + 3\sqrt{12}$

23. $5\sqrt{7} + 4\sqrt{63}$

24. $5\sqrt{3} + \sqrt{27} - \sqrt{12}$

25. $4\sqrt{2} - \sqrt{8} + \sqrt{50}$

26. $\sqrt{75} - 4\sqrt{3} + 2\sqrt{27}$

27. $\sqrt{12} + \sqrt{18} + \sqrt{50}$

28. $\sqrt{63} - \sqrt{28} + \sqrt{24}$

29. $\sqrt{50a} + \sqrt{8a}$

30. $\sqrt{32a} - \sqrt{18a}$

31. $3\sqrt{9x} - 5\sqrt{4x}$

32. $2\sqrt{4x^2y} + 3\sqrt{25x^2y}$

33. $2\sqrt{8a} + 4\sqrt{50a} - 7\sqrt{2a}$

34. $3\sqrt{48b} - 2\sqrt{12b} + \sqrt{3b}$

35. $\sqrt{50a} + 3\sqrt{12a} - \sqrt{18a}$

36. $4\sqrt{25x^2y} + 3\sqrt{81x^2y} - 2\sqrt{2y}$

37. We can find the height, h, of the given figure by finding b from the formula $b = \sqrt{c^2 - s^2}$. If $c = 10$ units and $s = 6$ units, find h.

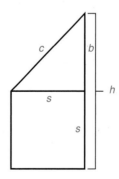

38. Use exercise 37 to find the height of the figure if $c = 13$ feet and $s = 5$ feet.

39. The figure is made up of 9 equal squares in which each square has an area of 7.29 square units. What are the dimensions of the figure?

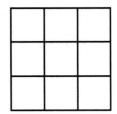

Directions   Perform the indicated operations and simplify. Assume that all variables represent nonnegative real numbers. See example 8–4 C.

40. $3\sqrt[3]{4} + 5\sqrt[3]{4}$

41. $7\sqrt[5]{2} - 4\sqrt[5]{2} + 3\sqrt[5]{2}$

42. $9\sqrt[4]{3} + 6\sqrt[4]{3} + 2\sqrt[4]{3}$

43. $\sqrt[3]{16} + \sqrt[3]{54}$

44. $\sqrt[3]{24} - \sqrt[3]{81}$

45. $\sqrt[3]{81} + 2\sqrt[3]{250}$

46. $\sqrt[3]{8a^2} + \sqrt[3]{27a^2}$

47. $\sqrt[4]{16x^3} + \sqrt[4]{81x^3}$

48. $\sqrt[4]{625a} - \sqrt[4]{81a}$

49. $\sqrt[3]{64x^2y} - \sqrt[3]{27x^2y}$

50. $\sqrt[3]{x^6y} + 2x^2\sqrt[3]{y}$

51. $3a\sqrt[3]{b^2} - \sqrt[3]{a^3b^2}$

52. $\sqrt[3]{16a^2b} + \sqrt[3]{54a^2b}$

53. Why are $3\sqrt{2x}$ and $2\sqrt{3x}$ not like radicals?

✪ Explore to Learn

In general $\sqrt{a} + \sqrt{b} \neq \sqrt{a + b}$.

a. Give a numeric example that illustrates this. (That is, find values of a and b that illustrate this concept.)

b. Find an example in which this statement is false—that is, find values for a and b for which $\sqrt{a} + \sqrt{b} = \sqrt{a + b}$.

Review Exercises

Directions   Multiply the following expressions. See section 3–2.

1. $3x(2x - y)$

2. $2a^2(a^2 - b^2)$

3. $(x - 1)(x - 1)$

4. $(y - 1)(y + 1)$

5. $(2x + 1)(2x - 1)$

6. $(x + 3y)(x + 2y)$

7. $(x - y)^2$

8. $(a + 2b)^2$

9. Ana can mow the lawn in one third the time it would take Ramon. Together they can mow the lawn in 2 hours. How long would it take for Ramon to mow the lawn alone? See section 5–7.

10. A liquid that is pure alcohol is to be mixed with a mixture containing 40% alcohol to obtain 30 gallons that is 70% alcohol. How many gallons of each must be used? See section 2–7.

8–5   Further Operations with Radicals

Multiplication of Radical Expressions

In section 8–2, we learned the procedure for multiplying two radicals. We now combine those ideas along with the *distributive property*, $a(b + c) = ab + ac$, to perform multiplication of radical expressions containing more than one term.

● **Example 8–5 A**
Multiplying radical expressions

Perform the indicated operations and simplify. Assume that all variables represent nonnegative real numbers.

1. $\sqrt{3}(3 + \sqrt{3}) = \sqrt{3} \cdot 3 + \sqrt{3}\sqrt{3}$ Distributive property
    $\qquad\qquad\quad = 3\sqrt{3} + \sqrt{9}$ $\sqrt{3}\sqrt{3} = \sqrt{9}$
    $\qquad\qquad\quad = 3\sqrt{3} + 3$ $\sqrt{9} = 3$

2. $\sqrt{3}(\sqrt{6} + \sqrt{21}) = \sqrt{3}\sqrt{6} + \sqrt{3}\sqrt{21}$ Distributive property
    $\qquad\qquad\quad = \sqrt{18} + \sqrt{63}$ Multiply radicands
    $\qquad\qquad\quad = \sqrt{9 \cdot 2} + \sqrt{9 \cdot 7}$ Factor $18 = 9 \cdot 2$; $63 = 9 \cdot 7$
    $\qquad\qquad\quad = 3\sqrt{2} + 3\sqrt{7}$ Simplify radicals

3. $(\sqrt{2} + \sqrt{3})(\sqrt{2} + 5\sqrt{3})$

    <u>Note</u> In this example, we are multiplying two binomials. Therefore, as we did in chapter 3, we will *multiply each term in the first parentheses by each term in the second parentheses*.

    $= \sqrt{2}\sqrt{2} + \sqrt{2} \cdot 5\sqrt{3} + \sqrt{3}\sqrt{2} + \sqrt{3} \cdot 5\sqrt{3}$ Distributive property
    $= \sqrt{4} + 5\sqrt{6} + \sqrt{6} + 5 \cdot \sqrt{9}$ Multiply radicands
    $= 2 + 5\sqrt{6} + \sqrt{6} + 5 \cdot 3$ $\sqrt{4} = 2, \sqrt{9} = 3$
    $= 2 + 5\sqrt{6} + \sqrt{6} + 15$
    $= 17 + 6\sqrt{6}$ Combine like terms

4. $(3 - \sqrt{2})(3 + \sqrt{2}) = 9 + 3\sqrt{2} - 3\sqrt{2} - \sqrt{4}$ Distributive property
    $\qquad\qquad\qquad\quad = 9 + 3\sqrt{2} - 3\sqrt{2} - 2$ Simplify radicals
    $\qquad\qquad\qquad\quad = 9 - 2$ Combine like terms
    $\qquad\qquad\qquad\quad = 7$ Subtract

    In this example, we observe that when we simplified, there were no longer any radicals in the answer.

5. $(\sqrt{a} - \sqrt{b})(\sqrt{a} + \sqrt{b})$
    $= \sqrt{a}\sqrt{a} + \sqrt{a}\sqrt{b} - \sqrt{b}\sqrt{a} - \sqrt{b}\sqrt{b}$ Distributive property
    $= \sqrt{a^2} + \sqrt{ab} - \sqrt{ab} - \sqrt{b^2}$ Multiply radicands
    $= a + \sqrt{ab} - \sqrt{ab} - b$ $\sqrt{a^2} = a$ and $\sqrt{b^2} = b$
    $= a - b$ Combine like terms

6. $(\sqrt{3} + 2\sqrt{2})^2$

$$= (\sqrt{3} + 2\sqrt{2})(\sqrt{3} + 2\sqrt{2})$$
$$= \sqrt{3}\sqrt{3} + \sqrt{3} \cdot 2\sqrt{2} + \sqrt{3} \cdot 2\sqrt{2} + 2\sqrt{2} \cdot 2\sqrt{2} \quad \text{Distributive property}$$
$$= \sqrt{9} + 2\sqrt{6} + 2\sqrt{6} + 4\sqrt{4} \quad \text{Multiply radicands}$$
$$= 3 + 2\sqrt{6} + 2\sqrt{6} + 4 \cdot 2 \quad \text{Simplify radicals}$$
$$= 3 + 2\sqrt{6} + 2\sqrt{6} + 8 \quad \text{Multiply}$$
$$= 11 + 4\sqrt{6} \quad \text{Combine like terms}$$

☛ *Quick check* Simplify. $2(\sqrt{3} + \sqrt{5})$; $\sqrt{2}(\sqrt{14} + \sqrt{6})$; and $(\sqrt{2} + \sqrt{3})(\sqrt{2} - 2\sqrt{3})$

Conjugate Factors

The type of factors that we are multiplying in examples 4 and 5 are called **conjugate factors**. The conjugate is used to rationalize the denominator of a fraction when the denominator contains two terms where one or both terms contain a square root. The idea of conjugate factors is derived from the factorization of the difference of two squares. When multiplying conjugate factors, we can simply write our answer as the square of the second term subtracted from the square of the first term.

In examples 4 and 5, we could have performed the multiplication as follows:

4. $(3 - \sqrt{2})(3 + \sqrt{2}) = (3)^2 - (\sqrt{2})^2 = 9 - 2 = 7$;
5. $(\sqrt{a} - \sqrt{b})(\sqrt{a} + \sqrt{b}) = (\sqrt{a})^2 - (\sqrt{b})^2 = a - b$.

The conjugate of a given factor is found by writing the original factor and changing the sign of the second term.

● **Example 8–5 B**

Forming a conjugate

Form the conjugates of the given expressions.

1. $\sqrt{7} + 2$ The conjugate is $\sqrt{7} - 2$.
2. $\sqrt{11} - \sqrt{6}$ The conjugate is $\sqrt{11} + \sqrt{6}$.
3. $-5 - 2\sqrt{3}$ The conjugate is $-5 + 2\sqrt{3}$.
4. $\sqrt{a} + \sqrt{b}$ The conjugate is $\sqrt{a} - \sqrt{b}$.

☛ *Quick check* Form the conjugate. $6 - 3\sqrt{2}$

Rationalizing the Denominator

If we wish to rationalize the denominator of the fraction

$$\frac{1}{3 - \sqrt{2}}$$

we recall from example 8–5 A number 4 that when we multiplied $3 - \sqrt{2}$ by $3 + \sqrt{2}$, there were no radicals left in our product. This result is precisely what we want to occur in our denominator. Therefore, to rationalize this fraction, we apply the fundamental principle of fractions and multiply the numerator and the denominator by $3 + \sqrt{2}$, the conjugate of the denominator.

$$\frac{1}{3 - \sqrt{2}} = \frac{1}{3 - \sqrt{2}} \cdot \frac{3 + \sqrt{2}}{3 + \sqrt{2}}$$

$3 + \sqrt{2}$ is the conjugate of the denominator

$$= \frac{1(3 + \sqrt{2})}{(3)^2 - (\sqrt{2})^2}$$

(first term)^2 − (second term)^2

$$= \frac{3 + \sqrt{2}}{9 - 2}$$

No radicals remain in the denominator

$$= \frac{3 + \sqrt{2}}{7}$$

Denominator is rationalized

● **Example 8–5 C**

Using conjugates to rationalize a denominator

Rationalize the denominators.

1. $\dfrac{2}{\sqrt{7} + 2} = \dfrac{2}{\sqrt{7} + 2} \cdot \dfrac{\sqrt{7} - 2}{\sqrt{7} - 2}$

Multiply by the conjugate of the denominator

$$= \frac{2(\sqrt{7} - 2)}{(\sqrt{7})^2 - (2)^2}$$

$(x + y)(x - y) = x^2 - y^2$

$$= \frac{2\sqrt{7} - 4}{7 - 4}$$

Simplify in numerator and denominator

$$= \frac{2\sqrt{7} - 4}{3}$$

Subtract in denominator

2. $\dfrac{5}{\sqrt{11} - \sqrt{6}} = \dfrac{5}{\sqrt{11} - \sqrt{6}} \cdot \dfrac{\sqrt{11} + \sqrt{6}}{\sqrt{11} + \sqrt{6}}$

Multiply by the conjugate of the denominator

$$= \frac{5(\sqrt{11} + \sqrt{6})}{(\sqrt{11})^2 - (\sqrt{6})^2}$$

$(x + y)(x - y) = x^2 - y^2$

$$= \frac{5(\sqrt{11} + \sqrt{6})}{11 - 6}$$

Simplify radicals

$$= \frac{5(\sqrt{11} + \sqrt{6})}{5}$$

Subtract in denominator

$$= \sqrt{11} + \sqrt{6}$$

Reduce (by 5) to lowest terms

3. $\dfrac{\sqrt{3}}{5 - 2\sqrt{3}} = \dfrac{\sqrt{3}}{5 - 2\sqrt{3}} \cdot \dfrac{5 + 2\sqrt{3}}{5 + 2\sqrt{3}}$

Multiply by the conjugate of the denominator

$$= \frac{\sqrt{3}(5 + 2\sqrt{3})}{(5)^2 - (2\sqrt{3})^2}$$

$(x + y)(x - y) = x^2 - y^2$

$$= \frac{5\sqrt{3} + 2\sqrt{9}}{5^2 - 2^2(\sqrt{3})^2}$$

Simplify radicals

$$= \frac{5\sqrt{3} + 2 \cdot 3}{25 - 4 \cdot 3}$$

Simplify radicals

$$= \frac{5\sqrt{3} + 6}{25 - 12}$$

Perform operations

$$= \frac{5\sqrt{3} + 6}{13}$$

Subtract in denominator

☛ *Quick check*   Rationalize the denominators. $\dfrac{3}{7 + \sqrt{2}}$ and $\dfrac{2}{\sqrt{11} - 3}$ ●

MASTERY POINTS

Can you
- Multiply radical expressions containing more than one term?
- Form conjugate factors?
- Multiply conjugate factors?
- Rationalize a denominator that has two terms in which one or both terms contain a square root?

Exercise 8–5

Directions

Perform the indicated operations and simplify. Assume that all variables represent positive real numbers. See example 8–5 A.

☞ *Quick Check*

Examples $2(\sqrt{3} + \sqrt{5})$ $\sqrt{2}(\sqrt{14} + \sqrt{6})$

Solutions $= 2\sqrt{3} + 2\sqrt{5}$ Distributive property

$= \sqrt{2}\sqrt{14} + \sqrt{2}\sqrt{6}$ Distributive property

$= \sqrt{28} + \sqrt{12}$ Product property

$= \sqrt{4 \cdot 7} + \sqrt{4 \cdot 3}$ Factor $28 = 4 \cdot 7$; $12 = 4 \cdot 3$

$= 2\sqrt{7} + 2\sqrt{3}$ $\sqrt{4} = 2$

Example $(\sqrt{2} + \sqrt{3})(\sqrt{2} - 2\sqrt{3})$

Solution $= \sqrt{2}\sqrt{2} - \sqrt{2} \cdot 2\sqrt{3} + \sqrt{3}\sqrt{2} - \sqrt{3} \cdot 2\sqrt{3}$ Distributive property

$= 2 - 2\sqrt{6} + \sqrt{6} - 2 \cdot 3$ $\sqrt{2}\sqrt{2} = 2$; $\sqrt{3}\sqrt{3} = 3$

$= 2 - \sqrt{6} - 6$ Combine like radicals

$= -4 - \sqrt{6}$ $2 - 6 = -4$

Directions   Perform the indicated operations and simplify. Assume that all variables represent positive real numbers. See example 8–5 A.

1. $3(\sqrt{2} + \sqrt{3})$

2. $5(2\sqrt{6} + \sqrt{2})$

3. $\sqrt{2}(\sqrt{3} + \sqrt{7})$

4. $\sqrt{5}(\sqrt{7} - \sqrt{3})$

5. $3\sqrt{2}(2\sqrt{3} - \sqrt{11})$

6. $\sqrt{6}(\sqrt{2} + \sqrt{3})$

7. $\sqrt{5}(\sqrt{15} - \sqrt{10})$

8. $\sqrt{14}(\sqrt{21} + \sqrt{10})$

9. $2\sqrt{7}(\sqrt{35} - 3\sqrt{14})$

10. $\sqrt{a}(\sqrt{ab} + \sqrt{b})$

11. $\sqrt{a}(3\sqrt{a} + \sqrt{b})$

12. $(5 + \sqrt{3})(4 - \sqrt{3})$

13. $(3 + \sqrt{2})(4 + \sqrt{2})$

14. $(5 - \sqrt{5})(5 - \sqrt{5})$

15. $(3 - 4\sqrt{a})(4 - 3\sqrt{a})$

16. $(7 + 2\sqrt{y})(6 + 5\sqrt{y})$

17. $(\sqrt{3} + \sqrt{2})(\sqrt{3} - \sqrt{2})$

18. $(\sqrt{7} + \sqrt{5})(\sqrt{7} - \sqrt{5})$

19. $(2 + \sqrt{6})(2 - \sqrt{6})$

20. $(5 - \sqrt{3})(5 + \sqrt{3})$

21. $(2 + \sqrt{5})^2$

22. $(3 - \sqrt{7})^2$

23. $(\sqrt{x} + \sqrt{y})^2$

24. $(\sqrt{a} - \sqrt{b})^2$

25. $(\sqrt{x} - \sqrt{y})(\sqrt{x} + \sqrt{y})$

26. $(2\sqrt{a} - \sqrt{b})(2\sqrt{a} + \sqrt{b})$

27. $(x\sqrt{y} + \sqrt{z})(x\sqrt{y} - \sqrt{z})$

28. $(a\sqrt{b} + c)(a\sqrt{b} - c)$

29. $(2\sqrt{x} + y)^2$

30. $(3\sqrt{a} + \sqrt{b})^2$

Directions	☞ *Quick Check*
Form the conjugates of the given expressions. See example 8–5 B.	**Example** $6 - 3\sqrt{2}$ **Solution** $6 + 3\sqrt{2}$ First term remains the same, change the sign of the second term

Directions   Form the conjugates of the given expressions. See example 8–5 B.

31. $11 - \sqrt{3}$

32. $-5\sqrt{7} - \sqrt{2}$

33. $\sqrt{a} + 3\sqrt{b}$

34. $a\sqrt{b} - \sqrt{c}$

Directions	☞ *Quick Check*
Simplify the following expressions, leaving all denominators rationalized. Assume that all variables represent positive real numbers and that no denominator is equal to zero. See example 8–5 C.	**Examples** $\dfrac{3}{7 + \sqrt{2}}$ $\qquad$ $\dfrac{2}{\sqrt{11} - 3}$ **Solutions** $= \dfrac{3}{7 + \sqrt{2}} \cdot \dfrac{7 - \sqrt{2}}{7 - \sqrt{2}}$ Multiply by the conjugate $= \dfrac{2}{\sqrt{11} - 3} \cdot \dfrac{\sqrt{11} + 3}{\sqrt{11} + 3}$ Multiply by the conjugate $= \dfrac{3(7 - \sqrt{2})}{(7)^2 - (\sqrt{2})^2}$ $(x + y)(x - y) = x^2 - y^2$ $= \dfrac{2(\sqrt{11} + 3)}{(\sqrt{11})^2 - (3)^2}$ $(x + y)(x - y) = x^2 - y^2$ $= \dfrac{3(7 - \sqrt{2})}{49 - 2}$ Simplify denominator $= \dfrac{2(\sqrt{11} + 3)}{11 - 9}$ Simplify denominator $= \dfrac{3(7 - \sqrt{2})}{47}$ Simplify denominator $= \dfrac{2(\sqrt{11} + 3)}{2}$ Simplify denominator $= \dfrac{21 - 3\sqrt{2}}{47}$ Multiply in numerator $= \sqrt{11} + 3$ Reduce fraction

Directions   Simplify the following expressions, leaving all denominators rationalized. Assume that all variables represent positive real numbers and that no denominator is equal to zero. See example 8–5 C.

35. $\dfrac{1}{\sqrt{2} + 3}$

36. $\dfrac{1}{\sqrt{3} - 2}$

37. $\dfrac{7}{2 + \sqrt{7}}$

38. $\dfrac{6}{3 - \sqrt{6}}$

39. $\dfrac{3}{\sqrt{6} - \sqrt{3}}$

40. $\dfrac{5}{\sqrt{11} + \sqrt{6}}$

41. $\dfrac{8}{\sqrt{3} - 1}$

42. $\dfrac{12}{\sqrt{5} - 1}$

43. $\dfrac{10}{\sqrt{6} - 2}$

44. $\dfrac{1}{\sqrt{a} + b}$

45. $\dfrac{3}{2\sqrt{3} - \sqrt{5}}$

46. $\dfrac{4}{2\sqrt{3} - \sqrt{6}}$

47. $\dfrac{\sqrt{2}}{\sqrt{2} + 1}$

48. $\dfrac{\sqrt{5}}{\sqrt{5} - 2}$

49. $\dfrac{\sqrt{15}}{\sqrt{5} - \sqrt{3}}$

50. $\dfrac{\sqrt{14}}{\sqrt{7} + \sqrt{2}}$

51. $\dfrac{1 + \sqrt{5}}{1 - \sqrt{5}}$

52. $\dfrac{\sqrt{3} - \sqrt{7}}{\sqrt{3} + \sqrt{7}}$

53. $\dfrac{\sqrt{a} + b}{\sqrt{a} - b}$

54. $\dfrac{3}{4 - 2\sqrt{2}}$

55. $\dfrac{\sqrt{3}}{3\sqrt{3} + 4}$

56. $\dfrac{12}{6 - 2\sqrt{5}}$

57. $\dfrac{10}{5 - 2\sqrt{5}}$

58. $\dfrac{12}{2\sqrt{5} - 2}$

59. Explain how to determine the conjugate of a two-term expression involving square roots.

✺ Explore to Learn

One of the examples in this section shows that $\dfrac{3}{7 + \sqrt{2}} = \dfrac{21 - 3\sqrt{2}}{47}$. Calculate the left side of this equation with a calculator. Calculate the right side of this equation also with a calculator. What is true about each value? Could you have predicted this before you did the calculation?

Review Exercises

Directions Perform the indicated operations and leave your answer with only positive exponents. Assume that no variable is equal to zero. See sections 3–1 and 3–3.

1. $(2^2)^3$

2. $2^2 \cdot 2^3$

3. $3^{-2} \cdot 3^4$

4. $\dfrac{3^2}{3^5}$

5. $(x^{-2})^3$

6. $(2a^2b)^3$

7. $\dfrac{x^2 y^5}{x^3 y^2}$

8. $x^{-4} \cdot x^{-5}$

9. A commuter train travels 90 miles in the same time that an express train travels 150 miles. If the express train averages 30 miles per hour faster than the commuter train, how fast is each train traveling? See section 5–7.

10. A chemist mixes pure acid with a 25% acid solution. How many liters of each must be used to obtain 100 liters that are 40% acid? See section 2–7.

8–6 Fractional Exponents

Fractional Exponents

In this section, we are going to develop the idea of a fraction used as an exponent. Consider the example

$$(a^{\frac{1}{2}})^2 = a^{\frac{1}{2} \cdot 2} = a^1$$

When we raise a power to a power, we multiply the exponents. In the previous section, we observed that if a represented a nonnegative real number, then

$$(\sqrt{a})^2 = a$$

Therefore, for our properties of exponents and our procedures for radicals to be consistent, the following statement must be true:

$$a^{1/2} = \sqrt{a}$$

We generalize as follows:

Definition of $a^{1/n}$

$$a^{1/n} = \sqrt[n]{a}$$

where n is a natural number greater than 1. Whenever n is even, a represents only nonnegative real numbers.

Concept

The expression $a^{1/n}$ represents the nth root of a.

● **Example 8–6 A**

Using the definition of $a^{1/n}$

Rewrite the following in radical notation and simplify where possible.

1. $5^{1/2} = \sqrt{5}$

2. $(64)^{1/2} = \sqrt{64} = 8$

3. $a^{1/3} = \sqrt[3]{a}$

4. $(-8)^{1/3} = \sqrt[3]{(-8)} = -2$

Once we become acquainted with fractional exponents, the process of changing the fractional exponent to radical form for simplification will become unnecessary.

Consider the following problem:

$$(\sqrt[3]{a})^2 = (a^{1/3})^2 = a^{2/3}$$

We observe that *when a number is raised to a fractional exponent, the numerator of the fractional exponent indicates the power to which the base is to be raised, and the denominator indicates the root to be taken.*

Definition of $a^{m/n}$

If a is any real number, m is any integer, and n is any positive integer, then if $\dfrac{m}{n}$ is reduced to lowest terms,

$$a^{m/n} = (\sqrt[n]{a})^m = \sqrt[n]{a^m}$$

provided that $\sqrt[n]{a}$ is a real number.

Concept

The expression $a^{m/n}$ represents the nth root of a raised to the mth power.

We calculate $a^{m/n}$ *by first finding the nth root of a and then raising the resulting number to the mth power.*

● **Example 8–6 B**

Using the definition of $a^{m/n}$

Simplify. Assume that all variables represent nonnegative real numbers.

Numerator is the power ↓

Denominator is the index ↓

1. $(16)^{3/4} = (\sqrt[4]{16})^3$
   $= 2^3$ 4th root of 16 is 2
   $= 8$ Standard form

2. $(32)^{2/5} = (\sqrt[5]{32})^2 = 2^2 = 4$

3. $(-27)^{2/3} = (\sqrt[3]{-27})^2 = (-3)^2 = 9$

4. $(x^6)^{1/3} = \sqrt[3]{x^6} = x^2$

☞ **Quick check** Simplify. $(49)^{1/2}$ and $(8)^{2/3}$

Operations with Fractional Exponents

We can extend the properties and definitions involving integer exponents to expressions that involve fractional exponents.

● **Example 8–6 C**

Using the properties of exponents with fractional exponents

Perform the indicated operations and simplify. Assume that all variables represent positive real numbers.

1. $5^{\frac{1}{2}} \cdot 5^{\frac{1}{3}} = 5^{\frac{1}{2}+\frac{1}{3}}$ Multiply like bases, add exponents
   $= 5^{\frac{3}{6}+\frac{2}{6}}$ Least common denominator is 6
   $= 5^{\frac{5}{6}}$ Add numerators

2. $\dfrac{2^{\frac{1}{2}}}{2^{\frac{1}{4}}} = 2^{\frac{1}{2}-\frac{1}{4}} = 2^{\frac{2}{4}-\frac{1}{4}} = 2^{\frac{1}{4}}$

3. $(-27)^{-\frac{2}{3}} = \dfrac{1}{(-27)^{\frac{2}{3}}} = \dfrac{1}{(\sqrt[3]{-27})^2} = \dfrac{1}{(-3)^2} = \dfrac{1}{9}$

4. $(2^3 x^9 y^{15})^{\frac{1}{3}} = (2^3)^{\frac{1}{3}}(x^9)^{\frac{1}{3}}(y^{15})^{\frac{1}{3}} = 2^{3 \cdot \frac{1}{3}} \cdot x^{9 \cdot \frac{1}{3}} \cdot y^{15 \cdot \frac{1}{3}}$
   $= 2^1 x^3 y^5 = 2x^3 y^5$

5. $x^{\frac{2}{3}} \cdot x^{\frac{3}{4}} = x^{\frac{2}{3}+\frac{3}{4}} = x^{\frac{8}{12}+\frac{9}{12}} = x^{\frac{17}{12}}$

6. $(y^{\frac{1}{2}})^{\frac{4}{3}} = y^{\frac{1}{2} \cdot \frac{4}{3}} = y^{\frac{2}{3}}$

7. $\dfrac{z^{\frac{1}{2}}}{z^{\frac{2}{3}}} = z^{\frac{1}{2}-\frac{2}{3}} = z^{\frac{3}{6}-\frac{4}{6}} = z^{-\frac{1}{6}} = \dfrac{1}{z^{\frac{1}{6}}}$

<u>Note</u> In example 7, $z^{1/6}$ in the denominator is simply another form of $\sqrt[6]{z}$. Therefore, if we want our answer in a rationalized form, we would proceed as follows:

$$\frac{1}{z^{1/6}} = \frac{1}{\sqrt[6]{z}} \cdot \frac{\sqrt[6]{z^5}}{\sqrt[6]{z^5}} = \frac{\sqrt[6]{z^5}}{z} = \frac{z^{5/6}}{z}$$

☞ **Quick check** Simplify. $y^{2/3} \cdot y^{3/2}$; $\dfrac{a^{3/4}b^{5/3}}{a^{1/4}b}$; and $(36)^{-1/2}$

Section 8–1 introduced the idea of *n*th root. Just as with square roots, most *n*th roots are irrational. These can be approximated using a calculator. Many calculators have a key marked $\boxed{\sqrt[x]{y}}$ to compute *n*th roots. However, others do not. In this case one must use a key like $\boxed{x^y}$, using the definition $x^{1/n} = \sqrt[n]{x}$. This idea requires that if there is no $\boxed{\sqrt[x]{y}}$ key, then both the $\boxed{x^y}$ and $\boxed{1/x}$ keys must be used. On the TI–81/82 the $\boxed{1/x}$ key is marked $\boxed{x^{-1}}$, and this calculator has a special key for cube roots.

● **Example 8–6 D**

Using a calculator to find nth roots

Compute the following irrational numbers to three decimal places with a calculator.

1. $\sqrt[3]{21.6} \approx 2.785$ 21.6 $\boxed{\sqrt[x]{y}}$ 3 $\boxed{=}$ or 21.6 $\boxed{x^y}$ 3 $\boxed{1/x}$ $\boxed{=}$
                                              or 21.6 $\boxed{\sqrt[3]{}}$ $\boxed{=}$

TI–81/82 $\boxed{\text{MATH}}$ 4 21.6 $\boxed{\text{ENTER}}$
Casio 7700GE/9700GE $\boxed{\text{SHIFT}}$ $\boxed{\sqrt[3]{}}$ 21.6 $\boxed{\text{EXE}}$

2. $\sqrt[4]{862.5} \approx 5.419$ 862.5 $\boxed{\sqrt[x]{y}}$ 4 $\boxed{=}$ or 862.5 $\boxed{x^y}$ 4 $\boxed{1/x}$ $\boxed{=}$
TI–81/82   862.5 $\boxed{\wedge}$ 4 $\boxed{x^{-1}}$ $\boxed{\text{ENTER}}$
Casio 7700GE/9700GE   4 $\boxed{\text{SHIFT}}$ $\boxed{\sqrt[x]{}}$ 862.5 $\boxed{\text{EXE}}$

● **Example 8–6 E**

Problem solving

At a height of *h* feet above the sea on level ground, the distance *d* in miles that a person can see an object is found by using the equation $d = 1.2h^{1/2}$. How far can someone see who is at the top of the Sears Tower, which is 1,454 feet high? Round the answer to the nearest mile.

$$d = 1.2h^{1/2} \qquad \text{Original equation}$$
$$d = 1.2(1{,}454)^{1/2} \qquad \text{Substitute}$$
$$d = 1.2\sqrt{1{,}454} \qquad \text{Change to radical form}$$
$$d = 1.2(38.131352) \qquad \text{Square root}$$
$$d \approx 46 \qquad \text{Multiply and round}$$

Therefore, a person on top of the Sears Tower could see approximately 46 miles.

MASTERY POINTS

Can you
- Express fractional exponents in radical form?
- Express radicals in fractional exponent form?
- Apply the properties and definitions involving integer exponents to fractional exponents?
- Compute decimal approximations to *n*th roots?

Exercise 8–6	☞ *Quick Check*

Directions
Simplify the given expressions. See examples 8–6 A, B, and C.

Examples $(49)^{1/2}$ 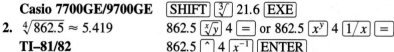 $8^{2/3}$ 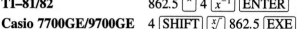 $(36)^{-1/2}$

Solutions $= \sqrt{49}$ Square root $= (\sqrt[3]{8})^2$ Numerator is the power Denominator is the index $= \dfrac{1}{(36)^{1/2}}$ $a^{-n} = \dfrac{1}{a^n}$
          $= 7$ $= \dfrac{1}{\sqrt{36}}$ Square root
                                  $= (2)^2$ Cube root of 8 is 2 $= \dfrac{1}{6}$ Simplify
                                  $= 4$

Directions Simplify the given expressions. See examples 8–6 A, B, and C.

1. $(36)^{1/2}$
2. $(25)^{1/2}$
3. $(a^6)^{1/3}$

4. $(b^{12})^{1/3}$
5. $(8)^{1/3}$
6. $(32)^{1/5}$

7. $(-27)^{1/3}$
8. $(-8)^{1/3}$
9. $(27)^{2/3}$

10. $(16)^{3/4}$
11. $(9)^{3/2}$
12. $(16)^{3/2}$

Directions
Perform the indicated operations and simplify. Assume that all variables represent positive real numbers. See example 8–6 C. Express answers with only positive exponents.

☛ **Quick Check**

Examples $y^{\frac{2}{3}} \cdot y^{\frac{3}{2}}$

Solutions $= y^{\frac{2}{3}+\frac{3}{2}}$ Multiply like bases, add exponents

$= y^{\frac{4}{6}+\frac{9}{6}}$ Least common denominator is 6

$= y^{\frac{13}{6}}$ Add numerators

$\dfrac{a^{\frac{3}{4}} b^{\frac{5}{3}}}{a^{\frac{1}{4}} b}$

$= a^{\frac{3}{4}-\frac{1}{4}} b^{\frac{5}{3}-1}$ Divide like bases, subtract exponents

$= a^{\frac{2}{4}} b^{\frac{5}{3}-\frac{3}{3}}$ Common denominators

$= a^{\frac{1}{2}} b^{\frac{2}{3}}$ Subtract numerators

Directions Perform the indicated operations and simplify. Assume that all variables represent positive real numbers. See example 8–6 C. Express answers with only positive exponents.

13. $(25)^{-1/2}$
14. $(9)^{-1/2}$
15. $(16)^{-3/4}$
16. $(27)^{-2/3}$
17. $(-8)^{-1/3}$
18. $(-27)^{-1/3}$
19. $2^{1/2} \cdot 2^{3/2}$
20. $3^{1/3} \cdot 3^{2/3}$
21. $2^{1/3} \cdot 2^{1/2}$
22. $5^{1/5} \cdot 5^{1/2}$
23. $x^{1/4} \cdot x^{3/4}$
24. $b^{1/3} \cdot b^{2/3}$
25. $c^{1/2} \cdot c^{1/4}$
26. $x^{1/4} \cdot x^{1/3}$
27. $\dfrac{2^{3/2}}{2^{1/2}}$
28. $\dfrac{3^{4/3}}{3^{1/3}}$

29. $\dfrac{2^{1/2}}{2^{1/3}}$
30. $\dfrac{7^{3/4}}{7^{2/3}}$
31. $\dfrac{a^{4/5}}{a^{1/5}}$
32. $\dfrac{x^{3/4}}{x^{1/2}}$
33. $\dfrac{y^{2/3}}{y^{1/2}}$
34. $\dfrac{x^{5/6}}{x^{2/3}}$
35. $(a^{2/3})^{1/2}$
36. $(c^{1/2})^{1/2}$
37. $(x^{1/2})^{4/3}$
38. $(y^{2/3})^{3/4}$
39. $(c^{-1/4})^{-2/3}$
40. $(y^{-1/2})^{-2/5}$

41. $(a^{-2/3})^{-1/2}$
42. $(b^{-1/2})^{-1/2}$
43. $(x^{1/4})^{-2/3}$
44. $(x^{1/3})^{-3/4}$
45. $(y^{-3/4})^{1/3}$
46. $(c^{-2/5})^{1/2}$
47. $(16a^4)^{3/4}$
48. $(x^3 y^{12})^{1/3}$
49. $(8a^6 b^3)^{2/3}$
50. $(27x^3 y^{12})^{2/3}$
51. $\dfrac{b^{3/4} c^{1/2}}{b^{1/4} c^{1/4}}$
52. $\dfrac{xy^{3/4}}{x^{1/2} y^{1/4}}$
53. $\dfrac{ab}{a^{1/2} b^{1/3}}$
54. $\dfrac{xy^{3/4}}{x^{2/5} y^{1/2}}$

Directions	**Example** $\sqrt[3]{8{,}256}$
Compute the following irrational numbers to three decimal places with a calculator.	**Solution** 8,256 $\boxed{\sqrt[x]{y}}$ 3 $\boxed{=}$ or 8,256 $\boxed{x^y}$ 3 $\boxed{1/x}$ $\boxed{=}$ or 8,256 $\boxed{\sqrt[3]{}}$ $\boxed{=}$
	TI–81/82 $\boxed{\text{MATH}}$ 4 8,256 $\boxed{\text{ENTER}}$
	Casio 7700GE/9700GE $\boxed{\text{SHIFT}}$ $\boxed{\sqrt[3]{}}$ 8256 $\boxed{\text{EXE}}$
	Answer ≈ 20.211

Directions   Compute the following irrational numbers to three decimal places with a calculator.

55. $\sqrt[3]{538}$

56. $\sqrt[3]{6{,}230}$

57. $\sqrt[4]{295}$

58. $\sqrt[4]{174}$

59. $4\sqrt[5]{81{,}000}$

60. $10\sqrt[5]{100}$

61. $5\sqrt[6]{0.24}$

62. $3\sqrt[6]{600}$

Directions   Solve the following problems. See example 8–6 E.

63. Find the number whose principal fourth root is 3.

64. Find the number whose principal cube root is -2.

65. The formula for approximating the velocity V in miles per hour of a car based on the length of its skid marks S (in feet) on dry pavement is given by
$$V = 4.9\, S^{1/2}$$
If the skid marks are 100 feet long, what was the approximate velocity?

66. The formula for approximating the velocity V in miles per hour of a car based on the length of its skid marks S (in feet) on wet pavement is given by
$$V = 3.46\, S^{1/2}$$
If the skid marks are 81 feet long, what was the approximate velocity of the car?

67. At a height of h feet above the sea on level ground, the distance d in miles that a person can see an object is found by using the equation
$$d = 1.2\, h^{1/2}$$
How far can someone see who is in a tower 400 feet above the ground?

68. How can you find the principal fourth root of a number on a calculator using only the square root key?

69. How can you find the principal eighth root of a number on a calculator using only the square root key?

70. Why must $a^{1/2} = \sqrt{a}$?

☸ **Explore to Learn**

a. Fill in the following table of values using a calculator.

b. Based on the values in the table, and assuming $a > 0$, what must be the restrictions on a for $a < \sqrt[n]{a}$ to be true?

c. Based on the values in the table, and assuming $a > 0$, what must be the restrictions on a for $a > \sqrt[n]{a}$ to be true?

n	$\sqrt{n}$	$\sqrt[3]{n}$	$\sqrt[4]{n}$
0.1			
0.5			
1			
2			
10			

Review Exercises

Directions   Perform the indicated operations. Assume that all radicands are nonnegative. See sections 8–1 and 8–2.

1. $(\sqrt{7})^2$

2. $(\sqrt{x})^2$

3. $(\sqrt{x+1})^2$

Directions   Perform the indicated operations. See section 3–2.

4. $(x+1)^2$

5. $(x-2)^2$

Directions   Find the solution set for the following equations. See section 4–7.

6. $x + 6 = x^2$

7. $x + 2 = x^2 - 9x + 18$

8. $x + 1 = x^2 + 2x + 1$

9. Two electrical voltages have a total of 156 volts (V). If one voltage is 32 V more than the other, find the two voltages. See section 2–7.

10. How many red balls are there in a bowl if four times the square of the number of balls is 196? See section 4–7.

8–7   Equations Involving Radicals

Radical Equations

An equation in which the unknown quantity appears under a radical symbol is called a **radical equation**. Examples of radical equations are

$$\sqrt{x} = 5; \quad \sqrt{x+2} = 7; \quad 4 + \sqrt{x+2} = x$$

In this section, we will consider radical equations containing only square roots.

Solving radical equations involves squaring both sides of an equation to eliminate the square roots. When we square both sides of an equation, we use the squaring property of equality.

Squaring Property of Equality

If P and Q are algebraic expressions and if
$$P = Q$$
then all solutions of $P = Q$ are also solutions of the equation
$$P^2 = Q^2$$

Concept

If each side of an equation is squared, the solution(s) of the original equation are also solution(s) of the resulting equation.

Extraneous Solutions

This property implies that there *may be* solutions of the equation $P^2 = Q^2$ that are not solutions of the original equation $P = Q$. If such solutions exist, they are called **extraneous solutions** (roots). Thus, all possible solutions must be checked in the original equation.

Solving a Radical Equation
1. Rewrite the equation (if necessary) so that a radical is by itself on one side of the equation.
2. Square each side of the equation and combine like terms.
3. Repeat steps 1 and 2 if a radical remains in the equation.
4. Solve the resulting equation.
5. Check all possible solutions in the original equation.

● **Example 8–7 A**

Solving an equation containing a square root

Find the solution set.

1.

$$\sqrt{x + 2} = 7$$

$(\sqrt{x + 2})^2 = (7)^2$ Square both sides

$x + 2 = 49$ Squaring a square root gives the radicand

$x = 47$ Subtract 2 from both sides

Check :

$\sqrt{x + 2} = 7$ Original equation

$\sqrt{(47) + 2} = 7$ Substitute into original equation

$\sqrt{49} = 7$ Simplify the radicand

$7 = 7$ True

The solution set is $\{47\}$.

2. $4 + \sqrt{x + 2} = x$

$\sqrt{x + 2} = x - 4$ Isolate the radical by subtracting 4

$(\sqrt{x + 2})^2 = (x - 4)^2$ Square both sides

$x + 2 = (x - 4)(x - 4)$ Simplify

$x + 2 = x^2 - 4x - 4x + 16$ Multiply

$x + 2 = x^2 - 8x + 16$ Subtract x and 2

$0 = x^2 - 9x + 14$ Solve the resulting quadratic equation

$0 = (x - 7)(x - 2)$ Factor

$x - 7 = 0$ or $x - 2 = 0$ Set each factor equal to 0 and solve

$x = 7$ or $x = 2$ Check for extraneous roots

Check :

for 7   for 2

$4 + \sqrt{x + 2} = x$ $4 + \sqrt{x + 2} = x$ Original equation

$4 + \sqrt{(7) + 2} = (7)$ $4 + \sqrt{(2) + 2} = (2)$ Substitute

$4 + \sqrt{9} = 7$ $4 + \sqrt{4} = 2$ Simplify

$4 + 3 = 7$ $4 + 2 = 2$ Simplify

$7 = 7$ True $6 = 2$ False

Therefore, 2 is an *extraneous root* and the solution set is $\{7\}$.

3.
$$\sqrt{x + 1} = x + 1$$
$$(\sqrt{x + 1})^2 = (x + 1)^2 \qquad \text{Square both sides}$$
$$x + 1 = (x + 1)(x + 1) \qquad \text{Simplify}$$
$$x + 1 = x^2 + x + x + 1 \qquad \text{Multiply}$$
$$x + 1 = x^2 + 2x + 1 \qquad \text{Combine like terms}$$
$$0 = x^2 + x \qquad \text{Solve the resulting quadratic equation}$$
$$0 = x(x + 1) \qquad \text{Factor}$$
$$x = 0 \text{ or } x + 1 = 0 \qquad \text{Set each factor equal to 0 and solve}$$

Therefore, $x = 0$ or $x = -1$.

Check:

for 0			for -1	
$\sqrt{x + 1} = x + 1$			$\sqrt{x + 1} = x + 1$	Original equation
$\sqrt{(0) + 1} = (0) + 1$			$\sqrt{(-1) + 1} = (-1) + 1$	Substitute
$\sqrt{1} = 1$			$\sqrt{0} = 0$	Simplify
$1 = 1$	True		$0 = 0$	True

Therefore, 0 and -1 are solutions of the equation. The solution set is $\{0, -1\}$.

4. $\sqrt{3x + 4} = \sqrt{x + 14}$

Note There are two square roots in this equation, but we can eliminate both square roots by squaring both sides.

$$(\sqrt{3x + 4})^2 = (\sqrt{x + 14})^2 \qquad \text{Square both sides}$$
$$3x + 4 = x + 14 \qquad \text{Solve for } x$$
$$2x + 4 = 14 \qquad \text{Subtract } x \text{ from both sides}$$
$$2x = 10 \qquad \text{Subtract 4 from both sides}$$
$$x = 5 \qquad \text{Divide both sides by 2}$$

Check:

$$\sqrt{3x + 4} = \sqrt{x + 14} \qquad \text{Original equation}$$
$$\sqrt{3(5) + 4} = \sqrt{(5) + 14} \qquad \text{Substitute}$$
$$\sqrt{15 + 4} = \sqrt{19} \qquad \text{Simplify}$$
$$\sqrt{19} = \sqrt{19} \qquad \text{True}$$

The solution set is $\{5\}$.

5.
$$x = \sqrt{x^2 - 3x - 6}$$
$$(x)^2 = (\sqrt{x^2 - 3x - 6})^2 \qquad \text{Square both sides}$$
$$x^2 = x^2 - 3x - 6 \qquad \text{Simplify}$$
$$0 = -3x - 6 \qquad \text{Subtract } x^2$$
$$3x = -6 \qquad \text{Add } 3x$$
$$x = -2 \qquad \text{Divide by 3}$$

Check :

$$x = \sqrt{x^2 - 3x - 6}$$ Original equation

$$(-2) = \sqrt{(-2)^2 - 3(-2) - 6}$$ Substitute

$$-2 = \sqrt{4 - 3(-2) - 6}$$ Exponents

$$-2 = \sqrt{4 + 6 - 6}$$ Multiply

$$-2 = \sqrt{4}$$ Addition and subtraction

$$-2 = 2$$ False

$x = -2$ does not check, and we conclude that there is no solution to this equation. The solution set is the empty set, $\varnothing$.

☞ *Quick check*   Find the solution set. $\sqrt{x + 5} = 6$ and $\sqrt{2x + 5} = \sqrt{x + 8}$ ●

● **Example 8–7 B**

Problem solving

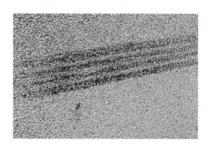

The formula for approximating the velocity V in miles per hour of a car based on the length of its skid marks S (in feet) on dry pavement is given by $V = 2\sqrt{6S}$. If the velocity is 48 mph, how long will the skid marks be?

$$V = 2\sqrt{6S}$$ Original equation

$$(48) = 2\sqrt{6S}$$ Substitute

$$24 = \sqrt{6S}$$ Divide both sides by 2 to isolate the radical

$$576 = 6S$$ Square both sides

$$96 = S$$ Divide both sides by 6

Therefore, the car would leave skid marks approximately 96 feet long.   ●

MASTERY POINTS

Can you
● Find the solution set of equations containing radicals?

Exercise 8–7	☞ *Quick Check*

Directions
Find the solution set.
See example 8–7 A.

Examples $\sqrt{x + 5} = 6$ $\sqrt{2x + 5} = \sqrt{x + 8}$

Solutions $(\sqrt{x + 5})^2 = (6)^2$ $(\sqrt{2x + 5})^2 = (\sqrt{x + 8})^2$ Square both sides

$\qquad\qquad x + 5 = 36$ $2x + 5 = x + 8$ Simplify

$\qquad\qquad\quad x = 31$ $x + 5 = 8$ Solve for x

$\qquad\qquad\qquad\qquad\qquad\qquad\qquad x = 3$

Check :

$\sqrt{(31) + 5} = 6$

$\sqrt{36} = 6$

$6 = 6$ True

The solution set is $\{31\}$.

Check :

$\sqrt{2(3) + 5} = \sqrt{(3) + 8}$ Substitute into original equation

$\sqrt{6 + 5} = \sqrt{11}$ Simplify

$\sqrt{11} = \sqrt{11}$ True

The solution set is $\{3\}$.

Directions Find the solution set. See example 8–7 A.

1. $\sqrt{x} = 4$

2. $\sqrt{x} = 5$

3. $\sqrt{x} = 9$

4. $\sqrt{x} = 7$

5. $\sqrt{x + 5} = 4$

6. $\sqrt{x - 3} = 5$

7. $\sqrt{x - 7} = 6$

8. $\sqrt{x + 3} = 7$

9. $\sqrt{2x + 1} = 5$

10. $\sqrt{2x + 6} = 6$

11. $\sqrt{3x + 1} = 4$

12. $\sqrt{5x - 4} = 6$

13. $\sqrt{x + 4} = 7$

14. $\sqrt{x + 2} = 9$

15. $\sqrt{x - 5} = 1$

16. $\sqrt{x - 4} = 2$

17. $\sqrt{x + 7} = 5$

18. $\sqrt{x + 8} = 4$

19. $\sqrt{x + 6} = 3$

20. $\sqrt{x + 10} = 5$

21. $\sqrt{2x + 1} = \sqrt{x + 5}$

22. $\sqrt{2x + 4} = \sqrt{3x - 2}$

23. $\sqrt{5x - 3} = \sqrt{2x + 9}$

24. $\sqrt{7x - 4} = \sqrt{3x + 20}$

25. $\sqrt{2x + 7} = \sqrt{4x + 1}$

26. $\sqrt{6x - 3} = \sqrt{4x + 5}$

27. $\sqrt{3x + 5} = \sqrt{5x + 1}$

28. $\sqrt{4x - 4} = \sqrt{x + 5}$

29. $\sqrt{5x - 7} = \sqrt{2x + 8}$

30. $\sqrt{4x - 15} = \sqrt{7x - 9}$

31. $\sqrt{5 + 2x} = \sqrt{2 + 3x}$

32. $\sqrt{9 - 5x} = \sqrt{15 - 7x}$

33. $x = \sqrt{x^2 - x + 3}$

34. $x = \sqrt{x^2 - 2x + 10}$

35. $x = \sqrt{x^2 - 2x - 8}$

36. $x = \sqrt{x^2 - 3x - 9}$

37. $2x = \sqrt{4x^2 - 2x + 14}$

38. $2x = \sqrt{4x^2 - 3x + 15}$

39. $x = \sqrt{x^2 - 16}$

40. $x = \sqrt{x^2 - 4}$

41. $\sqrt{x^2 + 1} = x + 2$

42. $\sqrt{x^2 + 3x} = x + 1$

43. $3 + \sqrt{x^2 + 3x} = x$

44. $\sqrt{x^2 + 12} - 2 = x$

45. $\sqrt{x + 6} = x$

46. $\sqrt{5x + 6} = x$

47. $\sqrt{2x + 8} = x$

48. $\sqrt{4x + 12} = x$

49. $6 + \sqrt{x - 4} = x$

50. $\sqrt{x + 2} - 2 = x$

51. $\sqrt{x + 4} + 8 = x$

52. $\sqrt{x + 6} = x$

53. $\sqrt{x + 7} = 2x - 1$

54. $\sqrt{2x - 1} + 2x = 7$

Directions Find the unknown number in problems 55–62.

55. The square root of the sum of a number and 6 is 5. Find the number.

56. The square root of the sum of a number and 9 is 7. Find the number.

57. The square root of the product of a number and 6 is 12. Find the number.

58. The square root of the product of a number and 9 is 6. Find the number.

59. A certain number is equal to the square root of the sum of that number and 12.

60. The square root of the product of a number and 12 is equal to the number increased by 3.

61. The square root of the sum of a number and 11 is 1 less than the number.

62. The square root of the product of a number and 4 is 3 less than the number.

63. At a height of h ft above the sea or level ground, the distance d in miles that a person can see an object is given by

$$d = \sqrt{\frac{3h}{2}}$$

How high must a person be to see an object 6 miles away?

64. The formula for approximating the velocity V in miles per hour of a car based on the length of its skid marks S (in feet) on dry pavement is given by

$$V = 2\sqrt{6S}$$

If the velocity is 36 mph, how long will the skid marks be?

65. On wet pavement, the formula in exercise 64 is given by

$$V = 2\sqrt{3S}$$

How long will the skid marks be if the car is traveling at 36 mph on wet pavement?

66. Find the number whose principal square root is 3.

67. Find the number whose principal square root is 10.

68. Find the number whose principal square root is 11.

69. Why must you check all solutions whenever the squaring property of equality is used?

Explore to Learn

The statement $\sqrt{a} + \sqrt{b} = \sqrt{a + b}$ was considered in earlier sections. Square each side of $\sqrt{a} + \sqrt{b} = \sqrt{a + b}$ and draw a conclusion about what must be true about the values of a and b for this statement to be true.

Review Exercises

Directions   Completely factor the following expressions. See sections 4–2 and 4–4.

1. $x^2 - 4$

2. $x^2 + 9x + 18$

3. $x^2 - 3x - 10$

4. $x^2 - 6x + 9$

Directions   Simplify. See section 8–1.

5. $\sqrt{81}$

6. $\sqrt{49}$

7. $\sqrt{121}$

8. Find the solution set for the equation $x^2 = 64$. See section 4–7.

Chapter 8
Lead-in Problem

Roger has a ladder that will extend to a length of 21 feet. For the ladder to be safe to climb on, it must be placed 7 feet away from the house. The roof is 20 feet above the ground. Will the ladder be able to reach the roof safely? If not, how far up will the ladder reach? (Leave your answer rounded to one decimal place.)

Solution

21 ft a ft

7 ft

$$a^2 + b^2 = c^2 \qquad \text{Pythagorean theorem}$$
$$a^2 + (7)^2 = (21)^2 \qquad \text{Substitute 21 for } c \text{ and 7 for } b$$
$$a^2 + 49 = 441 \qquad \text{Simplify}$$
$$a^2 = 392 \qquad \text{Isolate } a^2$$
$$a = \sqrt{392} \approx 19.8 \qquad \text{Round } \sqrt{392} \text{ to one decimal place}$$

The ladder will not be able to reach the roof safely. The ladder's maximum safe reach is approximately 19.8 feet.

Chapter 8 Summary

• Glossary

extraneous solution (page 520) a solution that does not satisfy the original equation.

like radicals (page 505) radicals having the same index and the same radicand.

perfect-square integer (page 484) an integer whose square root is an integer.

radical equation (page 520) an equation in which the unknown quantity appears under a radical symbol.

• New Symbols and Notation

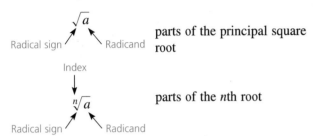

parts of the principal square root

parts of the nth root

$a^{1/2} = \sqrt{a}$ fractional exponent notation for the principal square root

$a^{1/n} = \sqrt[n]{a}$ fractional exponent notation for nth root

$a^{m/n} = (\sqrt[n]{a})^m = \sqrt[n]{a^m}$ fractional exponent notation

• Properties and Definitions

Definition of a square root (page 484)
For every pair of real numbers a and b, if $a^2 = b$, then a is called a square root of b.

Pythagorean theorem (page 486)

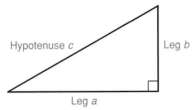

In a right triangle, the square of the length of the hypotenuse (the side opposite the right angle) is equal to the sum of the squares of the lengths of the two legs (the sides that form the right angle). If c is the length of the hypotenuse and a and b are the lengths of the legs, this property can be stated as

$$c^2 = a^2 + b^2$$

Definition of an nth root (page 487)
The nth root of number a, denoted by

$$\sqrt[n]{a}$$

is one of n equal factors such that $\sqrt[n]{a} = b$ and

$$\overbrace{b \cdot b \cdot b \cdots b}^{n \text{ factors}} = b^n = a$$

where n *is a natural number greater than 1.*

Product property for square roots (page 491)
For all nonnegative real numbers a and b,

$$\sqrt{a} \cdot \sqrt{b} = \sqrt{ab}$$

$\sqrt{a^2}$ property (page 492)
If a is any nonnegative real number, then

$$\sqrt{a^2} = a$$

Product property for radicals (page 493)

$$\sqrt[n]{a}\,\sqrt[n]{b} = \sqrt[n]{ab}$$

$\sqrt[n]{a^n}$ property (page 493)
If a is any nonnegative real number, then

$$\sqrt[n]{a^n} = a$$

Quotient property for square roots (page 498)
For any nonnegative real numbers a and b, where $b \neq 0$,

$$\sqrt{\frac{a}{b}} = \frac{\sqrt{a}}{\sqrt{b}}$$

Conditions necessary for a radical expression to be in simplest form, also called standard form (page 499)

1. The radicand contains no factors that can be written with an exponent greater than or equal to the index. ($\sqrt{a^3}$ violates this.)

2. The radicand contains no fractions. $\left(\sqrt{\dfrac{a}{b}} \text{ violates this.}\right)$

3. No radicals appear in the denominator. $\left(\dfrac{1}{\sqrt{a}} \text{ violates this.}\right)$

Quotient property for radicals (page 500)

$$\sqrt[n]{\frac{a}{b}} = \frac{\sqrt[n]{a}}{\sqrt[n]{b}} \qquad (b \neq 0)$$

Definition of $a^{1/n}$ (page 515)

$$a^{1/n} = \sqrt[n]{a}$$

where n is a natural number greater than 1. Whenever n is even, a represents only nonnegative real numbers.

Definition of $a^{m/n}$ (page 515)

If a is any real number, m is any integer, and n is any positive integer, then if $\dfrac{m}{n}$ is reduced to lowest terms,

$$a^{m/n} = (\sqrt[n]{a})^m = \sqrt[n]{a^m}$$

provided that $\sqrt[n]{a}$ is a real number.

Squaring property of equality (page 520)

If P and Q are algebraic expressions and if

$$P = Q$$

then all solutions of $P = Q$ are also solutions of the equation

$$P^2 = Q^2$$

• Procedures

[8–2]

Simplifying the principal square root (page 492)

1. If the number under the $\sqrt{}$ symbol (radicand) is a perfect square, write the corresponding square root.

2. If the number under the $\sqrt{}$ symbol has any perfect-square factors, write the corresponding square root as a coefficient of the radical. ($\sqrt{12} = \sqrt{4 \cdot 3} = 2\sqrt{3}$)

3. The square root is in simplest form when the radicand has no perfect-square integer or variable factors other than 1.

[8–3]

To rationalize a square root denominator (page 499)

1. Multiply the numerator and the denominator by the square root that is in the denominator. The radicand in the denominator will be a perfect-square integer.

2. Simplify the radical expressions in the numerator and the denominator.

3. Reduce the resulting fraction if possible.

To rationalize an nth root denominator (page 501)

1. Multiply the numerator and the denominator by a radical with the same index as the radical that we wish to eliminate from the denominator.

2. The exponent of each factor under the radical must be such that when we add it to the original exponent of the factor under the radical in the denominator, the sum will be equal to or divisible by the index of the radical.

3. Carry out the multiplication and reduce the fraction if possible.

[8–4]

To combine like radicals (page 505)

1. Perform any simplification within the terms.

2. Use the distributive property to combine terms that have like radicals.

[8–7]

Solving a radical equation (page 521)

1. Rewrite the equation (if necessary) so that a radical is by itself in one side of the equation.

2. Square each side of the equation and combine like terms.

3. Repeat steps 1 and 2 if a radical remains in the equation.

4. Solve the resulting equation.

5. Check all possible solutions in the original equation.

☀ Chapter 8 Error Analysis

1. Principal square root
   Example: $\sqrt{36} = -6$
   Correct answer: 6
   What error was made? (*see page 484*)

2. Rationalizing the denominator of an nth root
   Example: $\dfrac{x}{\sqrt[3]{x}} = \dfrac{x}{\sqrt[3]{x}} \cdot \dfrac{\sqrt[3]{x}}{\sqrt[3]{x}} = \dfrac{x\sqrt[3]{x}}{x} = \sqrt[3]{x}$
   Correct answer: $\sqrt[3]{x^2}$
   What error was made? (*see page 501*)

3. Addition in the radicand of a radical expression
Example: $\sqrt{9+4} = \sqrt{9} + \sqrt{4} = 3 + 2 = 5$
Correct answer: $\sqrt{13}$
What error was made? (*see page 492*)

4. Subtraction of radical expressions
Example: $\sqrt{3} - \sqrt{2} = \sqrt{3-2} = \sqrt{1} = 1$
Correct answer: $\sqrt{3} - \sqrt{2}$
What error was made? (*see page 506*)

5. Sums and differences of radical numbers
Example: $\sqrt{8} + 3\sqrt{2} - \sqrt{18} = 2\sqrt{2} + 3\sqrt{2} - \sqrt{18}$
$= 5\sqrt{2} - \sqrt{18}$
Correct answer: $2\sqrt{2}$
What error was made? (*see page 506*)

6. Multiplication of radical expressions
Example: $\sqrt{3}(\sqrt{3} + \sqrt{2}) = \sqrt{3} \cdot \sqrt{3} + \sqrt{2}$
$= 3 + \sqrt{2}$
Correct answer: $3 + \sqrt{6}$
What error was made? (*see page 509*)

7. Squaring a radical binomial
Example: $(\sqrt{2} - \sqrt{3})^2 = (\sqrt{2})^2 - (\sqrt{3})^2$
$= 2 - 3 = -1$
Correct answer: $5 - 2\sqrt{6}$
What error was made? (*see page 510*)

8. Fractional exponents to radical expressions
Example: $x^{3/2} = \sqrt[3]{x^2}$
Correct answer: $x\sqrt{x}$
What error was made? (*see page 516*)

9. Extraneous solutions of radical equations
Example: Find the solution set of $\sqrt{x+2} = x - 4$
$(\sqrt{x+2})^2 = (x-4)^2$
$x + 2 = x^2 - 8x + 16$
$x^2 - 9x + 14 = 0$
$(x - 7)(x - 2) = 0$
$x - 7 = 0$ or $x - 2 = 0$
$x = 7$ or $\quad x = 2 \quad$ {2,7}
Correct answer: {7}
What error was made? (*see page 521*)

10. Negative exponents
Example: $6^{-2} = -36$
Correct answer: $\dfrac{1}{36}$
What error was made? (*see page 202*)

Chapter 8 Critical Thinking

1. Given the numbers 33 and 27, determine a method by which you can multiply these numbers mentally.

2. There are two 5-gallon containers. One has 4 gallons of red fluid and the other 4 gallons of blue. One gallon of the blue fluid is poured into the container of red fluid, mixed well, and then 1 gallon of the now mixed 4-red to 1-blue fluid is poured back into the container of blue fluid. Is there more red fluid in the blue fluid container or more blue fluid in the red fluid container?

Chapter 8 Review

Directions Assume that all variables in problems 1–47 represent positive real numbers and that no denominator is equal to zero.

[8–1]

Directions Find the indicated value. Round to three decimal places if necessary.

1. $\sqrt{81}$
2. $\sqrt{12}$
3. $-\sqrt{28}$
4. $3\sqrt{15}$

Directions In the following right triangles find the length of the missing side. Round the answer to the nearest tenth if necessary. Assume that c represents the hypotenuse.

5. $a = 4$ in., $b = 5$ in.
6. $a = 9$ cm, $c = 15$ cm
7. $b = 10$ m, $c = 20$ m

[8–2]

Directions   Perform any indicated operations and simplify.

8. $\sqrt{40}$

9. $\sqrt{18a^2b^3}$

10. $\sqrt{2}\sqrt{14}$

11. $\sqrt{18}\sqrt{10}$

12. The length of the diagonal is the dimension used when describing the size of a square television picture tube. If the side of the television tube is 15 inches, what is the size of the picture tube rounded to the nearest inch?

[8–3]

Directions   Express the given radicals in simplest form with all denominators rationalized.

13. $\sqrt{\dfrac{16}{17}}$

14. $\sqrt{\dfrac{7}{18}}$

15. $\sqrt{\dfrac{a}{b}}$

16. $\sqrt{\dfrac{x}{y^3}}$

17. $\dfrac{a}{\sqrt{ab}}$

18. $\dfrac{2x}{\sqrt{xy}}$

19. An 18-foot ladder is placed against the wall of a house. If the bottom of the ladder is 8 feet from the house, how far from the ground is the top of the ladder? Round the answer to the nearest tenth of a foot.

[8–4]

Directions   Perform the indicated operations and simplify.

20. $3\sqrt{7} + 4\sqrt{7}$

21. $\sqrt{18} + 5\sqrt{2}$

22. $3\sqrt{20} - \sqrt{45}$

23. $2\sqrt{75} - \sqrt{3} + 5\sqrt{27}$

24. $\sqrt{50a} - 2\sqrt{8a}$

25. $7\sqrt{9x} - 2\sqrt{4x}$

[8–5]

Directions   Perform the indicated operations and simplify.

26. $\sqrt{3}(\sqrt{5} - \sqrt{7})$

27. $\sqrt{10}(\sqrt{14} + \sqrt{6})$

28. $(5 + \sqrt{7})(3 - \sqrt{7})$

29. $(6 - \sqrt{3})^2$

30. $(\sqrt{3} + \sqrt{5})^2$

31. $(2\sqrt{a} - \sqrt{b})(2\sqrt{a} + \sqrt{b})$

Directions   Express the given radicals in simplest form with all denominators rationalized.

32. $\dfrac{1}{\sqrt{3} - 2}$

33. $\dfrac{2}{\sqrt{6} + 4}$

34. $\dfrac{1}{\sqrt{a} + b}$

35. $\dfrac{x}{\sqrt{xy} + x}$

36. $\dfrac{a}{a + \sqrt{b}}$

37. $\dfrac{\sqrt{2} + 3}{\sqrt{2} - 3}$

[8–6]

Directions   Simplify the given expressions.

38. $(36)^{1/2}$

39. $(8)^{2/3}$

40. $(-8)^{1/3}$

41. $(32)^{-2/5}$

Directions Perform the indicated operations and simplify. For exercises 48–51 use a calculator and round to three decimal places.

42. $a^{2/5} \cdot a^{3/5}$

43. $b^{1/3} \cdot b^{3/4}$

44. $\dfrac{a^{1/2}}{a^{1/4}}$

45. $(a^{3/2})^{3/4}$

46. $(16a^4b^8)^{3/4}$

47. $\dfrac{a^2b^2}{a^{1/2}b^{3/2}}$

48. $\sqrt[3]{45}$

49. $\sqrt[4]{68}$

50. $5\sqrt[5]{25}$

51. $\dfrac{1}{2}\sqrt[3]{100}$

52. At a height of h feet above sea level on level ground, the distance d in miles that a person can see an object is found by using the equation

$$d = 1.2h^{1/2}$$

How far can someone see who is at the top of the Washington Monument 555 feet above the ground? Round the answer to the nearest tenth of a mile.

[8–7]

Directions Find the solution set.

53. $\sqrt{x} = 8$

54. $\sqrt{x-4} = 7$

55. $\sqrt{5x-3} = \sqrt{3x+5}$

56. $2x = \sqrt{4x^2 - x + 4}$

57. $x = \sqrt{x^2 - x - 5}$

58. $\sqrt{x^2 + 16} = x + 2$

59. $\sqrt{x+6} = x + 4$

60. $\sqrt{x-3} = x - 3$

61. The formula for approximating the velocity V in miles per hour of a car based on the length of its skid marks S (in feet) on dry pavement is given by

$$V = 2\sqrt{6S}$$

If the velocity is 50 mph, how long will the skid marks be?

62. On wet pavement, the formula in exercise 61 is given by

$$V = 2\sqrt{3S}$$

How long will the skid marks be if the car is traveling at 50 mph on wet pavement?

Chapter 8 Test

Directions Find the indicated root. Round to three decimal places where necessary.

1. $2\sqrt{80}$

2. $-\sqrt{100}$

3. $4\sqrt[3]{27}$

Directions Simplify the following expressions, leaving all denominators rationalized. Assume that all variables represent positive real numbers and that no denominator is equal to zero.

4. $\sqrt{60}$

5. $\sqrt{12} + 3\sqrt{3}$

6. $\sqrt{10}(\sqrt{5} - \sqrt{2})$

7. $\sqrt{14}\sqrt{7}$

8. $\sqrt[3]{27a^9b^8c^7}$

9. $\dfrac{6}{\sqrt{3}}$

10. $\sqrt{32x} + \sqrt{8x}$

11. $\sqrt[3]{\dfrac{a}{b^2}}$

12. $(a^{3/4})^{2/3}$

13. $\sqrt[5]{a^3}\,\sqrt[5]{a^4}$

14. $\dfrac{12}{\sqrt{10}}$

15. $\dfrac{x^{1/2}}{x^{1/3}}$

16. $(3 + \sqrt{x})^2$

17. $\sqrt[3]{54} - \sqrt[3]{16}$

18. $\dfrac{xy^2}{\sqrt[5]{y^2}}$

19. $\dfrac{4}{4 - \sqrt{2}}$

20. $(\sqrt{x} - \sqrt{2y})(\sqrt{x} + \sqrt{2y})$

21. $\dfrac{ab^{1/2}}{a^{1/2}b^{1/4}}$

Directions   Find the solution set.

22. $\sqrt{2x + 7} = 4$

23. $\sqrt{x + 2} = x$

24. $x = \sqrt{7x - 12}$

25. $x = \sqrt{x^2 + 9}$

Directions   In the following right triangles, if c represents the hypotenuse, find the length of the unknown side. Round the answer to the nearest tenth when necessary.

26. $a = 6$ ft, $b = 8$ ft

27. $a = 5$ cm, $c = 10$ cm

28. Under ideal conditions, the velocity v in meters per second of an object freely falling from a height h is given by $v = 4.4\sqrt{h}$. If the velocity is 66 meters per second, how far has the object fallen?

29. You have a ladder that will extend to a length of 24 feet. For the ladder to be safe to climb on, it must be placed 8 feet away from the house. How far up will the ladder reach? Round the answer to one decimal place.

30. A baseball diamond is a square that is 90 feet long on each side. How far is it from home plate to second base? Round the answer to the nearest foot.

Chapter 8 Cumulative Test

Directions   Perform the indicated operations and simplify. Assume that all variables represent positive real numbers and that no denominator is equal to zero.

1. $3[4(6 - 2) + (-5 + 4)]$

2. $\dfrac{8a^{-2}b^3c^0}{4a^{-5}b}$

3. $5\sqrt{12} + 2\sqrt{27}$

4. $\sqrt[3]{-8}$

5. $\dfrac{2x}{x - 1} - \dfrac{x - 3}{x - 1}$

6. $\dfrac{\sqrt{x}}{x - \sqrt{y}}$

7. $\sqrt[3]{81x^4y^6z}$

Directions   Factor completely.

8. $25c^2 - d^2$ **9.** $2x^2 + 7x - 4$ **10.** $6x^2 + 11x + 4$

Directions   Find the solution set for problems 11–16 and solve problems 17 and 18.

11. $2(4x - 3) = 5x - 7$

12. $x^2 = 9$

13. $3(x + 4) - 2(x - 3) = 12$

14. $\dfrac{1}{4}x + 2 = \dfrac{1}{2}x - 1$

15. $\dfrac{3x + 1}{9} + \dfrac{1}{12} = \dfrac{2x - 1}{3}$

16. $2x^2 + 3x + 1 = 0$

17. $-3 < 2x - 5 < 11$

18. $3x + 5 > x + 12$

19. Find the slope of the line passing through points (4,2) and (3,5).

20. Write $8x - 2y = 4$ in slope-intercept form and determine the slope and y-intercept.

21. Find the solution set for the system of linear equations.
$$5x - y = 4$$
$$x + 3y = 2$$

22. One number is four times a second number and their sum is 70. Find the numbers.

23. A punch machine can make 18 holes in 4 minutes. How many holes can the machine make in 5 hours?

24. The product of two consecutive positive even integers is 288. Find the integers.

25. The width of a rectangle is 6 feet less than its length. The perimeter of the rectangle is 96 feet. Find the dimensions.

A rock is dropped from the top of the Washington Monument. If the monument is 555 feet tall, how long will it take the rock to strike the ground?

9–1 Solving Quadratic Equations by Extracting the Roots

In section 4–7, we solved quadratic equations of the form

$$ax^2 + bx + c = 0, \ a \neq 0$$

by factoring. It was necessary that the quadratic expression $ax^2 + bx + c$ be factorable to use the method discussed. Let us review the procedures used in solving quadratic equations by factoring.

Solving a Quadratic Equation by Factoring

1. Write the quadratic equation in standard form with the leading coefficient positive.
2. Completely factor the quadratic expression.
3. Set each of the factors containing the variable equal to 0 and solve the linear equations.

Note For ease in factoring, we write the quadratic expression with the leading coefficient positive, $a > 0$.

● **Example 9–1 A**

Solving a quadratic equation by factoring

Find the solution set of each quadratic equation by factoring.

1. $x^2 - x - 12 = 0$

$(x - 4)(x + 3) = 0$ Factor the left side

$x - 4 = 0$ or $x + 3 = 0$ Set each factor equal to 0

$x = 4$ $x = -3$ Solve each equation for x

The solution set is $\{-3, 4\}$.

2. $3y^2 = 7y + 6$

$3y^2 - 7y - 6 = 0$ Write the equation in standard form

$(3y + 2)(y - 3) = 0$ Factor the left side

$3y + 2 = 0$ or $y - 3 = 0$ Set each factor equal to 0

$3y = -2$ $y = 3$ Solve each equation for y

$y = -\dfrac{2}{3}$ $y = 3$

The solution set is $\left\{-\dfrac{2}{3}, 3\right\}$.

Extracting the Roots

Given the quadratic equation $x^2 - 9 = 0$, factoring the left side and solving the resulting equations, we get

$$(x - 3)(x + 3) = 0$$
$$x - 3 = 0 \quad \text{or} \quad x + 3 = 0$$
$$x = 3 \quad \text{or} \quad x = -3$$

The solutions of the equation are 3 or -3.

We can obtain the same result if we write the equation in the form

$$x^2 = 9$$

Since 9 is positive, we can take the square root of each side of the equation. Then

$$x = \sqrt{9} = 3 \quad \text{or} \quad x = -\sqrt{9} = -3$$

and we obtain the same result. This development justifies the following method of solving a quadratic equation by **extracting the roots** using the **square root property**.

Square Root Property

If k is a nonnegative number and $x^2 = k$, then
$$x = \sqrt{k} \quad \text{or} \quad x = -\sqrt{k}$$

● **Example 9–1 B**

Solving a quadratic equation by extracting the roots

Find the solution set of the following quadratic equations by extracting the roots.

1. $x^2 = 25$

$x = \sqrt{25}$ or $x = -\sqrt{25}$ Extract the roots

$x = 5$ $x = -5$ $\sqrt{25} = 5$

The solution set is $\{-5, 5\}$.

2. $x^2 - 12 = 0$

$$x^2 = 12$$

			Add 12 to each side
$x = \sqrt{12}$	or	$x = -\sqrt{12}$	Extract the roots
$x = 2\sqrt{3}$		$x = -2\sqrt{3}$	$\sqrt{12} = 2\sqrt{3}$

The solution set is $\{-2\sqrt{3}, 2\sqrt{3}\}$.

3. $z^2 = -9$

Since -9 is negative and the property requires that k be a nonnegative number, we are not able to solve this equation in the set of real numbers. The equation has no real number solution so the solution set is the empty set, $\emptyset$.

4. $2x^2 = 98$

To extract the roots, the squared term must have coefficient 1.

$$2x^2 = 98$$
$$x^2 = 49$$

			Divide each term by 2
$x = \sqrt{49}$	or	$x = -\sqrt{49}$	Extract the roots
$x = 7$		$x = -7$	$\sqrt{49} = 7$

The solution set is $\{-7, 7\}$.

> **Quick check** Find the solution set of the equation $3x^2 = 24$ by extracting the roots.

Any equation that is written in the form

$$(x + q)^2 = k \text{ or } (px + q)^2 = k$$

can be solved by extracting the roots. Consider the following examples.

● **Example 9–1 C**

Solving a quadratic equation by extracting the roots

Find the solution set of the following quadratic equations by extracting the roots.

1. $(x - 2)^2 = 4$

			Extract the roots
$x - 2 = \sqrt{4}$	or	$x - 2 = -\sqrt{4}$	$\sqrt{4} = 2$
$x - 2 = 2$		$x - 2 = -2$	
$x = 2 + 2 = 4$		$x = 2 - 2 = 0$	Add 2 to each side

The solution set is $\{0, 4\}$.

2. $(y - 1)^2 = 24$

			Extract the roots
$y - 1 = \sqrt{24}$	or	$y - 1 = -\sqrt{24}$	$\sqrt{24} = 2\sqrt{6}$
$y - 1 = 2\sqrt{6}$		$y - 1 = -2\sqrt{6}$	
$y = 1 + 2\sqrt{6}$		$y = 1 - 2\sqrt{6}$	Add 1 to each side

The solution set is $\{1 - 2\sqrt{6}, 1 + 2\sqrt{6}\}$.

> **Quick check** Find the solution set of the equation $(x + 4)^2 = 9$ by extracting roots.

MASTERY POINTS

Can you
- Solve a quadratic equation by factoring?
- Solve quadratic equations of the form $x^2 = k$ and $(px + q)^2 = k$ by extracting the roots?

Exercise 9–1

Directions
Find the solution set of each quadratic equation by extracting the roots or by factoring. Express radicals in simplest form. See examples 9–1 A and B.

☞ *Quick Check*

Example $3x^2 = 24$

Solution $x^2 = 8$ Divide each side by 3

$x = \sqrt{8}$ or $x = -\sqrt{8}$ Extract the roots

$x = 2\sqrt{2}$ $x = -2\sqrt{2}$ $\sqrt{8} = \sqrt{4 \cdot 2} = 2\sqrt{2}$

The solution set is $\left\{-2\sqrt{2}, 2\sqrt{2}\right\}$.

Directions Find the solution set of each quadratic equation by extracting the roots or by factoring. Express radicals in simplest form. See examples 9–1 A and B.

1. $x^2 + 2x - 15 = 0$

2. $x^2 - 4x + 3 = 0$

3. $2y^2 - y - 6 = 0$

4. $5z^2 - 16z + 3 = 0$

5. $x^2 = 4$

6. $x^2 = 49$

7. $x^2 = 64$

8. $x^2 = 81$

9. $y^2 = 11$

10. $x^2 = 5$

11. $a^2 = 20$

12. $x^2 = 28$

13. $x^2 - 3 = 0$

14. $x^2 - 13 = 0$

15. $p^2 - 32 = 0$

16. $x^2 + 36 = 0$

17. $x^2 + 16 = 0$

18. $5x^2 = 75$

19. $3z^2 = 18$

20. $5x^2 = 15$

21. $2x^2 - 100 = 0$

22. $4x^2 - 64 = 0$

23. $7x^2 - 56 = 0$

24. $9a^2 - 162 = 0$

25. $\dfrac{3}{4}x^2 - 6 = 0$

26. $\dfrac{1}{5}x^2 - \dfrac{3}{5} = 0$

27. $\dfrac{1}{3}x^2 = \dfrac{2}{3}$

28. $\dfrac{2}{3}x^2 = 8$

29. $\dfrac{5}{2}x^2 - \dfrac{3}{5} = 0$

30. $\dfrac{4x^2}{3} - 3 = 0$

31. $\dfrac{1}{2}y^2 - \dfrac{3}{2} = 4$

32. $\dfrac{z^2}{4} - 6 = \dfrac{3}{4}$

Directions	$\blacktriangleright$ **Quick Check**
Find the solution set of each quadratic equation by extracting the roots. Express radicals in simplest form. See example 9–1 C.	**Example** $(x + 4)^2 = 9$

Solution

$$x + 4 = 3 \qquad \text{or} \qquad x + 4 = -3 \qquad \text{Extract the roots}$$
$$x = -4 + 3 \qquad\qquad x = -4 - 3 \qquad \text{Add } -4 \text{ to each side}$$
$$x = -1 \qquad\qquad\qquad x = -7$$

The solution set is $\{-7, -1\}$.

Directions   Find the solution set of each quadratic equation by extracting the roots. Express radicals in simplest form. See example 9–1 C.

33. $(x + 2)^2 = 4$

34. $(x + 6)^2 = 16$

35. $(x - 4)^2 = 25$

36. $(x - 3)^2 = 49$

37. $(x + 3)^2 = 6$

38. $(x - 1)^2 = 7$

39. $(x - 9)^2 = 18$

40. $(x + 8)^2 = 8$

41. $(x + 5)^2 = 32$

42. $(x - 10)^2 = 27$

Directions   Use a calculator to find the solution set for each of the following equations.

43. $x^2 = 5.76$

44. $y^2 = 10.89$

45. $x^2 - 3.61 = 0$

46. $z^2 - 4.84 = 0$

47. $(x - 1.3)^2 = 8.41$

48. $(y - 0.8)^2 = 7.29$

49. $(z + 0.2)^2 = 4.41$

50. $(x + 0.6)^2 = 2.89$

Directions	**Example** A square has an area of 16 square inches. Find the length of each side.
Solve by setting up a quadratic equation and extracting the roots.	**Solution** Use the formula $A = s^2$, where A is the area and s is the length of a side. Then $16 = s^2$ or $s^2 = 16$, and

$$s = \sqrt{16} \quad \text{or} \quad s = -\sqrt{16} \qquad \text{Extract the roots}$$
$$s = 4 \qquad\qquad\quad s = -4 \qquad\qquad \sqrt{16} = 4$$

Since a square cannot have a side that is -4 inches long, then $s = 4$ inches. The length of each side of the square is 4 inches.

Directions   Solve by setting up a quadratic equation and extracting the roots.

51. Find the length of each side of a square whose area is 25 square meters.

52. Given a square whose area is 45 square centimeters, how long is each side of the square?

53. A circle has an area of approximately 12.56 square feet. Find the approximate length of the radius r of the circle if $A \approx 3.14 r^2$.

54. Find the approximate length of the radius of a circle whose area is approximately 50.24 square yards. (Refer to exercise 53 for the formula.)

55. The square of a number less 81 is equal to zero. Find the number.

56. Four times the square of a number is 100. Find the number.

57. The square of a number is equal to nine times the number. Find the number.

58. If you subtract eight times a number from two times the square of the number, you get zero. Find the number.

59. The sum of the areas of two squares is 245 square inches. If the length of the side of the larger square is twice the length of the side of the smaller square, find the lengths of the sides of the two squares.

60. The length of the side of one square is three times the length of the side of a second square. If the difference in their areas is 128 square centimeters, find the lengths of the sides of the two squares.

61. The width of a rectangle is one-fourth of the length. If the area is 144 square meters, find the length of the rectangle. (*Hint: A = ℓw.*)

62. The length of a rectangle is three times the width. If the area of the rectangle is 147 square feet, find the dimensions of the rectangle.

63. The sum of the areas of two circles is 80π. Find the length of the radius of each circle if one radius is twice as long as the other.

Directions   Solve by using the relationship that exists for a square: The sum of the squares of two sides is equal to the square of a diagonal of the square.

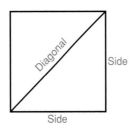

$$[(\text{side})^2 + (\text{side})^2 = (\text{diagonal})^2]$$

64. Find the length of the side of a square whose diagonal is 16 inches long.

65. Find the length of the side of a square whose diagonal is 10 centimeters long.

66. Find the length of the side of a square whose diagonal is 24 feet long.

67. Explain why $x^2 + 25 = 0$ has no real solution.

⊛ Explore to Learn

a. Example 9–1 A number 2 states that the solution set to the equation $3y^2 = 7y + 6$ is $\left\{-\dfrac{2}{3}, 3\right\}$. Write a paragraph that explains what this means. Show any calculations that might help clarify your statement.

b. Considering part (a), what is the solution set to the equation $3A^2 = 7A + 6$?

c. Considering parts (a) and (b), what is the solution set to the equation $300x^2 = 700x + 600$?

Review Exercises

Directions   Multiply the following. See section 3–2.

1. $(x - 2)^2$

2. $(3z + 2)^2$

Directions   Completely factor the following. See sections 4–2 and 4–3.

3. $x^2 + 18x + 81$

4. $9y^2 + 30y + 25$

Directions   Perform the indicated operations. See sections 5–2 and 5–4.

5. $\dfrac{3x}{x + 2} - \dfrac{x}{x^2 - 4}$

6. $\dfrac{x - 3}{x^2 - x - 2} \div \dfrac{x^2 - 9}{x + 1}$

7. One alloy of brass is 80% zinc and 20% copper. Another alloy of brass is 50% zinc and 50% copper. How many grams of each alloy should be used to make 400 grams of brass that is 70% zinc and 30% copper? See section 2–7.

8. A freight train passes through Junction City. If you triple the square of the number of cars and take away two, there are 430 cars. How many cars are in the freight train? See section 4–7.

9–2   Solving Quadratic Equations by Completing the Square

Building Perfect Square Trinomials

The methods we have used to solve quadratic equations thus far have applied to special cases of the quadratic equation. The method that we call **completing the square** involves transforming the quadratic equation

$$ax^2 + bx + c = 0$$

into the form

$$(x + q)^2 = k, \ k \geq 0$$

where q and k are constants. This latter equation can then be solved by extracting the roots, as we did in section 9–1.

Consider the following perfect square trinomials and their equivalent binomial squares.

$$x^2 + 10x + 25 = (x + 5)(x + 5) = (x + 5)^2$$
$$x^2 - 14x + 49 = (x - 7)(x - 7) = (x - 7)^2$$

In each of the perfect square trinomials on the left side,

    a. the coefficient of x^2 is 1.

    b. the third term, the constant, is the square of one-half of the coefficient of the variable x in the middle term.

We further observe that in the square of the binomial on the right side, the constant term in the binomial is one-half of the coefficient of the variable x in the middle term. That is,

    1. In the trinomial $x^2 + 10x + 25$, the constant term, 25, is the square of one-half of 10. Thus,

$$\frac{1}{2}(10) = 5 \text{ then}$$

$$\left[\frac{1}{2}(10)\right]^2 = (5)^2 = 25$$

                                                         └─── Constant term of the binomial

    2. In the trinomial $x^2 - 14x + 49$, the constant term, 49, is the square of one-half of -14. Thus,

$$\frac{1}{2}(-14) = -7 \text{ then}$$

$$\left[\frac{1}{2}(-14)\right]^2 = (-7)^2 = 49$$

                                                         └─── Constant term of the binomial

Now we can use these observations to "build" perfect square trinomials by completing the square and obtain the equivalent square of a binomial.

● **Example 9–2 A**

Forming a perfect square trinomial

Complete the square in each of the following expressions. Write the resulting expression as the square of a binomial.

1. $x^2 + 6x$

Since the coefficient of x is 6, the constant term is the square of one-half of 6.

$$\left[\frac{1}{2}(6)\right]^2 = (3)^2 = 9$$

The trinomial becomes

$$x^2 + 6x + 9$$

which factors into

$$(x + 3)(x + 3) = (x + 3)^2$$

2. $y^2 - 3y$

Since the coefficient of y is -3, the constant term is the square of one-half of -3.

$$\left[\frac{1}{2}(-3)\right]^2 = \left(-\frac{3}{2}\right)^2 = \frac{9}{4}$$

The trinomial becomes

$$y^2 - 3y + \frac{9}{4}$$

which factors into

$$\left(y - \frac{3}{2}\right)\left(y - \frac{3}{2}\right) = \left(y - \frac{3}{2}\right)^2$$

☞ *Quick check*   Complete the square in the expression $x^2 - 8x$. Write as the square of a binomial.   ●

Solutions by Completing the Square

> ### Solving a Quadratic Equation by Completing the Square
>
> To find the solution set of the quadratic equation $ax^2 + bx + c = 0$, $a \neq 0$, by completing the square, we proceed as follows:
> 1. If $a = 1$, proceed to step 2. If $a \neq 1$, divide each term of the equation by a and simplify.
> 2. Write the equation with the variable terms on the left side and the constant term on the right side.
> 3. Add to each side of the equation the square of one half of the numerical coefficient of the middle term of the original equation.
> 4. Write the left side as a trinomial and factor it as the square of a binomial. Combine like terms on the right side.
> 5. Extract the roots and solve the resulting linear equations.

The following examples show how we use the procedure to find the solution set of a quadratic equation by completing the square.

● **Example 9–2 B**

Solving a quadratic equation by completing the square

Find the solution set by completing the square.

1. $x^2 - 2x - 8 = 0$

We first isolate the terms containing the variable on the left side.

$$x^2 - 2x = 8 \qquad \text{\small Add 8 to each side}$$

Complete the square on the left side.

$$\left[\frac{1}{2}(-2)\right]^2 = (-1)^2 = 1 \qquad \text{\small Square one half of the coefficient of } x$$

Once we have found the value necessary to form a perfect square trinomial, we add it to both sides.

$$x^2 - 2x + 1 = 8 + 1 \qquad \text{\small Add 1 to each side}$$
$$(x - 1)^2 = 9 \qquad \text{\small Factor the left side and combine like terms on the right side}$$

The trinomial factors into the square of a binomial, and we solve by extracting the roots as we did in section 9–1.

$$
\begin{array}{llll}
x - 1 = \sqrt{9} & \text{or} & x - 1 = -\sqrt{9} & \text{\small Extract the roots} \\
x - 1 = 3 & & x - 1 = -3 & \text{\small } \sqrt{9} = 3 \\
x = 1 + 3 & & x = 1 - 3 & \text{\small Add 1 to each side} \\
x = 4 & & x = -2 & \text{\small Combine on right side}
\end{array}
$$

The solution set is $\{-2, 4\}$.

Note The original quadratic equation $x^2 - 2x - 8 = 0$ could have been solved by factoring.

$$x^2 - 2x - 8 = 0 \qquad \text{Original equation}$$
$$(x - 4)(x + 2) = 0 \qquad \text{Factor the quadratic}$$
$$x - 4 = 0 \quad \text{or} \quad x + 2 = 0 \qquad \text{Set each of the factors equal to 0}$$
$$x = 4 \quad \text{or} \qquad x = -2 \qquad \text{Solve for } x$$

2. $x^2 - 6x + 2 = 0$

Isolate the terms containing the variable on the left side.

$$x^2 - 6x = -2 \qquad \text{Add } -2 \text{ to each side}$$

Complete the square on the left side.

$$\left[\frac{1}{2}(-6)\right]^2 = (-3)^2 = 9 \qquad \begin{array}{l}\text{Square one half of the}\\\text{coefficient of } x\end{array}$$

$$x^2 - 6x + 9 = -2 + 9 \qquad \text{Add 9 to each side}$$

$$(x - 3)^2 = 7 \qquad \begin{array}{l}\text{Factor the left side and combine}\\\text{like terms on the right side}\end{array}$$

$$x - 3 = \sqrt{7} \qquad \text{or} \quad x - 3 = -\sqrt{7} \qquad \text{Extract the roots}$$

$$x = 3 + \sqrt{7} \qquad\qquad x = 3 - \sqrt{7} \qquad \text{Add 3 to each side}$$

The solution set is $\left\{3 - \sqrt{7}, 3 + \sqrt{7}\right\}$.

3. $4x^2 + 24x = 8$

To complete the square using the method we have described, it is necessary for the coefficient of x^2 to be 1. To get this, we divide each term of the equation by 4.

$$\frac{4x^2}{4} + \frac{24x}{4} = \frac{8}{4} \qquad \text{Divide each term by 4}$$

$$x^2 + 6x = 2 \qquad \text{Reduce}$$

Complete the square on the left side.

$$\left[\frac{1}{2}(6)\right] = (3)^2 = 9 \qquad \begin{array}{l}\text{Square one half of the}\\\text{coefficient of } x\end{array}$$

$$x^2 + 6x + 9 = 2 + 9 \qquad \text{Add 9 to each side}$$

$$(x + 3)^2 = 11 \qquad \begin{array}{l}\text{Factor the left side and combine}\\\text{like terms on the right side}\end{array}$$

$$x + 3 = \sqrt{11} \qquad \text{or} \quad x + 3 = -\sqrt{11} \qquad \text{Extract the roots}$$

$$x = -3 + \sqrt{11} \qquad\qquad x = -3 - \sqrt{11} \qquad \text{Subtract 3 from each side}$$

The solution set is $\left\{-3 + \sqrt{11}, -3 - \sqrt{11}\right\}$

➤ *Quick check*   Find the solution set of $x^2 + 10x + 5 = 0$ by completing the square.

MASTERY POINTS

Can you
- Complete the square of an expression in the form $x^2 + bx$?
- Solve a quadratic equation by completing the square?

Exercise 9–2	☛ *Quick Check*
Directions	**Example** $x^2 - 8x$
Complete the square of each of the following and factor as the square of a binomial. See example 9–2 A.	**Solution** Since the coefficient of x is -8, the constant term is the square of one-half of -8. $$\left[\frac{1}{2}(-8)\right]^2 = (-4)^2 = 16$$ The trinomial becomes $$x^2 - 8x + 16$$ which factors into $$(x - 4)(x - 4) = (x - 4)^2$$

Directions   Complete the square of each of the following and factor as the square of a binomial. See example 9–2 A.

1. $x^2 + 10x$

2. $x^2 + 4x$

3. $a^2 - 12a$

4. $y^2 - 18y$

5. $x^2 + 24x$

6. $b^2 + 16b$

7. $y^2 - 20y$

8. $x^2 - 22x$

9. $x^2 + x$

10. $x^2 + 5x$

11. $x^2 - 7x$

12. $y^2 - 9y$

13. $x^2 + 3x$

14. $b^2 + 7b$

15. $s^2 - 5s$

16. $x^2 - 3x$

Directions	☛ *Quick Check*
Find the solution set by completing the square. See example 9–2 B.	**Example** $x^2 + 10x + 5 = 0$
	Solution $x^2 + 10x = -5$ *Subtract 5 from both sides*
	Complete the square on the left side
	$$\left[\frac{1}{2}(10)\right]^2 = (5)^2 = 25$$ *Square one half of the coefficient of x*
	$x^2 + 10x + 25 = -5 + 25$ *Add 25 to each side*
	$(x + 5)^2 = 20$ *Factor the left side and combine on the right side*
	$x + 5 = \sqrt{20}$ or $x + 5 = -\sqrt{20}$ *Extract the roots*
	$x + 5 = 2\sqrt{5}$ or $x + 5 = -2\sqrt{5}$ $\sqrt{20} = 2\sqrt{5}$
	$x = -5 + 2\sqrt{5}$ or $x = -5 - 2\sqrt{5}$ *Subtract 5 from each side*
	The solution set is $\{-5 + 2\sqrt{5}, -5 - 2\sqrt{5}\}$

Directions   Find the solution set by completing the square. See example 9–2 B.

17. $x^2 + 8x + 7 = 0$

18. $x^2 + 12x + 11 = 0$

19. $a^2 - 4a - 12 = 0$

20. $y^2 - 10y + 9 = 0$

21. $x^2 - 4x = -3$

22. $x^2 + 14x = -13$

23. $u^2 - u - 1 = 0$

24. $a^2 + 3a - 1 = 0$

25. $x^2 - 5x + 2 = 0$

26. $n^2 - 3n - 2 = 0$

27. $y^2 - 4y = 81$

28. $b^2 - 12b = -25$

29. $h^2 + 21h + 10 = 0$

30. $3x^2 + 6x = 3$

31. $2x^2 + 4x - 8 = 0$

32. $3x^2 - 9x = -3$

33. $2y^2 + 8y + 2 = 0$

34. $3a^2 + 12a - 6 = 0$

35. $4x^2 - 4x = 8$

36. $2b^2 - 8b - 16 = 0$

37. $6a^2 - 12a = -6$

38. $6n^2 + 6n = 18$

39. $y^2 = 3 - y$

40. $x^2 + 1 = -3x$

41. $12 - 6a = 6a^2$

42. $4 - x^2 = 2x$

43. $-n^2 - 4 = -6n$

44. $1 - n^2 = 3n$

45. $2x^2 - 3 = 4x + 5$

46. $3a^2 + 5 = 12a + 8$

47. $(x + 3)(x - 2) = 1$

48. $(x - 5)(x - 3) = 4$

49. $x(x + 6) = -3$

50. $x(x - 2) = 5$

Directions	**Example**	A piece of lumber is divided into two pieces so that one piece is 5 inches longer than the other. If the product of their lengths is 104 square inches, what is the length of each piece?
Solve the following problems by setting up a quadratic equation and completing the square.	**Solution**	Let x represent the length of the shorter piece. Then $x + 5$ is the length of the longer piece.

the product of the lengths   is   104 square inches

$$x(x + 5) = 104$$

The equation is $x(x + 5) = 104$.

$$x^2 + 5x = 104 \qquad \text{Multiply on the left side}$$

$$\left[\frac{1}{2}(5)\right]^2 = \left(\frac{5}{2}\right)^2 = \frac{25}{4} \qquad \text{Square one half of the coefficient of } x$$

$$x^2 + 5x + \frac{25}{4} = 104 + \frac{25}{4} \qquad \text{Add } \frac{25}{4} \text{ to each side}$$

$$\left(x + \frac{5}{2}\right)^2 = \frac{416}{4} + \frac{25}{4} \qquad \text{Factor left side}$$

$$\left(x + \frac{5}{2}\right)^2 = \frac{441}{4} \qquad \text{Combine on the right side}$$

$$x + \frac{5}{2} = \sqrt{\frac{441}{4}} \quad \text{or} \quad x + \frac{5}{2} = -\sqrt{\frac{144}{4}} \qquad \text{Extract the roots}$$

$$x = -\frac{5}{2} + \frac{21}{2} \qquad\qquad x = -\frac{5}{2} - \frac{21}{2} \qquad \sqrt{441} = 21$$

$$x = \frac{-5 + 21}{2} = \frac{16}{2} = 8 \qquad x = \frac{-5 - 21}{2} = \frac{-26}{2} = -13$$

Combine like terms

The solution set of the equation is $\{8, -13\}$. Since we want the length of lumber, -13 is not an appropriate answer. Therefore, $x = 8$ and $x + 5 = 13$. The two pieces have lengths 8 inches and 13 inches.

Directions   Solve the following problems by setting up a quadratic equation and completing the square.

51. A metal bar is to be divided into two pieces so that one piece is 4 inches shorter than the other. If the sum of the squares of the two lengths is 208 square inches, find the two lengths.

52. To find the total surface area of an automobile cylinder, we use the formula $A = 2\pi r^2 + 2\pi rh$, where π is approximately equal to $\frac{22}{7}$. If the area A of the cylinder is approximately 88 square inches and the height h is 7 inches, find the approximate value of radius r. (Round to two decimal places.)

53. One surface of a rectangular solid has a width w that is 8 millimeters shorter than its length ℓ. If the area A of the surface is 105 square millimeters, what are its dimensions?

54. The length of a rectangular-shaped piece of paper is 7 inches longer than its width. What are the dimensions of the paper if it has an area of 78 square inches?

55. The perimeter of a rectangle is 52 inches and its area is 153 square inches. What are its dimensions? The formula for the perimeter of a rectangle is $P = 2\ell + 2w$. If we substitute 52 in place of P, we have $52 = 2\ell + 2w$. Divide each side of the equation by 2. Then $26 = \ell + w$. We can use this fact to establish the unknowns.
Width of rectangle: w
Length of rectangle: $26 - w$
Equation: $153 = w(26 - w)$

56. The perimeter of a rectangle is 38 centimeters and its area A is 88 square centimeters. What are its dimensions? (See exercise 55.)

57. The perimeter of a rectangle is 18 meters and its area is $19\frac{1}{4}$ square meters. What are its dimensions?

58. The area of a rectangular piece of sheet metal is 117 square inches. If the sum of the length ℓ and the width w is 22 inches, what are the dimensions of the metal plate?

59. A rectangular lot has an area of 84 square rods. If the sum of the length ℓ and the width w is 20 rods, what are the dimensions?

60. If two metal bars are the same length and if the length of one is increased by 3 centimeters and the second is decreased by 3 centimeters, the product of these two lengths is 27 square centimeters. Find the original lengths.

61. Two rectangular metal surfaces have the same width. If the width of one is increased by 6 inches and the other is increased by 8 inches, the product of the two widths is 99 square inches. Find the original widths.

62. If P dollars is invested at r percent compounded annually, at the end of two years it will grow to an amount $A = P(1 + r)^2$. At what rate will $200 grow to $224.72 in two years?

63. What is a perfect square trinomial?

64. In a perfect square trinomial, what is the relationship between the constant and the coefficient of the variable in the middle term?

✵ Explore to Learn

In an example in the exercises it was found that the solution set to the equation $x^2 + 10x + 5 = 0$ is $\{-5 + 2\sqrt{5}, -5 - 2\sqrt{5}\}$.

a. Compute the decimal approximations to the elements of this set.

b. Show by computation that each element is a solution to the equation.

Review Exercises

Directions   Evaluate the following expressions for the given values. See sections 1–8 and 8–1.

1. $\sqrt{a + b}$; $a = 2$ and $b = 7$

2. $\sqrt{b^2 - 4ac}$; $a = -1$, $b = 5$, and $c = 5$

3. Find the solution set of the system of linear equations
$$2x - y = 4$$
$$3x + 2y = 6$$
by any method.
See section 7–3.

4. Graph the equation $4x - 3y = -12$.
See section 6–2.

5. If 2 dozen oranges cost $2.48, how many dozen oranges can you buy for $11.16? Set up a proportion. See section 2–8.

9–3   Solving Quadratic Equations by the Quadratic Formula

Identifying a, b, and c in a Quadratic Equation

We have found solution sets of quadratic equations by factoring, extracting the roots, and by completing the square. Even though the solution set of *any* such quadratic equation can be found by completing the square, a general formula, which is called the **quadratic formula**, can be derived that will enable us to find the solution set in an easier fashion.

To use the quadratic formula, the equation must be written in standard form,

$$ax^2 + bx + c = 0, \ a \neq 0$$

and we must be able to identify the coefficients a, b, and c. In identifying a, b, and c, we note that

1. a is the coefficient of x^2
2. b is the coefficient of x
3. c is the constant term

● **Example 9-3 A**

Identifying the values of a, b, and c in a quadratic equation

Write each quadratic equation in standard form and identify the values of a, b, and c.

1. $3x^2 - 2x + 1 = 0$

The equation is in standard form.

$$3x^2 - 2x + 1 = 0$$

Constant term, $c = 1$
Coefficient of x, $b = -2$
Coefficient of x^2, $a = 3$

2. $3x^2 - 4 = x$

The equation must be written in standard form.

$$3x^2 - x - 4 = 0 \qquad \text{Subtract } x \text{ from each side}$$

$$a = 3, \ b = -1, \ c = -4$$

3. $4x(x - 3) = 2x - 1$

The equation must be written in standard form.

$$4x^2 - 12x = 2x - 1 \qquad \text{Multiply on left side}$$
$$4x^2 - 14x + 1 = 0 \qquad \text{Subtract } 2x \text{ from each side and add 1 to each side}$$

$$a = 4, \ b = -14, \ c = 1$$

☞ **Quick check**   Write the quadratic equation $2x^2 = 4 - 5x$ in standard form and identify the values of a, b, and c.   ●

Solving Quadratic Equations Using the Quadratic Formula

To derive the quadratic formula, we solve the equation $ax^2 + bx + c = 0$ by completing the square.

$$ax^2 + bx + c = 0$$

$$x^2 + \frac{b}{a}x + \frac{c}{a} = 0 \qquad \text{Divide each term of the equation by } a$$

$$x^2 + \frac{b}{a}x = -\frac{c}{a} \qquad \text{Subtract } \frac{c}{a} \text{ from each side}$$

$$\left[\frac{1}{2}\left(\frac{b}{a}\right) \right]^2 = \left(\frac{b}{2a}\right)^2 = \frac{b^2}{4a^2} \qquad \text{Square one half of the coefficient of } x$$

$$x^2 + \frac{b}{a}x + \frac{b^2}{4a^2} = -\frac{c}{a} + \frac{b^2}{4a^2}$$ Add $\frac{b^2}{4a^2}$ to each side

$$\left(x + \frac{b}{2a}\right)^2 = \frac{b^2}{4a^2} - \frac{c}{a}$$ Write left side as the square of a binomial and change the order of terms on the right side

$$\left(x + \frac{b}{2a}\right)^2 = \frac{b^2}{4a^2} - \frac{4ac}{4a^2}$$ Subtract fractions on the right side

$$\left(x + \frac{b}{2a}\right)^2 = \frac{b^2 - 4ac}{4a^2}$$

$$x + \frac{b}{2a} = \sqrt{\frac{b^2 - 4ac}{4a^2}} \quad \text{or} \quad x + \frac{b}{2a} = -\sqrt{\frac{b^2 - 4ac}{4a^2}}$$ Extract the roots

$$x + \frac{b}{2a} = \frac{\sqrt{b^2 - 4ac}}{\sqrt{4a^2}} \quad \text{or} \quad x + \frac{b}{2a} = -\frac{\sqrt{b^2 - 4ac}}{\sqrt{4a^2}}$$

$$x + \frac{b}{2a} = \frac{\sqrt{b^2 - 4ac}}{2a} \quad \text{or} \quad x + \frac{b}{2a} = -\frac{\sqrt{b^2 - 4ac}}{2a}$$

$$x = \frac{-b}{2a} + \frac{\sqrt{b^2 - 4ac}}{2a} \quad \text{or} \quad x = \frac{-b}{2a} - \frac{\sqrt{b^2 - 4ac}}{2a}$$ Subtract $\frac{b}{2a}$ from each side

$$x = \frac{-b + \sqrt{b^2 - 4ac}}{2a} \quad \text{or} \quad x = \frac{-b - \sqrt{b^2 - 4ac}}{2a}$$

The symbol $\pm$ means plus or minus. An example of its use is to simplify the statement "$x + y$ or $x - y$" into "$x \pm y$." Using this symbol, and noting that a cannot be zero because then the original equation is not a quadratic, these results can be combined into the **quadratic formula**.

The Quadratic Formula

If $ax^2 + bx + c = 0$ and $a \neq 0$, then $x = \dfrac{-b \pm \sqrt{b^2 - 4ac}}{2a}$.

This formula, developed in the sixteenth century, can be used to solve any quadratic equation.[1]

Note When writing the quadratic formula, be sure that the fraction bar extends all the way beneath the numerator.

$$\frac{-b \pm \sqrt{b^2 - 4ac}}{2a}$$

Fraction bar

A common mistake is to write this as

$$-b \pm \frac{\sqrt{b^2 - 4ac}}{2a}$$

[1] Quadratic equations were solved in ancient civilizations, often using geometric constructions, but nonpositive roots were rejected as not being part of the real world. Algebraic formulas like the quadratic formula for solving quadratic equations were developed by Rafael Bombelli (ca. 1526–1573).

Solving a Quadratic Equation by the Quadratic Formula

1. Write the equation in standard form with the leading coefficient positive ($a > 0$), if it is not already in this form.
2. Identify a (coefficient of x^2), b (coefficient of x), and c (the constant).
3. Substitute the values of a, b, and c into the quadratic formula.
$$x = \frac{-b \pm \sqrt{b^2 - 4ac}}{2a}$$
4. Simplify the resulting expressions.

Note For ease of computation, we write the quadratic expression with the leading coefficient positive, $a > 0$. The formula can be used for $a < 0$.

● **Example 9–3 B**

Solving a quadratic equation using the quadratic formula

Find the solution set using the quadratic formula.

1. $x^2 - 2x - 8 = 0$
   The equation is already in standard form where $a = 1$, $b = -2$, and $c = -8$.

$$x = \frac{-(-2) \pm \sqrt{(-2)^2 - 4(1)(-8)}}{2(1)}$$

   Replace a with 1, b with -2, and c with -8 in the quadratic formula

$$x = \frac{2 \pm \sqrt{4 - (-32)}}{2}$$

   Simplify by performing indicated operations

$$x = \frac{2 \pm \sqrt{36}}{2}$$

$$x = \frac{2 \pm 6}{2}$$

   $\sqrt{36} = 6$

$$x = \frac{2 + 6}{2} = \frac{8}{2} = 4 \quad \text{or} \quad x = \frac{2 - 6}{2} = \frac{-4}{2} = -2$$

   The solution set is $\{-2, 4\}$.

In example 9–2 B number 1, we saw that this equation could also be solved be completing the square or be factoring.

2. $x^2 = 4 - x$
   Write the equation in standard form: $x^2 + x - 4 = 0$. Then $a = 1$, $b = 1$, and $c = -4$.
   Substitute these values into the quadratic formula.

$$x = \frac{-(1) \pm \sqrt{(1)^2 - 4(1)(-4)}}{2(1)}$$

   Replace a with 1, b with 1, and c with -4

$$x = \frac{-1 \pm \sqrt{1 + 16}}{2}$$

$$x = \frac{-1 \pm \sqrt{17}}{2}$$

   Then $x = \dfrac{-1 + \sqrt{17}}{2} \quad \text{or} \quad x = \dfrac{-1 - \sqrt{17}}{2}$.

   The solution set is $\left\{ \dfrac{-1 - \sqrt{17}}{2}, \dfrac{-1 + \sqrt{17}}{2} \right\}$.

3. $x^2 - 7 = 0$

The equation can be written $x^2 + 0x - 7 = 0$, so $a = 1$, $b = 0$, and $c = -7$. Substitute these values into the quadratic formula.

$$x = \frac{-(0) \pm \sqrt{(0)^2 - 4(1)(-7)}}{2(1)}$$ Replace *a* with 1, *b* with 0, and *c* with −7

$$x = \frac{\pm\sqrt{28}}{2}$$

$$x = \pm\frac{2\sqrt{7}}{2}$$

$$x = \pm\sqrt{7}$$ Simplify by reducing

Thus $x = \sqrt{7}$ or $x = -\sqrt{7}$.

The solution set is $\left\{-\sqrt{7}, \sqrt{7}\right\}$.

<u>Note</u> The original quadratic equation could have been solved by extracting the roots.

$$x^2 - 7 = 0$$ Original equation
$$x^2 = 7$$ Add 7 to each side
$$x = \sqrt{7} \quad \text{or} \quad x = -\sqrt{7}$$ Extract the roots

4. $\dfrac{x^2}{8} + \dfrac{x}{2} = \dfrac{5}{4}$

When there are fractions, finding the solution will be easier if we multiply both sides by the LCD, 8, of all the fractions.

$$8\left(\frac{x^2}{8} + \frac{x}{2}\right) = \left(\frac{5}{4}\right)8$$ The LCD is 8

$$x^2 + 4x = 10$$ Multiply

$$x^2 + 4x - 10 = 0$$ Standard form

$$x = \frac{-(4) \pm \sqrt{(4)^2 - 4(1)(-10)}}{2(1)}$$ Replace *a* with 1, *b* with 4, and *c* with −10

$$= \frac{-4 \pm \sqrt{16 + 40}}{2}$$

$$= \frac{-4 \pm \sqrt{56}}{2}$$

Then
$$x = \frac{-4 + \sqrt{56}}{2} \qquad \text{or} \qquad x = \frac{-4 - \sqrt{56}}{2}$$

$$= \frac{-4 + 2\sqrt{14}}{2} \qquad\qquad = \frac{-4 - 2\sqrt{14}}{2}$$

$$= \frac{2(-2 + \sqrt{14})}{2} \qquad\qquad = \frac{2(-2 - \sqrt{14})}{2}$$

$$= -2 + \sqrt{14} \qquad\qquad\quad = -2 - \sqrt{14}$$

and the solution set is $\left\{-2 - \sqrt{14}, -2 + \sqrt{14}\right\}$.

Note Even though the quadratic formula can be used to solve any of the quadratic equations in this chapter, you might prefer to solve equations that factor, by factoring. Also look to see if the problem is in a form where you could easily extract the roots. Review examples 9–3 B numbers 1 and 3 for a comparison of solutions.

☞ *Quick check*   Find the solution set using the quadratic formula.

a. $2x^2 - 3x = 3$ b. $3 - \dfrac{2}{3}x^2 - \dfrac{7}{3}x = 0$

Problem Solving

Many useful formulas in the physical world have a second-degree term and are solved using the methods for quadratic equations. The following example illustrates this and some more application problems that require a quadratic equation to solve.

● **Example 9–3 C**

Problem solving

Solve the following problems.

1. The position of a particle moving on a straight line at time t in seconds is given by

$$s = 3t^2 - 5t \quad (t > 0)$$

where s is the distance from the starting point in feet. How many seconds will it take to move the particle 8 feet in a positive direction? We want t when $s = 8$ feet.

$(8) = 3t^2 - 5t$	Replace s with 8
$3t^2 - 5t - 8 = 0$	Write equation in standard form
$(3t - 8)(t + 1) = 0$	Factor the left side
$3t - 8 = 0$ or $t + 1 = 0$	Set each factor equal to 0
$t = \dfrac{8}{3}$ $t = -1$	Solve each equation

Since we do not want a negative answer, the particle will move 8 feet in $\dfrac{8}{3}\left(\text{or } 2\dfrac{2}{3}\right)$ seconds.

2. Find two consecutive whole numbers whose product is 156.

Let n represent the lesser whole number. Then $n + 1$ is the next consecutive whole number.

product of consecutive whole numbers	is	156
$n \cdot (n + 1)$	$=$	156

$n(n + 1) = 156$	
$n^2 + n = 156$	Multiply on the left side
$n^2 + n - 156 = 0$	Add -156 to each side

We will use the quadratic formula.

$$n = \frac{-(1) \pm \sqrt{(1)^2 - 4(1)(-156)}}{2(1)}$$ Replace *a* with 1, *b* with 1, and *c* with −156

$$n = \frac{-1 \pm \sqrt{1 + 624}}{2}$$

$$n = \frac{-1 \pm \sqrt{625}}{2}$$

$$n = \frac{-1 \pm 25}{2}$$ $\sqrt{625} = 25$

$$n = \frac{-1 + 25}{2} = 12 \text{ or } n = \frac{-1 - 25}{2} = -13$$

We reject the negative number since n is a whole number. So $n = 12$ and $n + 1 = 13$. The two consecutive whole numbers are 12 and 13.

Note The quadratic equation $n^2 + n - 156 = 0$ could have been solved by factoring: $(n + 13)(n - 12) = 0$.

3. The sum of a number and its reciprocal is $\frac{25}{12}$. Find the number and its reciprocal.

Let n represent the number. Then $\frac{1}{n}$ is the reciprocal of the number.

the sum of a number and its reciprocal   is $\frac{25}{12}$

$$n + \frac{1}{n} \qquad\qquad = \qquad \frac{25}{12}$$

$$n + \frac{1}{n} = \frac{25}{12}$$

$$12n \cdot n + 12n \cdot \frac{1}{n} = 12n \cdot \frac{25}{12}$$ Multiply each term by $12n$ (the LCD) to clear the fractions

$$12n^2 + 12 = 25n$$ Reduce in each term

$$12n^2 - 25n + 12 = 0$$ Write in standard form

$$(3n - 4)(4n - 3) = 0$$ Factor the left side

$$3n - 4 = 0 \text{ or } 4n - 3 = 0$$ Set each factor equal to 0 and solve

$$n = \frac{4}{3} \qquad\qquad n = \frac{3}{4}$$ Solve each equation

The number is $\frac{3}{4}$ and its reciprocal is $\frac{4}{3}$ or the number is $\frac{4}{3}$ and its reciprocal is $\frac{3}{4}$.

MASTERY POINTS

Can you
- Identify the values of a, b, and c in any quadratic equation?
- Use the quadratic formula

$$x = \frac{-b \pm \sqrt{b^2 - 4ac}}{2a}$$

to solve any quadratic equation?

Exercise 9–3

Directions
Write each quadratic equation in standard form and identify the values of a, b, and c, $a > 0$. See example 9–3 A.

☛ *Quick Check*

Example $2x^2 = 4 - 5x$

Solution Add $5x - 4$ to each side to get the equation in standard form

$$2x^2 + 5x - 4 = 0$$

Then $a = 2$, $b = 5$, and $c = -4$.

Directions   Write each quadratic equation in standard form and identify the values of a, b, and c, $a > 0$. See example 9–3 A.

1. $5x^2 - 3x + 8 = 0$

2. $4x^2 + x - 2 = 0$

3. $-6z^2 - 2z + 1 = 0$

4. $-3x^2 + x + 9 = 0$

5. $4x^2 = 2x - 1$

6. $y^2 = 5y + 3$

7. $x^2 = -3x$

8. $4x - 3x^2 = 0$

9. $5x^2 = 2$

10. $-8y^2 = -3$

11. $p(p+3) = 4$

12. $2x(x - 9) = 1$

13. $(x + 3)(x - 1) = 6$

14. $(z - 4)(2z + 1) = -6$

15. $8m^2 - (m + 3) = 2m - 1$

16. $3x^2 - (2x - 5) = x - 6$

Directions

Find the solution set, using the quadratic formula. If you feel that one of the other methods would have been easier, find the solution set for the equation a second time using that method. See example 9–3 B.

☞ *Quick Check*

Example $2x^2 - 3x = 3$

Solution Write the equation in standard form by subtracting 3 from both sides.

$$2x^2 - 3x - 3 = 0$$

Then $a = 2$, $b = -3$, and $c = -3$.

$$x = \frac{-(-3) \pm \sqrt{(-3)^2 - 4(2)(-3)}}{2(2)}$$

Replace *a* with 2, *b* with −3, and *c* with −3

$$x = \frac{3 \pm \sqrt{9 + 24}}{4}$$

$$x = \frac{3 \pm \sqrt{33}}{4}$$

$$x = \frac{3 + \sqrt{33}}{4} \text{ or } x = \frac{3 - \sqrt{33}}{4}$$

The solution set is $\left\{ \dfrac{3 - \sqrt{33}}{4}, \dfrac{3 + \sqrt{33}}{4} \right\}$

Directions Find the solution set, using the quadratic formula. If you feel that one of the other methods would have been easier, find the solution set for the equation a second time using that method. See example 9–3 B.

17. $x^2 - 3x + 2 = 0$

18. $y^2 + 6y + 9 = 0$

19. $a^2 - 2a + 1 = 0$

20. $x^2 + 10x + 24 = 0$

21. $x^2 - 25 = 0$

22. $2x^2 - 8 = 0$

23. $5x^2 - 10 = 0$

24. $3x^2 - 9 = 0$

25. $-x^2 = -3x$

26. $x^2 = 4x$

27. $5x^2 - 9x = 0$

28. $2x^2 = 7x$

29. $x^2 - 9x + 4 = 0$

30. $a^2 - 5a = 6$

31. $x^2 + 2x - 6 = 0$

32. $y^2 + y - 1 = 0$

33. $a^2 + 1 = 8a$

34. $2x^2 = 7x - 6$

35. $3y^2 = 5y + 6$

36. $4t^2 = 8t - 3$

37. $3a^2 = -9a - 2$

38. $x^2 - 9x = 6$

39. $3r^2 = r + 10$

40. $3a^2 + 5a = 4$

41. $x^2 + 8x + 16 = 0$

42. $2v^2 + 5v = -2$

43. $4x^2 + 25 = 20x$

44. $4x^2 - 7 = 12x$

45. $4x^2 + 12x + 9 = 0$

46. $4t^2 = 9t + 6$

47. $3a^2 - 2a - 7 = 0$

48. $4x^2 = 8 - 2x$

49. $3x^2 = 18 - 6x$

50. $9x^2 + 4 = 12x$

51. $3r^2 - 3r = 8$

| **Directions**
 Find the solution set of the following quadratic equations. See example 9–3 B. | ☞ *Quick Check* |

Example $3 - \dfrac{2}{3}x^2 - \dfrac{7}{3}x = 0$

Solution $9 - 2x^2 - 7x = 0$ Multiply each side by the LCD, 3
Write the equation in standard form.

$$2x^2 + 7x - 9 = 0$$

Then $a = 2$, $b = 7$, $c = -9$.

$$x = \frac{-(7) \pm \sqrt{(7)^2 - 4(2)(-9)}}{2(2)}$$ Replace *a* with 2, *b* with 7, and *c* with −9

$$x = \frac{-7 \pm \sqrt{49 + 72}}{4}$$

$$x = \frac{-7 \pm \sqrt{121}}{4}$$

$$x = \frac{-7 \pm 11}{4}$$

$$x = \frac{-7 + 11}{4} = \frac{4}{4} = 1 \text{ or } x = \frac{-7 - 11}{4} = -\frac{18}{4} = -\frac{9}{2}$$

The solution set is $\left\{-\dfrac{9}{2}, 1\right\}$.

Alternate solution

$$9 - 2x^2 - 7x = 0$$ Multiply each side by the LCD, 3
$$2x^2 + 7x - 9 = 0$$ Standard form
$$(2x + 9)(x - 1) = 0$$ Factor
$$2x + 9 = 0 \quad \text{or} \quad x - 1 = 0$$ Set each of the factors equal to 0
$$2x = -9 \qquad\qquad x = 1$$ Solve for *x*
$$x = -\frac{9}{2}$$

The solution set is $\left\{-\dfrac{9}{2}, 1\right\}$

Directions   Find the solution set of the following quadratic equations. See example 9–3 B.

52. $a^2 + a = \dfrac{15}{4}$

53. $y^2 - y = \dfrac{3}{5}$

54. $2x^2 - \dfrac{7}{2} + \dfrac{x}{2} = 0$

55. $\dfrac{2}{3}x^2 - x = \dfrac{4}{3}$

56. $\dfrac{3}{4}x^2 - \dfrac{1}{2}x - 4 = 0$

57. $\dfrac{1}{3}x^2 - \dfrac{3}{2} = \dfrac{1}{2}x$

58. $\dfrac{2}{3}y^2 - \dfrac{4}{9}y = \dfrac{1}{3}$

59. $\dfrac{2a}{3} = \dfrac{2}{9} - a^2$

Directions   Solve the following problems using methods for solving quadratic equations. See example 9–3 C number 1.

60. Use the formula $s = vt + \dfrac{1}{2}at^2$, where s is the distance traveled, v is the velocity, t is the time, and a is the acceleration of an object. Find t when (a) $s = 8$, $v = 3$, $a = 4$; (b) $s = 5$, $v = 4$, $a = 2$.

61. The distance s through which an object will fall in t seconds is $s = \dfrac{1}{2}gt^2$ feet, where $g = 32$ ft$/$sec^2. Find t (correct to tenth of a second) when (a) $s = 64$, (b) $s = 96$, (c) $s = 120$.

62. If a certain projectile is fired vertically into the air, the distance in feet above the ground in t seconds is given by $s = 160t - 16t^2$. Find t when (a) $s = 0$; (b) $s = 1{,}600$; (c) $s = 160$.

63. In a certain electric circuit, the relationship between i (in amperes), E (in volts), and R (in ohms) is given by $i^2R + iE = 8{,}000$. Find i ($i > 0$) when (a) $R = 2$ and $E = 80$, (b) $R = 4$ and $E = 60$.

64. A triangular-shaped plate has an altitude that is 5 inches longer than its base. If the area of the plate is 52 square inches, what is the length of the base b and the altitude h if the area of a triangle, A, is given by $A = \dfrac{1}{2}bh$?

65. The area of a triangle is 135 square inches. If the altitude is one-third the base, what are the lengths of the altitude and base? $\left(\text{Area } = \dfrac{1}{2} \text{ times base times altitude.}\right)$

66. A metal strip is shaped into a right triangle. In any right triangle, $c^2 = a^2 + b^2$, where c is the longest side, or hypotenuse, and a and b are the lengths of the other two sides, called legs. Find x when $a = x$, $b = x + 14$, and $c = x + 16$. (*Hint:* Substitute for a, b, and c in the above relationship and solve for x.)

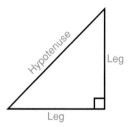

67. The hypotenuse of a right triangle is 10 millimeters long. One leg is 2 millimeters longer than the other. What are the lengths of the two legs? (Refer to exercise 66 for information about the hypotenuse and legs of a right triangle.)

68. The lengths of the legs of a right triangle are consecutive integers. If the hypotenuse is 5 centimeters long, what are the lengths of the legs of the triangle?

69. One leg of a right triangle is 2 inches longer than the other leg. If the hypotenuse is 4 inches long, what are the lengths of the legs of the triangle?

70. A 15-foot ladder is leaning against a building. If the base of the ladder is 6 feet from the base of the building, how high up the building does the ladder reach?

71. Joe leans a 50-foot ladder against his house. If the top of the ladder is 45 feet above the ground, how far out is the foot of the ladder from the house?

Directions   See example 9–3 C–2 and 3.

72. Find two consecutive whole numbers whose product is 210.

73. Find two consecutive negative even integers whose product is 224.

74. Find two consecutive odd positive integers whose product is 143.

75. The sum of a number and its reciprocal is $\dfrac{50}{7}$. What is the number?

76. The sum of a number and its reciprocal is $\dfrac{61}{30}$. What is the number?

77. In the quadratic equation $ax^2 + bx + c = 0$, why is it necessary that $a \neq 0$?

Explore to Learn

When applying the quadratic equation, if $b^2 - 4ac$ is zero then each root simplifies to $-\dfrac{b}{2a}$.

a. Make up a problem with values of a, b, and c so that this happens.
b. What can be said about this problem? In particular refer to sections 4–4 and 4–7.

Review Exercises

Directions   Perform the indicated operations. See sections 2–2 and 3–2.

1. $(4x^2 + 2x - 1) - (x^2 + x - 6)$

2. $(5y - 2)(y + 7)$

3. $(4z - 3)(4z + 3)$

4. $(3x - 5)^2$

Directions   Find the solution set of the following quadratic equations. See sections 9–1 and 9–3.

5. $3y^2 = 12$

6. $2x^2 - 7x - 4 = 0$

7. $x^2 - x = 10$

Directions   Subtract as indicated. See section 5–4.

8. $\dfrac{x + 2}{x^2 - 4} - \dfrac{x - 1}{x^2 - x - 6}$

9–4   Complex Numbers and Complex Solutions to Quadratic Equations

Complex Numbers

Recall that given the equation $x^2 = k$, we placed the restriction $k \geq 0$. Suppose that $k < 0$ as in the equation $x^2 = -9$. Extracting the roots as we did in section 9–1, we obtain

$$x = \sqrt{-9} \quad \text{or} \quad x = -\sqrt{-9}$$

But $\sqrt{-9}$ is not a real number since there is no real number whose square is -9. Thus, in the set of real numbers, the equation $x^2 = -9$ does not have a solution. We introduce a new set of numbers called the *complex numbers*. To define this set, we are going to assume that $\sqrt{-1}$ exists and refer to it by the symbol i.

Throughout the 3,000 year history of solving quadratic equations, equations like $x^2 + 9 = 0$ have appeared in certain situations. Since this is the same as the equation $x^2 = -9$, mathematicians felt the need to deal with an expression like $\sqrt{-9}$. Several hundred years ago it became apparent that by recognizing the expression $\sqrt{-1}$ as a valid expression, these equations could be solved. In 1777 the mathematician Leonhard Euler first used the following definition in a memoir presented in St. Petersburg, Russia. Euler chose the letter i to represent this expression because it stands for the word imaginary.

Definition of i

$$i = \sqrt{-1} \text{ and } i^2 = -1$$

Although the expression $\sqrt{-1}$, or i, may seem strange, it has turned out to have many, many applications in the sciences and engineering, where for example it often helps represent the phase of an electrical current. We can use this definition to define the square root of any negative number.

Definition of $\sqrt{-b}$

For any positive real number b, we define
$$\sqrt{-b} = i\sqrt{b}$$

We can now write

1. $\sqrt{-9} = i\sqrt{9} = 3i$
2. $\sqrt{-49} = i\sqrt{49} = 7i$

Now let us define a new set of numbers that includes the real numbers as a subset. These new numbers are called **complex numbers** and are composed of a real part denoted by a and an imaginary part denoted by b.

Definition of a Complex Number

A complex number is any number that can be written in the form $a + bi$, where a and b are real numbers and i represents $\sqrt{-1}$.

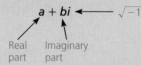

$a + bi$ is called the **standard form** of a complex number.
If $b \neq 0$, $a + bi$ is an imaginary number.
If $b = 0$, $a + 0i = a$ is a real number.

● **Example 9–4 A**

Identifying the real and imaginary parts of a complex number

The following are complex numbers.

1. $2 + 3i$, where $a = 2$ and $b = 3$

2. $4 - 2i$, where $a = 4$ and $b = -2$

Note We consider $4 - 2i$ to be in standard form since it is considered to be the same as $4 + (-2)i$.

3. $5i$ since $5i = 0 + 5i$, where $a = 0$ and $b = 5$

4. $\sqrt{-4}$ since $\sqrt{-4} = i\sqrt{4} = 2i = 0 + 2i$, where $a = 0$ and $b = 2$

5. $2 + \sqrt{-5}$ since $2 + \sqrt{-5} = 2 + i\sqrt{5} = 2 + \sqrt{5} \cdot i$, where $a = 2$ and $b = \sqrt{5}$

6. 7 since $7 = 7 + 0i$, where $a = 7$ and $b = 0$

☛ *Quick check*   Write $3 + \sqrt{-49}$ as a complex number $a + bi$.   ●

From example 9–4 A number 6, we can see that all real numbers are complex numbers, so the set of real numbers is a subset of the set of complex numbers. Figure 9–1 shows the relationship among the various sets of numbers that we have studied.

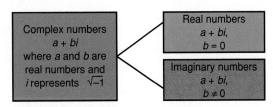

Figure 9–1

Addition and Subtraction of Complex Numbers

We add and subtract complex numbers in the same manner that we add and subtract algebraic expressions. That is, we combine the like terms.

Addition and Subtraction of Complex Numbers

1. Combine the real parts.
2. Combine the imaginary parts.
3. Leave the result in the form $a + bi$ or $a - bi$.

● **Example 9–4 B**

Addition and subtraction of complex numbers

Combine the following complex numbers. Write the answer in the form $a + bi$ or $a - bi$.

1. $(2 + 3i) + (4 - 5i)$

$\qquad = (2 + 4) + (3i - 5i)$ Commutative and associative properties

$\qquad = 6 + (-2i)$ Combine like terms

$\qquad = 6 - 2i$ Write in the form $a - bi$

2. $(3 + i) - (4 - 3i)$

$= 3 + i - 4 + 3i$ Definition of subtraction

$= (3 - 4) + (i + 3i)$ Commutative and associative properties

$= -1 + 4i$ Combine like terms

☞ *Quick check*   Subtract $(4 - 2i) - (6 + 5i)$

Multiplication of Complex Numbers

Multiplication of Two Complex Numbers

1. Multiply the numbers as if they are two binomials.
2. Substitute -1 for i^2.
3. Combine the like terms and leave the result in the form $a + bi$ or $a - bi$.

● **Example 9–4 C**

Multiplication of complex numbers

Multiply the following complex numbers. Write the answer in the form $a + bi$ or $a - bi$.

1. $2i(3 + 4i)$

$= 2i(3) + 2i(4i)$ Distributive property

$= 6i + 8i^2$ Multiply

$= 6i + 8(-1)$ $i^2 = -1$

$= -8 + 6i$ Write in the form $a + bi$

2. $(1 - 4i)(2 + 5i)$

$= 1(2 + 5i) - 4i(2 + 5i)$ Distributive property

$= 1(2) + 1(5i) + (-4i)(2) + (-4i)(5i)$

$= 2 + 5i - 8i - 20i^2$ Multiply

$= 2 + 5i - 8i - 20(-1)$ $i^2 = -1$

$= 2 + 5i - 8i + 20$

$= (2 + 20) + (5i - 8i)$ Combine like terms

$= 22 + (-3i)$

$= 22 - 3i$ Write in the form $a - bi$

Note In this example, we should be reminded once again of the FOIL method for multiplying. Remember, whenever i^2 appears, it must be replaced with -1.

3. $(5 + 2i)(5 - 2i)$

$= 5(5 - 2i) + 2i(5 - 2i)$ Distributive property

$= 25 - 10i + 10i - 4i^2$

$= 25 - 4(-1)$ Combine like terms

$= 25 + 4$

$= 29$

☞ *Quick check*   Multiply $(2 - 3i)^2$

Division of Two Complex Numbers

We call the polynomials $a + b$ and $a - b$ *conjugates* of one another and know that $(a + b)(a - b) = a^2 - b^2$. In like fashion, $5 + 2i$ and $5 - 2i$ are called **complex conjugates** of one another. We found in example 9–4 C number 3 that

$$(5 + 2i)(5 - 2i) = 5^2 + 2^2 = 25 + 4 = 29$$

Product of Complex Conjugates

The product of a complex number and its conjugate yields a real number

$$(a + bi)(a - bi) = a^2 + b^2$$

We use this property of complex numbers to perform division of two complex numbers.

Division of One Complex Number by Another Complex Number

1. Write the division as a fraction.
2. Multiply the numerator and the denominator by the conjugate of the denominator.
3. Multiply and simplify in the numerator. Use the special product property to simplify the denominator to a real number.
4. Write the result in the form $a + bi$ or $a - bi$.

● **Example 9–4 D**

Division of two complex numbers

Perform the indicated division. Write the answer in the form $a + bi$ or $a - bi$.

1. $\dfrac{4i}{2 + i}$

$$= \frac{4i}{2 + i} \cdot \frac{2 - i}{2 - i}$$ Multiply the numerator and the denominator by the conjugate of $2 + i$, which is $2 - i$

$$= \frac{4i(2 - i)}{(2 + i)(2 - i)}$$

$$= \frac{8i - 4i^2}{2^2 + 1^2}$$ Distributive property product of complex conjugates is a real number

$$= \frac{8i - 4(-1)}{4 + 1}$$ $i^2 = -1$

$$= \frac{4 + 8i}{5}$$

$$= \frac{4}{5} + \frac{8}{5}i$$ Divide each term of the numerator by 5

2. $\dfrac{4 + 3i}{3 - 5i}$

$= \dfrac{4 + 3i}{3 - 5i} \cdot \dfrac{3 + 5i}{3 + 5i}$ Multiply the numerator and the denominator by the conjugate of $3 - 5i$, which is $3 + 5i$

$= \dfrac{(4 + 3i)(3 + 5i)}{(3 - 5i)(3 + 5i)}$

$= \dfrac{12 + 20i + 9i + 15i^2}{3^2 + 5^2}$ Product of complex conjugates is a real number

$= \dfrac{12 + 29i + 15(-1)}{9 + 25}$

$= \dfrac{-3 + 29i}{34}$

$= -\dfrac{3}{34} + \dfrac{29}{34}i$ Divide each term of the numerator by 34

> ⚑ **Quick check**   Perform the indicated division $\dfrac{3 - 2i}{1 + i}$. Write the answer in the form $a + bi$ or $a - bi$.   ●

Quadratic Equations with Complex Solutions

When solving quadratic equations of the form $ax^2 + bx + c = 0$, the solutions can be determined by the quadratic formula

$$x = \dfrac{-b \pm \sqrt{b^2 - 4ac}}{2a}$$

In our work thus far, we obtained rational or irrational solutions and the radicand, $b^2 - 4ac$, was always positive or zero. When $b^2 - 4ac < 0$ (negative), we obtain complex solutions of the quadratic equation.

● **Example 9–4 E**

Solving quadratic equations that have complex solutions

Find the solution set of the following quadratic equations.

1. $(x - 3)^2 = -4$
This is in the form $(x + q)^2 = k$.

$\begin{aligned} x - 3 &= \pm\sqrt{-4} &&\text{Extract the roots} \\ x - 3 &= \pm 2i &&\sqrt{-4} = 2i \\ x &= 3 \pm 2i &&\text{Add 3 to each side} \end{aligned}$

The solution set is $\{3 - 2i, 3 + 2i\}$.

Note We could have solved the equation by expanding the left side, writing the quadratic equation in standard form, and using the quadratic formula.

2. $x^2 - 3x = -7$
Write the quadratic equation in standard form and use the quadratic formula.

$$x^2 - 3x + 7 = 0 \qquad \text{Add 7 to each side}$$

Now $a = 1$, $b = -3$, and $c = 7$.

$$x = \frac{-(-3) \pm \sqrt{(-3)^2 - 4(1)(7)}}{2(1)}$$

Replace a with 1, b with -3, and c with 7

$$= \frac{3 \pm \sqrt{9 - 28}}{2}$$

Perform indicated operations

$$= \frac{3 \pm \sqrt{-19}}{2}$$

$$= \frac{3 \pm i\sqrt{19}}{2}$$

The solution set is $\left\{ \dfrac{3 - i\sqrt{19}}{2}, \dfrac{3 + i\sqrt{19}}{2} \right\}$.

3. $(x + 1)(2x - 3) = -8$

$2x^2 - x - 3 = -8$ Multiply on the left side

$2x^2 - x + 5 = 0$ Write in standard form

Then $a = 2$, $b = -1$, and $c = 5$.   Use the quadratic formula

$$x = \frac{-(-1) \pm \sqrt{(-1)^2 - 4(2)(5)}}{2(2)}$$

Replace a with 2, b with -1, and c with 5

$$= \frac{1 \pm \sqrt{1 - 40}}{4}$$

Perform indicated operations

$$= \frac{1 \pm \sqrt{-39}}{4}$$

$$= \frac{1 \pm i\sqrt{39}}{4}$$

$\sqrt{-39} = i\sqrt{39}$

The solution set is $\left\{ \dfrac{1 - i\sqrt{39}}{4}, \dfrac{1 + i\sqrt{39}}{4} \right\}$.

☞ *Quick check*   Find the solution set. $y^2 - y = -5$ ●

The expression $b^2 - 4ac$ determines the type of solutions a quadratic equation will have. We call $b^2 - 4ac$ the **discriminant**.

Type of Solutions of a Quadratic Equation

If a, b, and c are rational numbers, and the discriminant $b^2 - 4ac$ is

Zero	Positive	Negative
$b^2 - 4ac = 0$	$b^2 - 4ac > 0$	$b^2 - 4ac < 0$
One rational repeated solution, namely $-\dfrac{b}{2a}$	If $b^2 - 4ac$ is a. a perfect square, *two* distinct *rational* solutions b. not a perfect square, *two* distinct *irrational* solutions	*Two* distinct *complex* (nonreal) solutions

● **Example 9–4 F**

Using the discriminant to determine the type of solutions to a quadratic equation

Determine the type of solution(s) that the following quadratic equations yield by using the discriminant.

1. $x^2 - x - 6 = 0$
Since $a = 1$, $b = -1$, and $c = -6$, then

$$b^2 - 4ac = (-1)^2 - 4(1)(-6) \qquad \text{Replace } a \text{ with 1, } b \text{ with } -1, \text{ and } c \text{ with } -6$$
$$= 1 + 24$$
$$= 25$$

Then $b^2 - 4ac > 0$ and, since 25 is a perfect square, we would obtain *two distinct rational* solutions.

2. $3y^2 + 2y - 2 = 0$
Since $a = 3$, $b = 2$, and $c = -2$, then

$$b^2 - 4ac = (2)^2 - 4(3)(-2) \qquad \text{Replace } a \text{ with 3, } b \text{ with 2, and } c \text{ with } -2$$
$$= 4 + 24$$
$$= 28$$

Then $b^2 - 4ac > 0$ and, since 28 is *not* a perfect square, we would obtain *two distinct irrational* solutions.

3. $x^2 - 10x + 25 = 0$
Since $a = 1$, $b = -10$, and $c = 25$, then

$$b^2 - 4ac = (-10)^2 - 4(1)(25)$$
$$= 100 - 100$$
$$= 0$$

Then $b^2 - 4ac = 0$ and we would obtain *one rational* solution.

4. $2y^2 + 2y + 7 = 0$
Since $a = 2$, $b = 2$, and $c = 7$, then

$$b^2 - 4ac = (2)^2 - 4(2)(7)$$
$$= 4 - 56$$
$$= -52$$

Then $b^2 - 4ac < 0$ and we would obtain *two distinct complex* solutions.

☞ **Quick check**   Determine the type of solutions $3x^2 + 2x = 4$ would yield by using the discriminant.   ●

MASTERY POINTS

Can you
- Write a complex number in standard form?
- Add, subtract, and multiply complex numbers?
- Rationalize the denominator of an indicated quotient of two complex numbers?
- Find the complex solutions of a quadratic equation?
- Determine the type of solutions of any quadratic equation?

Exercise 9–4	☞ *Quick Check*
Directions Write the following complex numbers in the form $a + bi$ or $a - bi$. See example 9–4 A.	**Example** $3 + \sqrt{-49}$ **Solution** $= 3 + i\sqrt{49}$ $\qquad = 3 + 7i$

Directions   Write the following complex numbers in the form $a + bi$ or $a - bi$. See example 9–4 A.

1. 9

2. -5

3. $4i$

4. $10i$

5. $\sqrt{-25}$

6. $\sqrt{-29}$

7. $4 + 2\sqrt{-4}$

8. $-3 - \sqrt{-10}$

Directions Add or subtract as indicated. Write the answer in the form $a + bi$ or $a - bi$. See example 9–4 B.	☞ *Quick Check* **Example** $(4 - 2i) - (6 + 5i)$ **Solution** $= 4 - 2i - 6 - 5i$ Definition of subtraction $\qquad = (4 - 6) + (-2i - 5i)$ Combine like terms $\qquad = -2 + (-7i)$ $\qquad = -2 - 7i$

Directions   Add or subtract as indicated. Write the answer in the form $a + bi$ or $a - bi$. See example 9–4 B.

9. $(1 + 2i) + (3 - i)$

10. $(5 + 4i) + (3 + 5i)$

11. $(5 - i) - (4 + 3i)$

12. $(1 - 5i) - (2 - i)$

13. $(3 + \sqrt{-1}) + (2 - 3\sqrt{-1})$

14. $(1 - \sqrt{-4}) - (3 + \sqrt{-9})$

15. $(-5 - \sqrt{-7}) - (4 + \sqrt{-7})$

16. $(2 + \sqrt{-11}) + (3 - \sqrt{-11})$

Directions Find the indicated products. Write the answer in the form $a + bi$ or $a - bi$. See example 9–4 C.	☞ *Quick Check* **Example** $(2 - 3i)^2$ **Solution** $= (2 - 3i)(2 - 3i)$ $\qquad = 2(2 - 3i) - 3i(2 - 3i)$ Distributive property $\qquad = 4 - 6i - 6i + 9i^2$ $\qquad = 4 - 12i + 9(-1)$ Combine like terms $i^2 = -1$ $\qquad = 4 - 12i - 9$ $\qquad = -5 - 12i$

Directions   Find the indicated products. Write the answer in the form $a + bi$ or $a - bi$. See example 9–4 C.

17. $3i(2 + 4i)$

18. $4i(5 - 2i)$

19. $(3 + 2i)(4 + i)$

20. $(5 - i)(3 + 4i)$

21. $(5 - 4i)(5 + 4i)$

22. $(5 - 5i)(5 + 5i)$

23. $(4 + 7i)^2$

24. $(3 - 2i)^2$

Directions
Perform the indicated division. Write the answer in the form $a + bi$ or $a - bi$. See example 9–4 D.

📌 *Quick Check*

Example $\dfrac{3 - 2i}{1 + i}$

Solution $= \dfrac{(3 - 2i)(1 - i)}{(1 + i)(1 - i)}$ Multiply numerator and denominator by conjugate of $1 + i$

$= \dfrac{3 - 5i + 2i^2}{1^2 + 1^2}$ Multiply as indicated; product of complex conjugates is a real number

$= \dfrac{3 - 5i - 2}{2}$ $2i^2 = 2(-1) = -2$

$= \dfrac{1 - 5i}{2}$

$= \dfrac{1}{2} - \dfrac{5}{2}i$ Divide each term of the numerator by 2

Directions   Perform the indicated division. Write the answer in the form $a + bi$ or $a - bi$. See example 9–4 D.

25. $\dfrac{5i}{2 + 3i}$

26. $\dfrac{6i}{6 - 7i}$

27. $\dfrac{4 + 2i}{3 - 5i}$

28. $\dfrac{1 + i}{2 - i}$

29. $\dfrac{5 - i}{4 + 3i}$

30. $\dfrac{4 - 7i}{5 + 2i}$

Directions
Find the solution set of the following quadratic equations. See example 9–4 E.

📌 *Quick Check*

Example $y^2 - y = -5$

Solution $y^2 - y + 5 = 0$ Write in standard form

$a = 1, b = -1, c = 5$

Using quadratic formula,

$y = \dfrac{-(-1) \pm \sqrt{(-1)^2 - 4(1)(5)}}{2(1)}$ Replace a with 1, b with -1, and c with 5

$= \dfrac{1 \pm \sqrt{1 - 20}}{2}$

$= \dfrac{1 \pm \sqrt{-19}}{2}$

$= \dfrac{1 \pm i\sqrt{19}}{2}$ $\sqrt{-19} = i\sqrt{19}$

The solution set is $\left\{ \dfrac{1 - i\sqrt{19}}{2}, \dfrac{1 + i\sqrt{19}}{2} \right\}$.

Directions   Find the solution set of the following quadratic equations. See example 9–4 E.

31. $x^2 + x + 2 = 0$

32. $x^2 - 3x + 7 = 0$

33. $x^2 + x + 5 = 0$

34. $x^2 + 2x + 6 = 0$

35. $x^2 + 3x + 4 = 0$

36. $2x^2 + 2x + 5 = 0$

37. $x^2 + 2x + 10 = 0$

38. $5x^2 + 4x + 2 = 0$

39. $x^2 - 3x = -5$

40. $x^2 + 5x = -9$

41. $2y^2 + y + 4 = 0$

42. $3y^2 - 2y + 3 = 0$

43. $2y^2 - 3y + 4 = 0$

44. $4y^2 - 3y + 4 = 0$

45. $x^2 + 4 = 0$

46. $x^2 + 25 = 0$

47. $x^2 + 12 = 0$

48. $x^2 + 18 = 0$

49. $(x + 2)^2 = -16$

50. $(x - 5)^2 = -3$

51. $(x+3)(x-2) = -11$

52. $(2y-1)(3y-2) = -3$

Directions	☛ **Quick Check**
Determine the type of solution(s) that the following quadratic equations yield, using the discriminant. See example 9–4 F.	**Example** $3x^2 + 2x = 4$ **Solution** $3x^2 + 2x - 4 = 0$ Write in standard form $\qquad a = 3, b = 2, c = -4$ $\qquad b^2 - 4ac = (2)^2 - 4(3)(-4)$ Replace a with 3, b with 2, and c with -4 $\qquad\qquad\qquad = 4 + 48$ $\qquad\qquad\qquad = 52$ The two solutions are distinct and irrational since $52 > 0$ and 52 is *not* a perfect square.

Directions   Determine the type of solution(s) that the following quadratic equations yield, using the discriminant. See example 9–4 F.

53. $y^2 + 2y - 5 = 0$

54. $2x^2 + x + 3 = 0$

55. $4x^2 - 12x + 9 = 0$

56. $3y^2 + y - 1 = 0$

57. $9x^2 - 3x = 0$

58. $3y^2 + 5y + 2 = 0$

59. $(x + 4)(x + 3) = 1$

60. $(2x + 3)(x + 5) = -3$

Directions   Solve the following problems.

61. The impedance of an electrical circuit is the measure of the total opposition to the flow of an electric current. The impedance Z in a series RCL circuit is given by

$$Z = R + i(X_L - X_C)$$

Determine the impedance if $R = 20$ ohms, $X_L = 12$ ohms, and $X_C = 22$ ohms.

62. Use the formula in exercise 61 to find Z if $R = 18$ ohms, $X_L = 14$ ohms, and $X_C = 24$ ohms.

63. For what values of x does the expression $\sqrt{9 - x}$ represent a real number?

64. For what values of x does the expression $\sqrt{x + 10}$ represent a real number?

65. For what values of x does the expression $\sqrt{x + 16}$ represent an imaginary number?

66. For what values of x does the expression $\sqrt{4 - x}$ represent an imaginary number?

67. What is the discriminant?

✦ Explore to Learn

It can be instructive to look at a geometric interpretation of complex numbers. First, observe that every complex number $a + bi$ can be paired up with an ordered pair (a,b), and this ordered pair represents a point in the rectangular coordinate system (chapter 6). Thus it can be graphed (plotted). For example, the complex number $2 - 3i$ can be represented by plotting the point $(2,-3)$.

a. Plot the point corresponding to $2 - 3i$. Compute the value of $i(2 - 3i)$. Plot this point. Compute the value of $i[i(2 - 3i)]$ (that is, i times the last value plotted). Plot this point. Compute i times this previous value. Plot this point. Connect the points with straight lines in the order plotted, and connect the last point to the first point. Give a geometric interpretation to the figure.

b. Try the same process starting with some other complex number and see what geometric figure appears.

c. On the basis of these experiments give a geometric interpretation of multiplication by i.

Review Exercises

Directions   Graph the following equations by finding *three* ordered pair solutions and then graphing the points. See section 6–2.

1. $y = 4x - 3$

2. $y = 2 - 3x$

3. Solve the system of linear equations
$$2x - y = 1$$
$$x + y = 3$$
by graphing. The exact answer will be given. See section 7–1.

4. Find the equation of the line through the points $(1,-3)$ and $(4,5)$. Write the answer in standard form. See section 6–4.

Directions   Perform the indicated operation. See section 5–2.

5. $\dfrac{3x}{x - 2} \cdot \dfrac{x^2 - x - 2}{4x^2}$

9–5   The Graphs of Quadratic Equations in Two Variables—The Parabola

The Parabola

Recall that the graph of a linear (first-degree) equation in two variables, is a straight line. The graph of a quadratic (second-degree) equation in two variables, is *not* a straight line. For this reason, we will require a number of points to plot the graph. The same procedure we used to graph linear equations can be applied to graph quadratic equations. The graph of any quadratic equation of the form $y = ax^2 + bx + c$, $a \neq 0$ is, in fact, a special *curve*, called a **parabola**.

Consider the quadratic equation given by $y = x^2 + 2x - 8$. Since we do not know what the graph looks like, we need to choose a sufficient number of points to plot.

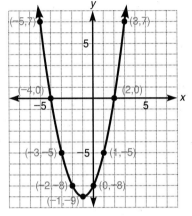

Figure 9–2

x	$y =$		$x^2 +$	$2x - 8$		(x,y)
-5	$y = (-5)^2 + 2(-5) - 8 = 7$					$(-5,7)$
-4	$y = (-4)^2 + 2(-4) - 8 = 0$					$(-4,0)$
-3	$y = (-3)^2 + 2(-3) - 8 = -5$					$(-3,-5)$
-2	$y = (-2)^2 + 2(-2) - 8 = -8$					$(-2,-8)$
-1	$y = (-1)^2 + 2(-1) - 8 = -9$					$(-1,-9)$
0	$y = (0)^2 + 2(0) - 8 = -8$					$(0,-8)$
1	$y = (1)^2 + 2(1) - 8 = -5$					$(1,-5)$
2	$y = (2)^2 + 2(2) - 8 = 0$					$(2,0)$
3	$y = (3)^2 + 2(3) - 8 = 7$					$(3,7)$

Plotting these points and passing a *smooth curve* through them, we have the graph of the quadratic equation $y = x^2 + 2x - 8$ (figure 9–2). This type of geometric figure is called a parabola.

Note We *do not* connect the points with straight line segments.

The *x*- and *y*-Intercepts of a Parabola

There are certain features of a parabola that we always wish to include in our graph, namely the *x*- and *y*-intercepts and the extreme (lowest or highest) point of the graph, called the **vertex**.

We observe from our example that when the curve crosses the *x*-axis, the value of *y* is zero, and when the curve crosses the *y*-axis, the value of *x* is zero. This is the same observation that we made with linear equations. We can generalize finding the *x*- and *y*-intercepts for any graph as follows:

1. To find any *x*-intercept(s), replace *y* by 0 in the equation and solve for *x*.
2. To find the *y*-intercept, replace *x* by 0 in the equation and solve for *y*.

● **Example 9–5 A**

Finding the x- and y-intercepts of a parabola

Find the x- and y-intercepts.

1. $y = x^2 - 4$

   a. Let $y = 0$

$$(0) = x^2 - 4 \qquad \text{Replace } y \text{ with } 0$$
$$0 = (x - 2)(x + 2) \qquad \text{Factor the right side, set each factor equal to 0}$$
$$\qquad\qquad\qquad\qquad\qquad \text{and solve for } x$$
$$x = 2 \text{ or } -2$$

The x-intercepts are 2 and -2. The points are $(2,0)$ and $(-2,0)$.

   b. Let $x = 0$

$$y = (0)^2 - 4 \qquad \text{Replace } x \text{ with } 0$$
$$y = -4$$

Therefore, the y-intercept is -4. The point is $(0,-4)$.

2. $y = x^2 - 6x + 9$

   a. Let $y = 0$

$$(0) = x^2 - 6x + 9 \qquad \text{Replace } y \text{ with } 0$$
$$0 = (x - 3)(x - 3) \qquad \text{Factor, set each factor equal to 0 and solve}$$
$$x = 3$$

Therefore, the x-intercept is 3. The point is $(3,0)$.

   b. Let $x = 0$

$$y = (0)^2 - 6(0) + 9 \qquad \text{Replace } x \text{ with } 0$$
$$y = 9$$

Hence, the y-intercept is 9. The point is $(0,9)$.

3. $y = -x^2 + 2x + 3$

   a. Let $y = 0$

$$(0) = -x^2 + 2x + 3$$
$$0 = x^2 - 2x - 3 \qquad \text{Multiply each side by } -1$$
$$0 = (x - 3)(x + 1) \qquad \text{Factor the right side and set each factor equal to 0}$$
$$x = 3 \text{ or } x = -1 \qquad \text{Solve each equation}$$

The x-intercepts are 3 and -1. The points are $(3,0)$ and $(-1,0)$.

   b. Let $x = 0$

$$y = -(0)^2 + 2(0) + 3$$
$$= 3$$

The y-intercept is 3. The point is $(0,3)$.

4. $y = x^2 + 1$

   a. Let $y = 0$

$$(0) = x^2 + 1$$

Then $x^2 = -1$ and $x = \pm\sqrt{-1}$. Since $\sqrt{-1}$ is not a real number, there are no real solutions for x. Hence, the graph has no x-intercepts.

   b. Let $x = 0$

$$y = (0)^2 + 1$$
$$= 1$$

The y-intercept is 1. The point is $(0,1)$.

Note From these examples, we see that the *y*-intercept is always the constant *c*. If the quadratic equation is in the standard form $y = ax^2 + bx + c$, the *y*-intercept will be the point $(0,c)$.

▶ *Quick check*   Find the *x*- and *y*-intercepts. $y = 2x^2 - 3x + 1$ ●

The Vertex and the Axis of Symmetry of a Parabola

We wish to find one remaining point of interest on the graph—the vertex. The vertex is the extreme point on the graph, that is, either the maximum or minimum value that the graph will attain. If our equation is in standard form, $y = ax^2 + bx + c$, it can be shown that the *x*-component of the vertex is given by $x = -\dfrac{b}{2a}$.

Once we have determined the *x*-component, we replace *x* with this value in the original equation and generate the corresponding *y*-value. Recall our original example: $y = x^2 + 2x - 8$. The value of *a* is 1 and *b* is 2. Therefore

$$x = -\frac{b}{2a} = -\frac{(2)}{2(1)} = -\frac{2}{2} = -1$$

We then substitute this value for *x* in the original equation and we obtain

$$y = (-1)^2 + 2(-1) - 8 = 1 + (-2) - 8 = -9$$

Hence, in this case, the vertex is the point with coordinates $(-1, -9)$. This means no matter what value *x* takes, *y* is *never* less than -9.

Note When the value of *a*, the coefficient of the squared term, is positive (as in this case), the parabola opens *upward* and the vertex is the *lowest* point of the graph. When *a* is negative, the parabola opens *downward* and the vertex is the *highest* point of the graph.

● **Example 9–5 B**

Finding the vertex of a parabola

Find the coordinates of the vertex of each parabola.

1. $y = x^2 - 4$

$$y = x^2 + 0x - 4 \qquad \text{Write equation in standard form}$$

Since $a = 1$ and $b = 0$, the vertex has coordinates

$$x = -\frac{b}{2a} = -\frac{(0)}{2(1)} = 0 \qquad \text{Replace } a \text{ with 1 and } b \text{ with 0}$$
$$y = (0)^2 - 4 = -4 \qquad \text{Replace } x \text{ with 0 in } y = x^2 - 4$$

The vertex is the point $(0, -4)$.

2. $y = x^2 - 6x + 9$

Since $a = 1$ and $b = -6$, the vertex has coordinates

$$x = -\frac{b}{2a} = -\frac{(-6)}{2(1)} = \frac{6}{2} = 3$$
$$\text{and } y = (3)^2 - 6(3) + 9$$
$$= 9 - 18 + 9$$
$$= 0$$

The vertex is the point $(3, 0)$.

3. $y = -x^2 + 2x + 3$

Since $a = -1$ and $b = 2$, so the vertex has coordinates

$$x = -\frac{b}{2a} = -\frac{(2)}{2(-1)} = \frac{2}{2} = 1$$
$$\text{and } y = -(1)^2 + 2(1) + 3$$
$$= -1 + 2 + 3$$
$$= 4$$

The vertex is the point $(1, 4)$.

4. $y = x^2 + 1$

Since $a = 1$ and $b = 0$, so the vertex has coordinates

$$x = -\frac{b}{2a} = -\frac{(0)}{2(1)} = -\frac{0}{2} = 0$$
$$\text{and } y = (0)^2 + 1$$
$$= 1$$

The vertex is the point $(0, 1)$.

➤ **Quick check**   Find the coordinates of the vertex. $y = 2x^2 - 3x + 1$

If the vertex is the point, (h, k), the vertical line, $x = h$, which passes through the vertex, is called the **axis of symmetry** of the graph of the parabola. The parabola is a symmetric curve and if we fold the graph along the axis of symmetry, the left half of the curve will coincide with the right half of the curve. For this reason, we choose two values of x that are greater than h and two values of x that are less than h when finding our arbitrary points to graph.

The Graph of a Parabola

To draw a reasonably accurate graph of any quadratic equation in two variables, we take the following steps.

Graphing the Parabola $y = ax^2 + bx + c$

1. Find the coordinates of the x- and y-intercepts.
   a. Let $x = 0$, solve for y to find the y-intercept.
   b. Let $y = 0$, solve for x to find the x-intercept(s).
2. Find the coordinates of the vertex.
   a. $x = -\dfrac{b}{2a}$
   b. Replace x with the value of $-\dfrac{b}{2a}$ in the original equation and solve for y.
3. Find the coordinates of four arbitrarily chosen points. Choose values of x such that, if the point (h, k) is the vertex of the parabola,
   a. two values are less than h and
   b. two values are greater than h.
4. Draw a smooth curve through the resulting points.

● Example 9–5 C

Graphing a parabola

Graph the following quadratic equations. Determine the equation of the axis of symmetry.

1. $y = x^2 - 4$

In our previous examples, we found the x- and y-intercepts and the vertex. We need only determine four more points and we will be ready to graph the function.

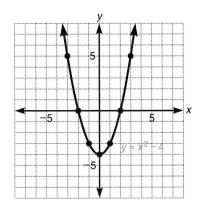

x	y	
2	0	} x-intercepts
−2	0	
0	−4	y-intercept and vertex
−1	−3	
1	−3	} Arbitrary points
−3	5	
3	5	

The axis of symmetry is the line $x = 0$ (y-axis).

2. $y = x^2 - 6x + 9$

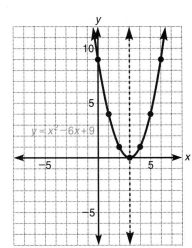

x	y	
3	0	x-intercept and vertex
0	9	y-intercept
1	4	
2	1	
4	1	} Arbitrary points
5	4	
6	9	

The axis of symmetry is the line $x = 3$.

3. $y = -x^2 + 2x + 3$

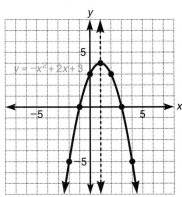

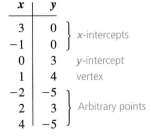

x	y	
3	0	} x-intercepts
−1	0	
0	3	y-intercept
1	4	vertex
−2	−5	
2	3	} Arbitrary points
4	−5	

The axis of symmetry is the line $x = 1$.

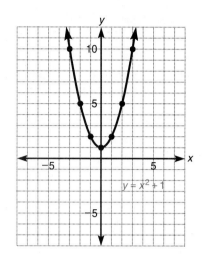

$y = x^2 + 1$

4. $y = x^2 + 1$

x	y	
0	1	y-intercept and vertex
-3	10	
-2	5	
-1	2	
1	2	Arbitrary points
2	5	
3	10	

The axis of symmetry is the line $x = 0$ (y-axis).

 Quick check Graph the equation and determine the equation of the axis of symmetry. $y = 2x^2 - 3x + 1$

A graphing calculator can be used to easily graph a second degree equation in the form $y = ax^2 + bx + c$. Example C in Graphing Calculators in the appendix shows how to do this. The algebraic methods of this section should still be used to find the exact values of intercepts and the coordinates of the vertex.

Quadratic equations are used in many physical situations. For example, if an object is thrown into the air, the graph of the distance the object travels versus the time it travels is a parabola.

● **Example 9–5 D**
Problem solving

A projectile is fired vertically into the air. Its distance s in feet above the ground in t seconds is given by $s = 160t - 16t^2$.

Note s is defined as a quadratic function of t. That is, the distance the object travels *changes with* time.

Find the highest point of the projectile (the vertex of the parabola) and the moment when the projectile will strike the ground. Graph the equation.

a. The vertex is the highest point since the parabola opens downward.

$$s = -16t^2 + 160t$$

where $a = -16$, $b = 160$. The t value of the vertex is

$$t = -\frac{b}{2a} = -\frac{(160)}{2(-16)} = -\frac{160}{-32} = 5 \qquad \text{Replace } a \text{ with } -16 \text{ and } b \text{ with } 160$$

The maximum height will be

$$s = -16(5)^2 + 160(5) = -400 + 800 = 400 \text{ feet}$$

The maximum height, $s = 400$ feet, is attained when $t = 5$ seconds.

b. The projectile will strike the ground when $s = 0$. Therefore, we set $s = 0$ and solve for t.

$$
\begin{aligned}
0 &= -16t^2 + 160t & \text{Replace } s \text{ with } 0 \\
&= -16t(t - 10) & \text{Set each factor equal to 0 and solve} \\
t &= 0 \text{ or } 10
\end{aligned}
$$

The value $t = 0$ seconds represents the time when the projectile was fired. Hence, the value $t = 10$ seconds represents the time when the projectile will strike the ground.

c. To graph the equation, we plot time, t, along the horizontal axis and distance, s, along the vertical axis.

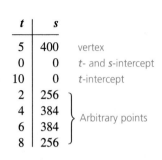

t	s	
5	400	vertex
0	0	t- and s-intercept
10	0	t-intercept
2	256	
4	384	Arbitrary points
6	384	
8	256	

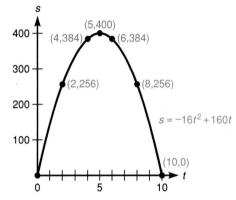

$s = -16t^2 + 160t$

Note We have used different scales on the t- and s-axes, and we do not represent values on the graph for negative time or distance, since they do not have any meaning in this example.

MASTERY POINTS

Can you
- Find the x- and y-intercepts of the graph of a quadratic equation?
- Find the coordinates of the vertex of the graph of a quadratic equation?
- Graph a quadratic equation in two variables?
- Determine the equation of the axis of symmetry?

Exercise 9–5

Directions
Find the x- and y-intercepts. If they do not exist, so state. See example 9–5 A.

☛ *Quick Check*

Example $y = 2x^2 - 3x + 1$

Solution   **a.** Let $y = 0$

$$(0) = 2x^2 - 3x + 1$$ Replace y with 0
$$0 = (2x - 1)(x - 1)$$ Factor the right side
$$2x - 1 = 0 \quad \text{or} \quad x - 1 = 0$$ Set each factor equal to 0 and solve

$$x = \frac{1}{2} \qquad\qquad x = 1$$

The x-intercepts are $\frac{1}{2}$ and 1. The points are $\left(\frac{1}{2},0\right)$ and $(1,0)$.

b. Let $x = 0$
Then $y = 1$. The y-intercept is 1. The point is $(0,1)$.

Directions   Find the x- and y-intercepts. If they do not exist, so state. See example 9–5 A.

1. $y = x^2 - 16$

2. $y = x^2 - 9$

3. $y = x^2 - 6x + 8$

4. $y = x^2 + 11x - 12$

5. $y = x^2 + 8x + 12$

6. $y = x^2 - 4x - 12$

7. $y = 5 - x^2$

8. $y = 7 - x^2$

9. $y = x^2 + 6x + 9$

10. $y = x^2 - 4x + 4$

11. $y = x^2 + 5$

12. $y = x^2 + 6$

13. $y = x^2 + x - 6$

14. $y = x^2 + 2x + 5$

15. $y = -x^2 + 4x - 3$

16. $y = 25 - x^2$

17. $y = 2x^2 + 3x + 1$

18. $y = 2x^2 - 7x + 6$

19. $y = -2x^2 - x + 6$

20. $y = -3x^2 + 7x + 6$

Directions
Find the coordinates of the vertex. Find the equation of the axis of symmetry. See example 9–5 B.

☞ Quick Check

Example $y = 2x^2 - 3x + 1$

Solution Now $a = 2$, $b = -3$, and $c = 1$

so $x = -\dfrac{b}{2a} = -\dfrac{(-3)}{2(2)} = \dfrac{3}{4}$ Replace a with 2 and b with -3

then $y = 2\left(\dfrac{3}{4}\right)^2 - 3\left(\dfrac{3}{4}\right) + 1$ Replace x with $\frac{3}{4}$

$= 2\left(\dfrac{9}{16}\right) - \dfrac{9}{4} + 1$

$= \dfrac{9}{8} - \dfrac{9}{4} + 1$

$= \dfrac{9}{8} - \dfrac{18}{8} + \dfrac{8}{8}$

$= -\dfrac{1}{8}$

The vertex is the point $\left(\dfrac{3}{4}, -\dfrac{1}{8}\right)$. The axis of symmetry is $x = \dfrac{3}{4}$.

Directions   Find the coordinates of the vertex. Find the equation of the axis of symmetry. See example 9–5 B.

21. $y = x^2 - 16$

28. $y = 7 - x^2$

34. $y = -x^2 + 4x - 3$

22. $y = x^2 - 9$

29. $y = x^2 + 6x + 9$

35. $y = -x^2 + 4x - 3$

23. $y = x^2 - 6x + 8$

30. $y = x^2 - 4x + 4$

36. $y = 25 - x^2$

24. $y = x^2 + 11x - 12$

31. $y = x^2 + 5$

37. $y = 2x^2 + 3x + 1$

25. $y = x^2 + 8x + 12$

32. $y = x^2 + 6$

38. $y = 2x^2 - 7x + 6$

26. $y = x^2 - 4x - 12$

33. $y = x^2 + x - 6$

39. $y = -2x^2 - x + 6$

27. $y = 5 - x^2$

40. $y = -3x^2 + 7x + 6$

Directions
Graph the following equations. Include the vertex and the points at which the graph crosses each axis. See example 9–5 C.

Quick Check

Example $y = 2x^2 - 3x + 1$

> *Note* $a = 2$, which is positive, so the parabola opens up.

Solution   The vertex and the x- and y-intercepts were found in the preceding examples and are listed in the following table.

x	y	
0	1	y-intercept
1	0	x-intercepts
$\frac{1}{2}$	0	
$\frac{3}{4}$	$-\frac{1}{8}$	Vertex
2	3	
-1	6	Arbitrary points
3	10	
-2	15	

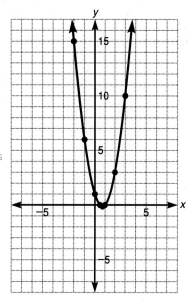

Directions   Graph the following equations. Include the vertex and the points at which the graph crosses each axis. The intercepts were found in exercises 1–20, and the vertex and axis of symmetry were found in exercises 21–40. See example 9–5 C.

41. $y = x^2 - 16$

45. $y = x^2 + 8x + 12$

49. $y = x^2 + 6x + 9$

50. $y = x^2 - 4x + 4$

42. $y = x^2 - 9$

46. $y = x^2 - 4x - 12$

51. $y = x^2 + 5$

43. $y = x^2 - 6x + 8$

47. $y = 5 - x^2$

52. $y = x^2 + 6$

48. $y = 7 - x^2$

44. $y = x^2 + 11x - 12$

53. $y = x^2 + x - 6$

56. $y = 25 - x^2$

59. $y = -2x^2 - x + 6$

57. $y = 2x^2 + 3x + 1$

54. $y = x^2 + 2x + 5$

60. $y = -3x^2 + 7x + 6$

58. $y = 2x^2 - 7x + 6$

55. $y = -x^2 + 4x - 3$

Directions   Graph each equation by plotting the variable for which the equation is solved along the vertical axis and by plotting the other variable along the horizontal axis. Graph the equation only in the regions for which the equation would have meaning. See example 9–5 D.

61. When a ball rolls down an inclined plane, it travels a distance $d = 6t + \dfrac{t^2}{2}$ feet in t seconds. Plot the graph showing how d depends on t. How long will it take the ball to travel 14 feet?

62. The output power P of a 100-volt electric generator is defined by $P = 100I - 5I^2$, where I is in amperes. Plot the graph showing how P depends on I for values of I from 0 to 5.

63. The current in a circuit flows according to the equation $i = 12 - 12t^2$, where i is the current and t is the time in seconds. Plot the graph of the relation given by the equation labeling the horizontal axis the t-axis.

64. If a projectile is fired vertically into the air with an initial velocity of 80 feet per second, the distance in feet above the ground in t seconds is given by $s = 80t - 16t^2$. Find the projectile's maximum height and when the projectile will strike the ground. Plot the graph showing how s depends on t.

65. Referring to exercise 64, if the initial velocity is 96 ft/sec, the equation is $s = 96t - 16t^2$. Find the maximum height and when the projectile will strike the ground. Plot the graph of this equation using the t-axis as the horizontal axis.

66. The distance s through which an object will fall in t seconds is $s = 16t^2$. Plot the graph showing the relation between s and t for the first 5 seconds.

67. An object is dropped from the top of the Empire State Building (1,250 feet tall), and the distance that the object is from the ground is given by the equation $s = 1,250 - 16t^2$. Plot the graph showing how s depends on t and determine when the object will strike the ground. (t is time in seconds. Round the answer to two decimal places.)

68. How can we use the axis of symmetry, $x = h$, to help graph a parabola?

Explore to Learn

Consider the graph in the figure. It shows the two x-intercepts, which are the solutions to the equation $0 = ax^2 + bx + c$, $\dfrac{-b - \sqrt{b^2 - 4ac}}{2a}$, and $\dfrac{-b + \sqrt{b^2 - 4ac}}{2a}$. The value of x at the vertex would be expected to be half-way between these two intercepts. This is the average of these two values. Compute this value (x in the figure) by adding the coordinates of the x-intercepts and dividing by 2.

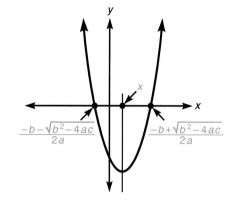

Chapter 9 Lead-in Problem

A rock is dropped from the top of the Washington Monument. If the monument is 555 feet tall, how long will it take the rock to strike the ground?

Solution   We use $s = 16t^2$, where s is the distance the rock fell and t is the time in seconds.

$555 = 16t^2$ Replace s with 555

$t^2 = \dfrac{555}{16}$ Divide each side by 16

$t = \sqrt{\dfrac{555}{16}} = \dfrac{\sqrt{555}}{4} \approx 5.9$ Extract the roots

or $t = -\sqrt{\dfrac{555}{16}} = -\dfrac{\sqrt{555}}{4} \approx -5.9$

Reject the negative value since time t must be positive. Thus, the rock will strike the ground in approximately 5.9 seconds.

Chapter 9 Summary

• Glossary

axis of symmetry (page 571)   the vertical line, $x = h$, which passes through the vertex, (h, k), of a parabola $y = ax^2 + bx + c$.

complex conjugates (page 561) $a + bi$ and $a - bi$ are complex conjugates of each other.

discriminant (page 563)   the radicand $b^2 - 4ac$ of the quadratic formula.

parabola (page 569)   the graph of any equation of the form $y = ax^2 + bx + c$, $a \neq 0$.

vertex (page 571)   the maximum or minimum point that the graph of a parabola will attain.

• New Symbols and Notation

$\pm$ plus or minus

$i = \sqrt{-1}$ and $i^2 = -1$

• **Properties and Definitions**

Square root property (page 534)

If k is a nonnegative number and $x^2 = k$, then

$$x = \sqrt{k} \text{ or } x = -\sqrt{k}$$

The quadratic formula (page 548)

If $ax^2 + bx + c = 0$ and $a \neq 0$, then

$$x = \frac{-b \pm \sqrt{b^2 - 4ac}}{2a}$$

Definition of i (page 558)

$$i = \sqrt{-1} \text{ and } i^2 = -1$$

Definition of $\sqrt{-b}$ (page 558)

For any positive real number b, we define

$$\sqrt{-b} = i\sqrt{b}$$

Definition of a complex number (page 558)

A complex number is any number that can be written in the form $a + bi$, where a and b are real numbers and i represents $\sqrt{-1}$.

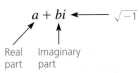

$a + bi$ is called the **standard form** of a complex number.
If $b \neq 0$, $a + bi$ is an imaginary number.
If $b = 0$, $a + 0i = a$ is a real number.

Product of complex conjugates (page 561)

The product of a complex number and its conjugate yields a real number

$$(a + bi)(a - bi) = a^2 + b^2$$

[9–1]

Solving a quadratic equation by factoring (page 533)

1. Write the quadratic equation in standard form with the leading coefficient positive.

2. Completely factor the quadratic expression.

3. Set each of the factors containing the variable equal to 0 and solve the linear equations.

[9–2]

Solving a quadratic equation by completing the square (page 541)

To find the solution set of the quadratic equation $ax^2 + bx + c = 0$, $a \neq 0$, by completing the square, we proceed as follows:

1. If $a = 1$, proceed to step 2. If $a \neq 1$, divide each term of the equation by a and simplify.

2. Write the equation with the variable terms on the left side and the constant term on the right side.

3. Add to each side of the equation the square of one half of the numerical coefficient of the middle term of the original equation.

4. Write the left side as a trinomial and factor it as the square of a binomial. Combine like terms on the right side.

5. Extract the roots and solve the resulting linear equations.

[9–3]

Solving a quadratic equation by the quadratic formula (page 549)

1. Write the equation in standard form with the leading coefficient positive ($a > 0$) if it is not already in this form.

2. Identify a (coefficient of x^2), b (coefficient of x), and c (the constant).

3. Substitute the values of a, b, and c into the quadratic formula.

$$x = \frac{-b \pm \sqrt{b^2 - 4ac}}{2a}$$

4. Simplify the resulting expressions.

[9–4]

Addition and subtraction of complex numbers (page 559)

1. Combine the real parts.

2. Combine the imaginary parts.

3. Leave the result in the form $a + bi$ or $a - bi$.

Multiplication of two complex numbers (page 560)

1. Multiply the numbers as if they are two binomials.

2. Substitute -1 for i^2.

3. Combine the like terms and leave the result in the form $a + bi$ or $a - bi$.

Division of one complex number by another complex number (page 561)

1. Write the division as a fraction.

2. Multiply the numerator and the denominator by the conjugate of the denominator.

3. Multiply and simplify in the numerator. Use the special product property to simplify the denominator to a real number.

4. Write the result in the form $a + bi$ or $a - bi$.

[9–5]

Graphing the parabola $y = ax^2 + bx + c$
(page 572)

1. Find the coordinates of the x- and y-intercepts.

   a. Let $x = 0$, solve for y to find the y-intercept.
   b. Let $y = 0$, solve for x to find the x-intercept(s).

2. Find the coordinates of the vertex.

   a. $x = -\dfrac{b}{2a}$

   b. Replace x with the value of $-\dfrac{b}{2a}$ in the original equation and solve for y.

3. Find the coordinates of four arbitrarily chosen points. Choose values of x such that, if the point (h, k) is the vertex of the parabola,

   a. two values are less than h and
   b. two values are greater than h.

4. Draw a smooth curve through the resulting points.

☀ Chapter 9 Error Analysis

1. Solving quadratic equations by completing the square
Example: Find the solution set of $x^2 + x - 8 = 0$ by completing the square.
$$x^2 + x = 8$$
$$x^2 + x + \frac{1}{4} = 8$$
$$\left(x + \frac{1}{2}\right)^2 = 8$$
$$x + \frac{1}{2} = \pm\sqrt{8} = \pm 2\sqrt{2}$$
$$x = -\frac{1}{2} \pm 2\sqrt{2} = \frac{-1 \pm 4\sqrt{2}}{2}\left\{\frac{-1 - 4\sqrt{2}}{2}, \frac{-1 + 4\sqrt{2}}{2}\right\}$$
Correct answer: $\left\{\dfrac{-1 - \sqrt{33}}{2}, \dfrac{-1 + \sqrt{33}}{2}\right\}$
What error was made? (*see page 541*)

2. Solving quadratic equations by quadratic formula
Example: Find the solution set of $x^2 + 2x - 5 = 0$ by quadratic formula. $a = 1$, $b = 2$, $c = -5$
$$x = -2 \pm \frac{\sqrt{(2)^2 - 4(1)(-5)}}{2(1)}$$
$$= -2 \pm \frac{\sqrt{24}}{2} = -2 \pm \frac{2\sqrt{6}}{2}$$
$$= -2 \pm \sqrt{6}\,\{-2 - \sqrt{6}, -2 + \sqrt{6}\}$$
Correct answer:
$\{-1 - \sqrt{6}, -1 + \sqrt{6}\}$
What error was made? (*see page 548*)

3. Finding the values of a, b, and c of a quadratic equation.
Example: Find the values of a, b, and c of $2x^2 - x = 4$. $a = 2$, $b = -1$, $c = 4$
Correct answer: $a = 2$, $b = -1$, $c = -4$
What error was made? (*see page 547*)

4. Extracting the roots
Example: Find the solution set of $x^2 = 25$ by extracting roots.
$$x^2 = 25$$
$$x = \sqrt{25} = 5\ \{5\}$$
Correct answer: $\{-5, 5\}$
What error was made? (*see page 534*)

5. Product of complex numbers
Example: $(2 + 3i)(1 - i) = (2)(1) - 2i + 3i - 3i^2$
$$= 2 - 2i + 3i - 3$$
$$= -1 + i$$
Correct answer: $5 + i$
What error was made? (*see page 560*)

6. Finding the y-intercept of a parabola
Example: Given the quadratic equation $y = 2x^2 - 2x - 3$, so the y-intercept is $(0, 3)$.
Correct answer: The y-intercept is $(0, -3)$.
What error was made? (*see page 569*)

7. Graph of a parabola
Example: The parabola that is the graph of $y = 2x^2 + x - 3$ opens *downward*.
Correct answer: opens *upward*
What error was made? (*see page 571*)

8. Squaring a radical binomial
   Example: $(\sqrt{6} + \sqrt{5})^2 = (\sqrt{6})^2 + (\sqrt{5})^2$
   $= 6 + 5 = 11$
   Correct answer: $11 + 2\sqrt{30}$
   What error was made? (*see page 510*)

9. Multiplying square roots
   Example: $\sqrt{-6} \cdot \sqrt{-6} = \sqrt{-6 \cdot -6} = \sqrt{36} = 6$
   Correct answer: -6
   What error was made? (*see page 558*)

10. Rationalizing the denominator of an *n*th root
    Example: $\dfrac{3}{\sqrt[4]{xy^2}} = \dfrac{3}{\sqrt[4]{xy^2}} \cdot \dfrac{\sqrt[4]{xy^2}}{\sqrt[4]{xy^2}} = \dfrac{3\sqrt[4]{xy^2}}{xy^2}$
    Correct answer: $\dfrac{3\sqrt[4]{x^3y^2}}{xy}$
    What error was made? (*see page 501*)

Chapter 9 Critical Thinking

1. The table shows three examples of three consecutive integers and a relation which is true in each case.

2	3	4	$3^2 = 2 \cdot 4 + 1$
5	6	7	$6^2 = 5 \cdot 7 + 1$
89	90	91	$90^2 = 89 \cdot 91 + 1$

   a. Generalize this pattern by describing it algebraically.
   b. Show why this pattern is always true.

2. In baseball the batting average is the number of hits divided by the number of times at bat (opportunities to hit). For example, 22 hits out of 80 times at bat would give an average of $\dfrac{22}{80} = 0.275$.
   (Batting averages are always given to the nearest thousandth.)

   A player may bat left- or right-handed, and some players can bat both ways. In this case a player is assigned both a left-handed (LH) and right-handed (RH) average as well as an overall average. It is possible for one player to have both a better RH and LH than another player, but have a worse overall average! This is an example of what is sometimes called Simpson's paradox. Give a numeric example of this. That is, create fictitious data for two players, A and B, so that player A has a better RH and LH than player B but has a worse overall average. (This situation has actually happened in professional baseball.)

Chapter 9 Review

[9–1]

Directions Find the solution set of the following quadratic equations by extracting the roots or by factoring.

1. $x^2 = 100$
2. $x^2 - 25 = 0$
3. $z^2 = 2$
4. $y^2 - 6 = 0$
5. $6x^2 = 24$
6. $8x^2 - 96 = 0$
7. $\dfrac{3}{4}x^2 = 12$
8. $\dfrac{2}{3}x^2 - 8 = 0$
9. $\dfrac{x^2}{3} - 8 = \dfrac{1}{4}$
10. $x^2 - x - 42 = 0$
11. $3y^2 - 7y + 4 = 0$

[9–2]

Directions   Find the solution set by completing the square.

12. $x^2 - 6x + 4 = 0$

13. $z^2 - 10z + 4 = 0$

14. $2a^2 - 8a = -2$

15. $3y^2 - 6y - 6 = 0$

16. $4 - x^2 = 5x$

17. $5 = 11y - y^2$

18. $x(x - 1) = 3$

19. $(x - 2)(x + 1) = 1$

20. $4x^2 - 3 = 8x - 1$

21. $\frac{1}{5}x^2 + \frac{2}{5}x = 2$

22. The length of a rectangle is 2 meters more than three times the width. Its area is 16 square meters. What are its dimensions? (Solve by completing the square.)

[9–3]

Directions   Find the solution set using the quadratic formula.

23. $x^2 - 2x - 5 = 0$

24. $x^2 - 8 = -4x$

25. $2y^2 - 3y = 5$

26. $3a^2 = 8 - 7a$

27. $2x^2 - 9 = 0$

28. $4x^2 = -7x$

29. $x^2 - \frac{2}{3}x = \frac{4}{3}$

30. $2x + \frac{3}{4} = \frac{3}{2}x^2$

31. A metal bar is to be divided into two pieces so that one piece is 3 inches longer than the other. If the sum of the squares of the two lengths is 117 square inches, find the two lengths. (Use the quadratic formula.)

[9–4]

Directions   Perform the indicated operations on the given complex numbers. Write the answer in standard form $a + bi$ or $a - bi$.

32. $(3 - 5i) + (2 + 4i)$

33. $(7 + i) - (3 - 8i)$

34. $(6 + i)(2 - 3i)$

35. $(-3 + 9i)(2 - i)$

36. $(6 - 4i)(6 + 4i)$

37. $(5 - 3i)^2$

38. $\dfrac{3i}{1 + i}$

39. $\dfrac{-4i}{2 - i}$

40. $\dfrac{2 - i}{3 - i}$

41. $\dfrac{5 + 2i}{4 + 3i}$

Directions   Find the solution set of the following quadratic equations.

42. $x^2 + 4x + 7 = 0$

43. $4y^2 - y = -5$

44. $(x + 1)(x - 1) = -8$

45. $(2x + 3)^2 = -4$

Directions   Determine the type of solutions the following quadratic equations will yield, using the discriminant.

46. $y^2 - 16y + 64 = 0$

47. $3x^2 - x - 2 = 0$

48. $5y^2 - 2y + 3 = 0$

49. $3x^2 + x - 3 = 0$

[9–5]

Directions   Find the x- and y-intercepts and the vertex of the graph of each quadratic equation. Find the equation of the axis of symmetry. Sketch the graph.

50. $y = x^2 - 4x - 12$

53. $y = 2 + x - 3x^2$

51. $y = 5x^2 - 6x + 1$

54. $y = 5x^2 - 2x$

52. $y = 8 - 2x - x^2$

55. $y = x - 3x^2$

56. $y = 4x^2 - 8$

58. $y = x^2 + 2$

57. $y = 9 - x^2$

59. $y = x^2 + 2x + 3$

Chapter 9 Test

Directions   Find the solution set.

1. $x^2 = 4x + 12$

2. $x^2 = 18$

3. $x^2 + 6x + 4 = 0$

4. $x^2 = 8x + 4$

5. $2x^2 + x - 4 = 0$

6. $9a^2 - 14 = 0$

7. $\frac{2}{3}x^2 + 3x + \frac{4}{3} = 0$

8. $y^2 + y + 4 = 0$

9. $(x + 5)(x + 2) = 4$

10. $y^2 = 2y - 6$

11. $(a + 5)^2 = 15$

12. $2x^2 + 11x - 21 = 0$

Directions   Find the solution set by completing the square.

13. $x^2 - 8x + 4 = 0$

14. $x^2 - 5x - 7 = 0$

Directions Determine the type of solution(s) that the following equations yield, using the discriminant.

15. $5x^2 - 2x + 7 = 0$

16. $9x^2 - 12x + 4 = 0$

Directions Perform the indicated operations. Write the answer in the form $a + bi$ or $a - bi$.

17. $(2 - \sqrt{-5}) - (1 + \sqrt{-5})$

18. $(3 + 4i)^2$

19. $\dfrac{3 + i}{2 - i}$

Directions Graph the following equations. Give the vertex, the equation of the axis of symmetry, and the points at which the graph crosses each axis.

20. $y = 4 - x^2$

21. $y = x^2 + 6x + 8$

23. When a ball on the moon is thrown upward with an initial velocity of 6 meters per second, its approximate height h after t seconds is given by $h = -0.8t^2 + 6t$. When will it reach its highest point? What is the highest point? When will it strike the ground?

24. A rectangle has an area of 65 square meters. If the length of the rectangle is 3 meters more than twice the width, find the dimensions of the rectangle.

22. A square has a side of length s. If the sum of the numerical values of the area and the perimeter of the square is 21, find s.

Final Examination

1. Insert the proper inequality symbol, $<$ or $>$, to make the statement $|-5| \quad |4|$ true.

Directions Perform the indicated operations and simplify the expression.

2. $38 - 10 \div 5 + 3 \cdot 4 - 2^3$

3. $-4[9 - 3(9 - 4) + 6]$

4. Evaluate the expression $a - b(2c - d)$ when $a = -4$, $b = 3$, $c = 5$, and $d = -6$.

Directions Simplify the following and leave the answers with only positive exponents. Assume that all variables represent nonzero numbers.

5. $x^5 \cdot x^{-3} \cdot x^4$

6. $\dfrac{2x^{-3}}{4x^2}$

7. $(3x^2 y^3)(-4xy^2)$

8. $(3xy^{-2})^{-3}$

9. $(7x^3 z^2)^0$

Directions Remove the grouping symbols and combine like terms.

10. $(4x^2 - y^2) - (2x^2 + y^2) + (5x^2 - 6y^2)$

11. $5x - (x - y) - 2x + y - (2x + 5y)$

Directions Perform the indicated operations and simplify.

12. $(y + 9)(y - 9)$

13. $(7z - 3w)^2$

14. $(x + 4)(5x^2 - 3x + 1)$

15. $\dfrac{5xy^2 - 3x^4 y + x^2 y^2}{xy}$

16. $(8y^2 - 2y - 3) \div (2y - 1)$

Directions Find the solution set of the following equations.

17. $2(x + 1) - 3(x - 3) = 4$

18. $x^2 - 11x - 12 = 0$

19. $3 + \dfrac{2}{x^2} - \dfrac{7}{x} = 0$

20. $\dfrac{3x}{6} - 2 = \dfrac{5x}{4}$

21. $8x^2 = 12x$

Directions Completely factor the following expressions.

22. $3x^2 - 6xy + 9x$

23. $a^2 - 4a - 21$

24. $4x^2 - 12x + 5$

25. $9a^2 - 64$

26. $6ax - 2ay + 3bx - by$

27. $x^2 - 10x + 25$

28. The product of two consecutive integers is 132. Find the integers.

Directions Perform the indicated operations and reduce to lowest terms.

29. $\dfrac{x^2 + 7x + 6}{x^2 - 4} \cdot \dfrac{x - 2}{x + 6}$

30. $\dfrac{3x}{4x - 8} \div \dfrac{9x}{x^2 - 4x + 4}$

31. $\dfrac{9}{x - 6} - \dfrac{5}{6 - x}$

32. $\dfrac{x - 2}{x + 5} + \dfrac{x + 4}{x^2 - 25}$

33. Simplify the complex fraction $\dfrac{5 + \dfrac{4}{y}}{4 - \dfrac{6}{y}}$.

34. Find the value of x if $15 : 6 = 8 : x$.

35. What is the ratio of 42 oz to 5 lb?

36. Find the equation of the line passing through points $(-2, 5)$ and $(1, -1)$. Write the answer in standard form.

37. Given the equation $2x - 3y = 9$, find the slope m and the y-intercept b of the line.

38. Find the solution set of the system of linear equations
$$x - 2y = 3$$
$$2x - 3y = -5.$$

39. The perimeter of a rectangle is 34 feet. If the length is 2 more than twice the width, what are the dimensions of the rectangle?

Directions   Simplify the following expressions by performing the indicated operations. Rationalize all denominators.

40. $\sqrt{27} - \sqrt{48}$

41. $\sqrt{3}(\sqrt{2} + \sqrt{3})$

42. $(4 + \sqrt{3})(4 - \sqrt{3})$

43. $(2 - \sqrt{7})^2$

44. $\dfrac{3}{3 - \sqrt{5}}$

45. $\sqrt[3]{-27}$

46. $(16)^{3/4}$

47. Find the solution set of the equation $\sqrt{x + 1} - 1 = x.$

48. Graph the linear equation $3x + 2y = 12$ using the x- and y-intercepts.

49. Sketch the graph of $y = x^2 - 5x - 6$ using the vertex, x- and y-intercepts, and four arbitrary points.

50. Find the solution set of the quadratic equation $4x^2 - 7x = 3$ using any method.

Directions   Perform the indicated operations on the following complex numbers. Leave your answer in the form of $a + bi$ or $a - bi$.

51. $(3 - 9i) - (2 + 10i)$

52. $(5 + 7i)^2$

53. $(1 - \sqrt{-5})(1 + \sqrt{-5})$

54. $\dfrac{3 + 4i}{2 - 3i}$

55. A druggist wishes to mix two different solutions, one containing 40% hydrogen peroxide and 60% water with another solution containing 20% hydrogen peroxide and 80% water to obtain 200 liters of a solution that is 30% hydrogen peroxide and 70% water. How many liters of each solution should she use?

56. Fumiko invested $31,000 in two stocks. She had an 11% profit on one investment and an 8% loss on the other investment. If her net profit was $370 for the year, how much did she have in each investment?

57. A photograph that is 8 inches by 12 inches is to be enlarged. If the enlargement calls for the longest side to be 32 inches, how many inches should the other side be?

58. Two lots are proportional in their lengths and widths. If the larger lot is 15 feet wide and 28 feet long and the length of the smaller lot is 24 feet, how wide is the smaller lot?

59. The area of a rectangle is 42 square feet. If the length is 1 foot longer than the width, what are the dimensions of the rectangle?

60. The area of a tennis court is 2,800 sq. ft. Find the length of the court if the length is $3\dfrac{1}{9}$ times the width.

61. Megan rows her boat 18 miles downstream in $2\frac{1}{2}$ hours. It takes her 6 hours to row back upstream. How fast can she row in still water? What is the speed of the current?

62. A particular projectile is distance d in feet from its starting point after t seconds of time has elapsed according to the formula $d = 2t^2 - 7t + 3$. How many seconds will it take to travel 12 feet?

63. A man lays out a rectangular garden that has a diagonal 13 yards long. If the length of the garden is 7 yards longer than the width, what are the dimensions of the garden?

64. Two airplanes leave from the same airport, one flying west and the other flying south. In 2 hours they are 500 miles apart. If one plane flies 50 miles per hour faster than the other, what is the speed of each plane?

Introduction to Graphing Calculators

Introduction

This book illustrates how to use basic scientific calculators where appropriate. This appendix is designed to show how to use the Texas Instruments TI-81, TI-82, TI-85, and the Casio fx-7700GE and Casio fx-9700GE graphing calculators on material found in this book. It should also help those with other brands or models to relate the examples to their particular calculator.

No textbook can fully support all of the variations of calculators that exist now or will appear in the future. The user of computing devices must also rely on the handbooks and manuals that accompany these sophisticated, complex devices.

Some Graphing Calculator Terminology

Multifunction Keys

Many keys have three functions. An example on the TI-81 is the x^2 key. On the calculator it looks like figure A–1. This means that this key serves three purposes:

Figure A–1

x^2	x^2	Square a value
$\sqrt{\ }$	2nd x^2	Compute the square root of a value
I	ALPHA x^2	The letter I

On the Casio, SHIFT is used instead of 2nd, but the idea is the same on all of these and other calculators.

To select the square root function we would show the keystrokes 2nd √, not 2nd x^2. This convention makes reading a sequence of keystrokes much more meaningful.[1] Thus, if 2nd (or SHIFT) or ALPHA precedes a keystroke, look for the symbol representing that key over the calculator's keys, not on the keys themselves.

Cursor Keys

Figure A–2

When manipulating graphs and some menus it is necessary to move a point, often called the cursor, around on the screen. All five of these calculators have four cursor keys for this purpose. They look something like figure A–2. We will refer to them often. They are also referred to as the up arrow, or right arrow, etc. keys.

Note The Casio also has one key that looks like →. This is used to assign values to memory locations—it is not a cursor key. See the section on using memory.

Menus

All of these calculators use menus to effect many of their capabilities. Without menus hundreds of keys might be necessary. A menu item is often selected with one of the function keys F1 ... F5. We use a colon notation to show which key is used and which menu is being selected. For example, F3:ZOOM means that there is a ZOOM menu item on the calculator's screen and it is selected with the F3 key. The symbol +:CAL would mean that the + key is to be used to select the CAL menu item.

TI-81/82/Casio The TI-81/82 and the Casios only show one menu on the screen at a time. To exit these menus you generally use 2nd QUIT on the TI-81/82 and EXIT on the Casios.

TI-85 This device sometimes shows submenus. An example is produced by the following sequence. To return from the zoom submenu to the graphing menu use EXIT. To return to the normal, or calculating, screen use EXIT again. The ▶ symbol means there are more menu choices; these can be seen by selecting MORE.

---

[1]The authors acknowledge that this convention is taken from excellent calculator manuals by John Paulling.

Keystroke	Effect	Screen display

Keystroke	Effect	Screen display				
GRAPH	Display graphing menu	y(x)=	RANGE	ZOOM	TRACE	GRAPH ▶
F3:ZOOM	Display the ZOOM submenu	y(x)= RANGE ZOOM TRACE GRAPH				
		BOX	ZIN	ZOUT	ZSTD	ZPREV ▶

It takes practice to learn the ins and outs of menus. When following examples in this text or in your calculator's handbook, always watch the screen to see what effect a certain keystroke has. Do not blindly follow keystrokes. If things get too confusing, begin again and gain more familiarity with the menus.

Standard Graphing Calculator Mode Settings

Calculators operate in different "modes." This is an undesirable but necessary characteristic of these devices in which the same keystrokes can have different effects.

The mode to which a calculator is set governs things like how numbers are displayed (fixed, scientific, engineering), the assumed base of numbers (2, 8, 10, 16), graph type (rectangular, polar, parametric), and many other characteristics.

In this book we always assume the calculator is set to graph equations in a rectangular coordinate system and that numbers are displayed in a normal format. The following table shows a standard configuration we will assume for each calculator, and how to make any necessary changes to the configuration.

TI-81	TI-82	TI-85
Use the MODE key. Use the up/down arrow keys to select, ENTER to change.	Use the MODE key. Use the up/down arrow keys to select, ENTER to change.	Use the 2nd MODE key. Use the up/down arrow keys to select, ENTER to change.
Norm Sci Eng **Float** 0123456789 **Rad** Deg **Function** Param **Connected** Dot **Sequence** Simul **Grid Off** Grid On **Rect** Polar	**Normal** Sci Eng **Float** 0123456789 **Radian** Degree **Func** Par Pol Seq **Connected** Dot **Sequential** Simul **FullScreen** Split	**Norm** Sci Eng **Float** 012345678901 **Radian** Degree **RectC** PolarC **Func** Pol Param DifEq **Dec** Bin Oct Hex **RectV** CylV SphereV **dxDer1** dxNDer

Casios

Choose the COMP menu when the machine is turned on (Press 1 for this). The modes are displayed. These should be 　　RUN / COMP G-type 　　: RECT/CONNECT angle 　　: Deg display 　　: Nrm2 M-D/Cpy : M-Disp	Use SHIFT SET UP and the up and down cursor keys to reset G-type (Graph type) and M-D/Cpy. Use SHIFT DISP F3:Nrm EXE to reset display to Nrm2. Use SHIFT DRG F1:Deg EXE to reset angle.

Note The display differs slightly between the Casio 7700GE and 9700GE models. When there is a difference, the 7700GE display will be used.

Numeric Calculating
on Graphing Calculators

Modern calculators can do much of the arithmetic in this book. The following table shows how to perform some typical basic calculations on the calculators mentioned above.

	Result	TI-81/82	TI-85	Casios
$\dfrac{6+9}{-3}$ See note 1.	-5	[(] 6 [+] 9 [)] [÷] [(−)] 3 [ENTER]	[(] 6 [+] 9 [)] [÷] [(−)] 3 [ENTER]	[(] 6 [+] 9 [)] [÷] [(−)] 3 [EXE]
$6+\dfrac{9}{-3}$	3	6 [+] 9 [÷] [(−)] 3 [ENTER]	6 [+] 9 [÷] [(−)] 3 [ENTER]	6 [+] 9 [÷] [(−)] 3 [EXE]
$(5+2)(8-3)$	35	[(] 5 [+] 2 [)] [(] 8 [−] 3 [)] [ENTER]	[(] 5 [+] 2 [)] [(] 8 [−] 3 [)] [ENTER]	[(] 5 [+] 2 [)] [(] 8 [−] 3 [)] [EXE]
$8+(-3)(2)^2$	-4	8 [+] [(−)] 3 [×] 2 [x^2] [ENTER]	8 [+] [(−)] 3 [×] 2 [x^2] [ENTER]	8 [+] [(−)] 3 [×] 2 [x^2] [EXE]
$\sqrt{13^2-12^2}$	5	[2nd] [√] [(] 13 [x^2] [−] 12 [x^2] [)] [ENTER]	[2nd] [√] [(] 13 [x^2] [−] 12 [x^2] [)] [ENTER]	[SHIFT] [√] [(] 13 [x^2] [−] 12 [x^2] [)] [EXE]
$\dfrac{1}{2}+\dfrac{1}{3}$	$0.8333\ldots$	1 [÷] 2 [+] 1 [÷] 3 [ENTER]	1 [÷] 2 [+] 1 [÷] 3 [ENTER]	1 [÷] 2 [+] 1 [÷] 3 [EXE]
$\dfrac{1}{2}+\dfrac{1}{3}$ on calculators with a fraction key (note 2) (Casios) or function (note 3) (TI-82/85)	$\dfrac{5}{6}$	1 [÷] 2 [+] 1 [÷] 3 [MATH] [1:▶Frac] [ENTER]	1 [÷] 2 [+] 1 [÷] 3 [2nd] [MATH] [F5:MISC] [MORE] [F1:▶Frac] [ENTER]	1 [$a^b/_c$] 2 [+] 1 [$a^b/_c$] 3 [EXE]

Note 1. Use the [(−)] key to enter a negative sign; use the [−] key for subtraction.

Note 2. Fraction key: Many calculators, including the Casios, have a key that computes numeric fractions exactly, up to certain limits. To enter $5\dfrac{1}{3} \div 2\dfrac{3}{8}$, enter 5 [$a^b/_c$] 1 [$a^b/_c$] 3 [÷] 2 [$a^b/_c$] 3 [$a^b/_c$] 8 [EXE] to see the result $2\dfrac{14}{57}$.

To see the answer as the improper fraction $\dfrac{128}{57}$, select [SHIFT] [d/c].

Note 3. Fraction function: The TI-82 and TI-85 will convert a decimal result into fractional form when possible, up to certain limits. To select the [▶Frac] feature, use [MATH] [1:▶Frac] on the TI-82 and on the TI-85 use [2nd] [MATH] [F5:MISC] [MORE] [F1:▶Frac] as shown in the example in the table. See the following discussion on adding this feature to the custom menu on the TI-85.

The TI-85's Custom Menu

If you want to use the ▶Frac feature on the TI-85 often, add it to the TI-85's custom menu. Do this as follows.

2nd CATALOG ·   This gets one near the bottom of the catalog menu.

Select F1 seven times.   Use PAGE↓ to find ▶Frac in the menu.

Select ▶Frac by using the down arrow key.

F3   Select the CUSTM menu item.

Select whichever of F1 to F5 corresponds to an empty slot in the custom menu. Use MORE if all are already filled.

EXIT EXIT

Now, to convert, say, 0.2 to fraction form, select CUSTOM and whichever of F1 to F5 contains the ▶Frac menu item. Assuming this is F1, use 0.2 CUSTOM F1 ENTER.

    Note Any menu item can be added to the custom menu. For example use the same steps as above to add the absolute value function, except omit the · and F1 keys because the abs function is the first item in the catalog. We recommend that you create the following custom menu to accomplish the operations in this book.

▶Frac	abs	round	$\sqrt[x]{\ }$	▶

Using Values Stored in Memory—Variables

Graphing calculators can store values in memory. The Texas Instruments use STO▶ for this purpose, and the Casios use →. For example, on the TI-81/85 models 12 + 8 STO▶ B ENTER will store the value 20 into a memory location called B. On the TI-82 it is 12 + 8 STO▶ ALPHA B ENTER. It is 12 + 8 → ALPHA B EXE on the Casios.

    Suppose one needed to compute the values of the five expressions shown in the table below. To compute them, one could store the value $\sqrt{23^2 - 15^2}$ into A and $\dfrac{12 + 15(-3)}{8 - 2}$ into B. Then the value of expression 2 in the following table, as an example, would be produced by 1 + ALPHA A − ALPHA B ENTER (or EXE). The table shows how to proceed, assuming A and B contain the values already described.

    Observe that, since the value in (5) is the value in (3), squared, one would store the value in (3) in another memory location, C. Then one simply uses ALPHA C to retrieve the value to square it. Different calculator models have different capacities for storing variables, but as a minimum the letters of the alphabet can always be used.

Expression	Using calculator variables	Value
1. $\quad 1 + \sqrt{23^2 - 15^2} + \dfrac{12 + 15(-3)}{8 - 2}$	1+A+B	12.936
2. $\quad 1 + \sqrt{23^2 - 15^2} - \dfrac{12 + 15(-3)}{8 - 2}$	1+A−B	23.936
3. $\quad 1 - \sqrt{23^2 - 15^2} + \dfrac{12 + 15(-3)}{8 - 2}$	1−A+B; store result in C	−21.936
4. $\quad 1 - \sqrt{23^2 - 15^2} - \dfrac{12 + 15(-3)}{8 - 2}$	1−A−B	−10.936
5. $\quad \left(1 - \sqrt{23^2 - 15^2} + \dfrac{12 + 15(-3)}{8 - 2}\right)^2$	C²	481.170

For Chapter 6—Basic Graphing Functions of a Graphing Calculator

Chapter 6 develops the idea of the *x-y* coordinate system. In addition to numeric calculations, all graphing calculators have a few graphing capabilities that we will use extensively:

1. Set the "range" for the graphing screen.
2. View the graphing screen.
3. Graph an equation that is solved for *y* in terms of *x*.
4. Trace and zoom.
5. Find an approximate value for an *x* or *y* intercept.

We will describe these capabilities in this introduction, and illustrate how they are accomplished with the graphing calculators mentioned above.

Set the Range for the Graphing Screen— the Set Mode Screen

Graphing calculators have a way to describe which part of the coordinate plane will be displayed. It is generally called setting the RANGE,[2] but on the TI-82 it is called WINDOW. Use the RANGE or WINDOW key (on the TI-85 select the RANGE choice in the menu displayed using the GRAPH key). A display similar to that in table A–1 appears. The Xmin and Xmax values refer to the range of *x*-values that will be displayed. The Ymin and Ymax values refer to the range of *y*-values that will be displayed. The Xscl and Yscl values refer to the tick marks that will appear on the axes. The Xres, if present, determines the number of *x*-values for which the expression will be calculated. It should be left at 1. Throughout the book, we will show the Xmin, Xmax, Xscl, Ymin, Ymax, Yscl values, in this order, in a box labeled RANGE. For the values shown in table A–1 we would write RANGE −10, 10, 1, −10, 10, 1 .

```
RANGE

Xmin = −10
Xmax = 10
Xscl = 1
Ymin = −10
Ymax = 10
Yscl = 1
Xres = 1
```

Table A–1

By entering numeric values and using the ENTER or EXE key to move down the list, the values in the RANGE may be changed. Note that the (−) (change sign) key is used to enter negative numbers. Do not use the subtract key, − . The values shown here are considered the standard zoom settings on the Texas Instruments calculators. To obtain these settings quickly on these devices use ZOOM 6:Standard (TI-81/82) and GRAPH F3:ZOOM F4:ZSTD EXIT CLEAR (TI-85).

Note The TI-81 terminology is used for the TI-81 and TI-82. For the Casio, we use fx 7700GE terminology.

To leave the set mode screen, use 2nd QUIT (TI-81/82) or EXIT (TI-85 and Casios).

View the Graphing Screen

Figure A–3 on the following page shows the screen's appearance for various settings of Xmax, Xmin, Ymax, Ymin. The values of Xscl and Yscl are one (1) except where labeled Xscl = 2, Yscl = 3. After setting these RANGE values and leaving the set mode screen, use the following steps to show the graphing screen and then return to the calculation screen.

TI-81/82

Y = CLEAR {Clears any equation in Y1; clear any others by moving down with the down arrow key and using CLEAR for each equation.}
GRAPH {observe the graphing screen} 2nd QUIT

TI-85

GRAPH F1:y(x) = CLEAR {Clears any equation in Y1; clear any others by moving down with the down arrow key and using CLEAR for each equation.}
EXIT F5:GRAPH {observe the graphing screen} EXIT .

Casio

MENU 1:COMP G↔T {observe the graphing screen} G↔T .

Observe that the distance between units is not the same on the screen. The calculator automatically makes horizontal units 1.5 or 1.7 times as long as vertical units.

Equal Distances on the Vertical and Horizontal Axes

To have horizontal and vertical distances be the same on the screen, the horizontal distance represented on the screen (Xmax − Xmin) must be either 1.5 or 1.7 times the vertical distance represented (Ymax − Ymin). The following table presents information related to a square display on each calculator.

[2]Not to be confused with the mathematical terms domain and range covered in this text.

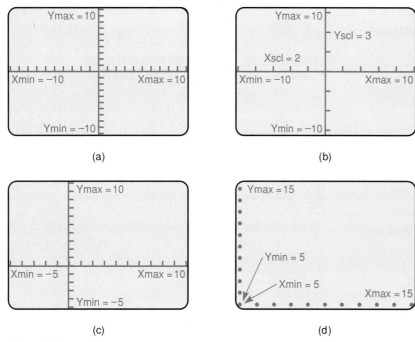

Figure A–3

	TI-81	TI-82	TI-85	Casios
Standard range settings	−10,10,1, −10,10,1	−10,10,1, −10,10,1	−10,10,1, −10,10,1	−10,10,1, −10,10,1
Square display settings	−15,15,1, −10,10,1	−15,15,1, −10,10,1	−17,17,1, −10,10,1	−17,17,1, −10,10,1
Keystrokes to get the square display settings; see note.	ZOOM 6:Standard ZOOM 5:Square	ZOOM 6:Z Standard ZOOM 5:Z Square	GRAPH F3:ZOOM F4:ZSTD MORE F2:ZSQR	Enter the square range settings directly using RANGE.

Note The values ±15, ±17 in the square display settings line are very accurate approximations to the exact values required.

Graph an Equation in Which y Is Solved for in Terms of x

If an equation is solved for the variable y in terms of the variable x, the graphing calculator can be used to view the graph of the equation.

Example A

Graph the straight line $y - 2x + 3 = 0$.

First solve the equation for y:

$$y - 2x + 3 = 0$$
$$y = 2x - 3$$

Assuming the calculator is set to the standard RANGE settings discussed above proceed as follows to obtain the graph. The figure shows what the display will look like.

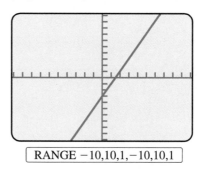

RANGE −10,10,1,−10,10,1

TI-81/82		TI-85	
$\boxed{Y=}$	Allows entry of up to four (ten) equations. See note 1.	$\boxed{\text{GRAPH}}$ $\boxed{\text{F1:y(x)=}}$	Allows entry of many equations. See note 1.
2 $\boxed{X\|T}$ $\boxed{-}$ 3	Enter the expression that represents y. See note 2.	2 $\boxed{\text{x-VAR}}$ $\boxed{-}$ 3	Enter the expression that represents y.
The display looks like $\begin{array}{l} : Y_1 = 2X - 3 \\ : Y_2 = \\ : Y_3 = \\ : Y_4 = \end{array}$		The display looks like $\boxed{\text{y1} = 2\text{x} - 3}$	
$\boxed{\text{ZOOM}}$ $\boxed{\text{6:Standard}}$		$\boxed{\text{EXIT}}$ $\boxed{\text{F3:ZOOM}}$ $\boxed{\text{F4:ZSTD}}$	
		When the graph is visible use the $\boxed{\text{CLEAR}}$ key to remove the menu from the bottom of the display. Use the $\boxed{\text{EXIT}}$ key to restore the menu.	

Note 1. If there are any equations already entered for Y_1 use the $\boxed{\text{CLEAR}}$ key before entering the equation. If there are any extra equations entered for Y_2, Y_3, or Y_4, move down with the down arrow key to that equation and use the $\boxed{\text{CLEAR}}$ key to clear that entry.

Note 2. On the TI-82 the key $\boxed{\text{X,T,}\theta}$ is used to enter the x variable, not $\boxed{X\|T}$.

Casios

Arrange to enter equations. Use the GRAPH mode. See note.	$\boxed{\text{MENU}}$ $\boxed{\text{6:Graph}}$
Enter equation into Y1.	2 $\boxed{\text{X,}\theta\text{,T}}$ $\boxed{-}$ 3 $\boxed{\text{F1:STO}}$ $\boxed{\text{F6:SET}}$
Enter the range.	$\boxed{\text{RANGE}}$ $\boxed{(-)}$ 10 $\boxed{\text{EXE}}$ 10 $\boxed{\text{EXE}}$ 1 $\boxed{\text{EXE}}$ $\boxed{(-)}$ 10 $\boxed{\text{EXE}}$ 10 $\boxed{\text{EXE}}$ 1 $\boxed{\text{EXE}}$
Leave the range setting mode.	$\boxed{\text{EXIT}}$
Construct the graph.	$\boxed{\text{F6:DRW}}$
Go back to compute mode.	$\boxed{\text{MENU}}$ $\boxed{\text{1:COMP}}$

Note If there are other equations in Y2, Y3, etc. clear them by storing a "blank" equation in them. To do this, press $\boxed{\text{AC/on}}$[3] $\boxed{\text{F1:STO}}$; position the cursor next to the equation to be cleared; select $\boxed{\text{F6:SET}}$. ●

For Chapter 7—Tracing, Zooming, and Graphing More than One Equation at a Time

A calculator's trace capability displays the x- and y-coordinates for a particular point. A zoom capability allows us to expand a graph around some particular point. We will discuss two methods of zooming.

Zooming by Expanding the Display by a Fixed Multiple

We will illustrate this by finding approximate values for the coordinates of the point where two straight lines intersect.

[3]The $\boxed{\text{AC/on}}$ key on the Casio is the $\boxed{\text{CLEAR}}$ key.

Example B

Graph the two straight lines $y = 2x - 3$ and $y = -\dfrac{1}{3}x + 2$. Then estimate the coordinates of the point where these two lines intersect.

To graph both lines using the standard RANGE settings proceed as follows for each calculator.

TI-81/82

$\boxed{Y=}$ $\boxed{\text{CLEAR}}$ 2 $\boxed{X\|T}$ $\boxed{-}$ 3   At this point the display is: $Y_1 = 2X - 3$

$\boxed{\text{ENTER}}$ $\boxed{\text{CLEAR}}$

$\boxed{(}$ $\boxed{(-)}$ 1 $\boxed{\div}$ 3 $\boxed{)}$ $\boxed{X\|T}$ $\boxed{+}$ 2   The display is: $Y_2 = (-1/3)X + 2$

At this point Y1 and Y2 are used. Y3 and Y4 are also available to store up to two more equations.

$\boxed{\text{ZOOM}}$ $\boxed{\text{6:Standard}}$ Standard settings

The graph shown appears. Select $\boxed{\text{TRACE}}$ and move the little box until it appears over the point of intersection by using the cursor keys.

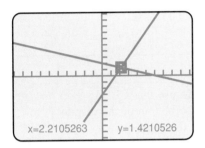

This will show that the coordinates at this point are approximately (2.1,1.3). To get a better picture expand the display. Use $\boxed{\text{ZOOM}}$ $\boxed{\text{2:Zoom In}}$ $\boxed{\text{ENTER}}$. This expands the graph around the point selected. It produces a new graph, shown.

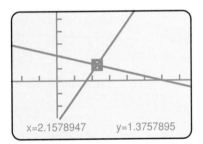

Tracing again will show that the coordinate of the point where the lines intersect is about (2.12,1.26). Repeatedly zooming and tracing will show that $x \approx 2.14$ and $y \approx 1.29$. Using methods shown in section 7–2 we could show that x is exactly $2\frac{1}{7}$ and y is exactly $1\frac{2}{7}$.

TI-85
$\boxed{\text{GRAPH}}$ $\boxed{\text{F1:y(x)=}}$ $\boxed{\text{CLEAR}}$ 2 $\boxed{\text{x-VAR}}$ $\boxed{-}$ 3
$\boxed{\text{ENTER}}$ $\boxed{\text{CLEAR}}$ $\boxed{(}$ $\boxed{(-)}$ 1 $\boxed{\div}$ 3 $\boxed{)}$ $\boxed{\text{x-VAR}}$
$\boxed{+}$ 2 $\boxed{\text{EXIT}}$

At this point the two equations are stored as y1 and y2. By using $\boxed{\text{ENTER}}$ and if necessary $\boxed{\text{CLEAR}}$ a practically unlimited number of equations can be stored. $\boxed{\text{F3:ZOOM}}$ $\boxed{\text{F4:ZSTD}}$ At this point the two equations are graphed. $\boxed{\text{F4:TRACE}}$ Use the right and left cursor keys to position the little box cursor over the point of intersection. The coordinates should be about (2.2,1.4).

Then zoom in as follows: $\boxed{\text{EXIT}}$ (redisplays the graph menu) $\boxed{\text{F3:ZOOM}}$ $\boxed{\text{F2:ZIN}}$ $\boxed{\text{ENTER}}$. After the screen is redrawn use $\boxed{\text{EXIT}}$ $\boxed{\text{EXIT}}$ to redisplay the menu and use $\boxed{\text{F4:TRACE}}$ and the cursor keys again to get a better approximation of the point of intersection. As noted above the actual values are $\left(2\frac{1}{7},1\frac{2}{7}\right)$.

Casios
Use the graph mode. When the calculator is turned on, select the GRAPH mode by using 6. Enter each equation as follows.
2 $\boxed{\text{X,θ,T}}$ $\boxed{-}$ 3 $\boxed{\text{F1:STO}}$; Now use the up/down cursor keys to place the cursor at Y1, then use $\boxed{\text{F6:SET}}$. Now, enter $\boxed{(}$ $\boxed{(-)}$ 1 $\boxed{\div}$ 3 $\boxed{)}$ $\boxed{\text{X,θ,T}}$ $\boxed{+}$ 2 $\boxed{\text{F1:STO}}$; now use the up/down cursor keys to place the cursor at Y2, then use $\boxed{\text{F6:SET}}$.
Reset the range: $\boxed{\text{RANGE}}$ $\boxed{(-)}$ 10 $\boxed{\text{EXE}}$ 10 $\boxed{\text{EXE}}$ 1 $\boxed{\text{EXE}}$ $\boxed{(-)}$ 10 $\boxed{\text{EXE}}$ 10 $\boxed{\text{EXE}}$ 1 $\boxed{\text{EXIT}}$ Then use $\boxed{\text{F6:DRAW}}$. Press $\boxed{\text{F1:Trace}}$ and use the cursor keys to move the blinking cursor along either line. Use the up/down cursor keys to select which line is being traced. Then use $\boxed{\text{F2:Zoom}}$ $\boxed{\text{F3:xf}}$ to redraw the expanded graph. Then use $\boxed{\text{EXIT}}$ $\boxed{\text{F1:TRCE}}$ to trace to the point of intersection again. Use $\boxed{\text{EXIT}}$ to remove any menu at the bottom of the display, which will enable you to see the x- and y-coordinates. As noted above the actual values are $\left(2\frac{1}{7},1\frac{2}{7}\right)$.

All these calculators have ways to change the amount by which the display is expanded when zooming. On the TI-81/82 it is $\boxed{\text{ZOOM}}$ (On the TI-82 $\boxed{\blacktriangleright}$ $\boxed{\text{MEMORY}}$) $\boxed{\text{4:Set Factors}}$; use $\boxed{\text{2nd}}$ $\boxed{\text{QUIT}}$ to return to the previous screen. On the TI-85 it is $\boxed{\text{GRAPH}}$ $\boxed{\text{F3:ZOOM}}$ $\boxed{\text{MORE}}$ $\boxed{\text{MORE}}$ $\boxed{\text{F1:ZFACT}}$; use $\boxed{\text{EXIT}}$ to return to the previous display. On the Casios use COMP mode ($\boxed{\text{MENU}}$ $\boxed{\text{1:COMP}}$), then select $\boxed{\text{SHIFT}}$ $\boxed{\text{Zoom}}$ $\boxed{\text{F2:FCT}}$; use $\boxed{\text{EXIT}}$ $\boxed{\text{MENU}}$ $\boxed{\text{1:COMP}}$ to return to the previous display.

Zooming with the Box Feature   Another zoom feature is the zoom box. The idea is to enclose that part of a graph that one wants expanded by a box. This automatically resets the RANGE and regraphs as required. The generic sequence is the same in all three calculators:

1. Select Zoom, then Box.
2. Use the cursor keys to place the cursor at one corner of the desired box. Usually this means placing the cursor to a point above and to the left of the point in which you are interested.
3. Use $\boxed{\text{ENTER}}$ or $\boxed{\text{EXE}}$ to mark the corner.

4. Use the cursor keys to place the cursor at a diagonally opposite corner of the desired box. As you move the point (probably down and to the right) a box will be drawn on the screen. Include the point of interest inside the box.
5. Use ENTER or EXE a second time to expand the display.

This feature is illustrated in example C.

For Chapter 9—Find an Approximate Value for an *x*-Intercept

An *x*-intercept of a graph is a point where a graph intersects the *x*-axis (horizontal axis), and a *y*-intercept is a point where a graph intersects the *y*-axis (vertical axis). To find approximations of these intercepts it is only necessary to graph the function, as in example A, and use the TRACE and ZOOM calculator features to find approximations to the coordinates at an intercept. Remember that the *y*-coordinate for an *x*-intercept is always 0, and the *x*-coordinate for a *y*-intercept is also always 0.

Example C illustrates how to graph a parabola and find decimal approximations of its *x*-intercepts.

Example C
Graph $y = 2x^2 - x - 3$ and find approximate values for its *x*-intercepts. Use the zoom box feature.

The figure shows the graph for the RANGE shown. It is clear that we need to expand the graph to see the right *x*-intercept better. The following steps show how to create the box shown in the second figure, which creates the third figure.

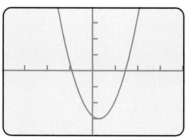

RANGE −4,4,1,−4,4,1

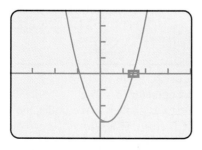

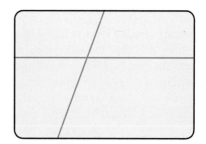

Enter the equation into the calculator; store it in Y1. Use example A as a guide for graphing equations. Clear all other equations.

TI-81/82
2 [X | T] [x^2] [−] [X | T] [−] 3

TI-85
2 [x-VAR] [x^2] [−] [x-VAR] [−] 3

Casio
2 [X,θ,T] [x^2] [−] [X,θ,T] [−] 3

Graph the equation after entering the equation and the range shown above. Next create the zoom box and expand the display. This is done as follows.

Step 1: Select box zoom

TI-81/82
[ZOOM] [1:Box]

TI-85
After graphing, the main graphing menu should be on the screen. If not, press [GRAPH]. Then select [F3:ZOOM] [F1:BOX].

Casios
[SHIFT] [F2:Zoom] [F1:BOX]

Step 2 (all calculators): Draw the box
Move the cursor so X ≈ 1.2 and Y ≈ 0.2. Then select [ENTER] or [EXE]. Now move the cursor right and down so that X ≈ 1.8 and Y ≈ −0.2. Then select [ENTER] or [EXE] again.

Next use the TRACE feature.

Step 1: Select the trace feature

TI-81/82
[TRACE]

TI-85
After using box zoom the zoom menu is still in the display. Use [EXIT] to get back to the main graphing menu, then select [F4:TRACE].

Casios
[SHIFT] [F1:Trace]

Step 2 (all calculators)

Use the right and left cursor (arrow) keys to move the trace cursor along the graph. Observe the *y*-coordinate. Move the cursor so it gets smaller. When it is 0, or when it changes sign, the cursor is at or near an intercept. It should be possible to verify that the *x*-coordinate at the *x*-intercept is approximately 1.5. To get better and better approximations of the value of *x* at the intercepts, repeat the zoom box process.

To find an approximation to the other intercept, reenter the original RANGE and use the zoom box feature at the other intercept. This should show that the intercept is -1.

The methods of chapter 9, or factoring, can be used to show that the exact values of the *x*-intercepts are -1 and 1.5. The TI-82/85 and Casio fx 9700 GE also have a built-in capability to find these numeric values automatically. On the TI-82 it is [2nd] [CALC] [2:root], and on the TI-85 it is in [GRAPH] [MORE] [F1:MATH] [F3:ROOT] (the cursor keys and enter key will also be needed in finding roots). On the Casio fx 9700 GE use [SHIFT] [G•SOLV] [F1:Root]; use the right cursor (once) to see both roots.

Summary and a Note

At this point you should have some idea about how to

- Set the RANGE for a graph.
- Graph an equation that is solved for *y* in terms of *x*.
- Use trace and zoom to expand a particular part of a graph.
- Find approximate values for the *x*- and *y*-intercepts for a particular graph.

These capabilities and others can be used throughout the text, wherever they are appropriate.

Note that we have not discussed ways in which to find the best RANGE settings for a graph. In fact the answer to that only comes from experience with the type of equation one is graphing. In fact, there might even be many RANGE settings that are useful for a particular graph—a wide view might be necessary to get an overall view of the shape of a graph, but an expanded view of some parts of the graph may also be necessary to best understand the graph.

Entering Algebraic Expressions into the Calculator

The following table can be of help in entering a mathematical expression into a specific calculator.

Algebraic notation	Generic calculator notation	TI-81/82 See note 1.	TI-85	Casios (COMP Mode)
$-5x^4$	$-5X^4$	[(−)] 5 [X|T] [^] 4	[(−)] 5 [x-VAR] [^] 4	[(−)] 5 [X,θ,T] [^] 4
$7x^3 - 2x^2$	$7X^3 - 2X^2$	7 [X|T] [MATH] 3 [−] 2 [X|T] [x^2] See note 2.	7 [x-VAR] [^] 3 [−] 2 [x-VAR] [x^2]	7 [X,θ,T] [^] 3 [−] 2 [X,θ,T] [x^2]
$\|-x + y\|$	abs(−X+Y)	[2nd] [ABS] [(] [(−)] [X|T] [+] [ALPHA] [Y] [)]	[CUSTOM] [F2:abs] [(] [(−)] [x-VAR] [+] [ALPHA] [Y] [)] See note 3.	[SHIFT] [MATH] [F3:NUM] [F1:Abs] [(] [(−)] [X,θ,T] [+] [ALPHA] [Y] [)]
$x - \dfrac{}{\sqrt{x+1}}$	X− √ (X+1)	[X|T] [−] [2nd] [√] [(] [X|T] [+] 1 [)]	[x-VAR] [−] [2nd] [√] [(] [x-VAR] [+] 1 [)]	[X,θ,T] [−] [SHIFT] [√] [(] [X,θ,T] [+] 1 [)]
$\dfrac{x+3}{x-3}$	(X+3)(X−3)	[(] [X|T] [+] 3 [)] [÷] [(] [X|T] [−] 3 [)]	[(] [x-VAR] [+] 3 [)] [÷] [(] [x-VAR] [−] 3 [)]	[(] [X,θ,T] [+] 3 [)] [÷] [(] [X,θ,T] [−] 3 [)]
	(X+3)(X−3)⁻¹	[(] [X|T] [+] 3 [)] [(] [X|T] [−] 3 [)] [x^{-1}]	[(] [x-VAR] [+] 3 [)] [(] [x-VAR] [−] 3 [)] [2nd] [x^{-1}]	[(] [X,θ,T] [+] 3 [)] [(] [X,θ,T] [−] 3 [)] [SHIFT] [x^{-1}]

Note 1. Remember that on the TI-82 the variable *x* is entered with [X,θ,T], not [X\|T].

Note 2. The TI-81/82 calculator has an x^3 operator built in. All calculators have an x^2 operator.

Note 3. Assumes the absolute value function was entered into the TI-85's custom menu as shown above under the TI-85's CUSTOM MENU.

Answers and Solutions

Chapter 1

Exercise 1–1
Answers to Odd-Numbered Problems
1. Readjust your schedule to allow adequate time to succeed.
3. Schedules will vary
5. You can write important facts on note cards, which can be reviewed whenever you have a few extra minutes.
7. the preface
9. Immediately correct any errors and learn the material that you missed on the test.

Exercise 1–2
Answers to Odd-Numbered Problems
1. $\frac{1}{2}$ 3. $\frac{5}{6}$ 5. $\frac{8}{9}$ 7. $\frac{7}{9}$ 9. 2 11. $\frac{20}{17}$ 13. $\frac{1}{2}$ 15. $\frac{49}{96}$ 17. $\frac{7}{12}$

19. $\frac{15}{28}$ 21. $\frac{2}{7}$ 23. $\frac{25}{17}$ or $1\frac{8}{17}$ 25. $\frac{51}{7}$ or $7\frac{2}{7}$ 27. $\frac{132}{7}$ or $18\frac{6}{7}$

29. 12 31. 12 33. $\frac{21}{32}$ 35. $\frac{1}{5}$ 37. $\frac{32}{21}$ or $1\frac{11}{21}$ 39. $\frac{20}{7}$ or $2\frac{6}{7}$

41. a. $187\frac{17}{48}$ in.^{3} b. $31\frac{39}{64}$ in.^{3} 43. 120 45. 126 47. 144

49. 385 51. 108 53. 60 55. $\frac{2}{3}$ 57. $\frac{7}{12}$ 59. $\frac{3}{5}$ 61. $\frac{13}{8}$ or $1\frac{5}{8}$

63. $\frac{17}{5}$ or $3\frac{2}{5}$ 65. $\frac{16}{15}$ or $1\frac{1}{15}$ 67. $\frac{7}{24}$ 69. $\frac{149}{270}$ 71. $\frac{11}{20}$

73. $\frac{57}{16}$ or $3\frac{9}{16}$ 75. $\frac{5}{3}$ or $1\frac{2}{3}$ 77. $86\frac{1}{2}$ ft 79. $10\frac{11}{12}$ pounds

81. $\frac{15}{32}$ 83. 34 plants 85. $2\frac{5}{8}''$ and $2\frac{3}{8}''$

Solutions to Trial Exercise Problems
7. $\frac{28}{36} = \frac{4 \cdot 7}{4 \cdot 9} = \frac{7}{9}$

18. $\frac{3}{4} \cdot 6 = \frac{3 \cdot 2 \cdot 3}{2 \cdot 2} = \frac{3 \cdot 3}{2} = \frac{9}{2}$ or $4\frac{1}{2}$

21. $\frac{6}{7} \div 3 = \frac{6}{7} \cdot \frac{1}{3} = \frac{6 \cdot 1}{7 \cdot 3} = \frac{2 \cdot 3}{7 \cdot 3} = \frac{2}{7}$

27. $7\frac{1}{3} \cdot 2\frac{4}{7} = \frac{22}{3} \cdot \frac{18}{7} = \frac{22 \cdot 3 \cdot 3 \cdot 2}{3 \cdot 7} = \frac{22 \cdot 3 \cdot 2}{7} = \frac{132}{7}$ or $18\frac{6}{7}$

33. $\dfrac{\frac{7}{8}}{\frac{4}{3}} = \frac{7}{8} \div \frac{4}{3} = \frac{7}{8} \cdot \frac{3}{4} = \frac{7 \cdot 3}{8 \cdot 4} = \frac{21}{32}$

35. $\frac{4}{5} \cdot \frac{2}{3} \cdot \frac{3}{8} = \frac{4 \cdot 2 \cdot 3}{5 \cdot 3 \cdot 8} = \frac{1 \cdot (4 \cdot 2 \cdot 3)}{5 \cdot (4 \cdot 2 \cdot 3)} = \frac{1}{5}$

42. $61\frac{1}{2} \div 14 = \frac{123}{2} \cdot \frac{1}{14} = \frac{123}{28} = 4\frac{11}{28}$ in.

45. $6 = 2 \cdot 3$
$14 = 2 \cdot 7$
$18 = 2 \cdot 3 \cdot 3$ LCD is $2 \cdot 3 \cdot 3 \cdot 7 = 126$

61. $1 + \frac{5}{8} = \frac{8}{8} + \frac{5}{8} = \frac{8 + 5}{8} = \frac{13}{8}$ or $1\frac{5}{8}$

67. $\frac{3}{8} - \frac{1}{12}$ (LCD is 24) $= \frac{3}{8} \cdot \frac{3}{3} - \frac{1}{12} \cdot \frac{2}{2} = \frac{9}{24} - \frac{2}{24} = \frac{7}{24}$

71. $\frac{7}{15} + \frac{5}{6} - \frac{3}{4} = \frac{7}{15} \cdot \frac{4}{4} + \frac{5}{6} \cdot \frac{10}{10} - \frac{3}{4} \cdot \frac{15}{15} = \frac{28}{60} + \frac{50}{60} - \frac{45}{60}$

(LCD is 60) $= \frac{28 + 50 - 45}{60} = \frac{33}{60} = \frac{11}{20}$

73. $8\frac{3}{16} - 4\frac{5}{8} = \frac{131}{16} - \frac{37}{8}$ (LCD is 16)

$= \frac{131}{16} - \frac{74}{16} = \frac{57}{16}$ or $3\frac{9}{16}$

77. $P = 2\ell + 2w = 2 \cdot 24\frac{1}{2} + 2 \cdot 18\frac{3}{4} = 2 \cdot \frac{49}{2} + 2 \cdot \frac{75}{4}$

$= 49 + \frac{75}{2} = \frac{98}{2} + \frac{75}{2} = \frac{98 + 75}{2} = \frac{173}{2}$

$= 86\frac{1}{2}$ ft

82. Since $\frac{3}{4}$ lived before 1940, therefore, $\frac{1}{4}$ lived after 1940.

$\frac{1}{4} \cdot \frac{16}{22} = \frac{1}{4} \cdot \frac{4 \cdot 4}{2 \cdot 11} = \frac{1 \cdot 2 \cdot 2}{2 \cdot 11} = \frac{2}{11}$

84. $41 \div 1\frac{1}{4} = \frac{41}{1} \div \frac{5}{4} = \frac{41}{1} \cdot \frac{4}{5} = \frac{164}{5} = 32\frac{4}{5}$. Since the number of plants is a whole number, 32 plants can be planted. Now, $32 \cdot 1\frac{1}{4} = \frac{32}{1} \cdot \frac{5}{4} = \frac{160}{4} = 40$ and $41 - 40 = 1$. So, the plants will be placed $\frac{1}{2}$ foot from each end.

Exercise 1–3
Answers to Odd-Numbered Problems
1. $\frac{2}{5}$ 3. $\frac{3}{20}$ 5. $\frac{1}{8}$ 7. $\frac{7}{8}$ 9. 19.019 11. 540.2927 13. 13.5585
15. 156.9876 17. 1.06964 19. 9.52816 21. 0.428412
23. 0.9100081809 25. 1.2 27. 40 29. 102 31. 2,500 33. 0.15
35. 0.65 37. $0.\overline{2}$ 39. $12.91 41. 13 cardinals 43. 32.61 seconds
45. 122.28 gallons 47. 1097.222 yd^2 49. $0.05 = \frac{1}{20}$

51. $0.12 = \dfrac{3}{25}$ **53.** $1.35 = \dfrac{27}{20}$ **55.** $3.25 = \dfrac{13}{4}$ **57.** $\dfrac{4}{5}$, 80%

59. $\dfrac{27}{50}$, 54% **61.** $\dfrac{23}{20}$, 115% **63.** 0.75, 75% **65.** 0.375, 37.5%

67. 2 **69.** 33.8 **71.** 550 **73.** $166.50 **75.** $23

77. $8.50 discount, $25.50 discount price **79.** 0.96 ounces

Solutions to Trial Exercise Problems

3. $0.15 = \dfrac{15}{100} = \dfrac{3 \cdot 5}{20 \cdot 5} = \dfrac{3}{20}$

13. $10.03 + 3.113 + 0.3342 + 0.0763 + 0.005 = 10.0300$
$$
\begin{array}{r}
10.0300 \\
3.1130 \\
0.3342 \\
0.0763 \\
0.0050 \\
\hline
13.5585
\end{array}
$$

19. $(7.006)(1.36) = $
$$
\begin{array}{r}
7.006 \\
1.36 \\
\hline
42036 \\
21018 \\
7006 \\
\hline
9.52816
\end{array}
$$

28. $21.681 \div 8.03 = 2.7$
$$
\begin{array}{r}
2.7 \\
8.03\overline{)21.681} \\
16\,06 \\
\hline
5\,621 \\
5\,621 \\
\hline
0
\end{array}
$$

33. $\dfrac{3}{20} = 20\overline{)3.000} = 0.15$
$$
\begin{array}{r}
.15 \\
20\overline{)3.000} \\
20 \\
\hline
100 \\
100 \\
\hline
\end{array}
$$

37. $\dfrac{2}{9} = 9\overline{)2.000} = 0.\overline{2}$ (repeating)
$$
\begin{array}{r}
.222\ldots \\
9\overline{)2.000} \\
18 \\
\hline
20 \\
18 \\
\hline
20
\end{array}
$$

39.
$$
\begin{array}{r}
14.36 \\
0.899 \;(89.9¢ = \$0.899) \\
\hline
12924 \\
12924 \\
11488 \\
\hline
12.90964 \approx \$12.91
\end{array}
$$

51. $12\% = 12.\% = 0.12$
$12\% = \dfrac{12}{100} = \dfrac{3 \cdot 4}{25 \cdot 4} = \dfrac{3}{25}$

59. $0.54 = 0.54. = 54\%$
$= \dfrac{54}{100} = \dfrac{27}{50}$

62. $2.40 = 2\dfrac{2}{5} = \dfrac{12}{5} = 240\%$

69. 26% of $130 = 0.26 \times 130 = $
$$
\begin{array}{r}
130 \\
0.26 \\
\hline
780 \\
260 \\
\hline
33.80 = 33.8
\end{array}
$$

77. discount $= 25\%$ of 34
$= 0.25 \times 34 = \$8.50$
price $= 34.00 - 8.50 = \$25.50$

Exercise 1–4

Answers to Odd-Numbered Problems

1. {Sun, Mon, Tues, Wed, Thur, Fri, Sat}

3. {January, February, March}

5. {Jan, March, May, July, Aug, Oct, Dec}

7. {a, l, g, e, b, r} **9.** {i, n, t, e, r, m, d, a} **11.** {3, 5, 7, 9}

13. {Sunday, Saturday} **15.** −$10, $150

17. −10 yards, 16 yards **19.** −14 points, 8 points

21.

23.

25.

27.

29. −6, −4, 0, 3, 6 **31.** −11, −7, −4, −1, 2 **33.** −2, 3, 4, 5, 11

35. $-1\dfrac{3}{4}, -\dfrac{1}{2}, 1\dfrac{1}{2}, 3, 4$ **37.** $4 < 8$ **39.** $9 > 2$ **41.** $-3 > -8$

43. $-10 < -5$ **45.** $0 < 4$ **47.** $0 > -6$ **49.** 2 **51.** 5 **53.** 4

55. $\dfrac{1}{2}$ **57.** $1\dfrac{1}{2}$ **59.** $-\dfrac{5}{8}$ **61.** −6 **63.** $|5| < |-7|$

65. $|0| < |-2|$ **67.** $|-8| > |-5|$ **69.** $|-6| > |-2|$

71. $7 > |-2|$ **73.** $|-4| < 6$ **75.** $|-27|$ **77.** $|-9|$

Solutions to Trial Exercise Problems

27.

For the location of $\sqrt{2}$, we use the approximation 1.414 from a calculator.

35. $-1\dfrac{3}{4}, -\dfrac{1}{2}, 1\dfrac{1}{2}, 3, 4$. The values represent an approximation of the coordinates.

40. $-2 > -4$, since −2 lies to the right of −4 on the number line.

46. $-3 < 0$, since −3 lies to the left of 0 on the number line.

58. $-\left|-2\dfrac{3}{4}\right| = -\left(2\dfrac{3}{4}\right) = -2\dfrac{3}{4}$

67. Since $|-8| = 8$ and $|-5| = 5$ and $8 > 5$, therefore, $|-8| > |-5|$.

Exercise 1–5

Answers to Odd-Numbered Problems

1. −13 **3.** 5 **5.** −5 **7.** −5 **9.** 0 **11.** −13.6 **13.** −11.1

15. $-\dfrac{1}{2}$ **17.** $\dfrac{1}{10}$ **19.** $-\dfrac{3}{4}$ **21.** 3 **23.** 10 **25.** −44 **27.** −22

29. −10 **31.** 0 **33.** 11 **35.** 15 **37.** 7° C **39.** ($1,031.98)

41. 2 mb **43.** 100.3 degrees **45.** −41° F **47.** $64

Solutions to Trial Exercise Problems

11. $(-8.7) + (-4.9) = -13.6$ The signs are the same so we add their absolute values $8.7 + 4.9 = 13.6$ and prefix this sum by their common sign.

15. $\left(-\dfrac{1}{6}\right)+\left(-\dfrac{1}{3}\right)=-\dfrac{1}{2}$ The signs are the same so we add their

absolute values $\dfrac{1}{6}+\dfrac{1}{3}=\dfrac{1}{6}+\dfrac{2}{6}=\dfrac{1+2}{6}=\dfrac{3}{6}=\dfrac{1}{2}$ and prefix this sum by their common sign.

21. $10+(-5)+(-2)=5+(-2)=3$ The numbers were added left to right.

23. $(-12)+(-10)+(8)+(24)=(-22)+(8)+(24)$
$=(-14)+(24)=10$

34.

| The sum of | increased | 10 |
| 15 and -18 | by | |

$$15+(-18)\quad+\quad 10\ =15+(-18)+10$$
$$=-3+10$$
$$=7$$

44. Let t represent the temperature at 1 P.M. To find the new temperature, we must *add* the rise in temperature to the original temperature.

temperature	is	temperature	rose	39°
at 1 P.M.		at 8 A.M.		
t	$=$	-13	$+$	39

$t=-13+39$
$t=26$

The temperature at 1 P.M. was 26° F.

Exercise 1–6
Answers to Odd-Numbered Problems

1. -1 **3.** 6 **5.** -12 **7.** -5 **9.** -4 **11.** 14 **13.** -6 **15.** 7
17. $-\dfrac{1}{4}$ **19.** $2\dfrac{5}{8}$ **21.** -9.4 **23.** -312 **25.** 301.8 **27.** -24
29. 11 **31.** -53 **33.** 16 **35.** -3 **37.** 10 **39.** -8 **41.** -25
43. $-38°$ C **45.** $-\$372.60$ **47.** $4\dfrac{1}{4}$ feet **49.** -22 **51.** -9
53. 61.3 **55.** $19\dfrac{1}{2}$ **57.** -13 **59.** -5 **61.** \$7.15 **63.** 38°
65. 14 years old **67.** \$243.37 **69.** 20,602 feet **71.** \$18.50

Solutions to Trial Exercise Problems

13. $(-6)+0=-6$ The sum of zero and a number is that number.
15. $7-0=7$ A number minus zero is that number.
26. $-12-(-10)-8=(-12)+(10)+(-8)=(-2)+(-8)=-10$
38. $12+3-16-10-(12+5)=12+3-16-10-17=$
$15-16-10-17=-1-10-17=-11-17=-28$
53. $41.7-(-19.6)=41.7+19.6=61.3$
65. Let a represent the age that Erin will be in the year 2000. We must find the *difference* between 2000 and 1986.

age in	is	the difference
the		between
year 2000		2000 and 1986
a	$=$	$2000-1986$

$a=2000-1986$
$a=14$

Erin will be 14 years old in the year 2000.

67. Let m represent the amount of money that Amy needs to borrow.

cost of	minus	assets	is	amount
stereo				to borrow
695.00	$-$	451.63	$=$	m

$695.00-451.63=243.37$

Amy needs to borrow \$243.37.

Exercise 1–7
Answers to Odd-Numbered Problems

1. 15 **3.** -28 **5.** -60 **7.** -36 **9.** 20 **11.** -105 **13.** -4.32
15. -13.769 **17.** -2.16 **19.** $\dfrac{9}{16}$ **21.** $\dfrac{1}{4}$ **23.** -120 **25.** 144
27. 0 **29.** 0 **31.** 1,050 **33.** \$1,399.20 **35.** $-\$92$
37. \$542.50 **39.** 700 gallons **41.** 2 **43.** -8 **45.** 2 **47.** -8
49. undefined **51.** 0 **53.** undefined **55.** -7 **57.** -5 **59.** -2
61. -4 **63.** 0 **65.** -3 **67.** undefined **69.** undefined **71.** $-1°$ C
73. 6 hours **75.** 9 seconds **77.** \$4 **79.** 25.6 mpg **81.** 4°
83. 32 books **85.** 12 minutes

Solutions to Trial Exercise Problems

11. $7\cdot(-1)(-3)(-5)=(-7)(-3)(-5)=(21)(-5)=-105$; negative answer because there were an odd number of negative factors.
27. $(-2)(0)(3)(-4)=0$ When zero is one of the factors, zero will be the answer.
30. Assets $(5)(-6)$, and his assets would change by (-30) dollars.
39. Let g represent the number of gallons of milk he sells in 4 weeks. Since there are 28 days in 4 weeks, we must *multiply* 28 by 25 to determine the gallons of milk sold.

total gallons	is	28 days	at	25 gallons
of milk he sells				per day
g	$=$	28	$\cdot$	25

$g=28\cdot25$
$g=700$

The grocer sells 700 gallons of milk.

51. $\dfrac{0}{-9}=0$, since $(-9)\cdot0=0$
59. $\dfrac{(-4)(-3)}{-6}=\dfrac{12}{-6}=-2$; odd number of negative factors
62. $\dfrac{(-4)(0)}{-8}=\dfrac{0}{-8}=0$, since $(-8)\cdot0=0$.
66. $\dfrac{(-2)(-4)}{(0)(4)}=\dfrac{8}{0}$ is undefined.
69. $\dfrac{(-6)}{(-3)(0)}=\dfrac{-6}{0}$ is undefined.
73. Let n represent the number of hours

$$\text{number of hours}=\dfrac{\text{number of miles}}{\text{rate of travel in miles per hour}}.$$

Hence $n=\dfrac{282}{47}=6$. The trip will take 6 hours.

85. Let m represent the number of minutes it took Hiroko to run 1 mile. Since there are 60 minutes in 1 hour, the race took 300 minutes + 12 minutes, which is 312 minutes. We must divide 312 minutes by 26 miles to determine the number of minutes per mile.

minutes per	is	number of	divided	number
mile		minutes	by	of miles
m	$=$	312	$\div$	26

$m=312\div26=12$

Hiroko ran 1 mile in 12 minutes.

Exercise 1–8

Answers to Odd-Numbered Problems

1. 16 **3.** −27 **5.** −36 **7.** −1 **9.** 1 **11.** 4 **13.** 26 **15.** 0
17. 3 **19.** 8 **21.** 1 **23.** 6 **25.** $\frac{12}{7}$ or $1\frac{5}{7}$ **27.** 11 **29.** 46
31. −10 **33.** 121 **35.** $\frac{5}{24}$ **37.** 96 **39.** −48 **41.** 78
43. −17 **45.** 38 **47.** 4 **49.** 12 **51.** −15.99 **53.** 38.47 **55.** 29
57. 33 **59.** $\frac{19}{288}$ **61.** $23\frac{1}{3}$°C **63.** $\frac{110}{7}$ or $15\frac{5}{7}$ square inches
65. $\frac{288}{41}$ or $7\frac{1}{41}$ inches **67.** \$374.50 **69.** 8 **71.** 3,328

Solutions to Trial Exercise Problems

2. $(-5)^4 = (-5)(-5)(-5)(-5) = (25)(-5)(-5) =$
$(-125)(-5) = 625$; positive since we have a negative number
to an even power
3. $(-3)^3 = (-3)(-3)(-3) = (9)(-3) = -27$; negative since we
have a negative number to an odd power
8. $-2^2 = -(2 \cdot 2) = -4 = -4$
17. $0(5 + 2) + 3 = 0(7) + 3 = 0 + 3 = 3$
25. $\frac{2}{3} \div \left(\frac{5}{6} - \frac{4}{9}\right) = \frac{2}{3} \div \left(\frac{15}{18} - \frac{8}{18}\right) = \frac{2}{3} \div \frac{7}{18}$

$= \frac{2}{3} \cdot \frac{18}{7} = \frac{2}{3} \cdot \frac{3 \cdot 6}{7} = \frac{12}{7}$ or $1\frac{5}{7}$
45. $4(2 - 5)^2 - 2(3 - 4) = 4(-3)^2 - 2(3 - 4) = 4(-3)^2 - 2(-1)$
$= 4(9) - 2(-1) = 36 - 2(-1) = 36 - (-2) = 38$
47. $\frac{5(3 - 5)}{2} - \frac{27}{-3} = \frac{5(-2)}{2} - \frac{27}{-3} = \frac{-10}{2} - \frac{27}{-3}$

$= (-5) - \frac{27}{-3} = (-5) - (-9) = 4$
54. $5[10 - 2(4 - 3) + 1] = 5[10 - 2(1) + 1] = 5[10 - 2 + 1]$
$= 5[8 + 1] = 5[9] = 45$
58. $\left(\frac{6 - 3}{7 - 4}\right)\left(\frac{14 + 2 \cdot 3}{5}\right) = \left(\frac{3}{3}\right)\left(\frac{14 + 6}{5}\right) = \left(\frac{3}{3}\right)\left(\frac{20}{5}\right) = 1 \cdot 4 = 4$
63. $\frac{22}{7} \cdot 3^2 - \frac{22}{7} \cdot 2^2 = \frac{22}{7} \cdot 9 - \frac{22}{7} \cdot 4 = \frac{198}{7} - \frac{88}{7} = \frac{198 - 88}{7}$

$= \frac{110}{7}$ or $15\frac{5}{7}$ square inches
69. Let p represent the total number of pieces of lumber. *Dividing*
the 16-foot board by 4 and the 12-foot board by 3 will give us the
number of pieces of lumber. If we *add* the number of pieces from
the 16-foot board to the number of pieces from the 12-foot board,
we will have the total number of pieces.

total number of pieces	is	number of pieces from the 16-foot board	combined with	number of pieces from the 12-foot board
p	=	$16 \div 4$	+	$12 \div 3$

$p = 16 \div 4 + 12 \div 3$
$\quad = 4 + 4$
$\quad = 8$

There will be 8 pieces of lumber.

Exercise 1–9

Answers to Odd-Numbered Problems

1. 9 **3.** 10 **5.** 0 **7.** 13 **9.** 0 **11.** 5 **13.** 61 **15.** −60 **17.** 48
19. 288 **21.** 25 **23.** 4 **25.** 21 **27.** 3 **29.** 20 **31.** −44
33. 0 **35.** 5 **37.** $6\frac{2}{3}$ **39.** 160 **41.** 288 **43.** 540.736 **45.** 108
47. 60 **49.** 41 **51.** 48 **53.** 54 **55.** 2,140 **57.** 28.6 **59.** 114
61. 256 **63.** 6 **65.** 2.016 **67.** 2.499 **69.** 9.45 **71.** $17\frac{431}{857}$
73. $41\frac{7}{13}$ rpm **75.** $12\frac{4}{33}$

Solutions to Trial Exercise Problems

8. $b^2 - d^2 = (\ )^2 - (\ )^2 = (3)^2 - (-3)^2 = 9 - 9 = 0$
17. $(3a + 2b)(a - c) = [3(\ ) + 2(\ )][(\ ) - (\ )] =$
$[3(2) + 2(3)][(2) - (-2)] = [6 + 6][4] = [12][4] = 48$
24. $(4a + b) - (3a - b)(c + 2d) = [4(\ ) + (\ )] - [3(\ ) - (\ )][(\ ) + 2(\ )] =$
$[4(2) + (3)] - [3(2) - (3)][(-2) + 2(-3)] = [8 + 3] - [6 - 3][(-2) +$
$(-6)] = [11] - [3][-8] = [11] - [-24] = 35$
29. $2c^2 - 3c + 6 = 2(-2)^2 - 3(-2) + 6 = 2(4) - (-6) + 6 =$
$8 + 6 + 6 = 14 + 6 = 20$
39. $I = prt; I = (\ )(\ )(\ ) = (1,000)(0.08)(2) = (80)2 = 160$
59. $A = \frac{I^2R - 120E^2}{R}; A = \frac{(\ )^2(\ ) - 120(\ )^2}{(\ )}$

$= \frac{(12)^2(100) - 120(5)^2}{(100)} = \frac{(144)(100) - 120(25)}{100}$

$= \frac{14,400 - 3,000}{100} = \frac{11,400}{100} = 114$
73. $V = \frac{vn}{N} = \frac{(90)(30)}{(65)} = \frac{90 \cdot 6 \cdot 5}{13 \cdot 5} = \frac{540}{13}$ or $41\frac{7}{13}$ rpm

Chapter 1 Review

1. See your instructor. **2.** See your instructor. **3.** $\frac{5}{7}$ **4.** $\frac{3}{4}$
5. $\frac{2}{3}$ **6.** $\frac{10}{7}$ or $1\frac{3}{7}$ **7.** $\frac{3}{5}$ **8.** $\frac{21}{20}$ or $1\frac{1}{20}$ **9.** $\frac{7}{8}$ **10.** $\frac{25}{8}$ or $3\frac{1}{8}$
11. $\frac{25}{3}$ or $8\frac{1}{3}$ **12.** $\frac{5}{8}$ acre **13.** $\frac{2}{5}$ cup **14.** 56 **15.** 120
16. $\frac{8}{7}$ or $1\frac{1}{7}$ **17.** $\frac{19}{24}$ **18.** $\frac{5}{6}$ **19.** $\frac{2}{9}$ **20.** $\frac{137}{20}$ or $6\frac{17}{20}$
21. $-\frac{7}{15}$ **22.** $\frac{7}{12}$ **23.** $\frac{7}{8}$ acre **24.** 263.51 **25.** 31.795
26. 1,355.09 **27.** 14.3 **28.** 1,152 **29.** \$98.59 **30.** ≈12.4 mpg
31. 0.6, 60% **32.** $\frac{1}{5}$, 20% **33.** $\frac{1}{10}$, 0.1 **34.** $1\frac{87}{100}$, 1.87
35. $\frac{11}{50}$, 22% **36.** $0.\overline{7}$, $77.\overline{7}$% **37.** 10 **38.** 68.4 **39.** 25
40. 78.72 **41.** {50, 51, 52, 53, 54, 55} **42.** {1, 2, 3, 4} **43.** {0}
44. {−3, −2, −1, 0, 1, 2, 3}

45. $-2, -\frac{1}{2}, 0, 3$

46. $-\frac{3}{4}, 1, \frac{3}{2}, \frac{5}{2}$

47. $-4, -1, \sqrt{2}, 4$

48. $-3, -2, \frac{1}{2}, \pi$

49. $<$ **50.** $<$ **51.** $>$ **52.** $<$ **53.** $>$ **54.** $>$ **55.** -4 **56.** 3
57. -6 **58.** 1 **59.** -6 **60.** 15 **61.** -4 **62.** -9 **63.** 3
64. 15 **65.** $-\frac{3}{4}$ **66.** $4\frac{1}{8}$ **67.** 3.4 **68.** -2.5 **69.** -9

70. 15 **71.** $-10°$ C **72.** \$1,081.85
73. a. $52,000 + (-3,000) + (-2,560) + (-3,300)$ **b.** \$43,140
74. $40°$ **75. a.** $9°, 8°, -5°, -6°$ **b.** Total rise was greater than the total fall by $6°$ **c.** $69° + (-11°) = 58°$ **76.** -21

77. $-\frac{112}{15}$ or $-7\frac{7}{15}$ **78.** 24 **79.** -144 **80.** 0 **81.** -7

82. 2 **83.** -6 **84.** undefined **85.** 0 **86.** -1.23 **87.** -1
88. 4,160 people **89.** 35 students **90.** \$3.40 **91.** 36.2 mpg
92. 25 **93.** -64 **94.** -16 **95.** -27 **96.** 98 **97.** -3 **98.** 20
99. 49 **100.** -9 **101.** 20 **102.** -14 **103.** 27 **104.** 1 **105.** -1
106. 72 **107.** -4 **108.** 4 **109.** 7

110. a. 3 **b.** $\frac{189}{4}$ or $47\frac{1}{4}$ **111.** 1,040

Chapter 1 Test

1. $\frac{5}{12}$ [1–2] **2.** 0 [1–7] **3.** $\frac{25}{8}$ or $3\frac{1}{8}$ [1–2] **4.** -5.76 [1–7]

5. -7 [1–7] **6.** 9 [1–2] **7.** 0 [1–7] **8.** 14 [1–8]

9. -25 [1–8] **10.** undefined [1–7] **11.** -5 [1–6] **12.** 27 [1–8]

13. $\frac{17}{20}$, 85% [1–3] **14.** 0.45, 45% [1–3] **15.** 46 [1–3]

16. 5.16 [1–3] **17.** $\{-1, 0, 1, 2\}$ [1–4] **18.** $>$ [1–4]

19. -48 [1–9] **20.** 20.6 [1–9] **21.** 74 wpm [1–7]

22. 225 ml [1–6] **23.** \$174 [1–6] **24.** \$43.55 [1–6]

25. \$101.62 [1–8]

Chapter 2

Exercise 2–1

Answers to Odd-Numbered Problems

1. 2 terms, $3x$ and $4y$ **3.** 3 terms, $4x^2$, $3x$, and -1

5. 1 term, $\frac{6x}{5}$ **7.** 3 terms, $8xy$, $\frac{5y}{2}$, and $-6x$

9. 3 terms, $5x^3$, $3x^2$, and -4 **11.** 3 terms, x, y, and z
13. 4 terms, a^2b, c, $-x^2y$, and z
15. 5 is the coefficient of x^2, 1 is understood to be the coefficient of x, -4 is the coefficient of z
17. 1 is understood to be the coefficient of x, -1 is understood to be the coefficient of y, -3 is the coefficient of z
19. -2 is the coefficient of a, -1 is understood to be the coefficient of b, 1 is understood to be the coefficient of c
21. polynomial, binomial, degree 1
23. not a polynomial because a variable is in the denominator
25. polynomial, monomial, degree 6
27. polynomial, trinomial, degree 7

29. $b - 3a$ **31.** $y + 5$ **33.** $x(y + z)$ **35.** $a - b$
37. (let x = the number) $x - 9$ **39.** $0.34x$ **41.** $\frac{1}{3}x$ or $\frac{x}{3}$
43. $x + 50$, where x is the amount in the savings account
45. $x - 100$, where x is the amount in the savings account
47. $x - 12$ **49.** $3x + 1$ **51.** $2(x + 4)$ **53.** $2x - 6$

55. a. $\frac{1}{3}(x + 5)$ or $\frac{x + 5}{3}$ **b.** $\frac{1}{2}x + 5$ or $\frac{x}{2} + 5$ **57.** $x(x - 3)$

59. $0.05(x - 10,000)$ **61.** $0.06x + (30,000 - x)$ **63.** $\frac{1}{7}y$ **65.** $\frac{2}{9}x$

67. $\frac{2}{5}x^2y$

Solutions to Trial Exercise Problems

8. $15x^2 + y$ has two terms. These are $15x^2$ and y.
17. The numerical coefficients are 1, -1, and -3.
23. Not a polynomial since x appears in the denominator.
38. Let x represent the number. Then 15% of x is $0.15x$.

46. $\frac{1}{2}x - 2x$ or $\frac{x}{2} - 2x$

49. Let x represent the number. Then the algebraic expression is $3x + 1$.
54. Let x represent the number. Then
  a. $2x - 6$
  b. $2(x - 6)$
60. Let the amount be represented by x. Then the algebraic expression is $0.25(12,000 - x)$.

Review Exercises

1. -6 **2.** -4 **3.** -8 **4.** 18 **5.** 0 **6.** -21 **7.** -25 **8.** 64
9. -2 **10.** 15 **11.** 22 **12.** 23

Exercise 2–2

Answers to Odd-Numbered Problems

1. like **3.** like **5.** unlike **7.** $9x$ **9.** $13a + 2b$ **11.** $14x$
13. $-3ab$ **15.** $7x^2 + 4x$ **17.** $13a - 5c - 2x^2$ **19.** $3x^2 - 3x$
21. $8x^2 + 2x + 1$ **23.** $3x + 8y$ **25.** $2x + 3y$ **27.** $2a + b + 5c$
29. $2x + 4y$ **31.** $-3x^2y + 15xy$ **33.** $70a + 9b$ **35.** $-8b + 10$
37. $3a - 9b$ **39.** $3x^2 + 3z$ **41.** $9x + 3$ **43.** $5a + 10$ **45.** $3x - 14$

47. $\frac{5}{6}x$ or $\frac{5x}{6}$ **49.** $\frac{13}{30}x - \frac{5}{6}y$ **51.** $-0.1x + 400$ **53.** $-2t_1 - 3t_2$

55. $6a_1 - 3a_2 + 2b_2$ **57.** $6x + 2y$ **59.** $6r_1 + 5r_2$
61. a. Since 12 cm $-$ 7 cm $=$ 5 cm, we might move 12 cm to the right and then 7 cm back. This is $3d_1 - d_2$.
  b. Since $14 - 8 = 6$, and we want -6 units, use $2d_1 - 2d_2$.
63. We want to move 9 cm to the left. This is -9 cm. Since $12 - 21 = -9$, use $3d_1 - 3d_2$.

65. $\frac{y}{10}$ dollars **67.** $(5n + 10d)$ cents **69. a.** $p + 12$ **b.** $p - 5$

71. $(258 - n + m)$ dollars **73.** $\frac{c}{50}$ **75.** $y + 2$ **77.** $(12f + t)$ inches

79. $(25,000n - 2,000)$ dollars **81.** $(9.95p + 12.99q)$ dollars

Solutions to Trial Exercise Problems

17. $3a + b + 2a - 5c - b - 2x^2 + 8a$
  $= (3a + 2a + 8a) + (b - b) - 5c - 2x^2$
  $= (3 + 2 + 8)a + (0) - 5c - 2x^2$
  $= 13a - 5c - 2x^2$

34. $(8xy + 9y^2z) - (13xy - 14yz)$
   $= 8xy + 9y^2z - 13xy + 14yz$
   $= (8xy - 13xy) + 9y^2z + 14yz$
   $= (8 - 13)xy + 9y^2z + 14yz$
   $= -5xy + 9y^2z + 14yz$

44. $-3(2a - 4) + 5(a - 3) - (a - 7)$
   $= -6a + 12 + 5a - 15 - a + 7$
   $= (-6a + 5a - a) + (12 - 15 + 7)$
   $= -2a + 4$

47. $\frac{1}{2}x + \frac{1}{3}x$ LCD of the fractions is 6.

$= \frac{3}{6}x + \frac{2}{6}x = \left(\frac{3}{6} + \frac{2}{6}\right)x = \frac{5}{6}x$ or $\frac{5x}{6}$.

60. a. $3d_1 - d_2 + d_1 + 5d_2 + 2d_1 - 2d_2 + 4d_1 + d_2 + 6d_1 + d_1 - 4d_2$
   $= 17d_1 - d_2$
   b. The robot's final position is as if it moved 17 steps right of distance d_1 and then left one step of distance d_2.

62. If you add or subtract even values you get only even values. Thus there is no combination of 4- and 6-cm moves that could produce any odd valued move. For example, a distance of 7 cm would be impossible to achieve.

67. Since 1 nickel = 5 cents, therefore, n nickels represent $5n$ cents. Since 1 dime = 10 cents, therefore, d dimes represent $10d$ cents. Hence, the amount of money she has in her purse is $(5n + 10d)$ cents.

76. The next greater odd integer is obtained by adding 2 to the odd integer z. Hence, the next greater odd integer is $z + 2$.

Review Exercises

1. $3x$ **2.** $6(a + 7)$

3. $(y - 2) \div 4$ or $\dfrac{y - 2}{4}$

4. (let x = the number) $5x$ **5.** (let x = the number) $x - 12$

6. (let x = the number) $\dfrac{x}{8} - 9$

Exercise 2–3

Answers to Odd-Numbered Problems

1. yes **3.** yes **5.** yes **7.** no **9.** yes **11.** no **13.** {16} **15.** {−3}
17. {−2} **19.** {−5} **21.** {−6} **23.** {−7} **25.** {14} **27.** {−1}
29. {4} **31.** {−5} **33.** {5} **35.** {6} **37.** {6} **39.** {8} **41.** {−6}
43. {5} **45.** {−7} **47.** {12} **49.** {16} **51.** {9} **53.** 26
55. 33 years old **57.** $735 **59.** $365
61. a. $50{,}000 = 40{,}000 + 10{,}000$
   $+ 2{,}000$
   $\underline{- 2{,}000}$
   $50{,}000 = 40{,}000 + 10{,}000$
   b. additive inverse
63. a. $50{,}000 = 40{,}000 + 10{,}000$
   $\underline{+ 5{,}000 \qquad\qquad 5{,}000}$
   $55{,}000 = 40{,}000 + 15{,}000$
   b. addition property of equality
65. a. $50{,}000 = \quad 40{,}000 + 10{,}000$
   $\underline{- 4{,}000 \qquad - 4{,}000}$
   $46{,}000 = \quad 36{,}000 + 10{,}000$
   b. addition property of equality

Solutions to Trial Exercise Problems

8. $3x + 2 = 5x - 1$; $x = \dfrac{3}{2}$

$3\left(\dfrac{3}{2}\right) + 2 = 5\left(\dfrac{3}{2}\right) - 1$

$\dfrac{9}{2} + 2 = \dfrac{15}{2} - 1$

$\dfrac{9}{2} + 2 = \dfrac{15}{2} - \dfrac{2}{2}$

$\dfrac{13}{2} = \dfrac{13}{2}$ (true)

Yes, $x = \dfrac{3}{2}$ is a solution.

23. $b + 7 = 0$
   $b + 7 - 7 = 0 - 7$
   $b = -7$
   $\{-7\}$
   Check: $(-7) + 7 = 0$
   $0 = 0$ (true)

28. $-y - 6 = -2y + 1$
   $-y + 2y - 6 = -2y + 2y + 1$
   $y - 6 = 1$
   $y - 6 + 6 = 1 + 6$
   $y = 7$
   $\{7\}$
   Check: $-(7) - 6 = -2(7) + 1$
   $-13 = -14 + 1$
   $-13 = -13$ (true)

38. $5(x + 2) = 4(x - 1)$
   $5x + 10 = 4x - 4$
   $5x - 4x + 10 = 4x - 4x - 4$
   $x + 10 = -4$
   $x + 10 - 10 = -4 - 10$
   $x = -14$
   $\{-14\}$

45. $3(z + 7) - (8 + 2z) = 6$
   $3z + 21 - 8 - 2z = 6$
   $z + 13 = 6$
   $z + 13 - 13 = 6 - 13$
   $z = -7$
   $\{-7\}$

58. Let b represent the original balance.

original balance	makes a deposit	of $42.50	equals	new balance
b	+	42.50	=	125.30

$b + 42.50 = 125.30$
$b + 42.50 - 42.50 = 125.30 - 42.50$ Subtract 42.50 from both sides.

$b = 82.80$

The original balance was $82.80.

Review Exercises

1. 16 **2.** −12 **3.** 1 **4.** 1 **5.** 1 **6.** 1

Exercise 2–4

Answers to Odd-Numbered Problems

1. $\{4\}$ **3.** $\{6\}$ **5.** $\{16\}$ **7.** $\{35\}$ **9.** $\{12\}$ **11.** $\{-3\}$ **13.** $\{-4\}$

15. $\{7\}$ **17.** $\{-4\}$ **19.** $\left\{\dfrac{7}{3}\right\}$ **21.** $\left\{\dfrac{3}{2}\right\}$ **23.** $\{0\}$ **25.** $\{0\}$ **27.** $\{15\}$

29. $\{-14\}$ **31.** $\{4\}$ **33.** $\{-7\}$ **35.** $\{11\}$ **37.** $\{-26\}$ **39.** $\left\{\dfrac{112}{3}\right\}$

41. 9 **43.** -63 **45.** $\$4.50$ **47.** $\$130$ **49.** 64

Solutions to Trial Exercise Problems

7. $\dfrac{1}{7}x = 5$

$7 \cdot \dfrac{1}{7}x = 7 \cdot 5$

$x = 35$
$\{35\}$

Check: $\dfrac{1}{7}(35) = 5$

$5 = 5$ (true)

15. $-4x = -28$

$\dfrac{-4x}{-4} = \dfrac{-28}{-4}$

$x = 7$
$\{7\}$

Check: $-4(7) = -28$
$-28 = -28$ (true)

23. $5x = 0$

$\dfrac{5x}{5} = \dfrac{0}{5}$

$x = 0$
$\{0\}$

Check: $5(0) = 0$
$0 = 0$ (true)

27. $\dfrac{x}{3} = 5$

$3 \cdot \dfrac{x}{3} = 3 \cdot 5$

$x = 15$
$\{15\}$

Check: $\dfrac{(15)}{3} = 5$

$5 = 5$ (true)

31. $2.6x = 10.4$

$\dfrac{2.6x}{2.6} = \dfrac{10.4}{2.6}$

$x = 4$
$\{4\}$

Check: $2.6(4) = 10.4$
$10.4 = 10.4$ (true)

38. $\dfrac{5}{7}x = 8$

$\dfrac{7}{5} \cdot \dfrac{5}{7}x = \dfrac{7}{5} \cdot 8$

$x = \dfrac{56}{5}$

$\left\{\dfrac{56}{5}\right\}$

Check: $\dfrac{5}{7}\left(\dfrac{56}{5}\right) = 8$

$8 = 8$ (true)

45. Let w represent Nancy's hourly wage.

30 hours	at	hourly wage	is	$135.0
30	·	w	=	135

$30w = 135$

$\dfrac{30w}{30} = \dfrac{135}{30}$ Divide both sides by 30.

$w = 4.5$

Nancy's hourly wage is $4.50.

Review Exercises

1. $5x - 2$ **2.** $2x + 1$ **3.** $12x + 2$ **4.** $3x - 5$ **5.** $10x - 1$ **6.** $5x + 1$

Exercise 2–5

Answers to Odd-Numbered Problems

1. $\{2\}$ **3.** $\{-2\}$ **5.** $\{36\}$ **7.** $\left\{\dfrac{16}{3}\right\}$ **9.** $\{4\}$ **11.** $\{0\}$ **13.** $\{3\}$

15. $\{0\}$ **17.** $\{1\}$ **19.** $\left\{\dfrac{8}{5}\right\}$ **21.** R **23.** R **25.** $\{-10\}$ **27.** $\left\{-\dfrac{9}{2}\right\}$

29. $\{18\}$ **31.** $\left\{-\dfrac{51}{8}\right\}$ **33.** $\{12\}$ **35.** $\varnothing$ **37.** $\varnothing$ **39.** $\{3\}$

41. $\left\{\dfrac{10}{7}\right\}$ **43.** $\left\{\dfrac{16}{11}\right\}$ **45.** $\left\{\dfrac{37}{10}\right\}$ **47.** $\left\{\dfrac{2}{5}\right\}$ **49.** $\{3\}$ **51.** $\{1\}$

53. a. $-7\dfrac{7}{9}°$ C **b.** $-32\dfrac{7}{9}°$ C **c.** $-16\dfrac{2}{3}°$ C **55.** $\dfrac{7}{3}$ **57.** $\{3.51\}$

59. $\{2.09\}$ **61.** $\{3.07\}$ **63.** $\{11.94\}$ **65.** 800 **67.** 458 **69.** 81
71. 620 **73.** $3,500$ **75.** $1,000$

Solutions to Trial Exercise Problems

20. $3x + 4x + 5 = 7x + 5$
Combine like terms on the left side.

$7x + 5 = 7x + 5$

Subtract $7x$ from each side

$7x + 5 - 7x = 7x + 5 - 7x$
$5 = 5$ (true)

This indicates that the equation is an identity and the solution set is the set R of real numbers.

27.
$$\frac{1}{2}x + 3 = \frac{3}{4}$$
$$\frac{1}{2}x + 3 - 3 = \frac{3}{4} - 3$$
$$\frac{1}{2}x = \frac{3}{4} - \frac{12}{4}$$
$$\frac{1}{2}x = -\frac{9}{4}$$
$$2 \cdot \frac{1}{2}x = 2 \cdot -\frac{9}{4}$$
$$x = -\frac{9}{2}$$
$$\left\{-\frac{9}{2}\right\}$$

35.
$$3(3x - 2) = 2x + 7x + 5$$
$$9x - 6 = 9x + 5$$
$$9x - 6 - 9x = 9x + 5 - 9x$$
$$-6 = 5 \qquad \text{(false)}$$
This indicates that the equation is a contradiction and the solution set is the empty set, $\varnothing$.

39.
$$3(2x - 1) = 4x + 3$$
$$6x - 3 = 4x + 3$$
$$6x - 4x - 3 = 4x - 4x + 3$$
$$2x - 3 = 3$$
$$2x - 3 + 3 = 3 + 3$$
$$2x = 6$$
$$\frac{2x}{2} = \frac{6}{2}$$
$$x = 3$$
$$\{3\}$$

43.
$$8 - 2(3x + 4) = 5x - 16$$
$$8 - 6x - 8 = 5x - 16$$
$$-6x = 5x - 16$$
$$-6x + 6x = 5x + 6x - 16$$
$$0 = 11x - 16$$
$$0 + 16 = 11x - 16 + 16$$
$$16 = 11x$$
$$\frac{16}{11} = \frac{11x}{11}$$
$$\frac{16}{11} = x$$
$$\left\{\frac{16}{11}\right\}$$

54. b. $W = 243, T = -3$
$$W = KT^4$$
$$(243) = K(-3)^4$$
$$243 = K \cdot 81$$
$$\frac{243}{81} = \frac{K \cdot 81}{81}$$
$$3 = K$$
Hence the value of K is 3.

58. $-3.11x + 5.33 = 9.08$
$$-3.11x = 9.08 - 5.33$$
$$-3.11x = 3.75$$
$$x = \frac{3.75}{-3.11}$$
$$x \approx -1.21$$
$$\{-1.21\}$$

76. Let the number be x. Then
$$0.18(100 + x) + 0.06x = 238.8$$
$$18 + 0.18x + 0.06x = 238.8$$
$$0.24x = 238.8 - 18$$
$$0.24x = 220.8$$
$$x = \frac{220.8}{0.24}$$
$$x = 920$$
The number is 920.

Review Exercises
1. 108 **2.** 144 **3.** 88 **4.** 360 **5.** 280 **6.** 92

Exercise 2–6
Answers to Odd-Numbered Problems

1. $h = 8$ **3.** $\ell = 10.5$ **5.** $h = 6$ **7.** $b = 12$ **9.** $b_2 = 8.5$
11. $r = 5$ **13.** $h = 6$ **15.** $h = 10$ **17.** $w = \dfrac{V}{\ell h}$ **19.** $P = \dfrac{I}{rt}$
21. $m = \dfrac{F}{a}$ **23.** $V = \dfrac{K}{P}$ **25.** $R = \dfrac{W}{I^2}$ **27.** $w = \dfrac{P - 2\ell}{2}$
29. $a = P - b - c$ **31.** $y = x - 4$ **33.** $x = \dfrac{y + 2}{3}$ **35.** $x = \dfrac{y + 5}{3}$
37. $x = \dfrac{3y + 2}{4}$ **39.** $a = \dfrac{by + c + 3}{y}$ **41.** $k = V - gt$
43. $b_1 = \dfrac{2A - b_2 h}{h}$ **45.** $a = \ell - dn + d$ **47.** $P = \dfrac{A}{1 + r}$
49. $f = \dfrac{T - g}{2}$ **51.** $q = \dfrac{D - R}{d}$ **53.** $x = \dfrac{y + 9}{6}$
55. $x = \dfrac{2y + 6}{3}$ **57.** $y = \dfrac{5x - 18}{3}$ **59.** $x = \dfrac{8y - 5}{9}$
61. $c = \dfrac{W - b^2 - R}{2b}$ **63.** $r = \dfrac{A - P}{Pt}$ **65.** $a = \dfrac{V + br^2}{r^2}$
67. $x = -6y$ **69.** $g = \dfrac{2vt - 2S}{t^2}$
71. $g = \dfrac{2s - 2vt}{t^2}$ **73.** $S = \dfrac{P + Cn + e}{n}$
75. $F = \dfrac{9}{5}C + 32$ or $= \dfrac{9C + 160}{5}$ **77.** $I = \dfrac{E}{R}$ **79.** $r = \dfrac{d}{t}$
81. $w = \dfrac{V}{\ell h}, \ell = \dfrac{V}{wh}, P = \dfrac{I}{rt}, t = \dfrac{I}{Pr}$ **83.** c **85.** c
87. c **89.** c **91.** c

Solutions to Trial Exercise Problems

35.
$$3x - y = 5$$
$$3x - y + y = 5 + y$$
$$3x = y + 5$$
$$\frac{3x}{3} = \frac{y + 5}{3}$$
$$x = \frac{y + 5}{3}$$

42.
$$V = k + gt, \text{ for } t$$
$$V - k = k + gt - k$$
$$V - k = gt$$
$$\frac{V - k}{g} = t$$
$$t = \frac{V - k}{g}$$

46.
$$\ell = a + (n-1)d, \text{ for } d$$
$$\ell - a = a + (n-1)d - a$$
$$\ell - a = (n-1)d$$
$$\frac{\ell - a}{n-1} = d$$

57. $\frac{5}{6}x - \frac{1}{2}y = 3$ LCD of the fractions is 6.

$$6 \cdot \frac{5}{6}x - 6 \cdot \frac{1}{2}y = 6 \cdot 3$$
$$5x - 3y = 18$$
Add $3y$ to each side.
$$5x = 3y + 18$$
$$5x - 18 = 3y$$
$$\frac{5x - 18}{3} = \frac{3y}{3}$$
$$y = \frac{5x - 18}{3}$$

61. $R = W - b(2c + b), \text{ for } c$
$$R = W - 2bc - b^2$$
$$R + 2bc = W - b^2$$
$$2bc = W - b^2 - R$$
$$c = \frac{W - b^2 - R}{2b}$$

75. $C = \frac{5}{9}(F - 32)$

Multiply both sides by 9.
$$9C = 9 \cdot \frac{5}{9}(F - 32)$$
$$9C = 5(F - 32)$$
$$9C = 5F - 160$$
$$9C + 160 = 5F$$
$$\frac{9C + 160}{5} = \frac{5F}{5}$$
$$F = \frac{9C + 160}{5}$$
$$\text{or } F = \frac{9C}{5} + \frac{160}{5}$$
$$F = \frac{9}{5}C + 32$$

81. $V = \ell wh$

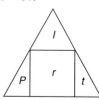

From this triangle, $w = \dfrac{V}{\ell h}$, $\ell = \dfrac{V}{wh}$

$I = Prt$

From this triangle,
$$P = \frac{I}{rt}, t = \frac{I}{Pr}$$

85. (a) translates to $y - 7 = 8$, (b) translates to $8y = y - 7$, (c) translates to $7 - y = 8y$, which is equivalent to $8y = 7 - y$ and this is the given equation. Hence (c) is the correct answer.

Review Exercises
1. -25 **2.** 25 **3.** -81 **4.** -27 **5.** x^4
6. (let $x =$ the number) x^2 **7.** ab **8.** xy

Exercise 2–7
Answers to Odd-Numbered Problems
1. 21 **3.** 28 **5.** 7 **7.** 15 **9.** 54 **11.** 24 **13.** 28 and 19
15. 84 and 12 **17.** 34 and 29 **19.** 108 and 12 **21.** 9 and 5
23. 30, 31, and 32 **25.** 15, 17, and 19 **27.** 20, 22, and 24
29. 14, 7, and 42 **31.** 9, 10, and 11 **33.** 34, 36, and 38
35. 24.7 inches **37.** $67\frac{1}{2}$ meters **39.** 8.5 cm, 8.5 cm, and 4 cm
41. Length is 19 ft, width is 16 ft.
43. Width is 9 inches, length is 26 inches.
45. Width is 18 ft, length is 23 ft. **47.** $13,000 at 8%; $7,000 at 6%
49. $11,000 at 8%; $4,000 at 6%
51. $8,000 at 10%; $10,000 at 8%
53. $8,000 at 9%; $22,000 at 7%
55. $8,000 at 11%; $17,000 at 18%
57. $7,700 at 13%; $13,300 at 9% **59.** $5,000 **61.** $14,000
63. 20 liters of 15% and 20 liters of 35% **65.** 8,800 pounds
67. $333\frac{1}{3}$ pounds **69.** 1,350 gallons

Solutions to Trial Exercise Problems
3. number   equation
$$x \qquad \frac{x}{4} + 6 = 13$$
Solution: $\frac{x}{4} + 6 = 13$
$$\frac{x}{4} = 7$$
$$x = 28$$
The number is 28.

30. first   second   third   equation
$$x \qquad 3 \cdot x \qquad x - 6 \quad (x) + (3 \cdot x) + (x - 6) = 44$$
Solution: $(x) + (3x) + (x - 6) = 44$
$$x + 3x + x - 6 = 44$$
$$5x - 6 = 44$$
$$5x = 50$$
$$x = 10$$
First is 10, second $(3x)$ is $3(10) = 30$, and the third $(x - 6)$ is $(10) - 6 = 4$.

55. number of   number of   equation
dollars at   dollars at
11% loss   18% profit

$$x \qquad 25,000 - x \quad (25,000 - x)(0.18) - x(0.11) = 2,180$$
Solution: $(25,000 - x)(0.18) - x(0.11) = 2,180$
$$(25,000)(0.18) - (0.18)x - (0.11)x = 2,180$$
$$4,500 - (0.29)x = 2,180$$
$$4,500 = (0.29)x + 2,180$$
$$2,320 = (0.29)x$$
$$8,000 = x$$
Therefore $8,000 was invested at 11% loss and $25,000 − $8,000 = $17,000 was invested at 18% profit.

59. Let the amount invested at 10% be x. Then the total amount at 9% would be $(x + 5,000)$. The equation is

$$(5,000)(0.08)(1) + x(0.1)(1) = (x + 5,000)(0.09)(1)$$
$$400 + 0.1x = 0.09x + 450$$
$$0.1x - 0.09x = 450 - 400$$
$$0.01x = 50$$
$$\frac{0.01x}{0.01} = \frac{50}{0.01}$$
$$x = 5,000$$

He must invest $5,000 more at 10%.

66. Let x gallons of 4% solution be mixed with 3,000 gallons of 10% solution to obtain $(x + 3,000)$ gallons of 8% solution. The equation is

$$(4\% \text{ of } x) + (10\% \text{ of } 3,000) = 8\% \text{ of } (x + 3,000)$$
$$0.04x + (0.1)(3,000) = 0.08(x + 3,000)$$
$$0.04x + 300 = 0.08x + 240$$
$$0.04x - 0.08x = 240 - 300$$
$$-0.04x = -60$$
$$\frac{-0.04x}{-0.04} = \frac{-60}{-0.04}$$
$$x = 1,500$$

The company should mix 1,500 gallons of 4% solution.

Review Exercises

1. 100 **2.** 84 **3.** 204 **4.** 108 **5.** 3,180 **6.** 16 **7.** 256 **8.** 40

Exercise 2–8
Answers to Odd-Numbered Problems

1. $\frac{3}{7}$; $3 : 7$ **3.** $\frac{5}{2}$; $5 : 2$ **5.** $\frac{3}{2}$; $3 : 2$ **7.** $\frac{5}{12}$; $5 : 12$ **9.** $\frac{60}{7}$; $60 : 7$

11. $\frac{1}{26}$; $1 : 26$ **13.** $\frac{2 \text{ lb}}{1 \text{ ft}^3}$; $2 \text{ lb} : 1 \text{ ft}^3$ **15.** $\frac{8 \text{ g}}{1 \text{ cm}^3}$; $8 \text{ g} : 1 \text{ cm}^3$

17. $\frac{60 \text{ miles}}{1 \text{ hr}}$; $60 \text{ miles} : 1 \text{ hr}$

19. A: $0.199 Brand B has lower unit price.
B: $0.183

21. A: $0.139 Brand A has lower unit price.
B: $0.149

23. A: $0.055 Brand A has lower unit price.
B: $0.057

25. A: $0.081 Brand A has lower unit price
B: $0.083

27. A: $0.29 Brand B has lower unit price.
B: $0.274

29. A: $0.341 Brand B has lower unit price.
B: $0.288

31. $\frac{13}{14}$ **33.** $\frac{21}{2}$ **35.** $\frac{9}{4}$ or $2\frac{1}{4}$ **37.** $\frac{1}{3}$ **39.** $x = \frac{5}{4}$ **41.** $p = \frac{100}{9}$

43. $x = \frac{16}{5}$ **45.** $a = \frac{12}{19}$ **47.** $x = 3$ **49.** $a = \frac{50}{13}$ **51.** 7 weeks

53. 36 grams of hydrogen **55.** 390 pounds **57.** $10\frac{1}{2}$ inches

59. 75 losses **61.** $x = \frac{30}{7}$ inches, $y = \frac{40}{7}$ inches **63.** 3,240 holes

65. 4 hours and 30 minutes **67.** $2\frac{1}{2}$ quarts **69.** 1,700,000 bottles

71. 15 inches **73.** 2.65 quarts **75.** 3.105 miles **77.** 2,996.4 grams
79. 12 kilograms

Solutions to Trial Exercise Problems

7. Since there are $3 \cdot 12 = 36$ inches in 3 feet, we have 15 inches to 36 inches written $\frac{15}{36} = \frac{5}{12}$ or $5 : 12$.

16. 300 miles to 10 gallons is written $\frac{300 \text{ miles}}{10 \text{ gallons}} = \frac{30 \text{ miles}}{1 \text{ gallon}}$, which we state as 30 miles per gallon or 30 mi: 1 gal.

28. Since 1 pound 4 ounces $= 20$ ounces,

the unit price for Brand A $= \dfrac{\$2.39}{20}$
$\approx \$0.120$

Unit price for Brand B $= \dfrac{\$1.89}{15}$
$= \$0.126$

So, Brand A has lower unit price.

30. $ME = \dfrac{\text{output}}{\text{input}} = \dfrac{375}{425} = \dfrac{15}{17}$

Therefore the mechanical efficiency is $\dfrac{15}{17}$

39. $\dfrac{9}{x} = \dfrac{36}{5}$, then $36 \cdot x = 9 \cdot 5$

$$36x = 45$$
$$x = \frac{45}{36} = \frac{5}{4}$$

Check: $\dfrac{9}{\frac{5}{4}} = \dfrac{36}{5}$ (true)

43. $6 : 15 = x : 8$, then $15 \cdot x = 6 \cdot 8$

$$15x = 48$$
$$x = \frac{48}{15} = \frac{16}{5}$$

Check: $6 : 15 = \dfrac{16}{5} : 8$

$$\frac{6}{15} = \frac{16}{5} \cdot \frac{1}{8}$$
$$\frac{2}{5} = \frac{2}{5} \quad \text{(true)}$$

46. $\dfrac{3}{4} : 4 = \dfrac{1}{2} : b$, then $\dfrac{3}{4} \cdot b = 4 \cdot \dfrac{1}{2}$

$$\frac{3}{4}b = 2$$
$$b = 2 \cdot \frac{4}{3} = \frac{8}{3}$$

Check: $\dfrac{3}{4} : 4 = \dfrac{1}{2} : \dfrac{8}{3}$

$$\frac{3}{4} \cdot \frac{1}{4} = \frac{1}{2} \cdot \frac{3}{8}$$
$$\frac{3}{16} = \frac{3}{16} \quad \text{(true)}$$

52. Let x represent the number of liters of gasoline to travel 1,428 kilometers.

Then $\dfrac{8 \text{ liters}}{84 \text{ km}} = \dfrac{x \text{ liters}}{1,428 \text{ km}}$ and

$$x \cdot 84 = 8 \cdot 1,428$$
$$84x = 11,424$$
$$x = \frac{11,424}{84} = 136$$

Therefore at the same rate of gasoline consumption, it would take 136 liters to travel 1,428 kilometers.

60. Let ℓ represent the length of the enlargement.

Then $\dfrac{10 \text{ in.}}{8 \text{ in.}} = \dfrac{\ell \text{ in.}}{36 \text{ in.}}$ and $8 \cdot \ell = 10 \cdot 36$

$$8\ell = 360$$
$$\ell = \frac{360}{8} = 45$$

Therefore the enlargement will be 45 inches long.

66. Let x represent the number of feet represented by $2\frac{5}{8}$ inches.

Then $\dfrac{\frac{1}{8} \text{ in.}}{1 \text{ ft}} = \dfrac{2\frac{5}{8} \text{ in.}}{x \text{ ft}}$ and $\dfrac{1}{8} \cdot x = 1 \cdot 2\frac{5}{8}$

$$\frac{1}{8}x = 2\frac{5}{8}$$
$$x = \frac{21}{8} \cdot 8 = 21$$

Therefore $2\frac{5}{8}$ inches represents 21 feet.

Review Exercises

1. -6 **2.** -6 **3.** 0 **4.** 14 **5.** -12 **6.** -10 **7.** 32 **8.** -16
9. 16 **10.** -64

Exercise 2–9
Answers to Odd-Numbered Problems

1. (number line: open circle at 2; marks 0, 2, 5)

3. (number line: closed circle at 1; marks 0, 1, 2)

5. (number line: open circle at 0; marks −5, 0)

7. (number line: open circle at 0; marks 0, 5)

9. (number line: closed circle at −4; marks −10, −5, −4)

11. (number line: open circles at 0 and ~3; marks −1, 0, 2, 5)

13. (number line: closed circles at 0 and 5; marks 0, 5)

15. (number line: open circle at 0, closed circle at 3; marks −1, 0, 3)

17. $x \le \dfrac{5}{2}$; (number line: closed circle at $\frac{5}{2}$; mark 0)

19. $x < 6$; (number line: open circle at 6; marks 0, 5, 6)

21. $x \ge 18$; (number line: closed circle at 18; marks 18, 20, 25)

23. $x \ge -9$; (number line: closed circle at −9; marks −10, −9, −5)

25. $x > -5$; (number line: open circle at −5; marks −5, 0)

27. $x > -\dfrac{5}{2}$; (number line: open circle at $-\frac{5}{2}$; marks −2, 3)

29. $x > \dfrac{16}{3}$; (number line: open circle at $\frac{16}{3}$; marks 5, 10)

31. $x \le 2$; (number line: closed circle at 2; marks 0, 2)

33. $x < 11$; (number line: open circle at 11; marks 5, 10, 11)

35. $x > 1$; (number line: open circle at 1; marks 0, 1, 5)

37. $x \ge 3$; (number line: closed circle at 3; marks 3, 5, 10)

39. $x > \dfrac{10}{7}$; (number line: open circle at $\frac{10}{7}$; marks 0, 1, 5)

41. $x > -10$; (number line: open circle at −10; marks −10, −5)

43. $x < \dfrac{16}{11}$; (number line: open circle at $\frac{16}{11}$; marks 0, 1)

45. $x \le \dfrac{37}{10}$; (number line: closed circle at $\frac{37}{10}$; marks 0, 3, 5)

47. $x \ge \dfrac{5}{2}$; (number line: closed circle at $\frac{5}{2}$; marks 0, 2)

49. $x \ge \dfrac{2}{5}$; (number line: closed circle at $\frac{2}{5}$; marks 0, 1)

51. $-2 < x < \dfrac{1}{2}$; (number line: open circles at −2 and $\frac{1}{2}$; marks −5, −2, 0)

53. $-\dfrac{4}{5} \le x \le \dfrac{1}{5}$; (number line: closed circles at $-\frac{4}{5}$ and $\frac{1}{5}$; marks −1, 0, 1)

55. $-2 < x \le \dfrac{5}{4}$; (number line: open circle at −2, closed circle at $\frac{5}{4}$; marks −2, 0, 1)

57. $-4 < x \le 1$; (number line: open circle at −4, closed circle at 1; marks −5, −4, 0, 1)

59. $-1 \le x < 7$; (number line: closed circle at −1, open circle at 7; marks −1, 0, 5, 7)

61. $-2 < x \le \dfrac{1}{3}$; (number line: open circle at −2, closed circle at $\frac{1}{3}$; marks −5, −2, 0)

63. (x = student's score) $x \ge 72$

65. (x = number of new lift trucks) $x \geq 8$
67. (P = selling price, C = cost) $P \geq 2C$
69. $x < 6$ **71.** $x \geq 12$ **73.** $x > 21$ **75.** $x \geq 7$
77. $3.3 \leq x \leq 4.0$ **79.** $x \geq 560$ **81.** $x > 18$ **83.** $5 < x < 27$
85. $8 < h < 12$ **87.** $8 < R < 14$

Solutions to Trial Exercise Problems

14. $1 \leq x < 4$

1 is included
4 is excluded.

22. $-4x < 12$

$\dfrac{-4x}{-4} > \dfrac{12}{-4}$

$x > -3$

35. $2x + (3x - 1) > 5 - x$
$2x + 3x - 1 > 5 - x$
$5x - 1 > 5 - x$
$5x + x > 5 + 1$
$6x > 6$
So, $x > 1$

43. $8 - 2(3x + 4) > 5x - 16$
$8 - 6x - 8 > 5x - 16$
$-6x > 5x - 16$
$-6x + 6x > 5x + 6x - 16$
$0 > 11x - 16$
$0 + 16 > 11x - 16 + 16$
$16 > 11x$
$\dfrac{16}{11} > \dfrac{11x}{11}$
$\dfrac{16}{11} > x$
$x < \dfrac{16}{11}$

56. $-2 < -x \leq 3$
$-2 < -1 \cdot x \leq 3$
$\dfrac{-2}{-1} > \dfrac{-1 \cdot x}{-1} \geq \dfrac{3}{-1}$
$2 > x \geq -3$
$-3 \leq x < 2$

66. A company will hire at least 2 new employees, but not more than 7. Let x be the number of new employees. Then the inequality would be $2 \leq x \leq 7$.

74. Two times a number plus 4 is greater than 6 but less than 14. Let x be the number. Then the inequality would be $6 < 2x + 4 < 14$.

Solving: $6 - 4 < 2x + 4 - 4 < 14 - 4$
$2 < 2x < 10$
$\dfrac{2}{2} < \dfrac{2x}{2} < \dfrac{10}{2}$
$1 < x < 5$

82. The perimeter of a rectangle must be less than 100 feet. If the length is known to be 30 feet, find all numbers that the width could be. Let x represent the width of the rectangle. Then the inequality would be $60 < 2 \cdot 30 + 2 \cdot x < 100$. Since the width of a real rectangle must be greater than zero and we already know the length to be 30, then the smallest value for the perimeter must be greater than 60 (two times the known length).

Solving: $60 < 2 \cdot 30 + 2x < 100$
$60 < 60 + 2x < 100$
$60 - 60 < 60 - 60 + 2x < 100 - 60$
$0 < 2x < 40$
$\dfrac{0}{2} < \dfrac{2x}{2} < \dfrac{40}{2}$
$0 < x < 20$

The width of the rectangle must be some real number greater than zero feet but less than 20 feet.

Review Exercises

1. -16 **2.** 16 **3.** -16 **4.** 16 **5.** x^5 **6.** (let x = the number) x^3
7. (let x = the number) x^2 **8.** xy

Chapter 2 Review

1. 3   **2.** 1   **3.** 2   **4.** 3   **5.** polynomial, degree is 2
6. polynomial, degree is 3   **7.** polynomial, degree is 2
8. not a polynomial because a variable is in the denominator   **9.** $5x$
10. $y - 7$ **11.** $z + 4$ **12.** (let x = the number) $2x + 6$
13. $2y - 7$ **14.** $\dfrac{1}{3}x + 10$ **15.** $0.25(x + 6{,}000)$ **16.** $\dfrac{c}{20}$ dollars
17. $(12f + t)$ inches   **18.** $-6a^2$ **19.** $-a^2 + a + 11$
20. $-11ab + 7b^2c + 11bc$ **21.** $-6y + 1$ **22.** $5ab + 3ac - 4bc$
23. $10a + 5$ **24.** $9x - 1$ **25.** $-x + 15$ **26.** $2a - 6$ **27.** $2x + 8$
28. yes   **29.** no   **30.** 48   **31.** 220   **32.** $\{7\}$ **33.** $\{21\}$ **34.** $\{-11\}$
35. $\{-6\}$ **36.** $\{4\}$ **37.** $\{19\}$ **38.** $\{-2\}$ **39.** $\{9\}$ **40.** $\{3\}$ **41.** $\{3\}$
42. $\{-7\}$ **43.** $\{-7\}$ **44.** $\{12\}$ **45.** $\{14\}$ **46.** $\{15\}$ **47.** $\{21\}$
48. $\left\{\dfrac{21}{4}\right\}$ **49.** $\{-18\}$ **50.** $\left\{-\dfrac{9}{2}\right\}$ **51.** $\{35\}$ **52.** $\{0\}$ **53.** $\{4\}$
54. $\{6\}$ **55.** $\{-8\}$ **56.** R **57.** $\left\{\dfrac{14}{3}\right\}$ **58.** $\left\{\dfrac{7}{3}\right\}$ **59.** $\{3\}$ **60.** $\left\{\dfrac{1}{2}\right\}$
61. $\varnothing$ **62.** $\left\{-\dfrac{1}{4}\right\}$ **63.** $\left\{\dfrac{7}{13}\right\}$ **64.** $\left\{-\dfrac{5}{2}\right\}$ **65.** $\{-14\}$
66. $\{-24\}$ **67.** $\left\{\dfrac{7}{4}\right\}$ **68.** $\{0\}$ **69.** $\left\{\dfrac{1}{5}\right\}$ **70.** $\left\{-\dfrac{7}{4}\right\}$ **71.** $\{2\}$
72. $a = \dfrac{F}{m}$ **73.** $I = \dfrac{E}{R}$ **74.** $P = \dfrac{k}{V}$ **75.** $g = V - k - t$
76. $b_2 = \dfrac{2A - b_1 h}{h}$ **77.** $x = \dfrac{4y}{3}$ **78.** b **79.** 36   **80.** 45
81. 97, 99, and 101   **82.** \$5,500 at 4% and \$6,500 at 5%
83. \$11,000 at 8% and \$9,000 at 7%
84. \$12,000 at 12% and \$13,000 at 19%
85. 89 pounds of 12% and 161 pounds of 40%   **86.** $3 : 7$ or $\dfrac{3}{7}$
87. $9 : 4$ or $\dfrac{9}{4}$ **88.** $2 : 5$ or $\dfrac{2}{5}$ **89.** 30 mi : 1 gal or $\dfrac{30 \text{ mi}}{\text{gal}}$
90. $2{,}193 : 881$ or $\dfrac{2{,}193}{881}$ **91.** $7 : 3$ or $\dfrac{7}{3}$

92. A: $0.158 Brand A has lower unit price.
B: $0.159
93. A: $0.115 Brand B has lower unit price.
B: $0.108
94. $x = 32$ **95.** $a = 3.6$ **96.** $y = \frac{54}{5}$ or $10\frac{4}{5}$ **97.** $p = \frac{2}{5}$
98. 35 feet   **99.** 12 qt of antifreeze and 4 qt of water

100. $x > 4$;

101. $x \le 3$;

102. $x > -7$;

103. $x < -4$;

104. $x < 2$;

105. $x > \frac{15}{7}$;

106. $x > \frac{9}{2}$;

107. $x \ge -\frac{6}{5}$;

108. $x \le \frac{2}{9}$;

109. $-\frac{11}{5} < x < \frac{3}{5}$;

110. $-1 < x \le \frac{1}{5}$;

111. $\frac{1}{2} < x < \frac{9}{4}$;

112. $-\frac{13}{3} \le x \le -\frac{1}{3}$;

113. $x \ge 71$
114. $x \ge 38$
115. $6 < x < 8$

Chapter 2 Test

1. yes, trinomial, degree is 2 [2–1]
2. yes, trinomial, degree is 7 [2–1]
3. not a polynomial since x appears in the denominator [2–1]
4. $2n + 2$ [2–1]   **5.** $2x^2$ [2–2]   **6.** $3a - 7$ [2–2]   **7.** $\{3\}$ [2–5]
8. R [2–5]   **9.** $\{24\}$ [2–5]   **10.** $\left\{\frac{28}{5}\right\}$ [2–8]   **11.** $\{11\}$ [2–5]
12. $\{0\}$ [2–5]   **13.** $\varnothing$ [2–5]   **14.** $\left\{\frac{14}{13}\right\}$ [2–5]   **15.** $r = \frac{A - P}{P}$ [2–6]
16. $y = \frac{3x + 5}{2}$ [2–6]   **17.** $x \ge -1$ [2–9]   **18.** $x > -6$ [2–9]

19. $-3 \le x \le 6$ [2–9]
20. A: $0.214, B: $0.216, Brand A has the lower unit price. [2–8]
21. $20.50 [2–8]   **22.** 25 liters of 10%, 75 liters of 2% [2–7]
23. 12 [2–7]   **24.** 40 shares of A, 80 shares of B [2–7]
25. $x \ge 83$ [2–9]   **26.** $6,500 at 12% profit,
$10,500 at 19% loss [2–7]

Chapter 2 Cumulative Test

1. 4 [1–6]   **2.** undefined [1–7]   **3.** -36 [1–8]   **4.** 38 [1–8]
5. 20 [1–8]   **6.** $4x^2y - 7xy^2$ [2–2]   **7.** $17a - 38$ [2–2]   **8.** 64 [1–9]
9. 234 [1–9]   **10.** 42 [1–7]   **11.** $x - y$ [2–1]   **12.** $\left\{\frac{5}{3}\right\}$ [2–5]
13. $\{0\}$ [2–5]   **14.** $\left\{-\frac{19}{2}\right\}$ [2–5]   **15.** $\left\{\frac{17}{16}\right\}$ [2–5]   **16.** $\left\{\frac{30}{11}\right\}$ [2–5]
17. $x \le -6$ [2–9]   **18.** $x < -7$ [2–9]   **19.** $x > \frac{8}{5}$ [2–9]
20. $-2 < x < 4$ [2–9]   **21.** $-1 \le x \le 6$ [2–9]
22. $b = P - a - c$ [2–6]   **23.** $y = \frac{x - az}{a}$ [2–6]   **24.** 4 [2–8]
25. $6,000 at 6%, $4,000 at 5% [2–7]   **26.** 14, 16, and 18 [2–7]
27. $x \le 6$ [2–9]   **28.** 30 liters [2–7]   **29.** 60 gallons [2–8]
30. A: $0.299 B: $0.287 Brand B has the lower unit price. [2–8]

Chapter 3

Exercise 3–1
Answers to Odd-Numbered Problems
1. a^5 **3.** $(-2)^4$ **5.** x^6 **7.** $(xy)^4$ **9.** $(x - y)^3$ **11.** $xxxx$
13. $(-2)(-2)(-2) = -8$ **15.** $5 \cdot 5 \cdot 5 = 125$
17. $(4y)(4y)(4y)(4y) = 256y^4$ **19.** $(x - y)(x - y)$ **21.** x^{11}
23. R^3 **25.** a^9 **27.** $5^5 = 3,125$ **29.** $4^7 = 16,384$
31. $(x - 2y)^{10}$ **33.** $(a - b)^{11}$ **35.** x^4y^4 **37.** $64x^3y^3z^3$ **39.** x^{15}
41. b^{25} **43.** $6x^4y^3$ **45.** a^7b^5 **47.** $30x^5$
49. $12a^7b^7c$ **51.** $12a^5b^3$ **53.** $-6a^3b^5$
55. a. 125 cubic units   **b.** 64 cubic units   **c.** 216 cubic units
57. $s = \frac{1}{2}gt^2$ **59.** $V = \frac{4}{3}\pi r^3$ **61.** $m^4 - 8$ **63.** $2x^2 - y^3$
65. $\frac{p^3}{q^2}$ **67.** $(x + y)^3$ and $x + y^3$

Solutions to Trial Exercise Problems
23. $R^2 \cdot R = R^2 \cdot R^1 = R^{2+1} = R^3$
27. $5^2 \cdot 5^3 = 5^{2+3} = 5^5 = 3,125$
43. $(2xy^2)(3x^3y) = 2 \cdot 3 \cdot xx^3y^2y = 6x^{1+3}y^{2+1} = 6x^4y^3$
55. a. $V = e^3$, then $V = (5)^3 = 5 \cdot 5 \cdot 5 = 25 \cdot 5 = 125$ cubic units
60. The cube of Theo's age is given as n^3. Since his mother is 6 years more than the cube of Theo's age, we add 6, giving $n^3 + 6$.

Review Exercises
1. $9a$ **2.** $8x$ **3.** $6ab$ **4.** $7xy$ **5.** $7a^2 + 5a$ **6.** $5x^2 + 5x$
7. $x^2y + 7xy^2$ **8.** $3ab^2 + 2a^2b$

Exercise 3–2
Answers to Odd-Numbered Problems
1. $2a^3b - 2ab^2c + 2abc^2$ **3.** $15ab^2 - 21ac^2$
5. $-15a^3b^2 + 5a^2b^3 - 20ab^4$ **7.** $3a^3b - 6a^2b^2 - 3ab^3$
9. $12a^2b^2 - 6ab^3$ **11.** $x^2 + 7x + 12$ **13.** $y^2 - 13y + 36$
15. $a^2 + 2a + 1$ **17.** $R^2 - 6R + 9$ **19.** $a^2 - 9$ **21.** $c^2 - 64$

23. $4a^2 - 1$ **25.** $6a^2 - 31a + 35$ **27.** $4x^2 - 49$
29. $3k^2 - 17kw - 6w^2$ **31.** $4x^2 - y^2$ **33.** $x^2 - 0.16$
35. $0.09b^2 - 0.16$ **37.** $x^2 + 10x + 25$ **39.** $4a^2 + 4ab + b^2$
41. $a^2 + 12ab + 36b^2$ **43.** $x^2 - 8x + 16$ **45.** $x^2 - 6xy + 9y^2$
47. $4a^2 - 9b^2$ **49.** $x^2 + 0.8x + 0.16$ **51.** $a^2 + a + 0.25$
53. $0.04y^2 - 0.4yz + z^2$ **55.** $0.25a^2 - ab + b^2$
57. $a^3 + 2a^2b - 7ab^2 + 4b^3$ **59.** $6x^3 + 21x^2 - 5x + 28$
61. $x^4 - x^3 - x^2 - 11x - 12$ **63.** $a^3 - 7a^2 + 4a + 12$
65. $2a^3 - 3a^2b - 2ab^2 + 3b^3$ **67.** $a^3 - 3a^2b + 3ab^2 - b^3$
69. $a^3 - 6a^2b + 12ab^2 - 8b^3$ **71.** $cx^2 - 4c^2x + 4c^3$

Solutions to Trial Exercise Problems

8. $(2x)(x - y + 5)(5y) = [(2x)(x - y + 5)](5y)$
$= [2x \cdot x - 2x \cdot y + 2x \cdot 5](5y) = [2x^2 - 2xy + 10x](5y)$
$= 2x^2 \cdot 5y - 2xy \cdot 5y + 10x \cdot 5y = 10x^2y - 10xy^2 + 50xy$

17. $(R - 3)^2$ is a special product.
$(R - 3)^2 = (R)^2 - [2 \cdot R \cdot (3)] + (3)^2 = R^2 - 6R + 9$

18. $(R + 2)(R - 2)$ is a special product.
$(R + 2)(R - 2) = (R)^2 - (2)^2 = R^2 - 4$

33. $(x - 0.4)(x + 0.4) = (x)^2 - (0.4)^2 = x^2 - 0.16$

49. $(x + 0.4)^2$ Use the special product: $(a + b)^2 = a^2 + 2ab + b^2$.
We have $(x + 0.4)^2 = (x)^2 + 2(x)(0.4) + (0.4)^2 = x^2 + 0.8x + 0.16$

53. $(0.2y - z)^2$ Use the special product: $(a - b)^2 = a^2 - 2ab + b^2$.
We have $(0.2y - z)^2 = (0.2y)^2 - 2(0.2y)(z) + (z)^2$
$= 0.04y^2 - 0.4yz + z^2$

57. $(a + 4b)(a^2 - 2ab + b^2)$
$= a^3 - 2a^2b + ab^2 + 4a^2b - 8ab^2 + 4b^3$
$= a^3 + 2a^2b - 7ab^2 + 4b^3$

63. $(a - 6)(a - 2)(a + 1) = [(a - 6)(a - 2)](a + 1)$
$= [a^2 - 2a - 6a + 12](a + 1) = [a^2 - 8a + 12](a + 1)$
$= a^3 + a^2 - 8a^2 - 8a + 12a + 12 = a^3 - 7a^2 + 4a + 12$

66. $(a + b)^3 = (a + b)(a + b)(a + b) = [(a + b)(a + b)](a + b)$
$= [a^2 + ab + ab + b^2](a + b) = [a^2 + 2ab + b^2](a + b)$
$= a^3 + a^2b + 2a^2b + 2ab^2 + ab^2 + b^3$
$= a^3 + 3a^2b + 3ab^2 + b^3$

Review Exercises

1. -5 **2.** 12 **3.** 3 **4.** -9 **5.** x^{12} **6.** a^{15} **7.** $27a^3b^3$ **8.** $8x^3$

Exercise 3–3
Answers to Odd-Numbered Problems

1. 1 **3.** 5 **5.** 1 **7.** $\dfrac{1}{R^5}$ **9.** $\dfrac{1}{9P^2}$ **11.** $\dfrac{9}{C^4}$ **13.** $\dfrac{y^3}{2}$ **15.** $\dfrac{2y^2}{x^4}$

17. $\dfrac{t^5}{r^2}$ **19.** $\dfrac{a^6}{b^6}$ **21.** $\dfrac{8}{27}$ **23.** $\dfrac{16x^4}{y^4}$ **25.** $\dfrac{27a^3}{b^3}$ **27.** y^2

29. b^2 **31.** $\dfrac{1}{R^4}$ **33.** 4 **35.** $\dfrac{1}{36}$ **37.** y^4 **39.** a **41.** $\dfrac{1}{y^6}$

43. a^3b^3 **45.** $4x^2y^2$ **47.** $\dfrac{1}{27a^3}$ **49.** x^3 **51.** $\dfrac{1}{a^6}$ **53.** x^2

55. $\dfrac{1}{a^2}$ **57.** $-\dfrac{1}{125}$ **59.** 27 **61.** $\dfrac{1}{a^6}$ **63.** $\dfrac{1}{x^{10}}$ **65.** a^6 **67.** 1

69. 1 **71.** $\dfrac{R^5}{S^9}$ **73.** See note on page 204.

Solutions to Trial Exercise Problems

3. $5a^0 = 5 \cdot 1 = 5$ **10.** $4z^{-2} = 4 \cdot \dfrac{1}{z^2} = \dfrac{4}{z^2}$

15. $2x^{-4}y^2 = 2 \cdot \dfrac{1}{x^4} \cdot y^2 = \dfrac{2y^2}{x^4}$ **32.** $\dfrac{3^4}{3^2} = 3^{4-2} = 3^2 = 9$

36. $\dfrac{x^4x^3}{x^2} = \dfrac{x^{4+3}}{x^2} = \dfrac{x^7}{x^2} = x^{7-2} = x^5$

48. $\dfrac{5^2a^3b}{5^3a^7b^3} = 5^{2-3}a^{3-7}b^{1-3} = 5^{-1}a^{-4}b^{-2} = \dfrac{1}{5^1} \cdot \dfrac{1}{a^4} \cdot \dfrac{1}{b^2} = \dfrac{1}{5a^4b^2}$

49. $x^{-4}x^7 = x^{-4+7} = x^3$
  Alternate: $x^{-4}x^7 = \dfrac{1}{x^4} \cdot x^7 = \dfrac{x^7}{x^4} = x^{7-4} = x^3$

Review Exercises

1. -10 **2.** 21 **3.** 4 **4.** -16 **5.** a^8
6. x^{12} **7.** x^6 **8.** $4a^2b^2$

Exercise 3–4
Answers to Odd-Numbered Problems

1. $8a^6$ **3.** $8x^6y^3$ **5.** $27x^6y^3$ **7.** $81x^2y^8$ **9.** $x^{16}y^{12}z^4$

11. $25a^{10}b^4c^8$ **13.** $\dfrac{1}{4a^4}$ **15.** $\dfrac{16}{x^4}$ **17.** $\dfrac{4x^2}{y^2}$ **19.** $\dfrac{b^9}{a^6c^3}$

21. $\dfrac{y^{12}}{27x^3}$ **23.** $\dfrac{y^{10}}{x^4z^6}$ **25.** $6x^7y^3$ **27.** $-6x^{10}y^4$ **29.** $6x^6y^3z^7$

31. $\dfrac{8x^3}{y^6}$ **33.** $\dfrac{9a^4}{b^6}$ **35.** $\dfrac{8x^6}{y^9}$ **37.** $\dfrac{4x^6y^6}{z^{10}}$ **39.** $\dfrac{b^8}{a^5}$ **41.** $\dfrac{R^2}{3S^4}$

43. $2x^2$ **45.** $\dfrac{3a^5}{2b^2}$ **47.** $\dfrac{4}{ab^9}$ **49.** $\dfrac{3S^3}{R^4}$ **51.** $a^{10}b^{17}$ **53.** $32a^{12}$

55. $32x^{14}y^{13}$ **57.** $\dfrac{y^2}{xz^4}$ **59.** $\dfrac{a^6b^{10}}{4}$ **61.** $\dfrac{b^6}{a^3c^3}$

63. See page 187.

Solutions to Trial Exercise Problems

13. $(2a^2)^{-2} = \dfrac{1}{(2a^2)^2} = \dfrac{1}{2^2(a^2)^2} = \dfrac{1}{2^2a^4} = \dfrac{1}{4a^4}$

25. $(3x^2)(2x^0y^2)(x^5y) = (3 \cdot 2)(x^2x^0x^5)(y^2y) = 6x^7y^3$

39. $\dfrac{a^{-2}b^3}{a^3b^{-5}} = \dfrac{b^3b^5}{a^2a^3} = \dfrac{b^{3+5}}{a^{2+3}} = \dfrac{b^8}{a^5}$

45. $\dfrac{6^{-1}a^3b^{-2}}{3^{-2}a^{-2}b^0} = \dfrac{3^2a^2a^3}{6^1b^21} = \dfrac{9a^{2+3}}{6b^2} = \dfrac{3a^5}{2b^2}$

51. $(a^2b^3)^3(ab^2)^4 = (a^2)^3(b^3)^3 \cdot a^4(b^2)^4 = a^6b^9a^4b^8 = a^{6+4}b^{9+8}$
$= a^{10}b^{17}$

57. $\left(\dfrac{xy^{-2}}{z^{-4}}\right)^{-1} = \dfrac{x^{-1}(y^{-2})^{-1}}{(z^{-4})^{-1}} = \dfrac{x^{-1}y^2}{z^4} = \dfrac{y^2}{x^1z^4} = \dfrac{y^2}{xz^4}$

Review Exercises

1. 35.34 **2.** 10.36 **3.** 16.72 **4.** 28.98 **5.** 18.81 **6.** 49.5

7. 10^8 **8.** 10^{12} **9.** 10^{-6} or $\dfrac{1}{10^6}$ **10.** 10^{-3} or $\dfrac{1}{10^3}$

Exercise 3–5
Answers to Odd-Numbered Problems

1. 2.55×10^2 **3.** 1.2345×10^4 **5.** 1.55×10^5 **7.** 8.55076×10^2
9. 1.0076×10^6 **11.** 1.2×10^{-4} **13.** 8.1×10^{-6} **15.** 7×10^{-4}
17. 9.4×10^{-11} **19.** 4.5×10^3 **21.** 5.85×10^6
23. 4.578×10^1 **25.** 2.985×10^{-8} **27.** $49,900,000$ **29.** 7.23
31. 0.0042 **33.** 0.00000147 **35.** 0.000789 **37.** 0.00000000482

39. 0.00000492 **41.** $1.22304 \times 10^{14} = 122{,}304{,}000{,}000{,}000$
43. $3.63226 \times 10^{-1} = 0.363226$
45. $1.76979 \times 10^{-7} = 0.000000176979$
47. $4.84481 \times 10^{8} = 484{,}481{,}000$ **49.** $1.4 \times 10^{3} = 1{,}400$
51. $4.6 \times 10^{3} = 4{,}600$ **53.** $5.849057 \times 10^{-3} = 0.005849057$
55. $1.52881 \times 10^{21} = 1{,}528{,}810{,}000{,}000{,}000{,}000{,}000$
57. $2 \times 10^{9} = 2{,}000{,}000{,}000$ **59.** 1.0×10^{22}
61. 9.45×10^{15} meters
63. a. approximately 9.78×10^{9} tons
   b. approximately 7.693×10^{9} tons

Solutions to Trial Exercise Problems

5. $155{,}000 = 1.55000. = 1.55000 \times 10^{5} = 1.55 \times 10^{5}$

11. $0.00012 = 0.0001.2 = 1.2 \times 10^{-4}$

27. $4.99 \times 10^{7} = 4.9900000 \times 10^{7} = 4.9900000. = 49900000.$
   $= 49{,}900{,}000$

35. 7.89×10^{-4} Move the decimal point four places to the left. We have $7.89 \times 10^{-4} = 0.000789.$

48. $(177{,}000) \div (0.15) = \dfrac{1.77 \times 10^{5}}{1.5 \times 10^{-1}} = \dfrac{1.77}{1.5} \times 10^{5-(-1)}$

   $= \dfrac{1.77}{1.5} \times 10^{6} = 1.18 \times 10^{6} = 1{,}180{,}000$

Review Exercises

1. $-\dfrac{11}{7}$ **2.** 5 feet and 11 feet **3.** 1 **4.** $27y^{3}$ **5.** $-8x^{9}y^{6}$

Exercise 3–6
Answers to Odd-Numbered Problems

1. $4x^{2}$ **3.** $-5x^{3}yz$ **5.** $3(a - b)$ **7.** $2a(b - c)$
9. $-4a^{2}b(x + y)$ **11.** $-4a^{2} + 2a$ **13.** $a^{2} - 3a + 2$

15. $5a^{2} - 3a + 4 - \dfrac{2}{a}$ **17.** $-x + y + 2y^{2}$ **19.** $10xy^{2} + 7 - 6y^{2}$

21. $a - c$ **23.** $x + 4$ **25.** $a - 3$ **27.** $x + 3$

29. $a + 9 + \dfrac{28}{a - 2}$ **31.** $a + 2 + \dfrac{4}{a + 3}$ **33.** $a + 3 + \dfrac{1}{a + 3}$

35. $2x - 17 + \dfrac{72}{x + 5}$ **37.** $3x - 5 + \dfrac{3}{2x - 3}$ **39.** $3a - 4 - \dfrac{4}{3a - 4}$

41. $x^{2} + 2x + 4$ **43.** $x^{2} + 2x + 3$ **45.** $b^{2} + 7b + 14 + \dfrac{6}{b - 1}$

47. $3a + 8$ **49.** $x^{2} - 1 + \dfrac{6x - 13}{x^{2} + 3x - 5}$

51. $y^{2} - 2y + 9 + \dfrac{-26y + 42}{y^{2} + 2y - 5}$ **53.** $-6x^{4} + 15x^{3} + 4x^{2} - 22x + 30$

Solutions to Trial Exercise Problems

5. $\dfrac{3(a - b)^{2}}{a - b} = \dfrac{3(a - b)^{2}}{(a - b)^{1}} = 3(a - b)^{2-1} = 3(a - b)^{1} = 3(a - b)$

   or $3a - 3b$ (*Note:* A quantity is treated as just one factor.)

15. $\dfrac{15a^{3} - 9a^{2} + 12a - 6}{3a} = \dfrac{15a^{3}}{3a} - \dfrac{9a^{2}}{3a} + \dfrac{12a}{3a} + \dfrac{6}{3a}$

   $= 5a^{2} - 3a + 4 - \dfrac{2}{a}$

17. $\dfrac{x^{2}y - xy^{2} - 2xy^{3}}{-xy} = \dfrac{x^{2}y}{-xy} - \dfrac{xy^{2}}{-xy} - \dfrac{2xy^{3}}{-xy} = -x + y + 2y^{2}$

21. $\dfrac{a(b - 1) - c(b - 1)}{b - 1} = \dfrac{a(b - 1)}{(b - 1)} - \dfrac{c(b - 1)}{(b - 1)} = a - c$

34. $(4a^{2} + 1 + 4a) \div (2a + 1)$

Arrange the dividend in descending powers of a

$$
\begin{array}{r}
2a + 1 \\
2a + 1 \overline{)\,4a^{2} + 4a + 1} \\
\underline{4a^{2} + 2a} \\
2a + 1 \\
\underline{2a + 1} \\
0
\end{array}
$$

$\dfrac{4a^{2} + 4a + 1}{2a + 1} = 2a + 1$

40. $(27a^{3} - 1) \div (3a - 1)$
   (*Note:* Insert zeros to hold positions where terms are missing.)

$$
\begin{array}{r}
9a^{2} + 3a + 1 \quad \text{Answer: } 9a^{2} + 3a + 1 \\
3a - 1 \overline{)\,27a^{3} + 0a^{2} + 0a - 1} \\
\underline{27a^{3} - 9a^{2}} \\
9a^{2} + 0a \\
\underline{9a^{2} - 3a} \\
3a - 1 \\
\underline{3a - 1} \\
0
\end{array}
$$

52. $\dfrac{600t + 10t^{2}}{600t} = \dfrac{600t}{600t} + \dfrac{10t^{2}}{600t} = 1 + \dfrac{t}{60}$

Review Exercises

1. $5x$ **2.** $3a^{2}$ **3.** $12ab$ **4.** $x^{3} + 2x^{2}$ **5.** $6a^{2} - 15a$
6. $3x^{3}y + 2x^{2}y^{2} - 7x^{2}y$

Chapter 3 Review

1. a^{12} **2.** a^{14} **3.** $4^{5} = 1{,}024$ **4.** $x^{4}y^{4}$ **5.** a^{15} **6.** $20a^{4}b^{5}$
7. $6x^{3}y^{7}$ **8.** $-15x^{5}$ **9.** $6a^{3}b^{5}$ **10.** $10x^{5}y^{5}$ **11.** $-6a^{6}b^{10}$
12. $4x^{3}$ **13.** $x^{2} + 6$ **14.** $x^{2}y^{3}$ **15.** $x^{2} + 10$ **16.** $15x^{2} - 10xy$
17. $-6a^{4}b + 9a^{3}b^{2} - 12a^{2}b^{3}$ **18.** $y^{2} - 10y + 25$
19. $x^{2} - x - 12$ **20.** $x^{2} + 10x + 25$ **21.** $a^{2} - 49$
22. $15x^{2} + 7xy - 2y^{2}$ **23.** $x^{3} + x^{2}y - 5xy^{2} - 2y^{3}$

24. $0.12x^{2} + 0.1x - 0.12$ **25.** $4a^{2} + 4ab + b^{2}$ **26.** $\dfrac{1}{b^{2}}$ **27.** $\dfrac{5}{a^{2}}$

28. a^{4} **29.** $\dfrac{1}{x^{3}}$ **30.** 3 **31.** $\dfrac{1}{a^{6}}$ **32.** x^{3} **33.** $\dfrac{a^{5}}{b^{5}}$

34. $\dfrac{4y^{2}z^{2}}{x^{2}}$ **35.** $\dfrac{b^{3}}{4a^{3}}$ **36.** $\dfrac{a^{9}}{b^{3}}$ **37.** $8a^{6}b^{9}$

38. $3^{12}x^{16}y^{20} = 531{,}441x^{16}y^{20}$ **39.** $\dfrac{y^{6}}{4x^{2}}$ **40.** $\dfrac{xy^{3}z^{3}}{2}$

41. $\dfrac{2a^{2}c^{3}}{b^{6}}$ **42.** $8x^{26}y^{25}$ **43.** $72a^{14}b^{3}$ **44.** $\dfrac{a^{9}}{c^{12}}$ **45.** $\dfrac{9a^{6}b^{4}}{c^{10}}$

46. $\dfrac{32x^{5}y^{20}}{z^{30}}$ **47.** 1.84×10^{3} **48.** 1.57×10^{-3} **49.** 1.07×10^{8}

50. 8.49×10^{11} **51.** 3.75×10^{1} **52.** 5.43×10^{-3} **53.** 504,000
54. 0.00639 **55.** 596 **56.** 0.00886 **57.** 0.000000735
58. 812,000,000 **59.** 2.67672×10^{5} **60.** 1.54818×10^{-8}
61. 7.2×10^{-3} **62.** 1.25×10^{-7} **63.** 5.2×10^{16}
64. 8.82×10^{12} BTUs **65.** $2a^{2} + 2ab$ **66.** $2a - 3 + 5a^{2}$

67. $5x - 3y^{3} + xy$ **68.** $2a + 3b - 6ab^{6}$ **69.** $4a + 1 + \dfrac{-2}{2a - 1}$

70. $3a - 2 + \dfrac{1}{a - 5}$ **71.** $x - 7$ **72.** $5x^{2} - x - 4$

Chapter 3 Test

1. 32 [3–1] **2.** a^{8} [3–1] **3.** $6a^{2} - 4ab$ [3–2] **4.** 5 [3–3]
5. a^{2} [3–3] **6.** $y^{2} - 14y + 49$ [3–2] **7.** x^{2} [3–3]

8. $27x^6y^{12}z^3$ [3–4]   **9.** $9x^2 - 4$ [3–2]   **10.** $\dfrac{27a^3}{b^3}$ [3–3]

11. $\dfrac{3a^2}{b^6}$ [3–3]   **12.** $-6x^3y^4$ [3–1]   **13.** $x^2 + 12x + 36$ [3–2]

14. $\dfrac{y^4}{4x^4}$ [3–4]   **15.** $\dfrac{8a^6}{b^9}$ [3–4]   **16.** $2a^2 - 5ab - 12b^2$ [3–2]

17. $\dfrac{y^5}{2x}$ [3–4]   **18.** 65,000 or 6.5×10^4 [3–5]   **19.** $3xy - 3xz$ [3–6]

20. $8a^3 - 6a^2 + 1$ [3–6]   **21.** $x^2 - x + 3$ [3–6]

22. $2a^2 + a - 1 - \dfrac{3}{2a - 1}$ [3–6]   **23.** $3x^2 - 4x - 5 + \dfrac{1}{x - 2}$ [3–6]

24. $5y^2$ [3–1]   **25.** 1.485×10^{30} [3–5]

Chapter 3 Cumulative Test

1. -2 [1–7]   **2.** undefined [1–7]   **3.** 0 [1–7]   **4.** 38 [1–8]
5. $x^2y + 2xy^2 - 3x^2y^2$ [1–9]   **6.** $9x^2 - 6xy + y^2$ [3–2]
7. $4a^4b^{10}$ [3–4]   **8.** -25 [1–7]   **9.** 0 [1–8]   **10.** $9x^2 - 4y^2$ [3–2]
11. $6x^3y^5$ [3–1]   **12.** 26 [1–8]   **13.** $x^3 - 2x - 1$ [3–2]

14. x^2 [3–3]   **15.** a^4 [3–3]   **16.** $x - 6 + \dfrac{1}{x - 2}$ [3–6]

17. $\{7\}$ [2–6]   **18.** $\left\{\dfrac{5}{3}\right\}$ [2–6]   **19.** $\left\{-\dfrac{19}{4}\right\}$ [2–6]

20. $x > -\dfrac{1}{3}$ [2–9]   **21.** $x > \dfrac{1}{6}$ [2–9]   **22.** $-8 \le x \le -1$ [2–9]

23. 97 [2–7]   **24.** 43.2 inches [2–8]
25. $18,500 at 8%; $11,500 at 7% [2–7]

Chapter 4

Exercise 4–1

Answers to Odd-Numbered Problems

1. $2(y + 3)$ **3.** $4(x^2 + 2y)$ **5.** $3(x^2y + 5z)$ **7.** $7(a - 2b + 3c)$
9. $3(5xy - 6z + x^2)$ **11.** $7(6xy - 3y^2 + 1)$ **13.** $2(4x - 5y + 6z - 9w)$
15. $5ab(4a - 12 + 9b)$ **17.** $3xy(x + 2)$ **19.** $2R^2(R^2 - 3)$
21. $x(2x^2 - x + 1)$ **23.** $3ab(5 + 6b - a)$ **25.** $xy(y + z + 1)$
27. $2L(L^2 - 9 + L)$ **29.** $5p(p + 2 + 3p^2)$ **31.** $(a + b)(x + y)$
33. $5(2a + b)(3x + 2y)$ **35.** $3(a + 4b)(x + 2y)$ **37.** $(b + 6)(8a - 1)$
39. $(a - 3)(a + 5)$ **41.** $(4x - 3)(2x - 3)$ **43.** $(a + b)(c + d)$
45. $(x + 3)(x + 4)$ **47.** $(x + 3)^2$ **49.** $(a + 3)(a + 5)$ **51.** $(x + 7)(x + 1)$
53. $(2a + b)(3x - 2y)$ **55.** $(2x + y)(2a - b)$ **57.** $(5x - 3y)(4x + z)$
59. $(a + 3b)(4x - 3y)$ **61.** $(2c - y)(a + 3b)$ **63.** $(c + 4y)(2a + 3b)$
65. $(3x + y)(2a + b)$ **67.** $(x - 2d)(3a + b)$ **69.** $(a + 5)(2a^2 + 3)$

71. $(4a^2 + 3)(2a - 1)$ **73.** $\pi r(s + r)$ **75.** $\dfrac{wx}{48EI}(2x^3 - 3\ell x^2 - \ell^3)$

77. $1, 34, 26, -2$ **79.** $9, 210, -86, \dfrac{-15}{4}$

Solutions to Trial Exercise Problems

11. $42xy - 21y^2 + 7 = 7 \cdot 6xy - 7 \cdot 3y^2 + 7 \cdot 1 = 7(6xy - 3y^2 + 1)$
25. $xy^2 + xyz + xy = xy \cdot y + xy \cdot z + xy \cdot 1 = xy(y + z + 1)$
33. $15x(2a + b) + 10y(2a + b) = 5(2a + b) \cdot 3x + 5(2a + b) \cdot 2y$
$= 5(2a + b)(3x + 2y)$
53. $6ax - 2by + 3bx - 4ay = 6ax + 3bx - 4ay - 2by = (6ax + 3bx) + (-4ay - 2by) = 3x(2a + b) - 2y(2a + b) = (2a + b)(3x - 2y)$

75. $Y = \dfrac{2wx^4}{48EI} - \dfrac{3\ell wx^3}{48EI} - \dfrac{\ell^3 wx}{48EI} = \dfrac{wx}{48EI} \cdot 2x^3 - \dfrac{wx}{48EI} \cdot 3\ell x^2$

$- \dfrac{wx}{48EI} \cdot \ell^3 = \dfrac{wx}{48EI}(2x^3 - 3\ell x^2 - \ell^3)$

Review Exercises

1. $x^2 + 9x + 18$ **2.** $a^2 + 5a - 24$ **3.** $x^2 + 8x + 12$ **4.** $a^2 - 2a - 24$
5. $a^2 - a - 6$ **6.** $b^2 - 6b + 8$ **7.** $x^2 - 7x + 10$ **8.** $x^2 - 10x + 24$
9. $2a^2 + 6a - 20$ **10.** $3a^2 - 18a - 48$ **11.** 18 ft by 27 ft

Exercise 4–2

Answers to Odd-Numbered Problems

1. -4 and 4   **3.** -5 and -4 **5.** -5 and 6   **7.** -4 and 3
9. -7 and 0   **11.** -1 and 8   **13.** 5 and 7   **15.** -3 and 6
17. -3 and 4   **19.** $(x + 1)(x + 6)$ **21.** $(x + 2)(x + 5)$
23. $(x + 1)(x + 14)$ **25.** $(b + 4)(b + 10)$ **27.** $(a + 6)(a + 3)$
29. $(x + 12)(x - 1)$ **31.** $(y + 15)(y - 2)$ **33.** $(x - 12)(x - 2)$
35. $(a + 8)(a - 3)$ **37.** $(x + 6)(x + 2)$ **39.** $(a - 6)(a + 4)$
41. $2(x + 5)(x - 2)$ **43.** $3(x - 8)(x + 2)$
45. will not factor, prime polynomial   **47.** $(y + 15)(y + 2)$
49. $4(x + 2)(x - 3)$ **51.** $5(a - 5)(a + 2)$ **53.** $(xy - 6)(xy + 3)$
55. $(xy + 12)(xy + 1)$ **57.** $3(xy + 3)(xy - 4)$ **59.** $(x + 2y)(x + y)$
61. $(a - 3b)(a + b)$ **63.** $(a - 3b)(a + 2b)$ **65.** $(x + 3y)(x - 5y)$

Solutions to Trial Exercise Problems

1. We want two integers whose product is -16 and their sum is 0. Since the sum is 0, the two integers must have the same absolute value and must be opposite in sign. The two integers are -4 and 4 because $(-4)(4) = -16$ and $(-4) + 4 = 0$.
5. We want two integers whose product is -30 and whose sum is 1. Since the product is negative and the sum is positive, the two integers must be opposite in sign and the integer with the greater absolute value must be positive. The two integers are -5 and 6 because $(-5)(6) = -30$ and $(-5) + 6 = 1$.
43. $3x^2 - 18x - 48 \qquad = 3(x^2 - 6x - 16)$
m and n are -8 and 2. $= 3(x - 8)(x + 2)$
52. $x^2y^2 - 4xy - 21 \qquad = (xy)^2 - 4(xy) - 21$
m and n are -7 and 3. $= (xy - 7)(xy + 3)$
59. We need to find m and n that add to $3y$ and multiply to $2y^2$. The values are $2y$ and y. The factorization is $x^2 + 3xy + 2y^2$
$= (x + 2y)(x + y)$.

Review Exercises

1. $x^2(a + b + c)$ **2.** $3x(x^2 + 4x - 2)$ **3.** $(3x + 5)(2x + 1)$
4. $(2x + 3)(3x - 2)$ **5.** $(4x + 1)(5x + 1)$ **6.** $(6x - 1)(2x + 3)$
7. $(x - 2)(3x - 5)$ **8.** $(7x - 3)(x - 9)$ **9.** $19,000 at 8%;
$18,000 at 6%

Exercise 4–3

Answers to Odd-Numbered Problems

1. $(3x + 1)(x + 3)$ **3.** $(5x + 1)(x + 2)$ **5.** $(3x + 4)(x + 2)$
7. $(5x + 3)(x + 2)$ **9.** $(2x + 3)(4x + 3)$ **11.** $(2x - 3)(x + 2)$
13. $(2x + 1)(x + 1)$ **15.** $(2R - 3)(R - 2)$ **17.** $(5x + 3)(x - 2)$
19. $(3x - 1)(3x - 1) = (3x - 1)^2$ **21.** will not factor, prime polynomial
23. $(2x + 3)(3x + 2)$ **25.** $(2x + 7)(2x + 3)$
27. will not factor, prime polynomial   **29.** $(3y - 8)(3y + 1)$
31. $(5x + 6)(2x - 1)$ **33.** $(2x - 5)(x - 2)$ **35.** $2(2x + 3)(x + 2)$
37. $2(2x + 3)(x + 1)$ **39.** $(2x + 3)(3x - 1)$ **41.** $2(x + 5)(x - 2)$
43. $(3x - 2)(2x + 3)$ **45.** $(2x + 3)^2$ **47.** $(7x - 1)(x - 5)$
49. $(3P + 1)(5P - 1)$ **51.** $2x(x - 5)(x + 2)$ **53.** $a(2a + 1)(a + 7)$
55. $(2x - 5)(4x + 3)$ **57.** $-16t(t - 2)$
59.   **61.** changes the sign of the middle term

Solutions to Trial Exercise Problems

35. $4x^2 + 14x + 12 \qquad = 2(2x^2 + 7x + 6)$

m and n are 3 and 4. $= 2[(2x^2 + 3x) + (4x + 6)]$

$\qquad\qquad\qquad\qquad = 2[x(2x + 3) + 2(2x + 3)]$

$\qquad\qquad\qquad\qquad = 2(2x + 3)(x + 2)$

51. $2x^3 - 6x^2 - 20x \qquad = 2x(x^2 - 3x - 10)$

m and n are -5 and 2. $= 2x(x - 5)(x + 2)$

Review Exercises

1. $x^2 - y^2$ **2.** $9a^2 - 4b^2$ **3.** $x^2 - 2xy + y^2$ **4.** $25a^2 - 16b^2$
5. $4a^2 + 4ab + b^2$ **6.** $16x^2 - 8xy + y^2$ **7.** $x^4 - 1$ **8.** $a^4 - 16$
9. 14 **10.** 48

Exercise 4–4

Answers to Odd-Numbered Problems

1. $(6)^2$ **3.** $(c)^2$ **5.** $(4x)^2$ **7.** $(2z^2)^2$ **9.** $(x + 1)(x - 1)$
11. $(a + 2)(a - 2)$ **13.** $(3 + E)(3 - E)$ **15.** $(1 + k)(1 - k)$
17. $(3b + 4)(3b - 4)$ **19.** $(b + 6c)(b - 6c)$ **21.** $(2a + 5b)(2a - 5b)$
23. $(5p + 9)(5p - 9)$ **25.** $8(x + 2y)(x - 2y)$ **27.** $5(r + 5s)(r - 5s)$
29. $2(5 + x)(5 - x)$ **31.** $(rs + 5t)(rs - 5t)$ **33.** $(x^2 - 3)(x^2 + 3)$
35. $(r^2 + 9)(r + 3)(r - 3)$ **37.** $(7x + 8y^2)(7x - 8y^2)$
39. $2(7xy + 5pc)(7xy - 5pc)$ **41.** $(x + 2)^2$ **43.** $(a + 8)^2$ **45.** $(x - 9)^2$
47. $(c - 7)^2$ **49.** $(a + 3)^2$ **51.** $(y - 3)^2$ **53.** $(2a - 3b)^2$ **55.** $(3c - 2d)^2$
57. $(2x + 3y)^2$ **59.** $(2x + 5y)^2$ **61.** $\dfrac{V}{8I}(h + 2v_1)(h - 2v_1)$

63. $\dfrac{1}{2}g(t_2 + t_1)(t_2 - t_1)$

Solutions to Trial Exercise Problems

17. $9b^2 - 16 = (3b)^2 - (4)^2 = (3b + 4)(3b - 4)$
25. $8x^2 - 32y^2 = 8(x^2 - 4y^2) = 8[(x)^2 - (2y)^2] = 8(x + 2y)(x - 2y)$
34. $x^4 - 1 = (x^2)^2 - (1)^2 = (x^2 + 1)(x^2 - 1) = (x^2 + 1)[(x)^2 - (1)^2]$
$\qquad = (x^2 + 1)(x + 1)(x - 1)$

Review Exercises

1. $(x + 6)(x + 2)$ **2.** $(7a + 9)(7a - 9)$ **3.** $(x - 4y)(3a + b)$
4. $2x(x + 3)(x + 4)$ **5.** $(2a + 3)(5a + 3)$ **6.** $(2a - 5)^2$
7. $(xy + 5)(xy + 3)$ **8.** $(x + 2y)^2$ **9.** 26 ft by 29 ft

Exercise 4–5

Answers to Odd-Numbered Problems

1. 4^3 **3.** 5^3 **5.** $(3x)^3$ **7.** $(a^2)^3$ **9.** $(2b^5)^3$ **11.** $(r + s)(r^2 - rs + s^2)$
13. $(2x + y)(4x^2 - 2xy + y^2)$ **15.** $(h - k)(h^2 + hk + k^2)$
17. $(a - 2)(a^2 + 2a + 4)$ **19.** $(x - 2y)(x^2 + 2xy + 4y^2)$
21. $(4x - y)(16x^2 + 4xy + y^2)$ **23.** $(3x - 2y)(9x^2 + 6xy + 4y^2)$
25. $(2a + 3b)(4a^2 - 6ab + 9b^2)$ **27.** $2(a + 2)(a^2 - 2a + 4)$
29. $2(x - 2)(x^2 + 2x + 4)$ **31.** $x^2(x + 3y)(x^2 - 3xy + 9y^2)$
33. $(x^2 + y)(x^4 - x^2y + y^2)$ **35.** $(a^3 - b)(a^6 + a^3b + b^2)$
37. $(x^4 - 3)(x^8 + 3x^4 + 9)$ **39.** $a^2(2b - a)(4b^2 + 2ab + a^2)$
41. $2(3r + s)(9r^2 - 3rs + s^2)$ **43.** $(xy - z)(x^2y^2 + xyz + z^2)$
45. $(a^5b^2 - 2c^3)(a^{10}b^4 + 2a^5b^2c^3 + 4c^6)$ **47.** $(ab + 2)(a^2b^2 - 2ab + 4)$
49. $(x^3y^4 + z^5)(x^6y^8 - x^3y^4z^5 + z^{10})$

Solutions to Trial Exercise Problems

27. $2a^3 + 16 = 2(a^3 + 8)$

$\qquad\qquad\qquad = 2[(a)^3 + (2)^3]$

Then $\qquad 2(\quad + \quad)[(\quad)^2 - (\quad)(\quad) + (\quad)^2]$

and $\qquad = 2(a + 2)[(a)^2 - (a)(2) + (2)^2]$

$\qquad\qquad\qquad = 2(a + 2)(a^2 - 2a + 4)$

44. $x^3y^9 - 1 = (xy^3)^3 - (1)^3$

Then $\qquad (\quad - \quad)[(\quad)^2 + (\quad)(\quad) + (\quad)^2]$

and $\qquad (xy^3 - 1)[(xy^3)^2 + (xy^3)(1) + (1)^2]$

$\qquad\qquad = (xy^3 - 1)(x^2y^6 + xy^3 + 1)$

Review Exercises

1. $(a - 2)(a - 5)$ **2.** $(3a + b)(2x - y)$ **3.** $(x + 2y)^2$ **4.** $(3a + b)(2a - b)$
5. $5a(a - 3)(a - 5)$ **6.** $(2x + 3)(3x - 4)$
7. \$7,500 at 10%; \$9,375 at 8%

Exercise 4–6

Answers to Odd-Numbered Problems

1. $(n + 7)(n - 7)$ **3.** $(7b + 1)(b + 5)$ **5.** $(xy + 4)(xy - 2)$
7. $(6 + y)(6 - y)$ **9.** $10(a - b)^2$ **11.** $4(a + 2b)(a - 2b)$
13. $(3a - b)(x + 2y)$ **15.** $(3x + 5)(2x - 1)$ **17.** $(2m - n)(3a + 2b)$
19. $(7b - 5)(b + 3)$ **21.** $(2x + 1)^2$ **23.** $(a - 2b)(3x + 4y)$ **25.** $(x - 4)^2$
27. $(a + 2)(a - 10)$ **29.** $(y + 3)(x^2 + 2)$ **31.** $(4x - 3)(x + 5)$
33. $6(x^2 - 4xy - 8y^2)$ **35.** $3xy(x + 5y)(m - 4n)$ **37.** $(3a - 5b)^2$
39. $3a(a^2 + 4)(a + 2)(a - 2)$ **41.** $3ab^3(a + b)^2$ **43.** $(3b + 7)(b - 13)$
45. $(3x + 2y)(a + 2b)$ **47.** $(6x - 1)(x + 2)$ **49.** $3x^2(x + 4)(x - 4)$
51. $(y + 3z)(y^2 - 3yz + 9z^2)$ **53.** $(x - y^3)(x^2 + xy^3 + y^6)$
55. $(a - 5)(a^2 + 5a + 25)$ **57.** $(x^9 + 3)(x^{18} - 3x^9 + 9)$
59. $(a^5 - b^5)(a^{10} + a^5b^5 + b^{10})$

Solutions to Trial Exercise Problems

35. $3x^2y(m - 4n) + 15xy^2(m - 4n) = 3xy(m - 4n)(x + 5y)$
$\qquad$ common factor of $3xy(m - 4n)$
37. $9a^2 - 30ab + 25b^2 = (3a)^2 - 2(3a)(5b) + (5b)^2 = (3a - 5b)^2$
39. $3a^5 - 48a = 3a(a^4 - 16) = 3a[(a^2)^2 - (4)^2] = 3a(a^2 + 4)(a^2 - 4)$
$\qquad = 3a(a^2 + 4)(a + 2)(a - 2)$

Review Exercises

1. $\{-3\}$ **2.** $\{3\}$ **3.** $\left\{-\dfrac{2}{3}\right\}$ **4.** $\left\{\dfrac{1}{4}\right\}$ **5.** $\{0\}$ **6.** $x + 7$ **7.** $x - 11$

8. $6(x^2 + x)$ **9.** $\dfrac{x^2 + 2x}{8}$ **10.** 13 **11.** 6 and 32

12. 22, 24, and 26

Exercise 4–7

Answers to Odd-Numbered Problems

1. $\{-5, 5\}$ **3.** $\{-6, 0\}$ **5.** $\{0, 7\}$ **7.** $\left\{-\dfrac{3}{2}, 3\right\}$ **9.** $\left\{-\dfrac{1}{2}, \dfrac{2}{3}\right\}$

11. $\left\{-\dfrac{1}{5}, \dfrac{1}{5}\right\}$ **13.** $\{7, 8\}$ **15.** $\left\{\dfrac{4}{3}, \dfrac{8}{5}\right\}$ **17.** $\{-5, 5\}$ **19.** $\{-4, 0\}$

21. $\left\{0, \dfrac{5}{3}\right\}$ **23.** $\left\{-\dfrac{3}{2}, 0\right\}$ **25.** $\{-5, 5\}$ **27.** $\{-3, 3\}$ **29.** $\{-8, 2\}$

31. $\{-7\}$ **33.** $\{-7, 2\}$ **35.** $\{-2, -1\}$ **37.** $\{-1, 12\}$ **39.** $\{-4, 8\}$

41. $\left\{-1, \dfrac{9}{2}\right\}$ **43.** $\left\{\dfrac{1}{3}, \dfrac{1}{2}\right\}$ **45.** $\left\{-\dfrac{3}{2}, \dfrac{4}{3}\right\}$ **47.** $\left\{-\dfrac{4}{3}, \dfrac{5}{2}\right\}$

49. $\{-1\}$ **51.** $\left\{-\dfrac{7}{3}, 4\right\}$ **53.** $\left\{-3, \dfrac{3}{2}\right\}$

55. -13 and -11 or 11 and 13 **57.** -18 and -16 or 16 and 18

59. -13 and -7 or 7 and 13 **61.** -2 and 0 or 4 and 6

63. -9 and -4 **65.** $-\dfrac{7}{3}$ and -6 or 2 and 7

67. $w = 4$ meters; $\ell = 6$ meters **69.** $w = 2$ feet; $\ell = 5$ feet

71. $w = 1$ inch or $w = 6$ inches **73. a.** $t = 2$ **b.** $t = 1$

75. a. 10 or 12 **b.** 12 or 20

77. a. $t = 2$ seconds **b.** $t = 4$ seconds

79. 1 second (on way up) and 5 seconds (on way down)

81. $t = 3$ seconds **83.** $b_2 = 8$ inches; $h = 7$ inches

85. $b_2 = 4$ meters; $b_1 = 8$ meters; $h = 4$ meters

87. $w = 7$ feet; $\ell = 8$ feet **89.** $w = 6$ feet; $\ell = 18$ feet

91. $w = 4$ inches, $\ell = 9$ inches

Solutions to Trial Exercise Problems

1. $(x + 5)(x - 5) = 0$
$x + 5 = 0$ or $x - 5 = 0$
$x = -5$ or $x = 5$
$\{-5, 5\}$

7. $(3x - 9)(2x + 3) = 0$
$3x - 9 = 0$ or $2x + 3 = 0$
$3x = 9$ or $2x = -3$
$x = 3$ or $x = -\dfrac{3}{2}$
$\left\{-\dfrac{3}{2}, 3\right\}$

23. $10a^2 = -15a$
$10a^2 + 15a = 0$
$5a(2a + 3) = 0$
$5a = 0$ or $2a + 3 = 0$
$a = 0$ or $2a = -3$
$a = -\dfrac{3}{2}$
$\left\{-\dfrac{3}{2}, 0\right\}$

27. $5y^2 - 45 = 0$
$5(y^2 - 9) = 0$
$5(y + 3)(y - 3) = 0$
$y + 3 = 0$ or $y - 3 = 0$
$y = -3$ or $y = 3$
$\{-3, 3\}$

33. $b^2 + 5b - 14 = 0$
$(b + 7)(b - 2) = 0$
$b + 7 = 0$ or $b - 2 = 0$
$b = -7$ or $b = 2$
$\{-7, 2\}$

38. $x^2 - 14x = 15$
$x^2 - 14x - 15 = 0$
$(x - 15)(x + 1) = 0$
$x - 15 = 0$ or $x + 1 = 0$
$x = 15$ or $x = -1$
$\{-1, 15\}$

45. $6x^2 + x - 12 = 0$
$(3x - 4)(2x + 3) = 0$
$3x - 4 = 0$ or $2x + 3 = 0$
$3x = 4$ or $2x = -3$
$x = \dfrac{4}{3}$ or $x = \dfrac{-3}{2}$
$\left\{-\dfrac{3}{2}, \dfrac{4}{3}\right\}$

46. $-6x = -3x^2 - 3$
$3x^2 - 6x + 3 = 0$
$3(x^2 - 2x + 1) = 0$
$3(x - 1)^2 = 0$
$x - 1 = 0$
$x = 1$
$\{1\}$

51. $3x^2 - 4x - 28 = x$
$3x^2 - 5x - 28 = 0$
$(3x + 7)(x - 4) = 0$
$3x + 7 = 0$ or $x - 4 = 0$
$3x = -7$ or $x = 4$
$x = \dfrac{-7}{3}$ or $x = 4$
$\left\{-\dfrac{7}{3}, 4\right\}$

61. Let x represent the lesser integer.
Then $x + 2$ is the next consecutive even integer.
Then $x(x + 2) = 2[x + (x + 2)] + 4$
$x^2 + 2x = 2(2x + 2) + 4$
$x^2 + 2x = 4x + 4 + 4$
$x^2 + 2x = 4x + 8$
$x^2 - 2x - 8 = 0$
$(x - 4)(x + 2) = 0$
$x - 4 = 0$ or $x + 2 = 0$
$x = 4$ or $x = -2$
$x + 2 = 6$ or $x + 2 = 0$
The consecutive even integers are 4 and 6 or -2 and 0.

64. Let x represent one integer.
Then $-3 - x$ is the other integer.
$x(-3 - x) = -70$
$-3x - x^2 = -70$
$0 = x^2 + 3x - 70 = (x + 10)(x - 7)$
$x + 10 = 0$ or $x - 7 = 0$
$x = -10$ or $x = 7$
$-3 - x = -3 - (-10) = 7$ or $-3 - x = -3 - 7 = -10$
The integers are -10 and 7.

69. Let w represent the width. Then $\ell = w + 3$. From "the area of a rectangle is numerically equal to twice the length," we get
$w(w + 3) = 2(w + 3)$
$w^2 + 3w = 2w + 6$
$w^2 + w - 6 = 0$
$(w + 3)(w - 2) = 0$
$w + 3 = 0$ or $w - 2 = 0$
$w = -3$ or $w = 2$
(*Note:* A geometric figure cannot have a negative width, so we ignore $w = -3$.)
The width is 2 feet and the length is $w + 3 = 5$ feet.

76. a. $P = 100I - 5I^2$, when $P = 480$.
$$480 = 100I - 5I^2$$
$$5I^2 - 100I + 480 = 0$$
$$5(I^2 - 20I + 96) = 0$$
$$5(I - 8)(I - 12) = 0$$
$$I - 8 = 0 \text{ or } I - 12 = 0$$
$$I = 8 \qquad\quad I = 12$$
Therefore $I = 8$ amperes or $I = 12$ amperes.

79. $s = v_0 t - 16t^2$, given $s = 80$ and $v_0 = 96$
$$80 = 96t - 16t^2$$
$$16t^2 - 96t + 80 = 0$$
$$16(t^2 - 6t + 5) = 0$$
$$16(t - 5)(t - 1) = 0$$
$$t - 5 = 0 \text{ or } t - 1 = 0$$
$$t = 5 \qquad\quad t = 1$$
So $t = 1$ second on the way up,
and $t = 5$ seconds on the way down.

84. Given $A = \frac{1}{2}h(b_1 + b_2)$, $A = 21$ square feet, $b_2 = 5$ feet, and

using "b_1 is 6 feet longer than the altitude h," we have
$b_1 = h + 6$.

Substitute $21 = \frac{1}{2}h[(h + 6) + 5]$

$$21 = \frac{1}{2}h(h + 11)$$

Multiply both sides by 2 to get $42 = h(h + 11)$
Then $42 = h^2 + 11h$
Add -42 to both sides to get $0 = h^2 + 11h - 42$
So $h^2 + 11h - 42 = 0$
Factor the left side: $(h + 14)(h - 3) = 0$
Then $h + 14 = 0$ or $h - 3 = 0$
So $h = -14$ or $h = 3$.
A trapezoid cannot have a negative altitude, so we ignore
$h = -14$. The altitude of the trapezoid is 3 feet.

87. Given $V = \ell wh$, $V = 224$ cubic feet and $h = 4$ feet, from "the
length is 1 foot longer than the width," $\ell = w + 1$.
Substituting, $224 = w(w + 1) \cdot 4$
Multiplying in the right side, $224 = 4w^2 + 4w$
Add -224 to both sides and interchange sides.
$$0 = 4w^2 + 4w - 224$$
$$4w^2 + 4w - 224 = 0$$
Factor 4 from each term to get $4(w^2 + w - 56) = 0$
Then $4(w + 8)(w - 7) = 0$ and $w + 8 = 0$ or $w - 7 = 0$
So $w = -8$ or $w = 7$.
The width of a box cannot be negative, so we discard $w = -8$.
The width is then 7 feet and the length is $w + 1 = 8$ feet.

Review Exercises

1. $\left\{\frac{2}{3}\right\}$ **2.** $\left\{\frac{5}{4}\right\}$ **3.** $\{-3, 3\}$ **4.** $\{-4, -2\}$ **5.** 18 **6.** -27 **7.** 0
8. undefined **9.** \$5,000 at 10%

Chapter 4 Review

1. $xy(x + z + yz)$ **2.** $a^3b(1 + b)$ **3.** $R^2(3R - 1 + 5R^2)$
4. $4y(y + 2 + 3y^2)$ **5.** $x^2(x^2 + 3x + 9)$ **6.** $4R^2S^2(4R - 3R^2S + 6)$
7. $5a^2b^2(2a^2b + 3 - 4a)$ **8.** $(a + b)(2 + x)$ **9.** $(x - 3z)(y + 4)$
10. $(3R + 1)(a + b)$ **11.** $(x - 3y)(2a - 3b)$ **12.** $(3a - b)(2x - y)$
13. $(x + 2y)(4a + 3b)$ **14.** $(a + 3b)(x - 4)$ **15.** $(x^2 + 4)(a - 2b)$
16. $(x - 7)(x - 2)$ **17.** $2a(a - 5)(a + 1)$ **18.** $(a + 12)(a + 2)$
19. $(x - 8)(x + 4)$ **20.** $(a - 18)(a + 2)$ **21.** $3(x - 5)(x + 2)$

22. $x(x - 3)(x + 2)$ **23.** $x(x - 7)(x + 3)$ **24.** $(ab + 3)(ab - 2)$
25. $(ab + 6)(ab + 4)$ **26.** $(ab - 6)(ab - 3)$ **27.** $(ab - 10)(ab + 2)$
28. $(2x + 1)(2x + 1)$ or $(2x + 1)^2$ **29.** $9(r - 2)(r - 2)$ or $9(r - 2)^2$
30. $(4x - 1)(x - 1)$ **31.** $(3a + 5)(3a - 2)$ **32.** $(4a - 3)(2a + 1)$
33. $(6x + 1)(4x + 3)$ **34.** $(4a - 3)(2a - 3)$ **35.** $(2a + 3)(a + 6)$
36. $(2a + 3)(2a - 3)$ **37.** $(6b + c)(6b - c)$ **38.** $(5 + a)(5 - a)$
39. $4(2x + y)(2x - y)$ **40.** $(3x + y^2)(3x - y^2)$
41. $(x^2 + 4)(x + 2)(x - 2)$ **42.** $(y^2 + 9)(y + 3)(y - 3)$ **43.** $(b + 6)^2$
44. $(c - 5)^2$ **45.** $(2x - 3)^2$ **46.** $(3x - 2)^2$
47. $(R + 2S)(R^2 - 2RS + 4S^2)$ **48.** $2(2x - 3)(4x^2 + 6x + 9)$
49. $(3a + 5b)(9a^2 - 15ab + 25b^2)$ **50.** $(xy - 1)(x^2y^2 + xy + 1)$
51. $2(x^3 + 5)(x^6 - 5x^3 + 25)$ **52.** $(4x^4 - y^5)(16x^8 + 4x^4y^5 + y^{10})$
53. $(ab^2 + c^3)(a^2b^4 - ab^2c^3 + c^6)$ **54.** $3x^3(4x - 1)$ **55.** $(a - 5)(a + 2)$
56. $(4a - 1)(a - 5)$ **57.** $(3y + 2)(3y - 2)$ **58.** $(2a + 3b)(3x - 2)$
59. $(b - 5)(b + 4)$ **60.** $(3x + 2)(3x + 5)$ **61.** $(a + 7)^2$
62. $3x^3(2x + 1)(2x - 1)$ **63.** $c(c + 4)(c + 5)$ **64.** $(4a - 1)^2$
65. $(b^2 + 1)(b + 1)(b - 1)$ **66.** $\{-3, 1\}$ **67.** $\{0, 8\}$ **68.** $\left\{-\frac{1}{5}, \frac{7}{3}\right\}$

69. $\left\{\frac{1}{7}, \frac{8}{5}\right\}$ **70.** $\left\{\frac{4}{3}, 9\right\}$ **71.** $\left\{-\frac{4}{5}, \frac{4}{5}\right\}$ **72.** $\left\{0, \frac{9}{4}\right\}$ **73.** $\{-1, 1\}$

74. $\{0, 64\}$ **75.** $\{-5, 5\}$ **76.** $\{-5, 6\}$ **77.** $\{1\}$ **78.** $\left\{-3, -\frac{1}{4}\right\}$

79. $\left\{-\frac{2}{5}, 2\right\}$ **80.** $\left\{-1, \frac{3}{4}\right\}$ **81.** $\{-3, 2\}$ **82.** 9 and 10
83. 8 feet by 13 feet **84.** 20 cattle **85.** $t = 6$

Chapter 4 Test

1. $(3a + b)(4x - 1)$ [4–1] **2.** $(x - 7)(x - 2)$ [4–2]
3. $(3y - 2)(2y + 3)$ [4–1] **4.** $(4x - 3)(x + 2)$ [4–3]
5. $(a + 6)(a + 2)$ [4–2] **6.** $3x^2(3x^3 - 4x + 2)$ [4–1]
7. $(z + 7)(z - 7)$ [4–4] **8.** $(x + 3y)(x^2 - 3xy + 9y^2)$ [4–5]
9. $(3x + 4)(2a + 3)$ [4–3] **10.** $(2a - b)(3x - y)$ [4–1]
11. $(b - 5)(b - 6)$ [4–2] **12.** $10(a - b)^2$ [4–4]
13. will not factor, prime polynomial [4–3]
14. $(6a + 1)(a + 3)$ [4–3] **15.** $(ab + 3)(ab - 4)$ [4–2]
16. $(2a + 3b)(2a - 3b)$ [4–4] **17.** $(2y - 5)^2$ [4–4]
18. $(a - b)(2c - 3d)$ [4–1] **19.** $(2a - b)(4a^2 + 2ab + b^2)$ [4–5]
20. $(x - 4y)(x + y)$ [4–2] **21.** $(d^2 + 9)(d + 3)(d - 3)$ [4–4]
22. $3(3x - y)(9x^2 + 3xy + y^2)$ [4–5] **23.** $(8x - 3)(x - 2)$ [4–3]
24. $a^2(a + 3b)(a^2 - 3ab + 9b^2)$ [4–5] **25.** $\{1, 6\}$ [4–7]
26. $\left\{-\frac{7}{3}, \frac{5}{2}\right\}$ [4–7] **27.** $\left\{0, \frac{1}{2}\right\}$ [4–7] **28.** $\{9\}$ [4–7]
29. 7 meters by 12 meters [4–7] **30.** -14 and -12, 12 and 14 [4–7]

Chapter 4 Cumulative Test

1. 33 [1–8] **2.** $8a^6b^3$ [3–4] **3.** $a^2 + 4ab + 4b^2$ [3–2] **4.** a^6 [3–1]
5. 12 [1–8] **6.** x^5y^4 [3–1] **7.** $5x - 7y$ [2–2] **8.** $9x^2 - 4y^2$ [3–2]
9. $\frac{8a^6}{b^3}$ [3–3] **10.** $\frac{1}{x^2}$ [3–3] **11.** $-2y$ [2–2] **12.** $\frac{x^6}{9y^4}$ [3–4]
13. $2x + 3 - \frac{3}{x - 4}$ [3–6] **14.** $\left\{\frac{9}{5}\right\}$ [2–5] **15.** $x < 12$ [2–9]
16. $\{7\}$ [2–5] **17.** $x \geq 3$ [2–9] **18.** $\{-3, 3\}$ [4–7] **19.** $\{2, 5\}$ [4–7]
20. $x < \frac{5}{2}$ [2–9] **21.** $2 \leq x \leq \frac{14}{3}$ [2–9] **22.** $x = 3y$ [2–6]
23. $x = \frac{5y + 2}{3}$ [2–6] **24.** $2ab(1 - 2ab - 4a^2b^4)$ [4–1]
25. $(2a + 3)^2$ [4–4] **26.** $(5c + 3d)(5c - 3d)$ [4–4]
27. $(2a + 3)(2a - 5)$ [4–3] **28.** $(x + 3)(x + 6)$ [4–2]
29. 14, 39 [2–7] **30.** 13, 15 [4–7] **31.** \$10,000 at 8%;
\$5,000 at 6% [2–7] **32.** 6 meters by 11 meters [4–7]

Chapter 5

Exercise 5–1

Answers to Odd-Numbered Problems

1. all real numbers except 0 **3.** all real numbers except 5

5. all real numbers except -3 **7.** all real numbers except $\dfrac{3}{4}$

9. all real numbers except $\dfrac{8}{3}$ **11.** all real numbers except 1 and 6

13. all real numbers except -2 and $\dfrac{4}{3}$

15. all real numbers except $-\dfrac{2}{3}$ and $\dfrac{2}{3}$

17. all real numbers except -3 and 3

19. all real numbers except $\dfrac{7}{3}$

21. all real numbers

23. all real numbers **25.** $\dfrac{3}{4}$ **27.** $\dfrac{2x}{5}$ **29.** $\dfrac{4x}{3}$ **31.** $-\dfrac{4}{3x^2}$

33. $\dfrac{4a}{5b}$ **35.** -5 **37.** $\dfrac{5}{4}$ **39.** $\dfrac{6}{x+3}$ **41.** $\dfrac{1}{a-b}$ **43.** $\dfrac{3}{5}$

45. $\dfrac{x-1}{2(x+1)}$ **47.** $\dfrac{x-3}{x+3}$ **49.** $\dfrac{x-5}{x-3}$ **51.** $\dfrac{2y+3}{4y-1}$

53. $\dfrac{1}{x^2+3x+9}$ **55.** $\dfrac{a-b}{a^2-ab+b^2}$ **57.** -4 **59.** $-\dfrac{2}{3}$

61. $-2x-2y$ **63.** $\dfrac{y-x}{y+x}$ **65.** $\dfrac{n-m}{m+n}$ **67.** $\dfrac{-4}{x+y}$

69. the set of all replacement values of the variable for which the rational expression is defined

71. answers will vary

Solutions to Trial Exercise Problems

6. $\dfrac{x+1}{2x-1}$ Set $2x - 1 = 0$, then $2x = 1$ and $x = \dfrac{1}{2}$. Domain is all real numbers except $\dfrac{1}{2}$.

9. $\dfrac{y+4}{8-3y}$ Set $8 - 3y = 0$, then $3y = 8$ and $y = \dfrac{8}{3}$. Domain is all real numbers except $\dfrac{8}{3}$.

12. $\dfrac{5s^2+7}{2s^2-s-3}$ Set $2s^2 - s - 3 = 0$ and factor. We have

$(2s - 3)(s + 1) = 0$. Then

$2s - 3 = 0$ or $s + 1 = 0$
$2s = 3$ or $s = -1$
$s = \dfrac{3}{2}$ $s = -1$

Domain is all real numbers except -1 and $\dfrac{3}{2}$.

16. $\dfrac{a-2}{4a^2-16}$ Set $4a^2 - 16 = 0$ and factor. We have
$4(a - 2)(a + 2) = 0$. This is true if and only if $a - 2 = 0$, $a = 2$ or $a + 2 = 0$, $a = -2$. Domain is all real numbers except -2 and 2.

21. $\dfrac{17q}{q^2+16}$

$q^2 + 16$, the sum of two squares, is not factorable. Now, $q^2 + 16$ never equals zero for any real values of q because $q^2 \geq 0$. Hence, the domain is all real numbers.

34. $\dfrac{15a^2x^3}{35ax^2} = \dfrac{3 \cdot 5 \cdot a \cdot a \cdot x \cdot x \cdot x}{7 \cdot 5 \cdot a \cdot x \cdot x}$

$= \dfrac{3 \cdot a \cdot x \cdot (5 \cdot a \cdot x \cdot x)}{7 \cdot (5 \cdot a \cdot x \cdot x)} = \dfrac{3ax}{7}$

41. $\dfrac{a+b}{a^2-b^2} = \dfrac{a+b}{(a+b)(a-b)} = \dfrac{1}{a-b}$

49. $\dfrac{x^2-3x-10}{x^2-x-6} = \dfrac{(x-5)(x+2)}{(x-3)(x+2)} = \dfrac{x-5}{x-3}$

58. $\dfrac{8b-8a}{a-b} = \dfrac{-8(a-b)}{(a-b)} = -8$

Review Exercises

1. $\left\{-\dfrac{9}{5}\right\}$ **2.** $w = \dfrac{P-2\ell}{2}$ **3.** -14 and -12 or 12 and 14

4. $x \leq 4$ **5.** $(4x + y)(4x - y)$ **6.** $(x + 1)(x - 17)$

7. $5(x - 2)(x + 1)$ **8.** $\$1,900$ at $7\dfrac{1}{2}\%$; $\$4,100$ at 9%

Exercise 5–2

Answers to Odd-Numbered Problems

1. $\dfrac{3}{5}$ **3.** $\dfrac{5}{6}$ **5.** $2a$ **7.** 10 **9.** $\dfrac{x}{4y}$ **11.** $\dfrac{3x}{4}$ **13.** $\dfrac{4}{35x}$

15. $\dfrac{35x}{4}$ **17.** $\dfrac{7c}{2b}$ **19.** $\dfrac{x}{6y^2}$ **21.** $\dfrac{16bcx}{3az}$ **23.** $\dfrac{2a}{5b^2x}$

25. $\dfrac{4}{x+y}$ **27.** $-\dfrac{3}{4}$ **29.** $-\dfrac{15}{4}$ **31.** $\dfrac{4}{a-5}$ **33.** $\dfrac{18(x-2)}{x+2}$

35. $\dfrac{24y(x-2)}{25}$ **37.** $r^2 - 5r + 4$ **39.** $-4x - 12$ **41.** $\dfrac{a-3}{a-5}$

43. $\dfrac{(x-3)(x+1)}{(x-1)(x+2)}$ **45.** $\dfrac{(2x-1)(x-1)}{(x-8)(x+7)}$ **47.** 1 **49.** $3x + 4$

51. $\dfrac{1}{(2x+1)^2}$ **53.** $\dfrac{2(a+6)}{3(a^2+3a+9)}$ **55.** $\dfrac{(z-7)(z+3)}{5(z^2-2z+4)}$

57. $y + 5$ **59.** $\dfrac{(x+10)(x+8)}{(x+4)(x+12)}$

Solutions to Trial Exercise Problems

10. $\dfrac{7a}{12b} \cdot \dfrac{9b}{28} = \dfrac{7a \cdot 9b}{12b \cdot 28} = \dfrac{7 \cdot a \cdot 3 \cdot 3 \cdot b}{2 \cdot 2 \cdot 3 \cdot b \cdot 2 \cdot 2 \cdot 7}$

$= \dfrac{3 \cdot a \cdot (7 \cdot 3 \cdot b)}{2 \cdot 2 \cdot 2 \cdot 2 \cdot (7 \cdot 3 \cdot b)} = \dfrac{3a}{16}$

20. $\dfrac{28m}{15n} \div \dfrac{7m^2}{3n^3} = \dfrac{28m \cdot 3n^3}{15n \cdot 7m^2} = \dfrac{4n^2}{5m}$

21. $\dfrac{24abc}{7xyz^2} \cdot \dfrac{14x^2yz}{9a^2} = \dfrac{3 \cdot 8 \cdot a \cdot b \cdot c \cdot 2 \cdot 7 \cdot x^2 \cdot y \cdot z}{7 \cdot x \cdot y \cdot z^2 \cdot 3 \cdot 3 \cdot a^2}$

$= \dfrac{8 \cdot b \cdot c \cdot 2 \cdot x(3 \cdot 7 \cdot a \cdot x \cdot y \cdot z)}{z \cdot 3 \cdot a(3 \cdot 7 \cdot a \cdot x \cdot y \cdot z)} = \dfrac{8 \cdot b \cdot c \cdot 2 \cdot x}{z \cdot 3 \cdot a} = \dfrac{16bcx}{3az}$

25. $\dfrac{x+y}{3} \cdot \dfrac{12}{(x+y)^2} = \dfrac{2 \cdot 2 \cdot 3(x+y)}{3(x+y)^2} = \dfrac{4}{x+y}$

30. $\dfrac{8y+16}{3-y} \cdot \dfrac{4y-12}{3y+6} = \dfrac{8(y+2) \cdot 4(y-3)}{-(y-3) \cdot 3(y+2)} = \dfrac{8 \cdot 4}{-3} = -\dfrac{32}{3}$

37. $\dfrac{r^2-16}{r+1} \div \dfrac{r+4}{r^2-1} = \dfrac{(r^2-16)(r^2-1)}{(r+1)(r+4)}$

$= \dfrac{(r-4)(r+4)(r+1)(r-1)}{(r+1)(r+4)} = \dfrac{(r-4)(r-1)}{1} = r^2 - 5r + 4$

39. $\dfrac{9 - x^2}{x + y} \cdot \dfrac{4x + 4y}{x - 3} = \dfrac{(3 - x)(3 + x) \cdot 4(x + y)}{(x + y)(x - 3)}$

$\quad = \dfrac{-1(x - 3)(x + 3) \cdot 4(x + y)}{(x + y)(x - 3)} = \dfrac{-1(x + 3) \cdot 4}{1}$

$\quad = -4(x + 3) = -4x - 12$

41. $\dfrac{a^2 - 5a + 6}{a^2 - 9a + 20} \cdot \dfrac{a^2 - 5a + 4}{a^2 - 3a + 2}$

$\quad = \dfrac{(a - 3)(a - 2) \cdot (a - 4)(a - 1)}{(a - 4)(a - 5) \cdot (a - 2)(a - 1)} = \dfrac{a - 3}{a - 5}$

47. $\dfrac{6r^2 - r - 7}{12r^2 + 16r - 35} \div \dfrac{r^2 - r - 2}{2r^2 + r - 10}$

$\quad = \dfrac{(6r^2 - r - 7)(2r^2 + r - 10)}{(12r^2 + 16r - 35)(r^2 - r - 2)}$

$\quad = \dfrac{(6r - 7)(r + 1)(2r + 5)(r - 2)}{(6r - 7)(2r + 5)(r - 2)(r + 1)} = 1$

49. $(3x^2 - 2x - 8) \div \dfrac{x^2 - 4}{x + 2}$

$\quad = \dfrac{(3x + 4)(x - 2)}{1} \cdot \dfrac{x + 2}{(x + 2)(x - 2)} = 3x + 4$

53. $\dfrac{10}{a^3 - 27} \cdot \dfrac{a^2 + 3a - 18}{15} = \dfrac{2 \cdot 5(a + 6)(a - 3)}{3 \cdot 5(a - 3)(a^2 + 3a + 9)}$

$\quad = \dfrac{2(a + 6)}{3(a^2 + 3a + 9)} = \dfrac{2a + 12}{3a^2 + 9a + 27}$

Review Exercises

1. $\dfrac{19}{12}$ **2.** $\dfrac{11}{24}$ **3.** $2(x + 5)(x - 5)$ **4.** $(x + 11)(x - 2)$ **5.** $(x + 4)^2$

6. $x = \dfrac{24}{5}$ **7.** $y = 15$ **8.** 7.89×10^{-5} **9.** 72 **10.** $\dfrac{98}{3}, \dfrac{14}{3}$

Exercise 5–3

Answers to Odd-Numbered Problems

1. $\dfrac{8}{x}$ **3.** $\dfrac{7}{p}$ **5.** $\dfrac{14x}{x + 2}$ **7.** $-\dfrac{2}{x}$ **9.** $\dfrac{x + 2}{x^2 - 1}$ **11.** $\dfrac{1}{a + 4}$

13. $\dfrac{1}{y + 3}$ **15.** $\dfrac{1}{a + 3}$ **17.** $\dfrac{1}{x - 5}$ **19.** $\dfrac{1}{a + 3}$ **21.** $\dfrac{1}{x + 2}$

23. $-\dfrac{1}{7}$ **25.** $\dfrac{9}{z}$ **27.** $-\dfrac{7}{x - 2}$ **29.** $\dfrac{9y}{y - 6}$ **31.** $\dfrac{-x + 4}{x - 5}$

33. $\dfrac{3y + 2}{2y - 3}$ **35.** $\dfrac{-x + 4}{2x - 5}$ **37.** $18x$ **39.** $48x^2$ **41.** $140y^3$

43. $64a^4$ **45.** $180a^3$ **47.** $3(x - 4)$ **49.** $18y^3(y - 4)$

51. $a(a + 1)(a - 1)$ **53.** $8(a + 2)(a + 1)$ **55.** $(a + 2)(a - 2)(a - 3)$

57. $(x - 1)(x + 1)^2$ **59.** $(a - 5)(a + 5)^2$ **61.** $(a - 2)(a - 3)(a + 3)$

63. $(a - 2)^2(a + 2)^2$ **65.** $(a + 3)(a - 2)(a + 4)$

67. the smallest (least) number or expression that is exactly divisible by each of the denominators

Solutions to Trial Exercise Problems

5. $\dfrac{5x}{x + 2} + \dfrac{9x}{x + 2} = \dfrac{5x + 9x}{x + 2} = \dfrac{14x}{x + 2}$

8. $\dfrac{3y - 2}{y^2} - \dfrac{4y - 1}{y^2} = \dfrac{(3y - 2) - (4y - 1)}{y^2}$

$\quad = \dfrac{3y - 2 - 4y + 1}{y^2} = \dfrac{-y - 1}{y^2}$

11. $\dfrac{3a - 8}{a^2 - 16} + \dfrac{4 - 2a}{a^2 - 16} = \dfrac{3a - 8 + 4 - 2a}{a^2 - 16}$

$\quad = \dfrac{a - 4}{a^2 - 16} = \dfrac{(a - 4)}{(a - 4)(a + 4)} = \dfrac{1}{a + 4}$

28. $\dfrac{1}{x - 7} - \dfrac{5}{7 - x} = \dfrac{1}{x - 7} + \dfrac{5}{x - 7} = \dfrac{6}{x - 7}$

33. $\dfrac{2y - 5}{2y - 3} - \dfrac{y + 7}{3 - 2y} = \dfrac{2y - 5}{2y - 3} + \dfrac{y + 7}{2y - 3} = \dfrac{3y + 2}{2y - 3}$

44. $4x^2 = 2^2 \cdot x^2$

$\quad 3x = 3 \cdot x$

$\quad 8x^3 = 2^3 \cdot x^3$

$\quad \text{LCD} = 2^3 \cdot 3 \cdot x^3 = 24x^3$

56. $y^2 - y - 12 = (y - 4)(y + 3)$

$\quad y^2 + 6y + 9 = (y + 3)^2$

$\quad \text{LCD} = (y + 3)^2(y - 4)$

Review Exercises

1. commutative property of addition **2.** $5(y + 2)(y - 2)$ **3.** $(x + 10)^2$

4. $(3y - 4)(y + 1)$ **5.** $\left\{\dfrac{8}{19}\right\}$ **6.** $\{-3,5\}$ **7.** $\$18{,}000$ at 10%

Exercise 5–4

Answers to Odd-Numbered Problems

1. $\dfrac{2x + 9}{12}$ **3.** $\dfrac{8x + 1}{48}$ **5.** $\dfrac{11}{3}$ **7.** $\dfrac{15y - 12}{(y + 4)(y - 5)}$ **9.** $\dfrac{10}{y - 2}$

11. $\dfrac{7x + 62}{4(x + 2)(x - 2)}$ **13.** $\dfrac{9x + 40}{x + 8}$ **15.** $\dfrac{x^2 + 4x}{(x + 1)(x - 1)}$

17. $\dfrac{9x + 22}{(x + 3)(x - 3)(x + 2)}$ **19.** $\dfrac{7y^2 - 13y}{(y - 3)^2(y + 1)}$

21. $\dfrac{2y^2 - 9y + 1}{(y - 5)(y - 3)(y + 2)}$ **23.** $\dfrac{2y - 15}{18}$ **25.** $\dfrac{25}{42y}$ **27.** $\dfrac{19a + 39}{60}$

29. $\dfrac{4x + 25}{18x}$ **31.** $\dfrac{-5x - 17}{(2x - 3)(x - 5)}$ **33.** $\dfrac{17}{7(y + 2)}$

35. $\dfrac{9x + 66}{x + 8}$ **37.** $\dfrac{-25x - 9}{3x + 1}$ **39.** $\dfrac{-6a + 3}{(a - 2)(a + 2)(a - 3)}$

41. $\dfrac{12y - 12}{(y - 6)(y + 4)(y - 3)}$ **43.** $\dfrac{-3a^2 + 11a - 3}{(a - 3)(a - 2)}$ **45.** $\dfrac{13}{12z}$

47. $\dfrac{52a}{45}$ **49.** $\dfrac{39b - 5}{36}$ **51.** $\dfrac{-z^3 + 9z^2 - 13}{(z - 2)(z + 2)(z - 1)(z + 1)}$

53. $\dfrac{bc + ac + ab}{abc}$ **55.** $\dfrac{2x + 2}{x(x + 2)}$ **57.** $\dfrac{Ir_1 - Ir_2}{Pr_1r_2}$ **59.** $\dfrac{3}{m}$ **61.** $\dfrac{h}{36}$

63. $\dfrac{48}{m}$ **65.** $w = \dfrac{A}{23}$ feet **67.** $b = \dfrac{2A}{9}$ rods **69.** $t = \dfrac{d}{55}$ hours

71. $\dfrac{b}{a}$ **73.** $b = \dfrac{9x + 20}{x^2 - x - 6}$ **75.** $y = \dfrac{x + 3}{5(x - 1)}$

Solutions to Trial Exercise Problems

6. $\dfrac{4}{x - 1} + \dfrac{5}{x + 3} = \dfrac{4(x + 3) + 5(x - 1)}{(x - 1)(x + 3)}$

$\quad = \dfrac{4x + 12 + 5x - 5}{(x - 1)(x + 3)} = \dfrac{9x + 7}{(x - 1)(x + 3)}$

11. $\dfrac{12}{x^2 - 4} + \dfrac{7}{4x - 8} = \dfrac{12}{(x + 2)(x - 2)} + \dfrac{7}{4(x - 2)}$

$\quad = \dfrac{12 \cdot 4}{4(x + 2)(x - 2)} + \dfrac{7(x + 2)}{4(x + 2)(x - 2)} = \dfrac{48 + (7x + 14)}{4(x + 2)(x - 2)}$

$\quad = \dfrac{7x + 62}{4(x + 2)(x - 2)}$

13. $5 + \dfrac{4x}{x + 8} = \dfrac{5(x + 8)}{x + 8} + \dfrac{4x}{x + 8} = \dfrac{5x + 40 + 4x}{x + 8} = \dfrac{9x + 40}{x + 8}$

25. $\dfrac{9}{14y} - \dfrac{1}{21y} = \dfrac{9 \cdot 3}{2 \cdot 3 \cdot 7 \cdot y} - \dfrac{1 \cdot 2}{2 \cdot 3 \cdot 7 \cdot y}$

$\quad = \dfrac{27}{42y} - \dfrac{2}{42y} = \dfrac{27 - 2}{42y} = \dfrac{25}{42y}$

27. $\dfrac{5a + 3}{12} - \dfrac{a - 4}{10} = \dfrac{5(5a + 3)}{60} - \dfrac{6(a - 4)}{60}$

$\quad = \dfrac{(25a + 15) - (6a - 24)}{60} = \dfrac{25a + 15 - 6a + 24}{60}$

$\quad = \dfrac{19a + 39}{60}$

34. $\dfrac{14}{5x-15} - \dfrac{8}{2x-6} = \dfrac{14}{5(x-3)} - \dfrac{8}{2(x-3)}$

$= \dfrac{14\cdot 2}{10(x-3)} - \dfrac{8\cdot 5}{10(x-3)} = \dfrac{28-40}{10(x-3)}$

$= \dfrac{-12}{10(x-3)} = -\dfrac{6}{5(x-3)}$

42. $\dfrac{2p}{p^2-9p+20} - \dfrac{5p-2}{p-5} = \dfrac{2p}{(p-5)(p-4)} - \dfrac{5p-2}{p-5}$

$= \dfrac{2p}{(p-5)(p-4)} - \dfrac{(5p-2)(p-4)}{(p-5)(p-4)} = \dfrac{2p-(5p^2-22p+8)}{(p-5)(p-4)}$

$= \dfrac{2p-5p^2+22p-8}{(p-5)(p-4)} = \dfrac{-5p^2+24p-8}{(p-5)(p-4)}$

52. $\dfrac{1}{f} = (u-1)\left(\dfrac{1}{R_1} + \dfrac{1}{R_2}\right) = (u-1)\left(\dfrac{R_2}{R_1R_2} + \dfrac{R_1}{R_1R_2}\right)$

$= (u-1)\left(\dfrac{R_2+R_1}{R_1R_2}\right)$

63. Let x be the other number. Since m is one of the numbers, then

$x \cdot m = 48$ and x (the other number) $= \dfrac{48}{m}$.

66. Using $A = \dfrac{1}{2}bh$, then $2A = bh$ (multiply each side by 2). Then

$h = \dfrac{2A}{b} = \dfrac{2(21)}{b} = \dfrac{42}{b}$ yards.

74. $\dfrac{5x}{x+3} \div y = \dfrac{x}{x-1}$; $\dfrac{5x}{x+3} = \dfrac{x}{x-1} \cdot y$; $\dfrac{5x}{x+3} \div \dfrac{x}{x-1} = y$;

$\dfrac{5x}{x+3} \cdot \dfrac{x-1}{x} = y$; $y = \dfrac{5(x-1)}{x+3}$

Review Exercises

1. $(x-7)^2$ **2.** $(2x-1)(x-5)$ **3.** $4(x+2)(x-2)$ **4.** x^2-81
5. $16x^2+24x+9$ **6.** $2x^2-15x-8$ **7.** 48   **8.** $12x^2$
9. $(x-3)^2(x+3)$ **10.** $\dfrac{13}{2x}$ **11.** $\dfrac{2x+1}{(x-2)(x-1)}$
12. 18 m, 6 m, 13 m

Exercise 5–5

Answers to Odd-Numbered Problems

1. $\dfrac{5}{6}$ **3.** $\dfrac{3}{2}$ **5.** $\dfrac{8}{9}$ **7.** $\dfrac{5}{2}$ **9.** $\dfrac{23}{30}$ **11.** $\dfrac{11}{2}$ **13.** $\dfrac{4x+1}{4x-3}$

15. $\dfrac{3a+1}{2-4a}$ **17.** $\dfrac{4a^2+3}{5a^2-3a}$ **19.** $\dfrac{ab-3}{ab+4}$ **21.** $\dfrac{y+x}{y-x}$ **23.** xy

25. $\dfrac{4b-5a^2}{a^2b(a-b)}$ **27.** 1   **29.** x **31.** $\dfrac{1}{x}$ **33.** $\dfrac{1}{a}$ **35.** $\dfrac{4}{x}$ **37.** $-\dfrac{y}{x}$

39. $\dfrac{2(x^2+y^2)}{(x+y)^2(x-y)^2}$ **41.** $\dfrac{2b+3}{a^3b^4}$ **43.** $\dfrac{3a-1}{17-a}$ **45.** $-\dfrac{5}{x+31}$

47. $\dfrac{-3a^2-5a+8}{a+29}$ **49.** $\dfrac{(y-3)(y+6)}{(y+1)(y-4)}$ **51.** $\dfrac{T_1}{T_2-T_1}$

53. $\dfrac{L_1L_2L_3}{L_2L_3+L_1L_3+L_1L_2}$

Solutions to Trial Exercise Problems

7. $\dfrac{7}{2+\dfrac{4}{5}} = \dfrac{7\cdot 5}{\left(2+\dfrac{4}{5}\right)\cdot 5} = \dfrac{7\cdot 5}{2\cdot 5 + \dfrac{4}{5}\cdot 5}$

$= \dfrac{35}{10+4} = \dfrac{35}{14} = \dfrac{5}{2}$

12. $\dfrac{\dfrac{6}{7} - \dfrac{5}{14}}{\dfrac{3}{14} - \dfrac{5}{7}} = \dfrac{\left(\dfrac{6}{7} - \dfrac{5}{14}\right)\cdot 14}{\left(\dfrac{3}{14} - \dfrac{5}{7}\right)\cdot 14} = \dfrac{\dfrac{6}{7}\cdot 14 - \dfrac{5}{14}\cdot 14}{\dfrac{3}{14}\cdot 14 - \dfrac{5}{7}\cdot 14}$

$= \dfrac{12-5}{3-10} = \dfrac{7}{-7} = -1$

15. $\dfrac{\dfrac{1}{a}+3}{\dfrac{2}{a}-4} = \dfrac{\left(\dfrac{1}{a}+3\right)\cdot a}{\left(\dfrac{2}{a}-4\right)\cdot a} = \dfrac{\dfrac{1}{a}\cdot a + 3\cdot a}{\dfrac{2}{a}\cdot a - 4\cdot a}$

$= \dfrac{1+3a}{2-4a} = \dfrac{3a+1}{2-4a}$

22. $\dfrac{\dfrac{3}{x^2}-\dfrac{4}{y}}{\dfrac{5}{x}+\dfrac{2}{y^2}} = \dfrac{\left(\dfrac{3}{x^2}-\dfrac{4}{y}\right)\cdot x^2y^2}{\left(\dfrac{5}{x}+\dfrac{2}{y^2}\right)\cdot x^2y^2} = \dfrac{\dfrac{3}{x^2}\cdot x^2y^2 - \dfrac{4}{y}\cdot x^2y^2}{\dfrac{5}{x}\cdot x^2y^2 + \dfrac{2}{y^2}\cdot x^2y^2}$

$= \dfrac{3y^2-4x^2y}{5xy^2+2x^2} = \dfrac{y(3y-4x^2)}{x(5y^2+2x)}$

26. $\dfrac{\dfrac{1}{a}-\dfrac{1}{b}}{\dfrac{1}{a^2}-\dfrac{1}{b^2}} = \dfrac{\dfrac{1}{a}\cdot a^2b^2 - \dfrac{1}{b}\cdot a^2b^2}{\dfrac{1}{a^2}\cdot a^2b^2 - \dfrac{1}{b^2}\cdot a^2b^2} = \dfrac{ab^2-a^2b}{b^2-a^2}$

$= \dfrac{ab(b-a)}{(b+a)(b-a)} = \dfrac{ab}{b+a} = \dfrac{ab}{a+b}$

37. $\dfrac{\dfrac{1}{x+y}-\dfrac{1}{x-y}}{\dfrac{1}{x+y}+\dfrac{1}{x-y}} = \dfrac{\left(\dfrac{1}{x+y}-\dfrac{1}{x-y}\right)\cdot(x+y)(x-y)}{\left(\dfrac{1}{x+y}+\dfrac{1}{x-y}\right)\cdot(x+y)(x-y)}$

$= \dfrac{\dfrac{1}{x+y}\cdot(x+y)(x-y) - \dfrac{1}{x-y}\cdot(x+y)(x-y)}{\dfrac{1}{x+y}\cdot(x+y)(x-y) + \dfrac{1}{x-y}\cdot(x+y)(x-y)}$

$= \dfrac{(x-y)-(x+y)}{(x-y)+(x+y)} = \dfrac{x-y-x-y}{x+y+x+y} = \dfrac{-2y}{2x} = -\dfrac{y}{x}$

46. $\dfrac{\dfrac{7}{b-7}+\dfrac{8}{b-5}}{\dfrac{6}{b^2-12b+35}} = \dfrac{\left(\dfrac{7}{b-7}+\dfrac{8}{b-5}\right)\cdot(b-5)(b-7)}{\dfrac{6}{(b-5)(b-7)}\cdot(b-5)(b-7)}$

The LCD is $(b-5)(b-7)$.

$= \dfrac{\dfrac{7}{b-7}\cdot(b-5)(b-7) + \dfrac{8}{b-5}\cdot(b-5)(b-7)}{6}$

$= \dfrac{7(b-5)+8(b-7)}{6} = \dfrac{7b-35+8b-56}{6} = \dfrac{15b-91}{6}$

Review Exercises

1. $\dfrac{6}{7}$ **2.** $\dfrac{x-7}{x+2}$ **3.** $2x^3+3x^2+x-21$ **4.** $\{3\}$ **5.** $\{-2\}$ **6.** $\{-3,1\}$

7. Domain is all real numbers except -7.
8. Domain is all real numbers except -2 and 2.
9. a. undefined   **b.** 0   **10.** 9 inches by 26 inches

Exercise 5–6

Answers to Odd-Numbered Problems

1. $\left\{\dfrac{8}{3}\right\}$ **3.** $\left\{\dfrac{14}{3}\right\}$ **5.** $\{-22\}$ **7.** $\left\{\dfrac{10}{9}\right\}$ **9.** $\left\{\dfrac{9}{2}\right\}$ **11.** $\left\{-\dfrac{5}{8}\right\}$

13. $\left\{\dfrac{19}{24}\right\}$ **15.** $\left\{-\dfrac{5}{27}\right\}$ **17.** $\{1\}$ **19.** $\left\{\dfrac{4}{37}\right\}$ **21.** $\left\{-\dfrac{13}{2}\right\}$ **23.** $\{36\}$

25. $\left\{\dfrac{47}{9}\right\}$ **27.** $\varnothing$ **29.** $\left\{\dfrac{11}{5}\right\}$ **31.** $\{40\}$ **33.** $\left\{-\dfrac{19}{3}\right\}$ **35.** $\left\{\dfrac{8}{3}\right\}$

37. $\left\{-\dfrac{2}{3}\right\}$ **39.** $\left\{-3, \dfrac{3}{2}\right\}$ **41.** $\left\{-4, \dfrac{2}{3}\right\}$ **43.** $\left\{-1, \dfrac{2}{3}\right\}$

45. $x = \dfrac{2y}{3y - 1}$ **47.** $c_1 = \dfrac{cc_2}{c_2 - c}$ **49.** $a = \dfrac{8b}{b - 8}$

51. $a = \dfrac{3b - 5}{4}$

53. $x = \dfrac{-2(y + 6)}{5}$

55. $D = \dfrac{P}{pL}$ **57.** $N = \dfrac{2P}{D - P}$

59. $f = \dfrac{Fa}{A}$ **61.** $T_2 = \dfrac{T_1 P_2 V_2}{P_1 V_1}$ **63.** $L_1 = \dfrac{L_2}{\alpha(t_2 - t_1) + 1}$

65. $R_1 = \dfrac{RR_2}{R_2 - R}$ **67.** equations whose solution sets are the same

69. Replace the variable in the original equation with the solution and see if the equation simplifies to a true statement.

71. The results are not the same because one is an expression and the other is an equation. The LCD is used to add the fractions in the expression. The LCD is used to clear the fractions in the equation.

Solutions to Trial Exercise Problems

2. $\dfrac{4x}{5} - \dfrac{2}{3} = 4$ The LCD is 15.
Multiply by 15 to get

$\dfrac{4x}{5} \cdot 15 - \dfrac{2}{3} \cdot 15 = 4 \cdot 15$

$\qquad 4x \cdot 3 - 2 \cdot 5 = 60$

$\qquad\quad 12x - 10 = 60$

$\qquad\qquad\quad 12x = 70$

$\qquad\qquad\qquad x = \dfrac{70}{12} = \dfrac{35}{6}$

The solution set is $\left\{\dfrac{35}{6}\right\}$.

6. $\dfrac{3R}{4} - 5 = \dfrac{5}{6}$ The LCD is 12.
Multiply by 12 to get

$\dfrac{3R}{4} \cdot 12 - 5 \cdot 12 = \dfrac{5}{6} \cdot 12$

$\qquad 3R \cdot 3 - 60 = 5 \cdot 2$

$\qquad\quad 9R - 60 = 10$

$\qquad\qquad\quad 9R = 70$

$\qquad\qquad\quad R = \dfrac{70}{9}$

The solution set is $\left\{\dfrac{70}{9}\right\}$.

15. $\dfrac{4}{6y} + 5 = \dfrac{1}{9y} + 2$ The LCD is $18y$
Multiply by $18y$ to get

$\dfrac{4}{6y} \cdot 18y + 5 \cdot 18y = \dfrac{1}{9y} \cdot 18y + 2 \cdot 18y$

$\qquad 4 \cdot 3 + 90y = 1 \cdot 2 + 36y$

$\qquad\quad 12 + 90y = 2 + 36y$

$\qquad\qquad\quad 90y = -10 + 36y$

$\qquad\qquad\quad 54y = -10$

$\qquad\qquad\qquad y = \dfrac{-10}{54} = \dfrac{-5}{27}$

The solution set is $\left\{-\dfrac{5}{27}\right\}$.

22. $\dfrac{R + 2}{10R} + \dfrac{4R - 1}{4R} = 2$ The LCD is $20R$
Multiply by $20R$ to get

$\dfrac{R + 2}{10R} \cdot 20R + \dfrac{4R - 1}{4R} \cdot 20R = 2 \cdot 20R$

$\qquad (R + 2) \cdot 2 + (4R - 1) \cdot 5 = 40R$

$\qquad\quad 2R + 4 + 20R - 5 = 40R$

$\qquad\qquad\qquad 22R - 1 = 40R$

$\qquad\qquad\qquad\qquad -1 = 18R$

$\qquad\qquad\qquad\qquad R = \dfrac{-1}{18}$

The solution set is $\left\{-\dfrac{1}{18}\right\}$.

34. $\dfrac{8}{a^2 - 6a + 8} = \dfrac{1}{a^2 - 16}$ Multiply both sides by the LCD, $(a - 2)(a + 4)(a - 4)$.

$\dfrac{8}{(a - 4)(a - 2)} \cdot (a - 2)(a + 4)(a - 4)$

$\qquad = \dfrac{1}{(a + 4)(a - 4)} \cdot (a - 2)(a + 4)(a - 4)$

$\qquad 8(a + 4) = 1(a - 2)$

$\qquad 8a + 32 = a - 2$

$\qquad\quad 8a = a - 34$

$\qquad\quad 7a = -34$

$\qquad\quad a = \dfrac{-34}{7}$

The solution set is $\left\{-\dfrac{34}{7}\right\}$.

37. $3x^2 + 4x + \dfrac{4}{3} = 0$ Multiply by 3 to obtain

$9x^2 + 12x + 4 = 0$ Factor the left side.

$\qquad (3x + 2)^2 = 0$

$\qquad\quad 3x + 2 = 0$

$\qquad\qquad 3x = -2$

$\qquad\qquad x = -\dfrac{2}{3}$ So

The solution set is $\left\{-\dfrac{2}{3}\right\}$.

42. $x^2 - \dfrac{5}{6}x = \dfrac{2}{3}$ Multiply by the LCD, 6, to obtain

$\qquad 6x^2 - 5x = 4$

$\qquad 6x^2 - 5x - 4 = 0$

$\qquad (2x + 1)(3x - 4) = 0$

$\qquad\quad 2x + 1 = 0 \text{ or } 3x - 4 = 0$

$\qquad\quad x = -\dfrac{1}{2} \text{ or } x = \dfrac{4}{3}$

The solution set is $\left\{-\dfrac{1}{2}, \dfrac{4}{3}\right\}$.

47. $\dfrac{1}{c} = \dfrac{1}{c_1} + \dfrac{1}{c_2}$ Multiply by the LCD cc_1c_2.

$\dfrac{1}{c} \cdot cc_1c_2 = \dfrac{1}{c_1} \cdot cc_1c_2 + \dfrac{1}{c_2} \cdot cc_1c_2$

$\qquad c_1c_2 = cc_2 + cc_1$

$\qquad c_1c_2 - cc_1 = cc_2$ Subtract cc_1 from both sides.

$\qquad (c_2 - c)c_1 = cc_2$ Factor c_1 in the left side.

$\qquad\qquad c_1 = \dfrac{cc_2}{c_2 - c}$ Divide both sides by $c_2 - c$.

52.
$$\frac{5}{m} + 4 = \frac{6a}{2m} + 3b \qquad \text{Multiply by the LCD } 2m.$$

$$\frac{5}{m} \cdot 2m + 4 \cdot 2m = \frac{6a}{2m} \cdot 2m + 3b \cdot 2m$$

$$10 + 8m = 6a + 6bm$$

$$8m = 6a - 10 + 6bm \quad \text{Subtract 10 from both sides.}$$

$$8m - 6bm = 6a - 10 \qquad \text{Subtract } 6bm \text{ from both sides.}$$

$$(8 - 6b)m = 6a - 10 \qquad \text{Factor } m \text{ in the left side.}$$

$$m = \frac{6a - 10}{8 - 6b} \qquad \text{Divide both sides by } 8 - 6b.$$

$$= \frac{2(3a - 5)}{2(4 - 3b)} = \frac{3a - 5}{4 - 3b}$$

58.
$$\frac{T_A}{T_B} = \frac{R_B}{R_A} \qquad \text{Multiply by the LCD } T_B R_A.$$

$$\frac{T_A}{T_B} \cdot T_B R_A = \frac{R_B}{R_A} \cdot T_B R_A$$

$$T_A R_A = R_B T_B$$

$$T_B = \frac{T_A R_A}{R_B}$$

62. Given $k = \frac{L_t - L_0}{L_0 t}$. Multiply both sides by $L_0 t$ to get

$kL_0 t = L_t - L_0$.
Add L_0 to both sides to get $L_t = kL_0 t + L_0$.
Factor L_0 in the right side to get $L_t = (kt + 1)L_0$. Divide both sides by $kt + 1$.

Then $\dfrac{L_t}{kt + 1} = L_0$ or $L_0 = \dfrac{L_t}{kt + 1}$.

Review Exercises

1. 14 **2.** $\dfrac{17}{3}$ **3.** 4 **4.** $y = -5x + 4$ **5.** $y = \dfrac{-2x + 6}{3}$

6. $y = \dfrac{x - 8}{4}$ **7.** $\dfrac{3y}{x^2}$ **8.** $\dfrac{y^4}{x^5}$ **9.** 30 m by 33 m

Exercise 5–7

Answers to Odd-Numbered Problems

1. $29\dfrac{1}{6}$ minutes **3.** $2\dfrac{14}{29}$ hours **5.** 12 hours **7.** $3\dfrac{1}{3}$ hours

9. 10 hours **11.** 110 miles at 55 mph, 210 miles at 60 mph

13. car A, 40 mph; car B, 50 mph **15.** $\dfrac{6}{7}$ mph **17.** 220 miles

19. $\dfrac{5}{9}$ **21.** 3 and 12 **23.** 6 **25.** 4 **27.** adult, 15 pills; child, 9 pills

29. $5. 46 **31.** $3\dfrac{9}{17}$ ohms **33.** 15 ohms **35.** 60 ohms

Solutions to Trial Exercise Problems

1. Let x represent the number of minutes required for both boys to mow the lawn. Then,

$$\frac{1}{50} + \frac{1}{70} = \frac{1}{x}$$

Multiply by the LCD of the fractions, that is, $350x$.

$$350x \cdot \frac{1}{50} + 350x \cdot \frac{1}{70} = 350x \cdot \frac{1}{x}$$

$$7x + 5x = 350$$

$$12x = 350$$

$$x = \frac{350}{12} = 29\frac{1}{6}$$

Together Jim and Kenny could mow the lawn in $29\dfrac{1}{6}$ minutes.

5. Let x represent the time required for the other painter to paint the house alone. Then $\dfrac{1}{6} + \dfrac{1}{x} = \dfrac{1}{4}$. Multiply both sides by the LCD, that is, $12x$.

$$12x \cdot \frac{1}{6} + 12x \cdot \frac{1}{x} = 12x \cdot \frac{1}{4}$$

$$2x + 12 = 3x$$

$$12 = x$$

Therefore the other painter would take 12 hours to paint the house alone.

13. Let x represent the average speed of car A.
Then $x + 10$ is the average speed of car B.

Using $t = \dfrac{d}{r}$, let t_A = time of car A, t_B = time of car B,

then $t_A = t_B$. Now $t_A = \dfrac{120}{x}$ and $t_B = \dfrac{150}{x + 10}$, so

$$\frac{120}{x} = \frac{150}{x + 10}. \qquad \text{Multiply both sides by the LCD, that is,}$$
$$x(x + 10).$$

$$x(x + 10) \cdot \frac{120}{x} = x(x + 10) \cdot \frac{150}{x + 10}$$

$$(x + 10)120 = 150x$$

$$120x + 1,200 = 150x$$

$$1,200 = 30x$$

$$40 = x$$

So $x + 10 = 50$. Therefore car B averaged 50 mph and car A averaged 40 mph.

20. Let x represent the numerator of the fraction. Then $x + 7$ is the denominator of the fraction. Then $\dfrac{x + 3}{(x + 7) - 1} = \dfrac{4}{5}$. Simplifying, we have the equation $\dfrac{x + 3}{x + 6} = \dfrac{4}{5}$. Multiply both sides by the LCD, that is, $5(x + 6)$.

$$5(x + 6) \cdot \frac{x + 3}{x + 6} = 5(x + 6) \cdot \frac{4}{5}$$

$$5(x + 3) = 4(x + 6)$$

$$5x + 15 = 4x + 24$$

$$x + 15 = 24$$

$$x = 9$$

$$x + 7 = 16$$

The original fraction is then $\dfrac{9}{16}$.

21. Let x represent the lesser of the two numbers. Then $4x$ is the greater of the two numbers.

Their reciprocals are then $\dfrac{1}{x}$ and $\dfrac{1}{4x}$ and we get the equation

$\dfrac{1}{x} + \dfrac{1}{4x} = \dfrac{5}{12}$. Multiply both sides by the LCD, that is, $12x$.

$$12x \cdot \frac{1}{x} + 12x \cdot \frac{1}{4x} = 12x \cdot \frac{5}{12}$$

$$12 + 3 = 5x$$

$$15 = 5x$$

then $x = 3$

$$4x = 12$$

Therefore the two numbers are 3 and 12.

To check, show $\dfrac{1}{3} + \dfrac{1}{12} = \dfrac{5}{12}$

$$\frac{4}{12} + \frac{1}{12} = \frac{5}{12}$$

$$\frac{5}{12} = \frac{5}{12}$$

28. Let x represent hourly wage of a journeyman electrician, then $\frac{3}{8}x$ is the hourly wage of an apprentice electrician.

$$x + \frac{3}{8}x = 29.70$$
$$8 \cdot x + 8 \cdot \frac{3}{8}x = 8 \cdot 29.70$$
$$8x + 3x = 237.60$$
$$11x = 237.60$$
$$x = \$21.60$$
$$\frac{3}{8}x = \$8.10$$

The journeyman earns \$21.60 per hour and the apprentice earns \$8.10 per hour.

30. Let R be the total resistance of the circuit. Then $\frac{1}{6} + \frac{1}{8} = \frac{1}{R}$.

Multiply by the LCD, that is, $24R$.

$$24R \cdot \frac{1}{6} + 24R \cdot \frac{1}{8} = 24R \cdot \frac{1}{R}$$
$$4R + 3R = 24$$
$$7R = 24$$
$$R = \frac{24}{7}$$

Therefore the total resistance in the parallel circuit is $\frac{24}{7}$ or $3\frac{3}{7}$ ohms.

32. Let $R_1 = $ the resistance in the unknown branch. Then $\frac{1}{R_1} + \frac{1}{30} = \frac{1}{12}$. Multiply by the LCD, that is, $60R_1$.

$$60R_1 \cdot \frac{1}{R_1} + 60R_1 \cdot \frac{1}{30} = 60R_1 \cdot \frac{1}{12}$$
$$60 + 2R_1 = 5R_1$$
$$60 = 3R_1$$
$$20 = R_1$$

Therefore the other branch has resistance of 20 ohms.

Review Exercises

1. $\{2\}$ **2.** $y = \frac{3x+6}{2}$ **3.** $8(y+2)(y-2)$ **4.** $(x+10)^2$

5. $(3y+2)(y-2)$ **6.** $\frac{5x^2+7x}{(x-1)(x+3)}$ **7.** $\frac{-2y^2-23y}{(2y+1)(y-5)}$

8. \$8,062. 50 at 9%; \$10,937. 50 at 7%

9. \$5,000 at 13% profit; \$12,500 at 9% loss

Chapter 5 Review

1. all real numbers except 0 **2.** all real numbers except -7

3. all real numbers except 9 **4.** all real numbers except $-\frac{2}{3}$

5. all real numbers except $\frac{3}{5}$ **6.** all real numbers except -4 and 3

7. all real numbers except -1 and 1 **8.** $\frac{3b}{a}$ **9.** $\frac{3xz}{2y^2}$

10. $\frac{x-7}{x+7}$ **11.** $\frac{x+3}{x+7}$ **12.** $\frac{6}{5}$ **13.** $-x-y$ **14.** $\frac{3p-2}{5p+1}$

15. $\frac{R-4}{3R-1}$ **16.** $\frac{5-n}{n+2}$ **17.** $\frac{9}{2}$ **18.** $\frac{9a}{b}$ **19.** $\frac{2x-y}{2x+y}$

20. $x-2$ **21.** $\frac{x(m-n)}{2}$ **22.** $\frac{1}{(y-1)^2}$ **23.** 2 **24.** $\frac{-1}{3(x+5)}$

25. $\frac{x-1}{x+1}$ **26.** $\frac{2a}{3}$ **27.** $\frac{8b}{21}$ **28.** $\frac{9}{2ab}$ **29.** $\frac{7(2x-1)}{3(3x+4)}$

30. $\frac{1}{(x-2)(x+6)}$ **31.** $\frac{10}{3}$ **32.** $\frac{1}{x-8}$ **33.** $\frac{3-b}{(b+3)(b-1)}$

34. $\frac{3(a+1)}{4(a-1)}$ **35.** $36x^2$ **36.** $42a^2b^2$ **37.** $2(x+3)(x-3)$

38. $(y+5)(y-5)(y+3)$ **39.** $4(z+2)(z-2)(z+1)$

40. $x(x+1)^2(3x-5)$ **41.** $\frac{3}{x}$ **42.** $\frac{4y}{y-2}$ **43.** $\frac{4x+4}{3x-1}$

44. $\frac{13-x}{2x-5}$ **45.** $\frac{1}{x-3}$ **46.** $\frac{x}{(x+2)^2}$ **47.** $\frac{68x}{105}$ **48.** $\frac{23}{48a}$

49. $\frac{47x-6}{(3x+1)(4x-3)}$ **50.** $\frac{4x^2-23x}{(x+4)(x-4)}$ **51.** $\frac{80a+48b-15ab}{20a^2b^2}$

52. $\frac{4a^2+4a+6}{a+1}$ **53.** $\frac{-10x^2-3}{x^2+1}$ **54.** $\frac{16x^2+5x+15}{24x^2}$

55. $\frac{13y^2-89y}{(y-9)(y+2)(y-2)}$ **56.** $\frac{2-2ax-36a}{(x+3)(x-3)}$ **57.** $\frac{h}{20}$

58. $w = \frac{84}{\ell}$ **59.** $\frac{16}{63}$ **60.** 1 **61.** $\frac{y-x}{y+x}$ **62.** $\frac{4+3x}{2-5x}$

63. $\frac{a^2-b^2}{a^2+b^2}$ **64.** $y-x$ **65.** $\frac{xy}{x+y}$ **66.** $\{-96\}$ **67.** $\left\{\frac{41}{48}\right\}$

68. $\left\{\frac{13}{24}\right\}$ **69.** $\left\{\frac{6}{5}\right\}$ **70.** $\left\{-\frac{39}{14}\right\}$ **71.** $\left\{\frac{7}{52}\right\}$ **72.** $\varnothing$ **73.** $\left\{-\frac{3}{2},\frac{3}{2}\right\}$

74. $\{-5,3\}$ **75.** $\left\{\frac{3}{2},6\right\}$ **76.** $x = \frac{a+b}{3}$ **77.** $x = \frac{y+3}{1-y}$

78. $y = \frac{4a-3b}{a}$ **79.** $y = 4b+5c$ **80.** $\ell = \frac{Aey}{F}$

81. $L = \frac{Wp-2E}{EF}$ **82.** $R = \frac{Mr}{M-2}$ **83.** $m = \frac{Fr}{v^2+gr}$

84. $5\frac{1}{3}$ hr, 16 hr **85.** $1\frac{5}{7}$ mph **86.** $\frac{2}{3}$ or 1

Chapter 5 Test

1. all real numbers except 4 [5–1] **2.** all real numbers except -2 and 0 [5–1] **3.** $24a^3$ [5–3] **4.** $(x+2)(x-2)^2$ [5–3]

5. $\frac{x+5}{x-2}$ [5–1] **6.** $\frac{2ax}{5y^2}$ [5–2] **7.** $\frac{1}{x-3}$ [5–3]

8. $\frac{x^2+5x-3}{9x^2}$ [5–4] **9.** $\frac{5}{2}$ [5–5]

10. $\frac{a+5}{2a-8}$ [5–4] **11.** $\frac{1}{a+4}$ [5–3] **12.** $\frac{b+2a}{a-2b}$ [5–5]

13. $\frac{y^2-y+6}{2(y+1)(y-1)}$ [5–4] **14.** $\frac{x+5}{2x+4}$ [5–2] **15.** $-\frac{3}{4}$ [5–1]

16. $\frac{(y+4)(y+1)}{(y+3)(y-1)}$ [5–2] **17.** a [5–5] **18.** $\frac{a+2}{a-1}$ [5–2]

19. $\frac{1}{h}$ [5–4] **20.** $\left\{-\frac{4}{7}\right\}$ [5–6] **21.** $\left\{-\frac{1}{2},\frac{4}{3}\right\}$ [5–6]

22. $y = \frac{7x}{6-4x}$ [5–6] **23.** $\frac{4}{5}, \frac{8}{5}$ [5–7] **24.** $11\frac{19}{31} \approx 11.6$ min [5–7]

Chapter 5 Cumulative Test

1. x^{11} [3–3] **2.** $\frac{1}{y^9}$ [3–3] **3.** $\frac{25x^4}{y^6}$ [3–4]

4. $10x^3 - 4x^2 + 11x - 9$ [2–2] **5.** $\left\{-\frac{26}{7}\right\}$ [2–5] **6.** $\left\{\frac{81}{4}\right\}$ [2–5]

7. $\left\{-\frac{3}{2},2\right\}$ [4–7] **8.** $x \geq -5$ [2–9] **9.** $2(x+3)(x-3)$ [4–4]

10. $3x^2(2x^3 - 12x + 3)$ [4–1] **11.** $(2x+5)(2x+3)$ [4–3]

12. $3(1+2x^3)(1-2x^3)$ [4–4] **13.** $49x^2 - 84x + 36$ [3–2]

14. $6x^3 - 19x^2 + 37x - 33$ [3–2] **15.** $15 - 22x^2 + 8x^4$ [3–2]

16. $x = \frac{120}{7}$ [2–8] **17.** 2,400 foot-pounds/min [2–8]

18. $\dfrac{a-6}{a-7}$ [5–1]   **19.** $\dfrac{y+4}{y+5}$ [5–1]

20. $y^2 - 5y + 6$ [5–2]   **21.** $\dfrac{2y+4x}{x^2 y^2}$ [5–4]

22. $\dfrac{3x+9y}{(x-y)(x+y)}$ [5–4]   **23.** $\dfrac{13x+56}{(x-7)(x+7)(x+2)}$ [5–4]

24. $\dfrac{3y-1}{4y+5}$ [5–5]   **25.** $\dfrac{y+x}{3x-4y}$ [5–5]   **26.** $\{2\}$ [5–6]

27. $\left\{-\dfrac{1}{3},6\right\}$ [5–6]   **28.** $q = \dfrac{fp}{p-f}$ [5–6]

29. $2\dfrac{2}{5}$ days [5–7]   **30.** $\dfrac{4}{9}$ [5–7]

Chapter 6

Exercise 6–1

Answers to Odd-Numbered Problems

1. $(1,2)$ and $(-1,-4)$ are solutions.
3. $(-1,2)$ and $(3,0)$ are solutions.   **5.** $(1,2)$ is a solution.
7. $(2,3)$ and $(0,0)$ are solutions.
9. $(-4,1)$ and $(-4,-4)$ are solutions.
11. $(-5,3)$ and $(-5,8)$ are solutions.   **13.** $(1,5),(-2,-4),(0,2)$
15. $(3,-3),(-4,11),(0,3)$ **17.** $(3,-5),(-2,10),(0,4)$
19. $(-2,-1),(3,0),\left(0,-\dfrac{3}{5}\right)$ **21.** $(1,-4),(-1,1),\left(0,-\dfrac{3}{2}\right)$
23. $(1,5),(-6,5),(0,5)$ **25.** $(7,-1),\left(-\dfrac{3}{5},-1\right),(0,-1)$
27. $(5,1),(-1,-2),(3,0)$ **29.** $(3,2),(-6,-4),(0,0)$
31. $(1,-2),(1,7),(1,0)$
33. a. $(75, \$170)$ **b.** $(300, \$620)$ **c.** $(1,000, \$2,020)$
35. a. $(2,236)$ **b.** $(12,216)$ **c.** $(0,240)$
37. a. $(10,50)$ **b.** $(15,65)$ **c.** $(8,44)$
39. a. $(3,165)$ **b.** $(8,440)$ **c.** $\left(\dfrac{26}{5},286\right)$ **41.** $(2,4)$ **43.** $(-1,3)$
45. $(-6,-1)$ **47.** $(0,4)$ **49.** $(5,0)$ **51.** $(-7,0)$ **53.** $\left(\dfrac{1}{2},3\right)$
55. $\left(\dfrac{3}{2},0\right)$

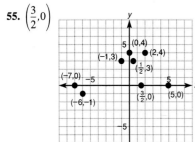

57. $B(2,-1)$ **59.** $D(0,6)$ **61.** $F(0,-6)$ **63.** $H(-4,4)$ **65.** $J(5,-5)$
67. III **69.** I **71.** IV **73.** III **75.** IV **77. a.** IV **b.** II **c.** III **d.** I
79. 0 **81.** $B, -1; D, 6; F, -6; H, 4; J, -5$
83. Figure is a horizontal line parallel to the x-axis having equation $y = 4$.

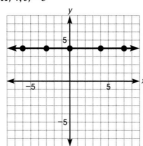

85. $(-5,5),(-2,2),(1,-1),(3,-3),(5,-5)$
The resulting figure is a line through the origin that lies in quadrants II and IV.

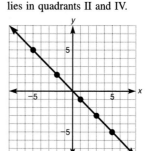

87.

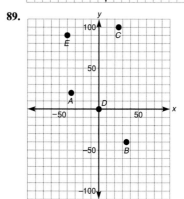

89.

91. A(Tues.,70), the temperature was 70°F on Tuesday. B(Wed.,80), the temperature was 80°F on Wednesday. C(Fri.,90), the temperature was 90°F on Friday. D(Sat.,85), the temperature was 85°F on Saturday.

93. $(0,4),(-4,0),\left(\dfrac{5}{2},\dfrac{13}{2}\right),(2,6)$

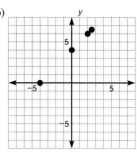

95. $(0,1), \left(-\frac{1}{2}, 0\right), (2,5), (-1,-1)$

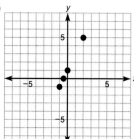

97. $(0,0), \left(\frac{1}{2}, 1\right), (2,4), (-1,-2)$

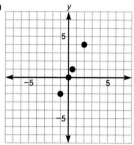

99. $(0,3), (-3,6), (3,0), \left(\frac{5}{2}, \frac{1}{2}\right)$

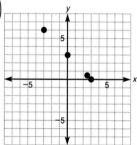

101. $(0,3), (3,-3), (-2,7), \left(\frac{3}{2}, 0\right)$

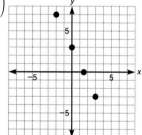

103. $(0,4), (-3,7), (2,2), (3,1)$

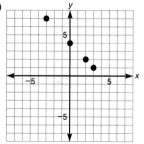

Solutions to Trial Exercise Problems

1. $y = 3x - 1; (1,2), (-1,-4), (2,3)$
Solution:

(a) $(1,2)$ (b) $(-1,-4)$ (c) $(2,3)$

$2 = 3(1) - 1$ $-4 = 3(-1) - 1$ $3 = 3(2) - 1$

$2 = 3 - 1$ $-4 = -3 - 1$ $3 = 6 - 1$

$2 = 2$ $-4 = -4$ $3 = 5$

(true)   (true)   (false)

Therefore $(1,2)$ and $(-1,-4)$ are solutions and $(2,3)$ is not a solution.

7. $3x = 2y; (2,3), (3,2), (0,0)$
Solution:

(a) $(2,3)$ (b) $(3,2)$ (c) $(0,0)$

$3(2) = 2(3)$ $3(3) = 2(2)$ $3(0) = 2(0)$

$6 = 6$ $9 = 4$ $0 = 0$

(true)   (false)   (true)

So $(2,3)$ and $(0,0)$ are solutions and $(3,2)$ is not a solution.

9. $x = -4; (-4,1), (4,2), (-4,-4)$
Solution:

Since we can write $x = -4$ as $x = 0 \cdot y - 4$, then $x = -4$ for *any* value of y. Therefore $(-4,1)$ and $(-4,-4)$ are solutions but $(4,2)$ is not since, substituting 4 for x, $4 = -4$ is false.

15. $y = 3 - 2x; x = 3, x = -4, x = 0$
Solution:

When $x = 3$, $y = 3 - 2(3) = 3 - 6 = -3$ $(3,-3)$

When $x = -4$, $y = 3 - 2(-4) = 3 + 8 = 11$ $(-4,11)$

When $x = 0$, $y = 3 - 2(0) = 3 - 0 = 3$ $(0,3)$

20. $x + 4y = 0; x = -4, x = 8, x = 0$
Solution:

When $x = -4$, $-4 + 4y = 0$

$\qquad\qquad\qquad 4y = 4$

$\qquad\qquad\qquad\quad y = 1\ (-4,1)$

When $x = 8$, $8 + 4y = 0$

$\qquad\qquad\qquad 4y = -8$

$\qquad\qquad\qquad\quad y = -2\ (8,-2)$

When $x = 0$, $0 + 4y = 0$

$\qquad\qquad\qquad 4y = 0$

$\qquad\qquad\qquad\quad y = 0\ (0,0)$

23. $y = 5; x = 1, x = -6, x = 0$
Solution:

We write $y = 5$ as $y = 0 \cdot x + 5$, so for *every* value of x, $y = 5$. Therefore $(1,5), (-6,5), (0,5)$.

28. $x = -3y + 1; y = -1, y = 2, y = 0$
Solution:

When $y = -1$, $x = -3(-1) + 1 = 3 + 1 = 4\ (4,-1)$

When $y = 2$, $x = -3(2) + 1 = -6 + 1 = -5\ (-5,2)$

When $y = 0$, $x = -3(0) + 1 = 0 + 1 = 1\ (1,0)$

31. $x = 1; y = -2, y = 7, y = 0$
Solution:

Since $x = 1$ can be written $x = 0 \cdot y + 1$, then $x = 1$ for *any* value of y, so we have $(1,-2), (1,7), (1,0)$.

36. Given $y = 240 - 2x$, when

a. $y = 200$, then $200 = 240 - 2x$

$\qquad\qquad\qquad\qquad -40 = -2x$

$\qquad\qquad\qquad\qquad\quad 20 = x \qquad (20,200)$

b. $y = 0$, then $0 = 240 - 2x$

$\qquad\qquad\qquad -240 = -2x$

$\qquad\qquad\qquad\quad 120 = x \qquad (120,0)$

c. $y = 210$, then $210 = 240 - 2x$

$\qquad\qquad\qquad\qquad -30 = -2x$

$\qquad\qquad\qquad\qquad\quad 15 = x \qquad (15,210)$

43.   48.   53.

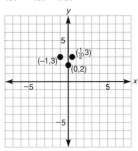

58. $(-2,0)$ **63.** $(-4,4)$ **64.** $(7,0)$ **68.** $(4,-1)$ lies in quadrant IV.

73. $\left(-\dfrac{5}{2}, -\dfrac{3}{4}\right)$ lies in quadrant III.

82. $(-3,5), (-3,3), (-3,0), (-3,-2), (-3,-5)$
They lie on a vertical
straight line.

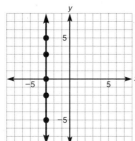

86.

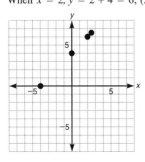

91. A (Tues.,70), the temperature was 70°F on Tuesday.
B (Wed.,80), the temperature was 80°F on Wednesday.
C (Fri.,90), the temperature was 90°F on Friday.
D (Sat.,85), the temperature was 85°F on Saturday.

93. When $x = 0$, $y = 0 + 4 = 4$ $(0,4)$
When $x = -4$, $y = -4 + 4 = 0$; $(-4,0)$
When $x = \dfrac{5}{2}$, $y = \dfrac{5}{2} + 4 = \dfrac{13}{2}$; $\left(\dfrac{5}{2}, \dfrac{13}{2}\right)$
When $x = 2$, $y = 2 + 4 = 6$; $(2,6)$

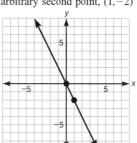

Review Exercises

1. $2x^2 + 4x$ **2.** $\dfrac{x+3}{2x+1}$ **3.** $\dfrac{x+10}{(x-2)(x+1)}$ **4.** -8 **5.** x^3 **6.** 0

7. stock A: 315 shares, stock B: 45 shares   **8.** $7\dfrac{1}{2}$ hours

Exercise 6–2
Answers to Odd-Numbered Problems

1. x-intercept, $(-2,0)$; y-intercept, $(0,4)$ **3.** x-intercept, $\left(-\dfrac{1}{3},0\right)$;
y-intercept, $(0,1)$ **5.** x-intercept, $(3,0)$; y-intercept, $(0,2)$

7. x-intercept, $\left(\dfrac{11}{2},0\right)$; y-intercept, $\left(0,\dfrac{11}{5}\right)$ **9.** x-intercept, $(0,0)$;
y-intercept, $(0,0)$ **11.** x-intercept, $(0,0)$; y-intercept, $(0,0)$

13. x-intercept, $(4,0)$; y-intercept, $(0,2)$ **15.** x-intercept, $(3,0)$;
y-intercept, $\left(0,-\dfrac{3}{2}\right)$

17. x-intercept, $(-2,0)$;
y-intercept, $(0,6)$

19. x-intercept, $(2,0)$;
y-intercept, $(0,-2)$

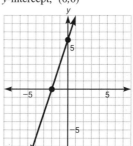

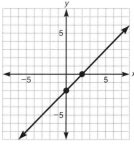

21. x-intercept, $(4,0)$;
y-intercept, $(0,-8)$

23. x-intercept, $(0,0)$;
y-intercept, $(0,0)$;
arbitrary second point, $(1,1)$

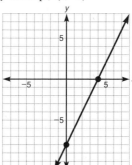

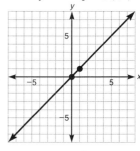

25. x-intercept, $(0,0)$;
y-intercept, $(0,0)$;
arbitrary second point, $(1,-2)$

27. x-intercept, $(0,0)$;
y-intercept, $(0,0)$;
arbitrary second point, $(1,2)$

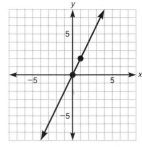

29. x-intercept, $(3,0)$;
y-intercept, $(0,-4)$

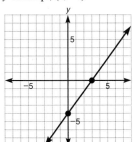

31. x-intercept, $(2,0)$;
y-intercept, $(0,5)$

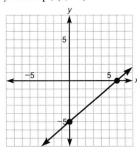

33. x-intercept, $(6,0)$;
y-intercept, $(0,-5)$

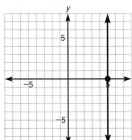

35. no x-intercept;
y-intercept, $(0,6)$

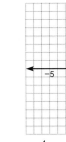

37. x-intercept, $(5,0)$;
no y-intercept

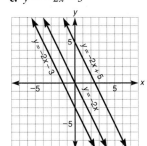

39. x-intercept, $(0,0)$; All real
numbers are y-intercepts.

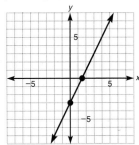

41. $y = 2x - 7$, $m = 2$, $b = -7$ **43.** $y = \dfrac{4}{3}x + 3$, $m = \dfrac{4}{3}$, $b = 3$

45. $y = -\dfrac{7}{3}x + \dfrac{10}{3}$, $m = -\dfrac{7}{3}$, $b = \dfrac{10}{3}$

47. $y = \dfrac{1}{5}x + \dfrac{7}{5}$, $m = \dfrac{1}{5}$, $b = \dfrac{7}{5}$

49. $y = \dfrac{8}{5}x - \dfrac{14}{5}$, $m = \dfrac{8}{5}$, $b = -\dfrac{14}{5}$

51. a. $y = -2x + 5$
b. $y = -2x$
c. $y = -2x - 3$

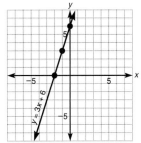

53. $y = 2x - 3$;
x-intercept $\left(\dfrac{3}{2},0\right)$;
y-intercept, $(0,-3)$

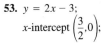

55. $3y - 2x = 6$;
x-intercept, $(-3,0)$;
y-intercept, $(0,2)$

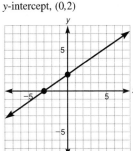

57.

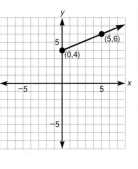

Solutions to Trial Exercise Problems

1. $y = 2x + 4$
Let $x = 0$,
then $y = 2(0) + 4 = 0 + 4 = 4$ $(0,4)$ is the y-intercept.
Let $y = 0$,
then $0 = 2x + 4$, $2x = -4$
$x = -2$ $(-2,0)$ is the x-intercept.

5. $2x + 3y = 6$
Let $x = 0$, then
$2(0) + 3y = 6$
$0 + 3y = 6$
$3y = 6$
$y = 2$ $(0,2)$ is the y-intercept.
Let $y = 0$, then
$2x + 3(0) = 6$
$2x + 0 = 6$
$2x = 6$
$x = 3$ $(3,0)$ is the x-intercept.

12. $y - 4x = 0$
Let $x = 0$, then
$y - 4(0) = 0$
$y - 0 = 0$
$y = 0$ $(0,0)$ is the y-intercept.
Let $y = 0$, then
$0 - 4x = 0$
$-4x = 0$
$x = 0$ $(0,0)$ is the x-intercept.

17. $y = 3x + 6$

x	y	
0	6	y-intercept
-2	0	x-intercept
-1	3	checkpoint

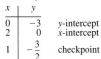

30. $3x - 2y = 6$

x	y	
0	-3	y-intercept
2	0	x-intercept
1	$-\dfrac{3}{2}$	checkpoint

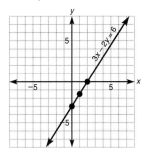

41. $y - 2x + 7 = 0$
$y - 2x = -7$ Add -7 to both sides.
$y = 2x - 7, m = 2, b = -7$ Add $2x$ to both sides.

43. $3y - 4x = 9$
$3y = 4x + 9$ Add $4x$ to both sides.
$y = \dfrac{4x + 9}{3}$ Divide both sides by 3.
$y = \dfrac{4}{3}x + 3, m = \dfrac{4}{3}, b = 3$ Divide by 3.

51. When $b = 5$, $y = -2x + 5$
When $b = 0$, $y = -2x$
When $b = -3$, $y = -2x - 3$

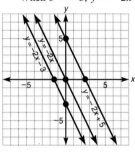

54. $x = y + 4$

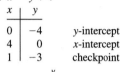

x	y	
0	-4	y-intercept
4	0	x-intercept
1	-3	checkpoint

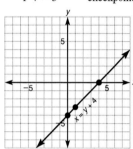

Review Exercises

1. $\dfrac{6}{5}$ **2.** $\dfrac{1}{x + 4}$ **3.** $\dfrac{x - 4}{x + 5}$ **4.** $-\dfrac{1}{5}$ **5.** $y = -3x + 6$ **6.** -4 or 4
7. 12.5% **8.** 33

Exercise 6–3

Answers to Odd-Numbered Problems

1. $m = -\dfrac{1}{2}$ **3.** $m = \dfrac{2}{3}$

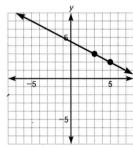

5. $m = 0$ **7.** slope is undefined.

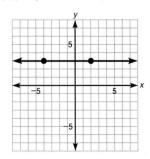

9. $m = \dfrac{2}{3}$ **11.** slope is undefined   **13.** $m = 1$ **15.** slope is undefined

17. $m = \dfrac{1}{7}$ **19.** $m = -\dfrac{3}{4}$ **21.** $m = \dfrac{10}{11}$

23. $(-5,3), (4,-2); m = -\dfrac{5}{9}$ **25.** $(3,5), (-2,-6); m = \dfrac{11}{5}$

27. $(6,5), (-3,-5); m = \dfrac{10}{9}$ **29.** $m = \dfrac{2}{3}$ **31.** $m = \dfrac{2}{3}$ **33.** $m = 4$

35. $m = \dfrac{15}{2}$

Solutions to Trial Exercise Problems

1. $(5,2), (3,3);$
$m = \dfrac{3 - 2}{3 - 5} = \dfrac{1}{-2} = -\dfrac{1}{2}$

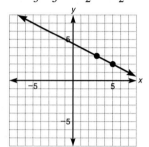

6. $(4,-4), (2,-4);$
$m = \dfrac{-4 - (-4)}{4 - 2} = \dfrac{0}{2} = 0$

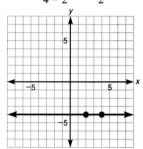

15. $(0,7), (0,-8); m = \dfrac{7 - (-8)}{0 - 0} = \dfrac{15}{0}$ slope is undefined

23. Using points $(-5,3)$ and $(4,-2)$, $m = \dfrac{3 - (-2)}{-5 - 4} = \dfrac{5}{-9} = -\dfrac{5}{9}$

29. $m = \dfrac{8}{12} = \dfrac{2}{3}$ **33.** $m = \dfrac{1,000}{250} = \dfrac{4}{1} = 4$

Review Exercises

1. -9 **2.** -13 **3.** $-\dfrac{1}{8}$ **4.** $\dfrac{b^5}{a^5}$ **5.** x **6.** 13 **7.** 55%
8. $y = -3x - 1$ **9.** $y = x - 2$

Exercise 6–4

Answers to Odd-Numbered Problems

1. $4x - y = 1$ **3.** $2x + y = -1$ **5.** $3x - 5y = -29$
7. $7x + 6y = 35$ **9.** $5x - 4y = 13$
11. $y = -3$ **13.** $x = 1$ **15.** $x - 2y = 0$
17. $3x + 8y = 7$ **19.** $x - y = -4$ **21.** $x + 7y = -6$
23. $8x - 3y = -24$ **25.** $8x + 5y = 0$
27. $y = -x + 2; m = -1; y$-intercept, 2
29. $y = -3x - 2; m = -3; y$-intercept, -2
31. $y = -\dfrac{2}{5}x + 2; m = -\dfrac{2}{5}; y$-intercept, 2
33. $y = \dfrac{7}{2}x + 2; m = \dfrac{7}{2}; y$-intercept, 2
35. $y = \dfrac{9}{2}x - 3; m = \dfrac{9}{2}; y$-intercept, -3
37. $y = \dfrac{8}{9}x - \dfrac{1}{9}; m = \dfrac{8}{9}; y$-intercept, $-\dfrac{1}{9}$

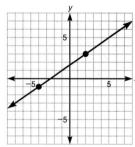

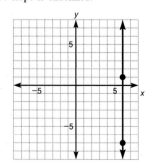

39. $m = 2; b = -4$

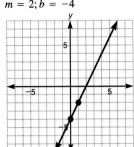

41. $m = -5; b = 2$

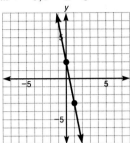

43. $m = \dfrac{2}{3}; b = -1$

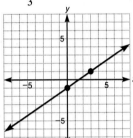

45. $m = -2; b = -3$

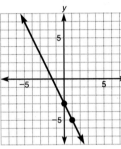

47. $m = \dfrac{5}{2}; b = -3$

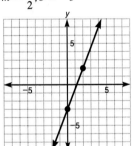

49. $m = -\dfrac{4}{3}; b = -3$

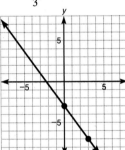

51.

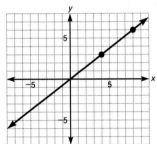

53.

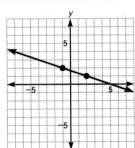

55.

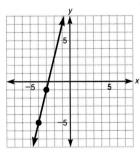

57.

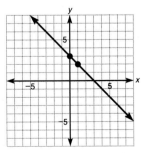

59.

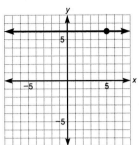

61.

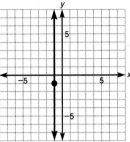

63. $y = 4x - 1$ **65.** $y = -\dfrac{5}{3}x + 2$ **67.** $y = -1$

69. $m_1 = -1, m_2 = -1$; parallel
71. $m_1 = -1, m_2 = 1$; perpendicular
73. $m_1 = 3, m_2 = -3$; neither

75. $m_1 = -5, m_2 = -\dfrac{10}{3}$; neither **77.** $m_1 = 4, m_2 = 4$; parallel

79. $m_1 = 0, m_2$ is undefined; perpendicular
81. $(-6,2), (7,-3); 5x + 13y = -4$ **83.** $(3,-2), (-4,-2); y = -2$
85. $(1,-4), (-6,4); 8x + 7y = -20$

Solutions to Trial Exercise Problems

1. $(1,3); m = 4$
Using $y - y_1 = m(x - x_1)$
$$y - 3 = 4(x - 1)$$
$$y - 3 = 4x - 4$$
$$y + 1 = 4x - 1$$
$$4x - y = 1$$

8. $(-3,0); m = -\dfrac{5}{8}$
$$y - 0 = -\dfrac{5}{8}[x - (-3)]$$
$$8y = -5(x + 3)$$
$$8y = -5x - 15$$
$$5x + 8y = -15$$

15. $(2,1)$ and $(6,3)$
$$m = \dfrac{3 - 1}{6 - 2} = \dfrac{2}{4} = \dfrac{1}{2}$$
Using $y - y_1 = m(x - x_1)$ and point $(2,1)$
$$y - 1 = \dfrac{1}{2}(x - 2)$$
$$2y - 2 = 1(x - 2)$$
$$x - 2y = 0$$

23. $(0,8)$ and $(-3,0)$
$$m = \dfrac{8 - 0}{0 - (-3)} = \dfrac{8}{3}$$
Using $y - y_1 = m(x - x_1)$ and point $(0,8)$
$$y - 8 = \dfrac{8}{3}(x - 0)$$
$$3y - 24 = 8x$$
$$8x - 3y = -24$$

29. $3x + y = -2$ $m = -3$
     $y = -3x - 2$ $b = -2$

31. $2x + 5y = 10$
$$5y = -2x + 10$$
$$y = \dfrac{-2x + 10}{5} \qquad m = -\dfrac{2}{5}$$
$$y = -\dfrac{2}{5}x + 2 \qquad b = 2$$

39. $y = 2x - 4$;
$m = 2$ and $b = -4$

$\left(\text{Note: } m = \dfrac{2}{1} = \dfrac{\text{rise}}{\text{run}}\right)$

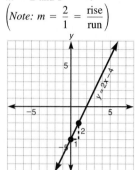

47. $5x - 2y = 6$
$\quad -2y = -5x + 6$
$\quad\quad y = \dfrac{-5x + 6}{-2}$
$\quad\quad y = \dfrac{5}{2}x - 3$

$m = \dfrac{5}{2}$ and $b = -3$

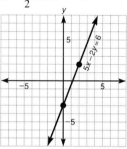

51. $(4,3); m = \dfrac{3}{4}$

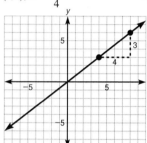

56. $(2,-3); m = 2 = \dfrac{2}{1}$

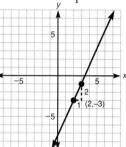

59. $(5,6); m = 0$

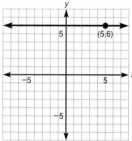

69. $x + y = 4$
$x + y = -7$
Then $y = -x + 4 \quad\quad m_1 = -1$
$\quad\quad y = -x - 7 \quad\quad m_2 = -1$
The slopes are equal, so
the lines are *parallel*.

74. $\quad x + 2y = 5$
$-6x + 3y = -1$
Now $x + 2y = 5 \quad\quad\quad -6x + 3y = -1$
$\quad\quad 2y = -x + 5 \quad\quad\quad 3y = 6x - 1$
$\quad\quad\quad y = -\dfrac{1}{2}x + \dfrac{5}{2} \quad\quad\quad y = 2x - \dfrac{1}{3}$

$m_1 = -\dfrac{1}{2} \quad\quad\quad\quad m_2 = 2$

The lines are *perpendicular* since $m_1 m_2 = -\dfrac{1}{2} \cdot 2 = -1$.

81. The line passes through points $(7,-3)$ and $(-6,2)$.

Then $m = \dfrac{2 - (-3)}{-6 - 7} = \dfrac{5}{-13} = \dfrac{-5}{13}$

and $y - (-3) = -\dfrac{5}{13}(x - 7)$

$\quad\quad y + 3 = -\dfrac{5}{13}(x - 7)$

$\quad 13y + 39 = -5x + 35$

$\quad 5x + 13y = -4$

Review Exercises

1. $0, -3, 9, \dfrac{9}{3}, \dfrac{0}{4}$ **2.** $\{-3\}$ **3.** $\left\{\dfrac{1}{7}\right\}$ **4.** $\{10\}$

5. $(3x - 2)(x + 2)$ **6.** does not factor   **7.** $8(y + 2x)(y - 2x)$

8. $41\dfrac{1}{2}$ ft and $58\dfrac{1}{2}$ ft   **9.** 100 pounds

Exercise 6–5
Answers to Odd-Numbered Problems

1.
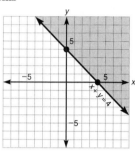

3.
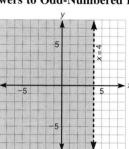

5.
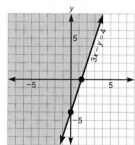

7.
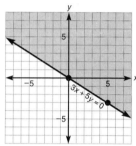

9.
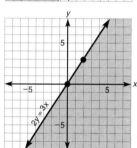

11.
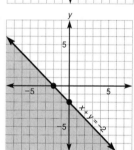

13.
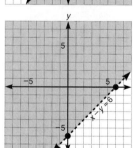

15.
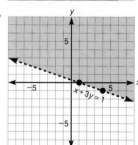

17.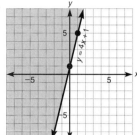

19.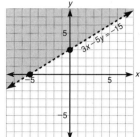

21.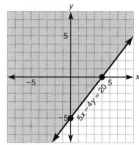

23.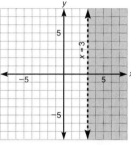

25.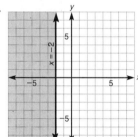

27.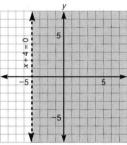

29.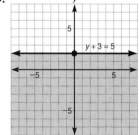

31.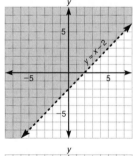

33.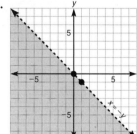

35.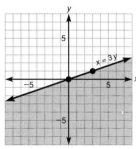

37.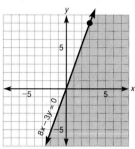

39.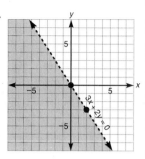

41. a. $x > 0$ and $y > 0$
b. $x < 0$ and $y > 0$
c. $x < 0$ and $y < 0$
d. $x > 0$ and $y < 0$

Solutions to Trial Exercise Problems

1. $x < 4$
Graph $x = 4$ in a dashed
line.
Using (0,0),
$0 < 4$ (true)
Shade half-plane
containing the origin.

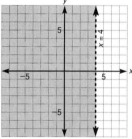

6. $2x - 4y > 8$
Graph $2x - 4y = 8$ in a
dashed line.
Using (0,0),
$2(0) - 4(0) > 8$
$0 - 0 > 8$
$0 > 8$ (false)
Shade half-plane that
does not contain the origin.

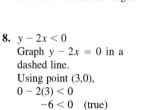

8. $y - 2x < 0$
Graph $y - 2x = 0$ in a
dashed line.
Using point (3,0),
$0 - 2(3) < 0$
$-6 < 0$ (true)
Shade half-plane
containing point (3,0).

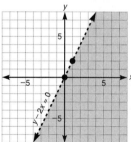

14. $x - y \geq -5$
Graph $x - y = -5$ in
a *solid* line.
Using (0,0),
$0 - 0 \geq -5$
$0 \geq -5$ (true)
Shade half-plane
containing the origin.

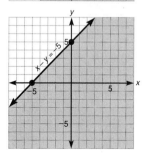

23. $x > 3$
Graph $x = 3$ in a dashed
line.
Using $(0,0)$,
$0 > 3$ (false)
Shade half-plane
not containing the origin.

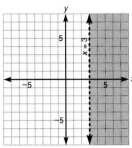

35. $x \geq 3y$
Graph $x = 3y$ in a solid
line.
Using point $(1,1)$,
$1 \geq 3$ (false)
Shade the half-plane that
does not contain the
point $(1,1)$.

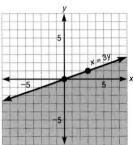

Review Exercises

1. all real numbers except 3   **2.** all real numbers except −5 and 2
3. all real numbers   **4.** $x + 3y = 7$ **5.** $3x - 4y = 8$ **6.** $x = -3$
7. $y = -3$ **8.** 11 and 44   **9.** $8\frac{2}{5}$ hours
10. 5 gallons of 2% milk and 15 gallons of skim milk

Exercise 6–6
Answers to Odd-Numbered Problems

1. yes, domain is $\{1,2,5,6\}$, range is $\{3,4,7,8\}$
3. yes, domain is $\left\{0,\frac{1}{2},3\right\}$, range is $\left\{0,\frac{1}{4},2\right\}$ **5.** no
7. yes, domain is $\{-2,-1,0,1\}$, range is $\{3\}$ **9.** no   **11.** yes
13. yes   **15.** no   **17.** no   **19.** yes   **21.** yes

23. a. $f(x) = x + 2$,
b. $\{(-3,-1),(-2,0),(-1,1),(0,2)\}$
c.

x	-3	-2	-1	0
y	-1	0	1	2

d.

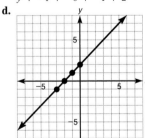

25. a. $f(x) = 2x - 2$,
b. $\{(-1,-4),(0,-2),(1,0),(2,2)\}$
c.

x	-1	0	1	2
y	-4	-2	0	2

d.

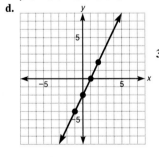

27. a. $f(x) = -2x + 4$,
b. $\{(0,4),(1,2),(2,0),(3,-2)\}$
c.

x	0	1	2	3
y	4	2	0	-2

d.

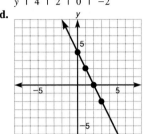

29. a. $f(x) = -\frac{2}{3}x + \frac{4}{3}$,
b. $\left\{\left(0,\frac{4}{3}\right),\left(1,\frac{2}{3}\right),(2,0),\left(3,-\frac{2}{3}\right)\right\}$
c.

x	0	1	2	3
y	$\frac{4}{3}$	$\frac{2}{3}$	0	$-\frac{2}{3}$

d.

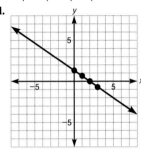

31. a. $f(x) = x$,
b. $\{(-1,-1),(0,0),(1,1),(2,2)\}$
c.

x	-1	0	1	2
y	-1	0	1	2

d.

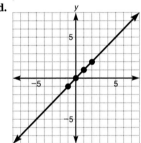

33.

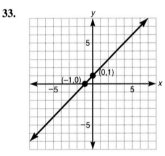

35.

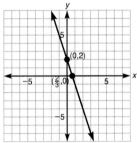

37.

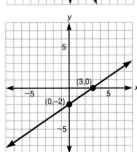

39.

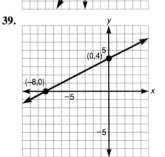

41.

43. $f(100) = 212$, $f(0) = 32$, $f(25) = 77$, $f(-10) = 14$
45. $h(2) = 55$, $h(3) = 78$, $h(4) = 101$
47. $g(2) = 0.14$, $g(3) = 0.21$, $g(1.2) = 0.084$

49. $g(N) = 0.7N, g(5) = 3.5, g(10) = 7, g(3) = 2.1$

51. $h(n) = \dfrac{n}{4} + 40, h(50) = 52.5, h(12) = 43, h(120) = 70$

53. $g(h) = 10.5h, g(40) = 420, g(48) = 504, g(28) = 294$

55. $f(25, 7) = 32, f(32, 11) = 43, f(146, 27) = 173$

Solutions to Trial Exercise Problems

5. No, it is not a function because the first component, -1, is repeated.

17. No, it is not a function because the graph is a vertical line and the value of x represented by the line is paired with every y-value.

29. a. $2x + 3y = 4; 3y = -2x + 4; y = -\dfrac{2}{3}x + \dfrac{4}{3};$

$$f(x) = -\dfrac{2}{3}x + \dfrac{4}{3}$$

b. $f(0) = -\dfrac{2}{3}(0) + \dfrac{4}{3} = \dfrac{4}{3}; f(1) = -\dfrac{2}{3}(1) + \dfrac{4}{3} = \dfrac{2}{3};$

$f(2) = -\dfrac{2}{3}(2) + \dfrac{4}{3} = 0; f(3) = -\dfrac{2}{3}(3) + \dfrac{4}{3} = -\dfrac{2}{3};$

$\left\{\left(0, \dfrac{4}{3}\right), \left(1, \dfrac{2}{3}\right), (2, 0), \left(3, -\dfrac{2}{3}\right)\right\}$

c.

x	0	1	2	3
y	$\dfrac{4}{3}$	$\dfrac{2}{3}$	0	$-\dfrac{2}{3}$

d.

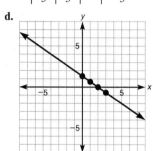

43. $f(C) = \dfrac{9}{5}C + 32$

$f(100) = \dfrac{9}{5}(100) + 32 = 9(20) + 32 = 180 + 32 = 212$

$f(0) = \dfrac{9}{5}(0) + 32 = 0 + 32 = 32$

$f(25) = \dfrac{9}{5}(25) + 32 = 9(5) + 32 = 45 + 32 = 77$

$f(-10) = \dfrac{9}{5}(-10) + 32 = 9(-2) + 32 = -18 + 32 = 14$

48. Using $F_f = \mu N$, where $\mu = 0.5, F_f = 0.5N$ and we define function h by $h(N) = 0.5N$. Then

$h(100) = 0.5(100) = 50$

$h(50) = 0.5(50) = 25$

$h(25) = 0.5(25) = 12.5$

Review Exercises

1. $y = -2x + 7$ **2.** $y = \dfrac{3x - 8}{4}$ **3.** $y = \dfrac{3}{5}$

4. Point of intersection is (2,0).

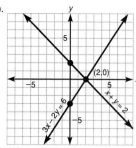

5. $x = 10$ **6.** $y = 2$ **7.** 26 teeth, 48 teeth

8. 8 ohms, 22 ohms   **9.** 4 ft, 8 ft

Exercise 6–7

Answers to Odd-Numbered Problems

1. O'Hare, approximately 48 million passengers

3. approximately 14 million passengers

5. Mexico, approximately 60 billion barrels

7. approximately 21 billion barrels   **9.** 51°F   **11.** 7° drop

13.

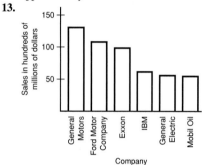

15.

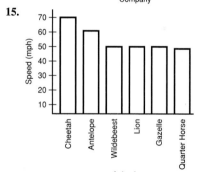

17.

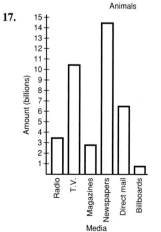

19.

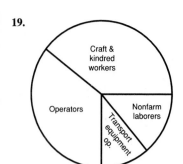

Blue-Collar Work Force cf 28,065,000

21.

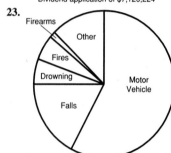

Dividend application of $7,120,224

23.

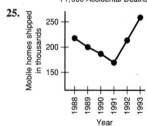

71,000 Accidental Deaths

25.

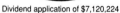

27.

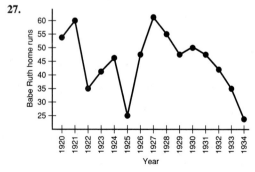

29.

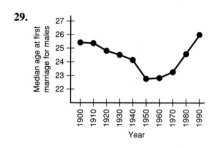

Solutions to Trial Exercise Problems

3. Reading from the vertical, Hartsfield International is approximately 42 million and Heathrow Airport is approximately 28 million. The difference between these values is approximately 14 million passengers.

7. Estimating from the circle graph, Mexico has approximately 60 billion barrels and the United States has approximately 39 billion barrels. The difference between these values is approximately 21 billion barrels.

11. The temperature at 4 PM was 51° F. The temperature at 6 PM was 44° F. The change was a 7° F drop in temperature.

Review Exercises

1. $x = 0$ **2.** $y = -1$ **3.** $y = -2x + 6$ **4.** $y = \dfrac{4x - 8}{3}$ **5.** 12

6. 14, 27, 41   **7.** 13, 18

Chapter 6 Review

1. $(-1,1), (0,4), (4,16)$

2. $(-2,-1), \left(0, \dfrac{1}{3}\right), (1,1)$

3. $(-7,-3), (0,-3), (5,-3)$ **4.** $(-3,15), (0,0), (3,-15)$

5. $(7,-2), (1,0), (-14,5)$ **6.** $\left(\dfrac{9}{4},-1\right), \left(\dfrac{7}{4},0\right), \left(\dfrac{1}{4},3\right)$

7. $(1,-8), (1,0), (1,2)$ **8.** $(-2,-3), (0,0), \left(\dfrac{2}{3},1\right)$

9. $(1,5)$
10. $(4,-4)$
11. $(-1,-6)$
12. $(0,-4)$
13. $\left(2,\dfrac{1}{2}\right)$
14. $\left(5,-\dfrac{2}{3}\right)$

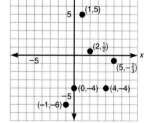

15. $(-1,-2)$ **16.** $(1,1)$ **17.** $(5,1)$ **18.** $(0,3)$ **19.** $(-2,1)$
20. $(3,-4)$

21. $(2,10), (0,4), \left(\frac{4}{3}, 8\right), \left(-\frac{2}{3}, 2\right)$　**22.** $(-2,9), \left(\frac{5}{2}, 0\right), (0,5), (3,-1)$

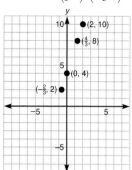

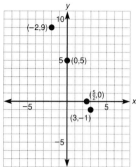

35.

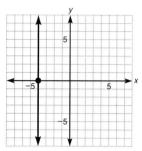

36.

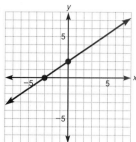

23. $(1,3), (-1,3), (0,2), (2,6)$　**24.** $(4,0), (3,-7), (0,-16), (-4,0)$

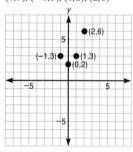

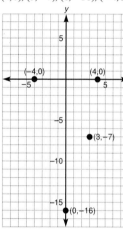

37.

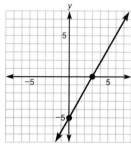

38. $m = -\frac{3}{5}$　**39.** m is undefined　**40.** $m = 0$　**41.** $m = 1$

42. $y = \frac{3}{4}x - 2; m = \frac{3}{4}; b = -2$

43. $y = -\frac{4}{3}x + \frac{2}{3}; m = -\frac{4}{3}; b = \frac{2}{3}$

44. $y = \frac{8}{3}x + \frac{1}{3}; m = \frac{8}{3}; b = \frac{1}{3}$

45. $y = 5x - 10; m = 5; b = -10$

46. $m = -1$ passing through $(0,-1)$ and $(-1,0)$; $x + y = -1$

47. $m = 1$ passing through $(2,0)$ and $(0,-2)$; $x - y = 2$

25. x-intercept, $\left(-\frac{5}{3}, 0\right)$; y-intercept, $(0,5)$　**26.** x-intercept, $(0,0)$;

y-intercept, $(0,0)$　**27.** x-intercept, (none); y-intercept, $(0,-2)$

28. x-intercept, $(6,0)$; y-intercept, (none)　**29.** x-intercept, $(2,0)$;

y-intercept, $\left(0, -\frac{4}{7}\right)$　**30.** x-intercept, $(0,0)$; y-intercept, $(0,0)$

48. $m = \frac{150}{1,050} = \frac{3}{21} = \frac{1}{7}$　**49.** $4x + y = 21$　**50.** $x - 3y = -12$

51. $x - 8y = -11$　**52.** $4x - 3y = -12$

53. $m = 3; b = 4$　**54.** $m = 2; b = -5$

31.

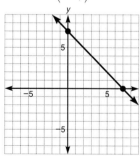

32.

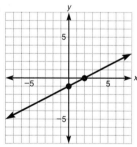

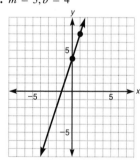

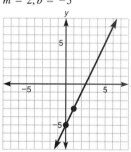

55. $m = \frac{3}{4}; b = 2$　**56.** $m = -\frac{4}{3}; b = 3$

33.

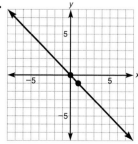

34.

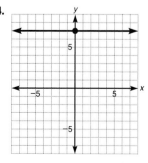

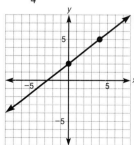

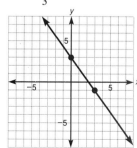

57.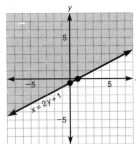

58.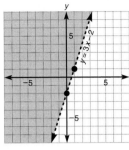

59.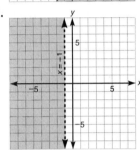

60.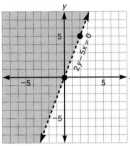

61.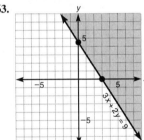

62.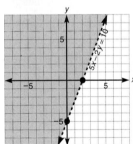

63.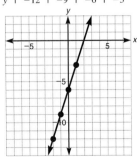

64.

70. a. $f(x) = 2x - 2$,
 b. $\{(-1,-4),(0,-2),(1,0),(2,2)\}$
 c.

x	-1	0	1	2
y	-4	-2	0	2

 d.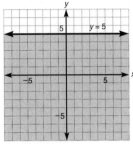

71. 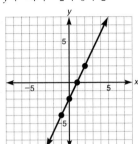 **72.**

73. $f(s) = 6s$, $f(3) = 18$, $f(14) = 84$, domain is positive real numbers

74. $17°$ C   **75.** $4°$ C   **76.** 800,000   **77.** 410,000   **78.** 1,860,000

79.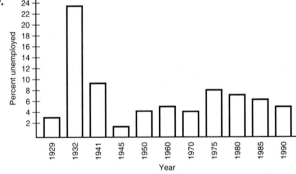

65. yes, domain is $\{2,3,4,5\}$, range is $\{4\}$
66. yes, domain is $\{2,3,4,5\}$, range is $\{2,3,4,5\}$ **67.** no   **68.** yes
69. a. $f(x) = 3x - 6$,
 b. $\{(-2,-12),(-1,-9),(0,-6),(1,-3)\}$
 c.

x	-2	-1	0	1
y	-12	-9	-6	-3

 d.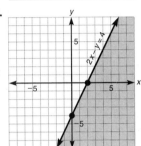

Chapter 6 Test

1. $-2, -1, 0$ [6–1]   **2.** x-intercept $(-6,0)$, y-intercept $(0,-2)$ [6–2]

3. $\dfrac{5}{4}$ [6–3]   **4.** Yes, domain is $\{-1,0,2,4\}$, range is $\{0,1,3\}$ [6–6]

5. $-\dfrac{1}{2}$, 2, perpendicular [6–4]   **6.** $\dfrac{2}{3}$, $-\dfrac{3}{2}$, perpendicular [6–4]

7. $y = 2x - 4$, $m = 2$, $b = -4$ [6–2]

8. $y = -\dfrac{3}{4}x + 3$, $m = -\dfrac{3}{4}$, $b = 3$ [6–2]   **9.** $y = 2x - 2$ [6–4]

10. $y = \dfrac{4}{5}x + \dfrac{7}{5}$ [6–4]   **11.** $y = -x + 4$ [6–4]

12. a. $(2,176)$ **b.** $\left(4\dfrac{1}{2},396\right)$ **c.** $(6,528)$ [6–1]   **13.** $\dfrac{9}{2}$ [6–3]

14. yes [6–6]   **15.** yes [6–6]

16. [6–2]

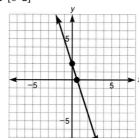

17. [6–2]

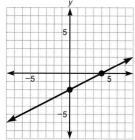

13. $-\dfrac{125}{a^9}$ [3–3] **14.** $3x^2 - 8x + 20 + \dfrac{-45}{x+2}$ [3–6]

15. x-intercept, $\left(\dfrac{7}{2},0\right)$; **16.** x-intercept, $(7,0)$;

y-intercept, $(0,-7)$ [6–2] y-intercept, none [6–2]

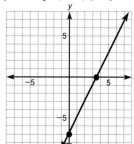

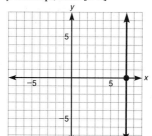

18. [6–6]

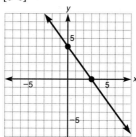

19. [6–5]

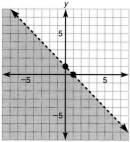

20. [6–6]

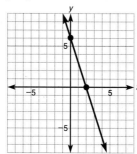

21. [6–5]

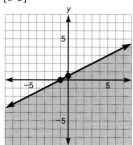

17. [6–5]

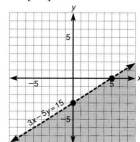

18. $m = \dfrac{7}{3}$ [6–3]
19. $3x - 4y = -13$ [6–4]
20. $3x + 5y = 27$ [6–4]
21. parallel [6–4]
22. neither [6–4]
23. 136 [2–7]
24. 12 [4–8]
25. legs are 5 inches and 12 inches [4–8]

22. [6–2]

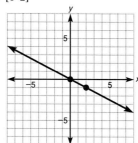

23. [6–2]

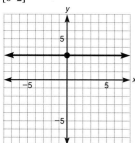

24. Army, 155 [6–7] **25.** Marines, 57 [6–7]

Chapter 6 Cumulative Test

1. 25 [1–9] **2.** $\left\{\dfrac{8}{3}\right\}$ [2–5] **3.** $\left\{\dfrac{60}{17}\right\}$ [6–6] **4.** $\left\{-\dfrac{5}{3}\right\}$ [2–5]

5. $\left\{-\dfrac{2}{3},2\right\}$ [6–6] **6.** $-1 < x \le 4$ [2–9]

7. $16y^2 - 24xy + 9x^2$ [3–2]

8. $\dfrac{5x^2 + 32x}{(x-7)(x+7)(x+6)}$ [5–3] **9.** $\dfrac{a^2 - 2a + 1}{(a+5)(a-2)}$ [5–3]

10. $\dfrac{6}{y+3}$ [5–2] **11.** $\dfrac{1}{x^9}$ [3–3] **12.** $\dfrac{1}{2^7} = \dfrac{1}{128}$ [3–3]

Chapter 7

Exercise 7–1

Answers to Odd-Numbered Problems

1. yes **3.** yes **5.** yes **7.** no **9.** yes **11.** no
13. solution: (3,2)

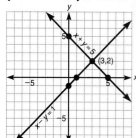

15. solution: $(3,-1)$

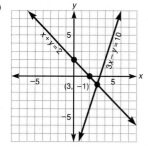

17. solution: $(-1, -5)$

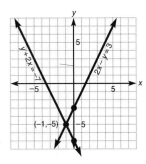

19. solution: $(2, 3)$

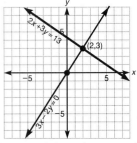

21. parallel lines: the system is inconsistent, no solution

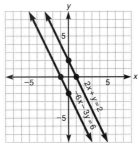

23. solution: $(-1, 0)$

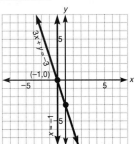

25. parallel lines: the system is inconsistent, no solution

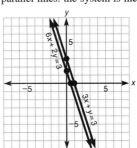

27. dependent: the solution is all ordered pairs satisfying the equation

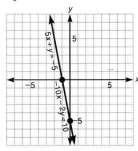

29. solution: $\left(\dfrac{12}{5}, \dfrac{6}{5}\right)$

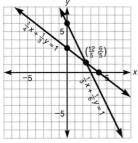

31. One that has no solution, when the graphs of the equations are parallel

33. The ordered pair that satisfies both equations. Graphically, it is the point of intersection.

Solutions to Trial Exercise Problems

5. $2x + y = 2$
$6x - y = 22 \qquad (3, -4)$
Let $x = 3$ and $y = -4$.
$2(3) + (-4) = 2 \qquad 6(3) - (-4) = 22$
$\qquad 6 - 4 = 2 \qquad\qquad 18 + 4 = 22$
$\qquad\qquad 2 = 2 \text{ (true)} \qquad 22 = 22 \text{ (true)}$
Therefore $(3, -4)$ is a solution.

10. $y = 4x - 3$
$y = -1 \qquad (-1, -1)$
Let $x = -1$ and $y = -1$.
$-1 = 4(-1) - 3 \qquad y = -1$
$-1 = -4 - 3 \qquad\quad -1 = -1 \text{ (true)}$
$-1 = -7 \text{ (false)}$
Therefore $(-1, -1)$ is *not* a solution.

15. $3x - y = 10$
$\quad x + y = 2$
Sketch the two lines. The two lines intersect at $(3, -1)$. So the solution is $(3, -1)$.

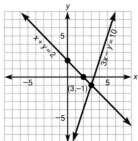

27. The system is *dependent* so all points on the line represent solutions. (*Note:* If we divide both sides of $-10x - 2y = 10$ by -2, we get equation $5x + y = -5$.)

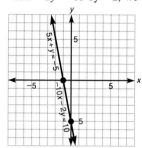

Review Exercises

1. $\left\{\dfrac{4}{5}\right\}$ **2.** $3x^2 - 10x + 30 + \dfrac{-89}{x+3}$ **3.** $y \le 5$ **4.** $-3 < x \le 3$

5. x^4 **6.** $x^5 - 2x^4 + x^3$ **7.** $4x^2 - 4xy + y^2$ **8.** 37 and 22

9. 10 pounds

Exercise 7–2

Answers to Odd-Numbered Problems

1. $\{(3,2)\}$ **3.** $\left\{\left(\dfrac{5}{3}, -5\right)\right\}$ **5.** inconsistent, $\varnothing$ **7.** $\left\{\left(\dfrac{5}{3}, 2\right)\right\}$

9. $\{(-2,3)\}$ **11.** $\{(-1,3)\}$ **13.** $\left\{\left(\dfrac{29}{10}, -\dfrac{3}{5}\right)\right\}$

15. dependent. The solution set is all the ordered pairs satisfying the equation.

17. $\{(1,-5)\}$

19. dependent. The solution set is all the ordered pairs satisfying the equation.

21. $\left\{\left(\dfrac{17}{21}, -\dfrac{11}{21}\right)\right\}$ **23.** $\left\{\left(\dfrac{22}{21}, \dfrac{13}{21}\right)\right\}$ **25.** $\{(5,-2)\}$ **27.** $\left\{\left(\dfrac{4}{5}, \dfrac{9}{5}\right)\right\}$

29. inconsistent, $\varnothing$ **31.** $\{(-3,7)\}$ **33.** $\{(7,3)\}$ **35.** $y = 5x$

37. $y = x + 20$ **39.** $x + y = 14{,}000$

41. y, because it is the variable whose opposite coefficients will be least in value

Solutions to Trial Exercise Problems

6. $\begin{aligned} -x + y &= 6 \\ x + 4y &= 4 \end{aligned}$ 　　　　 $\begin{aligned} -x + y &= 6 \\ x + 4y &= 4 \end{aligned}$

Add the equations. 　　　 $\begin{aligned} 5y &= 10 \\ y &= 2 \end{aligned}$

Substitute 2 for y in $-x + y = 6$. $-x + (2) = 6$

$-x = 4$ so $x = -4$

The solution is $(-4,2)$.

The solution set is $\{(-4,2)\}$.

10. $x - y = 3$ Multiply the first equation by 3.

$2x + 3y = 11$

$\begin{aligned} 3x - 3y &= 9 \\ 2x + 3y &= 11 \end{aligned}$

$5x \quad\;\; = 20$ Add the equations.

$x = 4$

Substitute 4 for x in $2x + 3y = 11$.

$2(4) + 3y = 11$

$8 + 3y = 11$

$3y = 3$

$y = 1$

The solution is $(4,1)$.

The solution set is $\{(4,1)\}$.

19. $\begin{aligned} -6x + 3y &= 9 \\ 2x - y &= -3 \text{ Multiply by 3.} \end{aligned}$ 　　 $\begin{aligned} -6x + 3y &= 9 \\ 6x - 3y &= -9 \end{aligned}$

Add: 　　　　　 $0 = 0$ (true)

dependent. The solution set is all ordered pairs satisfying the equation.

27. $\dfrac{1}{2}x + \dfrac{1}{3}y = 1$ Multiply by 6. → $3x + 2y = 6$

$\dfrac{2}{3}x - \dfrac{1}{4}y = \dfrac{1}{12}$ Multiply by 12. → $8x - 3y = 1$

(*Note:* We have just cleared denominators.)

$\begin{aligned} 3x + 2y &= 6 \text{ Multiply by 3.} \\ 8x - 3y &= 1 \text{ Multiply by 2.} \end{aligned}$ 　　 $\begin{aligned} 9x + 6y &= 18 \\ 16x - 6y &= 2 \end{aligned}$

Add: 　　　　 $25x \quad\;\; = 20$

$x = \dfrac{20}{25} = \dfrac{4}{5}$

Substitute $\dfrac{4}{5}$ for x in $\dfrac{1}{2}x + \dfrac{1}{3}y = 1$.

$\dfrac{1}{2}\left(\dfrac{4}{5}\right) + \dfrac{1}{3}y = 1$

$\dfrac{2}{5} + \dfrac{1}{3}y = 1$

$\dfrac{1}{3}y = \dfrac{3}{5}$

$y = \dfrac{9}{5}$

The solution is $\left(\dfrac{4}{5}, \dfrac{9}{5}\right)$.

The solution set is $\left\{\left(\dfrac{4}{5}, \dfrac{9}{5}\right)\right\}$.

34. $x + (0.4)y = 3.4$ Multiply by 10. 　 $10x + 4y = 34$

$(0.6)x - (1.4)y = 0.4$ Multiply by 10. 　 $6x - 14y = 4$

(*Note:* The coefficients are now integers.)

$10x + 4y = 34$ Multiply by 3. → 　 $30x + 12y = 102$

$6x - 14y = 4$ Multiply by -5. → 　 $-30x + 70y = -20$

$82y = 82$

$y = 1$

Substitute 1 for y in $x + 0.4y = 3.4$.

$x + (0.4)(1) = 3.4$

$x + 0.4 = 3.4$

$x = 3$

The solution is $(3,1)$. The solution set is $\{(3,1)\}$.

35. Since the length, y, is 5 times the width, x, then $y = 5x$.

38. Since each automobile travels 3 hours, then A travels $3x$ miles and B travels $3y$ miles. They are 500 miles apart after 3 hours, so $3x + 3y = 500$.

Review Exercises

1. $x = \dfrac{1}{3}$ **2.** $2x - 3y = -20$ **3.** $4x - 7y = -35$

4. $4(2x + y)(2x - y)$ **5.** $(3x - 2)^2$ **6.** $(5y + 4)(y - 2)$ **7.** $\{1\}$

8. 15 amperes and 30 amperes   **9.** $9\dfrac{1}{3}$ hours

Exercise 7–3

Answers to Odd-Numbered Problems

1. $\{(3,-1)\}$ **3.** $\{(-1,3)\}$ **5.** $\{(-5,-1)\}$ **7.** $\{(5,2)\}$

9. $\left\{\left(-\dfrac{5}{3}, -\dfrac{17}{3}\right)\right\}$ **11.** $\{(4,5)\}$ **13.** $\{(5,8)\}$ **15.** $\left\{\left(\dfrac{7}{5}, \dfrac{1}{5}\right)\right\}$

17. $\left\{\left(\dfrac{2}{13}, -\dfrac{14}{13}\right)\right\}$ **19.** inconsistent, $\varnothing$

21. dependent. The solution set is all the ordered pairs satisfying the equation.

23. $\{(-1,0)\}$ **25.** $\left\{\left(-\dfrac{11}{13}, -\dfrac{17}{13}\right)\right\}$ **27.** $\left\{\left(\dfrac{17}{20}, \dfrac{1}{10}\right)\right\}$ **29.** $\{(4,-2)\}$

31. dependent. The solution set is all the ordered pairs satisfying the equation.

33. $\left\{\left(\dfrac{13}{4}, \dfrac{7}{4}\right)\right\}$ **35.** inconsistent, $\varnothing$ **37.** $\left\{\left(5, \dfrac{21}{5}\right)\right\}$

39. $\left\{\left(\dfrac{28}{3}, 2\right)\right\}$ **41.** $\left\{\left(-\dfrac{7}{3}, -\dfrac{26}{3}\right)\right\}$ **43.** $\left\{\left(0, \dfrac{1}{2}\right)\right\}$

45. $\{(3,-4)\}$ **47.** $\left\{\left(\dfrac{12}{13}, -\dfrac{33}{13}\right)\right\}$ **49.** $\left\{\left(-\dfrac{76}{13}, \dfrac{60}{13}\right)\right\}$

51. $\left\{\left(\dfrac{20}{17}, -\dfrac{56}{17}\right)\right\}$ **53.** $x = 2y$ **55.** $x + y = 80$

57. Substitution, because there is an equation already solved for one variable in terms of the other variable.

Solutions to Trial Exercise Problems

3. $-2x + 5y = 17$
$y = 2x + 5$
Substitute $2x + 5$ for y in the first equation.
$-2x + 5(2x + 5) = 17$
$-2x + 10x + 25 = 17$
$8x + 25 = 17$
$8x = -8$
$x = -1$
Substitute -1 for x in $y = 2x + 5$.
$y = 2(-1) + 5$
$y = -2 + 5$
$y = 3$
The solution is $(-1,3)$.
The solution set is $\{(-1,3)\}$.

17. $5x - 3y = 4$
$x + 2y = -2$ $x = -2y - 2$
Substitute $-2y - 2$ for x in $5x - 3y = 4$.
$5(-2y - 2) - 3y = 4$
$-10y - 10 - 3y = 4$
$-13y - 10 = 4$
$-13y = 14$
$y = -\dfrac{14}{13}$
Substitute $-\dfrac{14}{13}$ for y in $x = -2y - 2$.
Then $x = -2\left(\dfrac{-14}{13}\right) - 2 = \dfrac{28}{13} - \dfrac{26}{13} = \dfrac{2}{13}$
The solution is $\left(\dfrac{2}{13}, \dfrac{-14}{13}\right)$.
The solution set is $\left\{\left(\dfrac{2}{13}, -\dfrac{14}{13}\right)\right\}$.

35. $-3x - 2y = 6$
$6x + 4y = 5$
Solving for y, $-3x - 2y = 6$. (Multiply by -1.)
$3x + 2y = -6$
$2y = -6 - 3x$
$y = -\dfrac{3}{2}x - 3$
Substitute $-\dfrac{3}{2}x - 3$ for y in $6x + 4y = 5$.
$6x + 4\left(-\dfrac{3}{2}x - 3\right) = 5$
$6x - 6x - 12 = 5$
$-12 = 5$ (false)
The system is *inconsistent* so there is no solution, $\varnothing$.

42. $y = 3x - 4$
$y = 3x + 5$
Then $3x - 4 = 3x + 5$
$-4 = 5$ (false)
The system is *inconsistent* so there is no solution, $\varnothing$.

45. $x - 3 = 0$
$3x + 2y = 1$
Solving $x - 3 = 0$ for x, we have $x = 3$.
Substitute 3 for x in $3x + 2y = 1$.
$3(3) + 2y = 1$
$9 + 2y = 1$
$2y = -8$
$y = -4$
The solution is $(3,-4)$.
The solution set is $\{(3,-4)\}$.

48. $\dfrac{2x}{3} - \dfrac{y}{2} = 1$ Multiply by 6. $4x - 3y = 6$

$3x + \dfrac{y}{4} = 2$ Multiply by 4. $12x + y = 8$

Now $4x - 3y = 6$ $4x - 3y = 6$
$12x + y = 8$ Multiply by 3. $\dfrac{36x + 3y = 24}{40x \qquad = 30}$
$x = \dfrac{30}{40} = \dfrac{3}{4}$

Substitute $\dfrac{3}{4}$ for x in the equation $3x + \dfrac{y}{4} = 2$.

Then $3\left(\dfrac{3}{4}\right) + \dfrac{y}{4} = 2$

$\dfrac{9}{4} + \dfrac{y}{4} = 2$ Multiply by 4.

$9 + y = 8$
$y = -1$
The solution is $\left(\dfrac{3}{4}, -1\right)$.

The solution set is $\left\{\left(\dfrac{3}{4}, -1\right)\right\}$.

53. The first current, x, is twice as much as the second current, y, so $x = 2y$.

Review Exercises

1. 16 ft and 26 ft   **2.** 4 yd and 9 yd   **3.** $b = \dfrac{2A - ch}{h}$ **4.** $\dfrac{2}{9}$

5. $\left\{\dfrac{5}{8}\right\}$ **6.** $-1 < x \le 1$

Exercise 7–4
Answers to Odd-Numbered Problems

1. $w = 4$ meters; $\ell = 14$ meters   **3.** $w = 14$ meters; $\ell = 36$ meters

5. $w = 11$ feet; $\ell = 20$ feet   **7.** \$2,500 at $3\frac{1}{2}$%; \$15,500 at 5%

9. \$18,200 at 5%; \$13,400 at 6%   **11.** \$4,400 at 7%; \$9,200 at 5%

13. 4 feet and 8 feet   **15.** 28 amperes and 52 amperes

17. 28 suits at \$125; 12 suits at \$185

19. 16 volts; 11 volts   **21.** 1st gear is 6 teeth; 2nd gear is 17 teeth

23. 12 ft by 29 ft   **25.** 8 ft by 24 ft   **27.** 345 chromium, 75 brass

29. 40 cc of 10% solution; 80 cc of 4% solution

31. 9 ounces of 80%; 3 ounces of 60%

33. 36 g of 60%, 54 g of 35%

35. $\dfrac{660}{7}$ centiliters of 3%; $\dfrac{40}{7}$ centiliters of 38%

37. 25 mph; 75 mph   **39.** Jane, $2\frac{1}{2}$ mph; Jim, 3 mph

41. wind, 69 mph; airplane in still air, 391 mph

43. $m = 3$, $b = -4$; $y = 3x - 4$

45. $m = -2$, $b = -1$; $y = -2x - 1$

47. $m = \dfrac{4}{5}$, $b = -4$; $y = \dfrac{4}{5}x - 4$

Solutions to Trial Exercise Problems

1. The perimeter P of a rectangle, by formula, is found by $P = 2\ell + 2w$, where ℓ is the length and w is the width of the rectangle. From "the perimeter of a rectangle is 36 meters," we get $2\ell + 2w = 36$. From "the length is 2 meters more than three times the width," we get $\ell = 3w + 2$.
We solve the system $2\ell + 2w = 36$
$$\ell = 3w + 2$$
Using substitution, substitute $3w + 2$ for ℓ in the first equation.

$2(3w + 2) + 2w = 36$	Then $\ell = 3w + 2$
$6w + 4 + 2w = 36$	$\ell = 3(4) + 2$
$8w + 4 = 36$	$\ell = 12 + 2$
$8w = 32$	$\ell = 14$
$w = 4$	

The solution of the system is $(\ell, w) = (14, 4)$. The rectangle has length $\ell = 14$ m and width $w = 4$ m.

6. Let x represent the amount invested at 8% interest and y be the amount invested at 6% interest. From "Phil has \$20,000, part of which he invests at 8% interest and the rest at 6%," we get $x + y = 20,000$. From "if his total income for 1 year from the two investments was \$1,460," we get $0.08x + 0.06y = 1,460$. Multiply by 100 to clear decimal points: $8x + 6y = 146,000$.
We solve the system $x + y = 20,000$
$$8x + 6y = 146,000$$
Multiply the first equation by -6. $(-6)x + (-6)y = -6(20,000)$

$$\begin{array}{r} -6x - 6y = -120,000 \\ \text{Add:} \quad 8x + 6y = 146,000 \\ \hline 2x = 26,000 \\ x = 13,000 \end{array}$$

Substitute 13,000 for x in $x + y = 20,000$: $13,000 + y = 20,000$, $y = 7,000$. Phil invested \$13,000 at 8% and \$7,000 at 6% interest.
Check: $(0.08)13,000 = \$1,040.00$
$$\underline{(0.06)7,000 = \$420.00}$$
Add to get income. $= \$1,460.00$

12. Let x represent the length of the longer piece of pipe and y be the length of the shorter piece of pipe. From "a piece of pipe is 19 feet long," we get $x + y = 19$. From "one piece is 5 feet longer than the other piece," we get $x = y + 5$.
Solve the system: $x + y = 19$
$$x = y + 5$$
Use substitution: Substitute $y + 5$ for x in $x + y = 19$.
$$(y + 5) + y = 19$$
$$2y + 5 = 19$$
$$2y = 14$$
$$y = 7$$
Substitute 7 for y in $x + y = 19$.
$$x + 7 = 19$$
$$x = 12$$
The pipe is cut into pieces 12 feet and 7 feet long.

21. Let x represent the number of teeth on the first gear and y be the number of teeth on the second gear. From "three times the number of teeth on a first gear is 1 more than the number of teeth on a second gear," we get $3x = y + 1$. From "five times the number of teeth on the first gear is 4 less than twice the number of teeth on the second gear," we get $5x = 2y - 4$.
Then $3x = y + 1$
$5x = 2y - 4$ is the system.
Rewrite the system.

$$\begin{array}{ll} 3x - y = 1 \text{ Multiply by } -2. \rightarrow & -6x + 2y = -2 \\ 5x - 2y = -4 & \underline{5x - 2y = -4} \\ \text{Add:} & -x = -6 \\ & x = 6 \end{array}$$

Substitute 6 for x in $3x = y + 1$. $3(6) = y + 1$
$$18 = y + 1$$
$$17 = y$$
The first gear has 6 teeth and the second gear has 17 teeth.

29. Let x represent the number of cubic centimeters (cc) of the 10% solution and y be the number of cc of the 4% solution.
Then $x + y = 120$ cc and $(0.10)x + (0.04)y = (0.06)120$.
Multiply the second equation by 100: $10x + 4y = 6(120)$, so $10x + 4y = 720$. Solve the system.

$$\begin{array}{ll} x + y = 120 \text{ Multiply by } -4. & -4x - 4y = -480 \\ 10x + 4y = 720 & \underline{10x + 4y = 720} \\ \text{Add:} & 6x = 240 \\ & x = 40 \end{array}$$

Substitute 40 for x in $x + y = 120$.
$$40 + y = 120$$
$$y = 80$$
Therefore the mechanic must have 40 cc of 10% solution and 80 cc of 4% solution.

37. Let x represent the speed of the faster car and y be the speed of the slower car. Then using "if they drive toward each other they will meet in 1 hour," $t = 1$ for each car, so $x + y = 100$. Using "if they drive in the same direction they will meet in 2 hours," $t = 2$ for each car and the faster car travels 100 miles farther, so $2x - 2y = 100$.
We now have the system $x + y = 100$
$$2x - 2y = 100$$
Multiply the first equation by 2.
$$2x + 2y = 200$$
$$\underline{2x - 2y = 100}$$
Add: $4x = 300$
$$x = 75$$
Then since $x + y = 100$, $75 + y = 100$, $y = 25$. Their speeds are 75 mph and 25 mph.

40. Let x represent the speed of the boat in still water and y be the speed of the current. Using $d = rt$ and (a) "a boat can travel 24 miles downstream in 2 hours," we get $2(x + y) = 24$, $2x + 2y = 24$; using $d = rt$ and (b) "and 16 miles upstream in the same length of time," we get $2(x - y) = 16$, $2x - 2y = 16$. We have the system of equations $2x + 2y = 24$

$$\underline{2x - 2y = 16}$$

Add: $4x \quad = 40$

$$x = 10$$

Then substituting 10 for x, $2(10) + 2y = 24$

$$20 + 2y = 24$$
$$2y = 4$$
$$y = 2$$

Therefore the boat travels at 10 mph in still water and the current is traveling at 2 mph.

43. Using $y = mx + b$, substitute
(a) 1 for x and -1 for y $\qquad -1 = (1)m + b$
(b) 2 for x and 2 for y $\qquad 2 = 2m + b$
Then $m + b = -1$ Multiply by -1. $\qquad -m - b = 1$
$\quad 2m + b = 2 \qquad\qquad\qquad\qquad \underline{2m + b = 2}$

$\qquad\qquad\qquad\qquad\qquad$ Add: $m \quad = 3$

Substitute 3 for m in $m + b = -1$.
$3 + b = -1$
$\quad b = -4$
The equation of the line is $y = 3x - 4$.

Review Exercises

1. $16x^4 y^6$ **2.** $\dfrac{y}{x^3}$ **3.** x^3 **4.** $x^2 - 5x + 4$ **5.** $12x^4 + 8x^3 - 12x^2$

6. $9y^2 - 4$ **7.** $16x^2 - 40xy + 25y^2$ **8.** $\{-1\}$ **9.** $\{-2,5\}$

Exercise 7–5
Answers to Odd-Numbered Problems

1.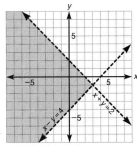

3.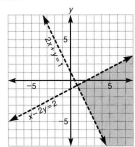

5.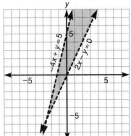

7.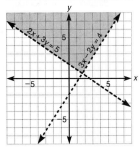

9.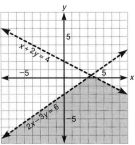

11.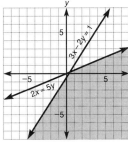

13.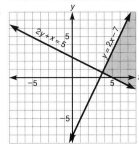

15.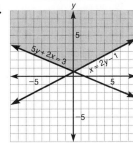

17.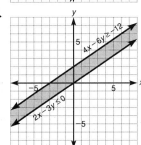

19.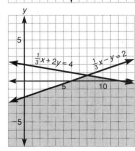

21.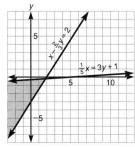

23.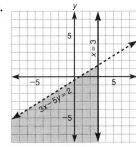

25.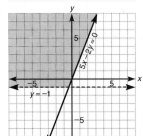

27.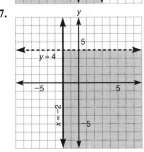

29.

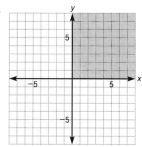

31. a. $x < 0$ and $y < 0$
b. $x > 0$ and $y < 0$

3. $(4,2)$

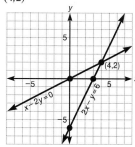

4. $(-2,6)$

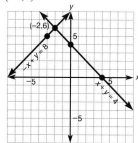

Solutions to Trial Exercise Problems

6. $5x - y < 5$
$\quad 2x + y \geq 1$

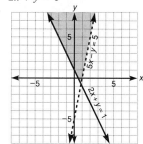

11. $3x - 2y \geq 1$
$\quad 2x \geq 5y$

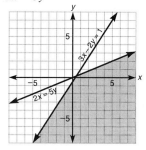

5. inconsistent

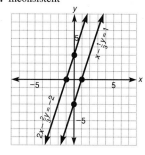

6. dependent; The solution is all the ordered pairs satisfying the equation.

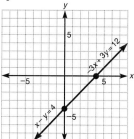

22. $\dfrac{3}{2}y - 4x \geq 2$

$\dfrac{1}{3}y > 1 - x$

(*Note:* Clear fractions.)
$3y - 8x \geq 4$
$y > 3 - 3x$

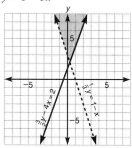

25. $5x - 2y \leq 0$
$\quad y > -1$

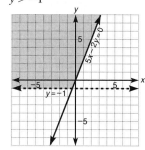

7. $\{(3,0)\}$ **8.** $\{(7,2)\}$ **9.** $\{(-1,1)\}$ **10.** $\{(1,-1)\}$ **11.** $\{(0,2)\}$
12. $\{(5,-3)\}$ **13.** $\{(5,2)\}$ **14.** inconsistent, $\varnothing$ **15.** $\{(2,1)\}$
16. dependent; The solution set is all the ordered pairs satisfying the equation.
17. $\{(5,3)\}$ **18.** $\{(5,-3)\}$ **19.** inconsistent, $\varnothing$
20. $\left\{\left(\dfrac{24}{11}, \dfrac{5}{11}\right)\right\}$ **21.** $\{(3,1)\}$ **22.** $\{(7,2)\}$
23. dependent; The solution set is all the ordered pairs satisfying the equation.
24. $\left\{\left(\dfrac{19}{11}, -\dfrac{17}{11}\right)\right\}$ **25.** 7 ft by 25 ft
26. \$11,000 at 14% profit; \$7,000 at 23% loss
27. 32 mph, 57 mph **28.** \$3,000 at $5\dfrac{1}{4}$%; \$5,200 at 10%
29. wind: 60 mph; plane: 300 mph
30. $m = 0$, $b = 2$. Equation of the line is $y = 2$.

31.

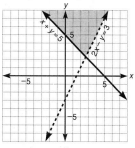

32.

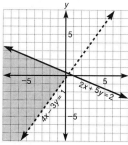

Review Exercises

1. $2x^2 + 3x + 7$ **2.** $4x^2 - 11x - 3$ **3.** $25y^2 - 20y + 4$
4. $49z^2 - 1$ **5.** a^4 **6.** xy^2 **7.** $-y^9$ **8.** $x(5x - 3)$
9. $(2x - 1)(2x + 5)$ **10.** $(3y - 5)^2$

Chapter 7 Review

1. $(3,1)$

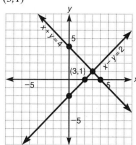

2. $(2,4)$

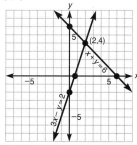

33.

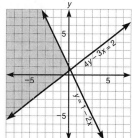

34.

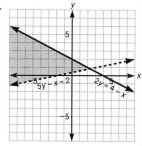

35.

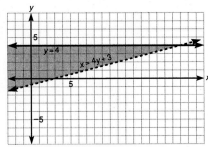

36.

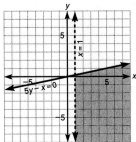

Chapter 7 Test

1. Solution: (2,3) [7–1]   **2.** Solution: (1,3) [7–1]

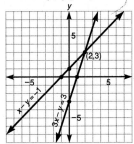   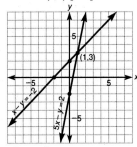

3. $\{(1,0)\}$ [7–2]   **4.** $\{(11,-7)\}$ [7–3]   **5.** $\{(1,-1)\}$ [7–3]

6. $\{(2,0)\}$ [7–3]   **7.** $\{(-2,10)\}$ [7–3]   **8.** $\left\{\left(\dfrac{1}{3},\dfrac{2}{3}\right)\right\}$ [7–3]

9. inconsistent, $\varnothing$ [7–3]   **10.** $\left\{\left(-3,-\dfrac{2}{5}\right)\right\}$ [7–3]

11. [7–5]   **12.** [7–5]

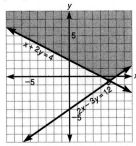   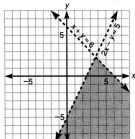

13. $12,000 at 7%, $8,000 at 5% [7–4]
14. boat: 10.5 mph; current: 1.5 mph [7–4]

15. $m = -\dfrac{1}{4}$, $b = 0$; equation of the line is $x + 4y = 0$ [7–4]

16. $11\dfrac{1}{2}$ in. by $6\dfrac{1}{2}$ in. [7–4]   **17.** $31\dfrac{1}{2}$ m by $28\dfrac{1}{2}$ m [7–4]

Chapter 7 Cumulative Test

1. $-\dfrac{1}{3}$ [1–9]   **2.** $\dfrac{y^2}{2x}$ [5–2]   **3.** $\dfrac{32-3x}{x(x-9)}$ [5–4]

4. $9x^2y - 14y^2$ [3–6]   **5.** $\dfrac{a-7}{7}$ [5–2]   **6.** $25x^2 - 30xy + 9y^2$ [3–2]

7. $2y + 6$ [3–6]   **8.** $\left\{\dfrac{15}{8}\right\}$ [5–6]   **9.** $\{-9,0\}$ [4–7]

10. $\left\{-\dfrac{11}{36}\right\}$ [5–6]   **11.** $\left\{-1,\dfrac{3}{7}\right\}$ [4–7]   **12.** $\{(4,2)\}$ [7–3]

13. $\{(1,-2)\}$ [7–3]   **14.** $\left\{\left(\dfrac{14}{17},\dfrac{1}{17}\right)\right\}$ [7–3]   **15.** inconsistent, $\varnothing$ [7–3]

16. $\left\{\left(\dfrac{5}{8},1\right)\right\}$ [7–3]   **17.** $\{(0,8)\}$ [7–3]

18. [7–5]

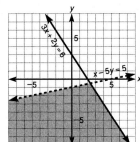

19. 42, 37 [2–7]   **20.** 5, 6 or $-6, -5$ [4–7]   **21.** 0 or 5 [4–7]

22. 1 or 2 [5–7]   **23.** $2\dfrac{2}{9}$ hours [5–7]   **24.** 7 in. by 13 in. [2–7]

25. 50 mph, 65 mph [7–4]

Chapter 8

Exercise 8–1
Answers to Odd-Numbered Problems

1. 10   **3.** 2   **5.** -12 **7.** -11 **9.** 11   **11.** -3 **13.** 4.243
15. 6.403   **17.** -7.211 **19.** 30   **21.** -6 **23.** 17.436   **25.** -77.460
27. 1.155   **29.** -0.839 **31.** 24 amperes   **33.** 13 units   **35.** 5 meters
37. 13 inches   **39.** 8 yards   **41.** 20 millimeters   **43.** 9.2 m
45. 8.7 ft   **47.** 9.2 cm   **49.** 17 ft   **51.** 20.1 ft   **53.** 2   **55.** 5
57. -2 **59.** 3   **61.** -3 **63.** 2   **65.** -3 **67.** 1   **69.** -1
71. 9 units   **73.** because no real number when squared is
equal to -16.

Solutions to Trial Exercise Problems

31. $I = \sqrt{\dfrac{\text{watts}}{\text{ohms}}} = \sqrt{\dfrac{(\ )}{(\ )}} = \sqrt{\dfrac{1,728}{3}} = \sqrt{576} = 24$

Answer: 24 amperes
53. $\sqrt[3]{8} = 2$, since $2 \cdot 2 \cdot 2 = 2^3 = 8$.
61. $-\sqrt[4]{81} = -3$, since $3 \cdot 3 \cdot 3 \cdot 3 = 3^4 = 81$ and the negative sign
indicates that we want the negative root.
67. $\sqrt[10]{1} = 1$, since $1^{10} = 1$. 1 raised to any power is 1 and the nth
root of 1 is 1 for any n.

Review Exercises

1. 3^2 **2.** $2^2 \cdot 3$ **3.** 2^3 **4.** $2^3 \cdot 5$ **5.** $2 \cdot 5^2$ **6.** 3^4 **7.** 2^6 **8.** 2^4
9. 12 teeth and 54 teeth   **10.** 30 ml of 60% and 90 ml of 80%

Exercise 8–2
Answers to Odd-Numbered Problems

1. 4   **3.** $2\sqrt{5}$ **5.** $3\sqrt{5}$ **7.** $4\sqrt{2}$ **9.** $4\sqrt{5}$ **11.** $7\sqrt{2}$ **13.** $a^3\sqrt{a}$
15. $2ab\sqrt{b}$ **17.** $3ab^2\sqrt{3ab}$ **19.** $3\sqrt{2}$ **21.** 15   **23.** $2\sqrt{15}$

25. $5\sqrt{15}$ **27.** $5\sqrt{3}$ **29.** $5x\sqrt{3}$ **31.** $4b\sqrt{3a}$ **33.** 13 feet
35. 24 mph **37.** $2\sqrt[5]{2}$ **39.** $2\sqrt[3]{3}$ **41.** $b^2\sqrt[3]{b^2}$ **43.** y^3
45. $b\sqrt[3]{4a^2}$ **47.** $2ab\sqrt[3]{2ab^2}$ **49.** $3ab^3\sqrt[3]{3a^2b^2}$ **51.** $b\sqrt[3]{b}$
53. a **55.** $3ab\sqrt[3]{2b}$ **57.** $3b\sqrt[4]{3a^3}$ **59.** $4a^5b^3\sqrt[3]{3b}$ **61.** 2 inches
63. 2 inches
65. because the square roots of negative numbers are not
real numbers.

Solutions to Trial Exercise Problems
15. $\sqrt{4a^2b^3} = \sqrt{4\cdot a^2\cdot b^2\cdot b} = \sqrt{4}\sqrt{a^2}\sqrt{b^2}\sqrt{b} = 2ab\sqrt{b}$
19. $\sqrt{6}\sqrt{3} = \sqrt{18} = \sqrt{9\cdot2} = \sqrt{9}\sqrt{2} = 3\sqrt{2}$
42. $\sqrt[3]{x^9} = \sqrt[3]{x^3\cdot x^3\cdot x^3} = \sqrt[3]{x^3}\sqrt[3]{x^3}\sqrt[3]{x^3} = x\cdot x\cdot x = x^3$
54. $\sqrt[3]{5a^2b}\sqrt[3]{75a^2b^2} = \sqrt[3]{375a^4b^3} = \sqrt[3]{125\cdot3\cdot a^3\cdot a\cdot b^3}$
$= \sqrt[3]{125}\sqrt[3]{a^3}\sqrt[3]{b^3}\sqrt[3]{3a} = 5ab\sqrt[3]{3a}$

61. $h = \sqrt[3]{\dfrac{12I}{b}} = \sqrt[3]{\dfrac{12(\ )}{(\ )}} = \sqrt[3]{\dfrac{12(2)}{(3)}} = \sqrt[3]{\dfrac{24}{3}}$
$= \sqrt[3]{8} = \sqrt[3]{2^3} = 2$ Answer: h = 2 inches

Review Exercises
1. $\dfrac{7}{8}$ **2.** $-4x^2y$ **3.** $\dfrac{2(y+5)}{y+1}$ **4.** 5 **5.** 3 **6.** 4 **7.** 6

8. x **9.** 20 liters **10.** $9\dfrac{1}{3}$ miles per hour

Exercise 8–3
Answers to Odd-Numbered Problems
1. $\dfrac{3}{5}$ **3.** $\dfrac{5}{7}$ **5.** $\dfrac{\sqrt{3}}{2}$ **7.** $\dfrac{8}{a}$ **9.** $\dfrac{\sqrt{2}}{2}$ **11.** $\dfrac{2\sqrt{7}}{7}$ **13.** $\dfrac{\sqrt{15}}{15}$
15. $\dfrac{2\sqrt{3}}{15}$ **17.** $\sqrt{2}$ **19.** $\dfrac{5\sqrt{2}}{2}$ **21.** $\sqrt{2}$ **23.** $3\sqrt{3}$ **25.** $\dfrac{\sqrt{10}}{2}$
27. $\dfrac{3\sqrt{10}}{2}$ **29.** $2\sqrt{2}$ **31.** $\dfrac{x\sqrt{y}}{y}$ **33.** $\dfrac{\sqrt{x}}{x}$ **35.** a^2
37. $\sqrt{41}$ meters **39.** 12 miles **41.** $\dfrac{1}{2}$ **43.** $\dfrac{3}{5}$ **45.** $\dfrac{a^2\sqrt[3]{3}}{b}$
47. $\dfrac{\sqrt[5]{a^4}}{b^2}$ **49.** $\dfrac{ab^2\sqrt[4]{bc}}{c^3}$ **51.** $\dfrac{\sqrt[5]{x^3y^2}}{z^3}$ **53.** $\dfrac{\sqrt[3]{20}}{5}$ **55.** $\dfrac{2\sqrt[4]{5}}{5}$
57. $\dfrac{x\sqrt[3]{y}}{y}$ **59.** $b\sqrt[3]{a}$ **61.** $\dfrac{a\sqrt[3]{bc^2}}{bc}$ **63.** $\dfrac{\sqrt[3]{a^2bc^2}}{bc}$ **65.** $\sqrt[3]{a^2b}$
67. 3 units
69. because the square roots of negative numbers are not
real numbers.

Solutions to Trial Exercise Problems
15. $\sqrt{\dfrac{4}{75}} = \dfrac{\sqrt{4}}{\sqrt{75}} = \dfrac{\sqrt{2^2}}{\sqrt{3\cdot5^2}} = \dfrac{2}{5\sqrt{3}}\cdot\dfrac{\sqrt{3}}{\sqrt{3}} = \dfrac{2\sqrt{3}}{5\cdot3} = \dfrac{2\sqrt{3}}{15}$

45. $\sqrt[3]{\dfrac{3a^6}{b^3}} = \dfrac{\sqrt[3]{3a^6}}{\sqrt[3]{b^3}} = \dfrac{a^2\sqrt[3]{3}}{b}$ **49.** $\sqrt[4]{\dfrac{a^4b^9}{c^{11}}} = \dfrac{\sqrt[4]{a^4b^9}}{\sqrt[4]{c^{11}}}$
$= \dfrac{ab^2\sqrt[4]{b}}{c^2\sqrt[4]{c^3}}\cdot\dfrac{\sqrt[4]{c^1}}{\sqrt[4]{c^1}} = \dfrac{ab^2\sqrt[4]{bc}}{c^2\sqrt[4]{c^4}} = \dfrac{ab^2\sqrt[4]{bc}}{c^2\cdot c} = \dfrac{ab^2\sqrt[4]{bc}}{c^3}$

55. $\sqrt[4]{\dfrac{16}{125}} = \dfrac{\sqrt[4]{16}}{\sqrt[4]{125}} = \dfrac{\sqrt[4]{2^4}}{\sqrt[4]{5^3}} = \dfrac{2}{\sqrt[4]{5^3}}\cdot\dfrac{\sqrt[4]{5^1}}{\sqrt[4]{5^1}} = \dfrac{2\sqrt[4]{5^1}}{\sqrt[4]{5^4}} = \dfrac{2\sqrt[4]{5}}{5}$

59. $\dfrac{ab}{\sqrt[3]{a^2}}\cdot\dfrac{\sqrt[3]{a^1}}{\sqrt[3]{a^1}} = \dfrac{ab\sqrt[3]{a}}{\sqrt[3]{a^3}} = \dfrac{ab\sqrt[3]{a}}{a} = b\sqrt[3]{a}$

61. $\sqrt[3]{\dfrac{a^3}{b^2c}} = \dfrac{\sqrt[3]{a^3}}{\sqrt[3]{b^2c}} = \dfrac{a}{\sqrt[3]{b^2c}}\cdot\dfrac{\sqrt[3]{b^1c^2}}{\sqrt[3]{b^1c^2}} = \dfrac{a\sqrt[3]{bc^2}}{\sqrt[3]{b^3c^3}} = \dfrac{a\sqrt[3]{bc^2}}{bc}$

67. $r = \sqrt[3]{\dfrac{3V}{4\pi}} = \sqrt[3]{\dfrac{3(\ )}{4(\ )}} = \sqrt[3]{\dfrac{3(113.04)}{4(3.14)}} = \sqrt[3]{\dfrac{339.12}{12.56}}$
$= \sqrt[3]{27} = \sqrt[3]{3^3} = 3$ Answer: 3 units

Review Exercises
1. $6x$ **2.** $4y$ **3.** $8ab$ **4.** $5xy$ **5.** x^2-9 **6.** x^2-y^2
7. $60\dfrac{1}{2}$ and $28\dfrac{1}{2}$ **8.** $4\dfrac{1}{2}$ days

Exercise 8–4
Answers to Odd-Numbered Problems
1. $9\sqrt{3}$ **3.** $10\sqrt{5}$ **5.** $6\sqrt{3}$ **7.** $3\sqrt{7}$ **9.** $3\sqrt{a}$ **11.** $8\sqrt{a}$
13. $7\sqrt{xy}$ **15.** $8\sqrt{a}+2\sqrt{ab}$ **17.** $4\sqrt{xy}+3\sqrt{y}$ **19.** $7\sqrt{2}$
21. $\sqrt{3}$ **23.** $17\sqrt{7}$ **25.** $7\sqrt{2}$ **27.** $2\sqrt{3}+8\sqrt{2}$ **29.** $7\sqrt{2a}$
31. $-\sqrt{x}$ **33.** $17\sqrt{2a}$ **35.** $2\sqrt{2a}+6\sqrt{3a}$ **37.** 14 units
39. 8.1 units by 8.1 units **41.** $6\sqrt[5]{2}$ **43.** $5\sqrt[3]{2}$
45. $3\sqrt[3]{3}+10\sqrt[3]{2}$ **47.** $5\sqrt[4]{x^3}$ **49.** $\sqrt[3]{x^2y}$ **51.** $2a\sqrt[3]{b^2}$
53. because the radicands are not the same.

Solutions to Trial Exercise Problems
14. $3\sqrt{x}+2\sqrt{y}-\sqrt{x} = 3\sqrt{x}-\sqrt{x}+2\sqrt{y} = 2\sqrt{x}+2\sqrt{y}$
18. $\sqrt{20}+3\sqrt{5} = \sqrt{2^2\cdot5}+3\sqrt{5} = 2\sqrt{5}+3\sqrt{5} = 5\sqrt{5}$
29. $\sqrt{50a}+\sqrt{8a} = \sqrt{2\cdot5^2\cdot a}+\sqrt{2^3\cdot a} = 5\sqrt{2a}+2\sqrt{2a} = 7\sqrt{2a}$
37. $h = b+s$, where s = 6 units and $b = \sqrt{c^2-s^2}$
$= \sqrt{(10)^2-(6)^2} = \sqrt{100-36} = \sqrt{64} = 8$. Then
$h = b+s = (8)+(6) = 14$. Answer: 14 units
43. $\sqrt[3]{16}+\sqrt[3]{54} = \sqrt[3]{2^4}+\sqrt[3]{2\cdot3^3} = 2\sqrt[3]{2}+3\sqrt[3]{2} = 5\sqrt[3]{2}$
50. $\sqrt[3]{x^6y}+2x^2\sqrt[3]{y} = x^2\sqrt[3]{y}+2x^2\sqrt[3]{y} = (1+2)x^2\sqrt[3]{y} = 3x^2\sqrt[3]{y}$

Review Exercises
1. $6x^2-3xy$ **2.** $2a^4-2a^2b^2$ **3.** x^2-2x+1 **4.** y^2-1 **5.** $4x^2-1$
6. $x^2+5xy+6y^2$ **7.** $x^2-2xy+y^2$ **8.** $a^2+4ab+4b^2$ **9.** 8 hours
10. 15 gallons of pure alcohol and 15 gallons of 40% alcohol

Exercise 8–5
Answers to Odd-Numbered Problems
1. $3\sqrt{2}+3\sqrt{3}$ **3.** $\sqrt{6}+\sqrt{14}$ **5.** $6\sqrt{6}-3\sqrt{22}$ **7.** $5\sqrt{3}-5\sqrt{2}$
9. $14\sqrt{5}-42\sqrt{2}$ **11.** $3a+\sqrt{ab}$ **13.** $14+7\sqrt{2}$
15. $12-25\sqrt{a}+12a$ **17.** 1 **19.** -2 **21.** $9+4\sqrt{5}$
23. $x+2\sqrt{xy}+y$ **25.** $x-y$ **27.** x^2y-z **29.** $4x+4y\sqrt{x}+y^2$
31. $11+\sqrt{3}$ **33.** $\sqrt{a}-3\sqrt{b}$ **35.** $\dfrac{-\sqrt{2}+3}{7}$ **37.** $\dfrac{-14+7\sqrt{7}}{3}$
39. $\sqrt{6}+\sqrt{3}$ **41.** $4\sqrt{3}+4$ **43.** $5\sqrt{6}+10$ **45.** $\dfrac{6\sqrt{3}+3\sqrt{5}}{7}$
47. $2-\sqrt{2}$ **49.** $\dfrac{5\sqrt{3}+3\sqrt{5}}{2}$ **51.** $\dfrac{-3-\sqrt{5}}{2}$
53. $\dfrac{a+2b\sqrt{a}+b^2}{a-b^2}$ **55.** $\dfrac{9-4\sqrt{3}}{11}$ **57.** $10+4\sqrt{5}$
59. The conjugate of a given expression is found by writing the
original expression and changing the sign of the second term.

Solutions to Trial Exercise Problems
5. $3\sqrt{2}(2\sqrt{3}-\sqrt{11}) = 3\sqrt{2}\cdot2\sqrt{3}-3\sqrt{2}\cdot\sqrt{11} = 6\sqrt{6}-3\sqrt{22}$
15. $(3-4\sqrt{a})(4-3\sqrt{a}) = 3\cdot4-3\cdot3\sqrt{a}-4\sqrt{a}\cdot4+4\sqrt{a}\cdot3\sqrt{a}$
$= 12-9\sqrt{a}-16\sqrt{a}+12\cdot a = 12-25\sqrt{a}+12a$

17. $(\sqrt{3} + \sqrt{2})(\sqrt{3} - \sqrt{2})$ Conjugates, therefore $= (\sqrt{3})^2 - (\sqrt{2})^2$
$= 3 - 2 = 1$

21. $(2 + \sqrt{5})^2 = (2 + \sqrt{5})(2 + \sqrt{5}) = 2 \cdot 2 + 2\sqrt{5} + 2\sqrt{5} + \sqrt{5}\sqrt{5}$
$= 4 + 4\sqrt{5} + 5 = 9 + 4\sqrt{5}$

26. $(2\sqrt{a} - \sqrt{b})(2\sqrt{a} + \sqrt{b})$ Conjugates, therefore
$= (2\sqrt{a})^2 - (\sqrt{b})^2 = 2^2(\sqrt{a})^2 - b = 4a - b$

38. $\dfrac{6}{3 - \sqrt{6}} = \dfrac{6}{3 - \sqrt{6}} \cdot \dfrac{3 + \sqrt{6}}{3 + \sqrt{6}} = \dfrac{6(3 + \sqrt{6})}{(3)^2 - (\sqrt{6})^2} = \dfrac{6(3 + \sqrt{6})}{9 - 6}$
$= \dfrac{6(3 + \sqrt{6})}{3} = 2(3 + \sqrt{6}) = 6 + 2\sqrt{6}$

51. $\dfrac{1 + \sqrt{5}}{1 - \sqrt{5}} = \dfrac{1 + \sqrt{5}}{1 - \sqrt{5}} \cdot \dfrac{1 + \sqrt{5}}{1 + \sqrt{5}} = \dfrac{(1)^2 + \sqrt{5} + \sqrt{5} + (\sqrt{5})^2}{(1)^2 - (\sqrt{5})^2}$
$= \dfrac{1 + 2\sqrt{5} + 5}{1 - 5} = \dfrac{6 + 2\sqrt{5}}{-4} = \dfrac{2(3 + \sqrt{5})}{-4} = \dfrac{-(3 + \sqrt{5})}{2}$
$= \dfrac{-3 - \sqrt{5}}{2}$

Review Exercises

1. 64 **2.** 32 **3.** 9 **4.** $\dfrac{1}{27}$ **5.** $\dfrac{1}{x^6}$ **6.** $8a^6b^3$ **7.** $\dfrac{y^3}{x}$ **8.** $\dfrac{1}{x^9}$

9. commuter train: 45 mph, express train: 75 mph
10. 20 liters of pure acid and 80 liters of 25% acid

Exercise 8–6

Answers to Odd-Numbered Problems

1. 6 **3.** a^2 **5.** 2 **7.** -3 **9.** 9 **11.** 27 **13.** $\dfrac{1}{5}$ **15.** $\dfrac{1}{8}$ **17.** $-\dfrac{1}{2}$

19. 4 **21.** $2^{5/6}$ **23.** x **25.** $c^{3/4}$ **27.** 2 **29.** $2^{1/6}$ **31.** $a^{3/5}$

33. $y^{1/6}$ **35.** $a^{1/3}$ **37.** $x^{2/3}$ **39.** $c^{1/6}$ **41.** $a^{1/3}$ **43.** $\dfrac{1}{x^{1/6}} = \dfrac{x^{5/6}}{x}$

45. $\dfrac{1}{y^{1/4}} = \dfrac{y^{3/4}}{y}$ **47.** $8a^3$ **49.** $4a^4b^2$ **51.** $b^{1/2}c^{1/4}$ **53.** $a^{1/2}b^{2/3}$

55. 8.133 **57.** 4.144 **59.** 38.349 **61.** 3.942 **63.** 81
65. 49 mph **67.** 24 miles
69. Take the square root 3 times, that is, the square root of the square root of the square root is the eighth root.

Solutions to Trial Exercise Problems

3. $(a^6)^{1/3} = a^{6 \cdot 1/3} = a^2$ **7.** $(-27)^{1/3} = \sqrt[3]{(-27)} = -3$
9. $(27)^{2/3} = (\sqrt[3]{27})^2 = (3)^2 = 9$

17. $(-8)^{-1/3} = \dfrac{1}{(-8)^{1/3}} = \dfrac{1}{\sqrt[3]{-8}} = \dfrac{1}{-2} = -\dfrac{1}{2}$

35. $(a^{2/3})^{1/2} = a^{2/3 \cdot 1/2} = a^{1/3}$ **39.** $(c^{-1/4})^{-2/3} = c^{(-1/4) \cdot (-2/3)} = c^{1/6}$

51. $\dfrac{b^{3/4}c^{1/2}}{b^{1/4}c^{1/4}} = b^{3/4 - 1/4}c^{1/2 - 1/4} = b^{2/4}c^{2/4 - 1/4} = b^{1/2}c^{1/4}$

Review Exercises

1. 7 **2.** x **3.** $x + 1$ **4.** $x^2 + 2x + 1$ **5.** $x^2 - 4x + 4$ **6.** $\{-2,3\}$
7. $\{2,8\}$ **8.** $\{-1,0\}$ **9.** 62 volts and 94 volts **10.** 7

Exercise 8–7

Answers to Odd-Numbered Problems

1. $\{16\}$ **3.** $\{81\}$ **5.** $\{11\}$ **7.** $\{43\}$ **9.** $\{12\}$ **11.** $\{5\}$ **13.** $\{9\}$
15. $\{36\}$ **17.** $\varnothing$ **19.** $\varnothing$ **21.** $\{4\}$ **23.** $\{4\}$ **25.** $\{3\}$ **27.** $\{2\}$

29. $\{5\}$ **31.** $\{3\}$ **33.** $\{3\}$ **35.** $\varnothing$ **37.** $\{7\}$ **39.** $\varnothing$ **41.** $\left\{-\dfrac{3}{4}\right\}$

43. $\varnothing$ **45.** $\{3\}$ **47.** $\{4\}$ **49.** $\{8\}$ **51.** $\{12\}$ **53.** $\{2\}$ **55.** 19
57. 24 **59.** 4 **61.** 5 **63.** 24 feet **65.** 108 feet **67.** 100
69. because there can be extraneous solutions.

Solutions to Trial Exercise Problems

17. $\sqrt{x} + 7 = 5$
$\sqrt{x} = -2$
$(\sqrt{x})^2 = (-2)^2$
$x = 4$
Check:
$\sqrt{4} + 7 = 5$
$2 + 7 = 5$
$9 = 5$ (false)
Therefore no solution and the solution set is $\varnothing$.

21. $\sqrt{2x + 1} = \sqrt{x + 5}$
$(\sqrt{2x + 1})^2 = (\sqrt{x + 5})^2$
$2x + 1 = x + 5$
$x + 1 = 5$
$x = 4$
Check:
$\sqrt{2(4) + 1} = \sqrt{(4) + 5}$
$\sqrt{8 + 1} = \sqrt{9}$
$\sqrt{9} = \sqrt{9}$
$3 = 3$ (true)
The solution set is $\{4\}$

41. $\sqrt{x^2 + 1} = x + 2$
$(\sqrt{x^2 + 1})^2 = (x + 2)^2$
$x^2 + 1 = (x + 2)(x + 2)$
$x^2 + 1 = x^2 + 2x + 2x + 4$
$x^2 + 1 = x^2 + 4x + 4$
$1 = 4x + 4$
$-3 = 4x$
$-\dfrac{3}{4} = x$

Check:
$\sqrt{\left(-\dfrac{3}{4}\right)^2 + 1} = \left(-\dfrac{3}{4}\right) + 2$

$\sqrt{\dfrac{9}{16} + 1} = -\dfrac{3}{4} + \dfrac{8}{4}$

$\sqrt{\dfrac{9}{16} + \dfrac{16}{16}} = \dfrac{5}{4}$

$\sqrt{\dfrac{25}{16}} = \dfrac{5}{4}$

$\dfrac{\sqrt{25}}{\sqrt{16}} = \dfrac{5}{4}$

$\dfrac{5}{4} = \dfrac{5}{4}$ (true)

The solution set is $\left\{-\dfrac{3}{4}\right\}$

45. $\sqrt{x + 6} = x$
$(\sqrt{x + 6})^2 = (x)^2$
$x + 6 = x^2$
$0 = x^2 - x - 6$
$0 = (x - 3)(x + 2)$
$x - 3 = 0$ or $x + 2 = 0$
$x = 3$ or $x = -2$
Check:
$\sqrt{(3) + 6} = (3)$
$\sqrt{9} = 3$
$3 = 3$ (true)
$\sqrt{(-2) + 6} = (-2)$
$\sqrt{4} = -2$
$2 = -2$ (false)
The solution set is $\{3\}$

49. $6 + \sqrt{x - 4} = x$
$\sqrt{x - 4} = x - 6$
$(\sqrt{x - 4})^2 = (x - 6)^2$
$x - 4 = (x - 6)(x - 6)$
$x - 4 = x^2 - 6x - 6x + 36$
$x - 4 = x^2 - 12x + 36$
$0 = x^2 - 13x + 40$
$0 = (x - 8)(x - 5)$
$x - 8 = 0$ or $x - 5 = 0$
$x = 8$ or $x = 5$
Check :
$6 + \sqrt{(8) - 4} = (8)$
$6 + \sqrt{4} = 8$
$8 = 8$ (true)
$6 + \sqrt{(5) - 4} = (5)$
$6 + \sqrt{1} = 5$
$7 = 5$ (false)
The solution set is $\{8\}$

59. Let x represent the number.
$\sqrt{x + 12} = x$
$(\sqrt{x + 12})^2 = (x)^2$
$x + 12 = x^2$
$0 = x^2 - x - 12$
$0 = (x - 4)(x + 3)$
$x - 4 = 0$ or $x + 3 = 0$
$x = 4$ or $x = -3$
Check :
$\sqrt{(4) + 12} = (4)$
$\sqrt{16} = 4$
$4 = 4$ (true)
$\sqrt{(-3) + 12} = (-3)$
$9 = -3$
$3 = -3$ (false)
Hence the number is 4.

Review Exercises

1. $(x + 2)(x - 2)$ **2.** $(x + 3)(x + 6)$ **3.** $(x + 2)(x - 5)$ **4.** $(x - 3)^2$
5. 9 **6.** 7 **7.** 11 **8.** $\{-8,8\}$

Chapter 8 Review

1. 9 **2.** 3.464 **3.** -5.292 **4.** 11.619 **5.** 6.4 in. **6.** 12 cm
7. 17.3 m **8.** $2\sqrt{10}$ **9.** $3ab\sqrt{2b}$ **10.** $2\sqrt{7}$ **11.** $6\sqrt{5}$
12. $15\sqrt{2}$ in. ≈ 21 in. **13.** $\dfrac{4\sqrt{17}}{17}$ **14.** $\dfrac{\sqrt{14}}{6}$ **15.** $\dfrac{\sqrt{ab}}{b}$
16. $\dfrac{\sqrt{xy}}{y^2}$ **17.** $\dfrac{\sqrt{ab}}{b}$ **18.** $\dfrac{2\sqrt{xy}}{y}$ **19.** 16.1 ft **20.** $7\sqrt{7}$
21. $8\sqrt{2}$ **22.** $3\sqrt{5}$ **23.** $24\sqrt{3}$ **24.** $\sqrt{2a}$ **25.** $17\sqrt{x}$
26. $\sqrt{15} - \sqrt{21}$ **27.** $2\sqrt{35} + 2\sqrt{15}$ **28.** $8 - 2\sqrt{7}$
29. $39 - 12\sqrt{3}$ **30.** $8 + 2\sqrt{15}$ **31.** $4a - b$ **32.** $-\sqrt{3} - 2$
33. $\dfrac{-\sqrt{6} + 4}{5}$ **34.** $\dfrac{\sqrt{a} - b}{a - b^2}$ **35.** $\dfrac{\sqrt{xy} - x}{y - x}$ **36.** $\dfrac{a^2 - a\sqrt{b}}{a^2 - b}$
37. $\dfrac{-11 - 6\sqrt{2}}{7}$ **38.** 6 **39.** 4 **40.** -2 **41.** $\dfrac{1}{4}$ **42.** a
43. $b^{13/12}$ **44.** $a^{1/4}$ **45.** $a^{9/8}$ **46.** $8a^3b^6$ **47.** $a^{3/2}b^{1/2}$ **48.** 3.557
49. 2.872 **50.** 9.518 **51.** 2.321 **52.** 28.3 mi **53.** $\{64\}$
54. $\{53\}$ **55.** $\{4\}$ **56.** $\{4\}$ **57.** $\varnothing$ **58.** $\{3\}$ **59.** $\{-2\}$ **60.** $\{3,4\}$
61. $104\dfrac{1}{6}$ ft **62.** $208\dfrac{1}{3}$ ft

Chapter 8 Test

1. 17.889 [8–1] **2.** -10 [8–1] **3.** 12 [8–1] **4.** $2\sqrt{15}$ [8–2]
5. $5\sqrt{3}$ [8–4] **6.** $5\sqrt{2} - 2\sqrt{5}$ [8–5] **7.** $7\sqrt{2}$ [8–2]
8. $3a^3b^2c^2 \sqrt[3]{b^2c}$ [8–2] **9.** $2\sqrt{3}$ [8–3] **10.** $6\sqrt{2x}$ [8–4]
11. $\dfrac{\sqrt[3]{ab}}{b}$ [8–3] **12.** $a^{1/2}$ [8–6] **13.** $a\sqrt[5]{a^2}$ [8–2] **14.** $\dfrac{6\sqrt{10}}{5}$ [8–3]
15. $x^{1/6}$ [8–6] **16.** $9 + 6\sqrt{x} + x$ [8–5] **17.** $\sqrt[3]{2}$ [8–4]
18. $xy\sqrt[5]{y^3}$ [8–3] **19.** $\dfrac{8 + 2\sqrt{2}}{7}$ [8–5] **20.** $x - 2y$ [8–5]
21. $a^{1/2}b^{1/4}$ [8–6] **22.** $\left\{\dfrac{9}{2}\right\}$ [8–7] **23.** $\{2\}$ [8–7] **24.** $\{3,4\}$ [8–7]
25. $\varnothing$ [8–7] **26.** $c = 10$ ft [8–1] **27.** $b \approx 8.7$ cm [8–1]
28. 225 meters [8–7] **29.** 22.6 feet [8–1] **30.** 127 feet [8–1]

Chapter 8 Cumulative Test

1. 45 [8–1] **2.** $2a^3b^2$ [3–3] **3.** $16\sqrt{3}$ [8–4] **4.** -2 [8–1]
5. $\dfrac{x + 3}{x - 1}$ [5–4] **6.** $\dfrac{x\sqrt{x} + \sqrt{xy}}{x^2 - y}$ [8–5] **7.** $3xy^2 \sqrt[3]{3xz}$ [8–2]
8. $(5c + d)(5c - d)$ [4–4] **9.** $(2x - 1)(x + 4)$ [4–3]
10. $(2x + 1)(3x + 4)$ [4–3] **11.** $\left\{-\dfrac{1}{3}\right\}$ [2–5]
12. $\{-3,3\}$ [4–7] **13.** $\{-6\}$ [2–5] **14.** $\{12\}$ [2–5]
15. $\left\{\dfrac{19}{12}\right\}$ [5–6] **16.** $\left\{-1,-\dfrac{1}{2}\right\}$ [4–7] **17.** $1 < x < 8$ [2–9]
18. $x > \dfrac{7}{2}$ [2–9] **19.** -3 [6–3]
20. $y = 4x - 2$; slope is 4; y-intercept is $(0,-2)$ [6–4]
21. $\left\{\left(\dfrac{7}{8},\dfrac{3}{8}\right)\right\}$ [7–2] **22.** 14, 56 [2–7] **23.** 1,350 [2–8]
24. 16, 18 [4–7] **25.** 21 feet by 27 feet [2–7]

Chapter 9

Exercise 9–1
Answers to Odd-Numbered Problems

1. $\{-5,3\}$ **3.** $\left\{-\dfrac{3}{2},2\right\}$ **5.** $\{-2,2\}$ **7.** $\{-8,8\}$
9. $\{-\sqrt{11}, \sqrt{11}\}$ **11.** $\{-2\sqrt{5},2\sqrt{5}\}$ **13.** $\{-\sqrt{3}, \sqrt{3}\}$
15. $\{-4\sqrt{2},4\sqrt{2}\}$ **17.** no real number solution, $\varnothing$ **19.** $\{-\sqrt{6}, \sqrt{6}\}$
21. $\{-5\sqrt{2},5\sqrt{2}\}$ **23.** $\{-2\sqrt{2},2\sqrt{2}\}$ **25.** $\{-2\sqrt{2},2\sqrt{2}\}$
27. $\{-\sqrt{2}, \sqrt{2}\}$ **29.** $\left\{-\dfrac{\sqrt{6}}{5}, \dfrac{\sqrt{6}}{5}\right\}$ **31.** $\{-\sqrt{11}, \sqrt{11}\}$
33. $\{-4,0\}$ **35.** $\{-1,9\}$ **37.** $\{-3 - \sqrt{6},-3 + \sqrt{6}\}$
39. $\{9 - 3\sqrt{2},9 + 3\sqrt{2}\}$ **41.** $\{-5 - 4\sqrt{2},-5 + 4\sqrt{2}\}$
43. $\{-2.4,2.4\}$ **45.** $\{-1.9,1.9\}$ **47.** $\{-1.6,4.2\}$
49. $\{-2.3,1.9\}$ **51.** 5 meters **53.** 2 feet **55.** 9 or -9 **57.** 0 or 9
59. 7 inches, 14 inches
61. 24 meters **63.** 4 units, 8 units **65.** $5\sqrt{2}$ centimeters
67. When trying to solve it by extracting the roots, we get $x^2 = -25$. Since no real number squared gives -25, there is no real solution.

Solutions to Trial Exercise Problems

11. $a^2 = 20$
Extract the roots.
$a = \sqrt{20}$ or $a = -\sqrt{20}$
$a = \sqrt{4 \cdot 5}$ or $a = -\sqrt{4 \cdot 5}$
$a = 2\sqrt{5}$ or $a = -2\sqrt{5}$
$\{-2\sqrt{5}, 2\sqrt{5}\}$

18. $5x^2 = 75$
Divide each side by 5.
$x^2 = 15$
Extract the roots.
$x = \sqrt{15}$ or $x = -\sqrt{15}$
$\{-\sqrt{15}, \sqrt{15}\}$

25. $\frac{3}{4}x^2 - 6 = 0$
Multiply each side by 4.
$3x^2 - 24 = 0$
Add 24 to each side.
$3x^2 = 24$
Divide each side by 3.
$x^2 = 8$
Then $x = \sqrt{8}$ or $x = -\sqrt{8}$
$x = 2\sqrt{2}$ or $x = -2\sqrt{2}$
$\{-2\sqrt{2}, 2\sqrt{2}\}$

33. $(x + 2)^2 = 4$
Extract the roots.
$x + 2 = \pm 2$
Add -2 to each side.
$x = -2 \pm 2$
So $x = -2 + 2$ or $x = -2 - 2$
$x = 0$ or -4
$\{-4, 0\}$

39. $(x - 9)^2 = 18$
Extract the roots.
$x - 9 = \pm\sqrt{18} = \pm 3\sqrt{2}$
Add 9 to each side.
$x = 9 + 3\sqrt{2}$ or $x = 9 - 3\sqrt{2}$
$\{9 - 3\sqrt{2}, 9 + 3\sqrt{2}\}$

43. $x^2 = 5.76$
Extract the roots.
$x = \sqrt{5.76}$ or $x = -\sqrt{5.76}$
$x = 2.4$ or $x = -2.4$
$\{-2.4, 2.4\}$

47. $(x - 1.3)^2 = 8.41$
Extract the roots.
$x - 1.3 = \pm\sqrt{8.41}$
$x - 1.3 = \pm 2.9$
$x = 1.3 \pm 2.9$
$x = 1.3 + 2.9$ or $x = 1.3 - 2.9$
$\{-1.6, 4.2\}$

58. Let n represent the number. Then n^2 is the square of the number and $8n$ is eight times the number.
The equation is
$$2n^2 - 8n = 0$$
$$2n(n - 4) = 0$$
$$2n = 0 \text{ or } n - 4 = 0$$
$$n = 0 \text{ or } n = 4$$
The number is 0 or 4.

61. Using $A = \ell w$, let $\ell =$ length of the rectangle. Then $\frac{1}{4}\ell =$ width of the rectangle.

The equation is $\ell \cdot \frac{1}{4}\ell = 144$
$$\frac{1}{4} \cdot \ell^2 = 144$$
$$\ell^2 = 576$$
$$\ell = \pm\sqrt{576} = \pm 24$$
Since length cannot be negative, then $\ell = 24$ meters

Review Exercises

1. $x^2 - 4x + 4$ **2.** $9z^2 + 12z + 4$ **3.** $(x + 9)^2$
4. $(3y + 5)^2$ **5.** $\dfrac{3x^2 - 7x}{(x + 2)(x - 2)}$ **6.** $\dfrac{1}{(x - 2)(x + 3)}$

7. $266\frac{2}{3}$ grams of 80% zinc and 20% copper, $133\frac{1}{3}$ grams of 50% zinc and 50% copper   **8.** 12 cars

Exercise 9–2

Answers to Odd-Numbered Problems

1. $x^2 + 10x + 25 = (x + 5)^2$ **3.** $a^2 - 12a + 36 = (a - 6)^2$
5. $x^2 + 24x + 144 = (x + 12)^2$ **7.** $y^2 - 20y + 100 = (y - 10)^2$
9. $x^2 + x + \dfrac{1}{4} = \left(x + \dfrac{1}{2}\right)^2$ **11.** $x^2 - 7x + \dfrac{49}{4} = \left(x - \dfrac{7}{2}\right)^2$
13. $x^2 + 3x + \dfrac{9}{4} = \left(x + \dfrac{3}{2}\right)^2$ **15.** $s^2 - 5s + \dfrac{25}{4} = \left(s - \dfrac{5}{2}\right)^2$
17. $\{-7, -1\}$ **19.** $\{-2, 6\}$ **21.** $\{1, 3\}$
23. $\left\{\dfrac{1 - \sqrt{5}}{2}, \dfrac{1 + \sqrt{5}}{2}\right\}$ **25.** $\left\{\dfrac{5 - \sqrt{17}}{2}, \dfrac{5 + \sqrt{17}}{2}\right\}$
27. $\{2 - \sqrt{85}, 2 + \sqrt{85}\}$ **29.** $\left\{\dfrac{-21 - \sqrt{401}}{2}, \dfrac{-21 + \sqrt{401}}{2}\right\}$
31. $\{-1 - \sqrt{5}, -1 + \sqrt{5}\}$ **33.** $\{-2 - \sqrt{3}, -2 + \sqrt{3}\}$
35. $\{-1, 2\}$ **37.** $\{1\}$ **39.** $\left\{\dfrac{-1 - \sqrt{13}}{2}, \dfrac{-1 + \sqrt{13}}{2}\right\}$
41. $\{-2, 1\}$ **43.** $\{3 - \sqrt{5}, 3 + \sqrt{5}\}$ **45.** $\{1 - \sqrt{5}, 1 + \sqrt{5}\}$
47. $\left\{\dfrac{-1 - \sqrt{29}}{2}, \dfrac{-1 + \sqrt{29}}{2}\right\}$ **49.** $\{-3 - \sqrt{6}, -3 + \sqrt{6}\}$
51. 12 inches; 8 inches   **53.** 15 millimeters by 7 millimeters
55. 17 inches by 9 inches   **57.** $5\frac{1}{2}$ meters by $3\frac{1}{2}$ meters
59. 14 rods by 6 rods   **61.** 3 inches
63. the square of a binomial

Solutions to Trial Exercise Problems

3. $a^2 - 12a$
Square one-half of the coefficient of a, -12.
$$\left[\frac{1}{2}(-12)\right]^2 = (-6)^2 = 36$$
Then $a^2 - 12a + 36 = (a - 6)^2$

9. $x^2 + x$
Square one-half of the coefficient of x, 1.
$$\left[\frac{1}{2}(1)\right]^2 = \left(\frac{1}{2}\right)^2 = \frac{1}{4}$$
So $x^2 + x + \dfrac{1}{4} = \left(x + \dfrac{1}{2}\right)^2$

11. $x^2 - 7x$

Square one-half of the coefficient of x, -7.

$$\left[\frac{1}{2}(-7)\right]^2 = \left(-\frac{7}{2}\right)^2 = \frac{49}{4}$$

So $x^2 - 7x + \frac{49}{4} = \left(x - \frac{7}{2}\right)^2$

23. $u^2 - u - 1 = 0$

Add 1 to each side.

$u^2 - u = 1$

Add $\left[\frac{1}{2}(-1)\right]^2 = \left(-\frac{1}{2}\right)^2 = \frac{1}{4}$ to each side.

$u^2 - u + \frac{1}{4} = 1 + \frac{1}{4}$

$\left(u - \frac{1}{2}\right)^2 = \frac{5}{4}$ Then

$u - \frac{1}{2} = \pm\sqrt{\frac{5}{4}} = \pm\frac{\sqrt{5}}{2}$

so $u = \frac{1}{2} \pm \frac{\sqrt{5}}{2} = \frac{1 \pm \sqrt{5}}{2}$

Then $u = \frac{1 + \sqrt{5}}{2}$ or $u = \frac{1 - \sqrt{5}}{2}$

$\left\{\frac{1 - \sqrt{5}}{2}, \frac{1 + \sqrt{5}}{2}\right\}$

30. $3x^2 + 6x = 3$

Divide each term by 3.

$x^2 + 2x = 1$

Then $\left[\frac{1}{2}(2)\right]^2 = (1)^2 = 1$, so add 1 to each side.

$x^2 + 2x + 1 = 1 + 1$

So $(x + 1)^2 = 2$. Then

$x + 1 = \pm\sqrt{2}$

so $x = -1 \pm \sqrt{2}$

$x = -1 + \sqrt{2}$ or $x = -1 - \sqrt{2}$

$\{-1 - \sqrt{2}, -1 + \sqrt{2}\}$

31. $2x^2 + 4x - 8 = 0$

Add 8 to each side to get $2x^2 + 4x = 8$.

Now divide each term by 2.

$x^2 + 2x = 4$

Add $\left[\frac{1}{2}(2)\right]^2 = (1)^2 = 1$ to each side. Then

$x^2 + 2x + 1 = 4 + 1$

$(x + 1)^2 = 5$

Extract the roots.

$x + 1 = \sqrt{5}$ or $x + 1 = -\sqrt{5}$

So $x = -1 \pm \sqrt{5}$ and we have

$\{-1 - \sqrt{5}, -1 + \sqrt{5}\}$

42. $4 - x^2 = 2x$

Add x^2 to each side.

$4 = x^2 + 2x$ or $x^2 + 2x = 4$

Then add $\left[\frac{1}{2}(2)\right]^2 = (1)^2 = 1$ to each side.

$x^2 + 2x + 1 = 4 + 1$

so $(x + 1)^2 = 5$

Extract the roots.

$x + 1 = \sqrt{5}$ or $x + 1 = -\sqrt{5}$

Then $x = -1 + \sqrt{5}$ or $x = -1 - \sqrt{5}$

$\{-1 - \sqrt{5}, -1 + \sqrt{5}\}$

47. $(x + 3)(x - 2) = 1$

Perform the indicated multiplication in the left side.

$x^2 + x - 6 = 1$

Add 6 to each side to get $x^2 + x = 7$.

Add $\left[\frac{1}{2}(1)\right]^2 = \left(\frac{1}{2}\right)^2 = \frac{1}{4}$ to each side.

Thus $x^2 + x + \frac{1}{4} = 7 + \frac{1}{4}$

and $\left(x + \frac{1}{2}\right)^2 = \frac{29}{4}$

Then $x + \frac{1}{2} = \pm\sqrt{\frac{29}{4}} = \pm\frac{\sqrt{29}}{2}$

so $x = -\frac{1}{2} \pm \frac{\sqrt{29}}{2} = \frac{-1 \pm \sqrt{29}}{2}$

Then $x = \frac{-1 + \sqrt{29}}{2}$ or $x = \frac{-1 - \sqrt{29}}{2}$

$\left\{\frac{-1 - \sqrt{29}}{2}, \frac{-1 + \sqrt{29}}{2}\right\}$

53. By "a surface of a rectangular solid has a width w that is 8 millimeters shorter than its length ℓ," we get $\ell =$ the length of the surface and $\ell - 8 =$ the width of the surface. Then given area $A = 105$ square millimeters, and using $A = \ell w$, $\ell(\ell - 8) = 105$, then $\ell^2 - 8\ell = 105$.

Add $\left[\frac{1}{2}(-8)\right]^2 = (-4)^2 = 16$ to both sides.

$\ell^2 - 8\ell + 16 = 105 + 16$

$(\ell - 4)^2 = 121$

$\ell - 4 = \pm\sqrt{121} = \pm11$

so $\ell - 4 = 11$ or $\ell - 4 = -11$

Then

$\ell = 4 + 11$ or $\ell = 4 - 11$

$\ell = 15 \qquad \ell = -7$

Since a rectangle must have positive length, -7 is ruled out. So the length $\ell = 15$ millimeters and the width $\ell - 8 = 7$ millimeters.

Review Exercises

1. 3   **2.** $3\sqrt{5}$ **3.** $\{(2,0)\}$

4.

5. 9 dozen

Exercise 9–3

Answers to Odd-Numbered Problems

1. $5x^2 - 3x + 8 = 0$; $a = 5$, $b = -3$, $c = 8$

3. $6z^2 + 2z - 1 = 0$; $a = 6$, $b = 2$, $c = -1$

5. $4x^2 - 2x + 1 = 0$; $a = 4$, $b = -2$, $c = 1$

7. $x^2 + 3x = 0$; $a = 1$, $b = 3$, $c = 0$

9. $5x^2 - 2 = 0$; $a = 5$, $b = 0$, $c = -2$

11. $p^2 + 3p - 4 = 0$; $a = 1$, $b = 3$, $c = -4$

13. $x^2 + 2x - 9 = 0$; $a = 1$, $b = 2$, $c = -9$

15. $8m^2 - 3m - 2 = 0$; $a = 8$, $b = -3$, $c = -2$ **17.** $\{1,2\}$

19. $\{1\}$ **21.** $\{-5,5\}$ **23.** $\{-\sqrt{2}, \sqrt{2}\}$ **25.** $\{0,3\}$ **27.** $\left\{0, \dfrac{9}{5}\right\}$

29. $\left\{\dfrac{9 - \sqrt{65}}{2}, \dfrac{9 + \sqrt{65}}{2}\right\}$ **31.** $\{-1 + \sqrt{7}, -1 - \sqrt{7}\}$

33. $\{4 - \sqrt{15}, 4 + \sqrt{15}\}$ **35.** $\left\{\dfrac{5 - \sqrt{97}}{6}, \dfrac{5 + \sqrt{97}}{6}\right\}$

37. $\left\{\dfrac{-9 - \sqrt{57}}{6}, \dfrac{-9 + \sqrt{57}}{6}\right\}$ **39.** $\left\{-\dfrac{5}{3}, 2\right\}$ **41.** $\{-4\}$

43. $\left\{\dfrac{5}{2}\right\}$ **45.** $\left\{-\dfrac{3}{2}\right\}$ **47.** $\left\{\dfrac{1 - \sqrt{22}}{3}, \dfrac{1 + \sqrt{22}}{3}\right\}$

49. $\{-1 - \sqrt{7}, -1 + \sqrt{7}\}$ **51.** $\left\{\dfrac{3 - \sqrt{105}}{6}, \dfrac{3 + \sqrt{105}}{6}\right\}$

53. $\left\{\dfrac{5 - \sqrt{85}}{10}, \dfrac{5 + \sqrt{85}}{10}\right\}$ **55.** $\left\{\dfrac{3 - \sqrt{41}}{4}, \dfrac{3 + \sqrt{41}}{4}\right\}$

57. $\left\{-\dfrac{3}{2}, 3\right\}$ **59.** $\left\{\dfrac{-1 - \sqrt{3}}{3}, \dfrac{-1 + \sqrt{3}}{3}\right\}$ **61. a.** 2 seconds

     b. $\sqrt{6}$ seconds $\approx$ 2.4 sec   **c.** $\dfrac{\sqrt{30}}{2}$ seconds $\approx$ 2.7 sec

63. a. $-20 + 20\sqrt{11}$ amperes   **b.** $\dfrac{-15 + 5\sqrt{329}}{2}$ amperes

65. $b = 9\sqrt{10}$ inches; $h = 3\sqrt{10}$ inches

67. 6 millimeters; 8 millimeters

69. $-1 + \sqrt{7}$ inches; $1 + \sqrt{7}$ inches   **71.** $5\sqrt{19}$ feet

73. $-16, -14$ **75.** 7 or $\dfrac{1}{7}$

77. If $a = 0$, then the equation becomes $bx + c = 0$ and it is no longer a quadratic equation.

Solutions to Trial Exercise Problems

7. $x^2 = -3x$

Add $3x$ to each side. Then
$x^2 + 3x = 0$ and
$a = 1$, $b = 3$, $c = 0$.

12. $2x(x - 9) = 1$

Perform the indicated multiplication.
$2x^2 - 18x = 1$
Add -1 to each side.
$2x^2 - 18x - 1 = 0$, so
$a = 2$, $b = -18$, $c = -1$.

18. $y^2 + 6y + 9 = 0$

Here $a = 1$, $b = 6$, and $c = 9$ so

$y = \dfrac{-6 \pm \sqrt{(6)^2 - 4(1)(9)}}{2(1)}$

$= \dfrac{-6 \pm \sqrt{36 - 36}}{2}$

$= \dfrac{-6 \pm \sqrt{0}}{2}$

$= \dfrac{-6}{2}$

$= -3$

$\{-3\}$

21. $x^2 - 25 = 0$

We can write this
$x^2 + 0x - 25 = 0$, so
$a = 1$, $b = 0$, $c = -25$.

Then $x = \dfrac{-0 \pm \sqrt{0^2 - 4(1)(-25)}}{2(1)}$

$x = \dfrac{\pm \sqrt{100}}{2}$

so $x = \pm \dfrac{10}{2}$ Then

$x = 5$ or $x = -5$

$\{-5,5\}$

26. $x^2 = 4x$

Add $-4x$ to each side and write the equation as
$x^2 - 4x + 0 = 0$
So $a = 1$, $b = -4$, $c = 0$,

and $x = \dfrac{-(-4) \pm \sqrt{(-4)^2 - 4(1)(0)}}{2(1)}$

$= \dfrac{4 \pm \sqrt{16}}{2}$

$= \dfrac{4 \pm 4}{2}$

Then $x = \dfrac{4 + 4}{2} = \dfrac{8}{2} = 4$ or $x = \dfrac{4 - 4}{2} = \dfrac{0}{2} = 0$

$\{0,4\}$

36. $4t^2 = 8t - 3$

Add $3 - 8t$ to each side.
$4t^2 - 8t + 3 = 0$
So $a = 4$, $b = -8$, $c = 3$,

and $t = \dfrac{-(-8) \pm \sqrt{(-8)^2 - 4(4)(3)}}{2(4)}$

$= \dfrac{8 \pm \sqrt{64 - 48}}{8}$

$= \dfrac{8 \pm \sqrt{16}}{8}$

so $t = \dfrac{8 \pm 4}{8}$

Then $t = \dfrac{8 + 4}{8} = \dfrac{12}{8} = \dfrac{3}{2}$ or $t = \dfrac{8 - 4}{8} = \dfrac{4}{8} = \dfrac{1}{2}$

$\left\{\dfrac{1}{2}, \dfrac{3}{2}\right\}$

54. $2x^2 - \dfrac{7}{2} + \dfrac{x}{2} = 0$

Multiply by the LCD, 2.
$4x^2 - 7 + x = 0$
Then write in standard form.
$4x^2 + x - 7 = 0$
Then $a = 4$, $b = 1$, and $c = -7$.

Thus $x = \dfrac{-1 \pm \sqrt{1^2 - 4(4)(-7)}}{2(4)}$

$= \dfrac{-1 \pm \sqrt{1 + 112}}{8}$

$= \dfrac{-1 \pm \sqrt{113}}{8}$

So $x = \dfrac{-1 + \sqrt{113}}{8}$ or $x = \dfrac{-1 - \sqrt{113}}{8}$

$\left\{\dfrac{-1 - \sqrt{113}}{8}, \dfrac{-1 + \sqrt{113}}{8}\right\}$

60. a. Using $s = vt + \dfrac{1}{2}at^2$, replace s with 8, v with 3, and a with 4.

$$8 = 3t + \frac{1}{2}(4)t^2$$
$$8 = 3t + 2t^2$$
$$2t^2 + 3t - 8 = 0$$
$$t = \frac{-3 \pm \sqrt{3^2 - 4(2)(-8)}}{2(2)}$$
$$= \frac{-3 \pm \sqrt{9 + 64}}{4}$$
$$= \frac{-3 \pm \sqrt{73}}{4}$$
$$t = \frac{-3 + \sqrt{73}}{4} \approx 1.39 \text{ or } t = \frac{-3 - \sqrt{73}}{4} \text{ (reject)}$$

The time is approximately 1.39.

66. Using $a^2 + b^2 = c^2$, replace a with x, b with $x + 14$, and c with $x + 16$.

$$x^2 + (x + 14)^2 = (x + 16)^2$$
$$x^2 + x^2 + 28x + 196 = x^2 + 32x + 256$$
$$2x^2 + 28x + 196 = x^2 + 32x + 256$$
$$x^2 - 4x - 60 = 0$$
$$x = \frac{-(-4) \pm \sqrt{(-4)^2 - 4(1)(-60)}}{2(1)}$$
$$= \frac{4 \pm \sqrt{16 + 240}}{2}$$
$$= \frac{4 \pm \sqrt{256}}{2}$$
$$= \frac{4 \pm 16}{2}$$

Then $x = \dfrac{4 + 16}{2} = \dfrac{20}{2} = 10$ or

$x = \dfrac{4 - 16}{2} = \dfrac{-12}{2} = -6$ (reject).

Thus, $x = 10$.

74. Let n represent the first odd positive integer. Then $n + 2$ is the next consecutive odd positive integer.
The equation is then $n(n + 2) = 143$
$$n^2 + 2n = 143$$
$$n^2 + 2n - 143 = 0$$
$$(n + 13)(n - 11) = 0$$
Then $n = -13$ and $n + 2 = -11$ or $n = 11$ and $n + 2 = 13$. Reject -13 and -11 since we want positive integers. Thus 11 and 13 are two consecutive odd positive integers.

Review Exercises

1. $3x^2 + x + 5$ **2.** $5y^2 + 33y - 14$ **3.** $16z^2 - 9$

4. $9x^2 - 30x + 25$ **5.** $\{-2,2\}$ **6.** $\left\{-\dfrac{1}{2}, 4\right\}$

7. $\left\{\dfrac{1 - \sqrt{41}}{2}, \dfrac{1 + \sqrt{41}}{2}\right\}$ **8.** $\dfrac{2x - 8}{(x + 2)(x - 2)(x - 3)}$

Exercise 9–4

Answers to Odd-Numbered Problems

1. $9 + 0i$ **3.** $0 + 4i$ **5.** $0 + 5i$ **7.** $4 + 4i$

9. $4 + i$ **11.** $1 - 4i$ **13.** $5 - 2i$ **15.** $-9 - 2i\sqrt{7}$

17. $-12 + 6i$ **19.** $10 + 11i$ **21.** $41 + 0i$ **23.** $-33 + 56i$

25. $\dfrac{15}{13} + \dfrac{10}{13}i$ **27.** $\dfrac{1}{17} + \dfrac{13}{17}i$ **29.** $\dfrac{17}{25} - \dfrac{19}{25}i$

31. $\left\{\dfrac{-1 - i\sqrt{7}}{2}, \dfrac{-1 + i\sqrt{7}}{2}\right\}$ **33.** $\left\{\dfrac{-1 - i\sqrt{19}}{2}, \dfrac{-1 + i\sqrt{19}}{2}\right\}$

35. $\left\{\dfrac{-3 - i\sqrt{7}}{2}, \dfrac{-3 + i\sqrt{7}}{2}\right\}$ **37.** $\{-1 - 3i, -1 + 3i\}$

39. $\left\{\dfrac{3 - i\sqrt{11}}{2}, \dfrac{3 + i\sqrt{11}}{2}\right\}$ **41.** $\left\{\dfrac{-1 - i\sqrt{31}}{4}, \dfrac{-1 + i\sqrt{31}}{4}\right\}$

43. $\left\{\dfrac{3 - i\sqrt{23}}{4}, \dfrac{3 + i\sqrt{23}}{4}\right\}$ **45.** $\{-2i, 2i\}$ **47.** $\{-2i\sqrt{3}, 2i\sqrt{3}\}$

49. $\{-2 - 4i, -2 + 4i\}$ **51.** $\left\{\dfrac{-1 - i\sqrt{19}}{2}, \dfrac{-1 + i\sqrt{19}}{2}\right\}$

53. two distinct irrational solutions
55. one rational solution **57.** two distinct rational solutions
59. two distinct irrational solutions
61. $z = 20 - 10i$ **63.** $x \le 9$ **65.** $x < -16$
67. the radicand $b^2 - 4ac$ of the quadratic formula

Solutions to Trial Exercise Problems

7. $4 + 2\sqrt{-4} = 4 + 2(2i) = 4 + 4i$

14. $(1 - \sqrt{-4}) - (3 + \sqrt{-9})$
$$= (1 - 2i) - (3 + 3i) = 1 - 2i - 3 - 3i$$
$$= (1 - 3) + (-2i - 3i)$$
$$= -2 + (-5i)$$
$$= -2 - 5i$$

23. $(4 + 7i)^2$
$$= 4^2 + 2(4)(7i) + (7i)^2 = 16 + 56i + 49i^2$$
$$= 16 + 56i + 49(-1)$$
$$= 16 + 56i - 49$$
$$= -33 + 56i$$

28. $\dfrac{1 + i}{2 - i} \cdot \dfrac{2 + i}{2 + i} = \dfrac{(1 + i)(2 + i)}{2^2 + 1^2} = \dfrac{2 + 3i + i^2}{4 + 1} = \dfrac{2 + 3i + (-1)}{5}$
$$= \dfrac{1 + 3i}{5} = \dfrac{1}{5} + \dfrac{3}{5}i$$

42. $3y^2 - 2y + 3 = 0$. Here $a = 3$, $b = -2$, and $c = 3$.
$$y = \frac{-(-2) \pm \sqrt{(-2)^2 - 4(3)(3)}}{2(3)} = \frac{2 \pm \sqrt{4 - 36}}{6}$$
$$= \frac{2 \pm \sqrt{-32}}{6}$$
$$= \frac{2 \pm i\sqrt{32}}{6}$$
$$= \frac{2 \pm 4i\sqrt{2}}{6}$$
$$= \frac{2(1 \pm 2i\sqrt{2})}{6}$$
$$= \frac{1 \pm 2i\sqrt{2}}{3}$$

The solution set is $\left\{\dfrac{1 - 2i\sqrt{2}}{3}, \dfrac{1 + 2i\sqrt{2}}{3}\right\}$.

59. $(x + 4)(x + 3) = 1$
$x^2 + 7x + 12 = 1$, and we have $x^2 + 7x + 11 = 0$. Then
$b^2 - 4ac = (7)^2 - 4(1)(11) = 49 - 44 = 5$; two distinct irrational solutions.

Review Exercises

1.
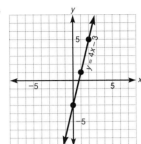

2.
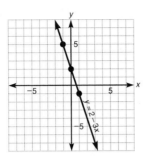

3. $\left(\dfrac{4}{3}, \dfrac{5}{3}\right)$

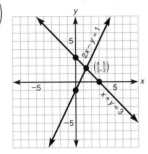

4. $8x - 3y = 17$

5. $\dfrac{3(x + 1)}{4x}$

41.
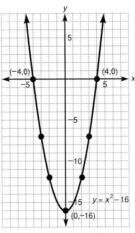

Exercise 9–5

Answers to Odd-Numbered Problems

1. x-intercepts, $(-4,0)$ and $(4,0)$; y-intercept, $(0,-16)$

3. x-intercepts, $(2,0)$ and $(4,0)$; y-intercept, $(0,8)$

5. x-intercepts, $(-6,0)$ and $(-2,0)$; y-intercept, $(0,12)$

7. x-intercepts, $(-\sqrt{5},0)$ and $(\sqrt{5},0)$; y-intercept, $(0,5)$

9. x-intercept, $(-3,0)$; y-intercept, $(0,9)$

11. x-intercepts, none; y-intercept, $(0,5)$

13. x-intercepts, $(-3,0)$ and $(2,0)$; y-intercept, $(0,-6)$

15. x-intercepts, $(1,0)$ and $(3,0)$; y-intercept, $(0,-3)$

17. x-intercepts, $(-1,0)$ and $\left(-\dfrac{1}{2},0\right)$; y-intercept, $(0,1)$

19. x-intercepts, $(-2,0)$ and $\left(\dfrac{3}{2},0\right)$; y-intercept, $(0,6)$

21. $(0,-16)$; $x = 0$ **23.** $(3,-1)$; $x = 3$ **25.** $(-4,-4)$; $x = -4$

27. $(0,5)$; $x = 0$ **29.** $(-3,0)$; $x = -3$ **31.** $(0,5)$; $x = 0$

33. $\left(-\dfrac{1}{2}, -\dfrac{25}{4}\right)$; $x = -\dfrac{1}{2}$ **35.** $(2,1)$; $x = 2$ **37.** $\left(-\dfrac{3}{4}, -\dfrac{1}{8}\right)$;

   $x = -\dfrac{3}{4}$ **39.** $\left(-\dfrac{1}{4}, \dfrac{49}{8}\right)$; $x = -\dfrac{1}{4}$

43.
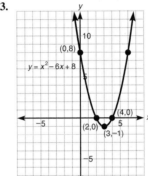

45.
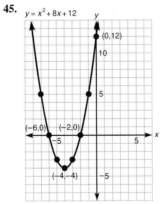

47.

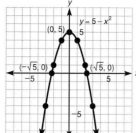

49.

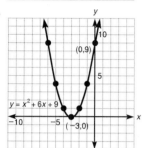

51.

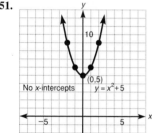

53.

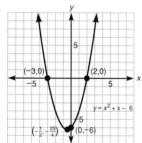

55.

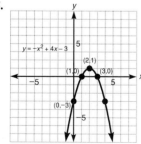

57.

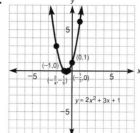

59.

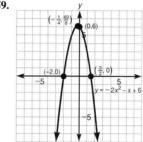

61.

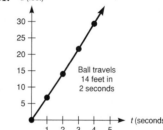

63.

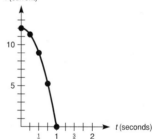

65.

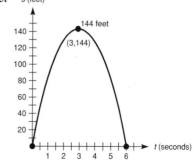

max. height is 144 feet, will strike
the ground in 6 seconds

67.

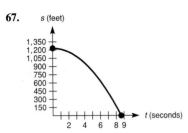

will strike the ground in approximately 8.84 seconds

Solutions to Trial Exercise Problems

3. $y = x^2 - 6x + 8$

Let $x = 0$, then $y = 0^2 - 6(0) + 8 = 8$.

Let $y = 0$, then $0 = x^2 - 6x + 8$. Factor the right side.

$0 = (x - 4)(x - 2)$

so $x = 4$ or $x = 2$

The y-intercept is $(0,8)$ and the x-intercepts are $(4,0)$ and $(2,0)$.

7. $y = 5 - x^2$

Let $x = 0$, then $y = 5 - 0^2 = 5$.

Let $y = 0$, then $0 = 5 - x^2$. Add x^2 to each side.

$x^2 = 5$

Extract the roots.

$x = \sqrt{5}$ or $x = -\sqrt{5}$ so the y-intercept is $(0,5)$ and the x-intercepts are $\left(-\sqrt{5},0\right)$ and $\left(\sqrt{5},0\right)$.

17. $y = 2x^2 + 3x + 1$

Let $x = 0$, then $y = 2(0)^2 + 3(0) + 1 = 1$.

Let $y = 0$, then $0 = 2x^2 + 3x + 1$. Factor the right side.

$0 = (2x + 1)(x + 1)$

then $2x + 1 = 0$ or $x + 1 = 0$

so $x = -\dfrac{1}{2}$ or $x = -1$

The y-intercept is $(0,1)$ and the x-intercepts are $(-1,0)$ and $\left(-\dfrac{1}{2},0\right)$.

23. $y = x^2 - 6x + 8$

Here $a = 1$ and $b = -6$

so $x = \dfrac{-b}{2a} = -\dfrac{-6}{2(1)} = 3$

then $y = (3)^2 - 6(3) + 8$

$= 9 - 18 + 8$

$= -1$

The vertex is at $(3,-1)$. Axis of symmetry is $x = 3$.

27. $y = 5 - x^2$

Here $a = -1$ and $b = 0$

so $x = \dfrac{-b}{2a} = \dfrac{-0}{2(-1)} = 0$

then $y = 5 - 0^2 = 5$

Therefore the vertex is at $(0,5)$. Axis of symmetry is $x = 0$.

37. $y = 2x^2 + 3x + 1$

Here $a = 2$ and $b = 3$ so $x = \dfrac{-b}{2a} = \dfrac{-3}{2(2)} = -\dfrac{3}{4}$. Then

$y = 2\left(-\dfrac{3}{4}\right)^2 + 3\left(-\dfrac{3}{4}\right) + 1$

$= 2\left(\dfrac{9}{16}\right) - \dfrac{9}{4} + 1$

$= \dfrac{9}{8} - \dfrac{18}{8} + 1$

$= -\dfrac{9}{8} + 1 = -\dfrac{1}{8}$

The vertex is at $\left(-\dfrac{3}{4},-\dfrac{1}{8}\right)$. Axis of symmetry is $x = -\dfrac{3}{4}$.

43. $y = x^2 - 6x + 8$

x	y	
0	8	y-intercept
4	0	x-intercepts
2	0	
3	-1	vertex
1	3	arbitrary points
5	3	
6	8	

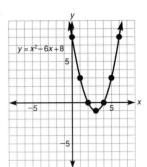

47. $y = 5 - x^2$

x	y	
0	5	y-intercept; vertex
$\sqrt{5}$	0	x-intercepts
$-\sqrt{5}$	0	
1	4	
2	1	arbitrary points
-2	1	
-1	4	

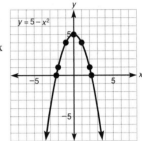

57. $y = 2x^2 + 3x + 1$

x	y	
0	1	y-intercept
$-\dfrac{1}{2}$	0	x-intercepts
-1	0	
$-\dfrac{3}{4}$	$-\dfrac{1}{8}$	vertex
1	6	arbitrary points
-2	3	
-3	10	

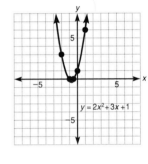

61. $d = 6t + \dfrac{t^2}{2}$ (*Note:* We must choose $t \geq 0$.)

t	d
0	0
1	$\dfrac{13}{2}$ or $6\dfrac{1}{2}$
2	14
3	$\dfrac{45}{2}$ or $22\dfrac{1}{2}$
4	32

$t = 2$ seconds
when $d = 14$ feet

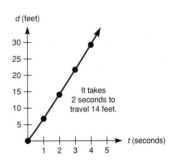

Chapter 9 Review

1. $\{-10,10\}$ **2.** $\{-5,5\}$ **3.** $\{-\sqrt{2},\sqrt{2}\}$ **4.** $\{-\sqrt{6},\sqrt{6}\}$
5. $\{-2,2\}$ **6.** $\{-2\sqrt{3},2\sqrt{3}\}$ **7.** $\{-4,4\}$ **8.** $\{-2\sqrt{3},2\sqrt{3}\}$

9. $\left\{-\dfrac{3\sqrt{11}}{2}, \dfrac{3\sqrt{11}}{2}\right\}$ **10.** $\{-6,7\}$ **11.** $\left\{1, \dfrac{4}{3}\right\}$

12. $\{3 - \sqrt{5}, 3 + \sqrt{5}\}$ **13.** $\{5 - \sqrt{21}, 5 + \sqrt{21}\}$

14. $\{2 - \sqrt{3}, 2 + \sqrt{3}\}$ **15.** $\{1 - \sqrt{3}, 1 + \sqrt{3}\}$

16. $\left\{\dfrac{-5 - \sqrt{41}}{2}, \dfrac{-5 + \sqrt{41}}{2}\right\}$ **17.** $\left\{\dfrac{11 - \sqrt{101}}{2}, \dfrac{11 + \sqrt{101}}{2}\right\}$

18. $\left\{\dfrac{1 - \sqrt{13}}{2}, \dfrac{1 + \sqrt{13}}{2}\right\}$ **19.** $\left\{\dfrac{1 - \sqrt{13}}{2}, \dfrac{1 + \sqrt{13}}{2}\right\}$

20. $\left\{\dfrac{2 - \sqrt{6}}{2}, \dfrac{2 + \sqrt{6}}{2}\right\}$ **21.** $\{-1 - \sqrt{11}, -1 + \sqrt{11}\}$

22. 2 meters by 8 meters   **23.** $\{1 - \sqrt{6}, 1 + \sqrt{6}\}$

24. $\{-2 - 2\sqrt{3}, -2 + 2\sqrt{3}\}$ **25.** $\left\{-1, \dfrac{5}{2}\right\}$

26. $\left\{\dfrac{-7 - \sqrt{145}}{6}, \dfrac{-7 + \sqrt{145}}{6}\right\}$ **27.** $\left\{-\dfrac{3\sqrt{2}}{2}, \dfrac{3\sqrt{2}}{2}\right\}$

28. $\left\{-\dfrac{7}{4}, 0\right\}$ **29.** $\left\{\dfrac{1 - \sqrt{13}}{3}, \dfrac{1 + \sqrt{13}}{3}\right\}$

30. $\left\{\dfrac{4 - \sqrt{34}}{6}, \dfrac{4 + \sqrt{34}}{6}\right\}$ **31.** 6 inches, 9 inches

32. $5 - i$ **33.** $4 + 9i$ **34.** $15 - 16i$ **35.** $3 + 21i$

36. $52 + 0i$ **37.** $16 - 30i$ **38.** $\dfrac{3}{2} + \dfrac{3}{2}i$ **39.** $\dfrac{4}{5} - \dfrac{8}{5}i$

40. $\dfrac{7}{10} - \dfrac{1}{10}i$ **41.** $\dfrac{26}{25} - \dfrac{7}{25}i$ **42.** $\{-2 - i\sqrt{3}, -2 + i\sqrt{3}\}$

43. $\left\{\dfrac{1 - i\sqrt{79}}{8}, \dfrac{1 + i\sqrt{79}}{8}\right\}$ **44.** $\{-i\sqrt{7}, i\sqrt{7}\}$

45. $\left\{\dfrac{-3 - 2i}{2}, \dfrac{-3 + 2i}{2}\right\}$

46. one rational solution   **47.** two distinct rational solutions

48. two distinct complex solutions

49. two distinct irrational solutions

50. x-intercepts, $(-2,0)$, $(6,0)$; y-intercept, $(0,-12)$; vertex, $(2,-16)$; $x = 2$

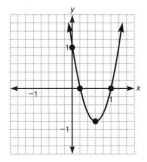

51. x-intercepts, $\left(\dfrac{1}{5},0\right)$, $(1,0)$; y-intercept, $(0,1)$; vertex, $\left(\dfrac{3}{5}, -\dfrac{4}{5}\right)$; $x = \dfrac{3}{5}$

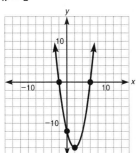

52. x-intercepts, $(-4,0)$, $(2,0)$; y-intercept, $(0,8)$; vertex, $(-1,9)$; $x = -1$

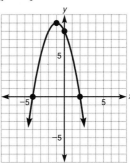

53. x-intercepts, $\left(-\dfrac{2}{3},0\right)$, $(1,0)$; y-intercept, $(0,2)$; vertex, $\left(\dfrac{1}{6}, \dfrac{25}{12}\right)$; $x = \dfrac{1}{6}$

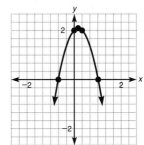

54. x-intercepts, $(0,0)$, $\left(\dfrac{2}{5},0\right)$; y-intercept, $(0,0)$; vertex, $\left(\dfrac{1}{5}, -\dfrac{1}{5}\right)$; $x = \dfrac{1}{5}$

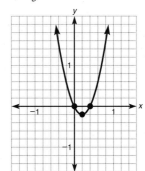

55. x-intercepts, $(0,0)$, $\left(\dfrac{1}{3},0\right)$; y-intercept, $(0,0)$; vertex, $\left(\dfrac{1}{6}, \dfrac{1}{12}\right)$; $x = \dfrac{1}{6}$

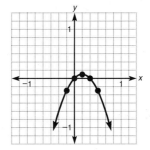

56. x-intercepts, $\left(-\sqrt{2},0\right)$, $\left(\sqrt{2},0\right)$; y-intercept, $(0,-8)$; vertex, $(0,-8)$; $x = 0$

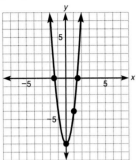

57. x-intercepts, $(-3,0)$, $(3,0)$; y-intercept, $(0,9)$; vertex, $(0,9)$; $x = 0$

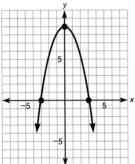

58. x-intercepts, none; y-intercept, $(0,2)$; vertex, $(0,2)$; $x = 0$

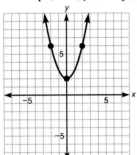

59. x-intercepts, none; y-intercept, $(0,3)$; vertex, $(-1,2)$; $x = -1$

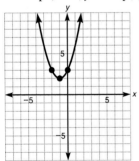

Chapter 9 Test

1. $\{-2,6\}$ [9–1]   **2.** $\{-3\sqrt{2},3\sqrt{2}\}$ [9–1]
3. $\{-3 - \sqrt{5}, -3 + \sqrt{5}\}$ [9–3]   **4.** $\{4 - 2\sqrt{5}, 4 + 2\sqrt{5}\}$ [9–3]
5. $\left\{\dfrac{-1 - \sqrt{33}}{4}, \dfrac{-1 + \sqrt{33}}{4}\right\}$ [9–3]   **6.** $\left\{-\dfrac{\sqrt{14}}{3}, \dfrac{\sqrt{14}}{3}\right\}$ [9–1]

7. $\left\{-4, -\dfrac{1}{2}\right\}$ [9–1]   **8.** $\left\{\dfrac{-1 - i\sqrt{15}}{2}, \dfrac{-1 + i\sqrt{15}}{2}\right\}$ [9–4]
9. $\{-6, -1\}$ [9–1]   **10.** $\{1 - i\sqrt{5}, 1 + i\sqrt{5}\}$ [9–4]
11. $\{-5 - \sqrt{15}, -5 + \sqrt{15}\}$ [9–1]   **12.** $\left\{-7, \dfrac{3}{2}\right\}$ [9–1]
13. $\{4 - 2\sqrt{3}, 4 + 2\sqrt{3}\}$ [9–2]   **14.** $\left\{\dfrac{5 - \sqrt{53}}{2}, \dfrac{5 + \sqrt{53}}{2}\right\}$ [9–2]
15. two distinct complex solutions [9–4]
16. one rational solution [9–4]
17. $1 - 2i\sqrt{5}$ [9–4]   **18.** $-7 + 24i$ [9–4]   **19.** $1 + i$ [9–4]
20.

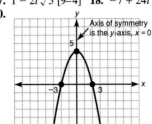

vertex, $(0,4)$; axis of symmetry, $x = 0$; x-intercepts, $(-2,0)$, $(2,0)$; y-intercept, $(0,4)$ [9–5]

21.

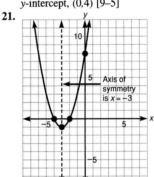

vertex, $(-3,-1)$; axis of symmetry, $x = -3$; x-intercepts, $(-4,0)$, $(-2,0)$; y-intercept, $(0,8)$ [9–5]
22. 3 [9–3]
23. Reaches the highest point in 3.75 sec. Highest point is 11.25 m. Strikes the ground in 7.5 sec. [9–5]
24. 13 m by 5 m [9–3]

Final Examination

1. > [1–4]   **2.** 40 [1–8]   **3.** 0 [1–8]   **4.** -52 [1–9]   **5.** x^6 [3–4]
6. $\dfrac{1}{2x^5}$ [3–4]   **7.** $-12x^3y^5$ [3–4]   **8.** $\dfrac{y^6}{27x^3}$ [3–4]   **9.** 1 [3–4]
10. $7x^2 - 8y^2$ [2–2]   **11.** $-3y$ [2–2]   **12.** $y^2 - 81$ [3–2]
13. $49z^2 - 42zw + 9w^2$ [3–2]   **14.** $5x^3 + 17x^2 - 11x + 4$ [3–2]
15. $5y - 3x^3 + xy$ [3–6]   **16.** $4y + 1 + \dfrac{-2}{2y - 1}$ [3–6]   **17.** $\{7\}$ [2–6]
18. $\{-1, 12\}$ [4–7]   **19.** $\left\{\dfrac{1}{3}, 2\right\}$ [6–5]   **20.** $-\left\{\dfrac{8}{3}\right\}$ [5–6]
21. $\left\{0, \dfrac{3}{2}\right\}$ [4–7]   **22.** $3x(x - 2y + 3)$ [4–1]
23. $(a - 7)(a + 3)$ [4–2]   **24.** $(2x - 1)(2x - 5)$ [4–3]
25. $(3a + 8)(3a - 8)$ [4–4]   **26.** $(2a + b)(3x - y)$ [4–6]
27. $(x - 5)^2$ [4–4]   **28.** 11 and 12 or -12 and -11 [4–8]
29. $\dfrac{x + 1}{x + 2}$ [5–2]   **30.** $\dfrac{x - 2}{12}$ [5–2]   **31.** $\dfrac{14}{x - 6}$ [5–3]

32. $\dfrac{x^2 - 6x + 14}{(x + 5)(x - 5)}$ [5–4] **33.** $\dfrac{5y + 4}{4y - 6}$ [5–5] **34.** $\dfrac{16}{5}$ [2–8]

35. $\dfrac{21}{40}$ or 21 : 40 [2–8] **36.** $2x + y = 1$ [6–4]

37. $m = \dfrac{2}{3};\ b = -3$ [6–4]

38. $\{(-19, -11)\}$ [7–3] **39.** 12 ft by 5 ft [2–7] **40.** $-\sqrt{3}$ [8–4]

41. $\sqrt{6} + 3$ [8–5] **42.** 13 [8–5] **43.** $11 - 4\sqrt{7}$ [8–5]

44. $\dfrac{9 + 3\sqrt{5}}{4}$ [8–5] **45.** -3 [8–1] **46.** 8 [8–6] **47.** $\{-1, 0\}$ [8–7]

48.

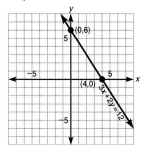

49.

x	y	
-1	0	} x-intercepts
6	0	
0	-6	y-intercept
$\dfrac{5}{2}$	$-\dfrac{49}{4}$	vertex
1	-10	}
2	-12	arbitrary points
3	-12	
4	-10	

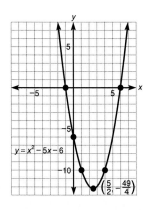

50. $\left\{\dfrac{7 - \sqrt{97}}{8}, \dfrac{7 + \sqrt{97}}{8}\right\}$ [9–3] **51.** $1 - 19i$ [9–4]

52. $-24 + 70i$ [9–4] **53.** 6 [9–4] **54.** $-\dfrac{6}{13} + \dfrac{17}{13}i$ [9–4]

55. 100 liters of 40%, 100 liters of 20% [2–7]

56. $15,000 at 11% profit; $16,000 at 8% loss [2–7]

57. $21\dfrac{1}{3}$ in. [2–8] **58.** $12\dfrac{6}{7}$ ft [2–8]

59. 7 ft by 6 ft [4–7] **60.** $93\dfrac{1}{3}$ ft by 30 ft [4–7]

61. still water, $5\dfrac{1}{10}$ mph; current, $2\dfrac{1}{10}$ mph [7–4] **62.** $4\dfrac{1}{2}$ sec [9–3]

63. 12 yards by 5 yards [9–3] **64.** 150 mph, 200 mph [9–3]

Photographs

Chapter 1

Table of Contents and Opener: Courtesy of Calgary Convention & Visitors Bureau. **page 4** Courtesy of Pat Martin. **page 9** Courtesy of the National Park Service. **page 34** Courtesy of the Florsheim Shoe Company. **page 42** Courtesy of the Jordan Information Bureau. **page 54** Courtesy of American Airlines. **page 69** Ice skaters on State Ice Rink 37, in downtown Chicago. Willy Schmidt/City of Chicago. **page 90** Courtesy of the National Park Service. **page 94** Courtesy of the University of Illinois Department of Chemistry.

Chapter 2

Table of Contents and Opener: Courtesy of Pat Martin. **page 107** Courtesy of Pat Martin. **page 110** Courtesy of OWI, Inc. **page 120** Courtesy of Schnuck Markets, Inc. **page 135** Courtesy of the Hertz Corporation. **page 140** Courtesy of Best Buy Co., Inc. **page 147** Courtesy of the Kentucky Department of Travel Development. **page 149** Courtesy of the American Plastics Council. **page 157** Courtesy of Kelloggs. **page 161** The historic Water Tower on North Michigan Avenue during the holidays. Willy Schmidt/City of Chicago. **page 180** Courtesy of Deere & Company. **page 184** Courtesy of Pat Martin.

Chapter 3

Table of Contents and Opener: Courtesy of Pat Martin. **page 211** Courtesy of NASA. **page 214** Courtesy of the American Society of Clinical Pathologists, Chicago. **page 223** Official U.S. Navy photo, by Lt. Cmdr. Art Leqare. **page 227** Courtesy of NASA. **page 230** Courtesy of Cardinal Pool, Champaign, Illinois. **page 229** Courtesy of Philips Petroleum Company.

Chapter 4

Table of Contents and Opener: Photograph provided by Easton. **page 236** Courtesy of NASA. **page 252** Courtesy of Pat Martin. **page 271** Courtesy of Pat Martin. **page 273** Brilliant fireworks display at Venetian Night '94, one of Chicago's free summer festivals on the lakefront. Peter J. Schultz/City of Chicago. **page 276** Courtesy of Pat Martin.

Chapter 5

Table of Contents and Opener: ©Nicholas DeSciose/Photo Researchers. **page 307** Courtesy of Greenview Companies. **page 310 #52** Courtesy of Lenscrafters, Champaign, Illinois. **page 310 #53** Courtesy of Pat Martin. **page 311** Courtesy of Kingery Printing Company. **page 325** Courtesy of Pat Martin. **page 327** Courtesy Buick Motor Division. **page 330** Courtesy of Pat Martin. **page 331** Courtesy of American Assoc. of Railroads/Wayne R. Gaylord. **page 341** Courtesy of Pat Martin.

Chapter 6

Table of Contents and Opener: Courtesy of the University of Illinois. **page 351** Courtesy of the American Society of Clinical Pathologists, Chicago. **page 356** Courtesy of Pat Martin. **page 380** Courtesy of the Coca-Cola Company. **page 393** Courtesy of Mr. Bulky Treats & Gifts. **page 414** Courtesy of Pat Martin. **page 415** Courtesy of Pat Martin.

Chapter 7

Table of Contents and Opener: Courtesy of the Chicago Board of Trade. **page 447** Courtesy of Pat Martin. **page 453** Courtesy of Cardinal Pool, Champaign, Illinois. **page 458** Courtesy of S&K Menswear. **page 466** Courtesy of Pat Martin. **page 467** Courtesy of the Illinois Department of Transportation, Bureau of Construction. **page 481** Courtesy of Pat Martin.

Chapter 8

Table of Contents and Opener: ©Patrick Donehue/Photo Researchers. **page 495** Courtesy of Zenith. **page 504** Courtesy of Pat Martin. **page 517** Sears Tower, the world's tallest building, in Chicago's Loop. Cheryl Tadin/City of Chicago. **page 523** Courtesy of Pat Martin. **page 530** Courtesy of the National Park Service. **page 531** Courtesy of Pat Martin.

Chapter 9

Table of Contents and Opener: Courtesy of the National Park Service. **page 538** Courtesy of Pat Martin. **page 580** ©Batman/Empire State Building. **page 588** Courtesy of NASA.

Index

WEEKLY SCHEDULE

Time	SUN	MON	TUES	WED	THURS	FRI	SAT
6 am – 7 am							
7 am – 8 am							
8 am – 9 am							
9 am – 10 am							
10 am – 11 am							
11 am – 12 pm							
12 pm – 1 pm							
1 pm – 2 pm							
2 pm – 3 pm							
3 pm – 4 pm							
4 pm – 5 pm							
5 pm – 6 pm							
6 pm – 7 pm							
7 pm – 8 pm							
8 pm – 9 pm							
9 pm – 10 pm							
10 pm – 11 pm							
11 pm – 12 am							
12 am – 6 am							
Special Notes							

WEEKLY SCHEDULE

Time	SUN	MON	TUES	WED	THURS	FRI	SAT
6 am – 7 am							
7 am – 8 am							
8 am – 9 am							
9 am – 10 am							
10 am – 11 am							
11 am – 12 pm							
12 pm – 1 pm							
1 pm – 2 pm							
2 pm – 3 pm							
3 pm – 4 pm							
4 pm – 5 pm							
5 pm – 6 pm							
6 pm – 7 pm							
7 pm – 8 pm							
8 pm – 9 pm							
9 pm – 10 pm							
10 pm – 11 pm							
11pm – 12 am							
12 am – 6 am							
Special Notes							

WEEKLY SCHEDULE

Time	SUN	MON	TUES	WED	THURS	FRI	SAT
6 am – 7 am							
7 am – 8 am							
8 am – 9 am							
9 am – 10 am							
10 am – 11 am							
11 am – 12 pm							
12 pm – 1 pm							
1 pm – 2 pm							
2 pm – 3 pm							
3 pm – 4 pm							
4 pm – 5 pm							
5 pm – 6 pm							
6 pm – 7 pm							
7 pm – 8 pm							
8 pm – 9 pm							
9 pm – 10 pm							
10 pm – 11 pm							
11pm – 12 am							
12 am – 6 am							
Special Notes							

WEEKLY SCHEDULE

Time	SUN	MON	TUES	WED	THURS	FRI	SAT
6 am – 7 am							
7 am – 8 am							
8 am – 9 am							
9 am – 10 am							
10 am – 11 am							
11 am – 12 pm							
12 pm – 1 pm							
1 pm – 2 pm							
2 pm – 3 pm							
3 pm – 4 pm							
4 pm – 5 pm							
5 pm – 6 pm							
6 pm – 7 pm							
7 pm – 8 pm							
8 pm – 9 pm							
9 pm – 10 pm							
10 pm – 11 pm							
11 pm – 12 am							
12 am – 6 am							
Special Notes							

WEEKLY SCHEDULE

Time	SUN	MON	TUES	WED	THURS	FRI	SAT
6 am – 7 am							
7 am – 8 am							
8 am – 9 am							
9 am – 10 am							
10 am – 11 am							
11 am – 12 pm							
12 pm – 1 pm							
1 pm – 2 pm							
2 pm – 3 pm							
3 pm – 4 pm							
4 pm – 5 pm							
5 pm – 6 pm							
6 pm – 7 pm							
7 pm – 8 pm							
8 pm – 9 pm							
9 pm – 10 pm							
10 pm – 11 pm							
11pm – 12 am							
12 am – 6 am							
Special Notes							

Geometric Figures: Plane and Solid

Plane

Parallelogram

Perimeter $P = 2a + 2b$

Area $A = ah$ (h is the altitude of the figure)

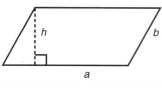

Rectangle

Perimeter $P = 2\ell + 2w$

Area $A = \ell w$

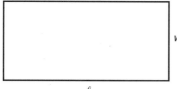

Square

Perimeter $P = 4s$

Area $A = s^2$

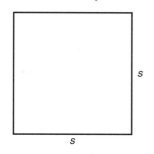

Trapezoid

Perimeter $P =$ sum of the lengths of the sides

Area $A = \dfrac{1}{2}h(b_1 + b_2)$

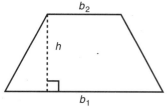

Triangle

Perimeter $P = a + b + c$

Area $A = \dfrac{1}{2}bh$

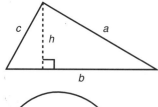

Circle

Circumference $C = 2\pi r = \pi d$

Area $A = \pi r^2$

(r is the radius and d is the diameter of the circle)

$d = 2r$ or $r = \dfrac{1}{2}d$

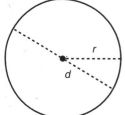